The 30th SIAR International Congress of Automotive and Transport Engineering

Ilie Dumitru · Dinu Covaciu ·
Laurențiu Racila · Adrian Rosca
Editors

The 30th SIAR International Congress of Automotive and Transport Engineering

Science and Management of Automotive and Transportation Engineering

Springer

Editors
Ilie Dumitru
Faculty of Mechanics
University of Craiova
Craiova, Dolj, Romania

Laurenţiu Racila
Faculty of Mechanics
University of Craiova
Craiova, Dolj, Romania

Dinu Covaciu
Department of Automotive and Transport Engineering
Transilvania University of Braşov
Brasov, Romania

Adrian Rosca
Faculty of Mechanics
University of Craiova
Craiova, Dolj, Romania

ISBN 978-3-030-32563-3 ISBN 978-3-030-32564-0 (eBook)
https://doi.org/10.1007/978-3-030-32564-0

This Springer imprint is published by the registered company Springer Nature Switzerland AG
The registered company address is: Gewerbestrasse 11, 6330 Cham, Switzerland

Preface

The International Congress **SMAT 2019 (Science and Management of Automotive and Transportation Engineering)** is held under the high patronage of Ministry of Education of Romania.

By the collaboration inside SIAR (Society of Automotive Engineers of Romania), this congress represents the effort of the academic peoples and researchers from the main Romanian universities centers, Bucharest, Brasov, Craiova, Cluj-Napoca, Pitesti and Timisoara.

The congress topics are focused on novel, most challenging issues on automotive and transportation, providing both field-related engineers and researchers within the industry and/or the academic environment the possibility to track down topical issues disseminated via the congress papers. Public presentation of papers enables the attendants to initiate and active participate in cross-fertilizing dialogs that can lead to new interdisciplinary projects and cooperation partnerships.

The main purpose of the SMAT 2019 Congress is to offer to all researchers and specialists in automotive field a large discussion environment, a place to present the newest achievements in scientific research and innovation activities, within the topics:

- Advanced Powertrain and Propulsion
- Road Vehicles and Environment
- Modern Transport Systems and Road Traffic
- Advanced Engineering Methods
- Materials and Technologies.

The congress aims to bring together academics, researchers, engineers and other specialists from the large field of automotive engineering and road transport, to foster some of their latest achievements and generate opportunities for further collaboration. Both plenary sessions and scientific sections will focus on the high quality of the submitted papers. Presentations will be mediated by leading moderators. We are glad to receive academics, researchers, engineers and other specialists within automotive engineering and road transport, students and anyone interested to attend our congress.

The congress topics underpin the latest trends and scientific development directions within the field of automotive and transport fields. The target audience will benefit from the most innovative solutions put forward by the submitted papers, within the field of automotive and transport, and further collaboration and professional networking opportunities that reside in this congress.

The congress has long historical progress, SMAT 2019 being the 30th edition. Over the time, the congress topics have been actualized permanently, one of the organizer's targets being the international visibility and the continued increase in the scientific content of the papers.

Many remarkable Romanian and international researchers from automotive and transports engineering will be present to the congress. During the congress acts, a scientific student contest will be organized and also a roundtable with social and economic representatives, about security, safety road and urban logistic.

The social, economic, technical and scientific impact of faculty of mechanics activities, a pillow of the University of Craiova, is finding in a large area of scientific and technical achievements of students, graduates and researchers of this faculty, all around the world.

Editors

Contents

Advanced Powertrain and Propulsion

The Calculus of the Pressures in the Characteristic Points of the Engine's Cycle for a GDI Engine 3
Vasile Blaga, Tudor Mitran, Sorin Moca, and Catalin Chioreanu

An Engine Mechanism Dynamic Analysis by Considering the Kinematic Elements as Deformable Ones 15
Nicolae Dumitru, Cristian Copilusi, Adrian Calangiu, and Ionut Geonea

Experimental Stand to Evaluate Engine Mass Air Flow (MAF) Sensor 26
Dragos Tutunea, Ilie Dumitru, Laurentiu Racila, Oana Otat, Ionut Geonea, Lucian Matei, and Cristina Rotea

Theoretical and Experimental Research on the Polluting Emissions for an Middle Class Vehicle Equipped with Engine that Uses the Double Energy Source, Gasoline and LPG 31
Valentin Nişulescu, Florian Ivan, Daniel Iozsa, and Gheorghe Bancă

Analysis of the Injector for Common Rail System, Including of the Electrical Command Signal 42
Cristian-Ioan Leahu

Influence of Tires Parameters on the Braking Process in Automobiles . . . 50
Robert Emil Simniceanu and Loreta Simniceanu

Mathematical Model for Stability Analysis of Vehicle-Driver System 54
Mario Trotea, Robert Emil Simniceanu, Augustin Constantinescu, and Stefanita Ciunel

A Study on the Influence of the Miller Cycle upon the Engine's Internal Aerodynamics . . . 58
Mihai Niculae, Adrian Clenci, Victor Iorga-Simăn, and Rodica Niculescu

3D Aerodynamic Simulation of the Heavy Duty Commercial Vehicle . . . 68
Nadica Stojanovic, Oday I. Abdullah, Josef Schlattmann, Ivan Grujic, and Danijela Miloraddovic

Research of Intelligent Control of Injection Systems for Subaru Competition Car . . . 82
Ioan-Aurel Chereceş, Doru-Laurean Băldean, and Adela-Ioana Borzan

Some Considerations on Optimal Placement of the Arm Hydraulic Cylinder to an Excavator . . . 88
Cristian-Ioan Leahu, Mircea Năstăsoiu, and Radu Tarulescu

Aspects Regarding Fuel Consumption and the Pollutant Products of Internal Combustion Engines for Commercial Vehicles . . . 95
Marius S. D. Capătă, Mădălin Florin Popa, and Nicolae Burnete

The Influence of the Hybridization Factor on the Energetic Performance in Test Cycles . . . 103
Cristian Renţea, Marius Băţăuş, Sorin Maican, Mircea Oprean, and Gheorghe Frăţilă

Road Vehicle and Environment

Investigating the Cyclic Variability of a Multi-cylinder Turbocharged Direct-Injection Spark-Ignition Engine Fueled with E10 . . . 113
Fazal Um Min Allah, Caio Henrique Rufino, Clayton Barcelos Zabeu, and Waldyr Luiz Ribeiro Gallo

CFD Study on Wheel Aerodynamic Performance in Side Wind Conditions for a Hatchback Vehicle . . . 124
Laurentiu Ilea, Daniel Iozsa, Cornelia Stan, and Claudiu Teodorescu

Technical Solutions for the Use of Internal Combustion Engine Combustion Gases . . . 131
Rareş Lucian Chiriac, Anghel Chiru, and Ovidiu Andrei Condrea

Evaluation of Tire Traction Performance on Dry Surface Based on Tire-Road Contact Stress . . . 138
Chen Liang, Liu Ji, Hoda Mousavi, and Corina Sandu

Performance, Combustion and Emissions Study of a DI Diesel Engine Running on Several Types of Diesel Fuels 153
Nicolae Vlad Burnete, Richard János Balint, Corneliu Adrian Măgherusan, and Dan Moldovanu

Research of Alternative Composite Solutions in Gas After-Treatment System for Diesel Cars 160
Bogdan-Manolin Jurchiş and Doru-Laurean Băldean

Study of Vehicles Energy Efficiency 165
Ion Lespezeanu, Laszlo Barothi, Leonard-Iulian Cucu, Cătălin-Dumitru Darie, Iulian Constantin Coropeţchi, and Ion Copae

Comparative Study Regarding the Operation of a Car's Internal Combustion Engine in Urban and Extra-Urban Environment 177
Ion Lespezeanu, Ramona-Monica Stoica, Radu Vilău, Daniel Sava, and Ion Copae

Marine Ship Equipped with Air Turbine – Electric Generator Aggregate 188
Radu Tarulescu, Stelian Tarulescu, Cristian-Ioan Leahu, and Dana Luca-Motoc

Influence of the Dynamic Vehicle Load on the Power Losses Required to Overcoming the Rolling Resistance 195
Nicolae Cordoş, Adrian Todoruţ, Călin Iclodean, and István Barabás

Research of Diesel Particle Filter Soot Impact and Regeneration Time Trend-Line 203
Emilian Vasile Borza, Doru-Laurean Băldean, and Bogdan-Manolin Jurchiş

Comparative Analysis of Three Distinct Porsche Models Concerning N.V.H. Aspects 209
Doru-Laurean Băldean

Mathematical Model for Study of Behavior of the Bodywork at Impact 214
Ana Maria Manea, Daniel Iozsa, Cornelia Stan, Claudiu Teodorescu, and Gheorghe Frăţilă

Aspects Regarding the Identification of Optimum Driver Comfort Level by Virtual Analysis of the Vertical Oscillations Generated by Road 221
Marian Florin Mitroi and Anghel Chiru

Research on the Use of Rheological Systems to Decrease the Vertical Oscillations Perceived by the Driver 231
Marian Florin Mitroi and Camelia Cerbu

On-Board Measurement of Emissions from Spark Ignition Engine Vehicle for Urban Routes 241
Stelian Ţârulescu, Radu Ţârulescu, and Cristian-Ioan Leahu

Modern Transport Systems and Road Traffic

Relationships Between Accident Severity and Weather and Roadway Adherence Factors in Crashes Occurred in Different Type of Collisions 251
Alin Drosu, Corneliu Cofaru, and Mihaela Virginia Popescu

Reconstruction of Rollover Collision Using PC Crash 11.1 and 3D Scanner 265
Laura Diana Grigorie, Dragos Sorin Dima, Alina Elena Romanescu, and Mihnea Marin

The Influence of Vehicle Low Impact Velocity over the Helmet Airbag Deployment and Cyclist Injuries 273
Ovidiu Andrei Condrea, Anghel Chiru, George Radu Togănel, and Daniel Dragos Trusca

Effects of Vehicle Impact Velocity on Cyclist Injuries and Kinematics for Rear Impacts 281
Ovidiu Andrei Condrea, Anghel Chiru, George Radu Togănel, and Daniel Dragos Trusca

Case Study Regarding the Reconstruction of a Real Vehicle-Bicycle Accident 289
Ovidiu Andrei Condrea, Anghel Chiru, George Radu Togănel, Daniel Dragos Trusca, and Rares Lucian Chiriac

Image Processing Applied in Road Traffic Using a Quadcopter 297
Carmen Gheorghe and Nicolae Filip

Road Traffic Studies for Urban Development 304
Georgiana Silvia Muresan, Carmen Gheorghe, and Nicolae Filip

Analysis of Occupant Injury Level in the Case of High-Speed Vehicle Slide-Spinning Collision 312
Laurentiu Claudiu Manea, Adriana Manea, Alexandru Ionut Radu, and Dragos Sorin Dima

Dynamic Considerations Regarding the Enrolling in the Curve with Small Radius of Railways, of the Locomotives 322
Iulian Turcanu and Marian Florin Mitroi

Methods of Analysis and Reconstruction of Traffic Events 332
Mihai Clinciu, Stefanita Ciunel, and Bebe Tica

Behavior of the Cranio-Cervical Biofidelity Complex Compared to the Cranio-Cervical Area of Anthropometric Test Devices 341
Mihai Clinciu and Stefanita Ciunel

Study of the Possibilities to Improve the Service Level of Traffic Light Intersections by Road Traffic Micro-simulation 349
Andrei Alexandru Boroiu, Elena Neagu, Alexandru Boroiu, and Sebastian Parlac

The Influence of the Color of the Reflective Vest on the Visibility of the Cyclist at Night 359
Bogdan Cornel Benea, Daniel Dragos Trusca, George Radu Togănel, and Alexandru Ionut Radu

Cross-Validation of Two Collision Modeling Software Based on the Reconstruction of a Real Collision - A Possible New Approach to Collision Reconstruction 366
Attila Iuliu Gönczi and Adrian Cipleu

Study of a Car-to-Pedestrian Collision in Case of Vehicle's Post-impact Braking 375
Bogdan Adrian Tolea, Alexandru Ionut Radu, Horia Beles, George Dragomir, and Sorin Moca

Study Regarding the Impact Between a Vehicle and a Two-Wheel Vehicle 383
Oana Oțăt, Ilie Dumitru, Victor Oțăt, Lucian Matei, and Dragoş Tutunea

Practical Use of Accelerometers in Road Traffic Studies 392
Dinu Covaciu, Daniela Florea, and Janos Timar

A New Method to Improve Computing Time for TSP Little Algorithm 400
Adrian Cipleu, Werner Ştefănescu, Ionel Vandici, and Attila Iuliu Gönczi

Experimental Research on Brake Behavior for Different Types of Commercial Vehicles 406
Mădălin Florin Popa, Marius S. D. Capătă, and Nicolae Burnete

Considerations Concerning the Study of Dynamic Equilibrium for Engine Axle and Bogies for Locomotives 414
Iulian Turcanu and Ioan Sebesan

Automotive Black Box and Development Platform Used for Traffic Risks Evaluation and Mitigation 426
Silviu Zamfir and Radu Drosescu

Experimental Studies of the Same Car Behavior in Different Pile-up Scenarios 439
Nicolae Ispas and Mircea Nastasoiu

Advanced Engineering Methods

Modeling of a Standardized Testing Track with the Use of Specialized Software 451
Daniela Voicu, Ramona-Monica Stoica, and Laszlo Barothi

Fuzzy Decision Algorithm for Driver Drowsiness Detection 458
Tiberiu Vesselenyi, Alexandru Rus, Tudor Mitran, Sorin Moca, and Csokmai Lehel

Contribution to Research the Applied Engineering Protocol to Implement a Fuzzy Regulator for Autonomous Driving of an Automotive Model Implemented in Virtual Reality 468
Florin Covaciu and Doru-Laurean Băldean

Study of Vehicle Dynamics Using Tensor Analysis 477
Radu Vilău, Ion Lespezeanu, Constantin-Iulian Pântece, Alexandru Vasile, and Ion Copae

Aerodynamic Study of an Electric Vehicle Prototype 485
Liviu Scurtu, Ancuta Jurco, Emilian Vasile Borza, Florin Mariasiu, Nicolae Vlad, and Silvan Morariu

Drag Coefficient Analysis for Side Mirrors of an Electric Vehicle Prototype 494
Liviu Scurtu, Adela-Ioana Borzan, Florin Mariasiu, Nicolae Vlad, Bogdan Varga, and Silvan Morariu

Design and CFD Simulation of an Automotive Air Heat Exchanger . . . 502
Monica Balcau and Liviu Scurtu

Application Model in AUTOSAR Software Development for Control Systems Design Through Fuzzy Methods 508
Cristian Spirleanu and Eugen Diaconescu

Sensors Used for Human Gait Monitoring 518
Daniela Tarnita and Alin Teofil Oncescu

Conceptual Design of an Urban Electric Microcar 525
Mehmet Murat Topaç, Merve Karaca, and Levent Bilal

New Approach for the Integration of CAD/CAPP/CAM Systems for Bearings Manufacturing 533
Dan George Prutica and Gheorghe Brabie

Design Optimization of the Towing Hook Used in Passenger Cars for Light Trailers 540
Mario Trotea, Augustin Constantinescu, and Loreta Simniceanu

A Study of the Mechanical Behavior of a Car Part Realized from Epoxy Resin Reinforced with Cotton Woven 550
Alexandru Bolcu, Nicolae Dumitru, Dumitru Bolcu, Cosmin Mihai Miritoiu, and Marius Marinel Stanescu

Modeling of Dynamic Behavior of a Spur Gear 559
Ionut Geonea, Laurentiu Racila, Ionica Valeriu, and Cosmin Mihai Miritoiu

Vibrations Influence on Automotive Drive 565
Valeriu Ionica, Dan Bagnaru, Ionut Geonea, and Leonard Ciurezu

Analysis Methodology of Experimental Results in Case of a Cockpit Module Under Impact Load 573
Paul Predoiu, Raluca Florentina Negrea, Dinel Popa, and Stefan Tabacu

Strain Analysis of a Steering Tie Rod 581
Ionela Alina Sandu, Ştefan Tabacu, Cătălin Ducu, and Constantin Onescu

Comparison Study Between a Tubular Type Camshaft and a Classic Type Using Finite Element Method 589
Sorin Vlad Igreţ and Radu Iulian Negrilă

Basic Subsystems Reliability Influence Study on the Main Characteristics of Automotive 597
Radu Iulian Negrila and Sorin Vlad Igreţ

Solicitations in the Rear Axle Support of a Karting Frame 607
Laurentiu Racila, Ilie Dumitru, Lucian Matei, Mihaela Racila, Dragoş Tutunea, and Ionut Geonea

Aspects Regarding the Stability Performances Calculation of the Wheeled Vehicles While Turning 614
Ion Preda, Gheorghe Ciolan, and Dinu Covaciu

Finite Element Analyses of the Influence of Tyre Functional Parameters on Contact Patch Stresses 624
Alexandra-Raluca Moisescu and Gabriel Anghelache

Construction of a Kart Chassis Through 3D Reconstruction Methods – Part 1 – Scanning and Alignment 636
Lucian Matei, Ilie Dumitru, Laurentiu Racila, Mihaela Racila, Dragoş Tutunea, and Oana Otat

Construction of a Kart Chassis Through 3D Reconstruction Methods – Part 2 – Reverse Engineering 643
Lucian Matei, Ilie Dumitru, Laurentiu Racila, Mihaela Racila, Dragoş Tutunea, and Oana Otat

Materials and Technologies

Generating Loci Through the Theory of Mechanisms 653
Iulian Popescu, Alina Elena Romanescu, and Ludmila Sass

Stereometric Analysis of Pt Schottky Contacts on Quaternary n-$Al_{0.08}In_{0.08}Ga_{0.84}N$ Thin Film 660
Ştefan Ţălu, Mihai Ţălu, Daniela Vintilă, Laura Diana Grigorie, and Alina Romanescu

Surface Roughness and Temperature in Dry Milling of an Austenitic Stainless Steel 670
Nicolae Craciunoiu, Daniela Tarata, Adrian Sorin Rosca, and Ionut Geonea

Complex Characterization of the Material 678
Ionela Alina Sandu, Constantin Onescu, Ştefan Tabacu, and Cătălin Ducu

Study About the Roughness Surface Quality for a Duralumin Alloy After Cutting Process 684
Ileana Cristina Pascu and Iulian Popescu

Valves Processing of Internal Power Engines by Ultrasonic Vibro-Cutting 691
Bebe Tică and C. Klement

Experimental Testing of Energy Absorbing Foams Used as Safety Structure Within Automobile's Cockpit Module 699
Raluca Florentina Negrea, Paul Predoiu, Stefan Tabacu, Catalin Ducu, and Denis Negrea

Virtual Modeling and Numerical Simulations of the Latitude Prosthesis - Human Elbow Assembly 706
Daniela Tarnita, C. Boborelu, D. Popa, R. Malciu, and D.-N. Tarnita

Author Index 713

Advanced Powertrain and Propulsion

The Calculus of the Pressures in the Characteristic Points of the Engine's Cycle for a GDI Engine

Vasile Blaga, Tudor Mitran(✉), Sorin Moca, and Catalin Chioreanu

The University of Oradea, Universitatii St. no. 1, 410087 Oradea, Romania
vblaga.ar@gmail.com, tudor_mitran@yahoo.com

Abstract. The GDI Engines are a viable option for the compliance with the more and more severe environmental protection standards. The main advantage of this type of engine is the operation with very lean mixtures at low and medium loads. The operation at low and medium loads takes place, mainly, in urban regime, exactly were a drastic reduction of the pollutant emissions produced by the road vehicles is necessary. The authors have developed an original model for the thermal calculus of a GDI Engine, depending on the rotational speed and on the environmental temperatures, that aims to be a useful instrument for then study of the thermal processes that take place in this type of engine. By developing a calculus algorithm of the state parameters in the characteristic points of engine's cycle, depending on the rotational speed and on the environmental temperature, one aims the development of a useful instrument to estimate the pollutant emissions and the identification of the measures that should be taken in order to reduce them. In this paper is presented the calculus algorithm for the pressures in the characteristic points of engine's cycle, excepting the ones at the intake port and the one at the end of the intake stroke. These are the subject of other paper.

Keywords: GDI Engine · The engine cycle · The working gas pressure

1 Introduction

GDI engines are currently the most used ICE for passenger cars. The operating conditions and the possibilities to develop the concept, including GCI engines, offer the possibility to build low consumption and low emission SI engines [1, 2].

There are different approaches for the study of the processes that take place in an ICE [3].

The authors have developed an original model of the thermal cycle for GDI Engine [4]. This is based on the theoretical cycle for SI engines, except the fact that the burning process takes place in two phases (Fig. 1):

- c-z – the phase of isochoric burning (the main phase of the burning process);

I. Dumitru et al. (Eds.): SMAT 2019, *The 30th SIAR International Congress of Automotive and Transport Engineering*, pp. 3–14, 2020.
https://doi.org/10.1007/978-3-030-32564-0_1

– z-u – the afterburning, considered to be a polytropic process of exponent $n_u < 1$; although the pressure inside the cylinder drops due to the increase of the instantaneous volume of the burning chamber, because the last quantities of fuel are burning, the temperature of the gases inside the cylinder rises ($T_u > T_z$);

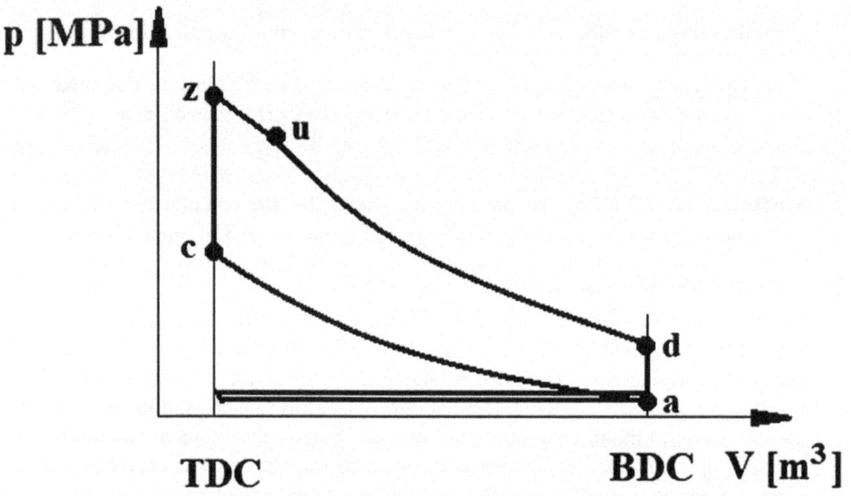

Fig. 1. The variation of the pressure during the burning process.

For the gas exchange process, for the compression process and for the expansion process, the simplifying assumptions were applied: the pressures during the intake stroke (p_a) and the pressure during the exhaust stroke (p_e) are considered to be constant and the compression processes, like the expansion one, is considered to be a polytropic one with constant exponent.

The determination of the pressure at the intake port (p_{ga}) and of the one at the end of the intake stroke (p_a) is the subject of another paper, that will be proposed for publishing. By knowing these values, all the initial values of the calculus of the thermal calculus of the engine's cycle are established.

The state parameters of the engine's cycle have been determined depending on the rotational speed and on the environmental temperature. Other theoretical studies revealed the fact that the environmental temperature has, practical, no influence. The calculus was developed for an air excess factor $\lambda = 1$.

2 The Calculus of the Pressures in the Characteristic Points of the Engine's Cycle

The relative pressure loss in the intake system:

$$\psi_o = \frac{\Delta p_a}{p_o} = \frac{p_o - p_a}{p_o} = 1 - \frac{p_a}{p_o}; \tag{1}$$

where:

- p_0 [Pa] is the environmental pressure;
- p_a [Pa] is the pressure at the end of the intake stroke;

The relative pressure loss in the exhaust system:

$$\psi_1 = \frac{\Delta p_r}{p_r} = \frac{p_r - p_o}{p_r} = 1 - \frac{p_o}{p_r}; \psi = \frac{p_a}{p_r}. \tag{2}$$

where

- p_r [Pa] is the pressure inside the cylinder at the exhaust stroke;

The polytropic exponent of the intake process:

$$n_a = 1 - \frac{k_a - 1}{\phi_a}; \quad \phi_a = 1,05; \ k_a = 1.458 - 0.525 \cdot 10^{-4} \cdot (T_0 + T_a) \tag{3}$$

where:

- φ_a is a coefficient that characterizes the deviation of the working process at the intake of the fresh charge from the adiabatic process;
- k_a is the adiabatic exponent of the intake process;
- T_0 [Pa] is the environmental temperature;
- T_a [Pa] is the temperature of the gases inside the cylinder at the end of the intake stroke;

The calculus of the parameters at the end of the intake stroke was presented in another paper [5].

The variation of the polytropic exponent of the intake process, calculated depending on the rotational speed and on the environmental temperature is presented in Fig. 2a. In Fig. 2b is presented the variation of the polytropic exponent of the intake process calculated depending on the rotational speed and in Fig. 2c is presented the variation of the polytropic exponent of the intake process calculated depending on the environmental temperature.

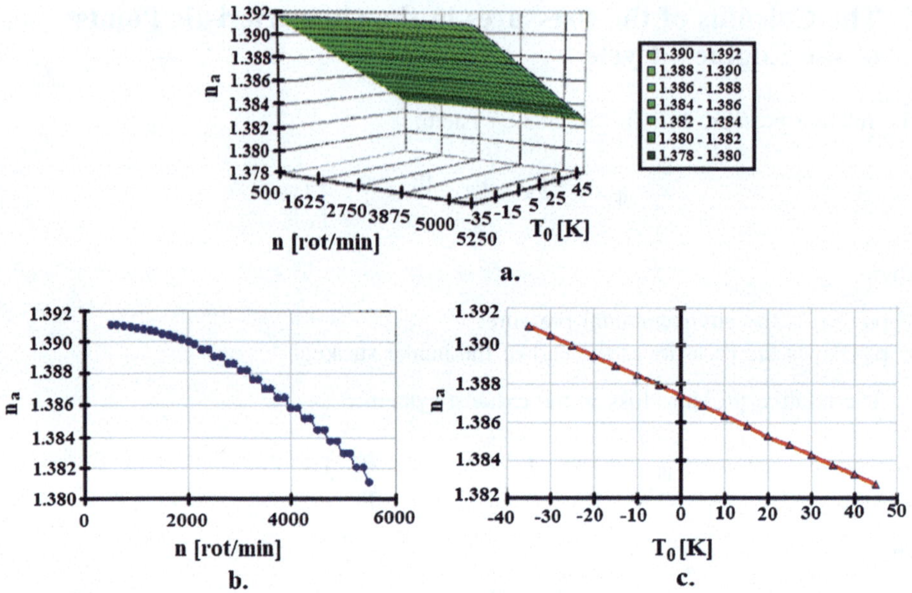

Fig. 2. The variation of the polytropic exponent n_a depending on: the rotational speed and on the environmental temperature (a), the rotational speed (b) and the environmental temperature (c).

The pressure at the end of the compression process (the end of the compression stroke):

$$p_c = p_a \cdot \varepsilon^{n_c} \; [Pa]; \tag{4}$$

where

- ε is the compression ratio;
- n_c is the polytropic exponent of the compression stroke (considered to be constant);

The polytropic exponent of the compression stroke (n_c) can be expressed as a function of the adiabatic exponent of the compression process (k_c) and of a coefficient (φ_c) that takes into consideration the heat exchanged by the working fluid with the cylinder's walls during this process:

$$n_c = 1 + \frac{k_c - 1}{\varphi_c}; \quad \varphi_c = 1{,}06; \tag{5}$$

The coefficient (φ_c) is a consequence of the energetic balance equation applied to the compression process.

$$\varphi_c = 1 - \frac{Q_{pc}}{\Delta U} = -\frac{L_{a-c}}{U_c - U_a} = \frac{L_{a-c}}{U_a - U_c} \tag{6}$$

where:

- L_{a-c} [J] is the absolute mechanical work in the compression process;
- $\Delta U = U_c - U_a$ [J] is the variation of the internal energy of the fresh charge during this process;
- U_c [J] is the internal energy of the working fluid at the end of the compression process (the end of the compression stroke – the point c in Fig. 1);
- U_a [J] is the internal energy of the working fluid at the end of the intake stroke (the point a in Fig. 1);

Because $Q_{pc} < 0$ and $\Delta U > 0$, the coefficient (φ_c) is super unit, so that $n_c < k_c$. In the conditions of the compression process, the adiabatic exponent (k_c) is calculated with the relation:

$$k_c = 1.458 - 0.525 \cdot 10^{-4} \cdot (T_0 + T_a) \quad (7)$$

where:

- T_c [K] is the temperature at the end of the compression stroke;

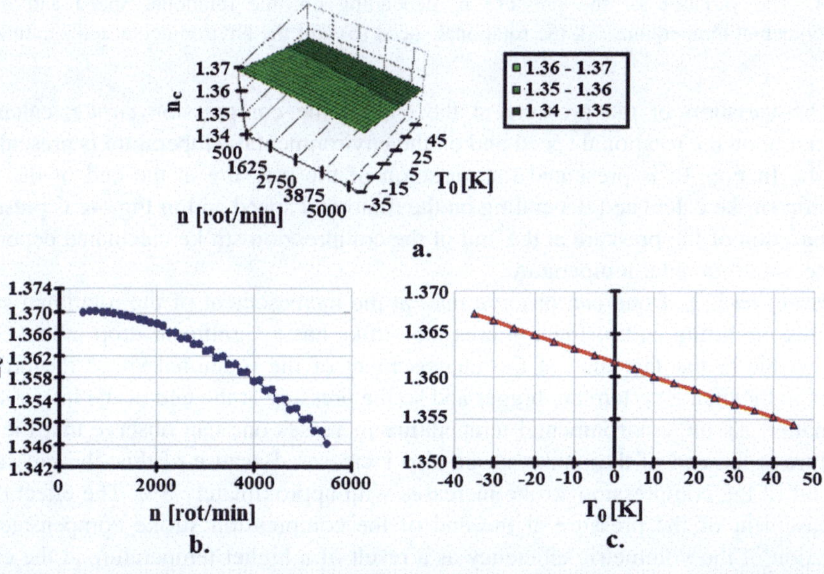

Fig. 3. The variation of the polytropic exponent n_c depending on: the rotational speed and on the environmental temperature (a), the rotational speed (b) and the environmental temperature (c).

The variation of the polytropic exponent of the compression process, calculated depending on the rotational speed and on the environmental temperature is presented in Fig. 3a. In Fig. 3b is presented the variation of the polytropic exponent of the compression process calculated depending on the rotational speed and in Fig. 3c is presented the variation of the polytropic exponent of the intake process calculated depending on the environmental temperature.

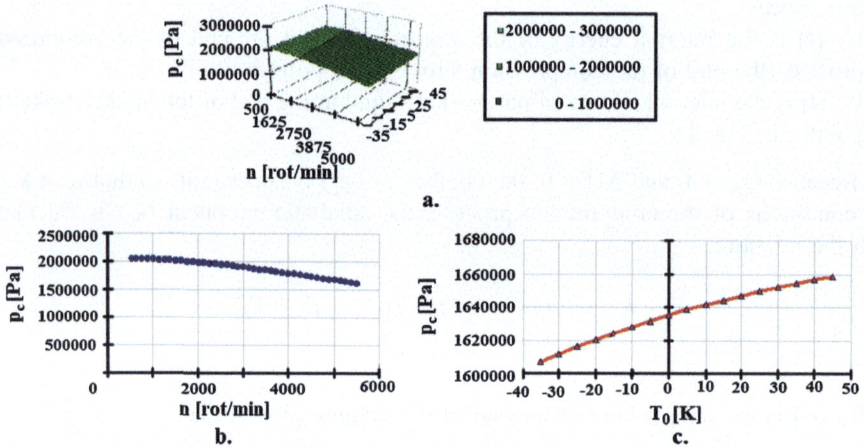

Fig. 4. The variation of the pressure p_c depending on: the rotational speed and on the environmental temperature (a), the rotational speed (b) and the environmental temperature (c).

The variation of the pressure at the end of the compression stroke, calculated depending on the rotational speed and on the environmental temperature is presented in Fig. 4a. In Fig. 4b is presented the variation of the pressure at the end of the compression stroke calculated depending on the rotational speed and in Fig. 4c is presented the variation of the pressure at the end of the compression stroke calculated depending on the environmental temperature.

From the Fig. 4 one can observe that, at the increasement of the rotational speed, the pressure at the end of the compression stroke has a significant drop (about 30%). This is due to the fact that, at the increasement of the rotational speed the pressure losses in the intake system are bigger and so the pressure at the end of the intake stroke is smaller. As the environmental temperature increases one can observe that the temperature at the end of the intake stroke also increases. Because of this, the pressure at the end of the compression stroke increases with approximately 3%. The effect of the increasement of the pressure at the end of the compression stroke compensates the reduction of the volumetric efficiency as a result of a higher temperature at the end of the intake stroke.

The pressure at the end of the isochoric burning phase (c-z in Fig. 1):

$$p_z = \alpha_{ai} \cdot p_c \ [Pa] \quad (8)$$

where:

- α_{ai} is the pressure raise ratio in the phase of the isochoric burning;

This ratio was determined by correlating the heat released in the phase of the isochoric burning with the parameters of the working fluid.

The variation of the pressure at the end of the isochoric burning, calculated depending on the rotational speed and on the environmental temperature is presented in Fig. 5a. In Fig. 5b is presented the variation of the pressure at the end of the isochoric burning calculated depending on the rotational speed and in Fig. 5c is presented the variation of the isochoric burning calculated depending on the environmental temperature.

As expected, from Fig. 5 one can observe that the pressure at the end of the isochoric burning phase decreases as the rotational speed increases because the pressure at the end of the compression stroke also drops.

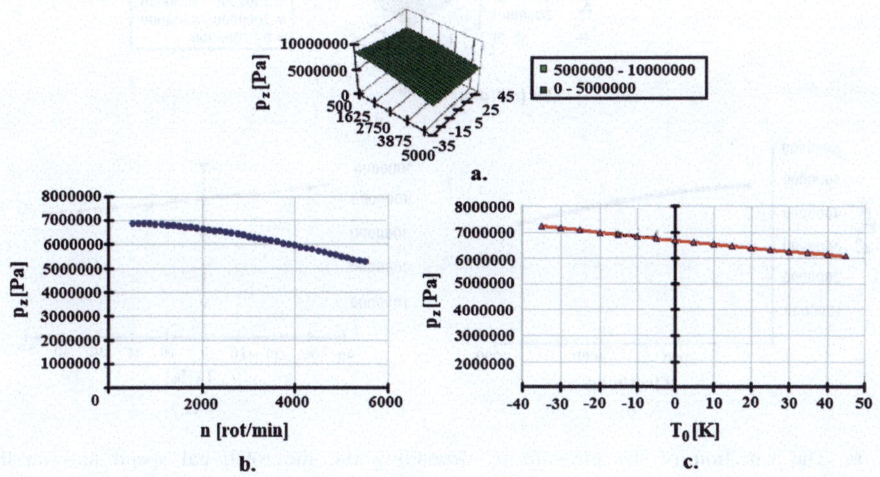

Fig. 5. The variation of the pressure p_z depending on: the rotational speed and on the environmental temperature (a), the rotational speed (b) and the environmental temperature (c).

The pressure at the end of the burning process (the point u in Fig. 1):

$$p_u = p_z \cdot \delta^{n_u} \; [Pa]; \tag{9}$$

where:

- $\delta = V_u/V_c$ is the volume increase ratio in the phase of the afterburning (z-u in Fig. 1). It was determined by correlating the heat released during this phase with the parameters of the working fluid (this methodology will be presented in another paper).
- V_u [m^3] is the volume of the burning chamber at the end of the burning process (the point u in the Fig. 1);
- V_c [m^3] is the minimum volume of the burning chamber (the volume in the point c from Fig. 1);
- n_u = 0,9 [6] is the mean adiabatic exponent of the afterburning;

In Fig. 6a is presented the variation of the pressure at the end of the burning process, calculated depending on the rotational speed and on the environmental temperature. In Fig. 6b is presented the variation of the pressure at the end of the burning process calculated depending on the rotational speed and in Fig. 6c is presented the variation of the pressure at the end of the burning process calculated depending on the environmental temperature.

The pressure at the end of the burning process drops, both at the rotational seed rising and at the environmental temperature increasement, because the pressure in the point z (see Fig. 1) also decreases (see Fig. 6).

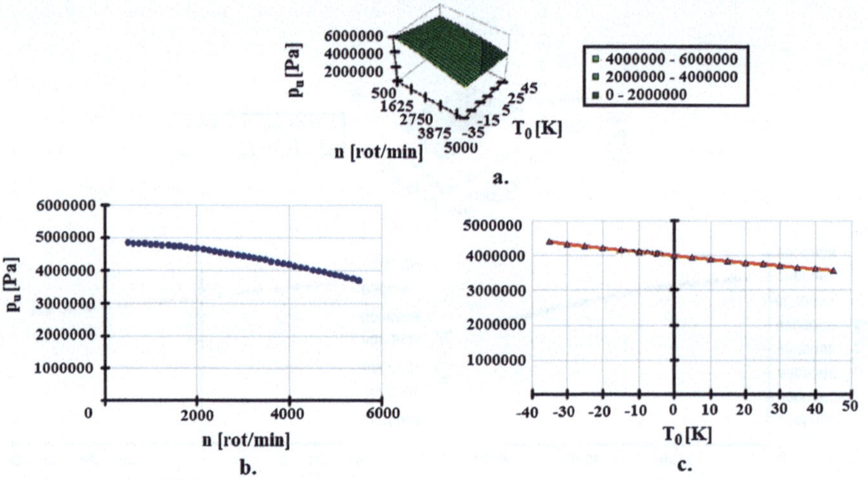

Fig. 6. The variation of the pressure p_u depending on: the rotational speed and on the environmental temperature (a), the rotational speed (b) and the environmental temperature (c).

The pressure at the end of the expansion process:

$$p_d = p_u \left(\frac{\delta}{\varepsilon}\right)^{n_d}; \tag{10}$$

where:

- n_d is the polytropic exponent of the expansion process;

$$n_d = 1 + \frac{k_d - 1}{\phi_d}; \tag{11}$$

The adiabatic exponent of the expansion process [6]:

$$k_d = 1{,}268 - \frac{0{,}035}{\lambda} + \frac{126}{T_u + T_d} \quad (12)$$

where;

- $\varphi_d = 0{,}92$ is the coefficient that characterizes the deviation of the working process from the adiabatic one [6];
- T_d [K] is the temperature at the end of the expansion stroke;

The variation of the polytropic exponent of the expansion process, calculated depending on the rotational speed and on the environmental temperature is presented in Fig. 7a. In Fig. 7b is presented the variation of the expansion exponent calculated depending on the rotational speed and in Fig. 6c is presented the variation of the expansion exponent calculated depending on the environmental temperature.

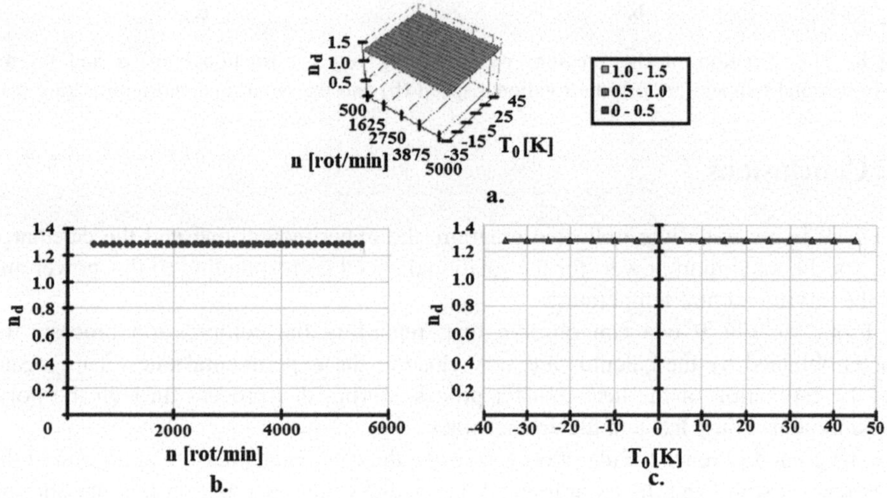

Fig. 7. The variation of the polytropic exponent n_d depending on: the rotational speed and on the environmental temperature (a), the rotational speed (b) and the environmental temperature (c).

The variation of the pressure at the end of the expansion stroke, calculated depending on the rotational speed and on the environmental temperature is presented in Fig. 7a. In Fig. 7b is presented the variation of the pressure at the end of the expansion stroke depending on the rotational speed and in Fig. 7c is presented the variation of the pressure at the end of the expansion stroke calculated depending on the environmental temperature.

One can observe a drop of the pressure at the end of the expansion stroke as the rotational speed and at the environmental temperature increase. This is due to a drop of the pressure at the end of the burning process as the rotational speed and the temperature of the environmental temperature increase.

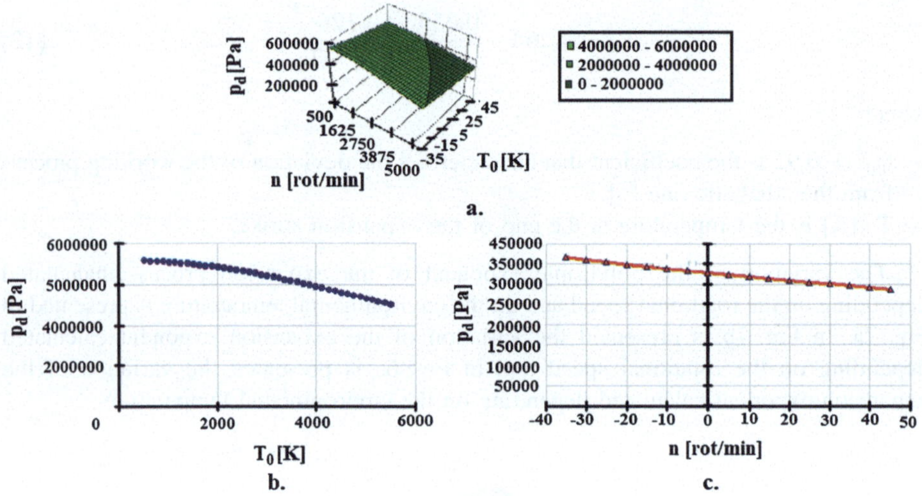

Fig. 8. The variation of the pressure p_d depending on: the rotational speed and on the environmental temperature (a), the rotational speed (b) and the environmental temperature (c).

3 Conclusions

In Fig. 9 is presented the indicated diagram, the experimental one and the calculated one, for the maximum power, for the rotational speed corresponding to this power and for the environmental temperature.

From the Fig. 9 one can observe that, regarding the compression process, the results obtained by the calculus are very close to the experimental ones. This means that the estimation of the heat transfer process during this process through the polytropic exponent (n_c) has s sufficient accuracy.

The great difference (of about 60% between the calculated pressure at the end of the compression stoke and the experimental one is due to the assumption that the burning process starts at the end of the compression stroke. In reality, as can be observed in Fig. 8, the burning starts before the piston reaches the TDC, at the end of the compression stroke, because the triggering of the spark plug takes place in advance regarding the TDC.

In exchange, the pressure calculated at the end of the isochoric phase is very close with the experimental one. The calculus for the determination of this pressure was developed based on the heat released through the burning of the fuel during the main phase of the burning process (c-z in Fig. 1). The calculus proves that, if the start of the burning process should take place at the end of the compression stroke, one can observe a drop of the mechanical work developed during an engine cycle due to the increasement of the mechanical work consumed for the compression of the working fluid. By triggering in advance (regarding the TDC) a part of the mechanical work that's necessary to compress the working fluid is furnished by the heat released through the burning of the first quantities of fuels.

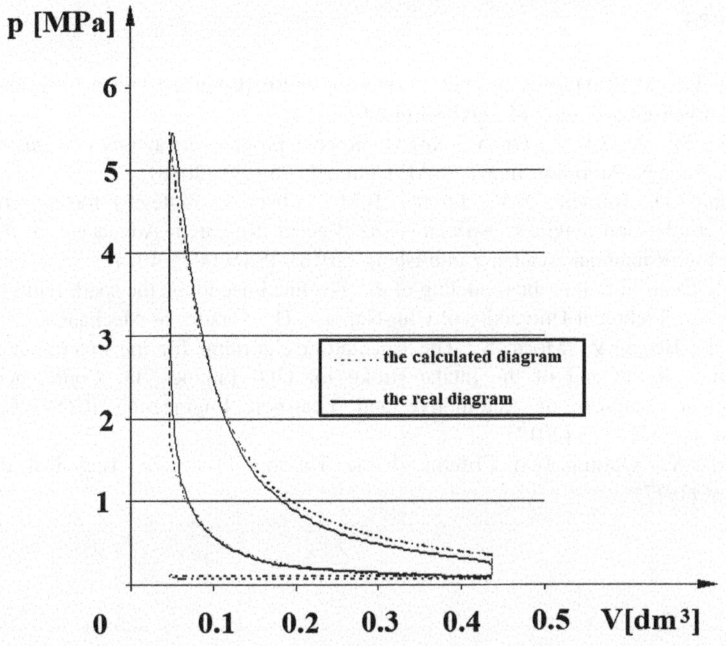

Fig. 9. Comparison between the calculated and the experimental data

Although the pressure calculated at the end of burning process is almost equal to the one experimentally determined, one can observe that the assumption of a burning process that includes an isochoric (main) phase of the burning process and an after-burning process of an exponent leads to an underestimation of the mechanical work produced during the burning process.

As a conclusion regarding the proposed model, by applying the simplifying assumptions, one can observe that the mechanical work developed during the burning phase is underestimated.

Also, from the Fig. 9 it results that, during the expansion process (u-d in the Fig. 1), the calculated pressure has a greater value than the measured one. This error can be reduced by determining with a greater accuracy the value of the polytropic exponent of the expansion process. This exponent estimates the heat exchange during the expansion process between the working fluid and the cylinder's walls.

The model proposed by the authors offers the possibility, after the validation in the conditions of an environmental temperature and at the maximum power rotational speed, to estimate the engine's parameters in any conditions of the engine's operation.

By knowing the influence of these parameters, one can adopt measures that should lead to limitation of the specific fuel consumption and, implicitly, at the reduction of the pollutant emissions.

References

1. Zhao, F., Lai, M.C., Harrington, D.L.: Automotive Spark-Ignited Direct-Injection Gasoline Engines. Elsevier Science Ltd., Oxford (1999)
2. Shuai, S., Ma, X., Li, Y., Qi, Y., Xu, H.: Recent progress in automotive gasoline direct injection engines. Automot. Innov. (SAE China) **1**, 95–113 (2018)
3. Moldovanu, D., Burnete, N.V., Jurchis, B.M.: Advanced strategies for investigation of internal combustion engines. Special Issue Recent Research Advances in Automotive Engineering. Romanian Academy publishing (2018). ISSN 0035-4074
4. Blaga, V.: Contribution to the modeling of the gasoline injection at the spark ignition engines. Ph.D. thesis, Technical University of Cluj-Napoca, The Faculty of Mechanics (2000)
5. Mitran, T., Blaga, V., Moca, S.: The calculation algorithm for the determination of the temperature at the end of the intake stroke for GDI Engines. In: Conference of 12th International Congress of Automotive and Transport Engineering (CONAT), Brasov, Romania, pp. 311–318 (2017)
6. Radcenko, V.: Optimization Criteria of the Thermal Processes. Technical Publishing, Bucharest (1977)

An Engine Mechanism Dynamic Analysis by Considering the Kinematic Elements as Deformable Ones

Nicolae Dumitru[1], Cristian Copilusi[1], Adrian Calangiu[2(✉)], and Ionut Geonea[1]

[1] Faculty of Mechanics, University of Craiova, Calea Bucuresti 107, Craiova, Romania
nicolae_dtru@yahoo.com

[2] FORD Company, Henri Ford 29, Craiova, Romania
acalangi@ford.com

Abstract. Through this paper is presented a methodology for optimal response analysis in dynamic conditions of an internal combustion engine mechanism. Thus, it will be analyzed a four stroke engine mechanism by considering this with rigid kinematic elements with the aid of MSC Adams and after this the same mechanism will be analyzed with kinematic deformable elements in ANSYS software environment. Also it will be presented an optimization algorithm of the proposed mechanism with the MSC Adams software when the mechanism elements are considered rigid ones and specific optimization algorithms for the mechanism. There are shown optimal modeling techniques of the mobile mechanical systems by considering the kinematic contact in accordance with the system dynamic response identification represented through kinematic and dynamic parameters variation diagrams.

The dynamic optimization algorithm represented through finite element method will be applied on the proposed engine mechanism connecting rod, by having as input data the kinematic and dynamic parameters obtained through a special interface ADAMS-ANSYS.

The obtained virtual models are designed with a flexible character in a parameterized mode. This means that the obtained results will be easily coupled with other results which were obtained through experimental tests.

Keywords: Dynamic analysis · Deformable elements · Finite element

1 Introduction

The heart of an internal combustion engine still remains in its own mechanism and an important factor on designing relays on materials. Our research addresses to the connecting rod component due to its complex state of loading. This component contributes to the transformation of a piston translational motion in a rotational one imposed to the crank shaft on high cyclic loads of the order of 10^5 to 10^9 cycles. Thus the component durability is of critical importance and the chosen component represents a topic

I. Dumitru et al. (Eds.): SMAT 2019, *The 30th SIAR International Congress of Automotive and Transport Engineering*, pp. 15–25, 2020.
https://doi.org/10.1007/978-3-030-32564-0_2

research under many aspects such as production technology, performance simulations, fatigue and materials.

Due to the modern techniques, it is well known that from a theoretical viewpoint, the mathematical models can help on an internal combustion engine design. But without their application on parameterized models and virtual simulations can go any further real results. Even though there are researches like [1, 2] where was designed a connecting rod in terms of maximum engine speed and loads applied at the crank and piston wrist pin ends by using finite element method in a static mode. Obviously the connecting rod two ends were designed to withstand different loads.

Therefore many research activities focuses on optimization process like in [3] where it was developed a mathematical model based on connecting rod weight and cost as objective functions. Constraints were imposed on the compression stress and on the bearing pressure at the crank shaft and the piston pin ends. The cost function was represented in some exponential form depending on geometric parameters. Other researches like [4, 5] presents methods to consider fatigue life as a constraint in optimal design of connecting rods. Optimization process were described as a methodology in [6, 7] by considering material properties as constraint functions and connecting rod weight as objective function. One research with an explicit design methodology was reported in [8] and incorporates an optimization technique that consists on primary engine components design by using finite element procedures for a 2D approach.

Our research objective is to develop an optimization process with modern software aid for internal combustion engine mechanism and moreover extended on the connecting rod component. The general method was reported in [9–11]. This research starts from a mathematical model presented in a second chapter of this research work and at the end a parameterized model was obtained and prepared for optimization process. Thus a four stroke engine mechanism 3D model was analyzed in an optimization process under MSC ADAMS environment and important results were achieved and presented in third chapter. These results give significant input data for a finite element analysis, using ANSYS, for the whole 3D model presented in fourth chapter. By having in sight equivalent von Misses stress, results after finite element analysis are presented in fifth chapter as input data for optimization process of the connecting rod component under ANSYS Mechanical APDL. Significant results after the optimization process were obtained and through this a method was validated and it can be applied on other mechanical mobile systems.

2 Mechanism Optimization Process

By using Multi-body Systems Method implemented through virtual design platform soft's, this will give the prototype possibility to submit into specific optimization processes. These can be done by modifying some geometrical characteristics and it can be virtually tested multiple constructive solutions by having in sight the system element characteristics, mass distribution, etc. The optimization process give positive results related to price comparative with the optimization analyses based on physical prototypes design and tests. The proposed optimization process is characterized by a kinematic and dynamic optimization algorithm included in MSC Adams software, which give the opportunity to parameterize the imported virtual model and to perform an analysis series by having the objective represented through design parameters identification which can have a major influence on the dynamic and kinematic behavior of the internal combustion engine mechanism. Thus, in MSC Adams software, an optimization process of a mobile mechanical system consists on crossing the following steps: parameterzing the proposed virtual model; defining the optimizing function objective; performing the design studies for main design variables identification, which have a significant influence onto objective functions; proper optimization process of the proposed mechanical system based on the main variables. The proposed internal combustion engine mechanism is shown in Fig. 1.

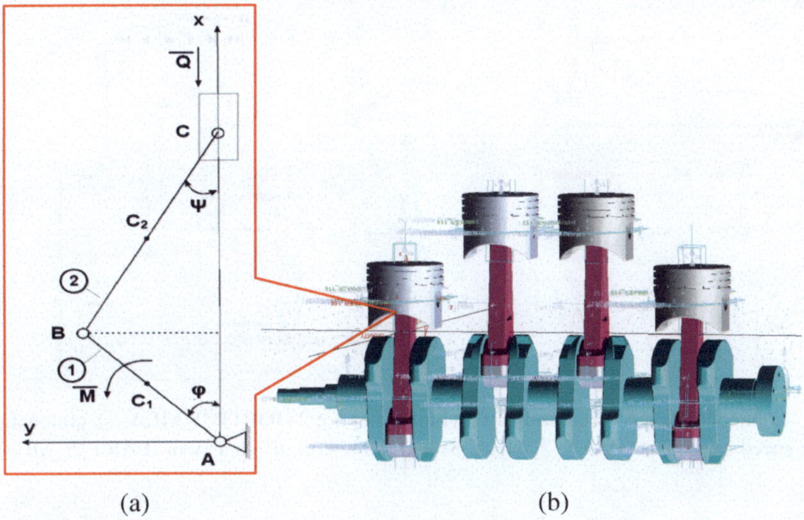

Fig. 1. (a) The kinematic scheme for a single cylinder (b) Virtual model of the internal combustion engine mechanism imported in MSC Adams

The design variables are represented by the coordinates of the corresponding points from Fig. 1, namely O, A and B. Thus, there will be obtained 9 design variables. The objective function is represented by decreasing the resultant connection forces which occur between connecting rod and crank shaft, as it can be remarked in Figs. 2 and 3.

The restriction function is represented by the connecting rod mechanism rotation angle or piston displacement. After main design variables identification, there will be started the proper optimization process based on an established number of iterations. The main task of this process is to research the optimal solution of the objective function by having in sight the variation domains of main design variables. Thus, there will be resulted the objective and restriction functions evolution, piston displacements after DV_4 variable modification, according with the design limits for three iterations, as it can be seen in Fig. 3.

The objective function is the resultant connection force decreasing between connecting rod and crank shaft (JOINT_2_MEA_1), and the restriction function is the connecting rod angular speed (PART_3_MEA_3).

By considering the obtained results plotted in Fig. 4, it can be remarked that the main design variables have been relatively modified from the initial mechanism phase. There are remarked small differences between initial and final variables values and we conclude that through minimal modifications brought through mechanism geometry it can be obtained an improved dynamic behavior of the analyzed mechanism. This certifies the optimization process. During successive iterations, the motion laws of the mechanism kinematic parameters were modified until the objective function is minimized. Thus, the final mechanism will decrease the objective function and in the same time there will be accomplished the defined design constraint.

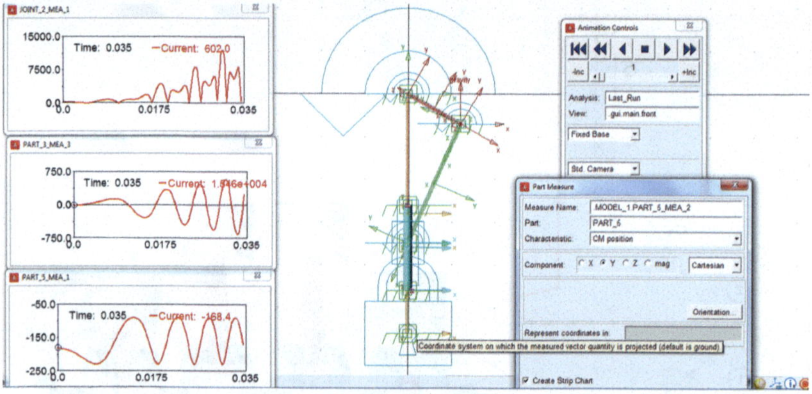

Fig. 2. Connection force variation from revolute joint no: 2 (JOINT_2_MEA_1), connecting rod angular speed on Z-axis (PART_3_MEA_3) and the piston displacement (PART_5_MEA_1).

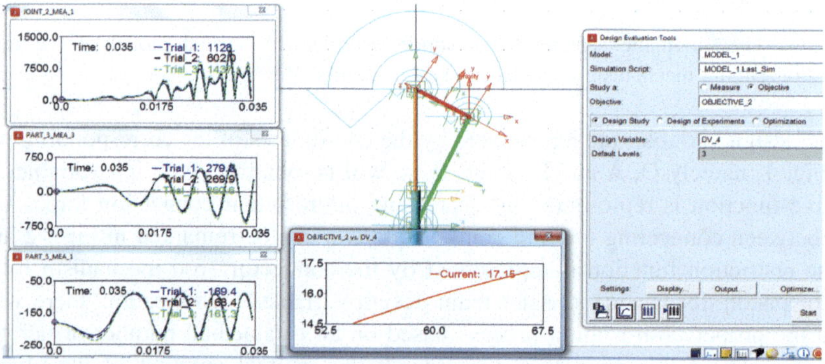

Fig. 3. Objective and restriction functions evolution, piston displacements after DV_4 variable modification, according with the design limits for three iterations.

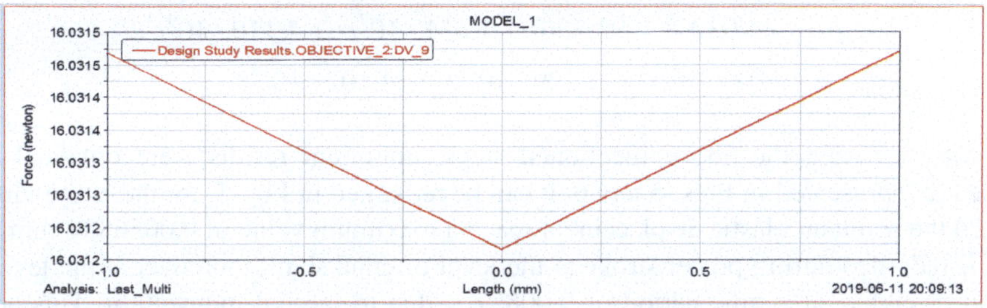

Fig. 4. Objective function for DV_9 variable evolution.

For each design variable it can be observed the behavior of the connection force variation during time between the connecting rod and crank shaft, but the connecting rod angular speed keeps the initial limits (in case of unoptimized mechanism). Also it can be remarked some variations of connection forces between piston and cylinder, during optimization process, but these gives insignificant modifications.

Another important argument was given by the design variables domains which these are characterized by a good evolution in the force interval of 14 to 18 N.

3 Mechanism Finite Element Modeling

For this stage it will be considered the same virtual model used in case of MSC Adams optimization process and for this it was created an importing interface for ANSYS V14.5. In order to perform the proposed mechanism dynamic analysis by using finite element method, it is necessary to perform the following steps

1. There will be imported the 3D model system, after this will be defined the kinematic joints by considering the elastic contact between elements, there will be introduced the material properties for each element, there will be meshed up through hexahedral finite element type the analyzed components and a structural transient setup analysis will be performed by introducing the contour conditions which are compatible with experimental tests.
2. Defining the pressure distribution laws for all four pistons and defining the proper parameters for dynamic analysis (direct method, integration step, control algorithm, etc.). Also there will be defined the local reference systems for monitoring the mechanical system motion in a dynamic mode.
3. There will be identified the variation laws during time of the mechanical system dynamic parameters or components, including stress, deformations, resultant displacements after combining the motions between rigid-solid with solid-deformable one. The input data are represented by the angular speed motion variation law in case of the crank shaft and the pressure distribution laws which were experimentally obtained [9]. These have the following mathematical expression:

$$p = -4.930 + 3203.386t - 1.631 \cdot 10^{7}t^{2} + 4.110 \cdot 10^{9}t^{3} \\ - 4.504 \cdot 10^{11}t^{4} + 2.229 \cdot 10^{13}t^{5} - 4.102 \cdot 10^{14}t^{6} \quad (1)$$

After crossing the above mentioned steps, important results were obtained and these are represented in Figs. 5 and 6. It can be remarked in Fig. 5, for the entire virtual model the resultant elastic displacement reach a maximum value of 0.0056834 mm and this is recorded during power stroke at the level of crankshaft, moreover at the level of counterweights, at a time period of 0.00857 s, due to inertial momentum. This argument is correlated with high values of equivalent von Misses stress obtained in case of wrist pins as it can be seen in Fig. 6. This reaches a maximum value of 19.902 MPa.

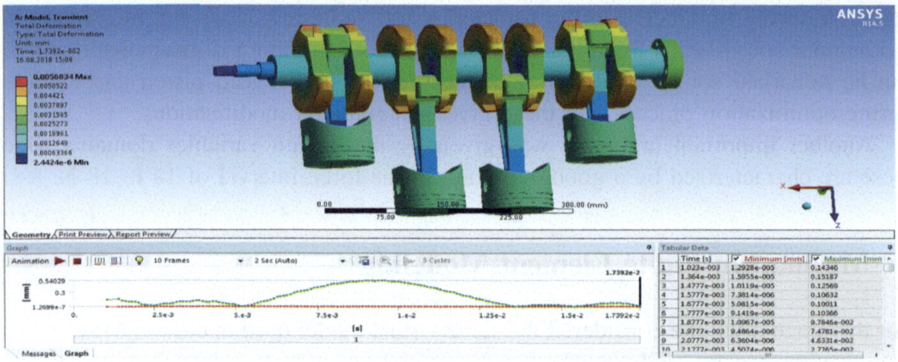

Fig. 5. Resultant elastic displacement variation laws [mm] vs. time [sec].

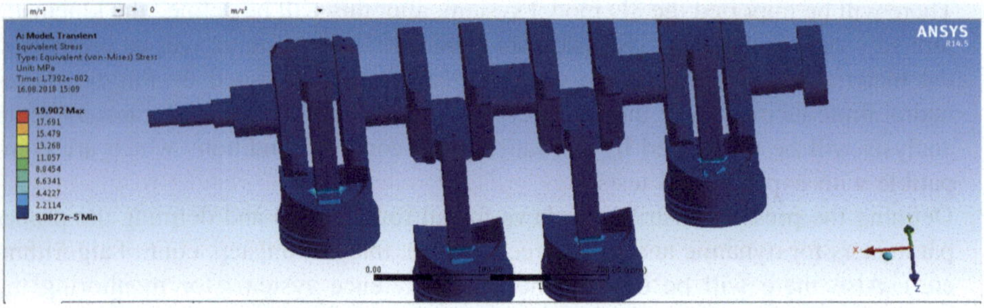

Fig. 6. Equivalent von Misses stress variation laws [MPa] vs. time [sec].

For the connecting rod dynamic mode optimization, a special attention was addressed to the connection forces distribution between connecting rod and crank shaft components. This is represented in Fig. 7. Thus, it can be remarked that in the connecting rod cap zone and crank shaft the connection forces reach a small value of 39342 N on x direction and −38532 N on y direction due to the stiffness of the analyzed component.

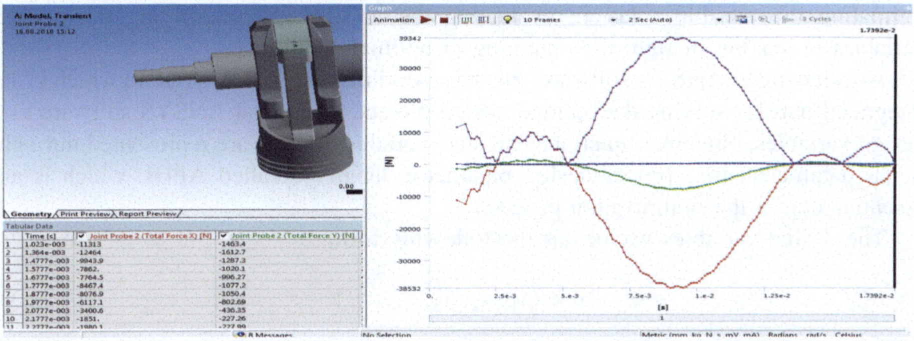

Fig. 7. Connecting force variation laws [N] from crank shaft bearing (connecting rod – crank shaft) vs. time [sec] – detailed view.

4 Connecting Rod Optimization Process

For the proposed internal combustion mechanism it is intended to perform a structural optimization process on the connecting rod component, with the aid of ANSYS Mechanical APDL software. This process will be performed only on the desired component. Thus, the reference elements which compose the optimization process are: objective function, design variables, constraints, approximation technique, optimization specific parameters (iterations number, tolerance). In general there are two optimization process types: form optimization and cross section optimization.

The structural optimization problem consists on accomplishing the following steps:

1. Formulating the functional requirements;
2. Choosing the proper topology;
3. Defining the design variables: x_1, x_2,…,x_n, and the answering ones: σ_1,…, σ_n;
4. Formulating the mathematical answer of the proposed system:

$$hj(x_1,\ldots,x_n) \geq 0\, j = 1, 2, .., t \tag{2}$$

5. Restriction identification with the following form: $g_i(x_1,\ldots,x_n) \geq 0 \quad i = 1, 2.$
6. Establishing criteria for evaluating the alternative solutions like: *f(x)*.
7. Choosing the initial solution x_0 for system analysis: $\sigma k_0 = K = 1,\ldots,t$.
8. Maintaining the restriction control.
9. Evaluating the *f(x)* function.
10. Evaluating the *F(x) = min* function.
11. Selective evaluation and elaborating the optimal solution.

Before the optimal solution evaluation it is necessary to formulate the optimal criteria through this was chosen. This criteria can be identified as the structure weight, production cost, but also can be defined any other parameter. Mathematically speaking, this criteria can be written as a function *f(x)* with the variables like: x_1, x_2,…, x_n. Thus, by crossing the above mentioned steps, the following problems will be solved: choosing the proper variables; system analysis through solving the proper mathematical equations for evaluating the system answer like: $\sigma_k = \sigma_k(x_1,\ldots, x_n)$ with: $k = 1, 2,\ldots, t$;

estimating the feasible solution; evaluating the f(x) criteria; modifying the proper variables by having in sight the obtaining of optimal solution.

As referring to ANSYS software, this has a optimization module (OPT) which is an integrated part for solving the optimal design project. In general ANSYS software use design variables, objective functions and state variables. These are represented through scalar parameters in a proper design parametric language called APDL which is an essential step in the optimization process.

The design variables vector has the following form:

$$x = (x_1, x_2, x_3, \ldots, x_n) \tag{3}$$

The design variables represent an important subject in case of a constraint with superior and inferior limits. In case of our model, there were defined the design variables like parameterized values of the connecting rod sectional elements (diameters, width and lengths, etc.). This component was analyzed by having in sight the elastic character and for this there were identified two major optimization problems: maximizing the connecting rod stiffness; minimizing the connecting rod volume and weight.

Thus, for the connecting rod optimization process, the following steps were accomplished: designing a parameterized connecting rod 3D model; contour conditions definition by creating a local coordinate cylindrical system for connecting rod loads modeling and applying constraints on the contact zone of the connecting rod – wrist pin – piston group; defining the inertial properties and material characteristics; identifying the variation diagrams for stress, deformations and displacements by monitoring the equivalent maximum von Misses stress; design variables definition and for each variable assigning the maximum and minimum limits; objective function definition, which in this case is represented by minimizing the connecting rod volume; state (restriction) function definition and this is represented by the maximum von Misses stress value.

In this case the design variables were represented by the interior and exterior radius for connecting rod closed hub and respectively the connecting rod cap. The aim of this optimization process is to obtain a minimum weight and an elastic form of the connecting rod with an optimal behavior from stress and deformation specific viewpoints. By following the above mentioned steps and running the optimal process of the proposed component, first set of results were obtained as von Misses stress distribution and resultant displacement which are shown in Figs. 8 and 9.

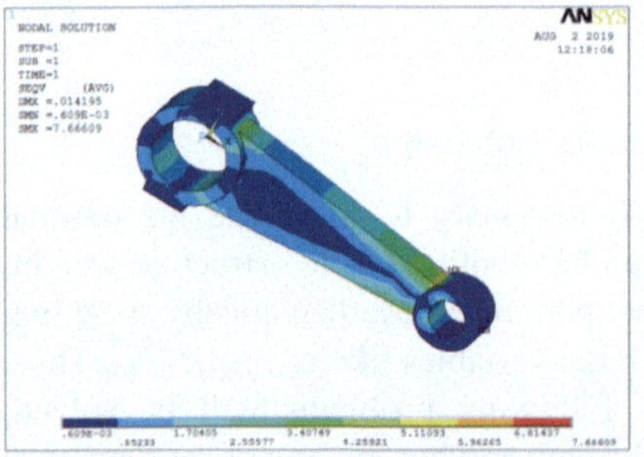

Fig. 8. Von Misses stress distribution.

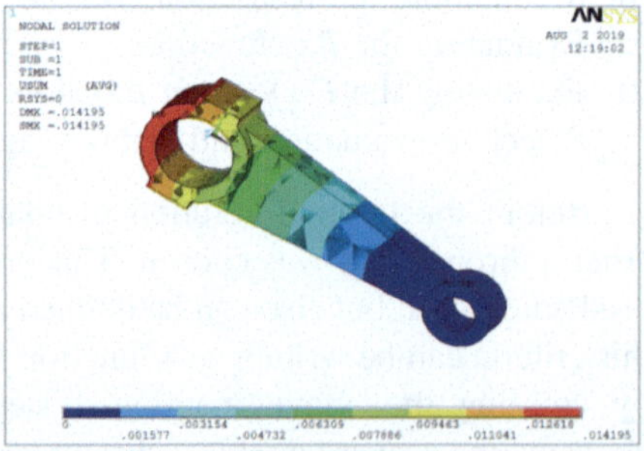

Fig. 9. Resultant displacement distribution.

By considering the reported results from Fig. 8 it can be observed a maximum von Misses stress value of 7.66609 MPa, before the optimization process. Also it was recorded a resultant displacement of 0.014195 mm in case of connecting rod cap segment as it can be seen in Fig. 9. These results represent input data for optimization process.

After optimization processing sequence, which was made for 30 iterations, important data were obtained and these are represented through a set of diagrams which are shown in Figs. 10, 11, 12, 13, 14 and 15. After connecting rod sectional optimization process, it can be remarked important equivalent von Misses stress reduced values from 7.66 MPa to 1.78 MPa according with Figs. 8 and 10. The same phenomena happens in case of elastic displacements which records reduced values from 0.014 mm to 0.0083 mm as it can be seen in Figs. 9 and 11.

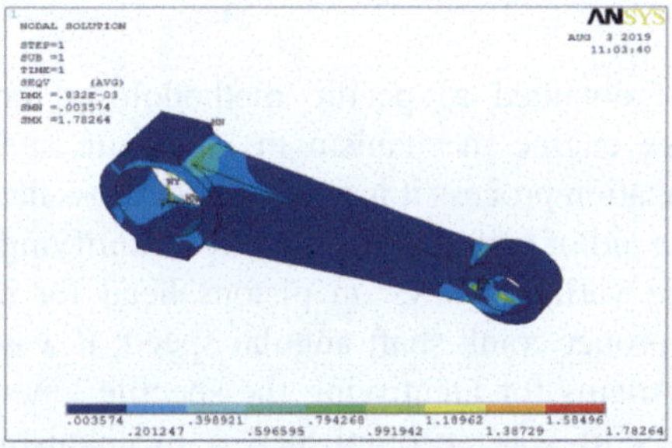

Fig. 10. Equivalent Von Misses stress distribution after optimization process.

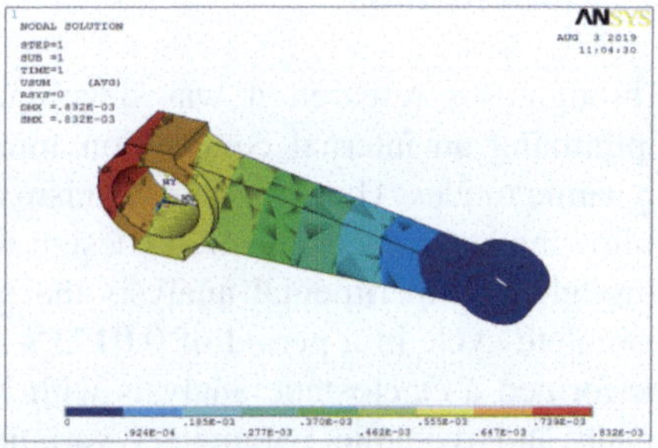

Fig. 11. Resultant displacement distribution after optimization process.

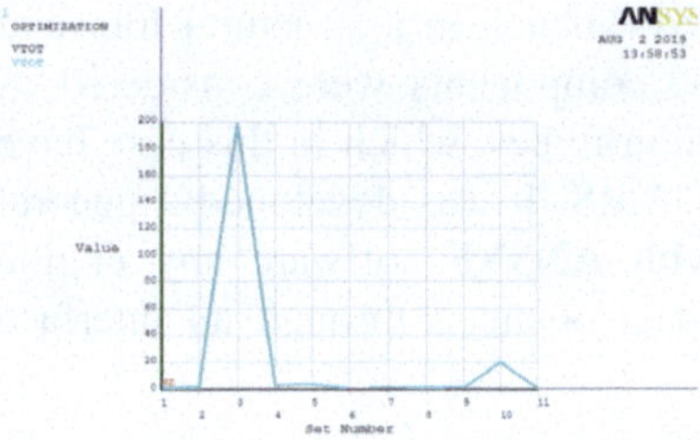

Fig. 12. Total volume depending on the design variables.

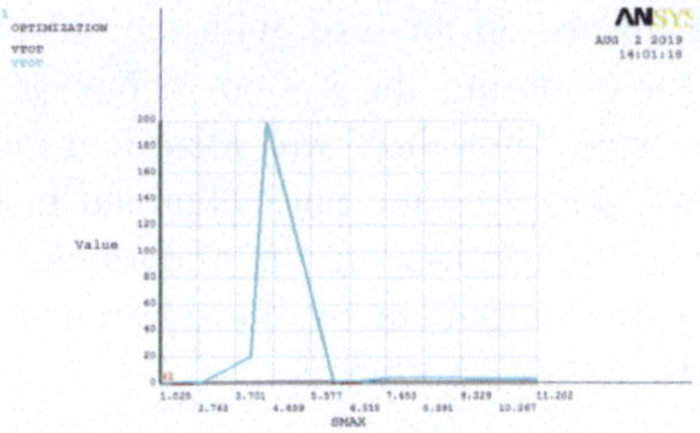

Fig. 13. Total volume depending on the maximum stress.

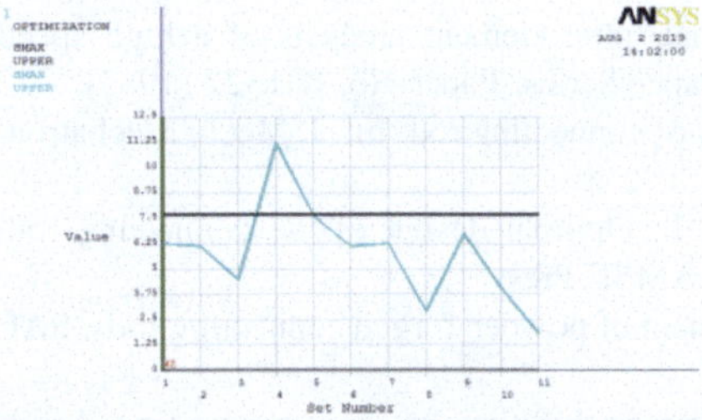

Fig. 14. Maximum stress depending on the design variables.

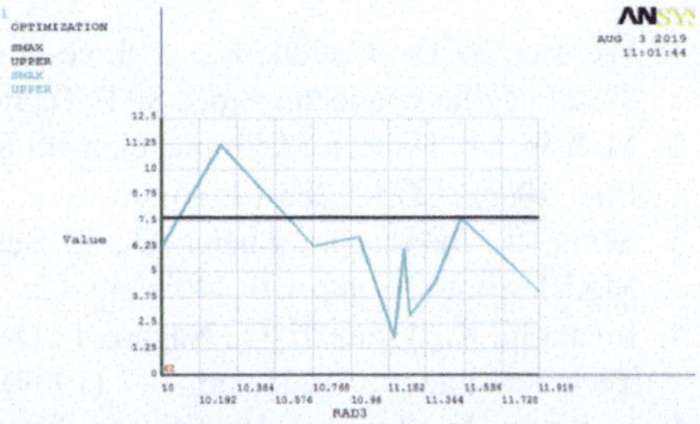

Fig. 15. Maximum stress depending on the connecting rod closed hub radius.

According with the obtained results, there are represented the objective function and state variable evolution as function of the iterations number and each design variable. Thus, in Figs. 12 and 14 it can be remarked the total volume evolution and obviously the connecting rod weight respectively the maximum equivalent von Misses stress evolution as function of the iteration number. Another results is reported in Fig. 13 which is represented the functional dependence between total volume and maximum stress during optimization process. A major remark for the obtained plots is the one regarding the von Misses stress variation in a large limit when there are made major radius values modifications in the area of closed hub and connecting rod cap zones, as it can be remarked in Fig. 15.

5 Conclusions

Through this research it was developed and obtained a specific methodology for optimizing an internal combustion four stroke engine mechanism in kinematic and dynamic modes. Thus, for a mechanism optimization process it is necessary to cross the following steps: mechanism 3D design with the aid of a modeling software; identifying through an experimental analysis the pressure variation laws on pistons head for a complete cycle in a period of 0.0173 s and a proper crank shaft angular speed; it was performed a cinetostatic analysis with MSC Adams for identifying the specific kinematic and dynamic parameters variation laws; it was accomplished a mechanism optimization process with the aid of MSC Adams by considering the kinematic elements as rigid ones with the aim of minimizing the joints connection forces.

By having in sight the same input data (angular speed and pressure piston head) it was performed an analysis with the aid of AnsysWorkbench in a structural transient mode for obtaining the system behavior when the components were considered as flexible ones. Thus there was identified the connection forces, which in this case have superior values like the ones obtained in MSC ADAMS. It was developed a general procedure for optimizing in a dynamic mode with ANSYS software any engine mechanism component by considering the input data obtained through an interface ADAMS-ANSYS software.

References

1. Webster, W.D., Coffell, R.: A three-dimensional finite element analysis of a high speed diesel engine connecting rod. SAE Technical Paper Series, Paper No. 831322 (1983)
2. Mohsin, M., Haneef, M.: Finite element analysis of connecting rod. Int. J. Mech. Mechatron. Eng. **10**(7), 1278–1283 (2016)
3. Serag, S., Sevien, L., Sheha, G., El-Beshtawi, I.: Optimal design of the connecting rod. Model. Simul. Control B **24**(3), 49–63 (1989). AMSE Press
4. Imahashi, K., Tsumuki, C., Nagare, I.: Development of powder-forged connecting rods. SAE Technical Paper 841221, pp. 1–7 (1989)
5. El-Sayed, M., Lund, E.H.: Structural optimization with fatiguelife constraints. Eng. Fract. Mech. **37**(6), 1149–1156 (1990)

6. Sarihan, V., Song, J.: Optimization of the wrist pin end of an automobile engine connecting rod with an interference fit. J. Mech. Des. Trans. ASME **112**, 406–412 (1990)
7. Athavale, S., Sajanpawar, P.R.: Studies on some modeling aspects in the finite element analysis of small gasoline engine. SAE Technical Paper Series, No. 911271 (1991)
8. Sonsino, C.M., Esper, F.J.: Fatigue design for PM components. European Powder Metallurgy Association EPMA (1994)
9. Norton, R.L.: Design of Machinery. McGraw Hill, New York (2008)
10. Juvinal, R.C.: Fundamentals of Machine Component Design. Willey, Hoboken (1999)
11. Dumitru, N., Ploscaru, C., Otat, O.: Sisteme mecanice mobile. Prototipare virtuala si analiza experimental. Universitaria Publishing house, Craiova (2018)

Experimental Stand to Evaluate Engine Mass Air Flow (MAF) Sensor

Dragos Tutunea(✉), Ilie Dumitru, Laurentiu Racila, Oana Otat, Ionut Geonea, Lucian Matei, and Cristina Rotea

Faculty of Mechanics, University of Craiova, Street Calea Bucuresti nr. 107, 200512 Craiova, Dolj, Romania
dragostutunea@yahoo.com

Abstract. In this paper an experimental setup was designed and realized in laboratory to test and measure the engine mass air flow sensor used in engine applications. The MAF sensor is installed usually between air filter and the intake manifold of the engine and measure the amount of air entering the engine or the air flow. A contaminated and failed mass air flow sensor shows a lower or higher air flow readings causing various drivability problems. The stand simulates the condition of function of MAF sensor by various air flows and temperature recording the flow by a pitot tube.

Keywords: Mass air flow · Air/fuel ratio · Engine Control Unit

1 Introduction

Nowadays the majority of engine cars are complex with a high number of sensors and actuators. This uplifts the efficiency, performance and safety of engines. The number of sensors used in the automotive industry are increasing to meet requirements of emissions regulations and optimum fuel consumption [1]. The Engine Control Unit (ECU) controls the main parameters of different subsystems of automobile engine gathering information's from sensor which detect coolant temperature, intake air temperature, exhaust gas oxygen concentration, manifold absolute pressure (MAP), mass air flow (MAF) rate, throttle position, crankshaft angular position/RPM and vehicle speed [2]. The MAF sensor is one of the most important sensor in the engine used in electronic fuel injection system to measure the air mass which is delivered to the engine. The signal allows ECU to make corrections in the quantity of air and fuel to be injected in the combustion chamber to have the correct Air-Fuel ratio [3]. The desired AFR ratio reduce the fuel consumption and emissions of combustion gases. The response time of hot-film/hot-wire mass air flow (MAF) sensors is fast but is affected by dust and dirt particles which pass through air filter destroying the thin hot-wire [4]. Hot-wire sensors utilize an electrical current to heat the assembly (wire and resistor) mounted in the intake air system. Variations in the flow of the intake air will change the values of amperage of current according to the cooling effects of the atmospheric air which pass to the heated wire. Than the MAF sensor converts to change of amperage in to a variable voltage that is send to ECU to indicate the value of air flowing into the

I. Dumitru et al. (Eds.): SMAT 2019, *The 30th SIAR International Congress of Automotive and Transport Engineering*, pp. 26–30, 2020.
https://doi.org/10.1007/978-3-030-32564-0_3

engine [5]. In this paper is evaluated the accuracy and performance of MAF sensors used in automotive applications.

2 Methods

The experimental stand was designed and built in the laboratory of Elements of Mechatronics of Faculty of Mechanics of Craiova. The scheme of principle of the stand is shown in Fig. 1. The stand use the fallowing equipment's:

- Termo Anemo Manometer MP 200 with Pitot tube;
- AX-3005DS Laboratory power supply;
- AX-18B multimeter;
- Experimental stand power supply 5 and 12 V;
- Digital thermometer −50 °C/+70 °C, TERM-722;
- Ammeter and voltmeter.

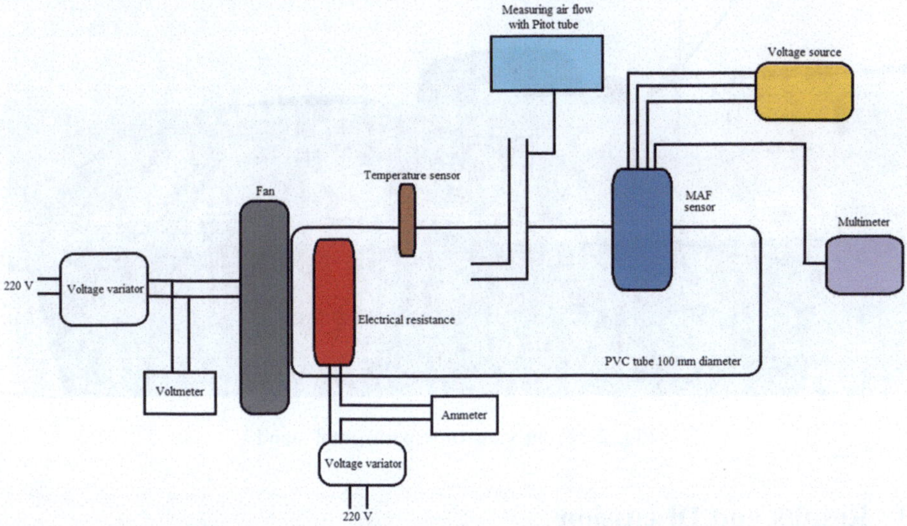

Fig. 1. Scheme of principle of MAF experimental stand

The testing procedure is composed by:

- The voltage supply (5 V, 12 V) is feed with 220 V;
- The laboratory power supply AX-3005DS is feed with 220 V;
- Feed the fan and electrical resistance with 220 V;
- The pins of MAF sensor are connected with voltage source;
- The voltage variators adjust the fan speed and temperature of the electrical resistance.

- Using the Thermo-Anemometer Manometer MP 200, the air flow inside the 100 mm diameter tube is measured by inserting a Pitot tube in the center at a distance of 200 mm;
- The variation of the air flow and its temperature is recorded by the MP 200 and the auxiliary temperature sensor;
- The air flow recorded by the MP 200 is correlated with the voltage generated by the pin of the MAF sensor.

  In the Figs. 2 and 3 is presented the final assembly of the experimental stand to evaluate engine mass air flow (MAF) sensor.

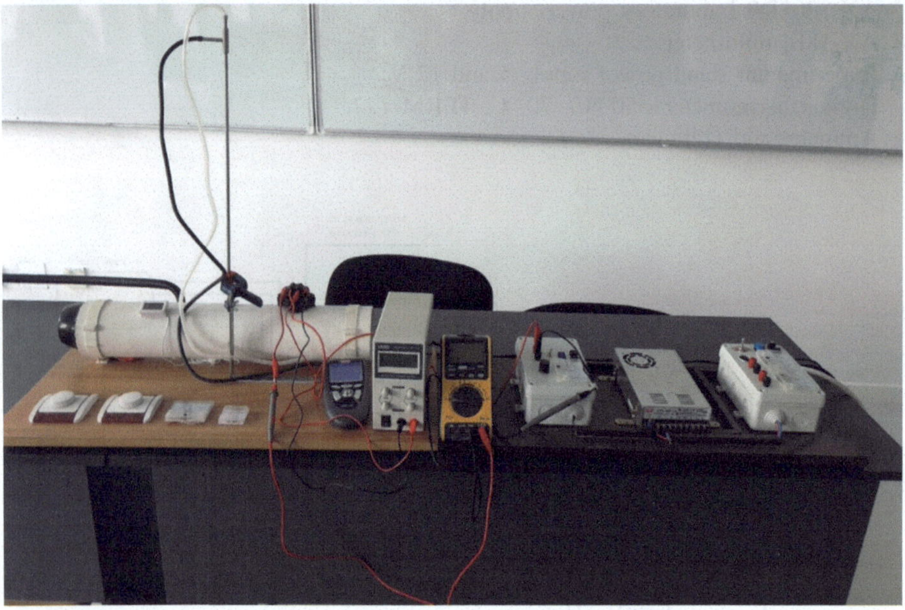

Fig. 2. Front view of experimental stand

3 Results and Discussion

In the Fig. 4 is presented the variation of the voltage generated at the MAF sensor pin function of the air flow recorded by the Pitot tube of Termo Anemo Manometer MP 200. The curve variation is similar as a polynomial function of third order. The variation of voltage is limited at lower air flow and higher at high air flows. The experimental stand can be used to evaluate different MAF sensors. When the engine is idling a reduce amount of air pass around the hot wire (a low electric current is necessary to keep the wire hot) but at high loads the air cools the hot wire and more electrical current is needed to maintain it hot.

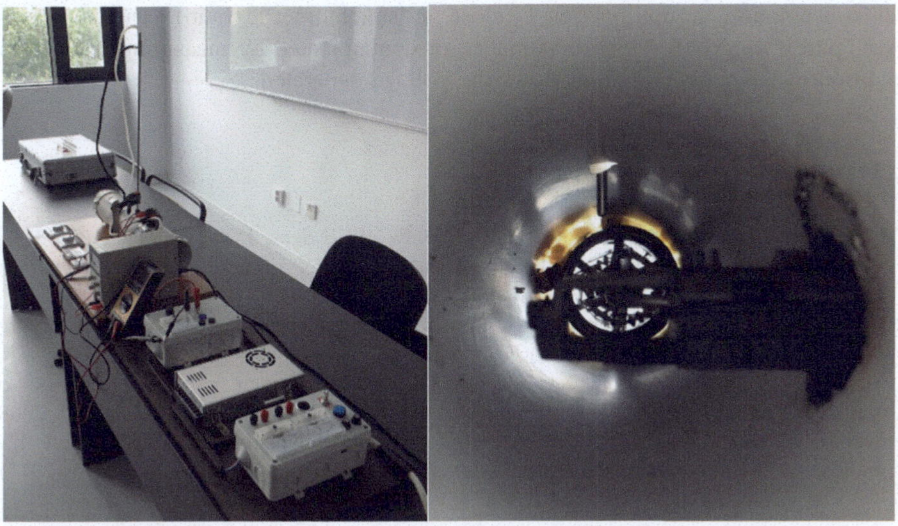

Fig. 3. Lateral view of experimental stand and MAF sensor

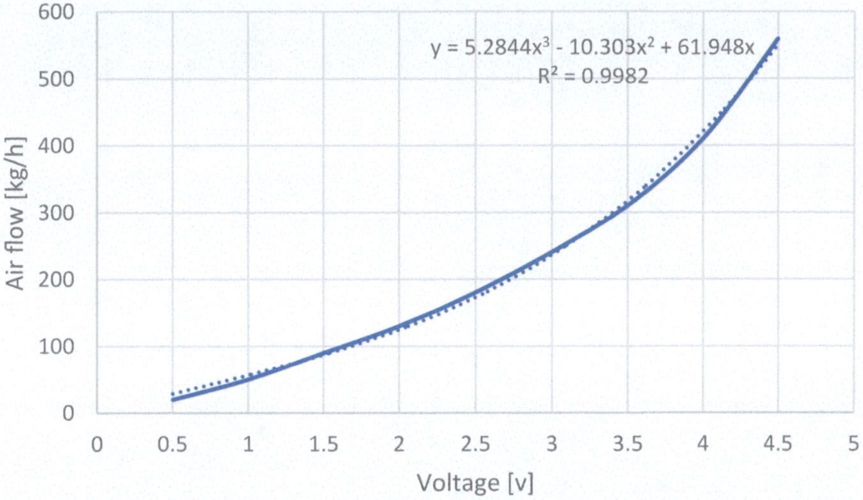

Fig. 4. Experimental values output of air flow (MP 200) and voltage of MAF sensor

4 Conclusion

In this paper an experimental stand was built to evaluate MAF sensors for automotive applications. The MAF sensor is one of the key components of electronic fuel injection system located between the air filter and the intake manifold of the engine. The ECU use the voltage air flow signal to calculate Air-Fuel Ratio to maintain the emissions and fuel consumption at optimal level (close to stoichiometric ratio). The stand can investigate if the MAF sensor is damaged or contaminated by measuring in different condition of temperature and air flows the voltage generated.

References

1. Fleming, W.J.: New automotive sensors—a review. IEEE Sens. J. **8**, 1900–1921 (2008)
2. Denton, T.: Book on Automobile Electrical and Electronic, 3rd edn. Elsevier Butterworth-Heinemann, Oxford (2004)
3. Powell, J.D., Fekete, N.P., Chang, C.F.: Observer-based air fuel ratio control. IEEE Control Syst. **18**(5), 72–83 (1998)
4. Xua, K.-J., Zhanga, J., Wanga, X.-F., Tengc, Q., Tanc, J., Zhanga, Y.-Y.: Improvements of nonlinear dynamic modeling of hot-film MAF sensor. Sens. Actuators, A **147**, 34–40 (2008)
5. Sazhin, O.: Novel mass air flow meter for automobile industry based on thermal flow microsensor I Analytical model and microsensor. Flow Measur. Instrum. **30**, 60–65 (2013)

Theoretical and Experimental Research on the Polluting Emissions for an Middle Class Vehicle Equipped with Engine that Uses the Double Energy Source, Gasoline and LPG

Valentin Nişulescu[1(✉)], Florian Ivan[2], Daniel Iozsa[1], and Gheorghe Bancă[1]

[1] Politehnica University of Bucharest, Bucharest, Romania
nisulescu.valentin@yahoo.com
[2] University of Piteşti, Piteşti, Romania

Abstract. Considering the severe conditions for reducing the chemical pollution caused by the vehicles imposed by regulations and rules worldwide, the liquefied petroleum gas becomes an attractive alternative fuel from an ecological and economical point of view. At present, there is almost no car manufacturer that does not include in the fabrication scheme spark ignition engines that use LPG as an alternative source of energy knowing the benefits of the GPL in terms of reducing pollutant emissions.

In the first part of the paper, using the AMESim simulation tool, for a middle - class vehicle, equipped with spark-ignition engine that uses as energy sources gasoline and LPG, the pollutant emissions of CO, HC and NOx were simulated. The simulation using the WLTC test cycle, was performed for both energy sources. Also using the simulation, the CO_2 was determined, based on which can be drawn conclusions regarding the fuel consumption.

The second part of the paper contains a comparative analysis between the simulated and measured values regarding the pollutant emissions and fuel consumption. The measured values, for the same vehicle, were obtained on the dynamometer using the WLTC test cycle.

Keywords: WLTC · LPG · Pollution · Economical · Ecological · Simulation

1 Introduction

Current studies indicates a significant increase regarding the interest of vehicle manufacturers but also of the users in terms of liquefied petroleum gas, at present the LPG production and distribution networks being highly developed in the community, extra-community and worldwide [1].

Also at european and worldwide level the legislative concerns of international authorities to reduce pollutant emissions from combustion processes in thermal engines are increasing in severity, determine the car manufacturers to quickly comply with these pollution limitation requirements, regardless of the main energy source or alternative used on the vehicle. The European trend also taking into account the

I. Dumitru et al. (Eds.): SMAT 2019, *The 30th SIAR International Congress of Automotive and Transport Engineering*, pp. 31–41, 2020.
https://doi.org/10.1007/978-3-030-32564-0_4

legislative limitations of chemical pollution, indicates the increase in the number of manufactured cars using gasoline engines at the expense of Diesel engines which may have a factory-fitted LPG system as an alternative energy source [2].

Currently, the European method for testing pollutant emissions for vehicles is represented by the WLTC Cycle (Worldwide harmonized Light-duty vehicles Test Cycle) Fig. 1, which starting with 1 september 2017 replaced the old test cycle NEDC (New European Driving Cycle) in vigor since 1992.

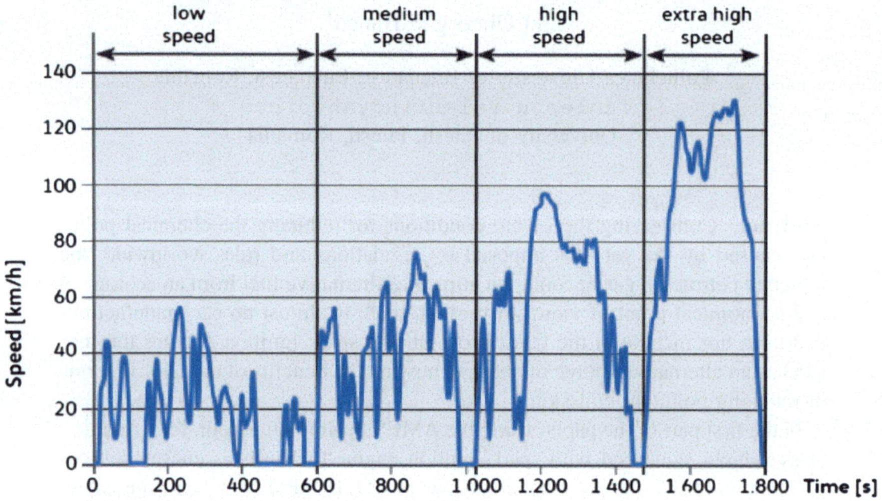

Fig. 1. WLTC cycle (worldwide harmonized Light-duty vehicles test cycle) [5]

After the introduction of the WLTC cycle, a one-year transition period was also possible for vehicles already type-approved on the old test cycle. In the case of the WLTC cycle, the test is performed on the dynamometer.

The technical characteristics of the WLTC test cycle are as follows:

- Driving time: 30 min;
- Stationary time proportion: 13;
- Route length: 23 km;
- Maximum speed: 131 km/h;
- Average speed: 47 km/h;
- Optional equipment taken into consideration;
- Shift points defined individually according to different number of gears in the different vehicles.

The speeds in the individual ranges are:

- Low speed: 56,5 km/h maximum speed and 18,9 km/h average speed
- Medium speed: 76,6 km/h maximum speed and 39,5 km/h average speed
- High speed: 97,4 km/h maximum speed and 56,6 km/h average speed
- Extra high speed: 131,3 km/h maximum speed and 92,0 km/h average speed.

2 Experimental

In the present study, starting from the WLTC test cycle were simulated and measured the CO, HC, NOx, CO_2 pollutant emissions of a medium-class vehicle which uses the dual energy source gasoline and LPG. The gasoline and LPG injection systems are the type - indirect injection, naturally aspirated engine, euro 6 pollution standard. The simulation was realized using the mathematical platform AMESim. The proposed AMESim model can be identified in the Fig. 2, with the choice of a middle class car having the following components:

- middle class vehicle;
- the driver component to ensure the control of the vehicle while running the WLTC cycle;
- manual gearbox with ratios parameterized according to the model;
- a spark-ignition engine, 4 cylinders, 1200 cm^3, indirect injection, naturally aspirated;
- three-way catalyst that simulates the exhaust gas treatment, system specific for a spark ignition engine, euro 6 standard;
- engine control unit to simulate the engine's operating strategy.

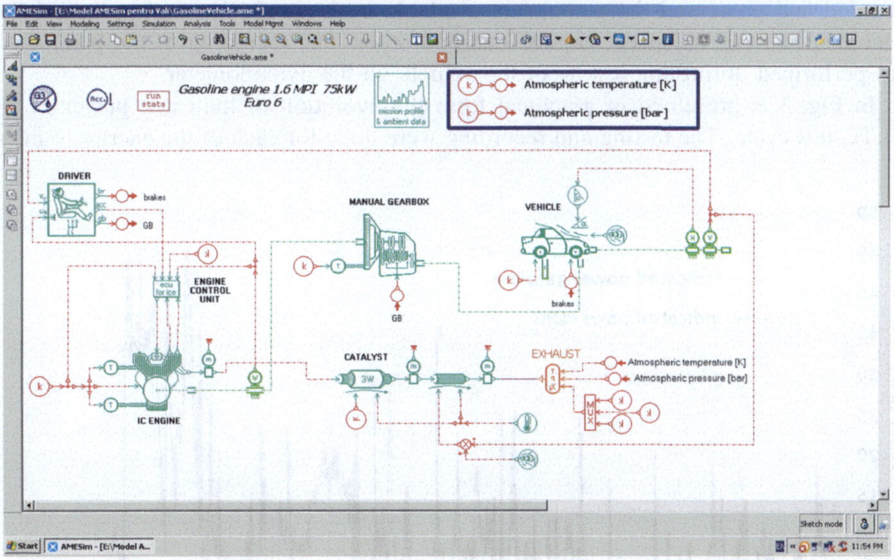

Fig. 2. AMESim model, spark ignition engine [4]

For the analysis of pollutant emissions and fuel consumption, the graphs resulting from the AMESim simulation were used, from where it extracted the final values for the masses of CO_2, CO, HC, NOx pollutants in order to perform a comparative analysis of these with the results obtained by direct measurement. The direct measurement of pollutant emissions was done on the dynamometer. The simulation and the measurement were performed for both energy sources; the resulting values are compared by making a percentage comparison, and indicated using graphs.

In terms of dynamic performances power and torque (on the WLTC cycle), these were determined during the test of pollutant emissions on the dynamometer by direct recording and processing of data from the engine control unit.

The control of the vehicle on the dynamometer was performed by the test engineer. The vehicle functions controlled by the test engineer were the vehicle speed and gears control. Before the test phase the vehicle was maintained for a period of 12 h at a constant temperature of 20 °C.

The measurement of the pollutant emissions on the dynamometer was performed in accordance with the Regulation No 83 of the Economic Commission for Europe of the United Nations (UN/ECE) - Uniform provisions concerning the approval of vehicles with regard to the emission of pollutants according to engine fuel requirements.

3 Results and Discussions

3.1 Indicated Power and Torque

In terms of dynamic performances, to obtain the indicated power and torque, were recorded and processed the parameters available in the engine control unit, such as: vehicle speed, engine speed, engine load and other. The recording of the parameters was performed during the testing of the vehicle on the dynamometer.

In Fig. 3 is presented in graphical form the evolution of indicated power in the WLTC test cycle. The testing and recording were done for each of the energy sources.

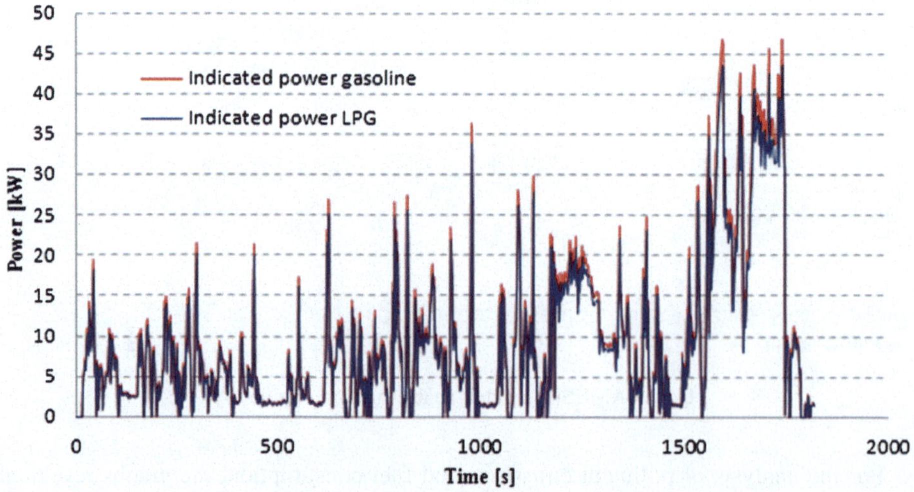

Fig. 3. The indicated power gasoline/LPG on the WLTC cycle

It can easily be noticed that the graphical form of power respects the speed profile of the test cycle which indicates the correct driving of the vehicle. After analyzing the values, there were differences between the two energy sources of about 7% in favor of the evolution using the gasoline. The gears shifting were carried out as described in test method in accordance with the Regulation No 83 of the Economic Commission for Europe of the United Nations (UN/ECE).

In Fig. 4 is presented in graphical form the evolution of indicated torque in the WLTC test cycle. And in this case the graphical form of torque respects the speed profile of the test cycle, with the difference of 7% higher in the case of gasoline. Generally from studies conducted so far, the power and torque indicated may differ by approximately 5% in relation to the real values obtained on the dynamometer [3].

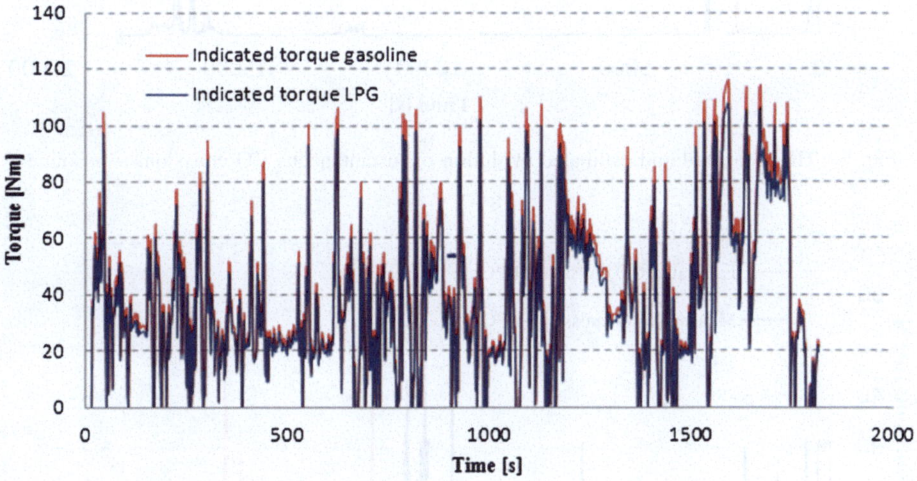

Fig. 4. The indicated torque gasoline/LPG on the WLTC cycle

3.2 CO Emissions

In Fig. 5 are displayed the instantaneous emissions as a result of modal measurement and estimation. The recordings on the dynamometer and modeling were realized at the 10 Hz frequency.

In the Fig. 6 is presented the evolution of CO emissions in the case of LPG test. After the analysis of the two graphs, it can be observed that the significant CO emissions are present in the cold start phase and also in the in the very high speed phase. Following the simulation and measurement, the cumulative CO emission values are as follows: gasoline estimated CO emissions 290.18 [mg/km]; gasoline measured CO emissions 346.41 [mg/km]; LPG estimated CO emissions 251.61 [mg/km]; LPG measured CO emissions 251.23 [mg/km].

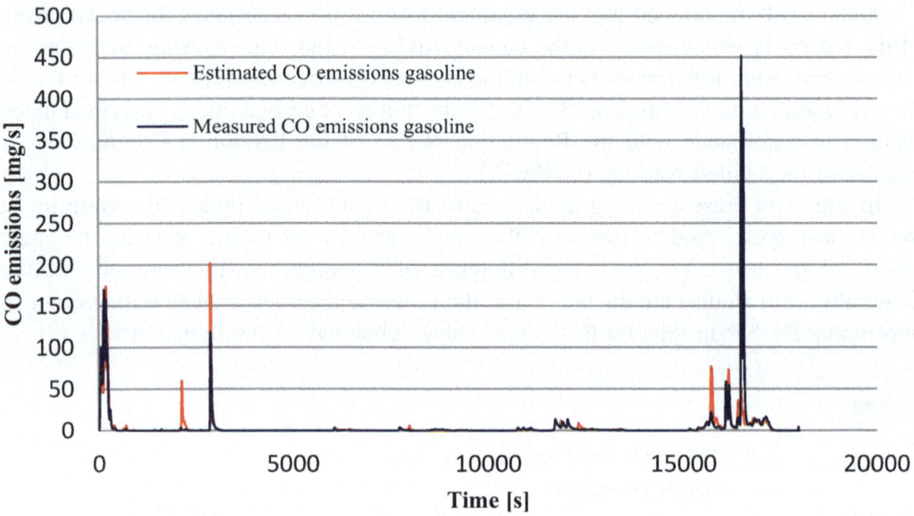

Fig. 5. The measured and estimated evolution of instantaneous CO emissions – gasoline

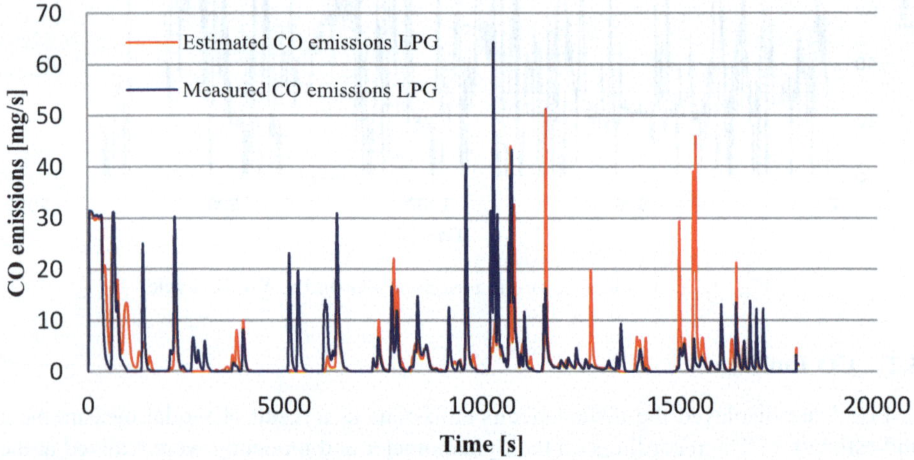

Fig. 6. The measured and estimated evolution of instantaneous CO emissions – LPG

3.3 HC Emissions

In the Fig. 7 is presented the evolution of HC emissions in the case of test using gasoline. It can be observed that the significant HC emissions are present in the cold start phase and remains at a high level until the engine is warmed up. From the analysis of the two graphs we can draw the concussion that even in the case of LPG the pollutant follows the same trend.

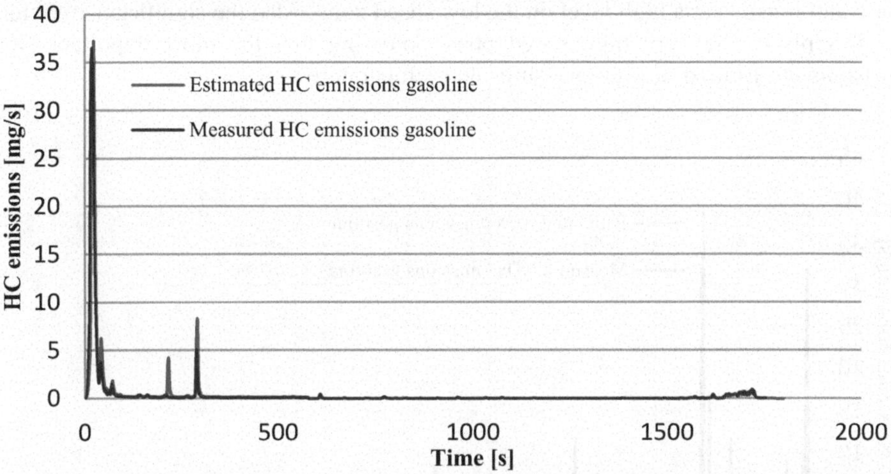

Fig. 7. The measured and estimated evolution of instantaneous HC emissions – gasoline

In the Fig. 8 is presented the evolution of HC emissions in the case of test using LPG. From simulation and measurement, the cumulative HC emission values are as follows: gasoline estimated HC emissions 32.56 [mg/km]; gasoline measured HC emissions 29.29 [mg/km]; LPG estimated HC emissions 24.29 [mg/km]; LPG measured HC emissions 29.55 [mg/km].

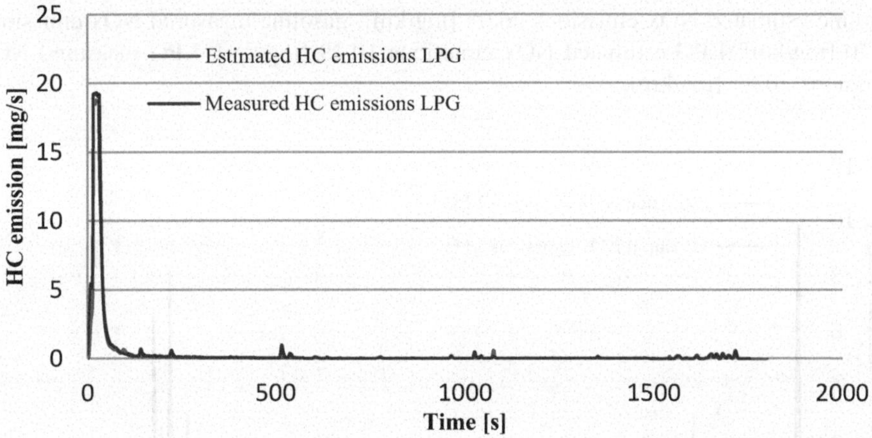

Fig. 8. The measured and estimated evolution of instantaneous HC emissions – LPG

3.4 NOx Emissions

In the Fig. 9 is presented the evolution of NOx emissions in the case of test using gasoline. It can be observed that the significant emissions are present after the cold start phase and remains at a high level on the low speed zone. Also the significant emissions are presents in the very high speed phase knowing that the most important NOx emissions are formed at high pressures and temperatures.

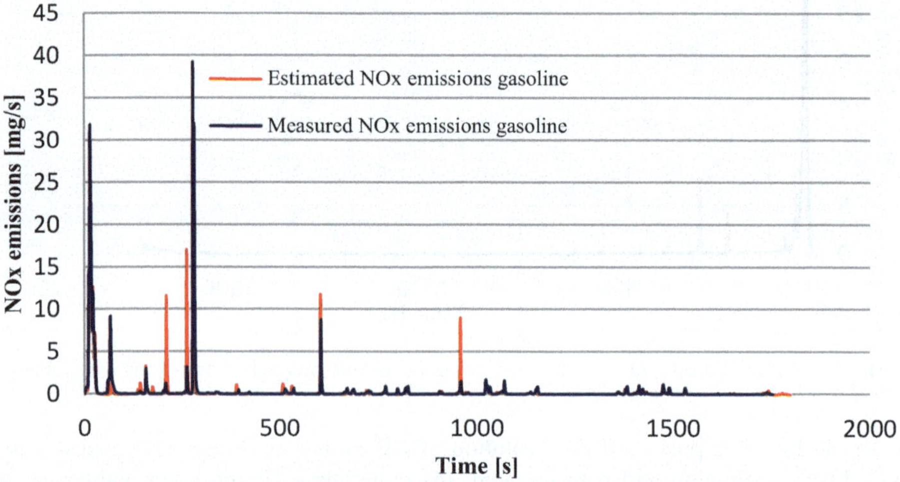

Fig. 9. The measured and estimated evolution of instantaneous NOx emissions – gasoline

In the Fig. 10 is presented the evolution of NOx emissions in the case of LPG test. From simulation and measurement, the cumulative HC emission values are as follows: gasoline estimated NOx emissions 26.08 [mg/km]; gasoline measured NOx emissions 27.70 [mg/km]; LPG estimated NOx emissions 10.29 [mg/km]; LPG measured NOx emissions 10.53 [mg/km].

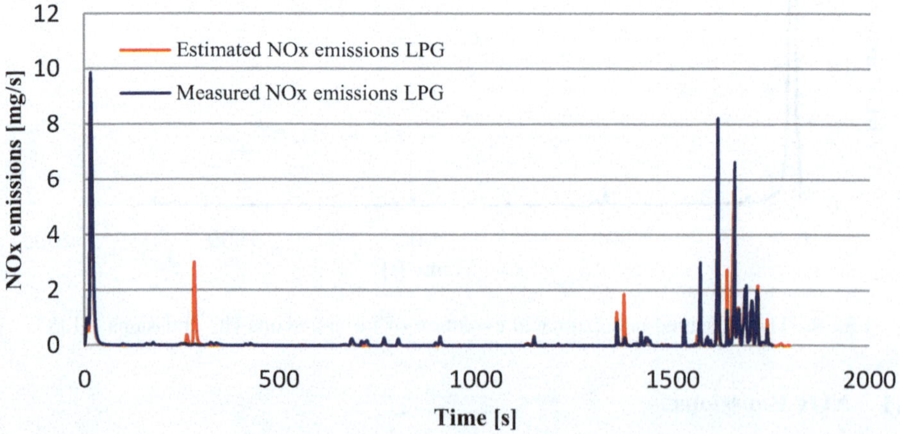

Fig. 10. The measured and estimated evolution of instantaneous NOx emissions – LPG

3.5 CO_2 Emissions

The evolution of CO_2 emissions in the case of gasoline test is presented in the Fig. 11. The variations of CO_2 indicate the results of fuel consumption for the two energy sources. It is noticeable that the graphical form of CO_2 emissions is dependent with the vehicle speed for the both energy sources, the simulation and measurement results where comparable.

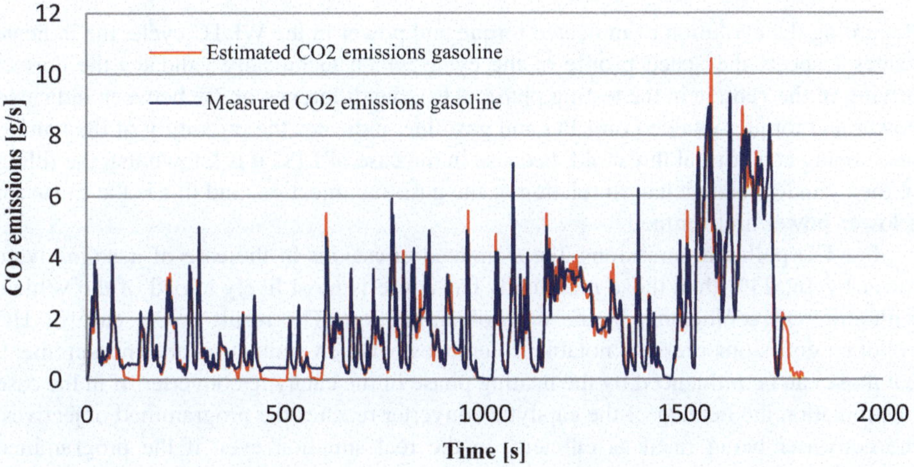

Fig. 11. The measured and estimated evolution of instantaneous CO_2 emissions – gasoline

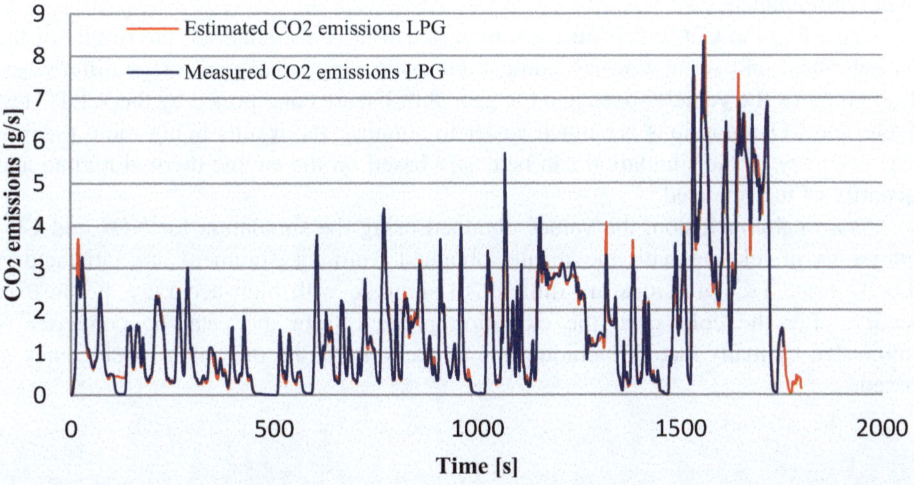

Fig. 12. The measured and estimated evolution of instantaneous CO_2 emissions – LPG

The evolution of CO_2 emissions in the case of LPG test is presented in the Fig. 12. After simulation and measurement, the cumulative CO_2 emission values are as follows: gasoline estimated CO_2 emissions 137.59 [g/km]; gasoline measured CO_2 emissions 143.13 [g/km]; LPG estimated CO_2 emissions 117.95 [g/km]; LPG measured CO_2 emissions 118.32 [g/km].

4 Conclusions

Regarding the evolution of indicated torque and power in the WLTC cycle, the indicate values respects the speed profile of the cycle which technically validates the correct driving of the vehicle in the testing phase. Also, the difference of 7% between indicated power and torque obtained on LPG and gasoline, indicates the proximity of the values, these being at a normal threshold, because in the case of LPG it is known that the filling of the cylinders is deficient in relation to the gasoline injection, and this is the cause for a lower power and torque.

For CO pollutant emissions, the estimated emissions in the case of gasoline were lower by 16, 23% than those measured. The cause is most likely a load of the vehicle with different coefficients in the case of simulation. The result in the case of HC pollutant emissions presents notable differences between simulation and measurement, but these can be influenced by the heating phase of the catalytic converter. If in the case of simulation the heating of the catalytic converter reaches the programmed objectives, the converter being rated as efficient, in the real situation even if the programmed objectives are achieved, the catalytic efficiency may be deficient with possible delays.

In the case of NOx pollutant emissions, for both energy sources, the results of the measurement and simulation are comparable, with small percentage differences. Considering the conditions for the formation of NOx emissions, under load, at high pressures and temperatures, it is understood why the results introduce small differences, because the real thermal regime is reached and the catalytic converter is in the area of maximum efficiency.

Regarding the CO_2 emissions, a similar situation is encountered, the results of the measurement and simulation are comparable, with very small percentage differences. Knowing that the vehicle speed and the gear shift list are conditioned by the WLTC test cycle, the CO_2 emissions are much easier to simulate the results being more precise. The accuracy of the simulation can be easily based on the engine thermal regime and quantity of fuel injected.

As a final conclusion, the values obtained using the simulation for NOx and CO_2 emissions in relation with the results obtained from measurement are satisfactory. The CO and HC emissions are difficult to simulate with high accuracy, because in reality, after the cold start, the oxidation process from the catalytic converter is influenced by many factors including the temperature based on which its efficiency is based.

References

1. Patil, P.N., Deshmukh, D.S., Patil, J.G.: A review on liquid petroleum gas as a fuel in spark ignition engine application. In: International Conference on Global Trends in Engineering, Technology and Management, ICGTETM-2016, pp. 124–128 (2016). ISSN 2231-5381
2. Mamidi, T., Suryawnshi, J.G.: Investigations on SI engine using liquefied petroleum gas (LPG) as an alternative fuel. Int. J. Eng. Res. Appl. **2**, 362–367 (2012). ISSN 2248-9622
3. Ivan, F., Lita, D., Busoi, A.: Methods and means of depollution of the engines for vehicles (2014). ISBN 978-606-25-0038-2
4. Help AMESim Rev12
5. https://www.group-training-online.com/logon.jsp. Accessed 17 May 2019

Analysis of the Injector for Common Rail System, Including of the Electrical Command Signal

Cristian-Ioan Leahu(✉)

Transilvania University of Brasov, 500036 Brasov, Romania
leahu.cristian@unitbv.ro

Abstract. Modern diesel engines are equipped with a Common Rail injection system. The injector is probably the most complex component in this system. To facilitate Common Rail system testing or research it can be individually mounted, separate from the engine, on an experimental stand. In this way, the shape and characteristics of the fuel jet can be analyzed more easily. Viewing and analyzing the electrical signal was done in the paper through the AVL Indicom program. In the paper is analyzed the control signal of an electromagnetic injector, during injection of a quantity of fuel from 2 to 40 mg, with differed strategy injection: one, two or three injections.

Keywords: Engine · Injector · Electric signal · Common Rail

1 Introduction

The injection system performance greatly influences the power, fuel consumption and pollutant emissions of diesel engines. Very important is the quality of the fuel jet sprayed in the cylinder. Injection spray quality can be appreciated by the following characteristics: spray dispersion angle, spray penetration, spray homogenous and spray droplet size (Fig. 1). These depend directly on the: injector nozzle geometry, the pressure, time and characteristic of injection [1, 2].

Modern diesel engines are equipped with a Common Rail injection system. The injector is probably the most complex component in a Common Rail system. The injectors used are of two types: piezoelectric and solenoid injector valve (Fig. 2) [3].

To facilitate Common Rail system testing or research it can be individually mounted, separate from the engine, on an experimental stand. In this way, the shape and characteristics of the fuel jet can be analyzed more easily, depending on the fuel pressure, form and duration of the injector control signal. It is also possible to analyze the action of the injector according to the pump parameters. So you can change the high pressure pump speed and fuel pressure at the pump outlet [4].

I. Dumitru et al. (Eds.): SMAT 2019, *The 30th SIAR International Congress of Automotive and Transport Engineering*, pp. 42–49, 2020.
https://doi.org/10.1007/978-3-030-32564-0_5

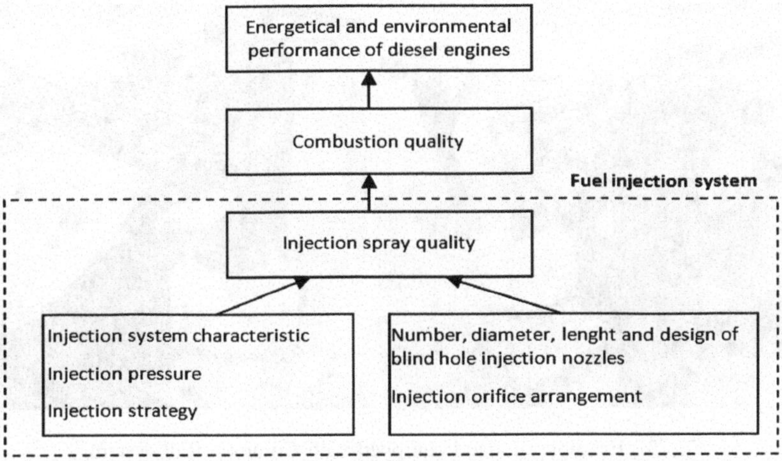

Fig. 1. The influence of the fuel injection system on performances diesel engines.

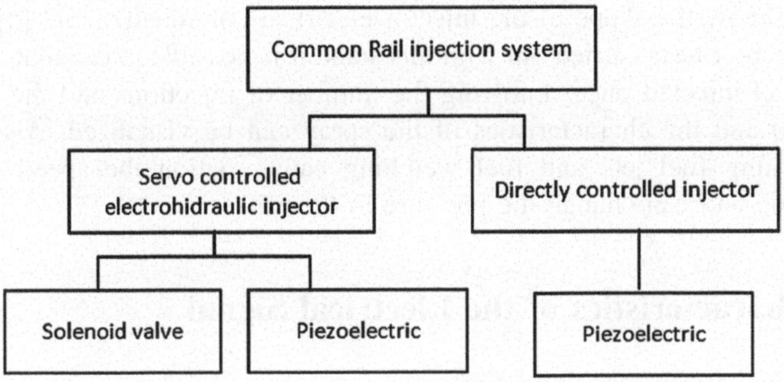

Fig. 2. The types of injectors of Common Rail system.

2 Example of Stand

A stand must contain the following components: fuel tank, high pressure pump (with low pressure pump included), common rail, injector, electronic control unit. It is practically mounted on a high pressure pump support along with the rail to which one or more injectors are connected (Fig. 3). The pump can be driven by an electric motor with variable speed.

By research or tests carried out with this stand it is possible to calculate the average amount of fuel injected once (knowing the number of injections and the mass of the injected fuel) and the characteristics of the spray can be visualized. Also, the stand besides counting fuel jets and fuel weighing can variation the speed of the high pressure pump and can change the pressure in the rail.

Fig. 3. Example of stand equipped with Common Rail system.

One of the biggest difficulties in construction such a Common Rail research stand is to provide the electrical signal needed to command the injectors. Therefore it is important to know the shape of the injector electrical command signal [5].

By research or tests carried out with this stand it is possible to calculate the average amount of fuel injected once (knowing the number of injections and the mass of the injected fuel) and the characteristics of the spray can be visualized. Also, the stand besides counting fuel jets and fuel weighing can variation the speed of the high pressure pump and can change the pressure in the rail.

3 The Characteristics of the Electrical Signal

3.1 Electrical Equipment and Software

In order to command the injector mounted on the test bench, a equipment is required that produce a electrical signal similar to the one made by the original electronic control unit (ECU) and to easy change the injection strategy. Such equipment is shown in Fig. 4.

Fig. 4. The equipment used to command a solenoid injector valve.

This equipment can be ordered via the F2IRE software. Whose interface is shown in Fig. 5. This program allows to control: injections advance, injections number, injection rate and the injected time. It is also possible to control the pressure in the rail [6, 7].

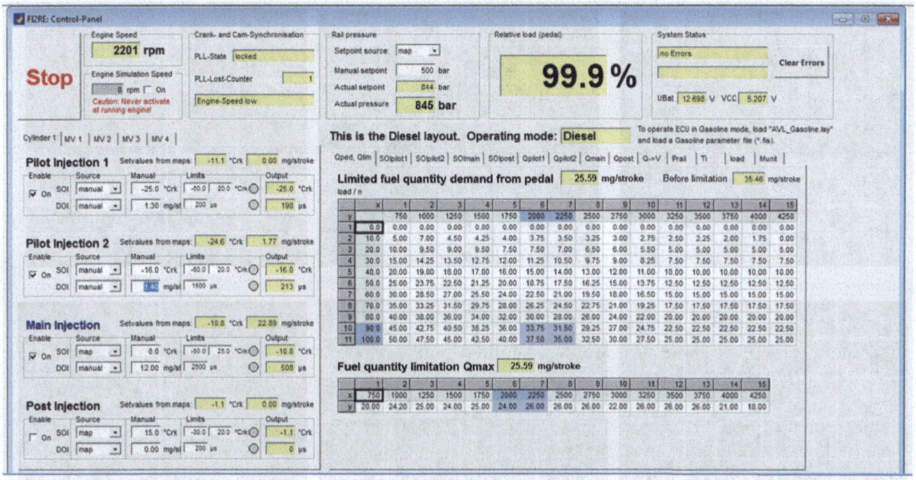

Fig. 5. FI2RE program interface

3.2 Analysis the Electrical Signal of the Solenoid Injector Valve

Viewing and analyzing the electrical signal was done through the AVL Indicom program (Fig. 6).

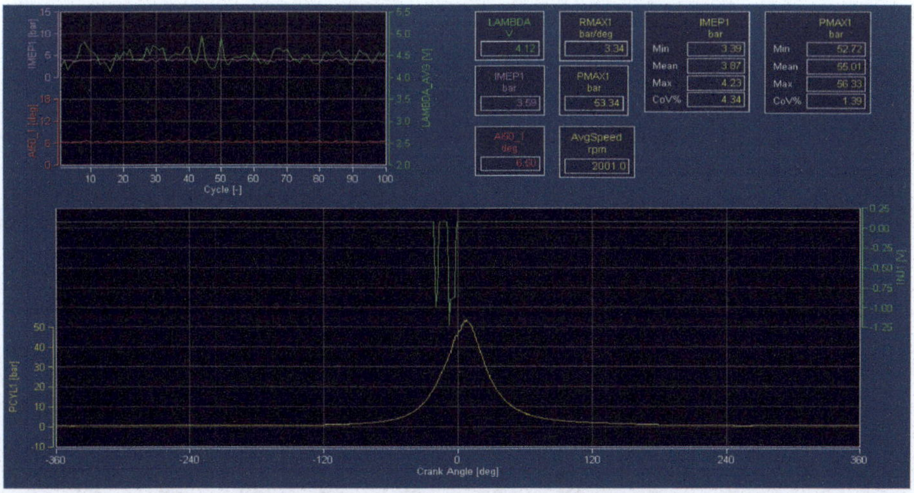

Fig. 6. AVL Indicom program interface

Several forms of the electric signal at different fuel injected quantities are shown in Fig. 7. During the measurements the pressure in the rail was maintained at 450 bar. Successively, the quantity of fuel injected (Q) changed, this varying the shape of the electrical signal.

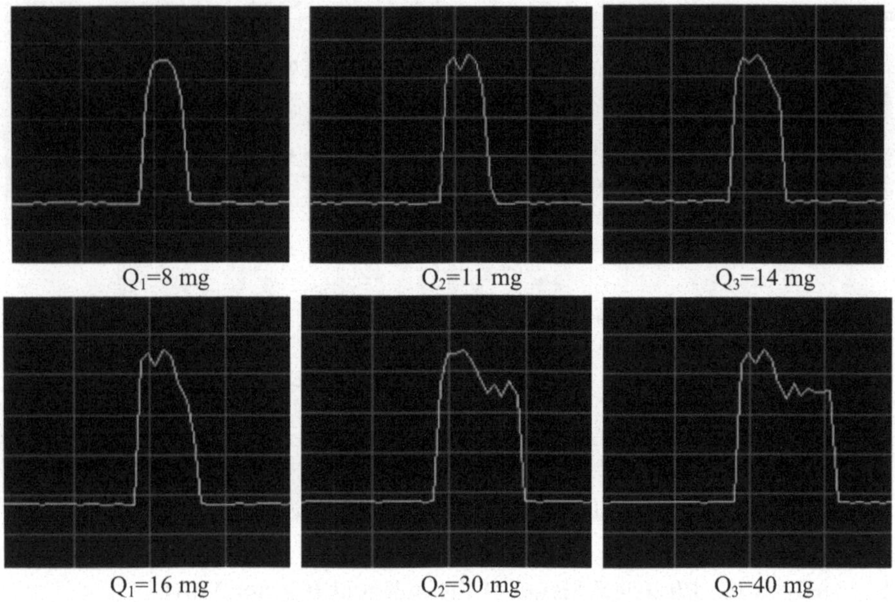

Fig. 7. Shapes of the electrical signal

As can be seen in Figs. 7 and 8, to increase the quantity Q from 8 mg to 16 mg, the duration of the injection increased from 556 μs to 772 μs.

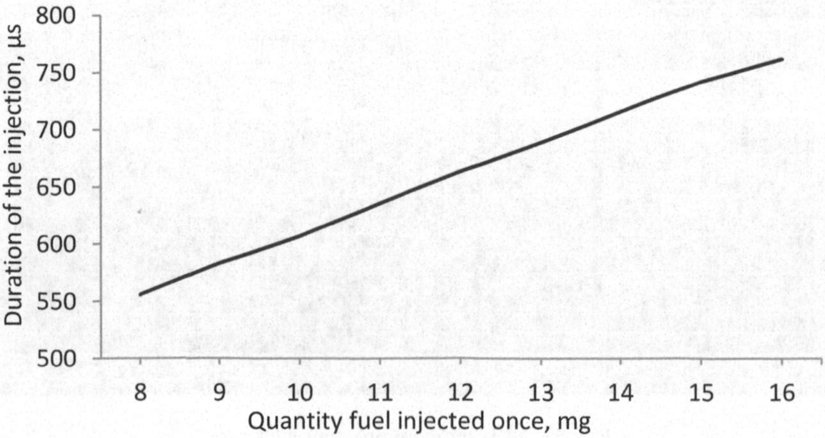

Fig. 8. Varying the duration injection in correlation with quantity fuel injected (Q).

It is noted that in order to increase the injected quantity (Q), both the injection duration and the signal value change. Also to injected 30 mg and 40 mg respectively, the duration of the injection is 1087 μs and 1277 μs. At the same time, the peak value of the electrical signal varied from 0,88 A to 0,58 A for 30 mg and 0,63 A to 0,90 A.

The following is the shapes of signals in the case of a split-injection consisting of a pilot and a main injection (Fig. 9). In this case, only the pilot injection was modified, more precisely the quantity of fuel injected (Q_{pilot}) changed.

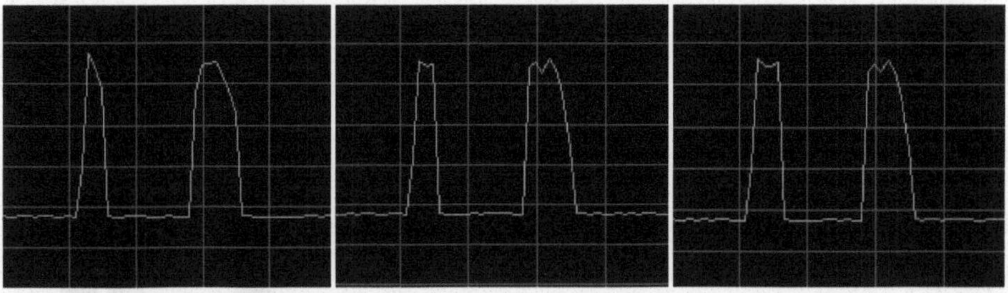

Q_{pilot}=2 mg and Q_{main}=12 mg Q_{pilot}=3 mg and Q_{main}=12 mg Q_{pilot}=5 mg and Q_{main} =12 mg

Fig. 9. Shapes of the electrical signal for two injection: one pilot and one main injections.

Finally, the form of signals for the split injection containing two pilot injections and a main injection is presented in Fig. 10. Successively, the quantity of fuel injected (Q_{pilot1}) changed, in case the first pilot injection.

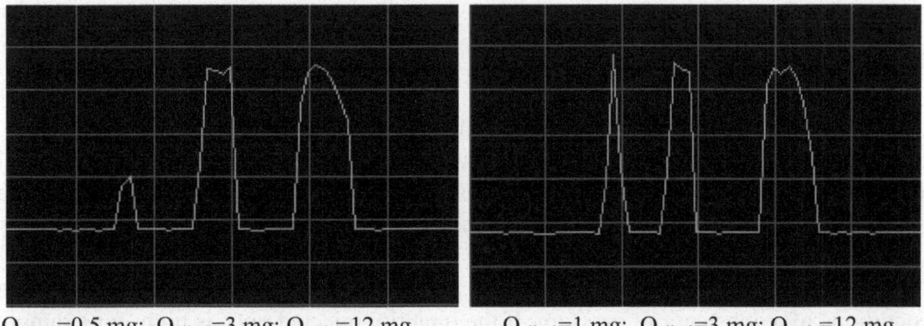

Q_{pilot1}=0,5 mg; Q_{pilot2}=3 mg; Q_{main}=12 mg Q_{pilot1}=1 mg; Q_{pilot2}=3 mg; Q_{main}=12 mg

Fig. 10. Shapes of the electrical signal for three injection: two pilot and one main injections.

Consider the injection strategy in which the following quantity of fuel are injected, as shown in Fig. 10: Q_{pilot1} = 1 mg; Q_{pilot2} = 3 mg; Q_{main} = 12 mg. The total duration of the effective fuel injection in the case these three injections is: 180 μs + 317 μs + 663 μs = 1160 μs. This means that a 16 mg dose was injected in 1160 μs, in contrast to the single injection when the duration of the fuel injected is 762 μs.

For example, in order to highlight the importance of the split injection, in Fig. 11 is presented the evolution of the emission of smoke depending on the engine speed at two injection strategy. It is observed that the smoke emission is lower at the speeds of 1600 rpm and 1800 rpm if the split injection is used. Once the speed increases by 2000 rpm the smoke emission is lower when there is only one injection. Also, according to [8], the temperature of exhaust gases decrease as the number of injections increased. The split injection is of special importance in the regeneration process of the particulate filter.

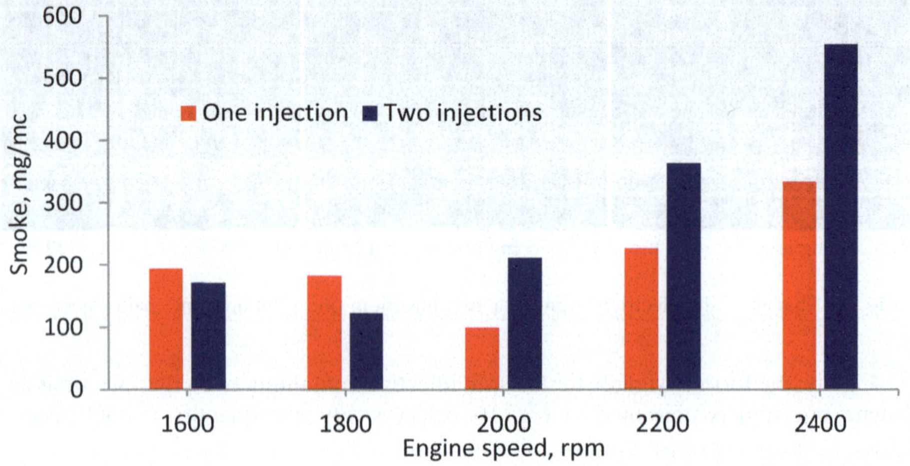

Fig. 11. Varying the smoke emission in correlation with diesel engine speed.

4 Conclusions

Common Rail system can be individually researched, separate from the engine, mounted on an experimental stand, in this way, the shape and characteristics of the fuel jet can be analyzed more easily. In this case it is important to know the shape of the injector electrical command signal, so that the injector mounted on the stand works similar to the conditions on the engine.

The injected quantity changes mainly by modified the injection duration, in the conditions of a constant rail pressure. If the quantity injected once is high, for example 30–40 mg then the maximum value of the electric signal also varies.

In the case of split injection, the duration of the fuel injected increases. For example, a single injection of 16 mg of fuel takes 762 μs. While divided injections (three injections per cycle) of 16 mg take 1074 μs (for two injections per cycle) and 1160 μs (three injections per cycle). Therefore, in the case of split injections, the injection duration may be reduced by an increase the pressure in the rail.

Due to the longer injection time, the engine may have reduced performances at high speeds (the duration of a cycle decreases with increasing speed) if split injection is used.

References

1. Merker, G.P., Schwarz, C., Teichmann, R.: Combustion Engines Development: Mixture Formation, Combustion, Emissions and Simulation. Springer, Berlin (2012)
2. Dragomir, G., Beles, H., Blaga, V., Stanasel, I., Negrea, V.D., Blaga, F.S.: Diesel engines pollution and functional-constructive performances compromise optimization, using fuzzy sets. In: Annals of DAAAM for 2008 & Proceedings of the 19th International DAAAM Symposium, Vienna, p. 437 (2008)
3. Basshuysen, R., Schäfer, F.: Internal combustion engine handbook: basics, components, systems, and perspectives. SAE International, Warrendale (2002)
4. Roma, C., Talabă, D., Abăitâncei, H., Leahu, C.I., Radu, S.: Analysis solicitations injection systems of high and very high pressure using multi domain simulation. In: AFASES – The 15th International Conference of Scientific Paper, Braşov, pp. 558–561 (2013)
5. Krogerus, T., Huhtala, K.: Diagnostics and identification of injection duration of common rail diesel injectors. Open Eng, **8**, 1–6 (2018). https://doi.org/10.1515/eng-2018-0001
6. Leahu, C.I., Tarulescu, S.: The influence of thermal regime on gasoline direct injection engine performance and emissions. In: IOP Conference Series: Materials Science and Engineering, vol. 147 (2016)
7. Tarulescu, S., Tarulescu, R., Soica, A.: Researches on combustion quality for a GDI experimental engine. In: AFASES – The 18th International Conference of Scientific Paper, Braşov, pp. 385–392 (2016)
8. Leahu, C.I., Tarulescu, S., Tarulescu, R.: The exhaust gas temperature control through an adequate thermal management of the engine. In: IOP Conference Series: Materials Science and Engineering, vol. 444 (2018)

Influence of Tires Parameters on the Braking Process in Automobiles

Robert Emil Simniceanu and Loreta Simniceanu(✉)

Faculty of Mechanical Engineering, University of Craiova, Craiova, Romania
lsimniceanu@yahoo.com

Abstract. The parameters of the braking process depend on the characteristics of the tires, and their modification has been noticed when the tread is worn. Experimentally, the increase in the braking distance (by default of the other parameters of the braking process) was found with the depth of the tread pattern. In this paper, mathematical relations are proposed to quantify the influence of the tread depth in the form of a 2nd order polynomial function. For the influence of the tire wear on the braking space, it is also proposed a dependence of the wear correction coefficient. Relationships are proposed to correct the theoretical relations for calculating the braking space and braking time with the value of the coefficient of influence thus determined.

Keywords: Automotive · Braking process · Tires parameters · Tires wear

1 Introduction

The performances of a vehicle, as well as aspects of traffic safety, depend on the condition and quality of the tires, the only contact surfaces between the vehicle and the road being the tire-rolling contact. They have to meet more features related to suspension, steering, speed, weight, noise, more than they have to deal with external forces exerted on them such as braking, acceleration, turns, shocks while running pits, edges, uneven road surfaces, wet, slippery, snow-covered or ice-covered roads. This is why tires are a very important element in car safety.

In the literature, analytical relations are proposed for determining the parameters of the braking process [1–3], used both in terms of vehicle performances, and for the reconstruction of traffic accidents [4, 5]. These analytical relationships do not take into account the direct influence of the physical parameters of the tires, but are taken into account by the adhesion coefficient. This does not take into account parameters such as tire pressure, tread wear, etc.

The parameters of the braking process depend on the characteristics of the tires and, with their wear, these parameters also change. In order to assess as accurately as the real processes of dynamic automotive and considering especially necessary for the activity of reconstruction of road accidents, the paper shows the influence of tire wear on the braking distance and quantify these influences as a correction coefficient braking distance and of braking time.

I. Dumitru et al. (Eds.): SMAT 2019, *The 30th SIAR International Congress of Automotive and Transport Engineering*, pp. 50–53, 2020.
https://doi.org/10.1007/978-3-030-32564-0_6

2 Experimental Considerations

Experimentally, the increase in the braking distance (by default of the braking time) with tread depth was found, as exemplified in the following graphs, with reference to different road conditions: dry road, wet road - Fig. 1, snow covered road - Fig. 2.

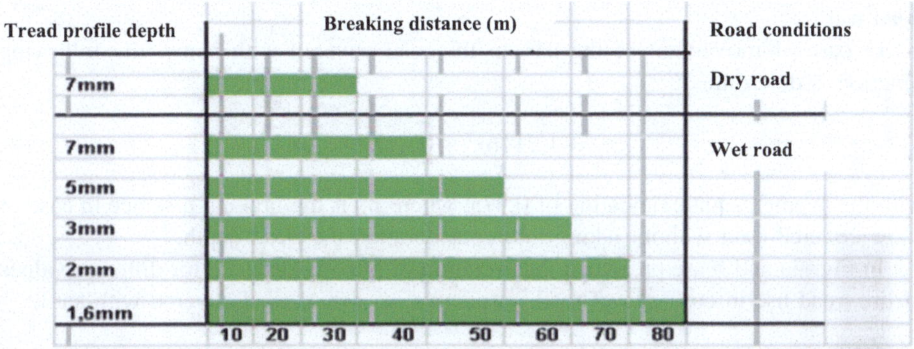

Fig. 1. Influence of the tread profile depth on the braking distance of the vehicle in cases of dry road and wet road

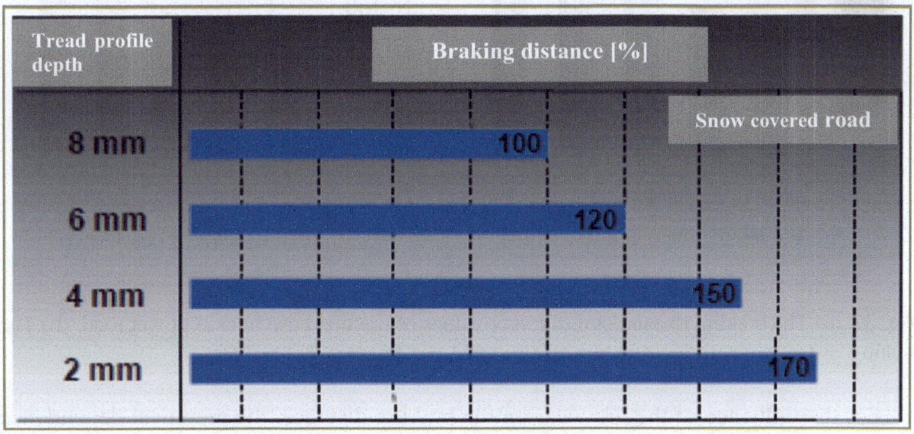

Fig. 2. Influence of the tread profile depth on the braking distance of the vehicle in case of snow covered road

3 Determination of the Correction Coefficients of the Braking Process Parameters

The influence of the tread depth on the braking distance is quantified as a polynomial function of the order 2.

In case of movement of the car on the wet road, it is determined the function given by the relationship:

$$s_{fu} = 0{,}9408x^2 + 1.5268x + s_{0u} \tag{1}$$

with the average square deviation of 0.9805, where s_{0u} is the braking distance in case of wet road without taking into consideration the tread depth and x is the value of tire tread use.

In case of movement of the car on the road covered with snow, the following function is determined:

$$s_{fz} = 1{,}7116x^2 + 0{,}677x + s_{0z} \tag{2}$$

with the average square deviation of 0.975, where s_{0z} is the braking distance in case of snow covered road without taking into consideration the tread depth.

In Fig. 3a and b are presented the variation of braking distance for different values of tire tread use in case of wet road, respective in case of road covered with snow.

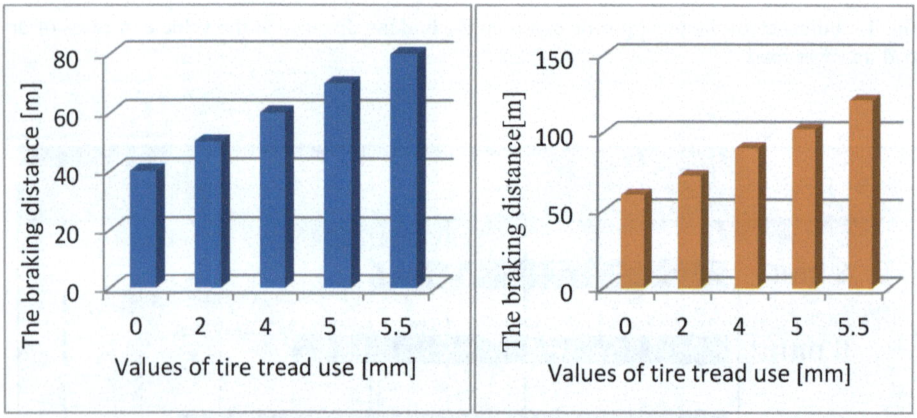

Fig. 3. (a) The braking distance for different values of tire tread use in case of wet road. (b) The braking distance for different values of tire tread use in case of snow road

For the influence of the tire wear on the braking distance, the graphical dependence of the correction coefficient applied to the theoretical brake area in relation to tire tread wear is proposed (Fig. 4). By applying this braking rate correction coefficient, significantly different values are obtained compared to uncorrected values (Fig. 5).

Using the tire wear correction coefficient according to the previously determined tire wear, the theoretical relations for calculating the braking distance and the braking time are corrected. The proposed calculations are:

$$s_f = \frac{v_2^2 - v_1^2}{2 \cdot \varphi \cdot g} \cdot c_u \tag{3}$$

$$t_f = \frac{v_2 - v_1}{\varphi \cdot g} \cdot c_u \tag{4}$$

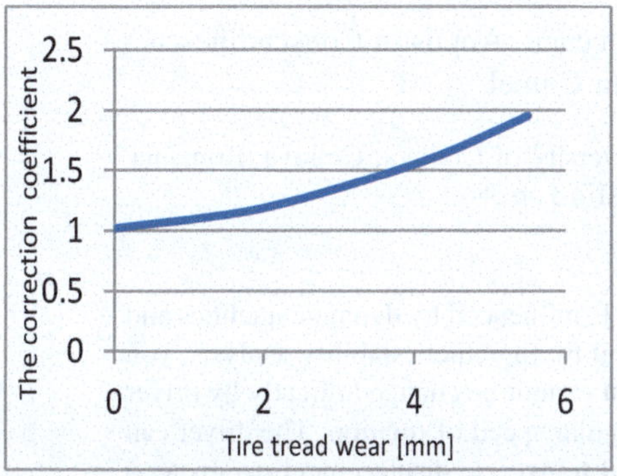

Fig. 4. The plot of the tire tread wear and brake distance coefficient correction variation

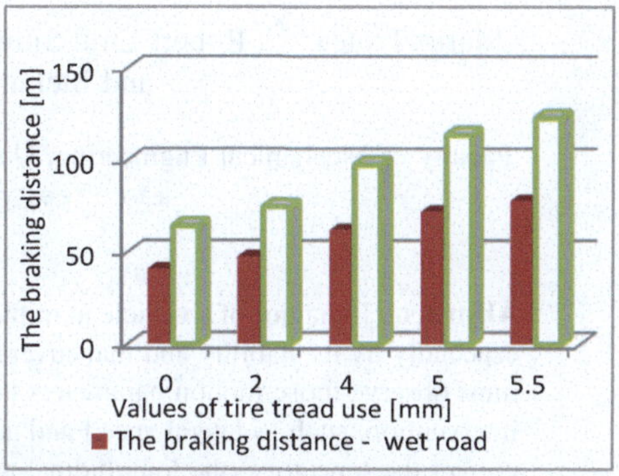

Fig. 5. The plot of the braking distance in case of wear tread corrected relations

4 Conclusions

The use of the correction coefficient of the space and the braking time is very important, producing significant percentage differences between corrected and uncorrected values. Thus, the braked area corrected for a 2 mm wear of the tread is increased by 25%, and in the case of 5–6 mm braking the increase of the brake distance by 100%, so for the accuracy of the theoretical determinations it is necessary to use these corrections.

References

1. Gillespie, Th.: Fundamentals of vehicle dynamics. SAE International, R-114 (1992)
2. Oţăt, V., Bolcu, D., Simniceanu, L.: Dinamica autovehiculelor, p. 380. Editura Universitaria Craiova (2005)
3. Simniceanu, L., Bogdan, M., Oţăt, V., Trotea, M.: Nonlinear differential system applied of a mechanical plan model of the automotives used for the nonlinear stability analysis (2017). http://iopscience.iop.org/article/10.1088/1757-899X/252/1/012023
4. Gaiginschi, R.: Reconstructia si expertiza accidentelor rutiere. Editura Tehnica, Bucuresti (2009)
5. Clinciu, M., Chiru, A., Ciunel, S.: Aspects regarding the reconstruction of traffic events. In: CONAT 2016 International Congress of Automotive and Transport Engineering, pp. 865–872. Springer (2016)

Mathematical Model for Stability Analysis of Vehicle-Driver System

Mario Trotea(✉), Robert Emil Simniceanu, Augustin Constantinescu, and Stefanita Ciunel

Faculty of Mechanical Engineering, University of Craiova, Craiova, Romania
mtrotea@yahoo.com

Abstract. Behavior of a vehicle in traffic is influenced by dynamic qualities and especially by its stability and maneuverability. In vehicle stability analysis, you must observe those motion parameters that cannot be changed directly by driver intervention, such as lateral speed and angular speed of rotation. The driver can correct the trajectory (the longitudinal and transverse displacement parameters) or the displacement velocity, and through them changes of the other motion parameters are made. The paper proposes a mathematical model for analyzing the movement of the car by taking into account the driver's intervention. To highlight the influence of driver intervention on vehicle stability, two mathematical models are analyzed: without taking into account the driver's action and analyzing the differences in the response of the vehicle in different driving situations.

Keywords: Automotive · Stability analysis · Mathematical model · Driver's action

1 Introduction

To analyze the behavior of a vehicle in different displacement situations, mechanical models and mathematical models with varying degrees of complexity, determined to respond to the elements of interest, are used. In the literature there are mechanical and mathematical models used especially in the study of automobile oscillations [1–3]. In analyzing the stability of an automobile, it is useful to consider those motion parameters that cannot be modified directly by the driver's intervention, such as lateral speed and angular speed of rotation [2, 4, 5]. The driver can correct the trajectory (parameter y) or the travel velocity v, and through them modify the other motion parameters [6–8]. The paper proposes a mathematical model of the behavior of the vehicle-driver system, taking into account the driver's action in analyzing the stability of this system.

2 The Mechanical and Mathematical Model of Automotives Moving

Previous work shows the mechanical model and the mathematical model appropriate for the study of the vehicle movement considering the next variables: y, v_y, φ, ω [2, 4, 5].

I. Dumitru et al. (Eds.): SMAT 2019, *The 30th SIAR International Congress of Automotive and Transport Engineering*, pp. 54–57, 2020.
https://doi.org/10.1007/978-3-030-32564-0_7

This mechanical model neglects the roll and pitch movements, and the travel speed is considered constant.

The driver's intervention in correcting the trajectory of the car is modeled in various forms [6–8]. Among the mathematical models in the literature, the present study uses the relationship proposed in [6]:

$$\dot{\delta}_s = -\left[\frac{k}{T_r}\left(y + \frac{L}{v_x}\dot{y}\right) + \frac{\delta_s}{T_r}\right] \tag{1}$$

where k is the feedback loop, Tr - response time and L is the driver's front visibility.

The analysis of the car or the car-driver system consists in the mathematical description of the vehicle's movement [2], taking into account the driver's intervention in correcting the trajectory, correlated with the possibilities and limits of the human factor. The aim is to emphasize the driver's significant influence on the car's trajectory and the stability of the vehicle until skidding occurs.

The mathematical equations describing the motion of the car form a system of nonlinear differential equations:

$$\dot{X} = f(X, t, \gamma) \tag{2}$$

where:

$X \in R^5$, $X = (v_y, \omega, y, \varphi, \delta_s)^T$

γ a symbol that includes the physical parameters of the system

3 Analysis of the Mathematical Model Solutions

The analysis focuses on two cases of movement. In the first case the vehicle moves in a straight line under the action of small external disturbances (amplitude Q = 0.01, ω = π, v = 20 m/s) and the case of higher amplitude oscillations (amplitude Q = 0.1, ω = 2,5, v = 10 m/s).

In the following graphs, the variation of the parameters $v_y(t), \omega(t), y(t), \phi(t)$ for the car-driver system (continuous line) and the model without taking into account the driver intervention (dotted line) is represented.

The case of vehicle moves in a straight line under the action of small external disturbances:

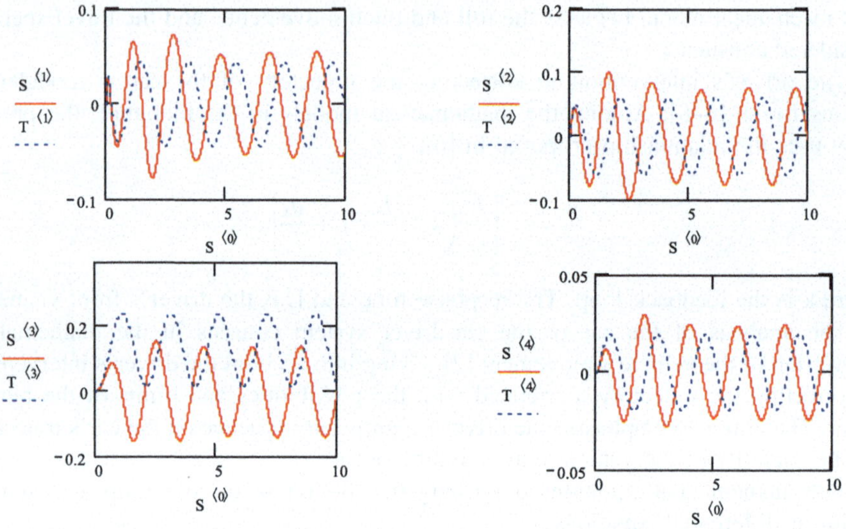

Fig. 1. The parameters variation in case car moving on straight line under the action of small external disturbances

The case of vehicle moves under the action of higher amplitude oscillations:

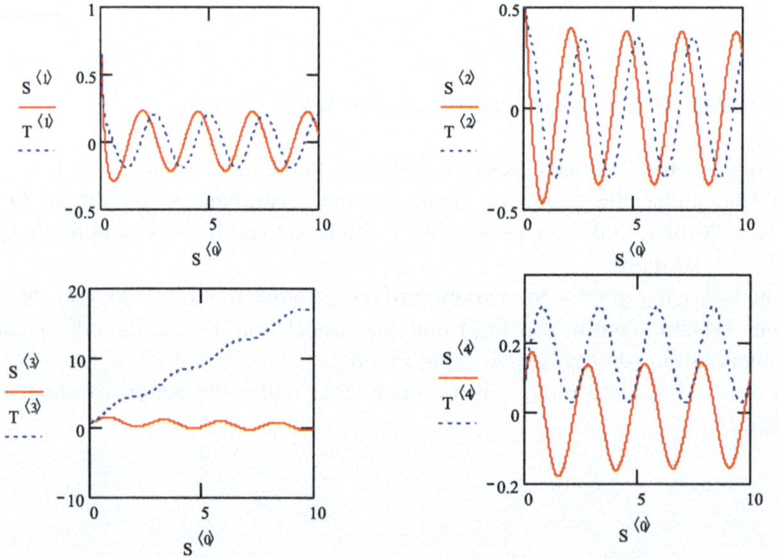

Fig. 2. The parameters variation in case of car moving on straight line under the action of small external disturbances

In the first case, it is noted that the response system is close to the two situations, the movement is regular and with a lower amplitude if the driver intervention is not taken into consideration, but there is no place along the line. Driver intervention causes the movement to be corrected and the vehicle to oscillate along the road axis (Fig. 1).

In the case of higher amplitude oscillations (amplitude Q = 0.1, v = 10 m/s), the displacement from the axle of the road is increasing in the case of the system without driver intervention, and if the action is taken into account the driver corrects the wheel's turning angle with values up to 0.135 rad, keeping the vehicle on the desired trajectory (Fig. 2).

4 Conclusions

Analyzing the results of the two models, comparing them and comparing them with experimental test results [3], it can be concluded that the introduction of the driver's command in the modeling of the movement leads to results much closer to the real ones, the studies of the dynamics of the vehicle aiming to obtain a model that simulates the real movement of the vehicle as accurately as possible.

References

1. Oţăt, V., Bolcu, D., Simniceanu, L.: Dinamica autovehiculelor, p. 380. Editura Universitaria, Craiova (2005)
2. Simniceanu, L., Bogdan, M., Oţăt, V., Trotea, M.: Nonlinear differential system applied of a mechanical plan model of the automotives used for the nonlinear stability analysis (2017). http://iopscience.iop.org/article/10.1088/1757-899X/252/1/012023
3. Neagoe, D., Bolcu, D.: Studiul stabilitatii si maniabilitatii autovehiculelor. Cercetari experimentale si teoretice. Universitaria Publishing, Craiova (2008)
4. Simniceanu, L., Otat, V., Trotea, M.: Determination to a plan model of the automotives moving used for the stability analysis. In: The 2015 International Conference on Mechatronics and Robotics Proceedings, MEROSTA 2015, Recent Advances in Mechanics, Mechatronics and Civil, Chemical and Industrial Engineering, Grecia, pp. 68–71 (2015). ISBN 978-1-61804-325-2
5. Simniceanu, L., Otat, V., Trotea, M., Bogdan, M.: The simulation of the vehicle motion based on generalized mathematical model of vehicle motion. In: Proceedings of the European Automotive Congress, EAEC-ESFA 2015, Bucuresti, vol. 1, pp. 741–747. Springer, Cham (2015)
6. Legouis, D.T., Laneville, A., Bourassa, P., Payre, G.: Characterization of dynamic vehicle stability using two models of the human pilot behavior. Veh. Syst. Dyn. **15**(1), 1–18 (1986)
7. Katzourakis, D.: Driver steering support interfaces near the vehicle's handling limits. http://resolver.tudelft.nl/uuid:87f43ef4-c94a-43f1-8c24-d487b387ebbb
8. Markkula, G.: Driver behavior models for evaluating automotive active safety From neural dynamics to vehicle dynamics. Chalmers University of Technology Goteborg, Sweden (2015). ISBN 978-91-7597-153-7

A Study on the Influence of the Miller Cycle upon the Engine's Internal Aerodynamics

Mihai Niculae[1], Adrian Clenci[1,2(✉)], Victor Iorga-Simăn[1], and Rodica Niculescu[1]

[1] University of Pitesti, Târgul din Vale street, 1, 110040 Pitești, Romania
{mihai.niculae, adrian.clenci}@upit.ro
[2] Le Cnam Paris, rue St. Martin, Paris, France

Abstract. Currently, the internal combustion engine is facing tough regulations regarding pollution and CO_2. In order to comply with the latest constraints concerning the CO_2 emission, the automotive manufacturers must increase the thermal efficiency of the combustion engine. One way which seems to be preferential for the manufacturers is to implement the Miller over-expanded cycle. One drawback of this cycle is that the flow inside the cylinders is steadier than in the case of the Otto cycle. Consequently, in the present study, in order to understand the air flow phenomena inside the cylinder, a 3D CFD simulation is used. The weight of the study is on the understanding of the internal aerodynamics in the case of the Miller cycle. Finally, solutions to improve the situation were also given.

Keywords: Computational Fluid Dynamics · Tumble · Turbulence · Aerodynamics · Miller cycle · Otto cycle

1 Introduction

1.1 Context

In the rush of the automotive manufactures to meet the current and future legislation concerning the CO_2 and pollutant emissions, new technologies are researched and implemented. Recently, the Miller over-expanded cycle, which is not at all a new concept, started to grow in popularity in the automotive world. The Miller cycle was patented in 1957 by Ralph [1] and in the past decades, it was implemented only on a narrow range of cars (mostly, hybridized vehicles such as Toyota, Ford etc.). Now, due to its features [3, 12], which are briefly presented in Sect. 1.2, many manufactures are starting to manifest the interest for it, [2].

As shown in [4] and [12], specific for the Miller cycle is that inside the cylinder, the flow is steadier than in the case of the Otto cycle.

Consequently, in the present study, the focus is on the air flow inside the cylinder for an engine operating under the Miller concept. More exactly, in order to understand the air flow phenomena occurring during the first two strokes of the engine, a 3D CFD simulation of the motored/driven engine was performed. Thus, the results concerning

I. Dumitru et al. (Eds.): SMAT 2019, *The 30th SIAR International Congress of Automotive and Transport Engineering*, pp. 58–67, 2020.
https://doi.org/10.1007/978-3-030-32564-0_8

the internal aerodynamics of the Miller engine are to be compared to the ones of a standard Otto engine.

1.2 Miller Concept

The Miller cycle represents one of the simplest alternatives to the Otto cycle that could be implemented without inflicting with the engine elements. The Miller concept can be defined as an engine cycle that dissociates the volumetric compression and expansion ratios by using different intake valve(s) closing moments (IVC), [2–4, 6].

The Miller concept is frequently done with the following methods, [3, 5]:

- extending the IVC moment to the compression stroke, long after the piston has passed the BDC (bottom dead center), a concept known as LIVC (late intake valve closing)
- advancing the IVC moment during the intake stroke, long before the piston has reached the BDC, a concept known as EIVC (early intake valve closing)

In this study, the focus is only on the EIVC strategy (Fig. 1). Since the intake valves are closing early, a smaller mass of air is trapped inside the cylinder during the intake phase in comparison with the Otto cycle (Fig. 2). After the IVC, the pressure starts to drop (see the phase IVC-2 in Fig. 1), and a kind of expansion occurs which, theoretically, generates a cooling effect and a calmer air movement inside the cylinder. Evidently, this phenomenon is characterized by mechanical work consumption, which, in theory is entirely recovered during the first part of the compression stroke (see the phase 2–3 in Fig. 1). During the phase IVC-2-3, the air acts like a spring, reducing the mechanical work consumption during first part of the compression stroke. It can be easily noticed that the actual compression (3–4) is reduced in respect to the expansion stroke (5–6), while in case of the Otto cycle (Fig. 1), the compression stroke (2″–3″) is similar with expansion stroke (4″–5″), [3, 5].

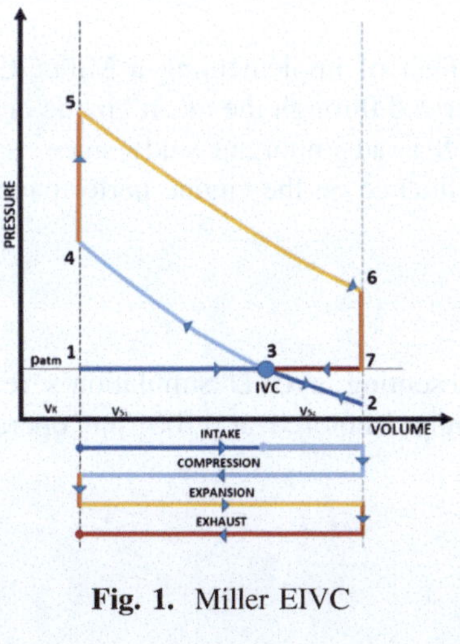

Fig. 1. Miller EIVC

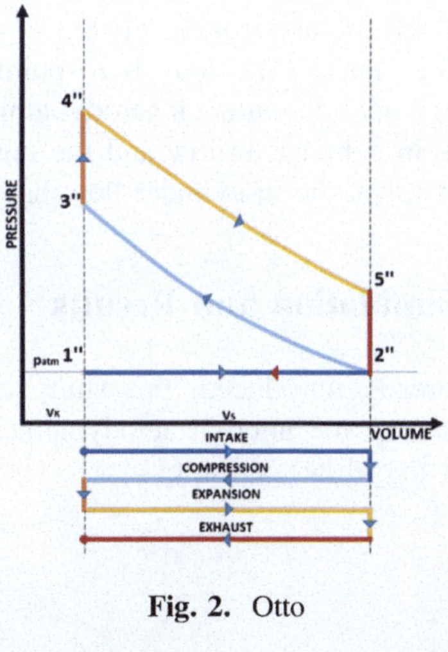

Fig. 2. Otto

1.3 Engine Internal Aerodynamics

As mentioned before, the focus of this study is on the internal aerodynamics of a Miller engine which was approached by a 3D CFD simulation of the motored/driven engine. Being about a non-combustion cycle, only the in-cylinder airflow is concerned. Generally speaking, inside the cylinder the mixture flow is either bulk/organized flow (large scale movements), which is essentially for its mixture properties, either turbulent, which is a chaotic and unstable motion of the fluid, [7–9]. The bulk flow consists in three types of motion: swirl squish and tumble, [10].

Swirl represents a circular movement of the mixture around the cylinder axis. This type of flow is generated either by the helical path of the intake duct or by the eccentric/side positioning of the intake port with respect to the symmetry plan of the cylinder, while at the same time, the design of the intake port creates a pressure drop which increases the pumping loses. Due to the pumping loses and the fact that this type of motion can inflict with the oil film that lies on cylinder wall in certain cases, other types of bulk motion are preferred in the SI engines, [10], such as tumble and/or squish.

Squish represents a radial movement that is obtained at the end of the compression stroke by step effect, when the mixture is literally squeezed between the piston head and the cylinder head, [10].

Tumble represents a rotational movement of the mixture in a perpendicular plan to the piston head. This type of flow is generated by the intake port and piston head geometry. Using tumble ports, the flow section is smaller which induces increased flow velocity and also less pumping work, [10, 11].

The tumble motion is created during the valve opening phase, while most of the flow goes on the upper part of the valve, generating a big tumble motion inside the cylinder. While the piston moves to the top dead center, due the angular momentum that is forcing the big tumble vortex to spin faster, it starts to break into small vortices. Thus, the tumble motion is more effective than swirl both in extracting energy from the piston movement and in transferring this kinetic energy of the mean motion into turbulence. After the combustion occurs, the vortices start to disappear due the heat generated by this process, [10].

The aim of this study is to point out the effect of implementing a Miller EIVC strategy upon the internal aerodynamics characterized through the mean kinetic energy of the in-cylinder airflow and the tumble ratio. It is an important study since as it is well-known, the in-cylinder flow has a direct influence on the engine performance.

2 Simulation and Results

As already introduced, this study is about presenting a CFD simulation's results concerning the internal aerodynamics of an engine motored at 1100 rpm operating under the Miller concept.

2.1 Simulation Description

Figure 3 shows the valve diagram for the two cases simulated: standard Otto (see the dash curves) and Miller (see continuous curves). For both cases we used a 70 mm piston stroke and a 72 mm bore with a volumetric compression ratio of 12.3.

Thus, two RANS (Reynolds Averaged Navier-Stokes) simulations were carried-out on the same geometry (Fig. 4) using AVL FIRE™ software. The computational domain was meshed using hexahedral cells. As the mesh is a moving and deformable one, the number of volume elements varies: 3.706.333 during the valve overlap and 1.510.648 at the end of compression stroke. This mesh size is the result of several prior CFD simulations showing a good trade-off between the need to have good results in a reasonable duration of simulation. For instance, these two single cycle 3D simulations (Otto and Miller) were performed each with a calculus step of 1° CA in about 30 h on an Intel Xeon machine (2.3 GHz, 2 CPU, 36 cores processor and 128 GB of RAM).

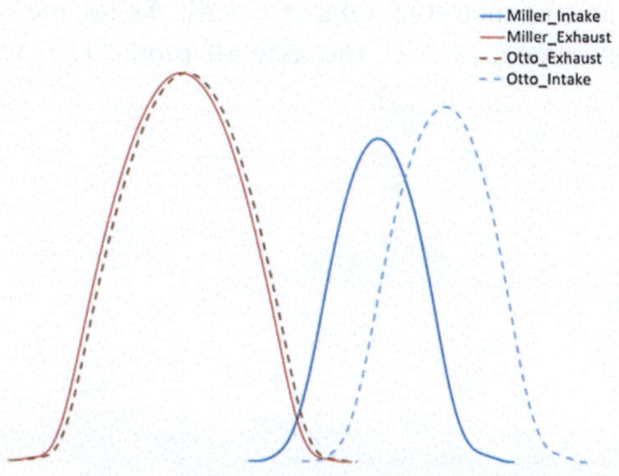

Fig. 3. Valve laws

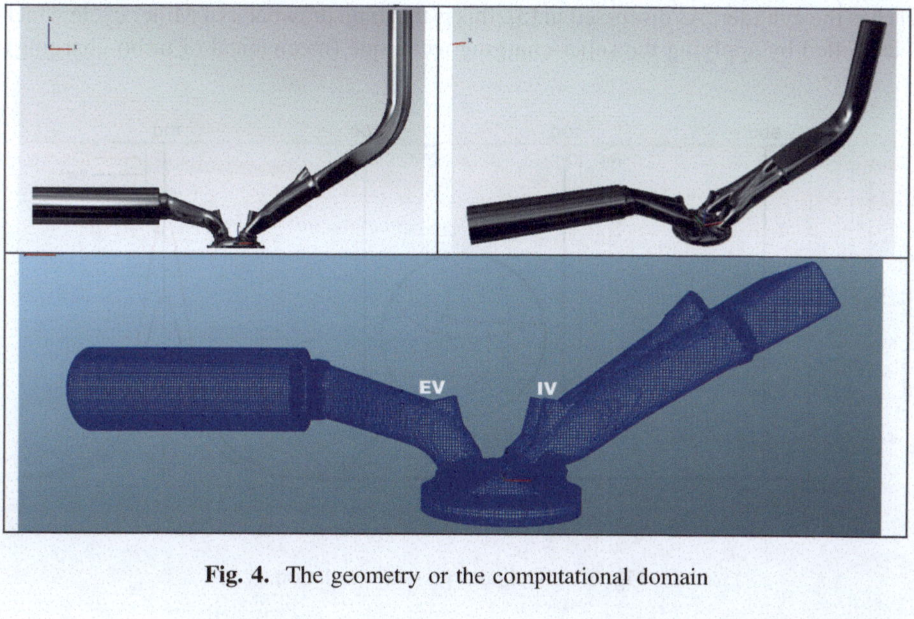

Fig. 4. The geometry or the computational domain

As for the boundary conditions (Fig. 5): a pressure inlet condition was applied at the entry surface in the intake geometry (see the upper right part of the pictures from Fig. 4) and a pressure outlet condition was applied at the exit surface of the exhaust geometry (see the lower left part of the pictures from Fig. 4). The rest of the geometry is an impenetrable adiabatic wall. As for the turbulence model, based on our previous experience, [4, 12], the k-zeta-f model [13] was chosen.

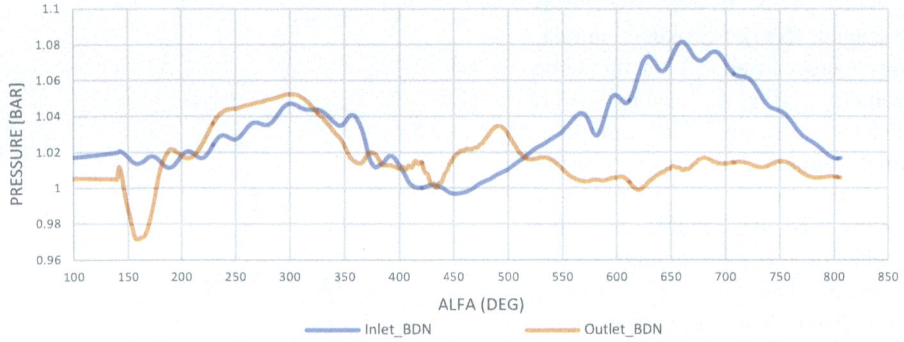

Fig. 5. Inlet-outlet pressure boundary conditions

2.2 Simulation Results. Discussion

In Fig. 6, the in-cylinder pressure is highlighted for both cycles, Otto and Miller. First thing that is easily noticed is the difference of peak pressure at the end of compression stroke, when the piston reaches TDC (top dead center). The difference of pressure is generated by the EIVC strategy specific to Miller cycle, which limits the mass of air that reaches the cylinder. As discussed in [3], this is the main drawback of Miller cycle which is cancelled by applying the super-charging technique (mechanical or turbo-charging).

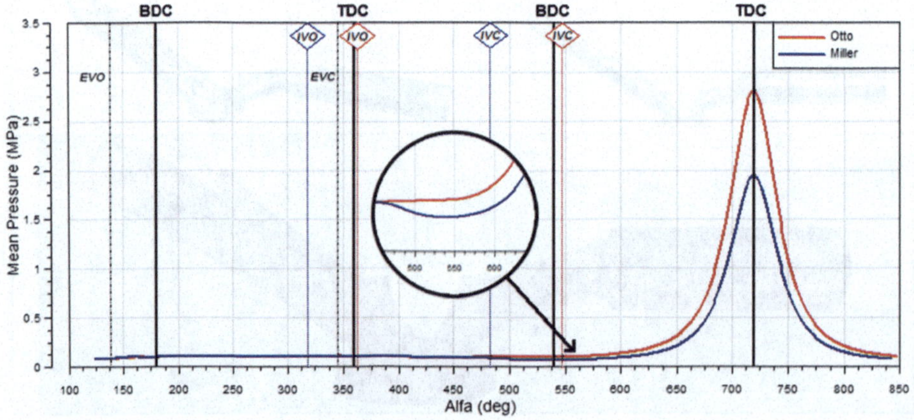

Fig. 6. In-cylinder pressure evolution

Going further with the analysis, as described previously, in Sect. 1.2, once the IVC occurs, the pressure starts to drop due to an expansion phenomenon that takes place during the reminder of the intake stroke. The effective compression is therefore delayed and actually starts when the in-cylinder pressure reaches the value recorded at IVC moment.

Tumble ratio is the ratio between the angular velocity distribution of the flow and the engine angular velocity, [14]. In Fig. 7 it can be observed that between EVO (exhaust valve opening) and IVO (intake valve opening), the level of tumble inside the cylinder is almost similar. At the IVO mark, the tumble ratio starts to behave differently, and is worth mentioning that the negative area of the chart specific to Otto cycle represents the fact that vortices are rotating in the opposite direction with respect to the Miller cycle case. Leaving aside the direction of rotation, if we are to speak in absolute values then one may notice that in case of the Otto cycle, the level of air motion is higher during the intake and compression strokes. In case of both cycles, on the last part of compression stroke (see the angular range between 650 and 700 °CA), the tumble ratio starts to drop. Consequently, even though the Otto absolute values observed in Fig. 7 are slightly higher than the ones of the Miller cycle, the effect of compression is clearly the destruction of the tumble motion for both cases. Certainly, for the Miller case, the effect is even more intense as it didn't start with a high intensity of tumble in the first place.

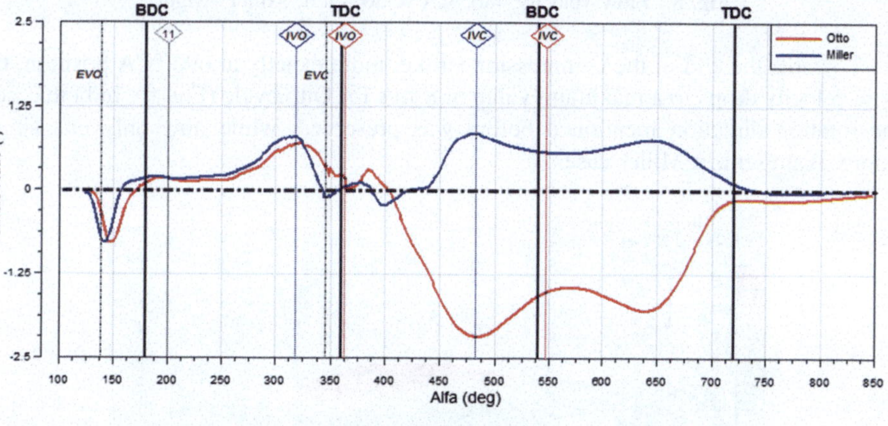

Fig. 7. Tumble ratio

The 486 °CA position represents the maximum level of tumble of both cycles, while the 690 °CA mark is a moment close to the end of compression stroke taken arbitrarily to show what's happening inside cylinder in the vicinity of the end of the compression (TDC). One may see in this arbitrarily angular position the moment of spark producing. As it is well known, the idea is to give enough energy to the airflow inside the cylinder prior to the spark ignition in order to ensure a fast propagation of the flame [10, 16].

To visualize and compare the air motion, the following pictures represent a side by side comparison of both cycles (Fig. 8). Since the comparison is made at 486 °CA, it means that in Miller case, the intake valves are fully closed, while in case of the Otto cycle, the valves are still open.

In case of the Otto cycle, it can be observed that a high velocity single vortex is located almost in the center of the cylinder. The red area represents a high velocity zone, generated by the reducing valve area due to the intake valve closing phase. In the right figure, the Miller cycle, there are two small vortices that are spinning in the opposite direction at a lower speed (if we refer to the left case).

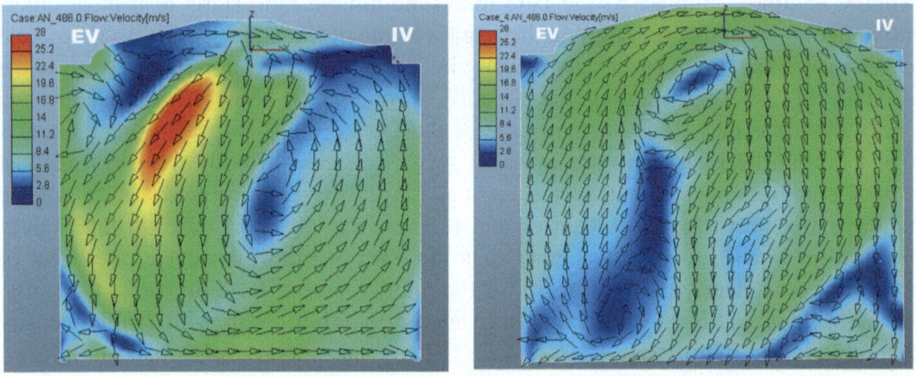

Fig. 8. Flow velocity 486 °CA (Otto – left; Miller – right)

Towards the end of the compression stroke, more exactly at 690 °CA position, the flow velocity drops to a maximum value of 8 m/s for Otto cycle (Fig. 9). In both cases, the rotation direction mentioned before was preserved, while now only one single vortex is present in Miller case.

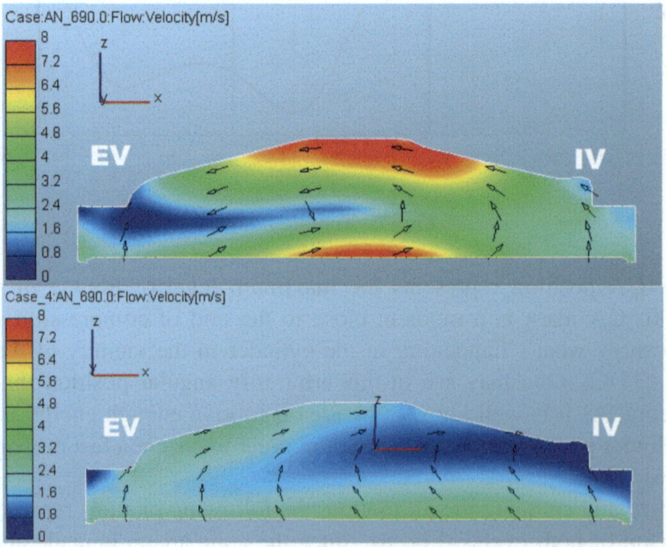

Fig. 9. Flow velocity 690 °CA (Otto – top; Miller – bottom)

Another parameter used to analyze the internal aerodynamics of an engine is the turbulence kinetic energy (TKE). According to [15], "TKE is the mean kinetic energy per unit mass associated with eddies in turbulent flow". As shown by Fig. 10, during the intake phase, in case of the Miller cycle, the TKE starts to increase fast, recording a peak just before the IVC moment. This is mainly due to the intensification of airflow velocity underneath the intake valves, which being in their closing phase feature a gradually decreasing flow area. However, the piston speed (W_p) plays also a role since it is the cause of airflow inside the cylinder. At this moment, piston speed still has a high value and it also contributes to the TKE peak mentioned before. Thus, two parameters (piston speed and flow area underneath the intake valves) are the cause of the TKE peak in the Miller case. Conversely, for the Otto case, the TKE peak is roughly in the same position, which could also be explained by the influence that piston speed has upon the airflow. Certainly, the peak value is smaller as the flow area underneath the intake valves is higher. Then, as seen in Fig. 10, the TKE evolution for the Otto case is above the one from the Miller case, which supports the previous explanations: for the Miller case, since the intake valves are already closed, it is only the piston speed, which supports the generation of TKE, while in the case of Otto cycle, two parameters contribute simultaneously to the obtaining of higher values: piston speed and flow area underneath the intake valves being in the closing phase, thus, increasing further the airflow velocity inside the cylinder, generating TKE. However, despite this significantly higher peak recorded prior to the IVC moment, at the end of compression, the Miller cycle features a lower TKE than the Otto cycle as it can be observed in Fig. 10. By proceeding further with the analysis of TKE evolution, one may notice that for both cases, the values recorded at the end of compression in the TDC position are obtained after a previous upward evolution. According to [16], these final peaks of TKE prior to TDC position are generated by the piston motion, which accelerates the flow at the end of compression stroke and the large vortexes (tumble motion – see Fig. 7) are converted into multiple small vortices, that rotates with high velocity (turbulence).

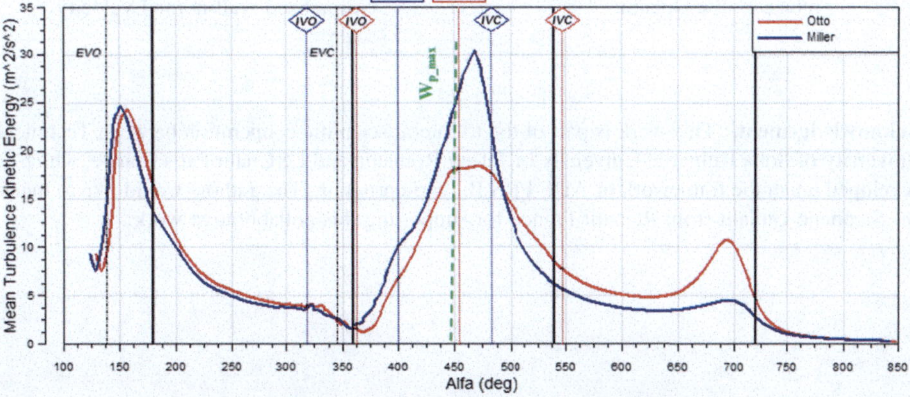

Fig. 10. Mean Turbulence Kinetic Energy

3 Conclusions and Future Works

The level of bulk flow, turbulence and flow speed at the end of the compression stoke plays a key role to the combustion phase. Just by applying the Miller cycle to a conventional Otto engine, the level of turbulence achieved is not satisfactory for the combustion phase. In order to have a good level of turbulence at the end of compression stroke it is necessary to generate and accelerate the motion during the intake phase and preserve it during the compression stroke. Our future work will focus on finding and testing ways to improve the Miller cycle in such way that in the end, a similar level of engine aerodynamics as in Otto cycle is obtained. For instance, as pointed out in [4] and [12] the following ideas could be exploited either singularly or together: higher piston stroke, tumble intake manifold, tumble intake ports, valve masking, adapted piston head geometry.

4 Notations and Abbreviations

BDC	Bottom Dead Center	k	Turbulent kinetic energy
CA	Crank angle	LIVC	Late Intake Valve Closing
CFD	Computational Fluid Dynamics	p_{atm}	Atmospheric pressure
CO_2	Carbon dioxide	RAM	Random Access Memory
CPU	Central Processing Unit	RANS	Reynolds Average Navier-Stokes
EIVC	Early Intake Valve Closing	SI	Spark Ignition
EV	Exhaust valves	TDC	Top Dead Center
EVC	Exhaust Valve Closing	TKE	Turbulence Kinetic Energy
EVO	Exhaust Valve Opening	V_k	Combustion chamber volume
f	Elliptic relaxation function	V_S	Cylinder volume
IV	Intake valves	W_p	Piston speed
IVC	Intake Valve Closing	W_{p_max}	Maximum piston speed
IVO	Intake Valve Opening	zeta	Normalized wall-normal velocity scale

Acknowledgement. This work is part of the tri-lateral scientific cooperation between Technical University of Sofia Bulgaria, University of Pitești Romania and Le Cnam Paris France, which is developed under the framework of AUF-FRS Bulgaria program. The authors would like to thank Dr. Stéphane Guilain from Renault France for supporting this collaborative work.

References

1. Ralph, M.: Miller cycle engine. Patent no. 2817322 (1957)
2. Hatamura, K., Hayakawa, M., Goto, T., Hitomi, M.: A study of the improvement effect of Miller-cycle on mean effective pressure limit for high-pressure supercharged gasoline engines. JSAE Rev. **18**(2), 101–106 (1997)
3. Clenci, A., Podevin, P., Descombes, G.: Etude thermodynamique de la détente prolongée. In: Colloque Francophone sur l'Energie, Environnement, Economie et Thermodynamique, COFRET 2008, Nancy, France, pp. 11–13 (2008)
4. Niculae, M., Clenci, A., Iorga-Siman, V., Niculescu, R.: An overview on the Miller-Atkinson over-expansion thermodynamic cycle. In: IManEE 2019 International Conference, Pitesti, p. 8 (2019)
5. Mo, H., Huang, Y., Mao, X., Zhuo, B.: Investigations on the potential of miller cycle for performance improvement of gas engine. Glob. J. Res. Eng. B Automot. Eng. **16**(1), 1–12, 2249–4596 (2016)
6. Clenci, A., Hara, V.: Analytical synthesis of intake cam in order to obtain Miller cycle technology. In: Mobility & Vehicle Mechanics International Conference, Kragujevac, Serbia (2002)
7. Vernet, J.: Detailed study of steady in-cylinder flow and turbulence using stereo-PIV, pp. 1–62. SCANIA, Sweden (2012)
8. Qi, Y.L., Dong, L.C., Liu, H., Puzinauskas, P.V., Midkiff, K.C.: Optimization of intake port design for si engine. Int. J. Automot. Technol. **13**(6), 861–872 (2011)
9. Aleiferis, P.G., Behringer, M.K., Malcolm, J.S.: Integral length scales and time scales of turbulence in an optical spark-ignition engine. Flow Turbul. Combust. **98**(2), 523–577 (2017)
10. Ryrholm, R., Eriksson, M.: Optical study of tumble spray and combustion, pp. 1–79, 1652–8557. Chalmers University of Technology (2015)
11. Kim, Y., Kim, M., Oh, S., Shin, W., Cho, S., Song, H.H.: A new physics-based modeling approach for a 0D turbulence model to reflect the intake port and chamber geometries and the corresponding flow structures in high-tumble spark-ignition engines. Energies **12**(10), 1898 (2019)
12. Niculae, M.: Stadiul actual al motoarelor ce utilizează ciclul Miller-Atkinson. Ph.D. report, Pitesti (2019)
13. Ogorevc, T., Sekavčnik, M., Katrašnik, T., Žun, I.: Three-dimensional numerical simulation of the exhaust stroke of a single-cylinder four-stroke ICE. Forsch. im Ingenieurwesen/Eng. Res. **73**(3), 147–160 (2009)
14. Bharadwaj, A.M., Madhu, K., Seemanthini, J., Vismay, K.G., Anand, S.M., Arvind, S.: Study of Swirl and Tumble Motion using CFD. Ijtarme **2**(1), 36–39 (2013)
15. Turbulence kinetic energy. https://en.wikipedia.org/wiki/Turbulence_kinetic_energy
16. Shirasaka, M.: The new four-cylinder horizontally-opposed gasoline engine from Subaru. MTZ Worldw. **72**(11), 26–33 (2010)

3D Aerodynamic Simulation of the Heavy Duty Commercial Vehicle

Nadica Stojanovic[1], Oday I. Abdullah[2,3], Josef Schlattmann[2], Ivan Grujic[1(✉)], and Danijela Miloraddovic[1]

[1] Faculty of Engineering, Department for Motor Vehicles and Motors, University of Kragujevac, 34000 Kragujevac, Serbia
ivan.grujic@kg.ac.rs
[2] System Technologies and Mechanical Design Methodology, Hamburg University of Technology, 21073 Hamburg, Germany
[3] Department of Energy Engineering, College of Engineering, University of Baghdad, Baghdad-Aljadria 47024, Iraq

Abstract. Owing to the fast development in the automotive field, the demands are focused on simultaneous improvement of performance, efficiency of fuel and comfort. In addition to these demands, there are other important points that should be taken into consideration such as reduction of the costs of maintenance and the environmental factors. All these requirements inherently contribute to develop the automotive field. Owing to the high prices of oil, the mechanical designers focus to reduce the fuel consumption as much as possible, and to achieve this objective, the resistance forces (drag force) that occur during vehicle operation should be reduced. In order to reduce the drag force, it is necessary for the vehicle to have a very good aerodynamic shape (low drag coefficient). The aim of this research paper is to analyze the typical designs of trucks that made by various manufacturers, and present the solutions that will reduce the effect of the shortcomings on the aerodynamic performance of the trucks. Numerical analysis was performed using Ansys software 14.5 [Fluid Flow (CFX)]. The results present air flow and the pressure distributions for different models of heavy duty trucks.

Keywords: Aerodynamic problem · Numerical analysis · Air resistance · Truck design

1 Introduction

Since the early years of the 20th century, when mass production of the commercial vehicles have started, it has been noticed that the air resistance plays a major role that affect the performance of vehicle motion. Although the fuel consumption was not a concern back years, today the manufacturers started to think seriously to develop a new design for the vehicle to reduce the aerodynamic drag. The main purpose of increasing the top speed of the vehicles is to compensate the lack of power of their engines [1]. However, the oil price varies (Fig. 1), so it is necessary for the automotive designers to reduce the fuel consumption of the vehicles as much as possible, or, more accurately, to reduce resistances that occur during the vehicle movement. Where, the air resistance is

I. Dumitru et al. (Eds.): SMAT 2019, *The 30th SIAR International Congress of Automotive and Transport Engineering*, pp. 68–81, 2020.
https://doi.org/10.1007/978-3-030-32564-0_9

proportional to square of vehicle speed. Therefore, when the vehicle speed increases, the drag coefficient increases too. This is mean that the fuel consumption increases too. Heavy commercial vehicles are known for being extremely inefficient compared to other ground vehicles, partly due to high aerodynamic drag. This is the result of their un-streamlined body shape [2]. Thus, the fuel consumption of the trucks is high, when the drag coefficient increases with increase of the speed [3]. Class 8 tractor-trailers (weight limit is higher than 14,969 kg) consume about 11-12% of the total US petroleum usage. At highway speeds, 65% of the energy expenditure for the Class 8 truck is to overcome the aerodynamic drag [4].

Generally, the heavy commercial vehicle's annual mileage can vary between 130,000 km and 160,000 km. Therefore, any reduction of aerodynamic drag will result in huge fuel savings and reduction of greenhouse gas emission. Although, the researchers made significant efforts in order to develop the different devices to save the fuel for the commercial vehicles over decade [2]. Owing to the importance of this topic, the researchers continue the investigation to find out a new approach to reduce the drag forces [5].

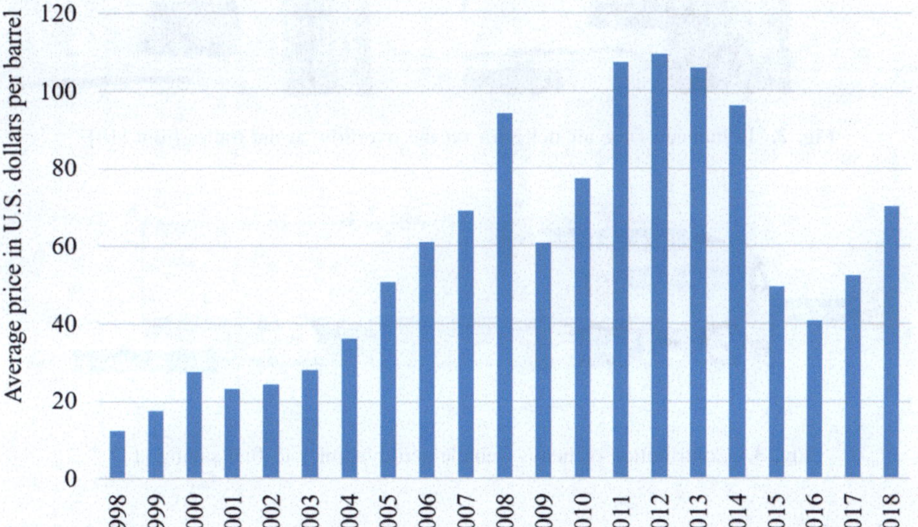

Fig. 1. Average annual OPEC crude oil price from 1998 to 2018 (in U.S. dollars per barrel) [6].

Trucks typically consist of two parts - manufacturers usually make chassis and cabin, while the trailers and the other parts of superstructure (e.g. containers, tanks…) are taken from other manufacturers. However, the heights of the cabin and the superstructure are different. By looking at the truck with the trailer, it can be seen that the trailer is higher than the truck cabin. Owing to the height difference between the trailer and truck cabin, the additional resistance will occur. In order to reduce this resistance, it should be added the air deflectors on the cabin. However, it will be easy to design them at the beginning, because of the relative motion between the cabin and the trailer [7]. The air resistance will be reduced by applying the air deflector on the roof and on the sides. Figure 2 shows the overflow of the air current of truck with and without the air deflector.

In this way, the entrance of the air between the cabin and the trailer is disabled. Beside the roof air deflectors, there are some other parts of the truck that contribute to the effect on the fuel consumption (reduction of the drag coefficient) as shown in Fig. 3.

It was found that the drag coefficient could be reduced up to 23% by using the trailer skirts, base plates and lengthened side extenders and air fairing [8]. Therefore, the main objective of many researches optimizes the external geometry of the truck to determine the aerodynamic influence of different parts to reduce the fuel consumption [9].

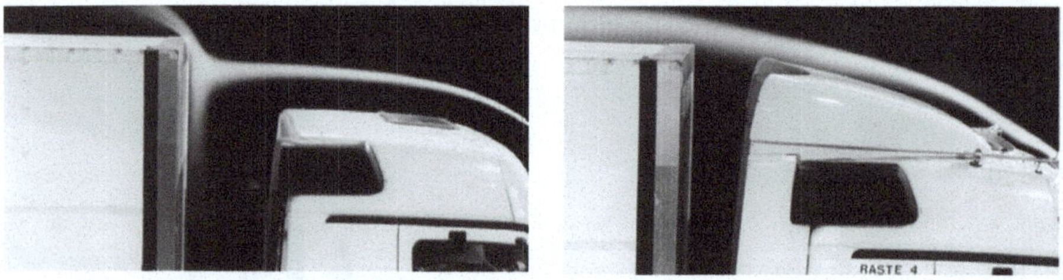

Fig. 2. Influence of the air deflector on the overflow at the trailer front [10]

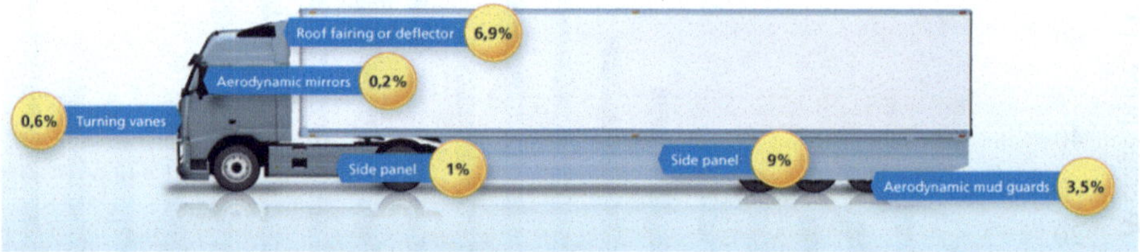

Fig. 3. Contribution of heavy vehicle aerodynamics to fuel saving [11]

In [12], the measurements showed that the drag coefficient of the truck is 0.608, while when the trailer is added the value of the drag coefficient is found 0.704. This means that, in case of the truck connected to the trailer, the drag coefficient increases by 15.8%. The effect of this increment cannot be neglected in the calculations [12].

The air resistance is proportional directly to the drag coefficient and it can be classified into three categories [13]:

- Body shape resistance: which depends on friction resistance on surfaces parallel to the air current, as well as on the shape of the front and the rear sides of the vehicle,
- Interrupted resistance: this type can be divided into two dominant components: the resistance due to the interruption between the car body and wheels and the resistance due to the existence of windows and channels on the roof and the glasses.

- Flow resistance: originating from the air flowing through the engine space, which it's illustrated in Fig. 4.

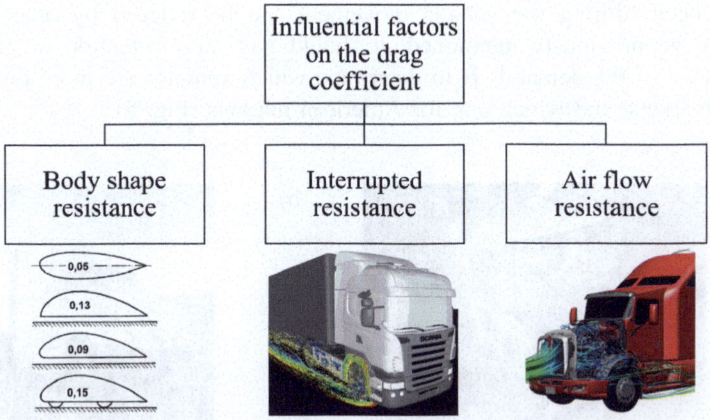

Fig. 4. Influential factors on the drag coefficient [13]

In this research paper the computational fluid dynamics (CFD) was used to achieve the numerical analysis. The computational fluid dynamics (CFD) can be applied for different categories: Turbulence models for Reynolds-Averaged Navier-Stokes (RANS) equations, Large Eddy Simulation (LES), Direct Numerical Simulation (DNS) and Lattice Boltzmann method (LBM). The LBM differs from the other methods by simulation of the movements of particles and it aims at recovering the hydrodynamics of the Navier-Stokes equation. RANS, on the other hand, uses the Navier-Stokes equation as a starting point and aims at solving them [8].

The Reynolds-averaged Navier-Stokes equations are solved using a finite volume method. The Renormalization Group (RNG) k-ε model was elected for closure of the turbulent quantities [14]. The external aerodynamics of a heavy truck is simulated using a validated computational fluid dynamics method and the external flow is presented by computer visualization.

Based on the available literature, the researchers applied different approaches and proposed many new designs to reduce the drag effect, but, unfortunately the value of drag still high. So, the researchers should continue the investigations in this specific area.

The purpose of the paper is to analyse the aerodynamic problems of different types of heavy-duty commercial vehicles. Three-dimensional models of the heavy duty commercial vehicles were built and analysed using a numerical approach. Ansys software 14.5 was used to find the solution of the aerodynamic problem of the selected vehicles. Finally, a comparison was made between the drag coefficients of each type of the vehicles in order to find the minimum value of drag coefficient, which means that it has the minimum resistance. Finally, the conventional design of the trucks assumes that aerodynamics is really important only for line haul where the average speeds are high.

1.1 3D Modelling of the Trucks

In order to reduce the consumption of fuel, it is necessary to modify for external shape of the vehicle, beside the modifications in the engine. Based this approach, the resistance that occur during the vehicle movement can be reduced by other ways. In addition to the previously mentioned, it should pay more attention on the market demands. One of the demands is to determine which vehicles are more desirable for European markets, and which one for American markets (Fig. 5).

a) b)

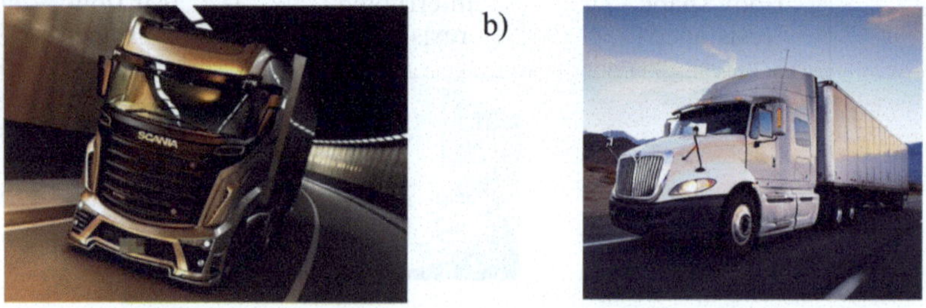

Fig. 5. Heavy duty vehicle that is specific for (a) European and (b) American market [15, 16]

It can be noticed that the models of the trucks which specifically desirable for European markets are those of new generation vehicles. Where, the influence of rear-view mirrors increases the drag coefficient with only 0.2%, as shown in Fig. 3. Because of that, the trucks without rear-view mirrors will be observed.

Figure 6 shows the three dimensional models of the heavy duty trucks that selected in this paper. These models of trucks have been built based on real dimensions (scale

M1

M2

M3

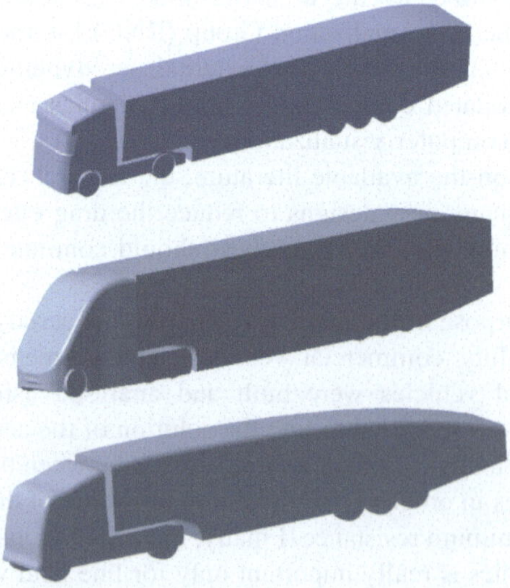

Fig. 6. 3D models of heavy duty trucks

1:1). The length of all trucks is approximately the same (around 18 m) while the surface area of the front side is 8.16 m^2 for the first model (M1), 8.7 m^2 for the second model (M2) and 7.95 m^2 for the third model (M3) truck.

Mesh quality is considered as an essential factor, which affects the error and the precision of the numerical results. It is very important to use the optimum mesh in order to obtain the results with high accuracy and reduce the numerical error. The tetrahedral elements were selected to build the finite element model of the heavy-duty vehicle. The corresponding number of elements and nodes for the finite element models are shown in Table 1.

Table 1. The mesh, number of elements and nodes of different heavy duty trucks.

	Mesh	*Elements*	*Nodes*
M1	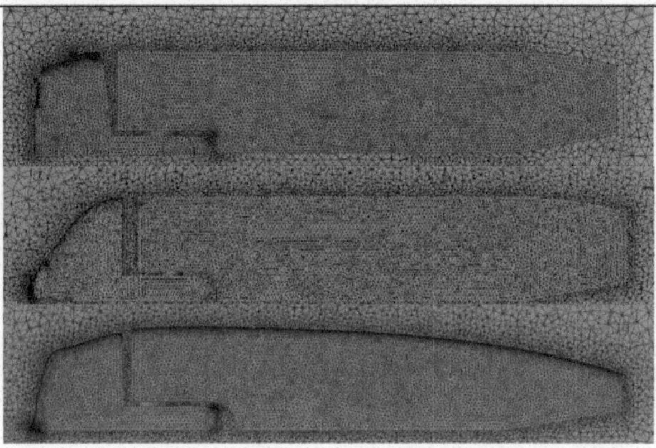	839014	157890
M2		738952	140490
M3		869709	162204

2 Mathematical Modelling and Boundary Conditions

Numerical analysis was performed using Ansys software Fluid Flow (CFX) module. The flowchart of the developed numerical approach of CFX problem is shown in Fig. 7. In all selected cases, the same initial conditions were applied for selected trucks. The initial conditions are:

- Vehicle speed = 27.8 m/s
- Atmospheric pressure = 101325 Pa
- Environmental temperature = 25 °C
- Air density = 1.225 kg/m^3

The influence of the lateral wind will not be considered in the analysis, only the drag coefficient will be determined. The applied model is k-ε model, where k is the turbulence kinetic energy which defined as the variance of the fluctuations in velocity and ε is the turbulence eddy dissipation - the rate at which the velocity fluctuations dissipate.

The values of k and ε obtain directly from the differential transport equations of the turbulence kinetic energy and turbulence dissipation rate are [17]:

$$\frac{\partial(\rho k)}{\partial t} + \frac{\partial}{\partial x_j}(\rho U_j k) = \frac{\partial}{\partial x_j}\left[\left(\mu + \frac{\mu_t}{\sigma_k}\right)\frac{\partial k}{\partial x_j}\right] + P_k + \rho\varepsilon + P_{kb} \quad (1)$$

$$\frac{\partial(\rho k)}{\partial t} + \frac{\partial}{\partial x_j}(\rho U_j k) = \frac{\partial}{\partial x_j}\left[\left(\mu + \frac{\mu_t}{\sigma_\varepsilon}\right)\frac{\partial \varepsilon}{\partial x_j}\right] + \frac{\varepsilon}{k}(C_{\varepsilon 1}P_k - C_{\varepsilon 2}\rho\varepsilon + C_{\varepsilon 1}P_{\varepsilon b}) \quad (2)$$

where $C_{\varepsilon 1}$, $C_{\varepsilon 2}$ and σ_ε are constants. P_{kb} and $P_{\varepsilon b}$ represent the influence of the buoyancy forces, which are described below. P_k is the turbulence production due to viscous forces. In addition to the independent variables, the following terms were used: density (ρ), the velocity vector (U_j), the molecular (dynamic) viscosity (μ) and the turbulence viscosity (μ_t).

The model constants are given as follows [17]:

- $C_{\varepsilon 1} = 1.44$,
- $C_{\varepsilon 2} = 1.92$,
- $\sigma_\varepsilon = 1.3$.

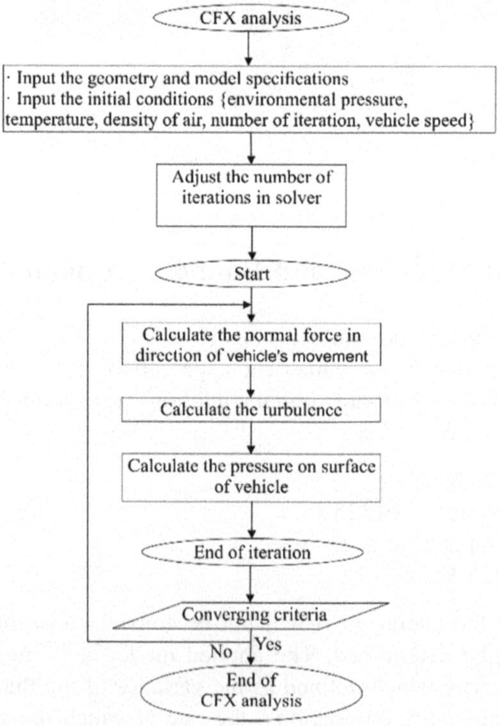

Fig. 7. Flowchart of the developed numerical approach for CFX problem

3 Results and Discussions

In this section, the drag coefficient, velocity streamlines surface and distribution of pressure over vehicle have been investigated deeply using computational Fluid Dynamics (CFD). It was selected three different models of heavy duty trucks to analyze the aerodynamic performance of selected trucks and finally make compression between them.

The drag coefficients for the models M1 and M3 that can often be seen in traffic is found to be 0.38. The minimum value of the drag coefficient of 0.31 was obtained for the vehicle M2.

Table 2. Values of the drag coefficient and velocity streamlines

	Drag co-efficient	Air flow velocity for different models of heavy duty vehicles
M1	0.38	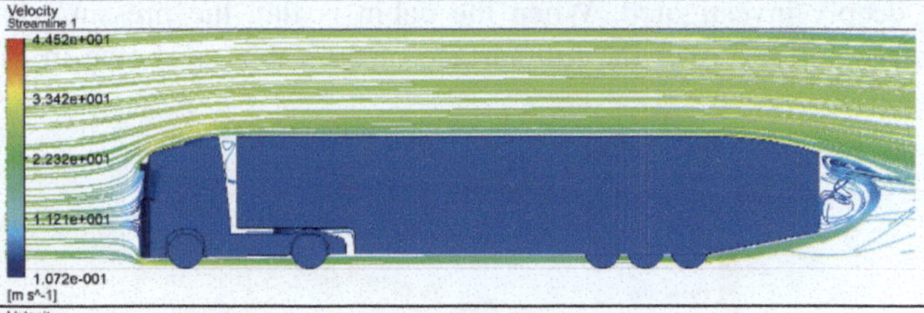
M2	0.31	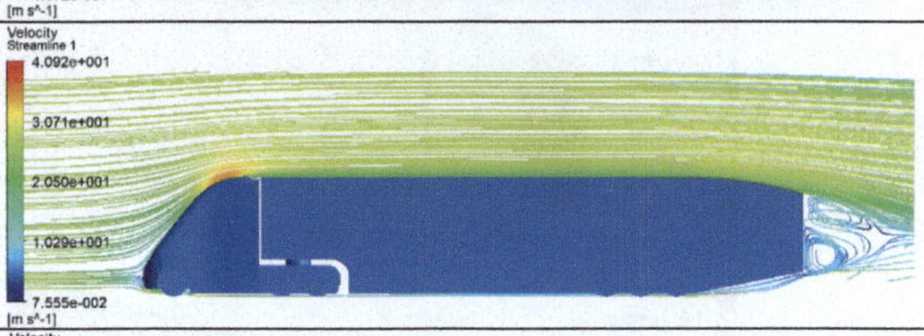
M3	0.38	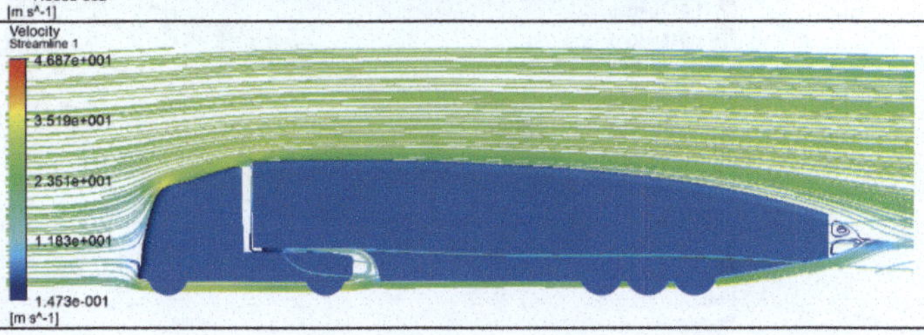

Aerodynamic drag is proportional to the square of velocity (Eq. 3), and hence the power needed to overcome the drag force, where it is proportional to the cube of velocity. This means that there is very strong relationship between the vehicle speed and the proportion of fuel used to overcome the drag. The form of pressure can be written as follows:

$$P_a = \frac{1}{2} c_x \rho A v^3 \quad (3)$$

where:

c_x - drag coefficient of the object,
ρ - density of the air,
A - frontal area of the object,
v - velocity of the air.

From the figures given in Table 2, it can be seen the critical zones, or more accurate, the high air currents turbulence. The turbulence has almost always occurred at the rear side of the trailer and between the cabin and the trailer. If the obtained results are taken into consideration later on in vehicle design, then this disadvantage can be removed or at least reduced to the minimum level. Also for such vehicles, it is important to pay more attention to the shape of the front area of the cabin. The effect of the dimensions and design of the surface on the normal vector of the air currents should be deeply investigated. When the cabin is flat, the pressure distribution is approximately uniform over the full surface of the cabin as shown in Fig. 8.

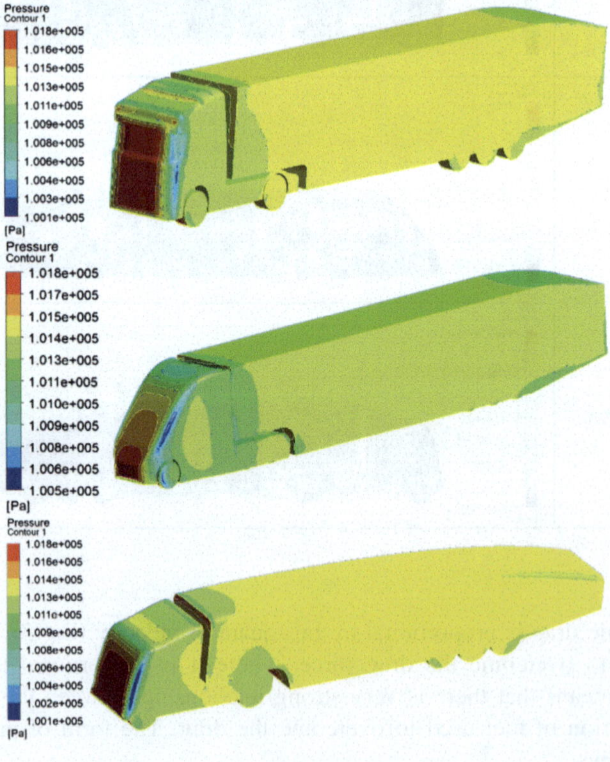

Fig. 8. Distribution of pressure over vehicle contour

The vehicle pushes the air forward during the movement, while at the rear side of the vehicle (in this case - behind the trailer), the air density decreases compared to the front end of the vehicle. Therefore, the vehicle should be created with the shape of the water drop, which is very efficient aerodynamic shape, where model (M2) represents this case. The critical zones for trucks models were determined and represented in Fig. 9 through turbulence kinetic energy. Based on the analysis of the selected models, it was found that the turbulence were occurred behind the trailer and at the vehicle sides.

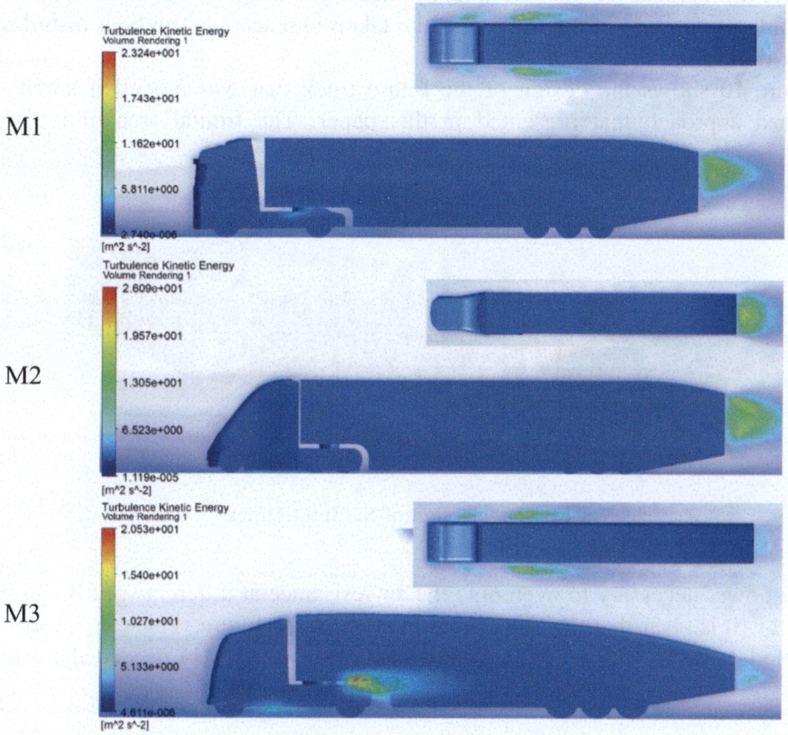

Fig. 9. Turbulence kinetic energy

4 Heavy Duty Vehicle of the Future

In order to reduce the drag coefficient for future heavy-duty vehicle, it should fulfil the following necessary requirements:

- The shape of the cabin must be similar to the truck model (M2), because the turbulence will not occur at lateral sides of the vehicle.
- The shape of the trailer should be as for the truck model (M3), because the turbulence behind the trailer is very small and can be neglected.

- The space between the cabin and the trailer should be small as much as possible.
- The height of the cabin and trailer should be the same in order to avoid the direct influence of the airflow on the front side of the trailer.

The shortcomings of the truck model (M2) occurred because the lateral spoilers were too close to the floor, and because of this the truck cannot be driven on all roads. Furthermore, it is necessary to find the optimal dimensions for the spoilers in order to avoid increase of the resistance. On the other hand, the vehicle should have the possibility of driving on different types of roads. These vehicles are used in the international transport of goods, so they must be driven on different roads of many countries. Therefore these trucks should be universal. Besides, during the design of such vehicles, the placing of all parts of the truck must be taken into account, without disturbance of comfort during the driving the vehicle.

Figure 10 shows the design of the future truck that was modelled based on the developed approach that presented in this paper. The frontal area of the truck is 8.43 m^2.

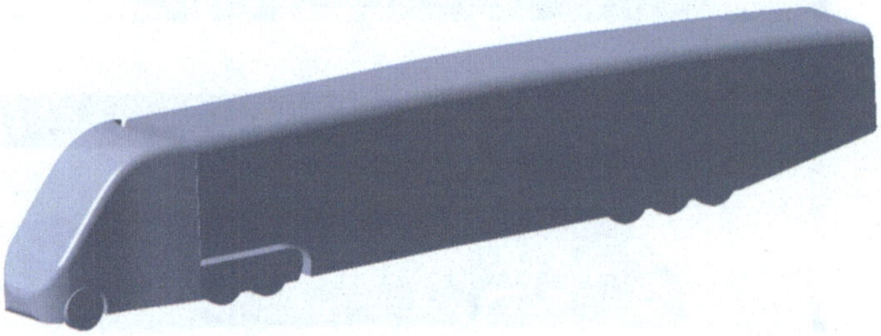

Fig. 10. Design of the future truck

The power necessary to overcome the air resistance at different speeds can be seen in Fig. 11. It is clear that the power requirement of the future vehicle is lower than all other models (M1, M2 and M3), while the truck model (M1) needed the maximum power.

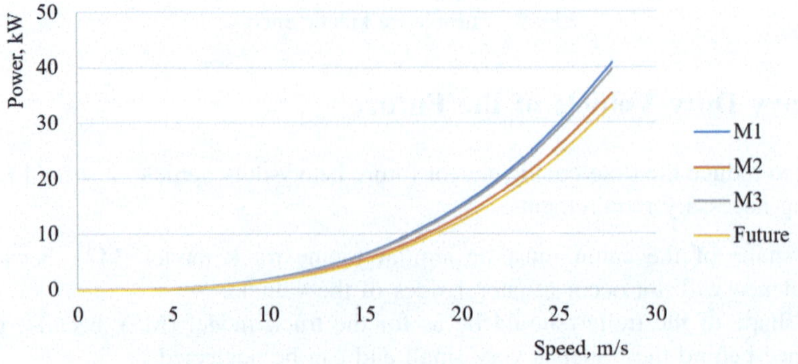

Fig. 11. Power to overcome aerodynamic drag

The drag coefficient for the future truck is 0.3. The air currents are shown in Fig. 12. The air turbulence occurred behind the trailer in this case also, which is illustrated in Fig. 13. However, the maximum values of turbulence are lower than typical models of trucks. In future research, such trailers should be studied in order to reduce the turbulence effect to minimum level.

As for the pressure values that occurred on the vehicle surface, they are the same as for the previous cases, Fig. 14. This is because the truck speed which assumed in the analysis was the same as in previous cases.

Fig. 12. Stremline on future vehicle

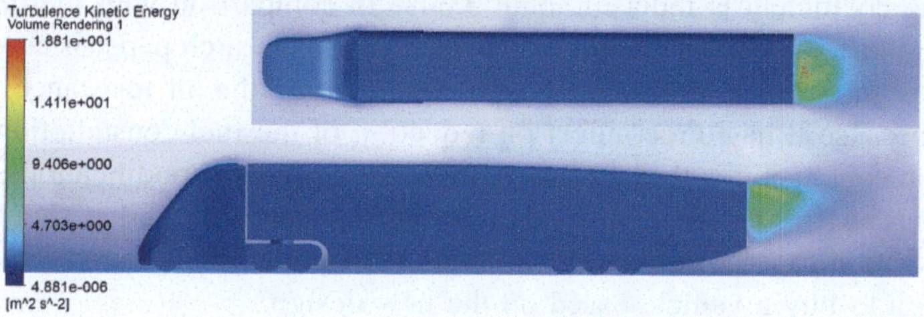

Fig. 13. Turbulence kinetic energy

Fig. 14. Distribution of pressure over vehicle contour

In market surveys, where the question was "Would you buy the vehicle which design is shown in Fig. 10?", there was no negative answer. But, the customers had other questions:

- What will be the cost of such truck?
- What will be the maintenance costs?
- What is fuel consumption for this truck?

The design of the truck is not very important factor, while the safety and profit is always an important factor.

5 Conclusions and Recommendations

Today, there is a strong tendency to reduce the fuel consumption, as much is possible. One of the ways is to reduce of the drag coefficient. This mission can be accomplished by alteration of the exterior design of the vehicle. This research paper presented the three existing models of trucks that were used as inputs to create the future truck as output of the analysis. After a deep study of aerodynamic behaviour of the trucks, the disadvantages of each truck were found. These disadvantages were taken into consideration to reduce their effects to minimum level for building the future design for truck.

The drag coefficient is reduced about 3.34% in comparison with the best existing model. Therefore, it can be said that the purpose of the research paper is accomplished. The new design of truck needs less power to overcome the air resistance. This is an advantage because it has direct effect on reduction of the fuel consumption.

Common conclusion is that there is more space for shape modifications, especially for the trailers. Also, the last update of the feasibility study of market should be taken into consideration to find out data on willingness of potential customers in the field of transportation to buy a vehicle based on the new design.

This research is preliminary research that will be followed by investigations of possibilities for optimization of trailers to reduce the turbulence as much as possible. The other important point is study the influence of implementation of spoilers on the trailer.

Acknowledgement. This paper was realized within the framework of the project "The research of vehicle safety as part of a cybernetic system: Driver-Vehicle-Environment", ref. no. TR35041, funded by the Ministry of Education, Science and Technological Development of the Republic of Serbia.

Also, the authors would like to thank the System Technology and Mechanical Design Methodology Group/Hamburg University of Technology to support this research paper.

References

1. Harun, C., Bavin, L., Israt, M., Hazim, M., Firoz, A.: Effect of various deflectors on drag reduction for trucks. Energy Procedia **110**, 561–566 (2017)
2. Pevitt, C., Chowdury, H., Moriaand, H., Alam, F.: A computational simulation of aerodynamic drag reductions for heavy commercial vehicles. In: Brandner, P.A., Pearce, B. W. (eds.) 18th Australasian Fluid Mechanics Conference, Launceston, Australia, pp. 369–372 (2012)
3. Xu, J., Zhou, S.: Flow field analysis of trucks and a design of an additional drag reduction device. Eng. Rev. **38**(1), 70–78 (2018)
4. Mccallen, R., Salari, K., Ortega, J., Castellucci, P., Pointer, W.D., Browand, F., Ross J., Storms, B.: DOE project on heavy vehicle aerodynamic drag, FY 2006 Annual Report (2007)
5. Cooper, K.R.: Full-Scale Wind Tunnel Tests of Production and Prototype, Second-Generation Aerodynamic Drag-Reducing Devices for Tractor Trailers. SAE Publication, SAE Paper No. 2006-01-3456 (2006)
6. Statista. https://www.statista.com/statistics/262858/change-in-opec-crude-oil-prices-since-1960/. Accessed 30 May 2019
7. Schütz, T.: Hucho – Aerodynamik des Automobils Strömungsmechanik, Wärmetechnik, Fahrdynamik, 6, KomfortATZ/MTZ-Fachbuch (2013)
8. Schoon, R., Pan, F.P.: Practical devices for heavy truck aerodynamic drag reduction, SAE, 2007-01-1781 (2007)
9. Harun, C., Hazim, M., Abdulkadir, A., Iftekhar, K., Firoz, A., Simon, W.: A study on aerodynamic drag of a semi-trailer truck. Procedia Eng. **56**, 201–205 (2013)
10. Wolf-Heinrich, H.: Aerodynamik des Automobils Strömungsmechanik, Wärmetechnik, Fahrdynamik, Komfort, 5th edn. Springer (2005)
11. Watts up with that. https://wattsupwiththat.files.wordpress.com/2009/11/semi-truck-savings.png. Accessed 02 Dec 2019
12. Bayındırlı, C., Erkan Akansu, Y., Sahir Salman, M.: The determination of aerodynamic drag coefficient of truck and trailer model by wind tunnel tests. Int. J. Automot. Eng. Technol. **5** (2), 53–60 (2016)
13. Janković, A.: Vehicle Dynamics (in Serbian), Kragujevac (2008)
14. Bayraktar, I., Baysal, O., Bayraktar, T.: Computational parametric study on external aerodynamics of heavy trucks. In: The Aerodynamics of Heavy Vehicles: Trucks, Buses, and Trains, pp. 485–501 (2004)
15. Diesel MB LP321. http://elizeudacarachata.blogspot.com/2012/11/scania-top-line-prata.html . Accessed 30 Apr 2019
16. Pluton Logistics. https://plutonlogistics.com/drumski-transport/kako-zaista-izgleda-posao-vozaca-kamiona-u-americi/. Accessed 30 Apr 2019
17. Ansys. https://users.encs.concordia.ca/home/m/m_mamu/ANSYS%20CFX%20documentation/cfx_thry.pdf. Accessed 30 Apr 2018

Research of Intelligent Control of Injection Systems for Subaru Competition Car

Ioan-Aurel Chereches, Doru-Laurean Băldean(✉), and Adela-Ioana Borzan

Technical University of Cluj-Napoca, 400641 Cluj-Napoca, Romania
doru.baldean@auto.utcluj.ro

Abstract. In the present research paper, an approach based on Engine Digital Scanning (EDS) was conducted in order to show the performances of more than ten engine operating regimes on the Subaru rally car, at various values regarding the intake manifold pressure, fuel injection duty and ignition advance. Main objective of the present work is to show the complex method of Engine Digital Scanning and to represent the experimental determinations and measurement values concerning racing car controls available on Subaru at Technical University from Cluj-Napoca. Specific objectives consist in experimental testing of ignition advance, engine speed, manifold pressure, fuel injection duty and highlighting the practical protocol implemented at Automotive Engineering and Transportation.

Keywords: Automotive · Diagnosis · Fuel-injection · Hydraulic · Racing

1 Introduction

In the present research paper, an approach based on Engine Digital Scanning (EDS) was conducted in order to show the performances of more than ten engine operating regimes on the Subaru rally car, at various values regarding the intake manifold pressure, fuel injection duty and ignition advance. Main objective of the present work is to show the complex method of Engine Digital Scanning and to represent the experimental determinations and measurement values concerning racing car controls available on Subaru at Technical University from Cluj-Napoca. Specific objectives consist in experimental testing of ignition advance, engine speed, manifold pressure, fuel injection duty and highlighting the practical protocol implemented at Automotive Engineering and Transportation.

2 Methodology, Materials and Input Data for Research

An experimental study of trend lines and actual values variation has been realized in order to highlight internal combustion engine's (ICE) operational indicators and to define the optimal conditions with higher intake pressure, different combustion and diminished pollution. By using modern technologies, measuring equipment for electronic system, and testing the vehicle researchers have looked to the graphs and

I. Dumitru et al. (Eds.): SMAT 2019, *The 30th SIAR International Congress of Automotive and Transport Engineering*, pp. 82–87, 2020.
https://doi.org/10.1007/978-3-030-32564-0_10

conducted a thorough analyze. The present research implements an innovative method based on digital measurement regarding operational parameters related to Subaru racing car. The Subaru car is a fully operational model, which was used and monitored in relation with ignition advance and fuel injection duty. An original approach of research was put into practice during the present research.

Main objective of this research is based on the methodology shown in Fig. 1.

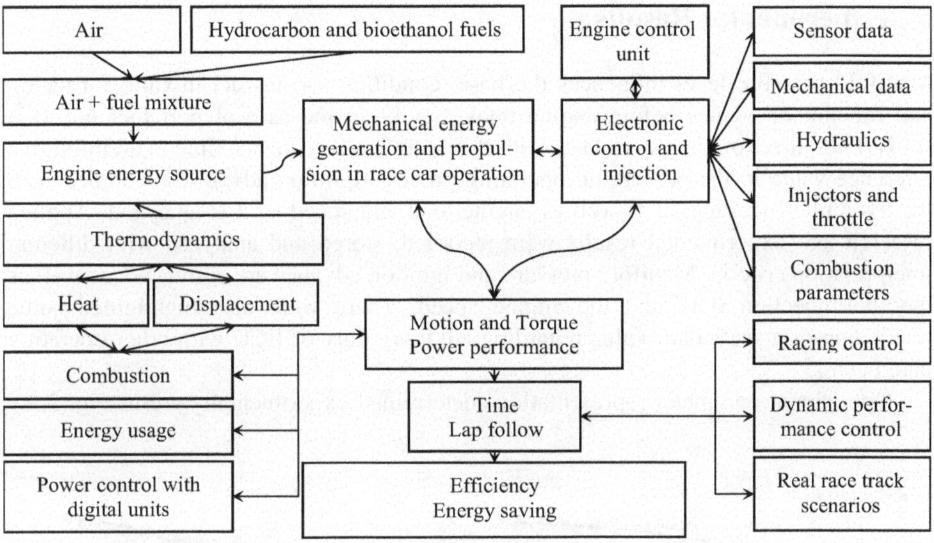

Fig. 1. Schematic of methodology in intelligent control of a race car.

The application of computer aided scanning in a racing car operating conditions represents an improvement of research capabilities, technologies and testing. The Motec interface allows to consider more parameters and to bring some new features in the study. There were considered some particular measurements regarding operational characteristics before [1–4], but the present research is one of the most applied in the practical sense of the Subaru research project. The latter one analyzes important aspects that define ignition and injection evaluating criteria by using electronic interface and comparative.

Data for Subaru competition car technical details are shown in the Table 1.

Table 1. Actual values regarding Subaru competition car

Parameters	Values	Parameters	Values
Model year	2008	Wheelbase, [mm]	2625
Vehicle body, [-]	CT9A	Length, [mm]	4490
Engine type, [-]	4G63T	Width, [mm]	1770
Gearbox	6 ratios	Height, [mm]	1450
Speed, [km/h]	(a) 200; (b) 250	Weight, [mm]	1350

Mathematical model for fuel injection duty in regard to the engine speed is given as follows:

$$F_{id} = -6 \cdot 10^{-6} \cdot x^2 + 0.0376 \cdot x + 45.324 \tag{1}$$

where F_{id} is fuel injection duty, in %, x – engine speed, in rpm.

3 Experimental Results

Manifold pressure level influences the basic condition for air-fuel mixture formation and for lambda values before engine intake stroke in the case of port fuel injection (PFI). Fuel injection duty correlates with the operational sequence time in addition to a reference value in the powertrain operating process. Ignition advance is coupled with the hydraulic-fuel charge, as well as engine load and speed so it is analyzed. Applied research and experimental results were recorded, stored and analyzed with different interpretation criteria. Manifold pressure and ignition advance are correlated, as well as the fuel injection duty and the engine speed. There were also highlighted some interaction of a secondary value regarding auxiliary duty of ECU with other operating parameters.

Operational parameter representations determined experimentally are in Fig. 2.

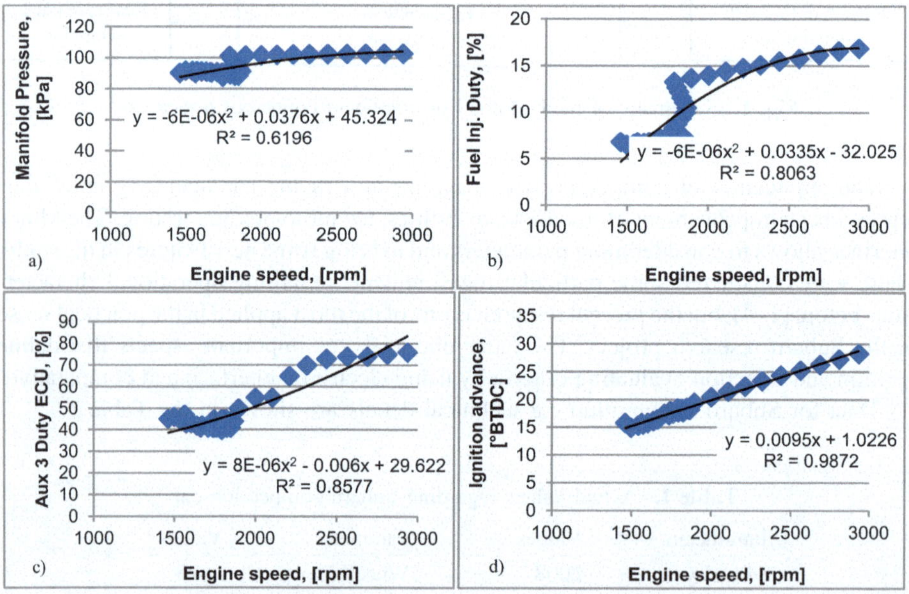

Fig. 2. Operational parameters of Subaru competition car versus engine speed a-manifold pressure; b-fuel injection duty; c-auxiliary injection duty of electronic control unit; d-ignition advance.

Operational fuel injection duty and manifold pressure at the intake representations measured experimentally on the Subaru competition car are given in Fig. 3.

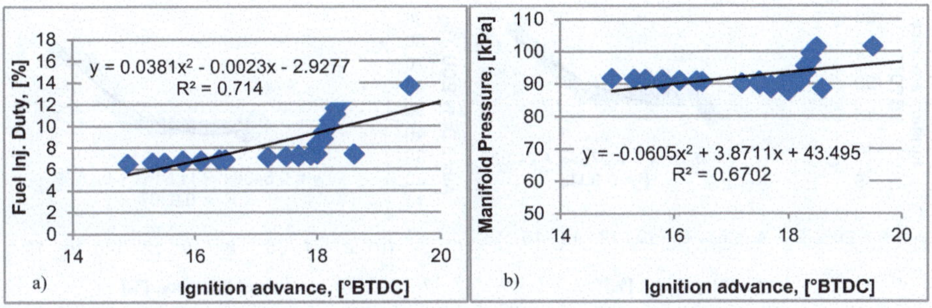

Fig. 3. Fuel injection and intake pressure versus ignition advance a-fuel injection correlated with ignition advance; b-manifold pressure correlated with ignition advance.

Operational parameters versus throttle position determined practically are in Fig. 4.

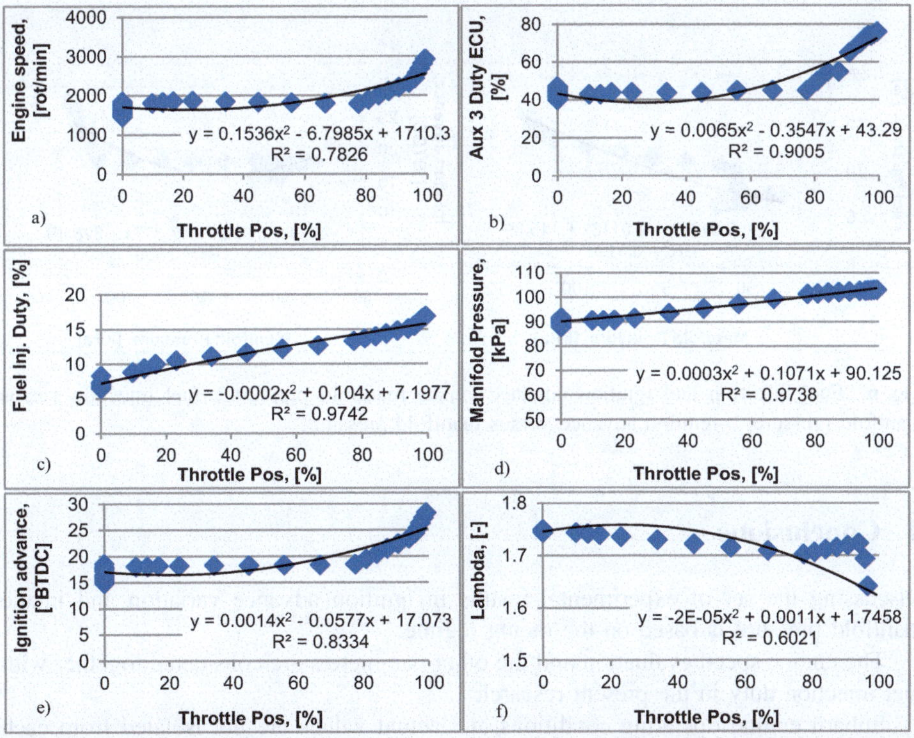

Fig. 4. Operational parameters versus throttle position a-engine speed; b-auxiliary duty from electronic control unit; c-fuel injection duty; d-manifold pressure; e-ignition advance; f-lambda.

Operational parameters versus fuel injection duty considered as an image of consumption determined practically are in Fig. 5.

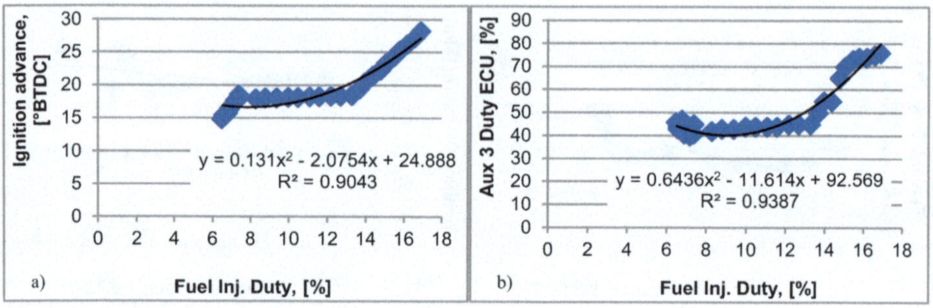

Fig. 5. Fuel injection and intake pressure versus ignition advance a-ignition advance versus fuel injection; b-auxiliary duty of electronic control unit correlated with fuel injection.

Actual values versus manifold pressure determined practically as a load expression are given in Fig. 6.

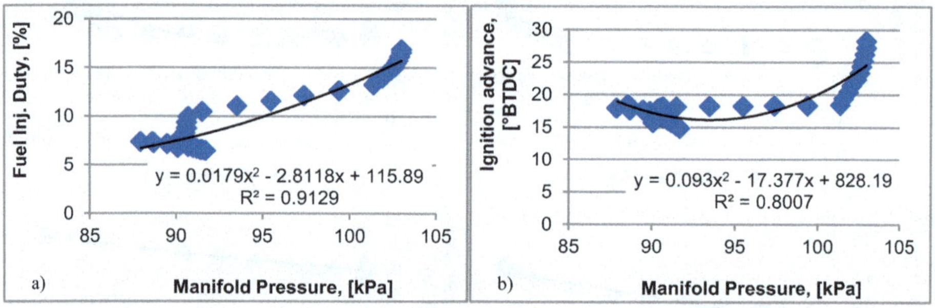

Fig. 6. Fuel injection and ignition advance versus manifold pressure a-fuel injection versus manifold pressure; b-ignition advance versus manifold pressure.

4 Conclusions

Discussing the set of experimental results in ignition advance variation and intake manifold pressure is based on the racing regime.

The engine speed evaluation and the other parameters are considered together with fuel injection duty in the present research.

Subaru engine operating conditions and actual values are not isolated from each other, but interacting with each other the whole environment. The engine, the operating system and the fuel injection duty, such as charge, should be studied in relation to the power-train structure and group of influences from the entire motor vehicle running on the race track and influencing and being influenced by the surrounding environment.

The fuel injection duty and ignition advance parameters have great influence on the vehicle's performance.

Ignition advance angle before top dead center increases with 20–30% when manifold pressure is in the interval of 100–105 kPa.

Research opportunities are thus emerging in order to steer up further studies of the outlined problems.

References

1. Van Basshuysen, R.: Gasoline Engine with Direct Injection. Processes, Systems, Development, Potential. Wiesbaden, Vieweg+Teubner, GWV Fachverlage GmbH (2009)
2. Ferenți, I., et al.: Analizing a thermal management solution of engine as embedded technology in rally vehicles/Analiza unei soluții de management termic al motorului ca tehnologie integrată în autovehiculul de raliu, 3 Mai 2017. https://stiintasiinginerie.ro/31-58. Accessed 3 June 2018
3. Ferenți, I., et al.: Artificial intelligence implemented in rally vehicles for increasing energetic efficiency in competitions/Inteligența artificială implementată în autovehicule de raliuri pentru creșterea eficienței energetice în competiții, 29 Mai 2018. https://stiintasiinginerie.ro/34-18. Accessed 3 June 2018
4. Toader, G.-V., et al.: Contributions to the experimental research of injection fuel supply system from 3.0i engine of BMW X5 road vehicles/ Contribuții la cercetarea experimentală a parametrilor sistemului de alimentare prin injecție de la motorul 3.0i al automobilului BMW X5, 4 Mai 2017. Accessed 5 June 2018

Some Considerations on Optimal Placement of the Arm Hydraulic Cylinder to an Excavator

Cristian-Ioan Leahu(✉), Mircea Năstăsoiu, and Radu Tarulescu

Transilvania University of Brasov, 500036 Brasov, Romania
leahu.cristian@unitbv.ro

Abstract. The development of hydraulic equipment is in the direction of increasing working pressures, increasing rotation speed and travel speeds, while ensuring satisfactory reliability and durability and reducing operating fluid losses. But, a high efficiency of the machine also implies a high-quality connection between the hydraulic and mechanical systems. A correct positioning of the hydraulic cylinders ensures the kinematic movement of the arms, boom and bucket. In addition, a correct positioning of the hydraulic cylinders implies the maximization of the heavy duty construction vehicles performance. In this sense, the arms excavators have a prolongation that permits the piston to have the extension stroke during digging. The length of this continuation of the arm must take account of the disadvantages of an oversized arm and of the advantage of reducing the force the cylinder must transmit. In order to highlight the importance of optimal cylinder positioning, in paper are analyzed the possibilities of placing the arm cylinder to an excavator.

Keywords: Excavator · Arm · Hydraulic cylinder · Optimal placement

1 Introduction

The construction and maintenance of roadways involves the use of heavy duty construction vehicles such as: excavators, bulldozers, backhoe, graders, well tractor scraper, trencher, loaders, telehandlers, feller bunchers, pile boring machine, dump truks, etc. In addition, some of these vehicles besides the construction domain are absolutely necessary in other fields, such as: industrial, forestry and agricultural.

All the vehicles listed fulfill their objective using a hydraulic system that contains cylinders. These represent the hydraulic elements of execution, which actuated in generally the elements of the mechanical system: booms, arms, bucket, blades and open-box bed [1].

The hydraulic system includes a pump driven by the internal combustion engine. It converts mechanical energy into hydraulic energy. This energy is transported by hydraulic oil through hoses and pipes. Hydraulic energy is transformed back into mechanical energy by hydraulic cylinders, which actuated the some mechanical system components [2]. In this way the hydraulic system and the mechanical system work together so that the heavy duty vehicles meets its main objective (mechanical digging, spreading, profiling, scarification, removal of weak soil, carry the material, lifting of materials, remove large trees, drilling, etc.).

I. Dumitru et al. (Eds.): SMAT 2019, *The 30th SIAR International Congress of Automotive and Transport Engineering*, pp. 88–94, 2020.
https://doi.org/10.1007/978-3-030-32564-0_11

The development of hydraulic equipment is in the direction of increasing working pressures, increasing rotation speed and travel speeds, while ensuring satisfactory reliability and durability and reducing operating fluid losses [3]. But, a high efficiency of the machine also implies a high-quality connection between the hydraulic and mechanical systems. A correct positioning of the hydraulic cylinders ensures the kinematic movement of the arms, boom and bucket. In addition, a correct positioning of the hydraulic cylinders implies the maximization of the heavy duty performance.

2 Aspects Considered in the Placement of Hydraulic Cylinders

In order to highlight the importance of correct cylinder positioning, the following are analyzed the possibilities of placing the arm cylinder to an excavator.

2.1 The Kinematic Movement

Two of the types of excavators commonly used in the construction and maintenance of roadways are those with a loader and digging bucket. Excavators with loader bucket are capable of excavating just above the excavator's support level. The other type of digging bucket (see Fig. 1) has a smaller bucket but can digging both above and below the level of heavy duty vehicle.

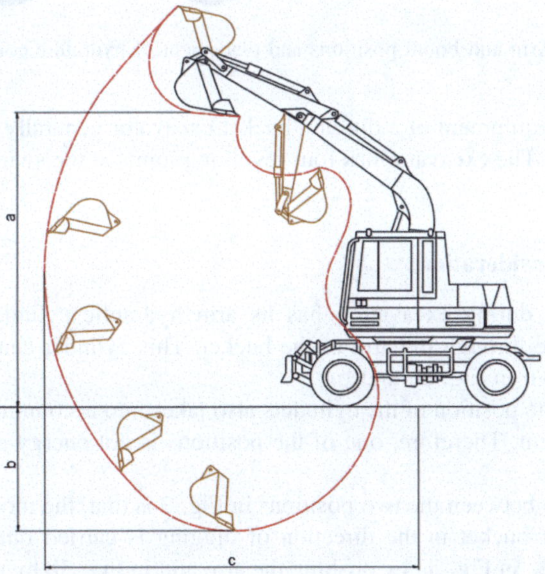

Fig. 1. The trajectory of the bucket to an excavator.

In Fig. 1, the distance a represents the maximum loading height of the bucket, the distance b represents the depth of excavation below the level of the machine and distance c is the maximum action range on the ground level.

The working equipment is controlled by cylinders and hydraulic actuators. The placement of the hydraulic components on the excavator work equipment is highlighted in Fig. 2.

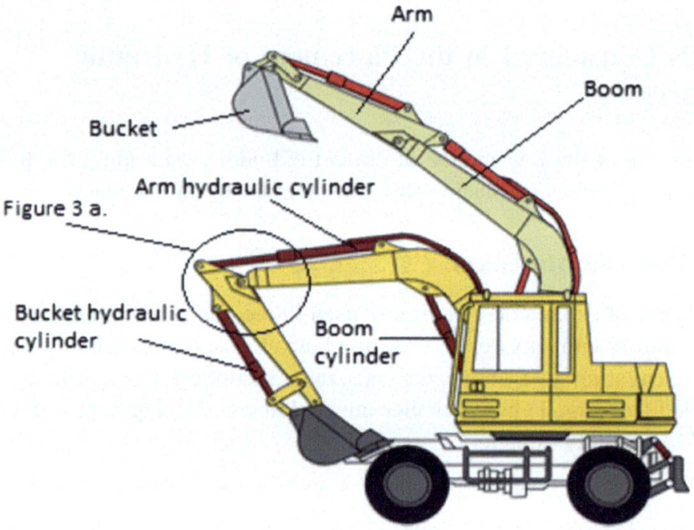

Fig. 2. Arm and boom positions and placement of hydraulic components.

The working equipment of a digging bucket excavator generally contains one arm, boom and bucket. The excavator has four revolving joints at the swing, boom, arm and bucket [1].

2.2 Energy Considerations

An essential role during excavation has its arm hydraulic cylinder. Practically, he transmits the force through the arm to the bucket. This cylinder can be placed in two positions, as shown in Fig. 3a and b.

However, to the position of the cylinders also takes into account the performance of the work equipment. Therefore, one of the positions is not energy efficient as will be seen below.

The difference between the two positions in Fig. 3 is that, the movement of the arm and implicitly the bucket in the direction of digging is carried out by the hydraulic cylinder as follows: in Fig. 3a by pushing the arm and in Fig. 3b by pulling the arm. In other words, in the left position during digging, the piston makes the extension stroke (see Fig. 4a) and in the right position the piston performs the retraction stroke (see Fig. 4b).

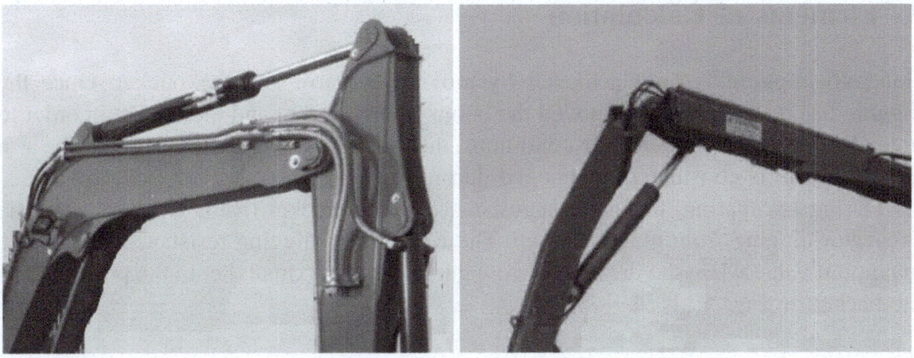

a. b.

Fig. 3. Possibilities of positioning the arm hydraulic cylinder.

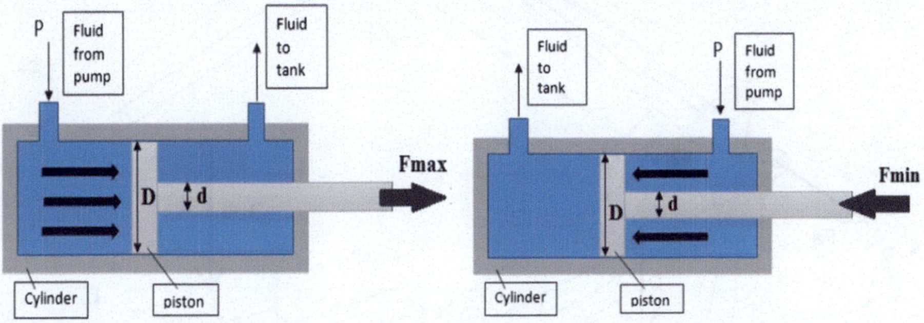

a. b.

Fig. 4. Double acting cylinder: a. Extension of the piston, b. retraction of the piston.

In the two strokes of the piston the force (F) transmitted by the piston rod is different because it differs the area (A) of the piston on which the oil pressure acts. For the extension stroke the piston area is effective: $A_{ex} = \frac{\pi \cdot D^2}{4}$. On the retraction stroke the effective area is diminished by the rod. For the retraction the piston ring area is effective: $A_{re} = \frac{\pi \cdot \left(D^2 - d^2\right)}{4}$. Therefore the extraction force is $F_{ex} = p \cdot \frac{\pi \cdot D^2}{4} \cdot \eta_{mec}$ and the retraction force $F_{re} = p \cdot \frac{\pi \cdot \left(D^2 - d^2\right)}{4} \cdot \eta_{mec}$.

It is necessary that during the digging the piston executes the stroke in which it to transmits the highest force, therefore the arm hydraulic cylinder is placed as shown in Fig. 3a.

3 Elements of Calculation

Generally digging is only achieved by moving the arm and the bucket. Once the digging has been completed, moved the boom to lift the arm and the bucket in order to unload the material. Under these conditions, the position of the joint O between the arm and the boom is considered unchanged during digging.

During excavating, the greatest resistance to the bucket (Rdb) teeth is when the excavator digging from the bottom up. The Rdb bucket digging resistance reaches the maximum value when the dredged furrow thickness is maximal, i.e. in the position 2 of the bucket, in Fig. 5 [4, 5].

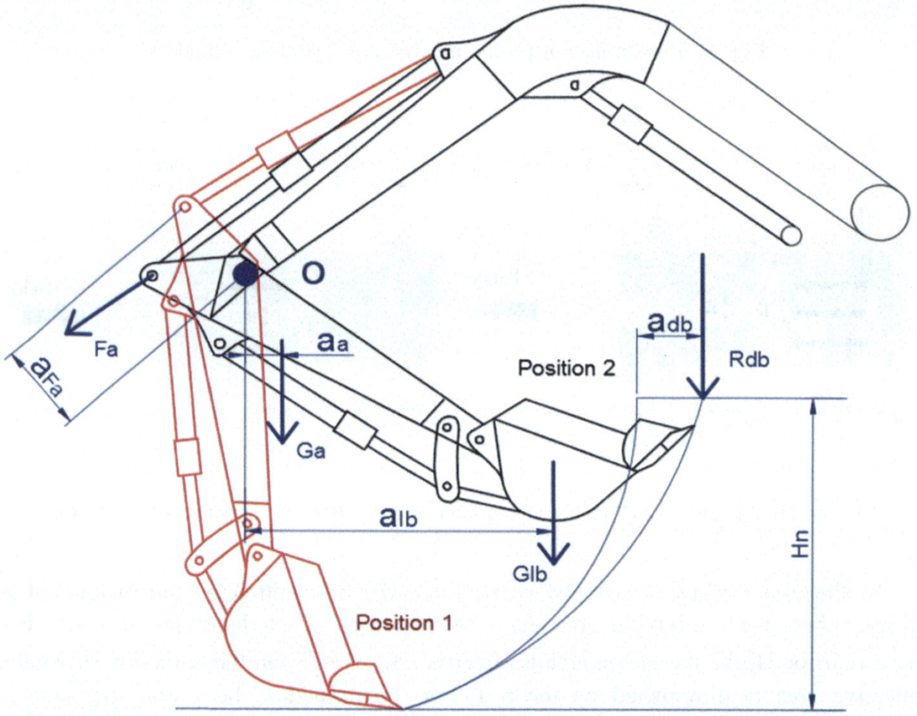

Fig. 5. The energy and dimensional parameters required for calculating optimal distance a_{Fa}.

As stated above, the drilling force depends directly on the Fa force transmitted by the arm hydraulic cylinder. The maximum force Fa is calculated when the cup is in the final position of digging, i.e. in position 2. If the equilibrium equation of the moments is calculated with respect to the O joint, it results [4]:

$$F_a = \frac{R_{db} \cdot a_{db} + G_a \cdot a_a + G_{lb} \cdot a_{lb}}{a_{Fa}} \quad (1)$$

were resistance R_{db} is,

$$R_{db} = c_1 \cdot \frac{q}{H_n \cdot c_2} \tag{2}$$

In relation (2), c_1 is the specific resistance to digging, q represents the capacity of the bucket, c_2 coefficient that depends on the earth category and H_n is the depth of digging. Also in relation (1), G_a and G_{lb} represent the weight of the arm and the loaded bucket. The parameters a_{db}, a_a, a_{lb} and a_{Fa} correspond to the distance between the forces and the O joint.

The distance a_{Fa} is chosen so that the force transmitted by the arm hydraulic cylinder has an optimum value. If it is considered an example of excavator and numerically replace parameters in relation 1, then it can be observed (see Fig. 6) that by changing the a_{Fa} distance from 0.2 to 0.9 m, the force F_a changes by up to 78%.

Modifying the force is necessary in order not to change the equilibrium of the moments with respect to the O joint.

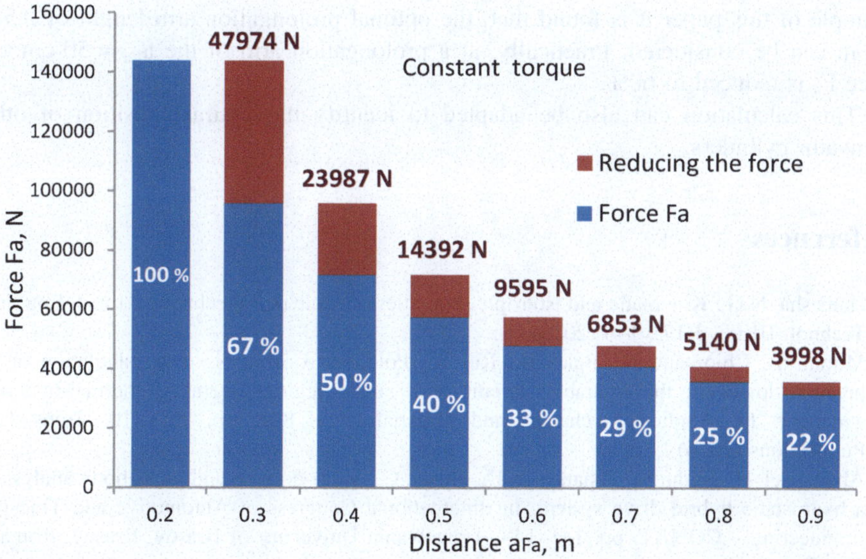

Fig. 6. The variation of the force Fa in relation to the distance a_{Fa}, in the conditions of a constant transmitting torque.

It is noted that by increasing a_{Fa} from 0.2 to 0.3 m an important decrease in the F_a force of 47974 N is obtained and from 0.3 to 0.4 a decrease of 23987 N. Thus, an optimal distance of 0.5 or 0.6 m can be considered. An even greater increase of the a_{Fa} distance over 0.6 m is not justifiable given the modest increase in F_a force and the constructive and functional complications that involve an excessive increase in the dimensions of the arm. In addition, for a distance of 0.6 m the force F_a can be generated by a pressure of 200 bar, which is a common value for excavators.

4 Conclusions

Hydraulic cylinders must be placed on working equipment so that they are able to transmit maximum forces during digging. In this sense, the arm has an extension (length a_{Fa}) that permits the piston to have the extension stroke during excavation. The length of this extension must take account, on the one hand, of the disadvantages of an oversized arm and, on the other hand, of the advantage of reducing the force the piston must transmit.

As shown in the paper, the values of the following parameters should be used when calculating the length of the prolongation arm: resistance to the bucket, weight of the arm and the loaded bucket, specific resistance to digging, capacity of the bucket, coefficient that depends on the earth category and depth of digging. According to the calculations in the paper there is a value of a length over which the reduction of force is insignificant in relation to the disadvantages introduced in the mechanism. In the example of this paper it is found that, the optimal prolongation arm length of 0.5 or 0.6 m can be considered. Practically, at a prolongation arm of the $a_{Fa} = 50$ cm, the force F_a is reduced to 60%.

This calculation can also be adapted to identify the optimal position of other excavator cylinders.

References

1. Hareesha, N.G.: Kinematic and isotropic properties of excavator mechanism. Int. J. Eng. Res. Technol.-IJERT **3**(17), 1–6 (2015)
2. Mitran, T., Chioreanu N., Beles H., Rus A., Polojintef-Corbu, N.: The calculation of the pressure losses in the hydraulic circuit of a single regime operating thermal-hydraulic generator. In: Applied Mechanics and Materials, vol. 822, pp. 305–310. Trans Tech Publications (2016)
3. Abaitancei, H., Radu, S., Tohaneanu, D., Husar, C.: Multi-domain and multi-body analysis of a hydrostatic hybrid drive system. In: International Congress of Automotive and Transport Engineering - CONAT, pp. 141–148. Transilvania University of Brasov, Brasov, Romania (2016)
4. Rosca, R., Dascalescu, D.S., Tenu, I., Rakosi, E., Manolache, G.: Autovehicule speciale si utilaje (Special vehicles and machinery). Politehnium Iasi Press, Iasi (2006)
5. Nastasoiu, M., Vasilache, L., Ispas, N., Soica, A., Nastasoiu, S.: Researches on the dimensional synthesis of the threepoint hitch couplers used at agricultural tractors. In: International Congress of Automotive and Transport Engineering - CONAT, pp. 47–54. Transilvania University of Brasov, Romania (2010)

Aspects Regarding Fuel Consumption and the Pollutant Products of Internal Combustion Engines for Commercial Vehicles

Marius S. D. Capătă[1(✉)], Mădălin Florin Popa[1], and Nicolae Burnete[2]

[1] Faculty of Automotive, Mechatronics and Mechanics, Technical University of Cluj Napoca, B-dul Muncii 103-105, 400461 Cluj Napoca, Romania
Dorin.capata@gmail.com

[2] Department of Automotive and Transportation, Technical University of Cluj Napoca, B-dul Muncii 103-105, 400461 Cluj Napoca, Romania

Abstract. The functioning regime of a vehicle represents the main element in the fuel's consumption and the pollutant products streamed into the atmosphere. If the human component can be acted on by training, in order to obtain performances in the exploitation of trucks, the functioning regime is, most of the times, imposed by the environmental and traffic conditions for the chosen route. This paper aims to analyze the vehicles' functioning regimes, underlining the dependency of the fuel consumption and the pollutant products on these factors. The analyses of the functioning parameters of the vehicles has been achieved with the help of the telematics systems supplied by the platform Dynafleet Online and SC Gerom Intertrans SRL-Ploiesti for commercial trucks that function in mixed regime internal-international. Through graphic representations the recorded values of fuel consumption and pollutant products are being underlined for usage in different atmospheric conditions and also for different types of road.

Keywords: Pollution · Fuel consumption · Road types · Atmospheric conditions

1 Introduction

An essential branch of the economy is represented by transportation. No matter if we refer to goods or persons, the terrestrial, maritime or air, transportation is at the basis of the industry's development, commerce, tourism and even education. Transportation has become a necessity that doesn't take into account the atmospheric condition, topography or road network, always finding ways for persons and goods to be moved from a place to another [1, 4].

It is well known that when the running conditions change, they determine changes in the functioning parameters of all the vehicle's components. If we refer to the climate changes, running in winter conditions imposes a more preventive type of driving, decreases the speed, increases the number of braking, increases idle rolling and the low gears are used more often [1, 2].

I. Dumitru et al. (Eds.): SMAT 2019, *The 30th SIAR International Congress of Automotive and Transport Engineering*, pp. 95–102, 2020.
https://doi.org/10.1007/978-3-030-32564-0_12

Functionally, the engines reach harder the optimum temperature, the rolling speed range increases and the transmission and braking systems are used more. An almost identical situation is met when the road type is change. The passing from highway to national or European road imposes speed reduction, the increase of brakes, acceleration and deceleration appear more often, the use of low gears appears [2, 3].

This paper aims to analyze the aspects related to the functioning regime of heavy-duty commercial vehicles and to underline the differences by comparing the data recorded in real time.

2 Methodology and the Interpretation of the Results

The development of the component systems of the heavy-duty commercial vehicles, the increase in the number of sensors and the interconnection od the sub-assemblies with the help of electronic architecture has facilitated the reading and recording certain functioning parameters. The transmission of these parameters and real time follow up has led to the development of new types of applications known as fleet monitoring systems. The usage of such systems brings a benefit to the transporters, the efficiency in their activity being a key element for profitability [3].

Each of the commercial vehicles producers has developed and has offered to his customers such systems for monitoring and administrating their fleet. Besides the localization of the vehicles in real time, the fleet monitoring systems offer their users the possibility to download different activity reports, their analyses improving the way of usage and the decrease of the exploitation costs.

The proposed study is based on collecting real time data, supplied by the monitoring platform dynafleet online, offered by Volvo Trucks, Fig. 1.

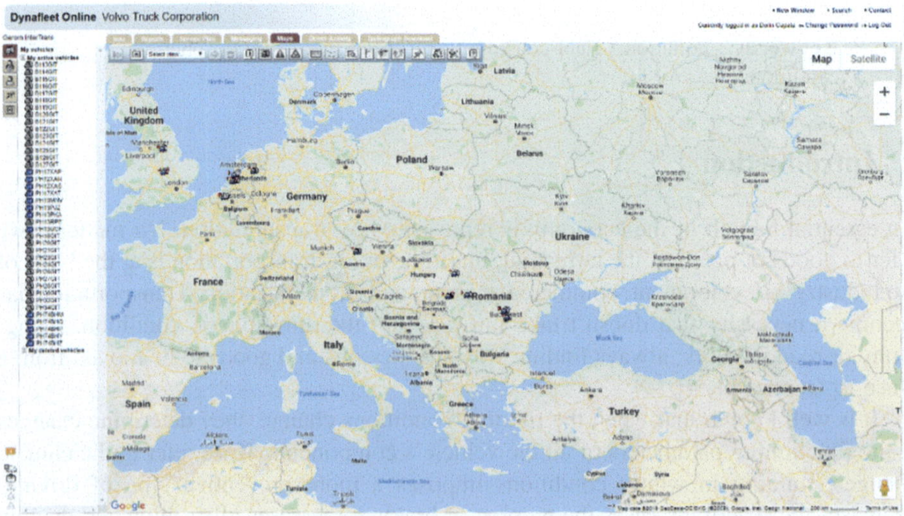

Fig. 1. The fleet monitoring system "dynafleet online" offered by Volvo Trucks [5].

For this study there have been chosen a number of 8 trucks, Volvo model FH4, produced in 2017, 460CP, Euro 6. By accessing the section reports of the fleet monitoring portal there have been used two of the variants, specifically the general data report and the environment report.

By using the general data report, it has been possible to record the data regarding the distance, fuel consumption, Ad Blue consumption, average speed, number of brakes and stops in 100 km, the percentage of idle rolling, the percentage of using cruise control, the percentage of using at maximum value the engine load as well as the total load of the truck divided in three categories: empty, average load and maximum capacity.

The environment report offers information regarding the main pollutant products such as the quantity of carbon monoxide (CO), carbon dioxide (CO_2), hydrocarbons (HC), nitrogen oxides (NO_x), particles (PM) and Sulphur oxides (SO_2).

The analyses of the data aimed two directions: the comparative study of the functioning regime in summer conditions (months April to September) – winter (moths October-March), respectively the comparative study of national and international rolling.

For the summer-winter comparative analyses the following have been recorded: consumption (fuel and AdBlue) and some of the data referring to functioning parameters (average speed, average number of brakes, idle rolling time, automatic pilot rolling time) for all the eight trucks that have been selected (Figs. 2 and 3).

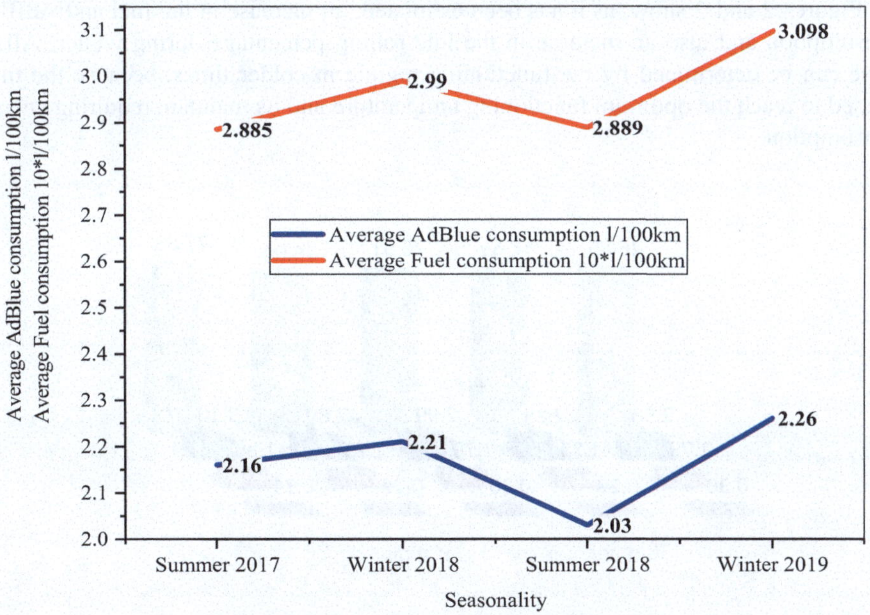

Fig. 2. Consumption data analyses

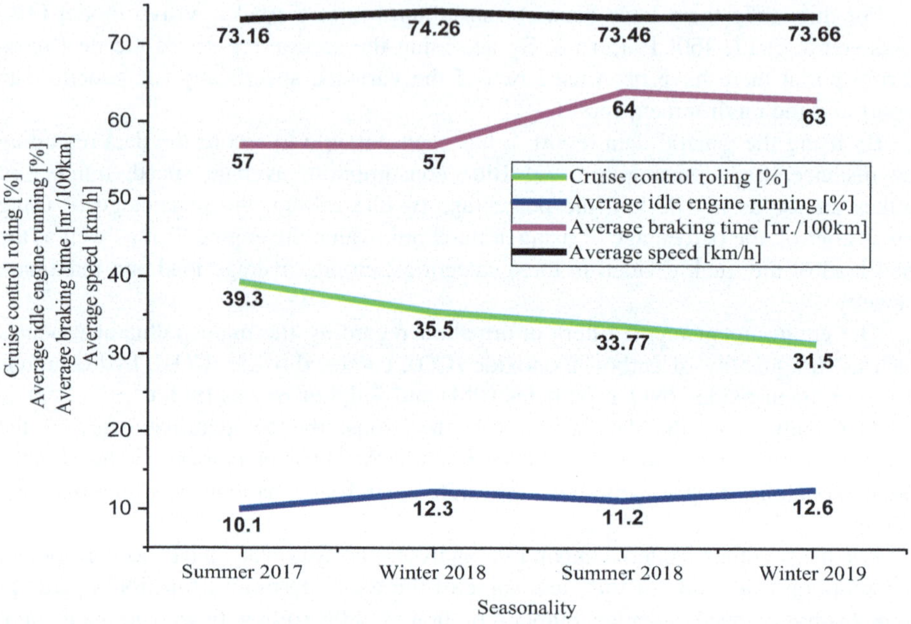

Fig. 3. Functional parameters

Figures 2 and 3 show, as it has been estimated, an increase in the fuel and AdBlue consumption and also an increase in the idle rolling percentage during winter. All of these can be determined by the functioning regime in colder times, because the time needed to reach the optimum functioning temperature and its maintain requiring energy consumption.

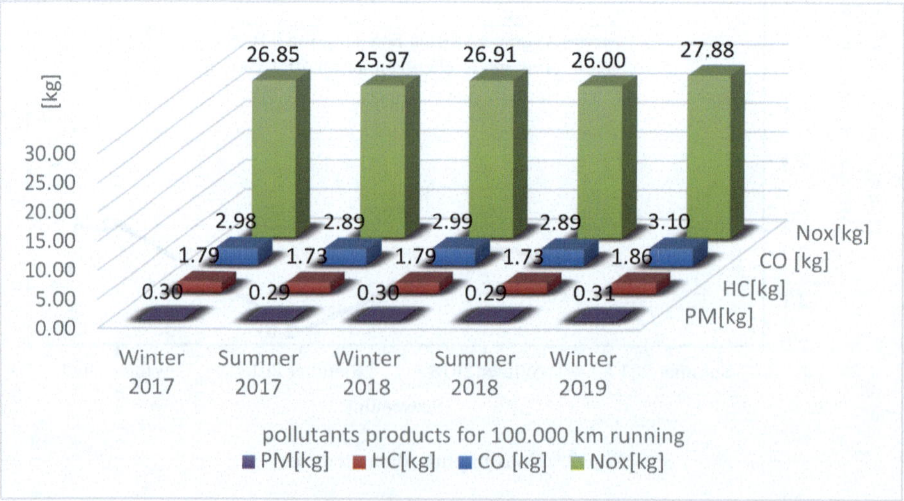

Fig. 4. The analyses of the main pollutants according to season

Surprisingly, the average speed as well as the average number of brakes are constant, the only parameters with noticeable changes being cruise control rolling.

From the point of view of the pollutant products, the analyses underlines increases in their values during winter time – Fig. 4, the predictability of the recorded values being supplied by the previously achieved analyses on fuel consumption.

For the comparative study, rolling on national/European (internal) road, in comparison to European/highway (international), there has been random chosen a number of 31 complete courses for all the selected trucks. Thus there have been avoided the alterations of results through changes in consumption or functioning parameters specific to each download/load, as well as the ones related to topography or congestion for different road sections. There have also been selected the courses with almost identical loads (values over 75% of the maximum admitted mass 20–24 tones), fuel consumption as well as the functioning regime being directly proportional to the truck's load.

Figures 5 and 6 present the recorded fuel and AdBlue consumption, while Figs. 7 and 8 present the values of the functioning parameters. Due to the specific of the chosen courses, for the values of the functioning parameters there have been pursued the average speed, the average number of brakes for 100 km, respectively the percentage of using the last gear.

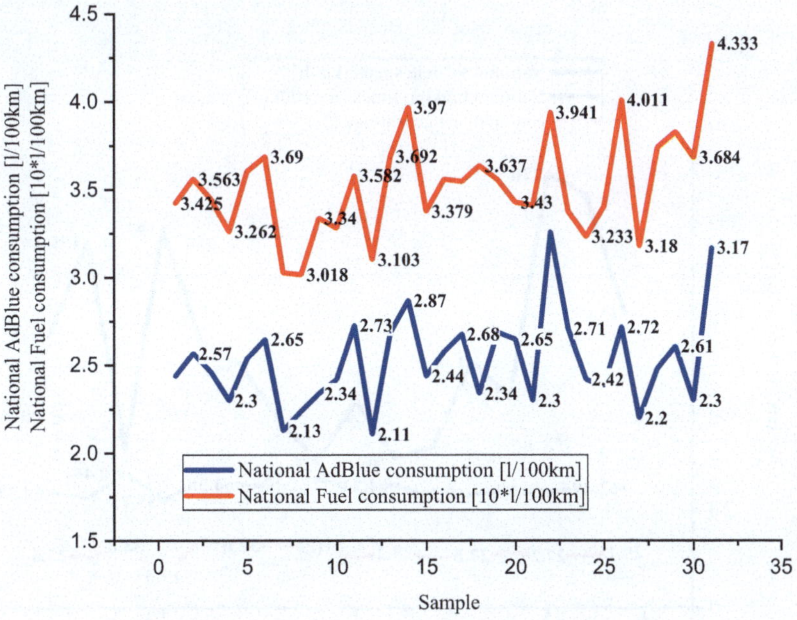

Fig. 5. Analyses of consumption data for national roads

By analyzing Figs. 5 and 6 it can be noticed a predictable increase in fuel and AdBlue consumption, for national roads, smaller values for fuel consumption appearing for national highways.

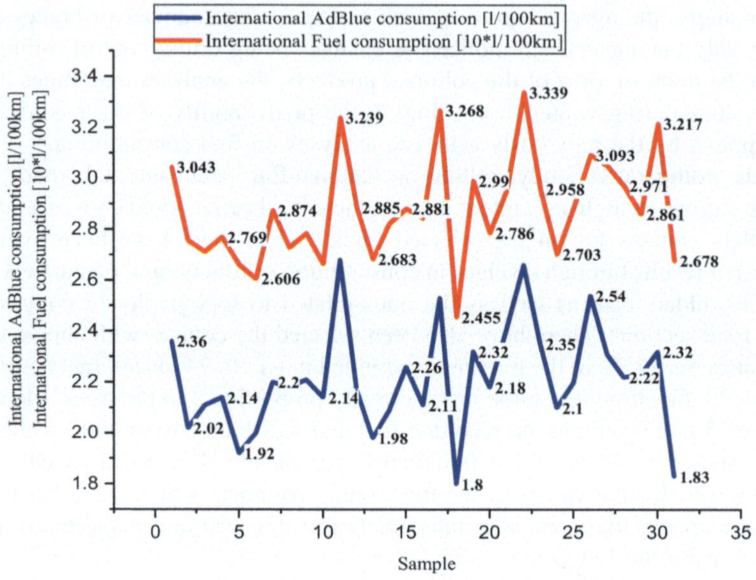

Fig. 6. Analyses for consumption data for international regime

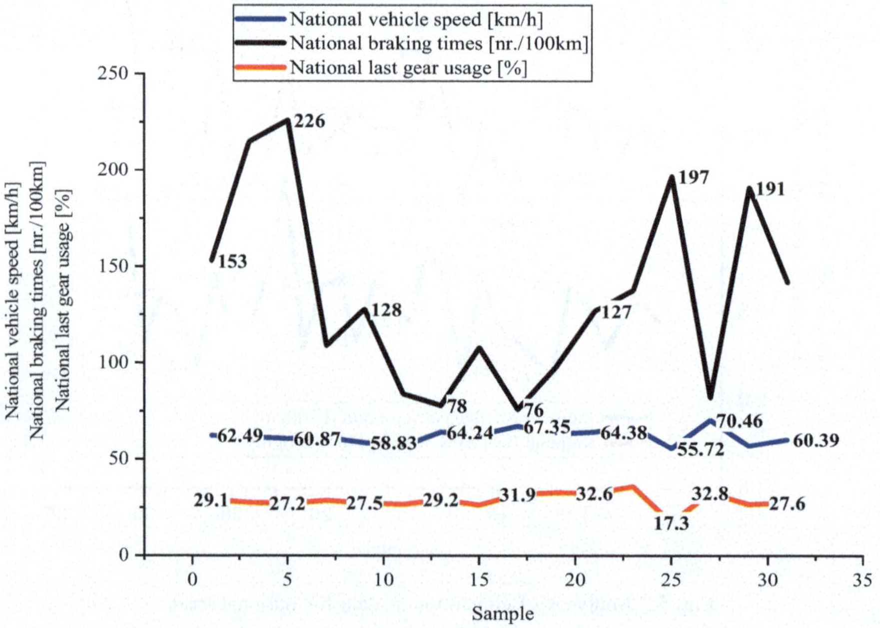

Fig. 7. Functional parameters in internal rolling regime

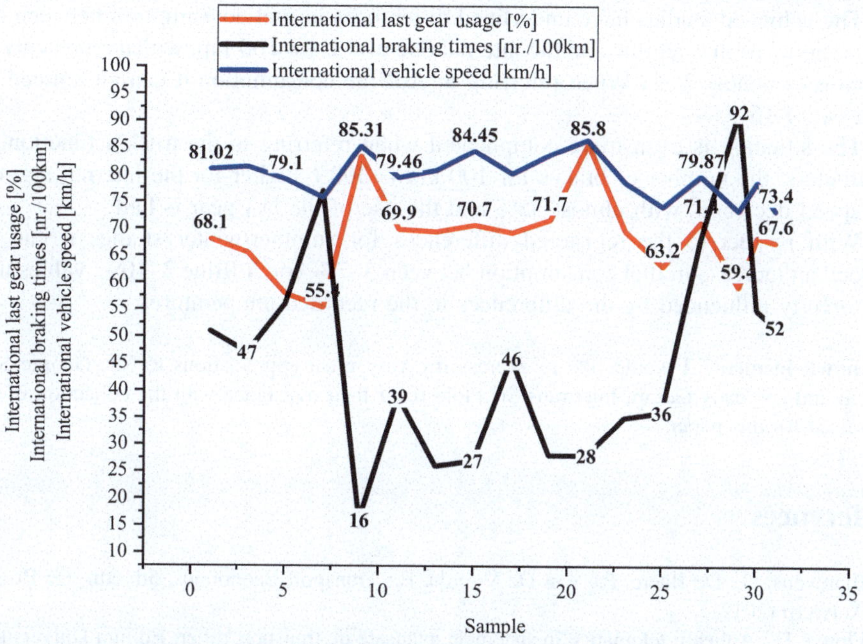

Fig. 8. Functional parameters in international rolling regime

With respect to the functioning parameters for trucks in the two rolling regimes, it can be noticed from Figs. 7 and 8 that there is a big difference between the registered values. The average number of brakes for 100 km has values 3 or 4 times bigger for national routes and also the use of the last gear decreases to half of the value registered internationally, while the average speed significantly reduces.

The national functioning parameters are predictable if we refer to the vehicle speed. Low vehicle speed and congested traffic have a direct and negative impact of braking and last gear usage.

The increase of fuel and automatically AdBlue consumption for internal traffic comes as a natural consequence as a result of the analyses of the graphs for functioning parameters. Repeated accelerating and braking as a result of driving conditions have a negative effect on fuel and Ad Blue consumption.

3 Conclusions of the Achieved Studies

The lack of infrastructure, overpopulation of roads and the limitation of driving speed have negative effects on the operation way of the trucks as well as on the drivers. The low efficiency within the internal transportation, meaning the number of mileage with respect to the number of driving hours overlaps to the increased exploitation costs and the negative impact on the environment.

The achieved studies have underlined differences in fuel consumption between the international rolling regime and the internal one of 7 L for 100 km, which represents an increase of almost 25%. When referring to AdBlue consumption it can be noticed an increase of 15%.

The situation is even more complicated when referring to the truck's functioning parameters: the number of brakes for 100 km is 300% higher for the internal regime, the speed decreases with almost 35% and the use of the last gear is half.

With respect to the registered differences for summer/winter studies it can be noticed an increase in fuel consumption between 5–7%, for AdBlue 3–10%, values that are directly influenced by the differences in the recorded temperatures.

Acknowledgement. I would like to express my very great appreciations to Mr. George Stocheciu and company Gerom Intertrans Srl Ploiesti for their assistance with the collection of the data used for this paper.

References

1. Blauwens, G., De Baere, P., Van De Voorde, E.: Transport Economics, 5th edn. De Boeck, Antwerp (2012)
2. Florea, D.: Aplicaţii telematice în sistemele avansate de transport rutier. Editura Universităţii Transilvania din Braşov (2004)
3. Cordos, N.: Elemente de telematica rutiera. Editura UT Press, Cluj Napoca (2015)
4. Todoruţ, I.-A., Barabás, I., Burnete, N.: Siguranţa autovehiculelor şi securitatea în transporturi rutiere. Editura U.T.PRESS, Cluj-Napoca (2012)
5. Dynafleer. https://www.dynafleetonline.com/AS. Accessed 18 June 2019

The Influence of the Hybridization Factor on the Energetic Performance in Test Cycles

Cristian Rențea(✉), Marius Bățăuș, Sorin Maican, Mircea Oprean, and Gheorghe Frățilă

University Politehnica of Bucharest, 313 Spl. Independentei, Bucharest, Romania
rentea.cristian@yahoo.com

Abstract. Reducing the fuel consumption and the pollutant emissions becomes more and more demanding, hence, in recent years, hybridization started to play an important part in the automotive industry.

While both hybrid and pure electric vehicles represent viable solutions for modern vehicle applications, hybrid electric vehicles (HEV) removes the disadvantage of using a single source of energy.

The aim of this paper is to investigate the energetic performance of a parallel HEV in different test cycles, such as Worldwide Harmonized Light Vehicles Test Cycle (WLTC), considering multiple levels of hybridization.

The study is done using complex models developed in a performant simulation environment.

The energetic performances are being compared in different cycles by analyzing the FCR (fuel consumption ratio) using three different values for the hybridization factor (HF).

A comparison of the test cycles is being realized using FCR parameter, also an analysis of the correlation between FCR and VSF (vehicle speed fluctuation) is done.

The simulation results obtained for the current application are compared with the ones obtained for a conventional vehicle as well as considering an electric vehicle.

Keywords: Parallel hybrid electric vehicle · Fuel consumption ratio · Test cycles · Hybridization factor

1 Introduction

Hybrid vehicles have a reduced fuel consumption and a lower pollutant emission level compared to the conventional vehicles, equipped with an internal combustion engine [1].

Based on the structure of the components in the powertrain, the hybrid electric vehicles are divided in three categories: serial, parallel and mixed hybrid respectively.

Another classification method of hybrid vehicles can be made taking into account the power output of the electric motor. Thus, in [2] the hybridization factor, which represents the ratio between the power of the electric motor and the total power for both thermal and electric sources of the vehicle, is defined.

I. Dumitru et al. (Eds.): SMAT 2019, *The 30th SIAR International Congress of Automotive and Transport Engineering*, pp. 103–110, 2020.
https://doi.org/10.1007/978-3-030-32564-0_13

$$HF = \frac{P_{em}}{P_{em} + P_{ICE}} \tag{1}$$

where P_{em} is the power of the electric motor and P_{ICE} represents the power of the internal combustion engine.

Considering the value of the hybridization factor and the energy storage capacity, hybrid electric vehicles can be classified according to the table below (Table 1).

Table 1. Levels of hybridization

Hybrid type	HF	Particularities
Minimal hybrid	0 … 0.1	Short range for purely electric mode
Mild hybrid	0.25 … 0.5	The energy provided by the purely electric mode cannot support a full standard driving cycle completion
Full hybrid	0.5… 0.7	The electric system fully supports the vehicle to follow a standard cycle

In case of a conventional vehicle hybridization, any excessive power supplied by the motor can cause an unbalance for the vehicle, thus it is not recommended [3]. Also in [4] is highlighted the need for a good balance between the powers of the electric motor and IC engine.

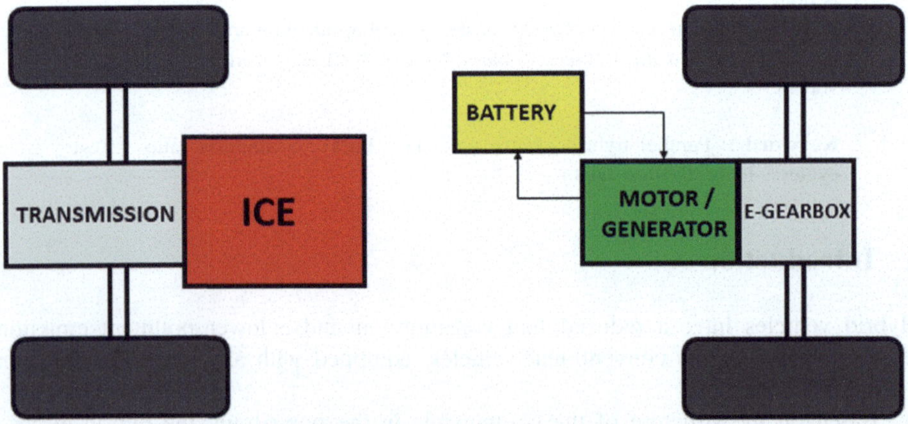

Fig. 1. Through the road parallel hybrid vehicle structure

For the purpose of this paper, the parallel hybrid structure, as shown in Fig. 1, is approached.

The parallel hybrid structure used is specific for a Through the Road Hybrid (TTR HEV). The TTR HEV is composed by an internal combustion engine, placed on the front axle, and an electric motor which provides power to the rear axle. The electric motor can also act as a generator, allowing the regenerative braking [3].

For the electric axle, the use of a single speed gearbox with a disengaging mechanism is driven by the commercial aspects for mass produced vehicles [4]. The employment of the disengaging mechanism prevent the over-speed of the electric motor at high vehicle speed while maximizing wheel torque during electric driving and the capability for electric boost in a wide range of vehicle speeds [4].

The paper investigates by means of simulation the influence of the hybridization factor on fuel economy considering five test cycles (US 06, Japanese 10–15, HWFET, NEDC and WLTC) with a big variation of cycle average speed (between 33.6 km/h and 83.4 km/h) and vehicle speed fluctuations (between 19.2% and 63%).

The simulation are done for a middle class passenger car with TTR HEV architecture for three hybridization factors in the mild hybrid range. The variations are obtained only by rescaling of the electric machine.

The transmission components play an important part for the parallel hybrid vehicle in the prospect of fuel consumption and pollutant emissions reduction, through increasing the engine operation efficiency.

The fuel consumption at constant speed and the vehicle speed fluctuation [5] are determined for the following test cycles: US 06, Japanese 10–15, HWFET, NEDC and WLTC.

2 Model Design

The model realized for this study (Fig. 2) was developed in LMS Imagine. Lab AMESim, a programme that contains more libraries of components (IFP Drive, Mechanical) that allow the study of the energetic performance for conventional powertrains and electric or hybrid electric vehicles [5–7].

The main submodels used for the HEV model are: a driver submodel (DRVDRVA00A-1) necessary to follow the cycle given by the mission profile and ambient data submodel (DRVMP2A-1), electric motor (DRVELMT0A), battery (DRVBAT03), internal combustion engine (DRVICE01E), control unit for a parallel hybrid vehicle (DRVCUPH1A), gearbox (DRVMGC01) and 1-D vehicle (DRVVEH4A) [6].

For the power electronic converter a special supercomponent that includes the standard DRVPET0B submodel was developed in order to completely interrupt the electric machine circuit when the torque command is null.

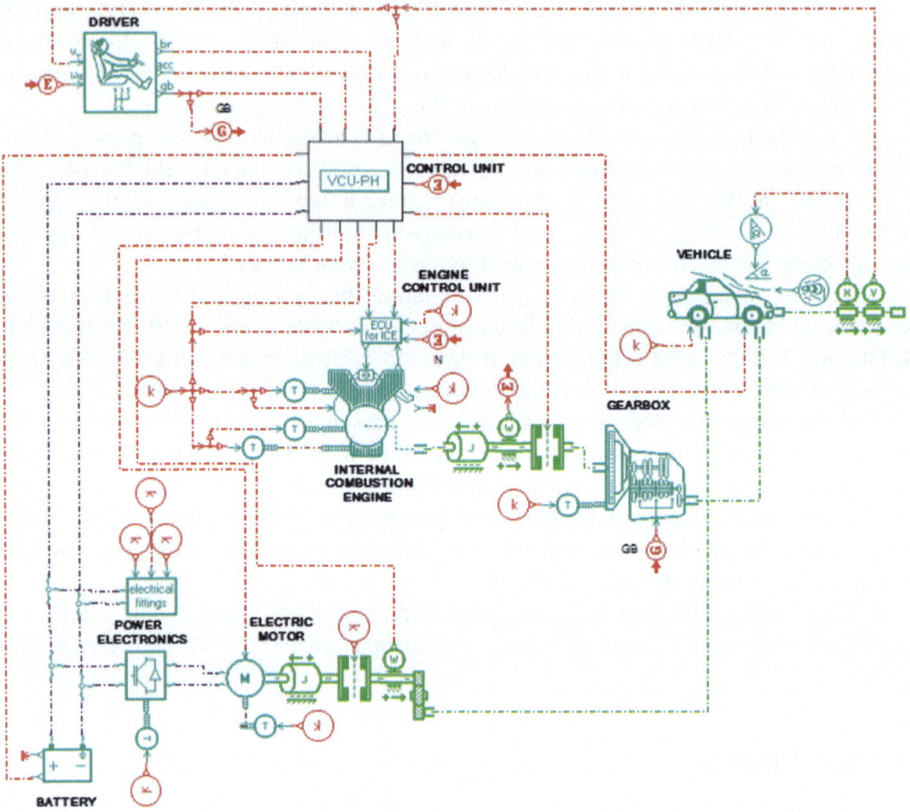

Fig. 2. The parallel hybrid model for the energetic performance study

The driver, the electric motor, the engine, the battery and the vehicle models send information to the control unit, which analyzes it to reduce the battery consumption. The electric machine is also used as a generator, to charge the battery when the driver during the regenerative braking.

By taking the information received into account, the control unit decides one of the following functioning modes [6]:

- Mode 1 - The IC engine is used to drive the vehicle, and if it is possible, is supplementary loaded with the generator to the optimum engine torque to regenerate the battery.
- Mode 2 - The electric motor is used to drive the vehicle, because the engine has a low rotary speed, but the flag for battery regeneration is kept.
- Mode 3 - The IC engine is used to drive the vehicle but battery regeneration take place only when the driver brakes (regenerative braking).
- Mode 4 - The electric motor is used to drive the vehicle and the battery energy regenerates only when the driver brakes.

3 Results

The three hybridization factors considered are: 0.22, 0.32 and 0.42. The simulations are done considering the IC engine, electric motor and transmission at nominal thermal regime. For each application and cycle an initial simulation of 10 repeating cycles is done to recover the stabilized final state of charge of the battery. This value is then used as initial state of the battery in order to eliminate the influence of the energy variation in the battery on the fuel consumption.

Figure 3 shows the fuel consumption of the vehicle for three different values of the hybridization factor, and five test cycles at the average speed of the cycle. The reference curve is considered the fuel consumption obtained at constant speed when the vehicle is driven only by the IC engine.

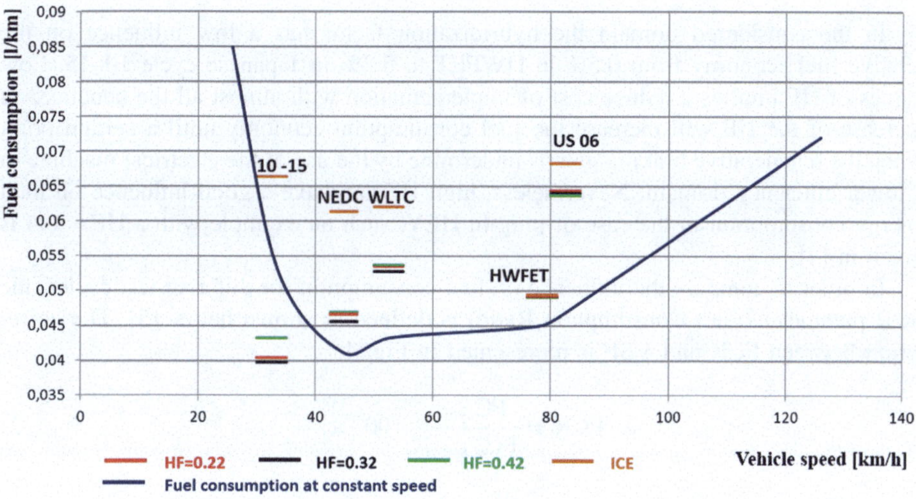

Fig. 3. Fuel consumption for different degrees of hybridization

By analyzing Fig. 3, it can be noticed that for the Japanese cycle (10–15) the fuel consumption obtained at constant speed when the vehicle uses the internal combustion engine as a single energy source is higher than the fuel consumption obtained in the test cycle when the vehicle has two sources of energy, this being caused by the low values for the driving speed. For a conventional vehicle the low resistances will imply the use of the IC engine in a domain with low efficiency. The electric motor has a large influence on fuel consumption for low speed cycles by reducing the time when internal combustion engine is used and by increasing the load on the IC engine during his operation.

One can observe that for the cycles with high Vehicle Speed Fluctuation (VSF) (over 50%) and low average speeds (less than 60 km/h) the hybridization has a great influence on the fuel consumption compared with the conventional powertrain.

Table 2 presents the relative fuel economy compared with a conventional gasoline vehicle that has the same parameters.

Table 2. The relative fuel economy

Cycle	Hybridization factor		
	0.22	**0.32**	**0.42**
10-15	38.2 %	40.3 %	35.1 %
NEDC	23.8 %	25.8 %	23.6 %
WLTC	13.5 %	14.8 %	13.4 %
HWFET	0.5 %	1.0 %	1.0 %
US 06	5.6 %	6.3 %	6.8 %

In the considered domain the hybridization factor has a low influence on the relative fuel economy from 0.5% in HWFET to 5.2% in Japanese cycle 10–15. Low values of HF implies a reduce cost of implementation with almost all the benefice, an increase of the HF will increase the fuel consumption economy until a certain point when the regenerative braking gain is undermine by the use of the electrical machine in a lower efficiency domain. Nevertheless, high HF can have a good influence on total energy consumption in the case of plug-in HEV, such an example with a HF = 0.4 is given in [4].

In order to compare the influence on fuel consumption for different test cycles, the FCR parameter (Fuel Consumption Ratio) is defined as shown below [5]. The correlation between FCR and VSF is represented in Fig. 4.

$$FCR = \frac{FC_{cycle}}{FC_{ct.speed}} \cdot 100\ [\%] \quad (2)$$

where: FC_{cycle} – fuel consumption obtained in the test cycle; $FC_{ct.speed}$ – fuel consumption obtained at constant speed movement equal to the average speed of cycle.

In Fig. 4, three regression lines are represented, one for the gasoline vehicle model, one for the electric vehicle model which have set as intercept (imposed point) the corresponding point of the constant speed movement (0, 100) and another for the HEV application (HF = 0.32) in which the corresponding point of the constant movement is not imposed. A linear dependency between FCR and VSF for the gasoline vehicle and another linear dependency between ECR and VSF for the electric vehicle can be observed (a coefficient of determination $R^2 = 0.785$, respectively $R^2 = 0.968$), whereas for the hybrid vehicle, there is no dependency (a coefficient of determination $R^2 = 0.0112$).

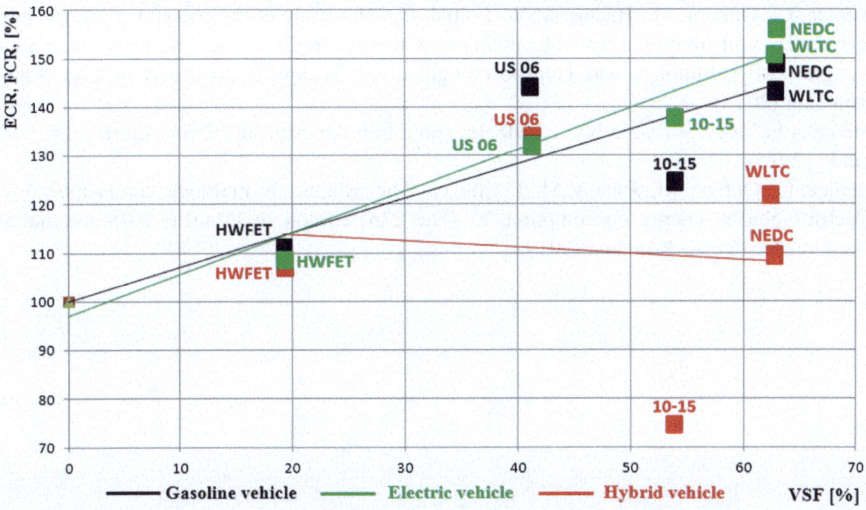

Fig. 4. Correlation between ECR, FCR and VSF parameters

4 Conclusions

When comparing using a relative parameter (FCR) the most disadvantageous cycles are US 06 and WLTC. The linear correlation between FCR and VSF observed in [5] for a vehicle equipped with a turbocharged diesel engine and here for the conventional configuration is not obtained for the HEV application. Also a linear correlation between the similar parameter Energy Consumption Ratio and VSF was obtained in [7] for an electric vehicle.

A high influence of the average cycle speed on fuel consumption for HEV vehicles is observed, but more investigations need to be done in order to also check the influences of the number of launches.

Low hybridization factor can allow for the maximum benefice/cost ratio while a high hybridization factor is reasonable only for plug-in HEV.

References

1. Umut Karaoglan, M., Sefa Kuralay, N., Ozgur Colpan, C.: The effect of gear ratıos on the exhaust emıssıons and fuel consumptıon of a parallel hybrid vehicle powertrain
2. Somà, A.: Trends and Hybridization Factor for Heavy - Duty Working Vehicles
3. Pisanti, C., Rizzo, G., Marano, V.: - Energy Management of Through-The-Road Parallel Hybrid Vehicles
4. Kern, D., Marschall, T.: The innovative transmission concept for the plug-in hybrid powertrain of the BMW 2 series active tourer. In: 14th International CTI Symposium Berlin (2015)

5. Renţea, C., Oprean, M., Băţăuş, M.V., Frăţilă, G.: The study of the energetic performance in test cycles considering a variable powertrain thermal regime. In: The IVth International Congress of Automotive and Transport Engineering, Technical University of Cluj Napoca, România (2018)
6. Siemens Industry Software NV – LMS Imagine. Lab Amesim. IFP Drive Library 14, User's guide (2015)
7. Renţea, C., Oprean, M., Băţăuş, M., Frăţilă, G.: The influence of multi-speed transmissions on electric vehicles energy consumption. In: The 23rd edition of IManEE 2019 International Conference, Piteşti, România (2019)

Road Vehicle and Environment

Investigating the Cyclic Variability of a Multi-cylinder Turbocharged Direct-Injection Spark-Ignition Engine Fueled with E10

Fazal Um Min Allah[1](✉), Caio Henrique Rufino[1], Clayton Barcelos Zabeu[2], and Waldyr Luiz Ribeiro Gallo[1]

[1] University of Campinas, Campinas, SP 13083-860, Brazil
fazaluminallah@hotmail.com
[2] Instituto Mauá de Tecnologia, Mauá, SP 09580-900, Brazil

Abstract. The innovative technologies have helped the spark-ignition engines to undergo a process of evolution in order to achieve lower pollutant emissions and better fuel economy. The downsized turbocharged direct-injection spark-ignition (DISI) engines help to achieve these objectives. Moreover, the addition as a fuel blend seems to be a promising fuel for internal combustion engines due to its lower pollutant emissions and sustainability characteristics. However, the in-cylinder cyclic variations during combustion affect the performance of internal combustion engines. In this paper, the in-cylinder cyclic variations of a turbocharged DISI engine are evaluated for a wide range of engine operating conditions (i.e., engine speed, engine load). The paper includes conventional statistical analysis as well as wavelet analysis. The in-cylinder pressure (indicated mean effective pressure and maximum in-cylinder pressure) and combustion durations (flame development duration, main combustion duration and combustion phasing) are used to analyze the cyclic variability. The relationships between combustion durations and indicated mean effective pressure are analyzed by using cross wavelet spectrum (XWS).

Keywords: Cyclic variations · Direct-injection · Wavelet analysis

1 Introduction

The internal combustion engines are widely used as power generation systems for vehicles and they have survived so far in a competitive market by undergoing through the processes of innovation and redesigning. Recently, the engines with lower pollutant emissions and higher fuel efficiency have become much more desirable due to the stringent environmental regulations and energy security problems. The downsized spark-ignition engines along with turbocharging, direct injection of fuel and variable valve timing can be employed to achieve lower fuel consumption and lower pollutant emissions [1–4]. Furthermore, the usage of ethanol fuel for direct injection spark-ignition (DISI) engines helps to overcome the threats of energy security and climate change [5]. The in-cylinder cyclic variations can cause drivability problems. The causes

I. Dumitru et al. (Eds.): SMAT 2019, *The 30th SIAR International Congress of Automotive and Transport Engineering*, pp. 113–123, 2020.
https://doi.org/10.1007/978-3-030-32564-0_14

of these variations are identifiable but it can be a difficult task to determine the degree of influence for each of them [6]. The in-cylinder cyclic variations can be characterized by different parameters (i.e., in-cylinder pressure, combustion durations and flame development etc.) [7]. Samuel et al. [8] evaluated the cyclic variations of a turbocharged DISI engine. They observed a higher level of cyclic variations for DISI engine as compared to that of the engine with throttle body injection (TBI). The unconventional approach adopted by Samuel et al. represented the coefficient of variation (CoV) as a function of crank position. The maximum values of CoV were more likely to be affected by engine operating conditions (i.e., engine speed and load) in turbocharged DISI engine. Hassaneen et al. [9] also compared the in-cylinder cyclic variations of turbocharged DISI engine and TBI engine. The in-cylinder combustion variations are strongly affected by spark timing in DISI engines. They concluded that the CoVs for maximum in-cylinder pressure (P_{max}) and indicated mean effective pressure (IMEP) were comparatively higher for turbocharged DISI engine as compared to that of TBI engine. These higher values of CoVs could be attributed to the non-homogeneities of air-fuel mixture in DISI engine. They also observed higher in-cylinder pressure, longer flame propagation duration and shorter rapid burn durations for turbocharged DISI engine. The experimental investigation carried out by Vaglieco et al. [10] demonstrated an improved performance of turbocharged DISI engine fueled with ethanol blends. Liu et al. [11] evaluated the effects of different injection strategies for a turbocharged DISI engine fueled with a gasoline-ethanol blend of E10. They concluded that the injection strategies strongly affect the in-cylinder cyclic variations.

Apart from the conventional statistical analysis of in-cylinder cyclic variations, Sen et al. [12] applied the wavelet analysis in order to identify the periodicities present within the time-series of IMEP at different engine loads. Their preference of wavelet analysis over short time Fourier transform (STFT) was due to the limitation of applying STFT for a time-series with multiple frequencies. Soon afterwards, several researches were published in which the wavelet analyses were performed for different parameters related to in-cylinder cyclic variations of spark-ignition and compression ignition engines [13–20]. Sen et al. [16] investigated the interdependencies of combustion durations and IMEP by applying cross wavelet analysis. Awad et al. [21], Litak et al. [22] and Yang et al. [23] have recently analyzed the in-cylinder cyclic variations by using wavelet analysis. However, the wavelet analysis is not yet applied to analyze the time-series of in-cylinder combustion parameters of a turbocharged DISI engine.

The present work aims at carrying out conventional statistical and wavelet analyses for in-cylinder cyclic variations of a turbocharged DISI engine. The experiments were carried out in order to acquire the data for 150 engine cycles of a turbocharged DISI engine. The wavelet analysis is applied to identify the periodicities in the IMEP time-series. The effects of engine operating conditions on cyclic variability are evaluated briefly. The interrelationships of IMEP and combustion durations are investigated by using cross wavelet spectra (XWS).

2 Materials and Methods

2.1 Experimental Setup

A 3-cylinder turbocharged direct injection spark ignition (DISI) engine with variable valve timing (VVT) is used to carry out experiments. The acquired data of 150 engine cycles is used for different engine operating conditions (i.e., engine load and engine speed). The test engine operates with a compression ratio of 10.4:1. The ethanol-gasoline (gasohol) blend is used as a fuel. The in-cylinder pressure profiles are obtained with a precision of 1 degree crank angle. The electronic control unit (ECU) was calibrated to operate with the optimum values of spark timing and air-fuel ratio. The spark timing is expressed as a function of engine speed for full engine load in Fig. 1.

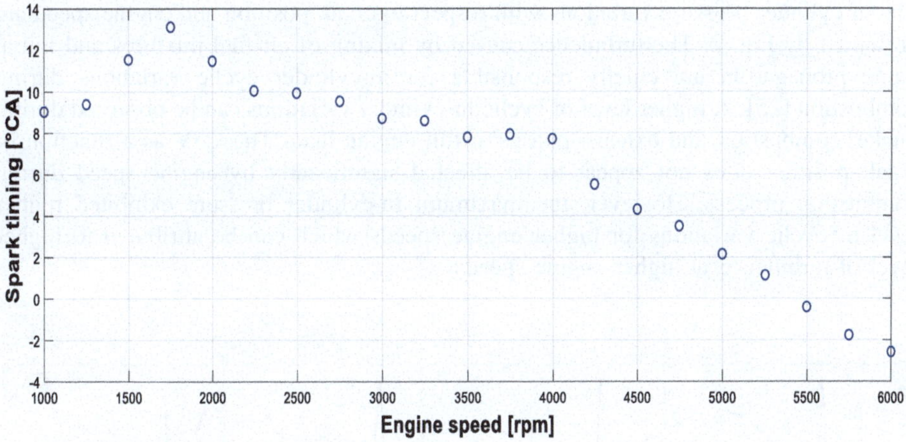

Fig. 1. Spark timing expressed as a function of engine speed at full engine load.

2.2 Wavelet Analysis

The conventional Fourier transform (FT) provides inadequate information about non-stationary signals. Therefore, the need for an appropriate analysis arises which can truly represent the signal in time-frequency domain. For a time series with N numbers of x variable localized along time for each time index of n, continuous wavelet transform (CWT) of a discrete time series is given by convolution of time localized values of variable x and analyzing wavelet (termed as mother wavelet). The mathematical representation of CWT is given by Eq. 1 where s and Ψ^* represent wavelet scale and complex conjugate of wavelet function respectively. The square modulus of the coefficient given by Eq. 1 gives the amount of signal energy for respective time and scale. The squared values of these coefficients are then plotted against respective time and scale in order to depict wavelet power spectrum [14]. The method adopted for the present work is briefly explained by Torrence and Campo [24]. The cross-wavelet transform of two time series is given by Eq. 2 where $W_{n1}(s)$ and $W_{n2}(s)$ define the wavelet power spectra of corresponding time series. The spectrum is given by square modulus of the coefficient given by Eq. 2 [14].

$$W_n(s) = \left(\frac{\partial t}{s}\right)^{\frac{1}{2}} \sum_{n=0}^{N-1} x_n \Psi^* \left(\frac{n' - n}{s}\right) \tag{1}$$

$$W_n^{12}(s) = W_{n1}(s) W_{n2}^*(s) \tag{2}$$

3 Results and Discussion

3.1 Effects of Engine Speed

The in-cylinder pressure variations with respect to crank position and engine speed are presented by Fig. 2. The turbulence caused by mixing of air-fuel mixtures and initial flame propagation are chiefly responsible for in-cylinder cyclic variations during combustion [25]. A higher level of cyclic in-cylinder variations can be observed during intake, combustion and exhaust phases at full engine load. The CoV as a function of crank position does not appear to be affected significantly by engine speed during combustion process. However, the maximum in-cylinder pressure exhibited higher level of cyclic variations for higher engine speeds which can be attributed to higher level of turbulence at higher engine speeds.

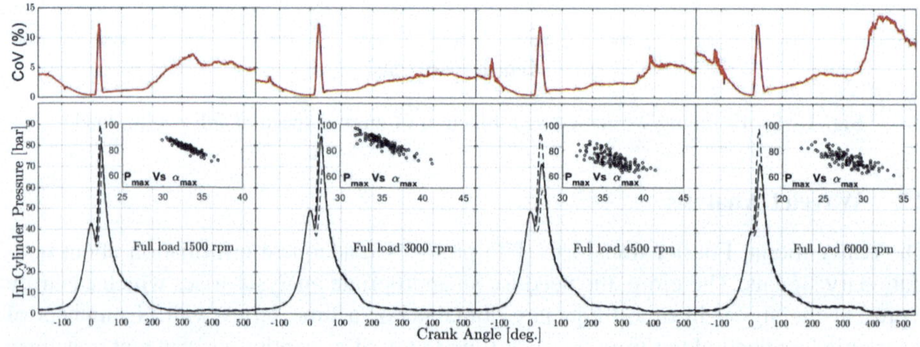

Fig. 2. In-cylinder pressure (maximum, mean and minimum) of 150 consecutive engine cycles and CoV of in-cylinder pressure expressed as functions of corresponding crank position for different engine speeds at full engine load.

Figure 3 presents IMEP and maximum in-cylinder pressure for 150 consecutive cycles. A higher level of cyclic variations can be observed for engine speed of 4500 rpm. There is a certain limit of rich air-fuel mixture beyond which the laminar flame speed starts to decrease [25]. The lowest value of relative air fuel ratio was

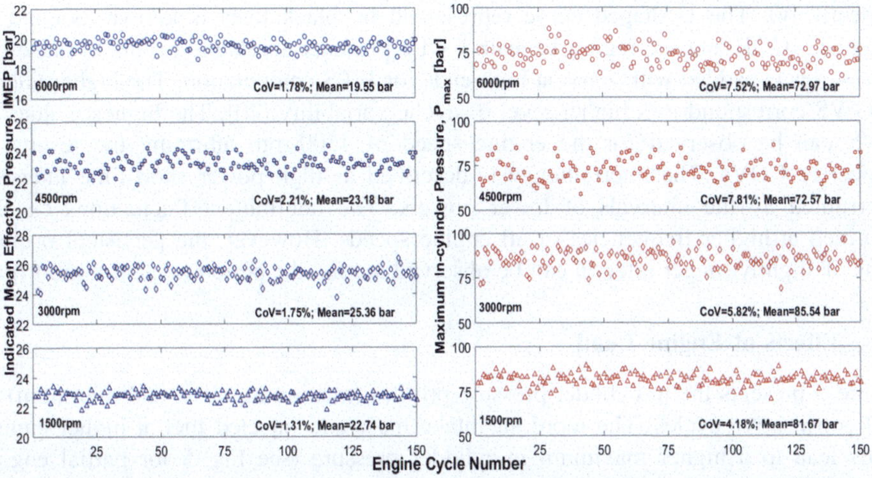

Fig. 3. IMEP and P_{max} of 150 consecutive engine cycles for different engine speeds at full engine load.

observed for the engine speed of 4500 rpm. The burn duration was delayed due to lower burning velocity. Therefore, the engine speed of 4500 rpm exhibited a higher value of CoV_{IMEP}.

The wavelet power spectra (WPS) and global wavelet spectra (GWS) for time series of IMEP is presented in Fig. 4. The details of the procedure adopted for WPS and GWS are given by Torrence and Campo [24]. The mean value of wavelet power spectra is given by GWS. The colors of the spectrum represent power level (i.e., red, blue and intermediate colors represent higher, lower and intermediate levels of power

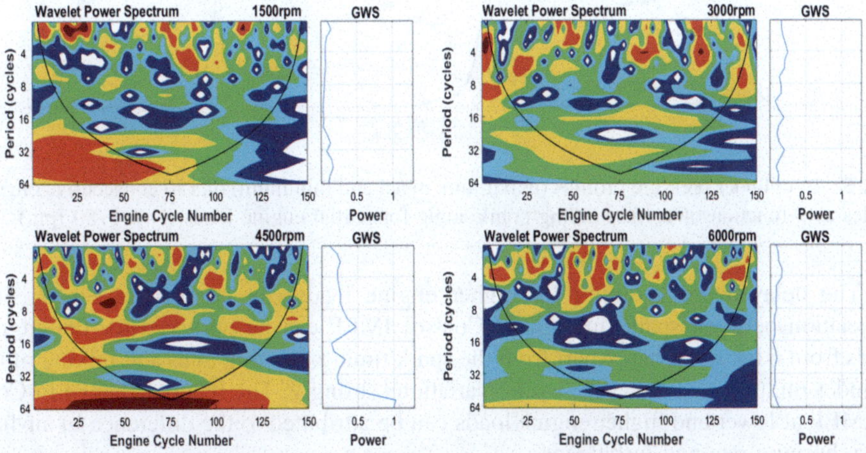

Fig. 4. WPS and GWS of IMEP time series for different engine speeds at full engine load.

respectively). The U-shaped curve represented by black lines is known as cone of influence (COI) and the region beneath COI represents unreliable results because the time series is padded with zeros at both ends for WPS computation. The higher values of GWS correspond to a higher level of cyclic variability [20]. The highest values of GWS can be observed for the engine speed of 4500 rpm affirming the results of respective COVs. The intermittency is described as high power short-time intervals interrupted by the intervals of lower powers. The intermittent fluctuations can be observed at higher frequencies of all engine speeds. However, the persistent oscillations of slightly longer interval can be observed around the period of 12 for 4500 rpm.

3.2 Effects of Engine Load

Figure 5 presents the in-cylinder pressure profiles (maximum, mean and minimum) of 150 consecutive cycles. The more air intake mass and injected fuel at higher engine loads lead to a higher maximum in-cylinder pressure (see Fig. 5 for partial engine loads). Figure 6 presents the CoVs for in-cylinder pressure as a function of crank position and engine load. The main sources of variation are the intensities of turbulence during intake, combustion and exhaust processes. The lower engine loads exhibited higher values of CoV during intake and exhaust processes.

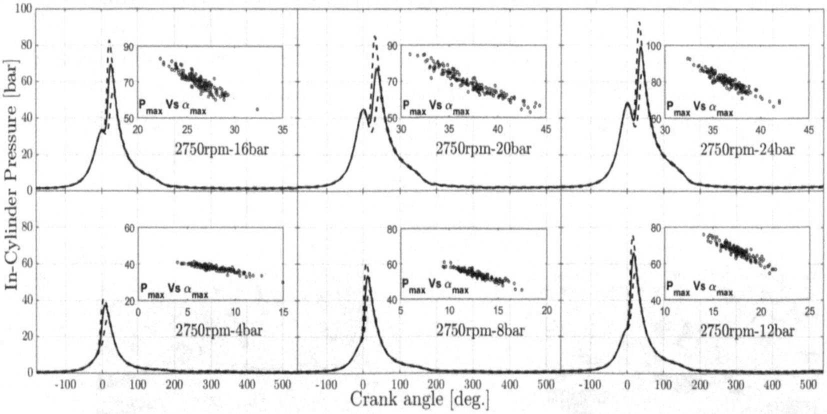

Fig. 5. In-cylinder pressure profiles (maximum, mean and minimum) of 150 consecutive engine cycles as a function of corresponding crank angle for partial engine loads (N = 2750 rpm).

The delayed spark timing for higher engine loads can be seen in Fig. 7. The correlation between spark timing and CoV of IMEP can be observed for the engine loads from 6 bar to 20 bar. Apart from the spark timing, burning rates and mixing of in-cylinder mixture also affect the cyclic variations strongly. The relative change in CoV of IMEP at lower and higher engine loads can be attributed to the difference in air-fuel ratio, burning rate and turbulence.

The effect of main combustion duration (CA10-90) can be seen in Fig. 8 where average values of CA10-90 and CoV of IMEP are plotted against corresponding engine

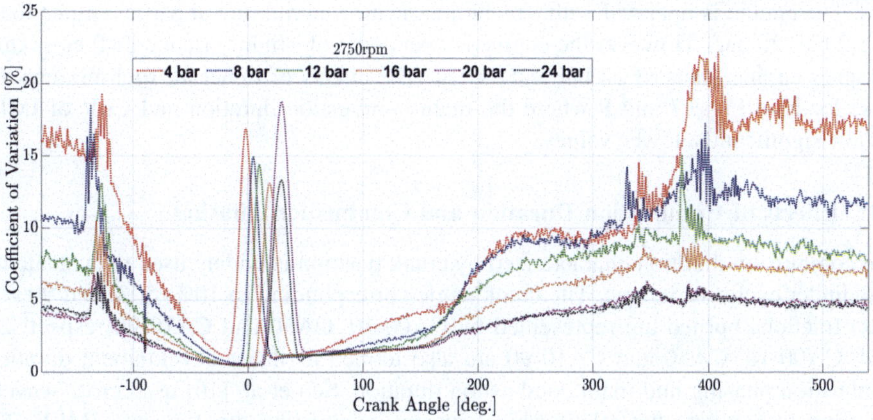

Fig. 6. CoV of in-cylinder pressure as a function of crank position and engine load (N = 2750 rpm).

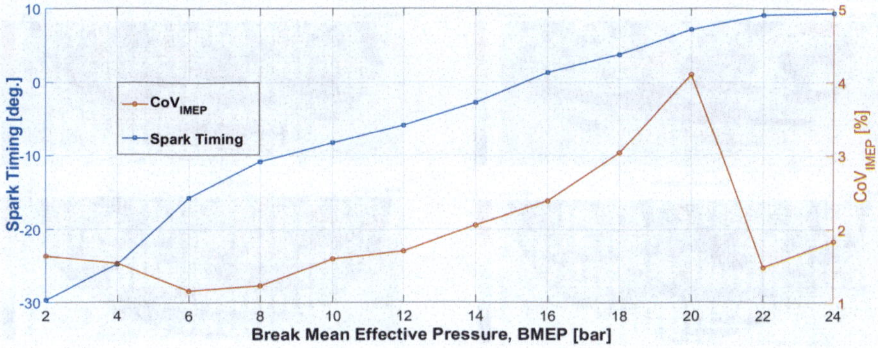

Fig. 7. CoV of IMEP as a function of crank position and engine loads (N = 2750 rpm).

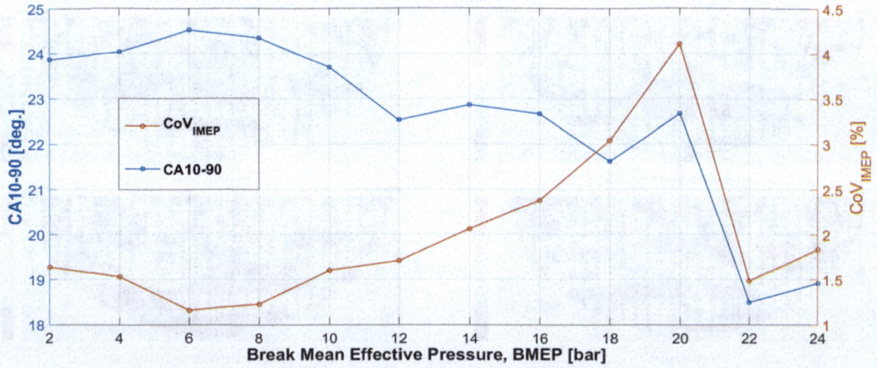

Fig. 8. CoV of IMEP as a function of crank position and engine loads (N = 2750 rpm).

load. The engine is operated with stoichiometric air-fuel mixture at partial engine loads (i.e., 2 bar–20 bar). However the engine is operated with slightly richer air-fuel mixture at higher engine loads of 22 bar and 24 bar. The effects of richer air-fuel mixtures for these loads in Figs. 7 and 8 where the main combustion duration and CoV of IMEP exhibit significantly lower values.

3.3 Effects of Combustion Duration and Combustion Phasing

The combustion duration characterized by crank position is widely used to evaluate the in-cylinder cyclic variations. The crank angle corresponding to 10%, 50% and 90% of mass fractions burned are represented by CA00-10, CA50 and CA10-90 respectively [25]. CA00-10, CA50 and CA10-90 are also termed as flame development duration, combustion phasing and main combustion duration. Sen et al. [16] used cross wavelet spectrum to evaluate the relationship between combustion duration and IMEP. The cross wavelet spectra of time series for in-cylinder combustion durations (CA00-10, CA50 and CA10-90) and IMEP are presented in Figs. 9, 10 and 11. The vectors

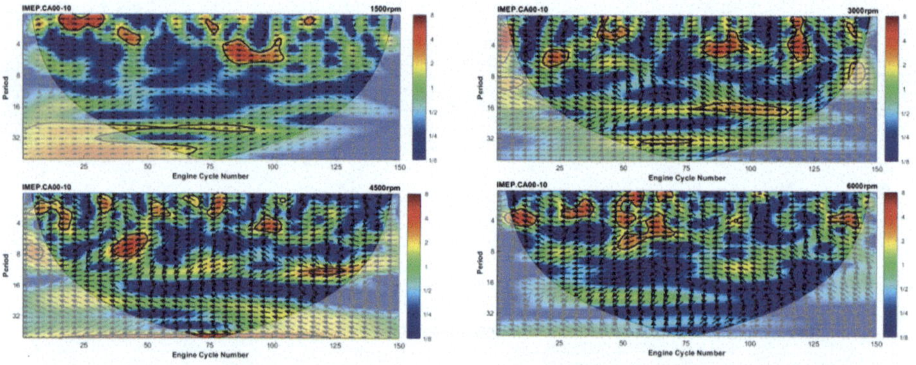

Fig. 9. XWS of CA00-10 and IMEP for different engine speeds at full engine load.

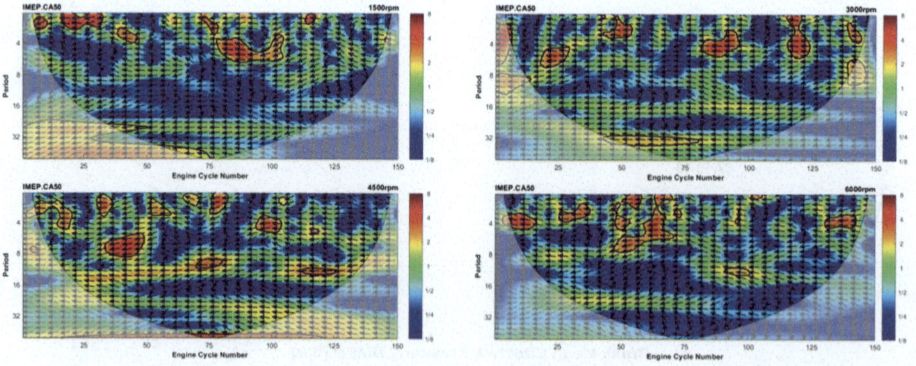

Fig. 10. XWS of CA50 and IMEP for different engine speeds at full engine load.

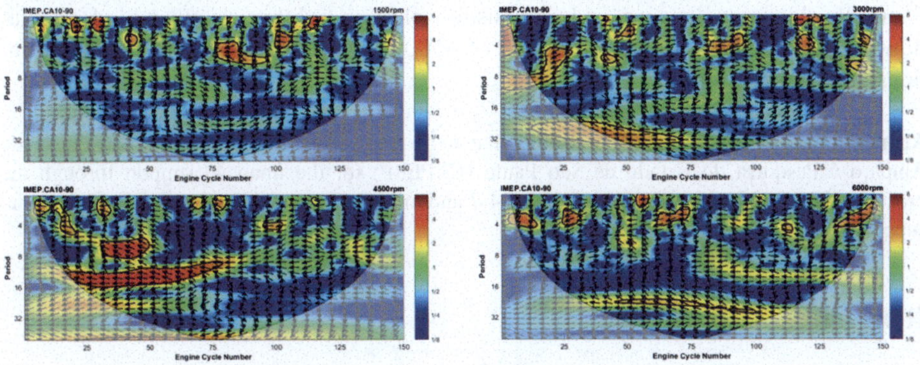

Fig. 11. XWS of CA00-90 and IMEP for different engine speeds at full engine load.

denoted by left (right) arrows indicate the anti-phase (in-phase) relationships. Anti-phase relationships can be observed for flame development duration and IMEP. The relationship between combustion phasing (CA50) and IMEP follows a similar kind of trend to that of CA00-10 because it is strongly dependent on flame development duration [26]. The faster flame development duration results in earlier CA50, consequently, yielding a higher value of IMEP. The spark timing is delayed for higher engine loads in order to avoid knock. The partial engine loads experience earlier spark timing and subsequently exhibit lower cyclic variability. The engine speeds of 1500 rpm, 3000 rpm and 6000 rpm encompasses more area of higher common power (represented by red color) for XWS of CA00-10 and CA50 as compared to that of CA10-90. However, the engine speed of 4500 rpm expectedly exhibited a period of high common power for longer time interval for XWS between CA10-90 and IMEP.

4 Conclusions

The in-cylinder pressure and combustion data of 150 engine cycles are obtained by carrying out experiments for a turbocharged DISI engine in order to evaluate the cyclic variation. The parameters such as CoV of IMEP and P_{max} are used to carry out congenital statistical analysis. The CoV as a function of crank position is also presented for different engine operating conditions. The intake, combustion and exhaust processes exhibited higher level of instabilities for in-cylinder pressure. The CoV of IMEP does not exceed the limit of 10% [25] (3%–5% for modern engines) for all tested engine operating conditions. The engine speed of 4500 rpm at full engine load exhibited higher CoV of IMEP due to the delayed combustion process. The results of wavelet analysis affirmed the trends followed by CoVs for different engine operating conditions. Higher in-cylinder cyclic variations are observed for higher engine loads due to the delayed spark timing and increased turbulence. The CoV of IMEP exhibited a strong relationship between spark timing and CoV of IMEP at partial engine loads. However, this trend appeared to be reversed for higher engine loads of 22 bar and 24 bar due to a richer air fuel mixture. The cross wavelet spectrum has shown that the flame development duration and combustion phasing exhibit a similar kind of relationship with IMEP. The engine speed of 4500 rpm exhibited slightly stronger relationship with CA1090.

Acknowledgements. The authors would like to express their gratitude to the Fundação de Amparo á Pesquisa do Estado de São Paulo (FAPESP) for the financial support through the processes Nrs. 2018/18859-1 and 2018/20394-7 and to PSA Group for the technical and financial support.

References

1. Leduc, P., Dubar, B., Ranini, A., Monnier, G.: Downsizing of gasoline engine: an efficient way to reduce CO_2 emissions. Oil Gas Sci. Technol. **58**(1), 115–127 (2003)
2. Lecointe, B., Monnier, G.: Downsizing a gasoline engine using turbocharging with direct injection. SAE Technical Paper 2003-01-0542 (2003)
3. Lake, T., Stokes, J., Murphy, R., Osborne, R., Schamel, A.: Turbocharging concepts for downsized DI gasoline engines. SAE Technical Paper 2004-01-0036 (2004)
4. Bandel, W., Fraidl, G.K., Kapus, P.E., Sikinger, H., Cowland, C.N.: The turbocharged GDI engine: Boosted synergies for high fuel economy plus ultra-low emissions. SAE Technical Paper 2006-01-1266 (2006)
5. Cohn, D.R., Bromberg, L., Heywood, J.B.: Direct injection ethanol boosted gasoline engines: biofuel leveraging for cost effective reduction of oil dependence and CO_2 emissions. MIT Plasma Science and Fusion Center (2005)
6. Han, S.B.: Cycle-to-cycle variations under cylinder-pressure-based combustion analysis in spark ignition engines. KSME Int. J. **14**(10), 1151–1158 (2000)
7. Stone, R.: Introduction to Internal Combustion Engines, 2nd edn. Palgrave Macmillan, UK (1992)
8. Samuel, S., Morrey, D., Whelan, I., Hassaneen, A.: Combustion characteristics and cycle-by-cycle variation in a turbocharged-intercooled gasoline direct-injected engine. SAE Technical Paper 2010-01-0348 (2010)
9. Hassaneen, A.E., Samuel, S., Whelan, I.: Combustion instabilities and nanoparticles emission fluctuations in GDI spark ignition engine. Int. J. Automot. Technol. **12**(6), 787–794 (2011)
10. Catapano, F., Sementa, P., Vaglieco, B.M.: Air-fuel mixing and combustion behavior of gasoline-ethanol blends in a GDI wall-guided turbocharged multi-cylinder optical engine. Renew. Energy **96**, 319–332 (2016)
11. Duan, X., Liu, J., Yuan, Z., Guo, G., Liu, Q., Tang, Q., Deng, B., Guan, J.: Experimental investigation of the effects of injection strategies on cycle-to-cycle variations of a DISI engine fueled with ethanol and gasoline engine blend. Energy **165**, 455–470 (2018)
12. Sen, A.K., Litak, G., Taccani, R., Radu, R.: Wavelet analysis of cycle-to-cycle variations in an internal combustion engine. Chaos Solitons Fractals **38**, 886–893 (2008)
13. Sen, A.K., Longwic, R., LitaK, G., Górski, K.: Analysis of cycle-to-cycle pressure oscillations in a diesel engine. Mech. Syst. Signal Process. **22**, 362–373 (2008)
14. Sen, A.K., Litak, G., Yao, B.-F., Li, G.-X.: Analysis of pressure fluctuations in a natural gas engine under lean burn conditions. Appl. Therm. Eng. **30**, 776–779 (2010)
15. Sen, A.K., Litak, G., Finney, C.E.A., Daw, C.S., Wagner, R.M.: Analysis of heat release dynamics in an internal combustion engine using multifractals and wavelets. Appl. Energy **87**, 1736–1743 (2010)

16. Sen, A.K., Zheng, J., Huang, Z.: Dynamics of cycle-to-cycle variations in a natural gas direct injection spark-ignition engine. Appl. Energy **88**, 2324–2334 (2011)
17. Longwic, R., Sen, A.K., Górski, K., Lotko, W., Litak, G.: Cycle-to-cycle variation of combustion process in a diesel engine powered by different fuels. J. VibroEng. **13**(1), 120–127 (2011)
18. Ceviz, M.A., Sen, A.K., Küleri, A.K., Öner, V.: Engine performance, exhaust emissions, and cyclic variations in lean-burn SI engine fueled by gasoline-hydrogen blends. Appl. Therm. Eng. **36**, 314–324 (2012)
19. Gęca, M., Wendeker, M., Litak, G.: Combustion variability and uniqueness in cylinders of a large power radial engine. J. VibroEng. **14**(2), 582–590 (2012)
20. Sen, A.K., Medina, A., Curto-Risso, P.L., Hernández, A.C.: Effect of ethanol addition on cyclic variability in a simulated spark ignition gasoline engine. Meccanica **49**, 2285–2297 (2014)
21. Abdalla, A.M., Ali, O.M., Awad, O.I., Tao, H.: Wavelet analysis of an SI engine cycle-to-cycle variations fuelled with the blending of gasoline-fusel oil at a various water content. Energy Convers. Manag. **183**, 746–752 (2019)
22. Sen, A.K., Litak, G., Wagner, R.M.: Cyclic heat release variability in a spark ignition engine under exhaust gas recirculation. In: MATEC Web of Conferences, vol. 252, p. 01108 (2019)
23. Yang, L.-P., Bodisco, T.A., Zar, A., Marwan, N., Chu-van, T., Brown, R.J.: Analysis of the non-linear dynamics of inter-cycle combustion variations in an ethanol fumigation-diesel dual-fuel engine. Nonlinear Dyn. **95**(3), 2555–2574 (2019). https://doi.org/10.1007/s11071-018-4708-x
24. Torrence, C., Campo, G.P.: A practical guide to wavelet analysis. Bull. Am. Meteor. Soc. **79**, 61–78 (1998)
25. Heywood, J.B.: Internal Combustion Engines Fundamentals. McGraw-Hill Inc., New York (1988)
26. Dahlander, P., Hemdal, S.: High-speed photography of stratified combustion in an optical GDI engine for different triple injection strategies. SAE Technical Paper 2015-01-0745 (2015)

CFD Study on Wheel Aerodynamic Performance in Side Wind Conditions for a Hatchback Vehicle

Laurentiu Ilea(✉), Daniel Iozsa, Cornelia Stan, and Claudiu Teodorescu

Autovehicule Rutiere Department, University Politehnica of Bucharest, 313 Splaiul Independentei Street, Bucharest, Romania
laurentiu.ilea@yahoo.com

Abstract. Compliance with legislation requirements and customer's needs are key elements considered when developing the systems for vehicles. For assessing the aerodynamic performance of passenger vehicles, both exterior body design and add-on elements are optimised for homologation conditions at 0° airflow. In order to develop less polluting vehicles, new methods are sought to simulate as close as possible real life usage. This will raise new challenges as some solutions that lead to an increased aerodynamic performance at 0-degree airflow, may as well affect it in side-wind conditions. The current paper identifies how elements like wheel cap and wheel fairing (optimised in previous research) will influence the aerodynamic performance of a hatchback body vehicle. Using the CFD method, the airflow behaviour was evaluated though the pressure, velocity and vortices magnitude representation, while the global performance was analysed through the Cd numerical value. By comparing the results between the two levels of equipment, at different airflow angles, it can be concluded that at high angles the aerodynamic elements start affecting the Cd parameter. Research of aerodynamic performance will be extended in order to develop an active wheel element, optimised for both head-on airflow and different side wind angles.

1 Introduction

The vehicle is the result of the human desire to move, to explore and to conquer faraway places. A functional tool in its beginnings, soon after a speed machine, a working horse for new borne industry, the vehicle has to offer as many purposes as human needs. With time, design and comfort were added to the set of qualities that sold a new model. The requests imposed by authorities, linked mainly to crash safety and, more recently, to emissions [1], should be also considered in addition to meeting customers' needs. The current paper has the purpose of assessing through CFD the aerodynamic performance of a hatchback vehicle and its influence on CO_2 emissions. The study is focused on a side wind analysis for two technical definitions: one standard definition with steel rim wheel and a second technical definition with full wheel caps and full rear wheel fairing, while using the same base vehicle. It is expected that the

I. Dumitru et al. (Eds.): SMAT 2019, *The 30th SIAR International Congress of Automotive and Transport Engineering*, pp. 124–130, 2020.
https://doi.org/10.1007/978-3-030-32564-0_15

second definition will have better results for the homologation process (as currently defined) [2], but with degradation in side wind due to increased projected surface and air detachment mechanism implications.

2 Vehicles Aerodynamic Performance Assessment

The vehicle development process is a complex, elaborated sequence of small steps that is meant to reunite all engineering branches so they converge towards a final product which successfully answers both customers and authorities' requests. The aerodynamic homologation process has the purpose of officialising the CdA value for each technical definition of the vehicle, so customer can be informed about the CO_2 emissions, in concordance to WLTP protocols [3]. The homologation is done in wind tunnels, with the vehicle perfectly aligned with airflow on X direction, at specific air speeds and ambient temperatures [2]. Although an official protocol, due to the desire to simulate more closely the real-life vehicle usage, changes will most likely soon be brought, changes like side wind tests – specified in Real Driving Emissions pilot protocol [4].

The assessment of aerodynamic performance is validated through a physical test, but the whole development process is mostly based on CFD tests. The current paper proposes to use the already wide available tool to anticipate the automotive industry need to test aerodynamic equipment not just for $\alpha = 0°$, but also for side wind situations that are often encountered in real life. The CFD tool has the advantage of allowing the analysis of speed and pressure, physical parameters that are invisible to the human eye. This method has a great advantage also from the time and costs point of view [5].

3 Vehicle Aerodynamics in Small Angles Cross Wind

The current analysis is done using Lattice-Boltzman mathematical model, in a classical aerodynamic CFD simulation setup that uses a hatchback base vehicle. The study is focusing on aerodynamic performance evaluation at nominal position ($\alpha = 0°$) and also small deviation angles ($0° < \alpha < 20°$). It will be considered a standard wheel configuration – 15 in. steel rim without wheel cap, as seen in Fig. 1(a) and an aerodynamic wheel configuration – 15 in. alloy rim equipped with full front and rear wheel caps and rear wheel fairing, as presented in Fig. 1(b).

The sensibility of vehicle aerodynamics in side winds is studied for α between 0° and 20° deviation from vehicle X axis, as graphically described in Fig. 2. These values were chosen due to high frequency of occurrence in real life, as some studies show [6].

The CFD results are integrated in a graphic that indicates the drag coefficient variation with deviation angle, as seen in Fig. 3. A first conclusion that can be highlighted is that aerodynamic performance has a significant variation with deviation angle. The Cd evolution can be divided in two main parts, for $\alpha < 10°$ and for $\alpha > 10°$. The Cd is decreasing with 0.020 counts until $\alpha = 4°$ and continues with a slight increase until $\alpha = 8°$. For $\alpha = 10°$ it can be noticed that there is a small degradation, followed by the constant increase of the Cd. If for $\alpha < 10°$ it can be considered that the Cd has in average approximately the same value, for $\alpha > 10°$ it is obvious from Fig. 3

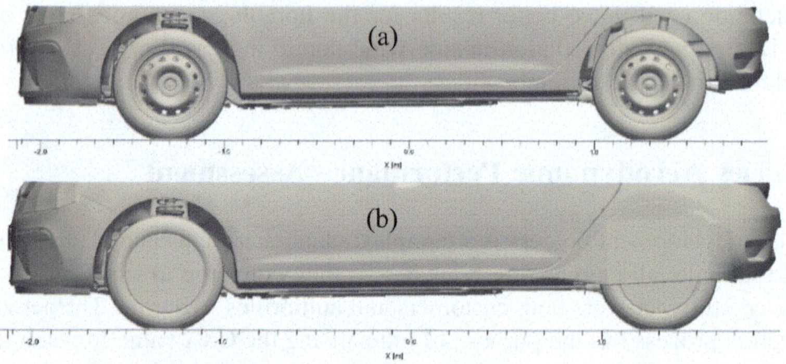

Fig. 1. Studied wheel configurations.

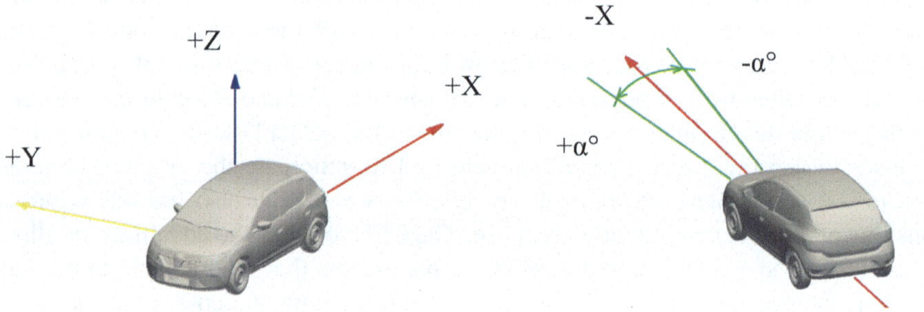

Fig. 2. Vehicle coordinates system definition.

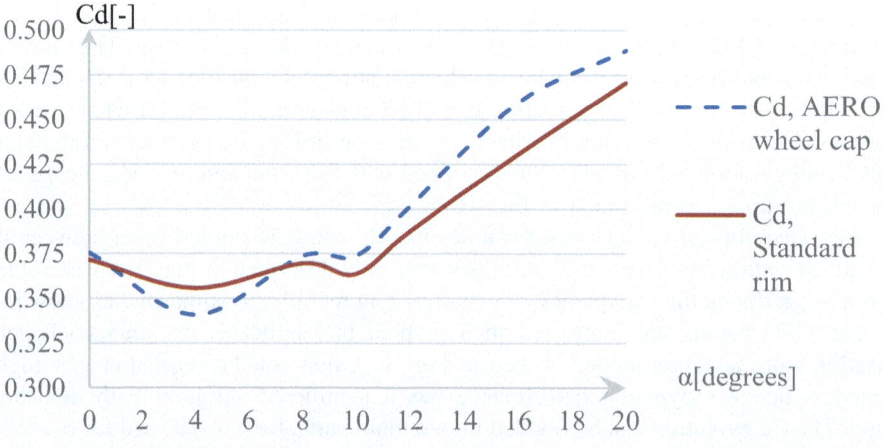

Fig. 3. Drag coefficient variation with α, for standard and aero wheels.

that the aerodynamic performance is degrading without any indication of tipping toward smaller values. This is an important indication that at $\alpha = 10°$ an inflexion point exist, an evidence that something fundamental is happening at this side wind value, for this body design in particular.

To understand what is changing at exactly $\alpha = 10°$ deviation angle, advantage must be taken of CFD tool that allows the user to record and plot the results under various forms. The shape and size of the air drag for a vehicle may be determined through the velocity magnitude [6], as seen in Fig. 4. The three views, along each axis, are indicating what are the main areas that contribute to local drag creation – front wheels drag, front pillar drag, exterior mirrors drag, rear wheels drag, rear drag.

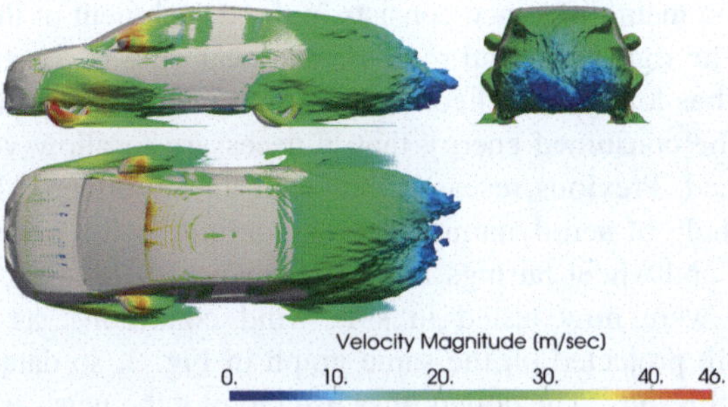

Fig. 4. Drag shape and size plotted with velocity magnitude, for $\alpha = 0°$.

To understand and evaluate Cd evolution it is necessary to analyse the drag zones that are developed under different deviation angles. This comparison is plotted in Fig. 5, where it can be easily observed how the local drag zones have different magnitude depending on α angle.

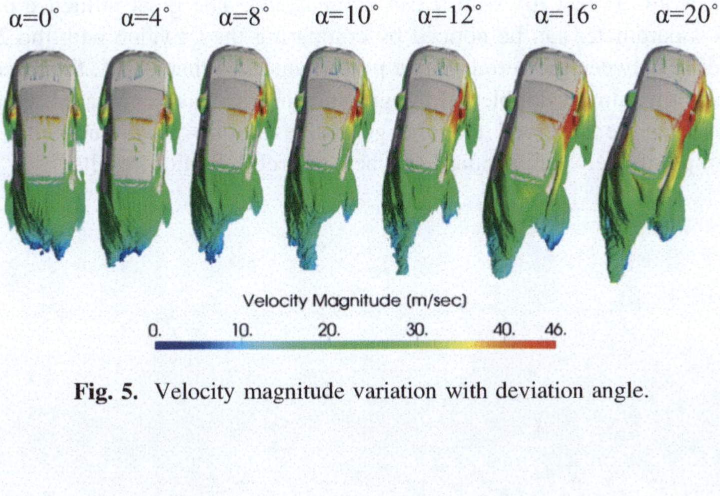

Fig. 5. Velocity magnitude variation with deviation angle.

For $\alpha < 10°$ it can be noticed that the rear wake is transformed from a an almost uniform shape into an asymmetric shape, with a longer, but thinner volume. In the same time, the drag associated with the right front pillar and exterior mirror, is intensifying. The overall effect on the global aerodynamic performance is the one noticed in Fig. 3. The Cd oscillation depending on α can, in this specific case, be explained by the different amount of energy consumed by each mentioned local drag area.

For $\alpha < 8°$ there is a beneficial effect due to reduced energy in the rear drag, which for vehicles, is the most important, although, the other areas are still stabile and almost at the same level of energy.

At $8° < \alpha < 8°$, an intensification of pillar and front wheel drag starts to appear. This is how the slight increase of global Cd can be explained.

At $\alpha \geq 10°$, the main difference consists in the detachment of the right pillar local drag shape. Also, the right front and rear wheels local drag is constantly intensifying. This phenomenon has a negative effect on the aerodynamic performance, which leads to an increase of the consumed energy that is necessary to allow vehicle to keep the same traveling speed. Previous research indicated an important Cd gain that can be obtained with the help of aerodynamic technical definition for wheels: full wheel cap on all four wheels and wheel fairings for the rear wheels.

These devices were now tested in side wind conditions, as explained above. The CFD results are projected on the same graph in Fig. 3, so data can be compared with standard wheel results. The dotted lines used to plot the new set of results indicate the same general behaviour (as for the standard wheel), however a bigger amplification of the curve can be observed. This implies that for $\alpha < 8°$, the gain is bigger than the one obtained with the standard technical definition of the vehicle. For $8° < \alpha < 20°$, the results are worse due to the negative effect of the wheel cap that does not allow the transversal air flow to pass, creating an additional resistance. The difference can be visualised in Fig. 6 by plotting the pressure deficit for both configurations, at different α angles.

As already mentioned, to have a better comprehension of vehicle's total aerodynamic resistance, it is necessary to relate to CdA parameter and not only to Cd. In Fig. 7, the evolution of CdA with α can be analysed. The great influence of the side wind on this parameter can be noticed by comparing the 0° value with the 20° value. The difference between the aerodynamic performances is that at 20°, the force resulted from the air flow, almost doubles. If the graphics in Fig. 7 are compared with the ones in Fig. 3, it can be observed that the general evolution is similar, with the same intersection points between standard and aero wheel definition results.

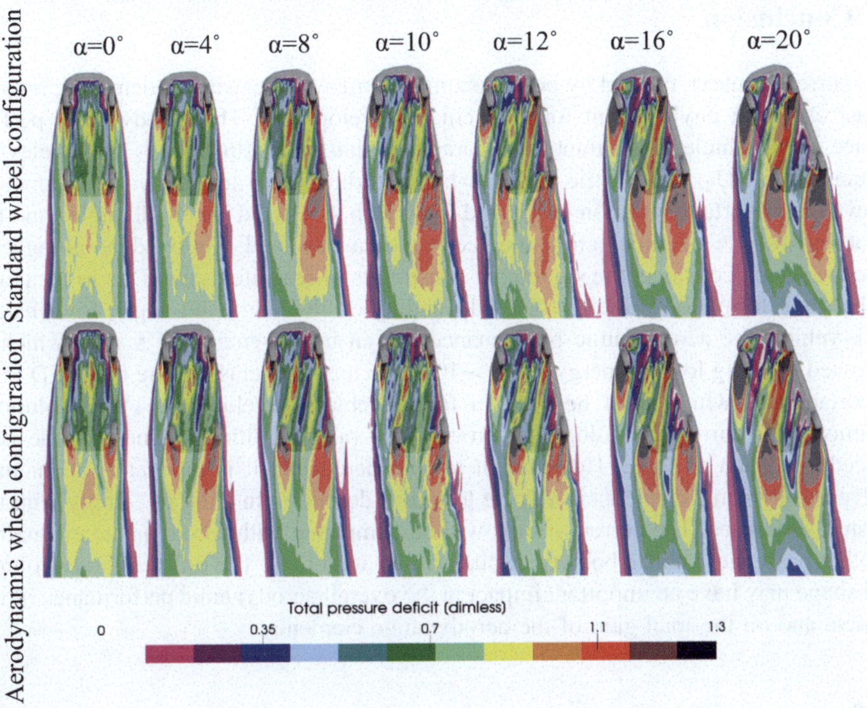

Fig. 6. Pressure deficit comparison between standard and aero wheel, for different α values.

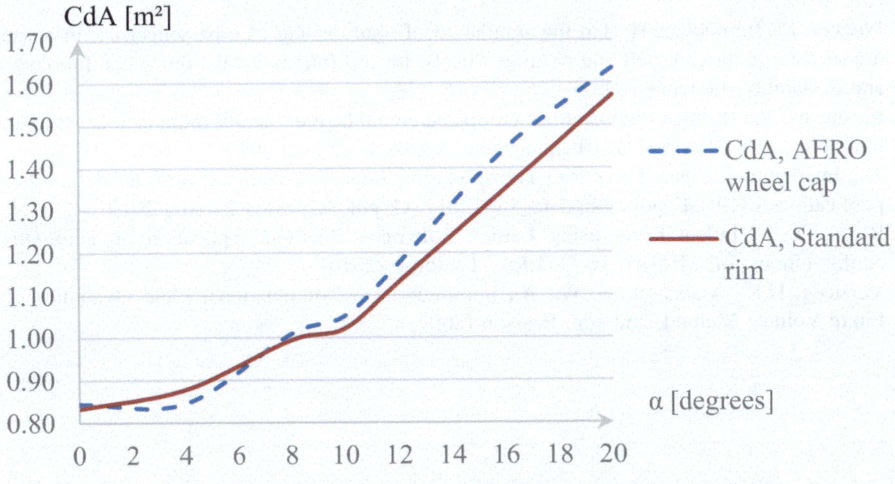

Fig. 7. CdA variation with α, for standard and aero wheels.

4 Conclusion

The current context, created by both customer desires and legislation demands, impose a fast changing environment for the vehicle development. The aerodynamic performance of a vehicle is an important parameter that can help both ICE vehicles by reducing the CO_2 and electric vehicles by extending their autonomy. CFD analysis shows the important variation of Cd and CdA with side wind for small angles and the great importance of anticipating the request of a new and improved homologation process, which considers the significant constraints of real-life usage. Due to local and global drag formation, which evolves depending on the side wind angle, for a hatchback vehicle the aerodynamic performance has an improvement for $\alpha < 8°$ which is followed by a big loss of energy for $\alpha > 10°$. The main conclusions are that CFD is an important tool which must be used in future vehicle development. Also, solutions optimised for current homologation process can raise significant issues for the new procedures, such as RDE. However, it is considered that it is mandatory to include active aerodynamic elements in vehicle technical definition in order to comply with the harsh CAFÉ rules. The current results will be completed with aerodynamic evaluation of other passenger vehicle body types like saloon and break. The different body volume and shape may have an important impact in the overall aerodynamic performance of the vehicle and on the total gain of the aerodynamic elements.

References

1. European Union. https://ec.europa.eu/clima/policies/transport/vehicles/cars_en. Accessed 22 Aug 2019
2. Mecker, E., Berneburg, H.: On the simulation of road driving of a passenger car in a wind tunnel using a moving belt and rotating wheels. In: 3rd International Conference Innovation and Reliability, Florence (1992)
3. European Union. https://publications.europa.eu/en/publication-detail/-/publication/7ac4d4cb-b9f7-11e7-a7f8-01aa75ed71a1/language-en. Accessed 22 Aug 2019
4. The International Council on Clean Transportation. https://www.theicct.org/sites/default/files/publications/EU-RDE_policy-update_18012017_vF.pdf. Accessed 22 Aug 2019
5. Ricot, D.: Turbulent flows using Lattice Boltzmann Method application on automotive configurations. In: MUSAF 16 ONERA, Toulouse (2016)
6. Versteeg, H.K., Malalasekera, W.: An Introduction to Computational Fluid Dynamics The Finite Volume Method, 2nd edn. Pearson (2007)

Technical Solutions for the Use of Internal Combustion Engine Combustion Gases

Rareş Lucian Chiriac(✉), Anghel Chiru, and Ovidiu Andrei Condrea

Transilvania University of Braşov, B-dul Eroilor nr. 29,
500036 Braşov, Romania
{rares-lucian.chiriac,achiru}@unitbv.ro

Abstract. Internal combustion engines have an efficiency of operating which can be exploit to increase its performance. The analysis of the energy balance of the internal combustion engine shows that there is an important reserve of exhaust gas energy, which can be capitalized in the direction of increasing the indicated yield. One solution is to use this energy to drive a turbine coupled to an electric generator. The paper presents an assessment of the energy that can be recovered from the exhaust gas energy of the internal combustion engine and technical solutions that can be applied.

Keywords: Energy balance · Electric generator · Indicated yield · Exhaust gases · Internal combustion engine

1 Introduction

The turbocharger is designes to introduce a large amount of air into the engine cylinders. It can also be used as a brake motor for large trucks with diesel engines. The turbocharger is a great potential to be developed.

Indeed, cost and development time must be reduced to meet both customer wishes and stringent emission standards. The current turbocharger simulation codes are mostly based on the search boxes (airflow and air mass) supplied by the manufacturers [1].

2 Objective

Internal combustion engines must meet certain constructive, energetic and ecological requirements. The thermal energy sources of the engine are limited, therefore areas where new constructive solutions are required must be optimized. The general methods of internal combustion engine efficiency (Fig. 1) are: variable distribution, direct injection optimization and fuel mixture formation, flue gas recirculation, heat transfer, supercharging. Interaction between the thermal engine sub-domains is influenced by each other, both directly and indirectly [2].

One direction of research is that of reducing the size of the engine. By reducing the size, we can also understand the implementation of new constructive solutions.

I. Dumitru et al. (Eds.): SMAT 2019, *The 30th SIAR International Congress of Automotive and Transport Engineering*, pp. 131–137, 2020.
https://doi.org/10.1007/978-3-030-32564-0_16

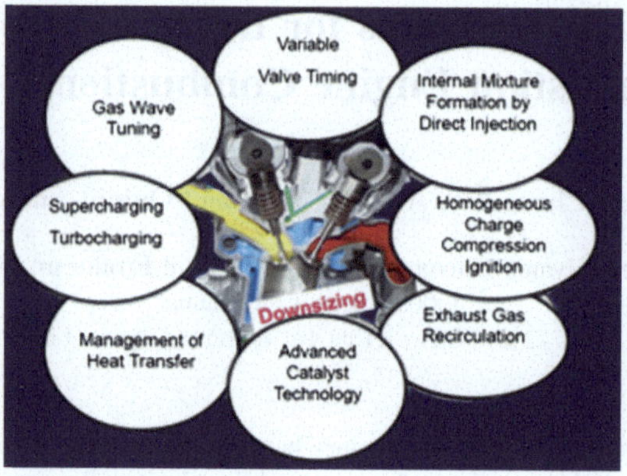

Fig. 1. Domains to reduce the dimensions of internal combustion engines

Studies show that it is possible to reduce the engine cylinder by 60% and still reach the torque curve of a classic thermal engine. Reducing the dimensions of internal combustion engines involves the attributes needed to use such a concept for premium vehicles [4].

Automotive propulsion systems to provide an overview of the overall energy system of the car are shown in Fig. 2 [2].

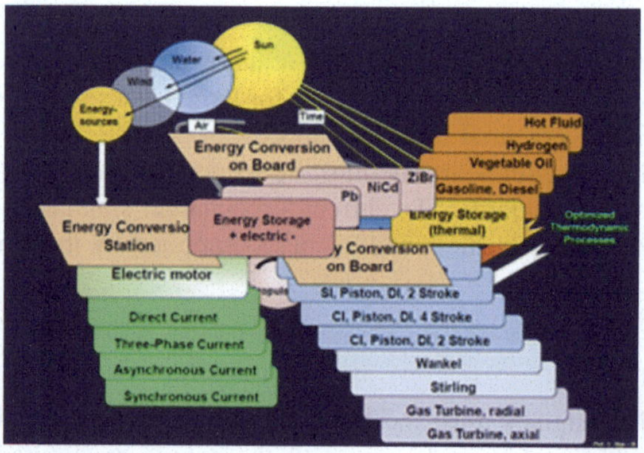

Fig. 2. Overview of processes, propulsion systems, and source of energy for ecological propulsion

From all the technical solutions applied to the development, the work analyzes the exhaust brake design requirements and considerations implemented for heavy-duty heavy duty vehicles. The exhaust brake is mounted on the exhaust galleries. It aims at taking the braking energy of the vehicle to turn it into electricity. The turbine is connected to a power generator via a shaft.

Stronger and faster vehicles have increased the load of the conventional braking system, which has reduced the service life of the braking system, particularly brake linings. This is not only due to higher braking speeds, but also due to frequent braking applications, especially in urban driving conditions and heavy-duty descent. In order to cope with the excessive heat loss resulting from continuous braking during repeated braking, there must be a system for dividing the load of the service brakes. This is done by using auxiliary brakes, such as the exhaust brakes. This solution has proven to be effective after intense use [5].

3 Energy Assessment

The configurations studied by the new exhaust brake system proposed by the automotive industry for internal combustion engines to recover energy from the engine and also some of it to turn it into green energy for automotive consumers but also to store go from the thermal balance equations. Basic energy, impulse and preservation mass equations are resolved along the average flow line. The mass and energy equations are solved for each volume, and the impulse equation is resolved for each volume limit. Figure 3 shows the schematic of the energy values and the flow of the heat flow [6] (Table 1).

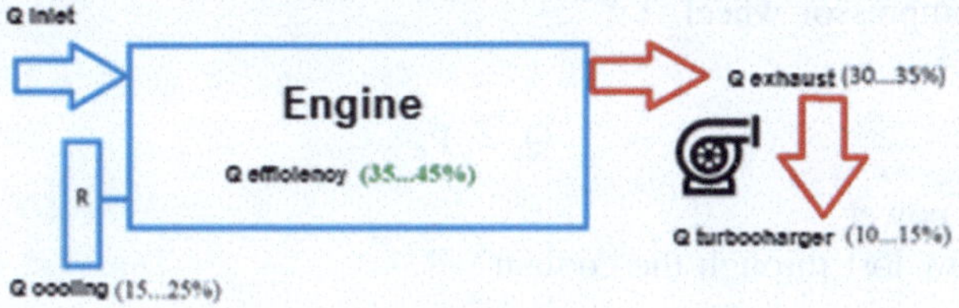

Fig. 3. Scheme of energy balance elements of internal combustion engines

Table 1. Indicative values for the thermal flows heat balance

Engine type	$\dot{Q}_e$[%]	$\dot{Q}_{ch}$[%]	$\dot{Q}_w$[%]	$\dot{Q}_n$[%]	$\dot{Q}_r$[%]
Spark ignition engine	21 … 29	14 … 28	28 … 48	1 … 40	8 … 19
Diesel engine	28 … 41	16 … 33	24 … 44	1 … 4	8 … 16

where:

$\dot{Q}$ – the heat flow obtained by burning the fuel consumed by the engine (available energy); $\dot{Q}_e$ – useful heat; $\dot{Q}_{ch}$ – the heat flow lost through the cooling fluid; $\dot{Q}_w$ – the heat flow lost through the combustion gases along the evacuation; $\dot{Q}_n$ – the heat loss due to incomplete chemical burn; $\dot{Q}_r$ – the residue of the heat balance.

Equation of Mass Continuity

$$\frac{dm}{dt} = \sum\nolimits_{boundaries} m_{flux} \tag{1}$$

Equation: Energy Conservation

$$\frac{d(me)}{dt} = p\frac{dV}{dt} + \sum\nolimits_{boundaries} m_{flux}H - h_{flux}A\left(T_{gas} - T_{wall}\right) \tag{2}$$

Equation: Engine power

$$P_{engine} = BMEPV_D\eta_{cycle} \tag{3}$$

Equation: The thermal balance of the engine is

$$\dot{Q} = \dot{Q}_e + \dot{Q}_{ch} + \dot{Q}_w + \dot{Q}_n + \dot{Q}_r \tag{4}$$

The residue of the heat balance can be further on capitalized through the turbocharger by producing also electrical energy with the help of a generator. After the use of the residue energy to drive the turbocharger and compress the air, the hybrid turbocharger can also produce electrical energy true the connected generator on the turbocharger compressor wheel.

Useful heat

$$\dot{Q}_e = P_e \tag{5}$$

P_e - engine power

The heat flow lost through the coolant

$$\dot{Q}_{ch} = \dot{m}_{lichid}c(t_e - t_i) \tag{6}$$

where:

$\dot{m}_{lichid}$ - cooling liquid flow;
c - medium specific heat of water;
t_e, t_i - inlet and outlet temperature of the coolant;

The heat flow lost through the combustion gases along the exhaust

$$\dot{Q}_w = \dot{m}_g c_{pmg} t_{ge} - \dot{m}_{air} c_{pma} t_a \tag{7}$$

where:

$\dot{m}_g$ - exhaust gas flow rate;
$\dot{m}_{air}$ - air flow rate;
c_{pmg}, c_{pma} - specific exhaust gases specific to air and air;
t_{ge} - outlet gas temperature;
t_a - inlet air temperature

Flow rate of heat loss due to incomplete chemical burns

$$\dot{Q}_{ch} = \sum V_i Q_l \tag{8}$$

where:

V_i - the amount of unbranched exhaust gas components
Q_l - Reduced combustion heat of gaseous components in exhaust gases

The residual flow rate of the heat balance.

$$\dot{Q}_r = \dot{Q} - \left(\dot{Q}_e + \dot{Q}_{ch} + \dot{Q}_w + \dot{Q}_n\right) \tag{9}$$

4 Results

The values recorded for the energy balance can be seen in Table 2 where they are presented according to the type of thermal engine (Otto and Diesel Engines).

Table 2. Values recorded for the energy balance

Engine type	$\dot{Q}_e$[%]	$\dot{Q}_{ch}$[%]	$\dot{Q}_w$[%]	$\dot{Q}_n$[%]	$\dot{Q}_r$[%]
The spark ignition engine, Otto	21 … 32	14 … 28	20 … 30	1 … 3	8 … 10
Compression ignition engine, Diesel	28 … 45	16 … 33	20 … 30	1 … 4	8 … 10

5 Design Solution

Brake performance is an important factor in the design of the diesel engine for cars. Many devices have been developed to achieve a higher braking force; however, most are complicated and expensive.

A schematic solution of the electric power take-off brake is shown in Fig. 4 where it is observed that there are four groups of components that together form the proposed solution, namely: 1 - the classic exhaust brake (T-Turbine), 2- the drive shaft 3 - gearbox with gears, 4 - alternator/generator.

Also, in the schematic diagram, one can be in the area marked 1, the exhaust gas outlet. Single-stage gearboxes are optional, but when used as a transmission solution. They are designed to reduce the rotational speed of the generator by a ratio of 1/4 relative to the rotational speed at which the energy generating exhaust brake operates [8].

It has been found that the predictions from the computerized simulation have agreed with the measured data, not only with regard to the braking force of the exhaust, but also the valve coil, outlet port pressure, and so on. The simulation program has

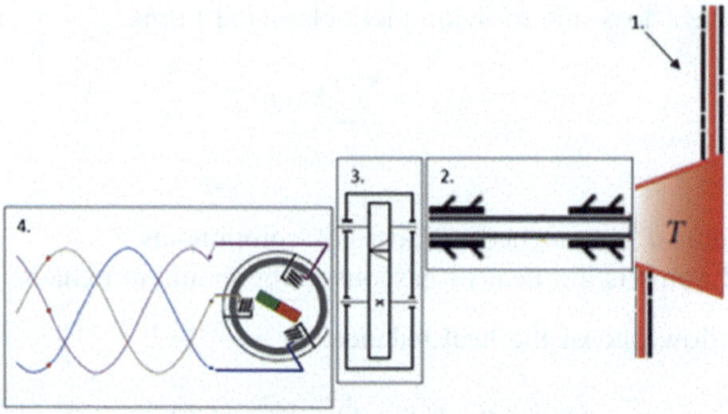

Fig. 4. Generator of electric power

proven to be a useful means of predicting the effect of different brake design parameters and has enabled an optimized exhaust brake system to be achieved at the design stage.

The initial exhaust brake force obtained on a specific turbo-diesel engine has reached the effective average pressure value of 0.56 MPa at rated speed. Changing the existing exhaust manifold configuration, which took better account of the pulsating pressure available there, increased the braking pressure to 0.74 MPa. A pressure effective braking of 0.90 MPa, corresponding to 10.3% of rated power, was achieved by adopting a specially designed valve spring [9].

6 Conclusions

Generating electricity generating brake is a reliable, practical and very efficient solution in many ways: production costs in relation to its applicability, maintenance, reduced construction geometry.

Acknowledgement. I would like to thank for the collaboration and support of Garrett, turbochargers and Siemens.

References

1. Chiriac, R.: New constructive solutions for hybrid turbochargers- as electrical energy generator for increasing the green supply of the vehicule. Editura Universității Transilvania, Brașov, România (2017)
2. Stan, C.: Congresul Internațional CONAT 2016, Brașov, România (2016)
3. Squatriglia, C.: Three Is the New Four as Engines Downsize (2012)
4. Lewis, A.: Ultra Boost for Economy: extending the limits of extreme engine downsizing. SAE Int. J. Eng. **7**, 387–417 (2012)
5. Khan, P: Development and Evaluation of Exhaust Brake Systems for Light Commercial Vehicle. SAE Technical Paper (2005)
6. Teo, A.: Effectiveness of series and parallel turbo compounding on turbocharged diesel engine. J. Mech. Eng. Sci **8**, 1448–1459 (2015)

7. Luft, M.: Losses of IC engine a chance for electrical energy recuperation. Sci. J. Sci. J. (2012)
8. Seuma, J.: Impact of Turbocharger Non-Adiabatic Operation on Engine Volumetric Efficiency and Turbo Lag (2012)
9. Akiba, K.: The Optimized Design of the Exhaust Brake of the Automotive Diesel Engine. SAE Technical Paper (1981)

Evaluation of Tire Traction Performance on Dry Surface Based on Tire-Road Contact Stress

Chen Liang[1], Liu Ji[1], Hoda Mousavi[2], and Corina Sandu[2](✉)

[1] School of Automotive and Traffic Engineering, Jiangsu University, Zhenjiang, Jiangsu, People's Republic of China
[2] TMVS Lab, Virginia Tech, Blacksburg, VA, USA
csandu@vt.edu

Abstract. Tire footprint properties have important effects on tire-road contact pressure distribution and, consequently, on the general performance of the tire when it is in contact with the road. Most of the research conducted in this area is related to the tire performance under static loading conditions. Past studies show that the slip ratio of the tire plays a significant role on tire footprint characteristics. In this study, to account for the effect of the tire footprint characteristics on the tire performance, a finite element tire model with detailed tread pattern (similar to the tire 205/55R16 used for testing and modeling) has been developed using ABAQUS software. The model has been validated using the experimental methods for both footprint dimensions and vertical stiffness values. The comparison of the results from experiment and simulation show a satisfying agreement within 3.6% error. The model has next been used to simulate tire-road interaction for both, static loading conditions and slippage. The results for the footprint properties of the tire with applied normal load of 4 kN for slip ratio of 10%, 15%, 20%, 35%, and 40% are presented. The footprint characteristics of the tire have been parametrized in order to describe the changes in the tire-road contact patch properties under different operational conditions. The parameters with the greatest effect on tire traction performance have been identified and the relationships between these parameters and the tire performance have been presented.

Keywords: Tire · Slip ratio · Footprint properties · Tread pattern

1 Introduction

Vehicle maneuver abilities, such as traction, steering, and braking are highly dependent on the characteristics and properties of the contact patch. Tire contact patch is the section of the tire in contact with road, which is also called the tire footprint. The study of the tire footprint characteristics is thus very important to understand what parameters affect the properties of the contact patch and consequently the performance of the tire.

Tire performance related to the tire-road contact phenomenon can be regarded as tire grounding performance. Researchers have made great efforts through the years to develop many methods for evaluating the grounding performance of tires. However, among the existing methods, those that employ experimental tests are costly and time

I. Dumitru et al. (Eds.): SMAT 2019, *The 30th SIAR International Congress of Automotive and Transport Engineering*, pp. 138–152, 2020.
https://doi.org/10.1007/978-3-030-32564-0_17

consuming. There are also simulation methods that implement comprehensive and complex models of tire-road interaction. These methods are computational expensive and costly due to the need for highly advanced computing facilities.

Considering the important and fundamental role of the tire contact patch properties on grounding performance of the tire, one can focus on the characteristics of the tire footprint in order to find a simple and effective way to evaluate tire grounding performances. A summary of studies on tire grounding performances that focused on the tire-road contact phenomenon is presented in this section.

Tamada and Shiraishi [1] used the heel/toe wear index as an evaluation index of the wear performance of the tire. The wear performance was evaluated by simulating the tire wear process under braking conditions. Sridharan and Sivaramakrishnan [2] analyzed the friction between the tread blocks and the road surface using finite element analysis method (FEA) and proposed that the tire grip performance can be improved by controlling the deformation of the tread pattern. Mashadi et al. [3] found the relations between the Magic Formula coefficients and tire physical properties, such as the inflation pressure, vertical load, and tire rim diameter using an FE model with a simple tread pattern. Wang et al. [4] used the maximum friction between the tire and the road surface as an evaluation index of the tire grip performance and studied the effect of the tread pattern stiffness on the tire grip performance through FE simulation. Their study showed that both the tread pattern deformed shape and the stiffness have a direct effect on tire-road contact pressure distribution. Roth et al. [5] proposed an optical method based on white light refraction and reflection for the measurement of tire footprint analysis. Footprint contact analysis of static and rolling tires are illustrated. Castillo Aguilar et al. [6] used a self-improved test bench to obtain the tread characteristics and the pressure distribution in the tire-ground contact patch.

Tire footprint characteristics are affected by tire geometrical and structural parameters, such as tire dimensions, tire tread pattern, tire compounds properties (including the material properties of different sections of the tire) and also by those parameters related to tire operational conditions, such as applied normal load, tire inflation pressure, camber and toe angles, slip ratio, etc. According to previously conducted studies [7, 8], tires show different grounding performance with different footprint characteristics. One of the important properties of the tire that affects the traction performance is the tire slip ratio. Some studies, such as [9, 10], showed that vehicles under wheel slip ratio between 15% and 20% can obtain the best traction performance on dry surfaces.

Many experimental methods are available to obtain the tire footprint and pressure distribution in the contact patch under static loading conditions, such as [11, 12]. However, obtaining these parameters for a tire under rolling conditions, especially when slip ratio exists, is really difficult and there are not enough studies to address this.

In this study, a simulation method has been developed to study the tire-road contact phenomenon not only under static loading conditions, but also when the tire is rolling with slip ratio from 2% to 35%. The footprint characteristics of the tire have been parametrized in order to describe the changes in the tire-road contact patch properties under different operational conditions. Next, the relationship between the tire traction performance and its footprint characteristics have been investigated. The parameters that affect the tire traction performance the most have been identified and the relationships between these parameters and tire performance have been presented.

2 FEA Modeling of Tire

The tire is a composite of various materials, including rubber compounds, steel cord, cord-rubber composite, etc. The steel cord with high modulus and low elongation is the predominant load-carrying part. The rubber compounds with low modulus and high elongation are able to sustain the inflation pressure and improve wear of the tire, as well as the handling performance of the vehicle. As rubber is a hyperplastic material, Neo-Hook material formulation is suitable to simulate its nonlinearly mechanical behavior and its incompressible property [13].

For this simulation, the rubber material properties were measured through uni-axial tension tests. The dumbbell shaped rubber specimen and the uni-axial tension test machine are shown in Fig. 1.

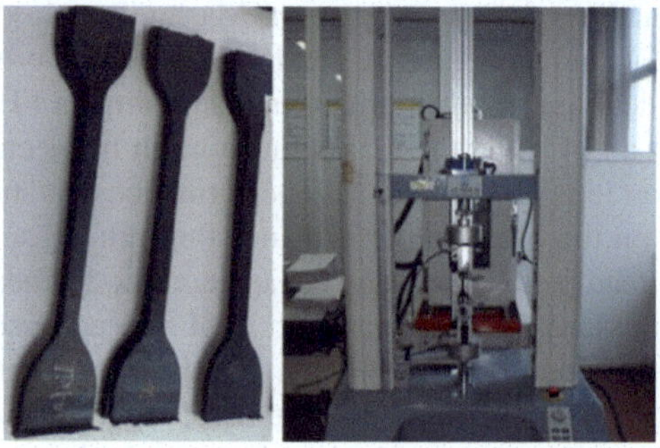

Fig. 1. Rubber specimen and uni-axial tension test machine

The reinforcement elements, usually called cords, are assumed to be linear elastic materials in this simulation. Rebar elements are selected to simulate the orthotropic elastic properties of the cord-rubber system. To characterize the mechanical property of the cords, the elastic modulus E and the Poisson's ratio ε are required. To obtain the elastic modulus E of the cords, several tests have been conducted using a uni-axial test machine. The rubber material properties and the reinforcement material properties employed in this simulation are given in Table 1.

Table 1. Constants of tire materials for simulation

Rubber material	Constants for the Neo-Hook model	
Component	C10 [N/mm^2]	D1
Tread	0.671	0.03
Inner	0.503	0.04
Apex	0.721	0.027
Sidewall	0.336	0.06
Undertread	0.839	0.024
Cords	Constants for linear elastic model	
Component	Elastic modulus [N/mm^2]	Poisson's ratio
Carcass	3300	0.4
Belt	84000	0.4

2.1 Description of FE Tire Model

A passenger car tire 205/55R16, shown in Fig. 2, has been selected for the simulation to study the tire-road contact patch. A 2D axisymmetric model for the tire cross-section without the tread pattern has been developed using AutoCAD. For developing the 3D model of the tire tread pattern, CATIA has been used. These two models have been modified using Hyperworks. Then, they were merged using ABAQUS. The complete 3D model of the tire has been developed in ABAQUS using the rotation of the merged section around the y axis. To simplify the model, only one repeated unit for tread pattern has been considered for the analysis.

In the tire model, the C3D8H and SFM3D4R elements were implemented to define the rubber material and the steel cord, respectively. The tire bead has been defined as an isotropic material by selecting the C3D8R element. The total number of nodes and elements of the 3D model developed are 205383 and 166439, respectively. The steps conducted for developing the model are shown in Fig. 3.

Fig. 2. PCR 205/55R16

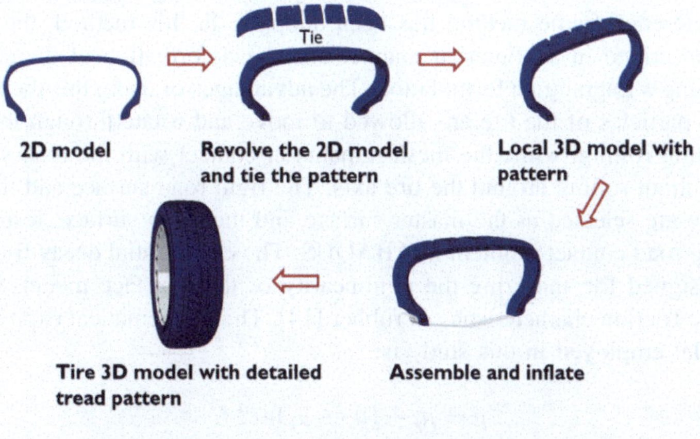

Fig. 3. The processes of building the FE tire model

3 Simulation Methodology

3.1 Static Loading Simulation

An analytical surface has been used to simulate the flat road surface. The coefficient of friction between the tread surface and the road surface has been defined using a Coulomb friction model. The tire inflation pressure has been considered to be 240 kPa. In this simulation, the rim has been assumed to be a fixed rigid body. To account for this, constrains have been applied to fix all the 6 degrees of freedom (DOFs) of the rim (three for the linear displacement and three for the rotations around x, y, and z axes). To simulate the tire deformation due to applied normal load, the location of the road surface has been moved toward the wheel center. The equivalent applied load of 4 kN has been used in this simulation, as illustrated in Fig. 4.

Fig. 4. Static loading simulation

3.2 Traction Simulation with Different Slip Ratios

To analyze the simulation of the tire under steady-state rolling conditions in ABAQUS, a moving reference frame method has been adopted. In this method, the rigid body rotation is described in an Eulerian manner and the deformation of the elements are described using a Lagrangian formulation. The advantages of using this method are that the material particles of the tire are allowed to move and rotate through the tire mesh thus simulating rolling, while the mesh remains in contact with the road surface continuously without rolling around the tire axis. The rigid road surface and the tire tread outer surface are selected as the master surface and the slave surface, respectively, to solve the tire/road contact problem in ABAQUS. The exponential decay friction model has been assigned for analyzing the nonlinearity of the interface model, which fully considers the friction characteristics of rubber [14]. The mathematical expression of the friction model employed in this study is:

$$\mu = \mu_k + (\mu_s - \mu_k)e^{-d_c \cdot s} \tag{1}$$

where μ_k is the kinetic friction coefficient, μ_s is the static friction coefficient, d_c is a user-defined decay coefficient, and s is the slip rate. In the simulation, the kinetic friction coefficient value used was 0.65, the static friction coefficient was considered to be 0.8, and the decay coefficient used was 0.5.

The slip ratio definition used in this study is presented in the following equation,

$$S = \left| \frac{V - r\omega}{V} \right| * 100\% \tag{2}$$

where r is the tire rolling radius, ω is rolling angular velocity, and V is longitudinal velocity of the wheel. As it can be seen from the Eq. (2), the slip ratio of the tire can be controlled by controlling the angular and longitudinal velocities of the tire. To change the angular velocity, a different torque may applied on the tire. This torque may produce traction or braking forces, as illustrated in Fig. 5. For the free rolling tire, the zero torque is applied. For a free rolling tire, one can assume that the slip ratio is negligible (close to zero) and the relationship between the longitudinal velocity, the tire rolling radius, and the angular velocity is as follows,

$$V = r\omega \tag{3}$$

In the simulation performed, in order to change the slip ratio on the contact patch, the angular velocity of the tire was changed.

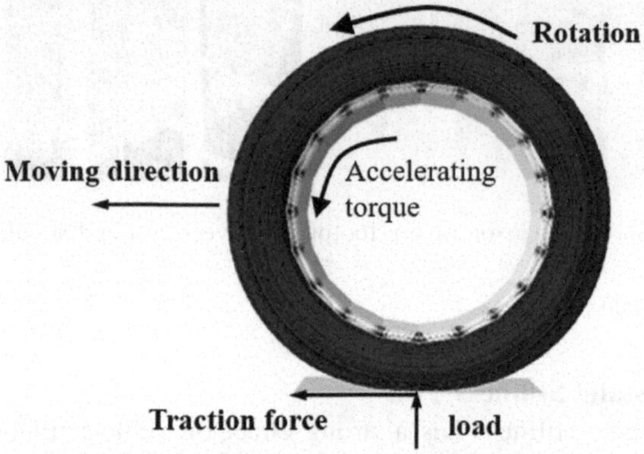

Fig. 5. Illustration of tire traction simulation

4 Validation of Tire FE Model and Simulation Methodology

4.1 Static Loading Footprint

In order to validate the effectiveness of the tire FE model, a static loading test was carried out using the tire stiffness tester MTM-2. The load used in the ink print test (conducted to obtain the tire footprint under static conditions) and in the corresponding simulation was 4.821 kN and the inflation pressure was 240 kPa. The comparison of the results for the tire footprint between the simulation and the data collected experimentally is shown in Fig. 6. The footprint geometry parameters were compared in Table 2 for both, simulation and test data. As shown in Table 2, the maximum error between the experimental test and the simulation results for the geometry parameters is 1.86%. According to the results shown, the FE tire model developed is a valid model capable to simulate the tire-road contact phenomenon and thus the simulation results obtained using this model are expected to provide accurate results of the tire footprint.

Table 2. The geometry parameters of footprint

	Test [mm]	Simulation [mm]	Error [%]
Length	147	148	0.67
Width	161	164	1.86

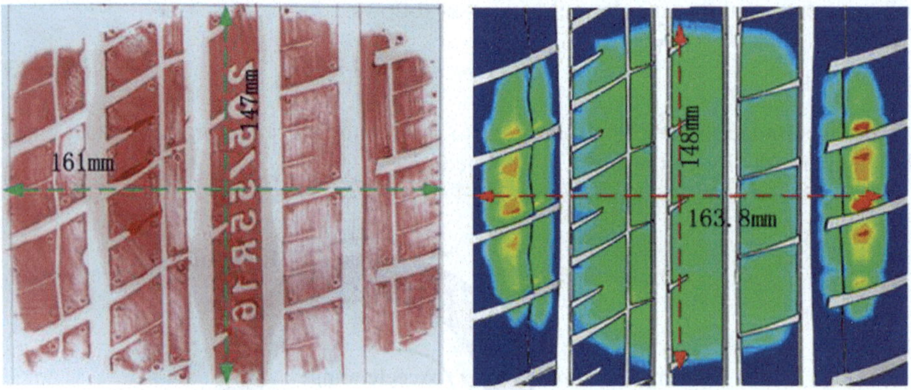

Fig. 6. Comparison of tire footprint between test and simulation

4.2 Vertical Static Stiffness

a. Tire Vertical Static Stiffness Test

The tire vertical static stiffness has a strong effect on vehicle riding performance. In order to further validate the FE tire model, an electro-hydraulic tire dynamic tester, shown in Fig. 7, was used to test the vertical static stiffness of the radial tire 205/55R16. For this machine, the wheel load can be adjusted up to 30 kN. Maximum circumferential speed could be around 120 km/h. The slip angle can be adjusted during the measuring process, in the range between −0.31 rad and +0.31 rad. The inclination angle may vary from −0.14 rad to +0.14 rad. There is a force sensor at the spin center of the tire wheel which can measure longitudinal force Fx, lateral force Fy, normal load Fz, and the moments round y axis and z axis. The Data Translation 9800+pc system was used to acquire the experimental data. Similar to our FEA simulation, to apply the static normal loading a steel plate was inserted under the tire to simulate the road surface.

Fig. 7. Tire drum testing equipment

Various wheel normal loads were generated by the hydraulic cylinders and were measured through the calibrated load cells in the vertical direction. The tire vertical deformation was measured by the laser distance sensor mounted on the equipment. Similar to the input for the simulation, the tire inflation pressure was set at 240 kPa and wheel load was set at 4 kN.

b. Comparison and Discussion

The comparison between the results of the experiment and the simulation using the FE tire model developed is shown in Fig. 8.

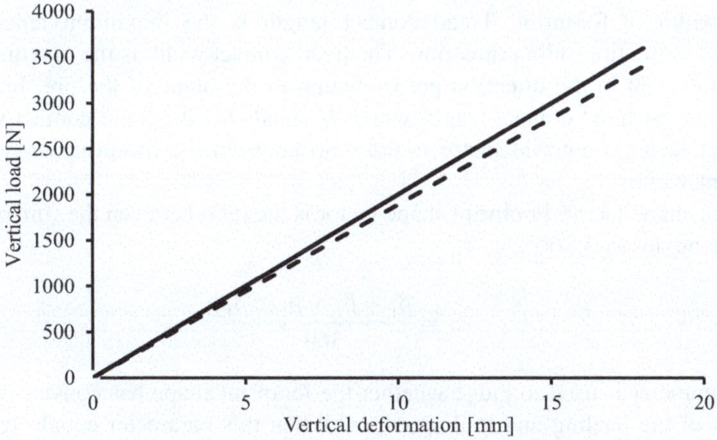

Fig. 8. Comparison of tire vertical static stiffness

As it can be seen from this figure, the vertical load of the tire and its vertical deformation are linearly dependent. Using the data obtained, the vertical stiffness of the tire can be calculated. The vertical stiffness obtained from the experiment and simulation are 206.2 N/mm and 198.8 N/mm respectively. There is a satisfactory agreement, the result of the simulation being within 3.6% error with the experimental data.

5 Footprint Characterization

This paper describes the characteristics of the shape of the contact area between the tire and the road surface as the geometric characteristics of footprint. The characteristics of the ground pressure distribution between the tire and the road surface are called mechanical characteristics of the tire ground pressure distribution. The detailed information about these characteristics, the reader can refer to published literature [15].

5.1 Geometric Characteristics of Footprint

The geometric characteristics considered in this study include the following parameters, defined as:

a. contact area. Contact area is the true area in contact between the tire tread and the pavement. It is shown as C-area in Fig. 9.
b. footprint area. Footprint area is the area obtained by vertically projecting the entire area between the tire tread and the pavement onto the surface of the road. It differs from the contact area as it also contains the projected area of the tread groove, shown as F-area in Fig. 9.
c. contact area ratio. Contact area ratio is the ratio between the contact area and the footprint area.
d. length/width of footprint. Tread contact length is the maximum length of the footprint in the tire rolling direction. The tread contact width is the maximum width of the footprint in the direction perpendicular to the plane of the tire. In Fig. 9, L represents the tread contact length while W stands for the tread contact width.
e. footprint factor. Footprint factor is the ratio between the footprint length and the footprint width.
f. footprint shape factor. Footprint shape factor is the ratio between the sum of the four exterior angles and 360°,

$$\varphi = \frac{\beta_1 + \beta_1 + \beta_1 + \beta_1}{360^\circ} \tag{4}$$

This parameter is used to judge whether the footprint shape has convex or concave profiles of the leading and trailing edges. When this parameter equals 1, the footprint shape is a rectangle. When this parameter is bigger than 1, the footprint shape is concave (the leading edge and trailing edge); otherwise, it means the footprint shape is convex (the leading edge and trailing edge).

g. footprint length factor. The footprint length factor is defined as,

$$\begin{cases} DOFds = \overline{AB}/DS70 \\ DOFus = \overline{CD}/US70 \\ DOFavg = (DOFds + DOFus)/2 \\ LFds = L/DS70 \\ LFus = L/US70 \\ LFavg = (LFds + LFus)/2 \end{cases} \quad (5)$$

where *US70* and *DS70* are the contact length at the 70% portion of the contact width, *DOFavg* is the second length factor, and *LFavg* is the third length factor.

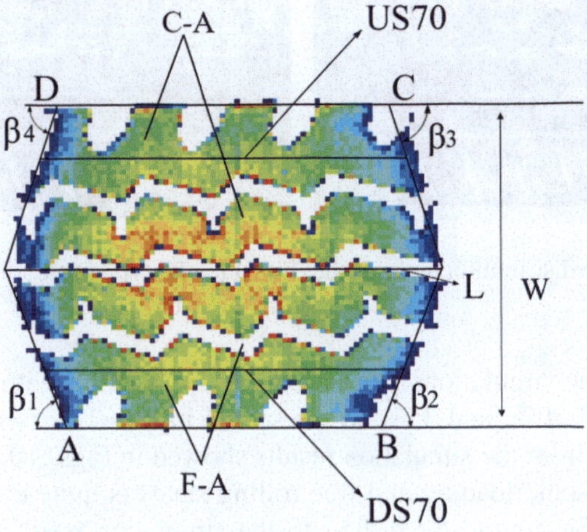

Fig. 9. Illustration of footprint parameters

5.2 Mechanical Characteristics of Footprint

The mechanical characteristics are mainly the following parameters:

a. average contact pressure. The average contact pressure is defined as,

$$\overline{p} = \text{Load}/\text{C-area} \quad (6)$$

b. contact pressure variance. The contact pressure variance is defined as,

$$\alpha = \sqrt{\frac{1}{\mathrm{n}-1}\sum_{i=1}^{n}(P_i - \bar{P})^2} \quad (7)$$

where n is the number of the nodes in the contact area and Pi is the contact pressure of the node i.

6 Results and Discussion

6.1 Simulation Results

Using the simulation method discussed in Sect. 3, the results presented in this section have been obtained. The tire static loading simulation and the tire steady-state free rolling simulation were performed first and the results are presented in Figs. 10 and 11, respectively.

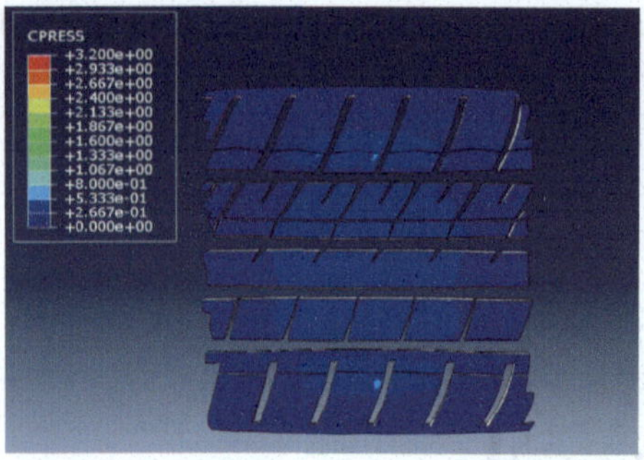

Fig. 10. Tire pressure distribution under static loading [MPa]

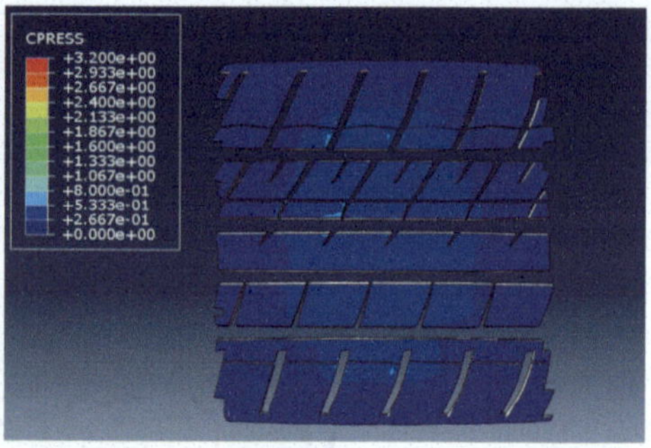

Fig. 11. Tire pressure distribution under free rolling [MPa]

The results of the simulations performed for tire traction with a slip ratio of 2%, 4%, 8%, 10%, 15%, 20%, and 35% are presented in Fig. 12.

It is can be seen from the simulation results showed in Figs. 10, 11, and 12, that the tire footprint under static loading and free rolling states is quite even compared to the tire footprint with slip ratios. According to the simulation results, the larger the slip ratio is, the greater the value of the maximum contact pressure of the footprint. This may be due to the buckling phenomenon of the tread pattern block edges during sliding.

To acquire traction force, the tire needs to roll with a certain slip ratio. The simulation result of tire traction force presented in Fig. 13 illustrates the changing law of the tire traction force with the increase in the slip ratio. It is shown that the tire traction force increases sharply with the slip ratio increase when the slip ratio is smaller than 8%. The traction force remains almost constant when the slip ratio is larger than 10%.

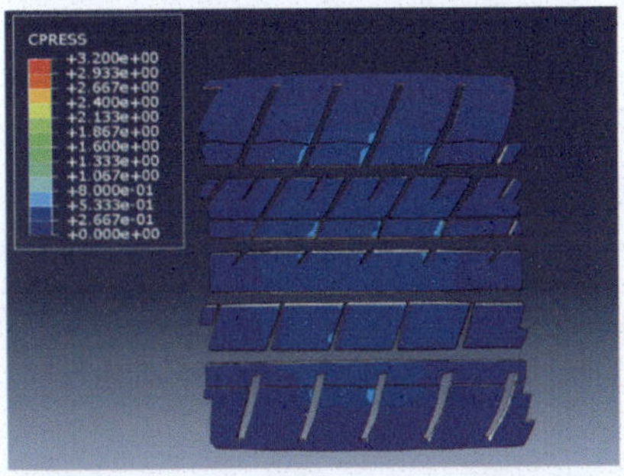

a. Slip ratio 2%

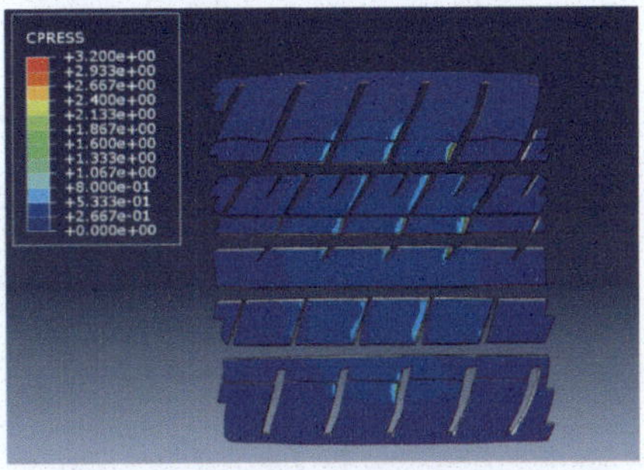

b. Slip ratio 4%

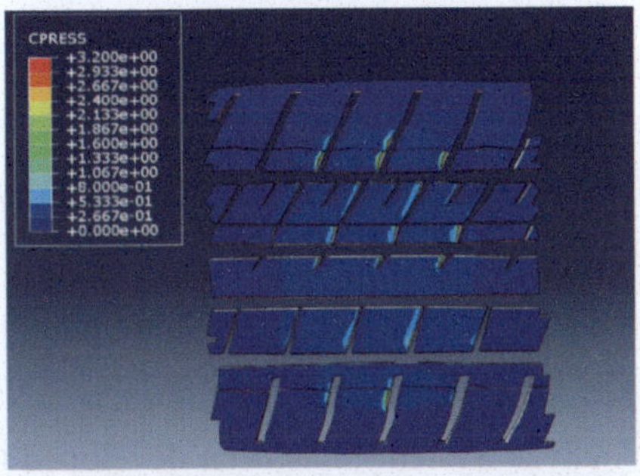

c. Slip ratio 8%

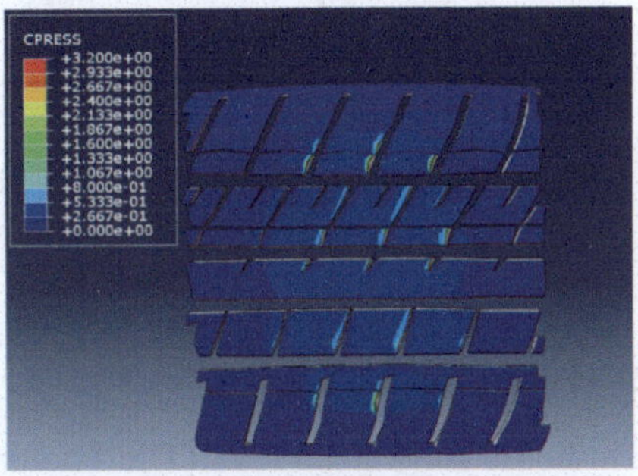

d. Slip ratio 10%

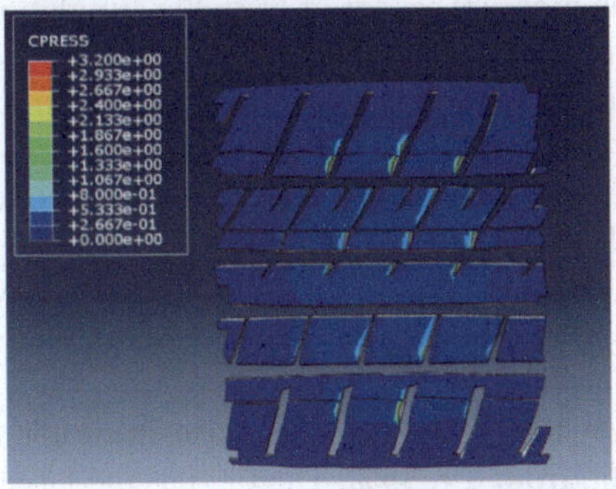

e. Slip ratio 15%.

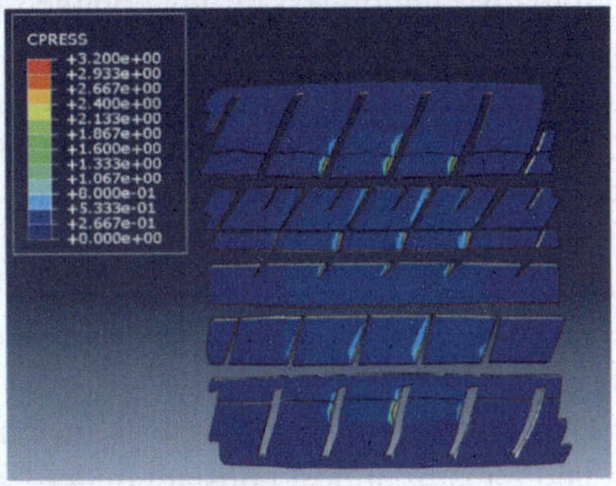

f. Slip ratio 20%

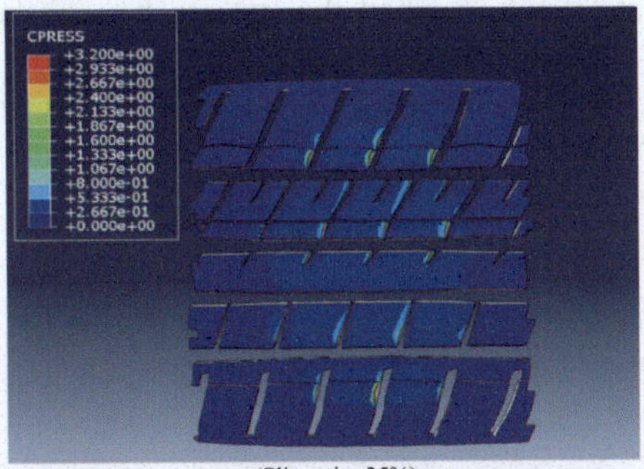

g. (Slip ratio=35%)

Fig. 12. Tire-road contact pressure distribution under different slip ratios [MPa]

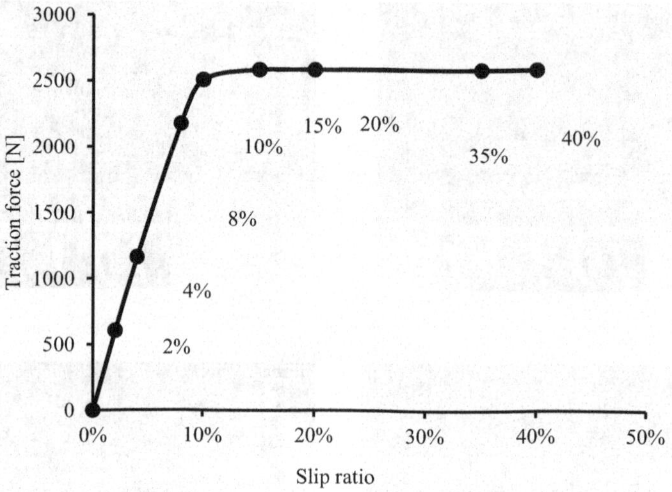

Fig. 13. Traction force under slip ratio

6.2 Footprint Analysis

An in-house software, TFAS (Tire Footprint Analyze System), has been developed to calculate footprint parameters described in Sect. 5. TFAS can read the tire ground pressure distribution test results obtained by the pressure sensor or from the simulation. Then, the software performs filtering noise reduction and boundary extraction processing on the data. After some calculations, the software will output the footprint parameters described in Sect. 5. Figure 14 lists some of the footprint parameters under static loading, free rolling state, and with different slip ratios. It is interesting to notice that the tire under free rolling has the smallest contact length, contact width, and contact area. It is useful for the tire rolling under the most economic state to save energy; the smallest tire contact area means that the tire can roll with the smallest rolling resistance.

The footprint contact length is increasing with the slip ratio increase from 0% to 8%, and remains almost constant under slip ratio from 8%-35%. The tire has the largest footprint contact width under static loading. It is noticed that the footprint contact width reaches the value of static loading when the tire reaches the largest traction force at about 8% slip ratio. It is also interesting to notice that the tire contact area increases with the slip ratio increasing from 0% to 35%. The tire contact pressure variance is also increasing with the slip ratio; this can explain why the maximum contact pressure is increasing with the slip ratio increase.

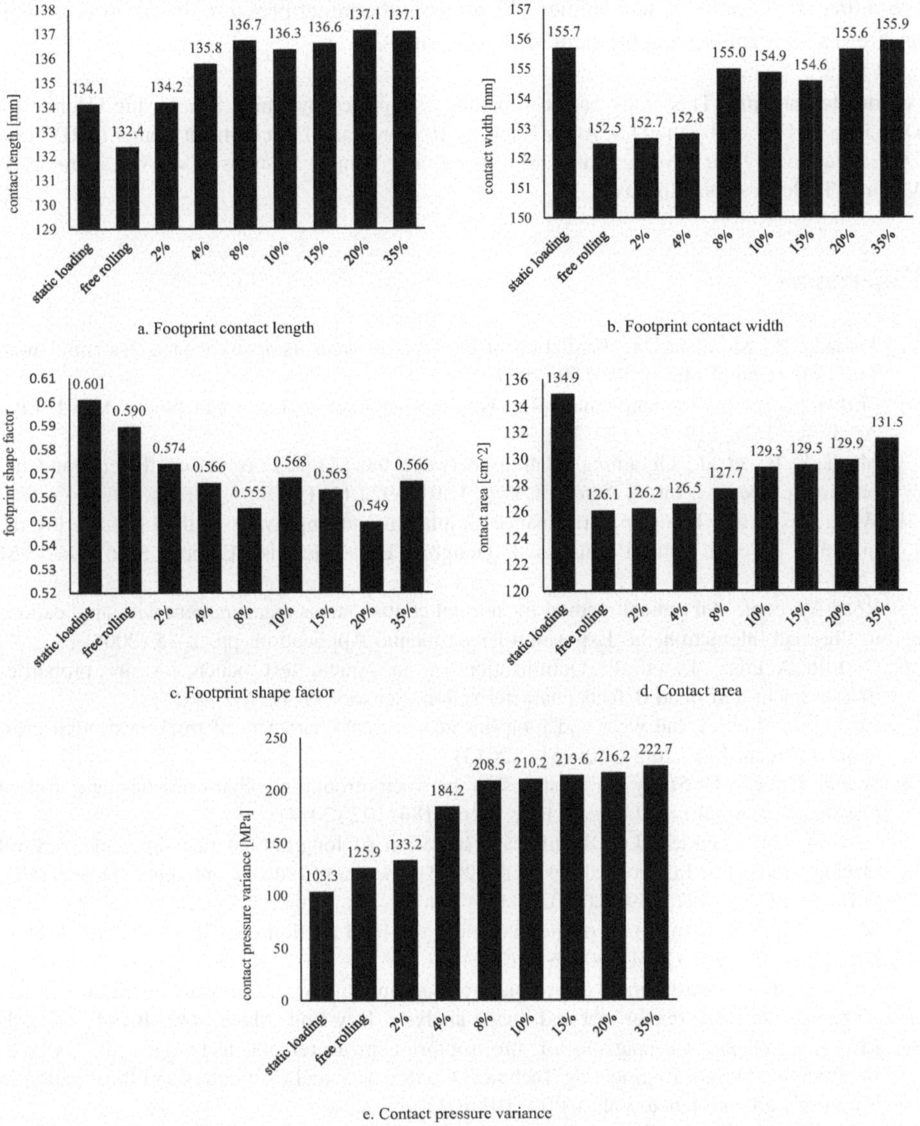

Fig. 14. Footprint parameters under different working state

7 Conclusion and Future Work

This paper presents results obtained as part of a more comprehensive study of tire performance on various surfaces. The paper focuses on the relationship between the tire traction performance and the tire-road contact pressure distribution. A finite element tire model has been used to study the relationships between tire traction forces and tire footprint parameters through simulations. In order to validate these simulation results, the future work includes test studies of tire-ground contact pressure distribution as well as tire traction force measurement with slip ratios.

Acknowledgments. This study has been partially supported by China Scholarship Council for Dr. Liang and Dr. Ji, by the U.S. Army Engineer Research and Development Center (ERDC) for Dr. Sandu, and by the Terramechanic, Multibody, and Vehicle Systems (TMVS) Laboratory at Virginia Tech for Mrs. Mousavi.

References

1. Tamada, R., Shiraishi, M.: Prediction of uneven tire wear using wear progress simulation. Tire Sci. Technol. **45**(2), 87–100 (2017)
2. Sridharan, K., Sivaramakrishnan, R.: Dynamic behavior of tyre tread block. Am. J. Eng. Appl. Sci. **5**(2), 119–127 (2012)
3. Mashadi, B., et al.: Obtaining relations between the Magic Formula coefficients and tire physical properties. Int. J. Automot. Eng. **1**, 911–922 (2015)
4. Wang, G., et al.: Tire grip performance simulation and improving method considering the influence of tread pattern stiffness. J. Donghua Univ. (English Edition) **33**(6), 947–951 (2016)
5. Roth, V., et al.: An optical method for normal contact stress measurement with applications to tire-road interaction. In: Exposition Experimental Application, pp. 1–15 (2004)
6. Castillo Aguilar, J., et al.: Optimization of an optical test bench for tire properties measurement and tread defects characterization. Sensors **17**(4), 707 (2017)
7. Liang, C., et al.: Tread wear and footprint geometrical characters of truck bus radial tires. Chin. J. Mech. Eng. **26**(3), 506–511 (2013)
8. Wang, G., et al.: Study on relationship between grounding characteristics and rolling resistance of radial tire. J. Mech. Eng. **8**(16), 186–192 (2014)
9. Carlson, C.R., Gerdes, J.C.: Nonlinear estimation of longitudinal tire slip under several driving conditions. In: Proceedings of the 2003 American Control Conference, Denver, CO, USA, vol. 6, pp. 4975–4980 (2003)
10. Müller, S., et al.: Estimation of the maximum tire-road friction coefficient. ASME. J. Dyn. Sys. Meas. Control **125**(4), 607–617 (2004)
11. Lu, R.S., et al.: Measurement of vehicle tire footprint pattern and pressure distribution using piezoresistive force sensor mat and image analysis. Key Eng. Mater. **437**, 467–471 (2010)
12. Els, P.S., et al.: Comparison of tire footprint measurement techniques. In: ASME. International Design Engineering Technical Conferences and Computers and Information in Engineering Conference, vol. 3:V003T01A027
13. Zhou, H., et al.: Analysis of radiation noise characteristics of tires with different tread patterns. Noise Vib. Control **32**(6), 175–178 (2012)
14. Wang, H., et al.: Effect of surface friction on tire–pavement contact stresses during vehicle maneuvering. J. Eng. Mech. **140**(4), [04014001(8)] (2013)
15. Lian, C., et al.: Experiment study of radial tire contact pressure distribution evaluation. Automot. Technol. **11**, 38–42 (2013)

Performance, Combustion and Emissions Study of a DI Diesel Engine Running on Several Types of Diesel Fuels

Nicolae Vlad Burnete, Richard János Balint, Corneliu Adrian Măgherusan(✉), and Dan Moldovanu

Technical University of Cluj-Napoca, Muncii Blvd. 103-105, 400641 Cluj-Napoca, Romania
{nicolae.vlad.burnete, dan.moldovanu}@auto.utcluj.ro, {richard.balint, adrian.magherusan}@student.utcluj.ro

Abstract. The present study aims to investigate the differences between several summer grade diesel fuels available on the local market with respect to their impact on engine performance, combustion and emission characteristics, by analyzing various combustion parameters (ignition delay, rate of heat release, maximum indicated pressure, maximum cylinder temperature etc.). The tests were performed on a turbocharged four-cylinder 1.9L DI Diesel engine operating at 0.1, 0.3, 0.5 and 0.7 load at constant 1700 rpm, using 7 different types of diesel fuel. The results showed peak difference percentages of 8.05% for BSFC, 32.29% for NO, 4.38% for peak in-cylinder pressure, 5.83% for in-cylinder temperature and 6.56% for ROHR. There were no significant differences in ignition delay.

Keywords: Engine performance · Combustion analysis · Diesel fuel · In-cylinder pressure

1 Introduction

Current diesel engine development trends and stricter emission limitations push the limits of diesel engine combustion process research. The combustion process in diesel engines favors a high engine energy efficiency and therefore a promising fuel consumption without power output penalties. In order to obtain more efficient engines that have a reduced environmental impact, a precise control of combustion phasing, rate of heat release, pressure gradient, combustion temperatures etc. [1–3] is required. Combustion and pressure trace analysis play an important role in achieving this desiderate [4, 5].

Diesel engine combustion is also influenced by the quality of the fuel used, meaning that, in order to achieve a clean and complete combustion process an adequate quality of the fuel is required. Considering that the engine cannot be optimized for every market available fuel, the latter must comply with strict quality standards, thus ensuring that the promised engine performance and reduced emission output are achieved.

I. Dumitru et al. (Eds.): SMAT 2019, *The 30th SIAR International Congress of Automotive and Transport Engineering*, pp. 153–159, 2020.
https://doi.org/10.1007/978-3-030-32564-0_18

The purpose of this effort is to investigate the differences between several summer grade diesel fuels available on the local market in order to highlight differences between engine performance and combustion process parameters.

2 Methodology

For this study a 1.9 l turbocharged direct injection diesel engine was used. The engine has a peak power of 63 kW at 3100 rpm and a peak torque of 215 Nm at 1900 rpm. Relevant engine data is presented in Table 1. Injection parameters (timing and quantity) were controlled by the engine ECU and since the load and speed did not change between fuels, the resulting values were found to be the same (Table 2). Engine speed and load were controlled via a Weinlich Steuerungen MP 80/6000 eddy-current brake. In-cylinder pressure was measured with a Kistler 6056A pressure sensor and the results were averaged over 21 cycles. For this study, seven regular summer grade diesel fuels that comply with the EN 590 regulations were tested with respect to engine performance, pollutant emissions and combustion process parameters. The emissions values were recorded every 5 s using a Testo 350 gas analyzer. For these tests engine speed was kept constant at 1700 rpm while the load varied (0.1, 0.3, 0.5 and 0.7).

Table 1. Engine data

Parameter	Value	M.U.
No. of cylinders	4	–
Valves per cylinder	2	–
Bore	79.5	mm
Stroke	95.5	mm
Compression ratio	19.5	–
Fueling system	Bosch VE VP 37	–
Charging	Garrett GT1544V	–

Table 2. Injection parameters and test conditions (averaged)

Load [-]	Timing [°CA] bTDC	Quantity [mg/cycle]	Charge air pressure [kPa]	Intake air temperature [°C]	EGR [%]
0.1	0.7	7.4	114.3	25	50
0.3	0.7	14.0	118.3	26	55
0.5	1.1	22.4	137.7	26	53
0.7	4.6	29.8	162.2	26	20

The pressure traces where smoothed using a Butterworth filter. A 1st Law analysis of the combustion process was performed based on the filtered pressure curves using the model presented in [1]. The pollutant emissions values recorded over 1 min were averaged and the then converted to brake specific quantities using Eq. (1) [6].

$$E_{p,i} = E_{V,i} \cdot \frac{M_i}{M_{exh}} \cdot \frac{\dot{m}_{exh}}{P_{eff}} \tag{1}$$

where $E_{p,i}$ is the exhaust emission in [g/kWh]; $E_{V,i}$ is the exhaust emission value as volume fraction in [ppm]; M_i is the mol mass of the component i in [km/kmol]; M_{exh} is the mol mass of the exhaust [kg/kmol]; $\dot{m}_{exh}$ is the exhaust mass flow [kg/h]; P_{eff} is the power output.

3 Results and Discussions

For brevity only 0.5 load results are shown here (Table 3), but the trends were similar for all other loads. For the performance comparison the fuel with the lowest break specific fuel consumption (BSFC) was chosen as a reference. The results show that at 0.5 load there is a difference between the fuel with the highest BSFC (Fuel 2) and the fuel chosen as reference (Fuel 7) of 3.85 g/kWh which represents 1.8%. For the other fuels, the difference in terms of BSFC compared to Fuel 7 were close to 0.9%. Since injection parameters did not vary from one fuel to another, a reduction in terms of fuel consumption would be noticed. For the other loads, BSFC varied with 3.64% at 0.3 load (Fuel 2 vs Fuel 7), 8.05 at 0.1 load (Fuel 4 vs Fuel 6) and with 2.32% at 0.7 load (Fuel 3 vs Fuel 7). With respect to emissions, at 0.5 load Fuel 6 led to the lowest carbon monoxide (CO) values, while Fuel 5 led to the lowest nitrogen oxide emissions (NO). However, at 0.1 load Fuel 4 showed the lowest amount of NO, while at 0.3 and 0.7 load, Fuel 1.

Table 3. Engine performance at 50% load

Fuel	Power [kW]	Torque [Nm]	BSFC [g/kWh]	CO [g/kWh]	NO [g/kWh]
Fuel 1	21.2	119.5	215.5	0.01455	2.145
Fuel 2	21.2	120.0	217.5	0.00755	2.320
Fuel 3	21.2	119.0	215.5	0.01721	2.460
Fuel 4	21.3	119.5	214.5	0.01448	2.686
Fuel 5	21.2	119.0	215.5	0.00105	2.143
Fuel 6	21.3	120.0	214.5	0.00104	2.259
Fuel 7	21.2	119.0	213.6	0.00265	2.414

The next step of the study was the analysis of the pressure traces, which are an average of 21 engine cycles. The traces for 0.5 load are presented in Fig. 1. The lowest and the highest values for the peak in-cylinder pressure were achieved using Fuel 7 and Fuel 1 respectively, which means a difference of 3.15%. Regarding the other loads, a difference of 2.05%, 4.38% and 3.94% was observed at 0.1, 0.3 and 0.7 load respectively. The lower pressure values with would lead to lower stresses on the engine pistons and crank mechanism components.

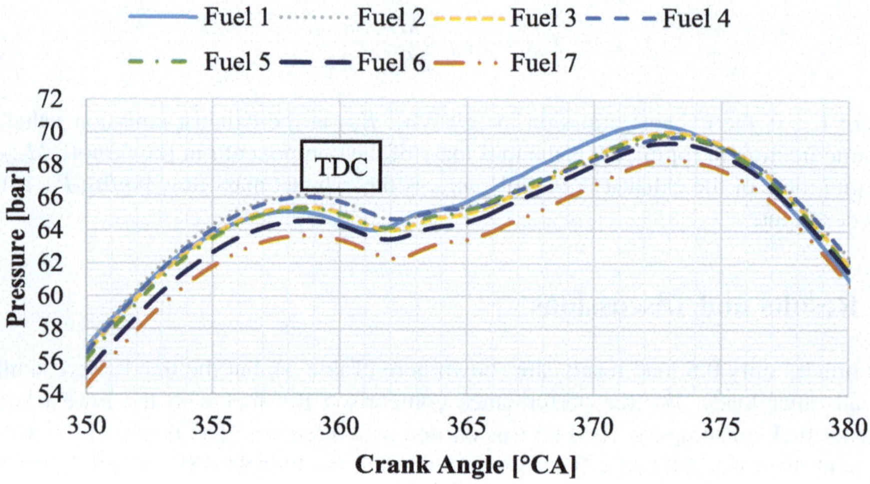

Fig. 1. In-cylinder pressure traces for all fuels at 0.5 load.

In-cylinder temperature traces are represented in Fig. 2. The results showed that between the highest and the lowest in-cylinder temperatures there was a difference of 64K at 0.1 load, 103K at 0.3 and 0.5 load and 42K at 0.7 load.

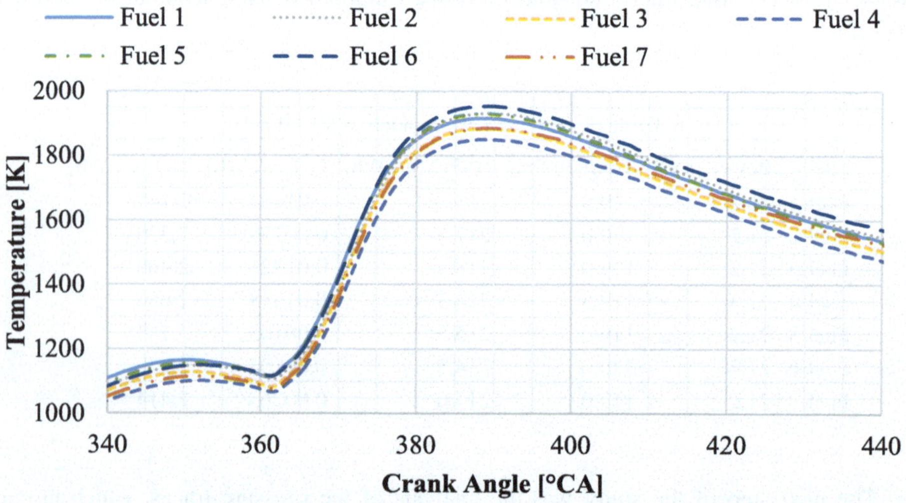

Fig. 2. In-cylinder temperature for all fuels at 0.5 load.

At 0.5 load, the NO_x emissions were the highest for Fuel 4, although it showed the lowest peak temperature. Considering that summer grade fuels contain at least 6.5% v/v biodiesel (BD), it is possible that Fuel 4 contains a higher percentage of BD, thus more embedded oxygen that favours NO formation.

Considering that during engine operation a cycle to cycle variability is to be expected and consequently a variable peak in-cylinder temperatures, combining this with a variable content of biodiesel this can lead, in the least favorable case (high temperature and high oxygen content), to very high amounts of NO emissions.

In Fig. 3 the rate of heat release is shown for all 7 fuels. The peak ROHR values showed a difference of 22.2% at load 0.1, 5.6% at load 0.3 and 6.2% at loads 0.5 and 0.7. At load 0.5 the lowest value was obtained for Fuel 1 with 37.01 J/°CA, which showed lower values starting at 13 °CA after TDC. As a result, in spite of Fuel 1 showing the highest peak pressure value, the in-cylinder pressure decreases faster than for the other fuels, due to the fact that the amount of heat released cannot compensate the heat losses through the cylinder walls. However, the in-cylinder temperature showed the 4th highest value, which can be explained by a somewhat higher lower heating value of the fuel compared to Fuels 3, 4 and 7 (here, it must me mentioned that a lower heating value of 43 000 kJ/kg was considered for all fuels). With respect to the ignition delay, no significant differences (under 0.5 °CA) were observed between the tested fuels, regardless of the engine load.

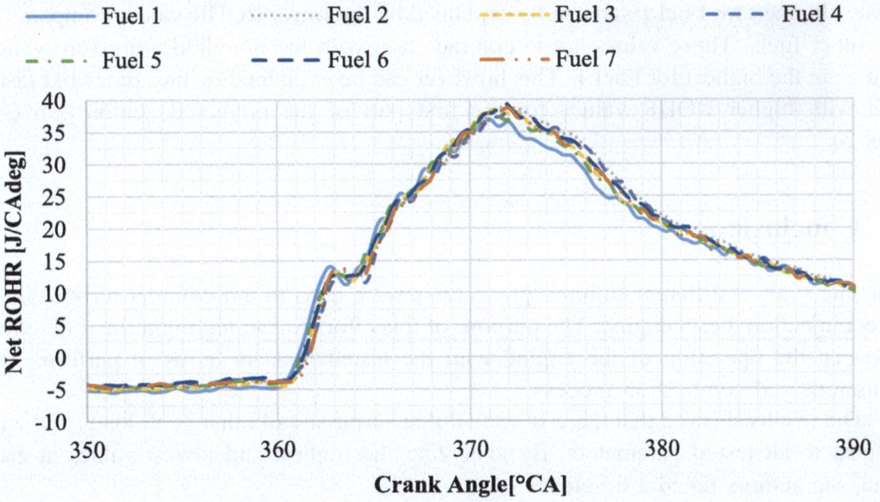

Fig. 3. Net Rate of Heat Release for all fuels at 0.5 load.

Figure 4 represents the accumulated heat released for all 7 fuels at 50% load. At all loads, the difference between the AHR is approximately 5.44%. Considering the same lower heating value for all fuels, this would result in a similar combustion efficiency of more than 90% for all fuels and at all loads.

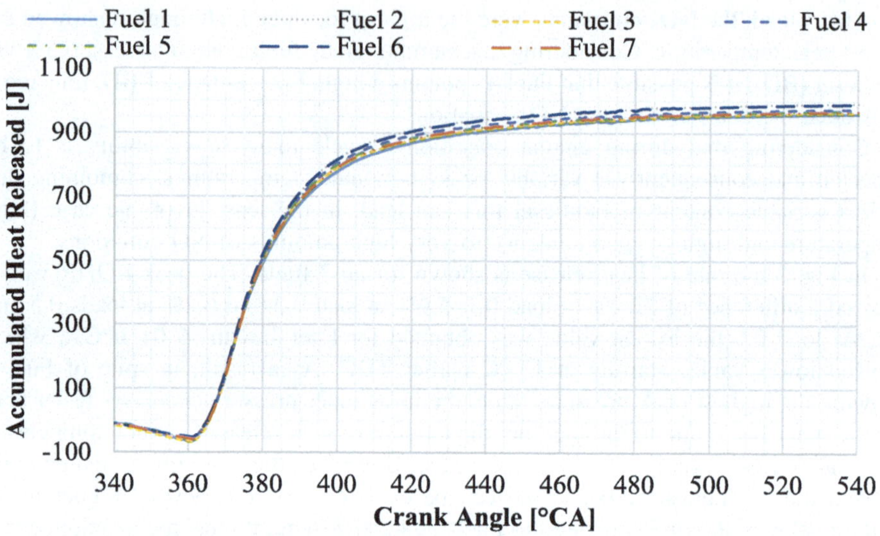

Fig. 4. Accumulated heat release for the 7 fuels at 0.5 load.

At 0.5 load the highest amounts of heat were released when using Fuels 2 and 6 with approximately 993 J for each of the fuels. The lowest value of approximately 962 J was obtained for Fuel 1 and can be explained by the lower ROHR values compared to the other fuels. These values are in contradiction with the in-cylinder pressure values that were the highest for Fuel 1. This however can be explained by the somewhat faster and with higher ROHR values for the first part of the premixed combustion (see Fig. 3).

4 Conclusions

For this study 7 different summer grade fuels were used in a 4-cylinder turbocharged direct injection diesel engine. The purpose of this effort was to highlight the differences between the operation of the engine with the tested fuels in terms of performance, emissions and combustion process.

The results showed that there is no fuel that surpasses all other at all loads and with respect to all tested parameters. By analyzing the highest and lowest values at each load, the authors noted a difference of:

- 3.95% for the BSFC,
- 9.71% for the CO_2 emissions,
- 35.45% for the NO emissions,
- 4.38% for peak in-cylinder pressure,
- 5.83% for in-cylinder temperature and
- 6.56% for the ROHR.

The main conclusion that can be drawn is that engine research alone cannot overcome all challenges with respect to performance and pollutant emissions. As a result, a concomitant development of both the engine and the fuel is required.

Table 4 summarizes the cost estimations for the considered fuels. The first column represents the price of the fuel per liter, while in the second column the cost of a full tank. In order to estimate the cost of a full tank, a volume of 65 L was considered, which represents an average fuel tank size of a large family car. The difference in percentage were calculated with respect to Fuel 7. Therefore, by analyzing the fuel consumption and price of the fuels, one can make an optimal choice between engine performance and fuel costs.

Table 4. Cost estimations for the tested fuels

Fuel	Price/l [RON]	Price/tank [RON]	Cost difference [%]
Fuel 1	5.72	371.8	+0.527
Fuel 2	5.66	367.9	−0.530
Fuel 3	5.7	370.5	+0.175
Fuel 4	5.72	371.8	+0.527
Fuel 5	5.66	367.9	−0.530
Fuel 6	5.75	373.75	+1.054
Fuel 7	5.69	369.85	–

Considering the above results, one can note that the fuel plays an important role in the internal combustion operation and strict fuel quality standards are required in order to ensure high engine performance with reduced emission output.

References

1. Burnete, N.V., Şerdean, F., Burnete, N., Moldovanu, D.: Combustion analysis for single cylinder engine running on various types of diesel, pp. 593–600 (2019). https://doi.org/10.1007/978-3-319-94409-8
2. Jörg, C., Schaub, J., Neumann, D., Pischinger, S.: Diesel combustion control via rate shaping. MTZ Worldw **79**, 16–21 (2018). https://doi.org/10.1007/s38313-018-0001-0
3. Steffens, C., Korfer, T., Hanses, G., Rosplesch, A., Kremer, F., Schaub, J.: Improvement of comfort aspects for high efficiency diesel engines. SAE Technical Paper Series, 1 (2013). https://doi.org/10.4271/2013-26-0119
4. Asad, U., Zheng, M.: Fast heat release characterization of a diesel engine. Int. J. Therm. Sci. **47**, 1688–1700 (2008). https://doi.org/10.1016/j.ijthermalsci.2008.01.009
5. Bueno, A.V., Velásquez, J.A., Milanez, L.F.: A new engine indicating measurement procedure for combustion heat release analysis. Appl. Therm. Eng. **29**, 1657–1675 (2009). https://doi.org/10.1016/j.applthermaleng.2008.07.023
6. VDMA. Exhaust emission legislation Diesel and gas engines (2017)

Research of Alternative Composite Solutions in Gas After-Treatment System for Diesel Cars

Bogdan-Manolin Jurchiş and Doru-Laurean Băldean(✉)

Technical University of Cluj-Napoca, 08544 Cluj-Napoca, Romania
{bogdan.jurchis,doru.baldean}@auto.utcluj.ro

Abstract. The present research paper elaborates on the possibilities and real applications in the field of testing and simulations of the influences of using different fuels with Diesel engines, as well as the effect that each fuel type or fuel blend has upon the brake specific fuel consumption, on the emissions of nitrous oxide, carbon oxide and hydrocarbons. There was used Diesel Standard fuel and its blends with Bio Diesel (from canola oil) and ethanol in different volumetric proportions. Particular research setting was created and results were gathered with precise digital equipment. Also tests and simulations were conducted in order to obtain the data by different means for an easier validation procedure. By using some proportions of Bio Diesel and ethanol slight improvements may be detected which encourage the further research into the matter. Comparable performances of brake specific fuel consumption were achieved at medium high throttle position between 60–80% both in testing and simulation with considered fuels.

Keywords: After-treatment · Diesel · Engine · Pollution · Vehicle

1 Introduction

Efforts for increasing the fuel efficiency and to reduce consumption by different technological improvements were made through time by Bennett [1] when applying additives and by other researchers when using vegetable oils to improve the operational characteristics [2]. Some researches regarding the possibility of perfecting the compression ignited engine supplied comparatively with canola oil and diesel standard were also developed at the Technical University of Cluj-Napoca [3]. Oils had been supplied to the thermal engines for a time now in order to improve the performance of their operation [4]. Combustion engines are one of the main sources of pollution and have a negative influence upon metropolitan low air quality. Most engines have oxides of nitrogen (usually tagged as NOx, consisting from nitric oxide, NO, and nitrogen dioxide, NO_2 – a gas compounds which may generate pulmonary sickness) in their exhaust emissions, as well as carbon oxide (CO), and other carbon derivatives. The later ones are products of partial combustion reactions between hydrocarbons (HC) and oxygen [5, 6]. The amount of chemical compounds depends on after treatment design and load [7]. The products of incomplete combustion reach into the exhaust pipeline changing the gas composition during the forced evacuation. The oxidation process of the carbon particles develops in the expansion stroke of the operation cycle. Multiple

I. Dumitru et al. (Eds.): SMAT 2019, *The 30th SIAR International Congress of Automotive and Transport Engineering*, pp. 160–164, 2020.
https://doi.org/10.1007/978-3-030-32564-0_19

reactions of chemicals and physical interferences lead to an important influence upon the gas dynamics inside the case of after-treatment equipment. These reactions and physical influences are determined and defined through simulation and through practical testing. Testing the oxidation catalysts is an engineering approach that creates some challenges in setting up the equipment for the practical effort.

Generally there are two types of catalysts used for their exhaust gas treatment capabilities regarding oxidation and reduction of toxic pollutants.

Usually catalyst monoliths are coated for optimal operation regarding after-treatment with specific metals (Pt, Pd, Rh) or with metals oxide, such as: Al_2O_3, SiO_2, CeO_2, ZrO_2, SnO_2, Cr_2O_3, MnOx, NiO, CuO, Co_3O_4, etc. Rare metals such as palladium and platinum when applied upon silica or alumina base are highly efficient catalysts for accelerated oxidation of CO and HC.

2 Materials and Method

Practical part of experiment and tests were developed by using the powertrain presented as a complex assembly of internal combustion engine 1, injection system 2, gas monitor system 3, after-treatment system 4, electronic control unit 5, engine speed sensor 6, and pressure regulator 7, working as a whole system as shown in corresponding (see Fig. 1). Getting through the testing procedure with a specific construction and energy supply system in comparison with alternative fuel types lead to the possibility of better measurements regarding brake specific fuel consumption and other actual values, such as nitrous oxides, carbon oxides, hydrocarbons and their trend-lines versus engine's load in study of the after-treatment process. Applying the computer aided design means and simulation will be analyzed each scenario. The present research is conducted with a reference structural model of power-train with after-treatment system. Based on the reference model, new supply solutions are proposed and tested comparatively in order to gain results concerning engine's emissions. Experimental testing method was applied using a multi-fuel supply system.

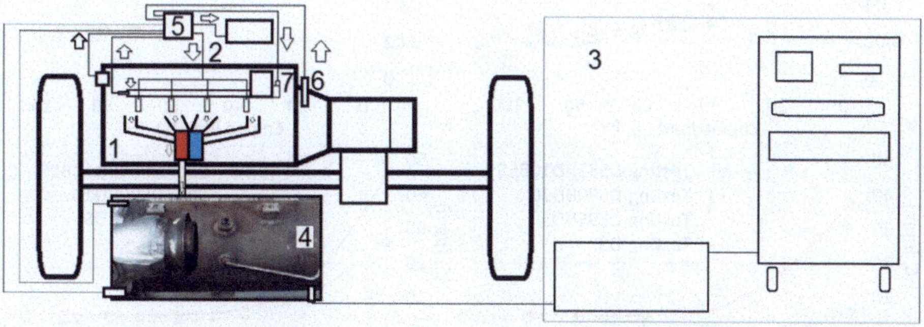

Fig. 1. Variation curves of specific fuel consumption and chemical pollutants vs. engine load.

3 Results and Discussions

During the applied experimentation and simulation first significant results were collected and stored concerning fuel specific consumption (see Fig. 2a, b), nitrous oxides (see Fig. 2c, d), carbon oxide (see Fig. 2e, f) and hydrocarbons emissions (see Fig. 2g, h). It has been observed an influence of positive sense when blending the DS with Bio Diesel and Ethanol, by lowering the nitrous oxides and even for the unburned hydrocarbons. Testing and simulation results may be analyzed in parallel.

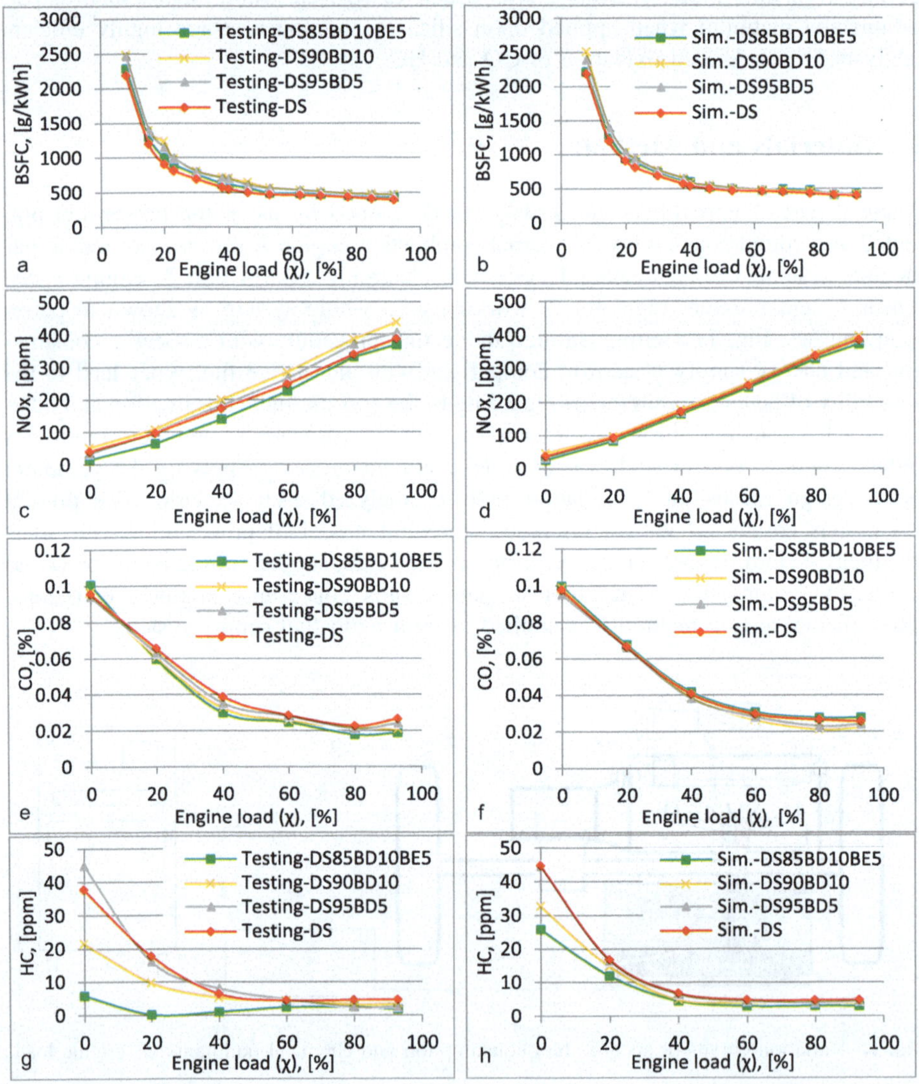

Fig. 2. Variation curves of specific fuel consumption and chemical pollutants vs. engine load.

The second set of experimental results were acquired regarding physical actual values of particulate matter, fuel specific consumption, exhaust gas temperature upstream of particulate filter, particulate filter differential pressure, voltage on the exhaust gas differential pressure (see Fig. 3). It was seen a positive influence at some regimes when blending the DS with Bio Diesel and Bio Ethanol, by decreasing the particulate matter during testing with 2–4% compared with theoretical estimations due to an improved combustion in the case of the tertiary blend (DS85BD10BE5).

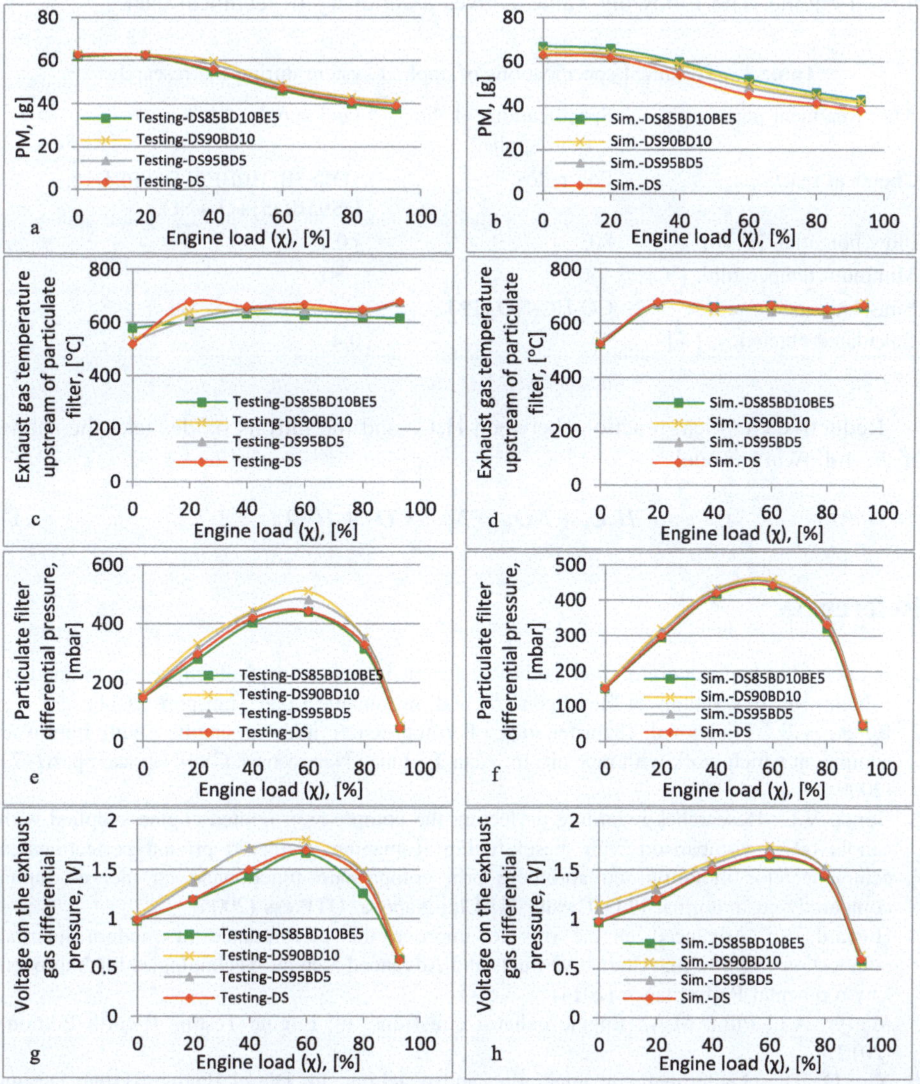

Fig. 3. Variation curves of particulate matter and exhaust gas characteristics vs. engine load.

4 Conclusions and Discussions

Also there were recorded some differences between testing and simulations even in regard to the exhaust gas temperature upstream of particulate filter. The achieved results were compared in parallel and if the after treatment system is working properly the specific fuel consumption decreases with engine load.

By implementing innovative alternatives regarding the fuel supply (with liquid composite blends) influence upon the chemical state of emissions from internal combustion engine. The following Table 1 gives a summary of technical data.

Table 1. Technical specifications of applied system during the research.

After-treatment parameter	Specifications of the system applied	Fuel supply data
Chemical reactions	Poli-redox	{DS85BD10BE5, DS90BD10, DS95BD5}+O_2+NO_x
Flow potential, [m^3/h]	80	<0.1
Minimum temperature, [°C]	450	−30
Emissions monitored	CO-HC-NO_x-PM	–
Calculated efficiency, [%]	93	0.4

Reductive chemical reactions between H_xC_y and the nitrous oxides take the shape of the following model:

$$H_xC_y + NO_x = N_z + CO_2 + H_2O \tag{1}$$

References

1. Bennett, J.: Application of fuel additives for maintenance of fuel efficiency in modern vehicles. In: Innovations in Fuel Economy and Sustainable Road Transport (2011)
2. Borza, V.E.: Functional characteristics of compression ignition engines with improved equipment which work with rape oil. In: Acta Technica Napocensis, Cluj-Napoca, pp. 67–73 (2009)
3. Borza, V.E.: Researches regarding perfecting the compression ignited engine supplied with canola oil in comparison with diesel fuel/in Romanian: Cercetări privind perfecţionarea echipamentelor motorului cu aprindere prin comprimare funcţionând cu ulei de rapiţă comparativ cu motorină, PhD Thesis, TU Cluj-Napoca, UTPress (2006)
4. Howard, K.: Advanced engine oils to improve the performance of modern internal combustion engines. In: Alternative Fuels and Advanced Vehicle Technologies for Improved Environmental Performance (2014)
5. Martyr, A.J., Plint, M.A.: Engine exhaust emissions. In: Engine Testing (Fourth Edition) (2012)
6. Xin, Q.: Diesel aftertreatment integration and matching. In: Diesel Engine System Design (2013). https://www.sciencedirect.com/topics/engineering/active-regeneration
7. Xin, Q.: Diesel engine air system design. In: Diesel Engine System Design (2013)

Study of Vehicles Energy Efficiency

Ion Lespezeanu(✉), Laszlo Barothi, Leonard-Iulian Cucu, Cătălin-Dumitru Darie, Iulian Constantin Coropețchi, and Ion Copae

Military Technical Academy "Ferdinand I", Bucharest, Romania
Ion_lespezeanu@yahoo.com

Abstract. This paper presents some issues regarding vehicles energy efficiency which implies simultaneous approach on the dynamic and economic parameters. Nowadays, the interest in improving the dynamic performances of the vehicles together with reducing the fuel consumption has grown, issue that is directly related to energy efficiency. In the sense of those presented, the paper approaches the criteria of appreciation of the dynamics and economy used in the specialized literature. Also, there are defined and established some appreciation values criteria of energy efficiency of vehicles which addresses both aspects. The study is based on the experimental data obtained from the test of a gasoline powered passenger car, data acquired from the on-board computer by a specialized tester. The paper proposes a study method of dynamic performances of vehicles related to energy efficiency.

Keywords: Energy efficiency · Fuel consumption · Dynamic parameters · Economic parameters

1 Introduction

Among the main requirements imposed on vehicles there are those related to their dynamics and economy; regularly the two requirements cannot be satisfied simultaneously on a desired level. In classic specialized literature, the study of dynamics and economy is made separately from each other, without interconnection appeal between them.

In classic sense of the specialized literature, references to dynamics belong only to the start, usually by speed, acceleration, time and starting distance. In broad sense, dynamism refers to the whole movement, and only to the fact that dynamics means any variation over time [1].

Similarly, in the classical sense of the specialized literature, references to economics belong only to fuel consumption, expressed in various forms [2]. In a broad sense other aspects should be targeted, such as referring to fuel input, engine efficiency, propulsion, etc. [3–6].

I. Dumitru et al. (Eds.): SMAT 2019, *The 30th SIAR International Congress of Automotive and Transport Engineering*, pp. 165–176, 2020.
https://doi.org/10.1007/978-3-030-32564-0_20

2 Tendency Referring to Energy Efficiency

Lately, the issue of vehicle's energy efficiency, which targets dynamism and economy, are in close interdependence with one another [1, 7–9]. In other words, there is no need for a high dynamics without interest in the energy effort to get it; for this reason, the main efforts are directed more towards improving the economy than towards the dynamics, obviously also due to the limitation of oil resources. This effort is explainable as the vehicles are transport means with low energy efficiency, wasting much energy introduced by fuel.

Efforts to improve energy efficiency aim simultaneously economy and dynamism, but in one sense: reducing fuel consumption by affecting dynamics within required limits. In this sense, the latest studies and researches in this field operate with the concept called ERFC (Emphasis on Reducing Fuel Consumption), with quantitative expression, which means what percentage from dynamism performances is sacrificed for fuel consumption reducing [7]. Criterion value of ERFC is stabilized with expression:

$$\mathrm{ERFC} = \frac{C_c - C_r}{C_c - C_p}\ [\text{-}] \tag{1}$$

Where: C_c – current fuel consumption; C_r – realized fuel consumption; C_p – potential fuel consumption (possible for reducing).

For example, from the desire to reduce fuel consumption in 2035 to 62.5% from the current value by sacrificing 50% of dynamism performance, from relation (1) is realized a relative fuel consumption of:

$$C_r = C_c - \mathrm{ERFC}(C_c - C_p) = 1 - 0{,}5(1 - 0{,}625) = 0{,}81\ [\text{-}] \tag{2}$$

As it is noticed from Fig. 1 too [7].

As it can be observed, the value of ERFC it is a criterion of energy efficiency, because the decrease of fuel consumption principally is based on decrease of the dynamics of the vehicle.

In order to not dramatically affect the dynamics, the construction companies will have to improve the manufacturing technologies and adopt new constructive measures. Specialists in this domain say that this will have a big impact upon the mass of the vehicle, meaning its significant reduction. For example, if the ERFC is 50% (Fig. 1), then the current vehicle must have 10% less weight. As a consequence of the above, Fig. 2 presents the forecast for 2035 on the dynamics and economy of an average middle class, according to the ERFC implementation rate compared to the current situation [7]; for dynamism the start time td from 0 km/h up to 100 km/h was targeted, and for economy fuel consumption at 100 km covered by C100.

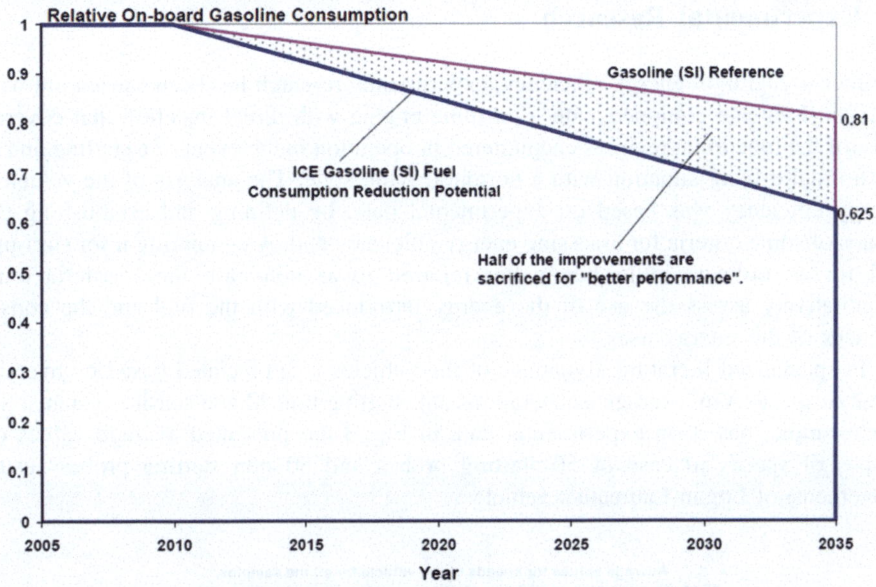

Fig. 1. Relative on-board gasoline consumption

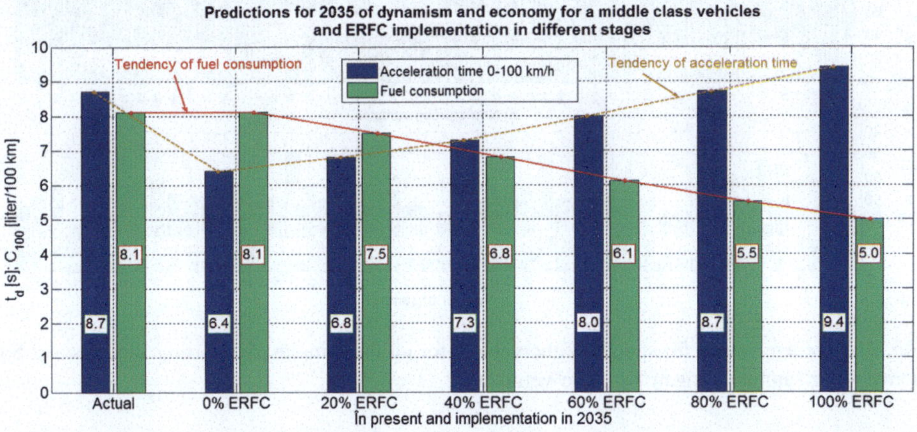

Fig. 2. Predictions for 2035 of dynamism and economy for a middle class vehicle and ERFC implementation in different stages

From Fig. 2 is considered that in 2035 the dynamism will decrees, and economy will increase with ERFC value increase. If the ERFC will not be implemented in 2035 then fuel consumption will remain the same with the actual one, and the vehicle's dynamics will increase by mass reduction. Instead, a 100% ERFC implementation will lead in 2035 to a start time of $t_{100} = 9.4$ s and to a fuel consumption of $C_{100} = 5$ L/100 km.

3 Experimental Research

In order to highlight energy efficiency, experimental research has been carried out with a Logan Laureate equipped with a gasoline engine with direct injection that covered most of the functional regimes encountered in operation in the event of a starting and in the normal driving situation with a normal driving style. The analysis of the vehicle's energy efficiency was based on experimental data, by defining and establishing the values of some criteria for assessing energy efficiency and by comparing it for start-ups and for an ordinary shift, hereinafter referred to as non-start; these criteria must quantitatively assess the use of the energy introduced with the fuel and the consequences of this energy use.

In specialized literature, dynamics of the vehicles is appreciated based on average value of speeds Vm, average accelerations am, starting time td and starting distance Sd. For example, based on experimental data in Fig. 3 are presented average values on probes of speeds in case of 50 starting probes and 50 non starting probes (usual movement) of Logan-Laureate's vehicle.

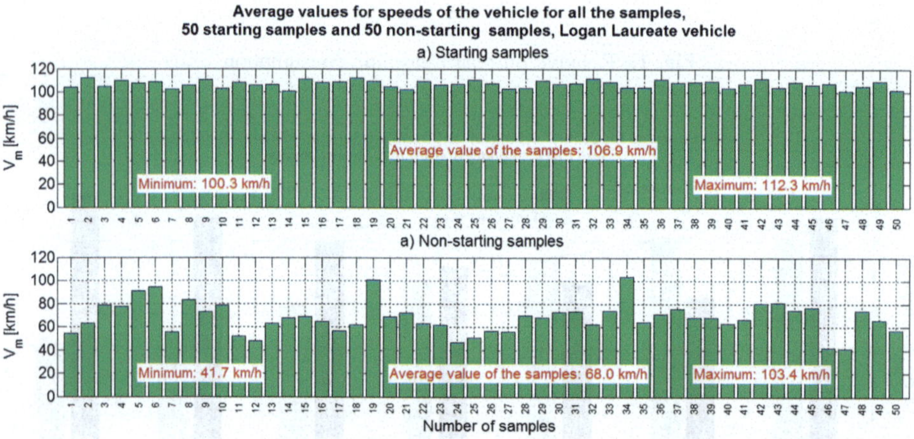

Fig. 3. Average values for speeds of the vehicle for all the samples, 50 starting samples and 50 non-starting samples, Logan Laureate vehicles

In addition, in specialized literature the economy of the vehicles is appreciated based on fuel consumption for 100 km travelled C100, hourly fuel consumption Ch etc. For example, based on the experimental data in Fig. 4, are shown the average values of fuel consumption for 100 km and hourly fuel consumption are shown for 50 starting probes and 50 normal probes for Logan-Laureate.

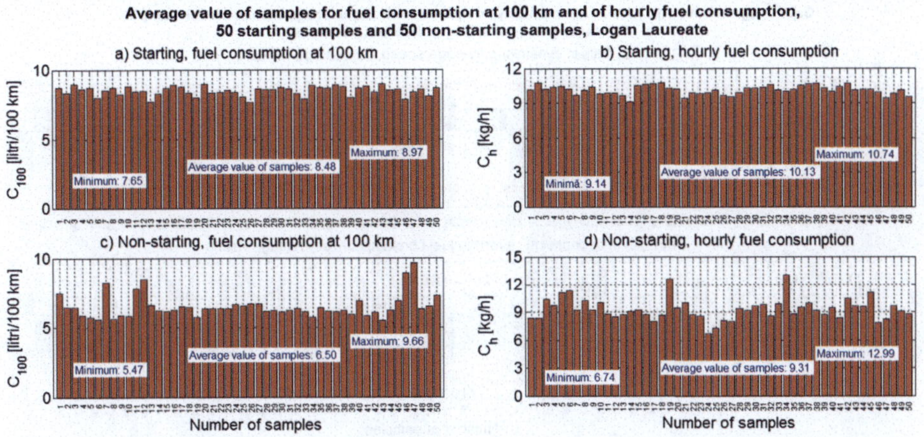

Fig. 4. Average values of samples for fuel consumption at 100 km and an hourly fuel consumption, 50 starting samples and 50 non-starting samples, Logan Laureate

The graphs from Figs. 3 and 4 confirm a known aspect: in comparison with the normal movement, the starting is generally characterized by high dynamics and low economy.

As mentioned in the specialized literature in the field of motor vehicles dynamics and economy are analyzed separately from each other. The practice of exploitation has shown that usually for a particular vehicle an increased dynamics is achieved with increased fuel consumption, as shown in Fig. 5 at the starting and in Fig. 6 at non-starting. Therefore, in this case, it is necessary to compare the loss/gain in the economy and the gain/loss in dynamics [1].

The graphs in Figs. 5 and 6 show the gain in dynamics and the loss in economy of 50 start and non-starting probes of the Logan Laureate by comparing with the average values of all the speeds V and respectively fuel consumption per 100 km, C_{100}. Because it operates with the notion of gain in dynamics, it follows that the positive values belong to the gain, and the negative ones of the loss; also because it deals with the notion of loss in economy, it results that the positive values belong to loss, and the negative ones to gain.

In Fig. 5, are also provided some examples of some experimental samples. For example, in P1 sample the loss in dynamics is 2.5% (negative) and the loss in economy is 2.7% (positive); therefore, there is both a loss of dynamism and a loss of economy, that is, an unfavorable, always unwanted situation. Instead, in P4 sample gains in dynamics is 2.7% (positive) and the loss in economy is 1.8% (positive); therefore, the gain in dynamics is greater than the loss in economy, thus a positive overall effect. In case of P9 sample, gained 3.2% dynamics, with a 2.9% economic gain, so a favorable and always desirable situation. In P14 sample dynamics lost was 6,2% (have negative value), with a gain of economy of 2,5% (have negative value); in this case the economy earning is less than loss dynamics, thus a negative overall effect.

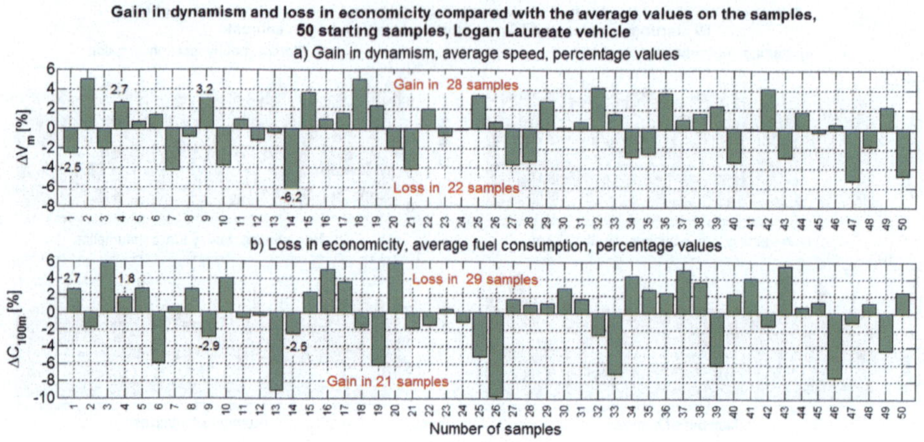

Fig. 5. Gain in dynamism and loss in economicity compared with the average values on the samples, 50 starting samples Logan Laureate

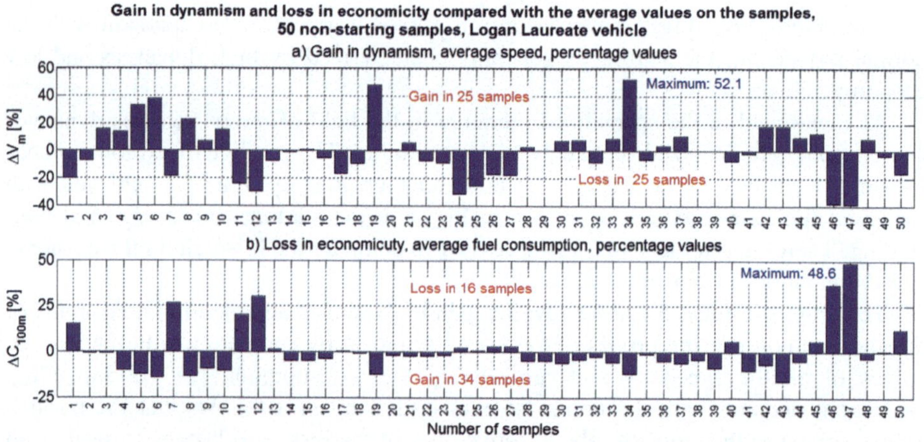

Fig. 6. Gain in dynamism and loss in economicity compared with the average values on the samples, 50 non-starting samples, Logan Laureate

Figures 5 and 6 also show the number of samples with gain and loss in dynamics and economy. The graphs in Fig. 6 show that in the case of non-starting the values on the samples are higher than in the case of starting.

4 Criteria for Assessing of Energy Efficiency

To assess at the same time the dynamics and economy of the vehicles, so for the purpose of establishing their energy efficiency, it is necessary to set specific energy efficiency criteria. In the next part there are presented some criteria for assessing the energy efficiency of vehicles, based on the experimental data obtained from the tests.

1. Ratio between kinetic energy of the vehicle W_{cin} and input energy from the fuel W_i:

$$k_c = \frac{W_{cin}}{W_i} \cdot 100[\%] \tag{3}$$

In Fig. 7 is presented the obtained results for the 50 samples in case of starting, and in Fig. 8 of non- starting probes.

The two energies from expression (3) are established with the following relations:

$$W_{cin} = \frac{m_a v^2}{2} [\mathrm{J}] \tag{4}$$

Respectively:

$$W_i = \frac{C_h Q_i S_p}{V} [\mathrm{J}] \tag{5}$$

where were noted: ma – mass of the vehicle; Ch – hourly fuel consumption of the engine; Qi – the lower calorific value of the fuel; Sp –travelled distance; v [m/s], V [km/h] – speed of the vehicle.

As can be seen from Fig. 7c, on the starting part, 25.85% from introduced energy by the fuel is actually used to drive the car; sample values of this report vary on range 24.32 ÷ 27.66. Similarly, for Fig. 8c, it is observed that on the non-starting part 18.77% from the energy introduced with the fuel is actually used for the movement of the vehicle; sample values of this report range are between 7.7 ÷ 30.77%.

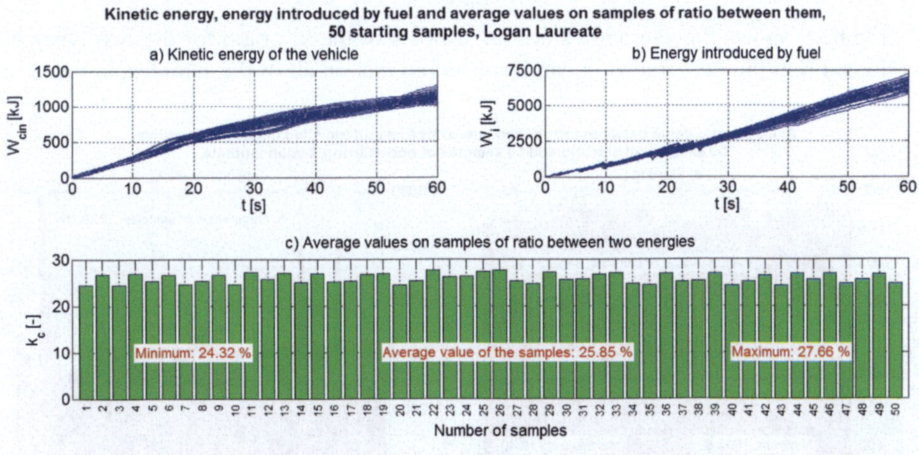

Fig. 7. Kinetic energy, energy introduced by fuel and average values on samples of ration between them

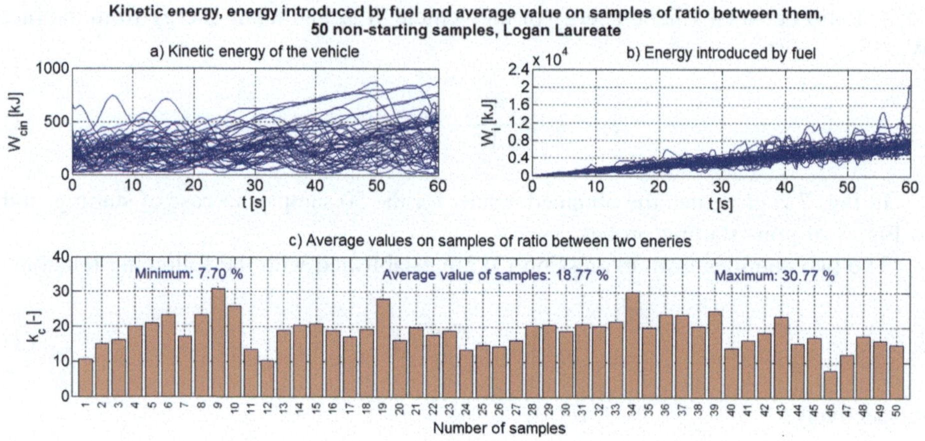

Fig. 8. Kinetic energy, energy introduced by fuel and average value on samples of ration between them

As can be seen, in the case of starting the fuel is used more efficiently and the distribution of the values on the samples is more concentrated.

Because relation (3) appeals to kinetic energy of vehicle, which consist of consumed energy for effective movement, as well as to the introduced energy with fuel, it appears that this criterion is the most important in assessing energy efficiency; this criterion has a dynamics component W_{cin} and one economic W_i.

2. Ratio between the introduced energy from the fuel W_i and travelled distance by the vehicle S_p:

$$k_{ws} = \frac{W_i}{S_p}\,[\mathrm{kJ/km}] \tag{6}$$

Figure 9 shows the average values on samples of the k_{ws} ratio for the two types of motions. From the two graphs, it can be observed that on all of the samples, in the case

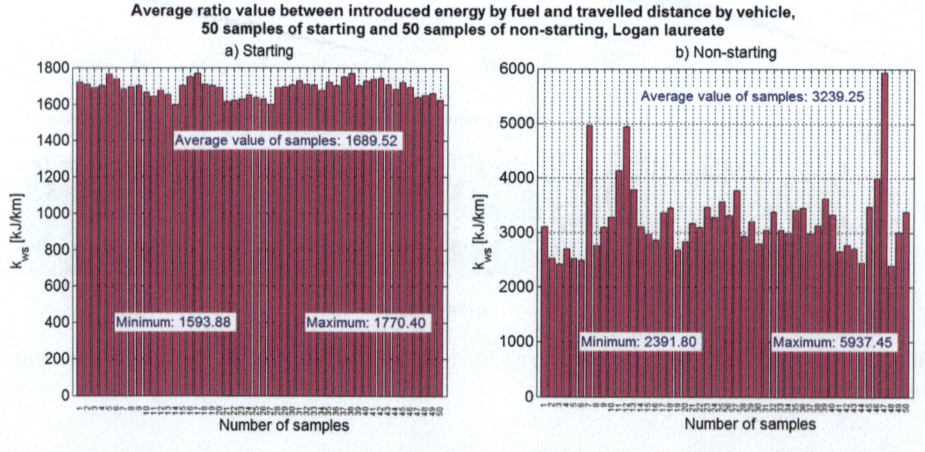

Fig. 9. Average ratio value between introduced energy by fuel and travelled distance by vehicle

of a regular trip for a kilometer, fuel consumption is almost 1.92 times higher than in case of starting.

3. Ratio between volumetric fuel consumption C_v and speed of the vehicle v:

$$k_{cv} = \frac{C_v}{v}[\mathrm{ml}/(\mathrm{m/s})] \tag{7}$$

Which shows the fuel consumption required to develop a travel speed of 1 m/s.

4. Ratio between volumetric fuel consumption C_v and acceleration of the vehicle a:

$$k_{ca} = \frac{C_v}{a}\left[\mathrm{ml}/\left(\mathrm{m/s}^2\right)\right] \tag{8}$$

Which is the fuel consumption required to develop a vehicle acceleration of 1 m/s^2.

Volumetric fuel consumption in milliliters (ml), results from the following relation:

$$C_v = \frac{C_{100}}{100} S_p \cdot 1000[\mathrm{ml}] \tag{9}$$

Where Sp is the travelled distance by the vehicle.

In Fig. 10 there are presented average values on samples of the last two appreciation criteria of energy efficiency of the vehicle.

From Fig. 10 results that for obtaining a speed of 1 m/s it is necessary a volumetric fuel consumption on all of the samples of approximate 1.18 bigger than in case of non-starting. Also, the graphs show that in order to obtain an acceleration of 1 m/s^2, volumetric fuel consumption on the samples is almost 1.19 times higher in the case of non-start. In addition, from Fig. 10 results that the energy effort to obtain an acceleration of 1 m/s^2 is much higher than that for achieving a speed of 1 m/s: more than 20.1 times than in case of starting and closer with 20.3 in case of non-starting.

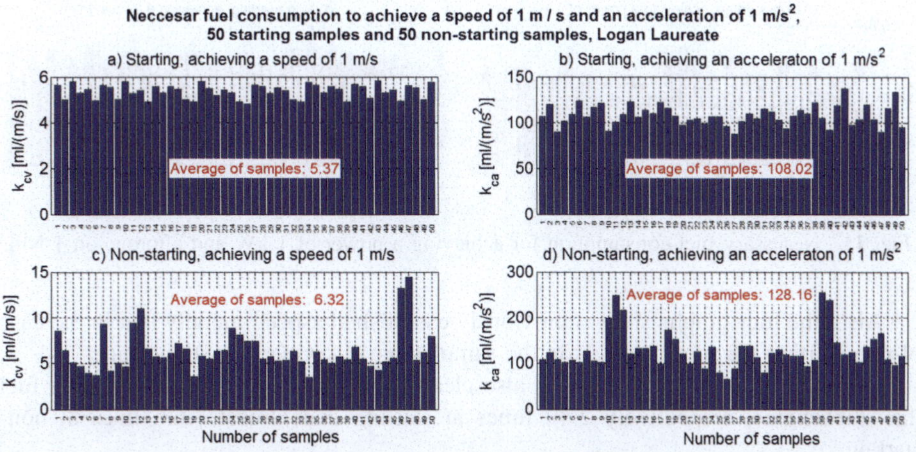

Fig. 10. Necessary fuel consumption to achieve a speed of 1 m/s and an acceleration of 1 m/s^2

5. Ratio between hourly fuel consumption C_h and engine's power P_e:

$$k_{mp} = \frac{C_h}{P_e}\,[\mathrm{kg/kWh}] \tag{10}$$

With values from Fig. 11, graphs from left side.

6. Ratio between hourly fuel consumption C_h and engine's torque M_e:

$$k_{mm} = \frac{C_h}{M_e}\,[\mathrm{kg/Nmh}] \tag{11}$$

With values from Fig. 11, graphs from right side.

The criterion from relation (10) expresses the hourly fuel consumption needed to develop engine power with 1 kW, and the one in relation (11) expresses the hourly fuel consumption required to develop a torque of 1 Nm. It can therefore be considered that the two criteria express the energy efficiency of the engine of a vehicle.

The graphs in Fig. 11 show that in order to obtain a power of 1 kW the fuel consumption on all of the samples is almost 1.63 times higher than in the case of non-starting. In addition, it is noted that in order to obtain engine torque of 1 Nm, fuel consumption on all of the samples is almost 1.12 times higher than in the case of non-starting.

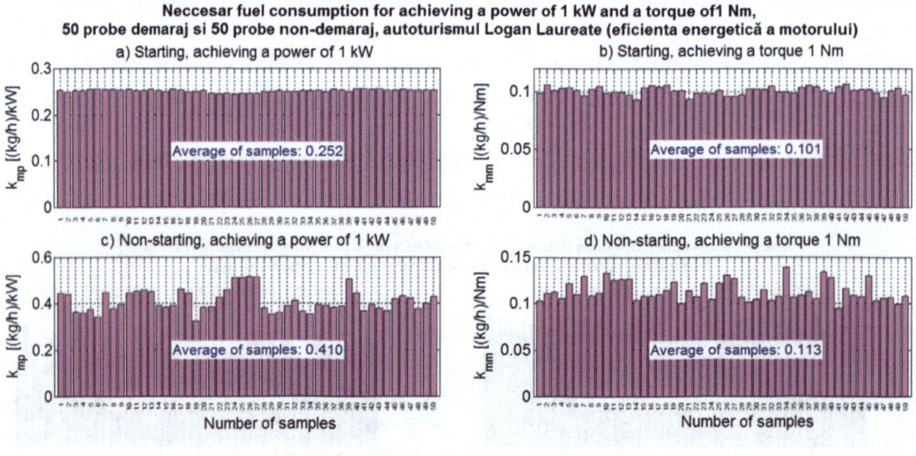

Fig. 11. Necessary fuel consumption for achieving a power of 1 kW and a torque on 1 Nm

Also, the four graphs show a previously established aspect, namely that in the case of non-starting the mean values on the samples have a wider variation range.

From these last four graphs it is also clear that the power unit requires more fuel than the engine, respectively 2.49 times at starting and almost 3.63 times at non-starting.

7. Ratio between fuel consumption for 100 km travelled C_{100} and power on the wheel P_r:

$$k_{pp} = \frac{C_{100}}{P_r}[(1/100\text{km})/\text{kW}] \tag{12}$$

With values from Fig. 12, graphs from left side.

8. Ratio between fuel consumption for 100 km travelled C_{100} and torque on the wheel M_r:

$$k_{pm} = \frac{C_{100}}{M_r}[(1/100\text{km})/\text{Nm}] \tag{13}$$

With values from Fig. 12, graphs from right side.

The criterion from relation (12) expresses the fuel consumption required to develop a power output from wheel of 1 kW, and the one in relation (13) expresses the fuel consumption needed to develop a torque of 1 Nm on the wheel. It can therefore be considered that the two criteria express the energy efficiency of the propulsion system of the vehicle.

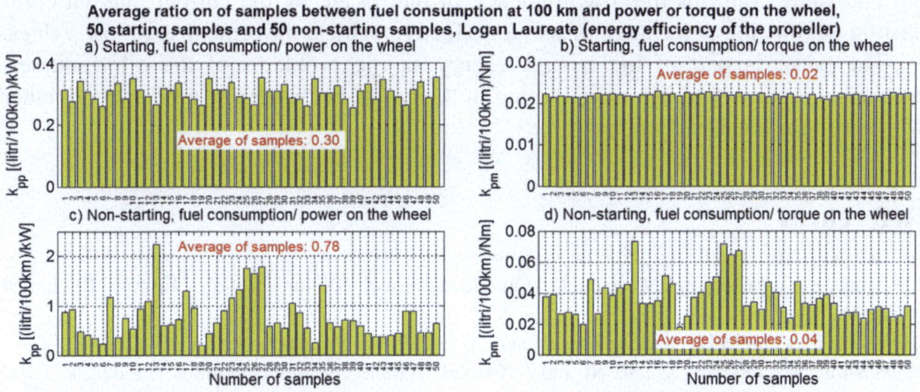

Fig. 12. Average ratio on of samples between fuel consumption at 100 km and power of torque at the wheel

The graphs from Fig. 12 show that in order to obtain a power output of 1 kW, fuel consumption on all of the samples is 2.6 times higher than in the case of non-starting. In addition, it is noted that in order to obtain a torque of 1 Nm on the wheel, fuel consumption on all of the samples is 2 times higher than in the case of non-starting.

From these last four graphs it is also evident that increasing the power by 1 unit requires higher fuel consumption than increasing the torque, respectively 15 times at starting and almost 19.5 times at non-starting.

All the criterias presented above give us important information regarding the relation between fuel consumption and dynamic parameters like speed, acceleration, torque, power etc. Depending on the analysis we want to perform, different criterias should be used. For example, if we want to study the energy efficiency of the engine we should use k_{mp} and k_{mm} but if the study subject is the transmission or the wheels, then k_{pp} and k_{pm} should be used because it takes into account the loss of power or torque from the engine to the wheels. For an advanced study of a vehicle's energy efficiency, a combination of all the criterias presented above should be used depending on what we are trying to highlight.

5 Conclusions

Based on the ones presented above it can be concluded that in comparison with normal movement, the starting is characterized with a high dynamicity and low economic, but increased energy efficiency.

Also, the graphs show that in the case of non-starting, the average values of the parameters have a wider variation domain than in the case of starting-up.

In the paper, there were defined some criteria for assessing the energy efficiency of vehicles and there were stabilized related values based on the experimental data. In similar ways other appreciation criteria of energy efficiency can be defined, which target simultaneously the dynamics and economic of the vehicle.

The study confirms the low efficiency of fuel usage by the current internal combustion engine, the percentage of use for effective vehicle movement with low values.

The paper is part of the current energy saving problems of the oil resources, specialists in automotive domain aiming in this sense better energy efficiency together with a certain level of dynamism.

References

1. Copae, I., Lespezeanu, I., Cazacu, C.: Dinamica autovehiculelor. Publisher Ericom, Bucharest (2006)
2. Jazar, R.: Vehicle Dynamics: Theory and Applications. Springer, New York (2008)
3. Mastinu, G., Ploechl, M.: Road and Off-Road Vehicle System Dynamics Handbook. CRC Press, Boca Raton (2014)
4. Rajamani, R.: Vehicle Dynamics and Control. Springer, New York (2006)
5. Vilău, R.: Considerații privind modelarea matematică a dispozitivelor pneumatice de frânare. In: Revista Tehnica Militară, DpA, DCD 1, pp. 10–12 (2005)
6. Bandivadekar, A.: On the Road in 2035. Report, Laboratory for Energy and the Environment, Massachusetts Institute of Technology (2008)
7. Blackmore, D.R., Thomas, A.: Fuel Economy of the Gasoline Engine. Wiley, New York (1977)
8. Cheah, H., Heywood, J.: Meeting US passenger vehicle fuel economy standards in 2016 and beyond. Energy Policy **39**, 454–466 (2011)
9. Fuel Economy Guide Model Year (2018). www.fueleconomy.gov. Accessed 22 Aug 2019

Comparative Study Regarding the Operation of a Car's Internal Combustion Engine in Urban and Extra-Urban Environment

Ion Lespezeanu(✉), Ramona-Monica Stoica, Radu Vilău, Daniel Sava, and Ion Copae

Military Technical Academy "Ferdinand I", Bucharest, Romania
ion_lespezeanu@yahoo.com

Abstract. This paper presents a comparative study of a compression ignition engine operating in different situations and at different loads and engine speeds. There are shown graphs of experimental results, obtained by using concepts and algorithms from mathematical statistics. A study regarding the influence of functional factors on engine dynamics and fuel savings is presented by using Correlation Analysis, Variance Analysis, Information Theory and Sensitivity Analysis. The specific characteristics of vehicle driving in urban and extra-urban environments are highlighted.

Keywords: Comparative study · Compression ignition engine · MANOVA · ANOVA · Energy efficiency

1 Introduction

Studying the functioning of diesel engines under different working regimes is easier today due to the electronic control of the engine parameters. The processes that occur during engine operation can be analyzed based on information obtained from the on-board computer, which implies an interdisciplinary approach to engine dynamics. As a result, the paper performs an analysis of the experimental data acquired from integrated sensors; the analysis of the experimental data offers the possibility to determine the dynamics of the engine and to make considerations on its specific fuel consumption.

2 Experimental Research

The experimental data were obtained from a car equipped with a supercharged diesel engine, with common rail injection system. For the acquisition of data regarding the operation of the engine and the vehicle a software developed by the manufacturer was used. The tests that were carried out allowed the acquisition and storage of a certain number of experimental data, of which 80 experimental samples were kept; of these, 40 tests were carried out in an urban environment and another 40 in an extra-urban one. The frequency of acquisition was 4000 values for each parameter, for a period of time of 400 s; which results in a measurement rate of 10 values/s.

I. Dumitru et al. (Eds.): SMAT 2019, *The 30th SIAR International Congress of Automotive and Transport Engineering*, pp. 177–187, 2020.
https://doi.org/10.1007/978-3-030-32564-0_21

3 Data Statistical Processing

Figures 1 and 2 show the variations of the vehicle's and engine's speed for 40 tests carried out in urban and extra-urban areas.

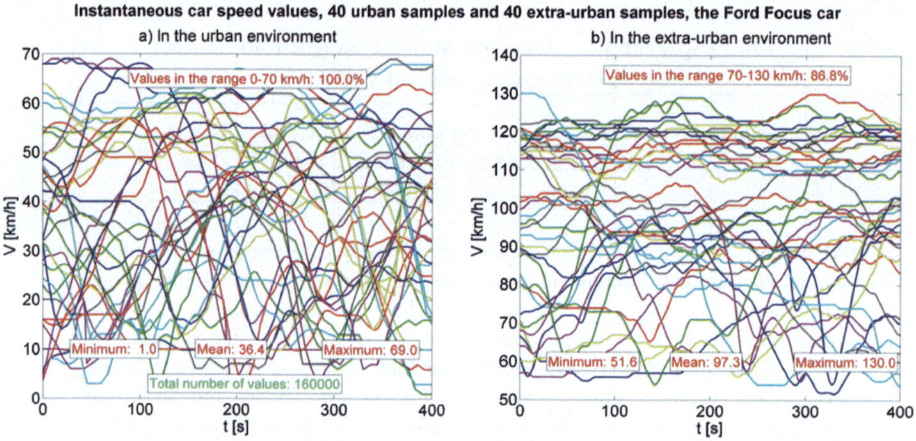

Fig. 1. Instantaneous car speed values

Likewise, by using first order statistical characteristics, in graphs were are also presented minimal, average and maximal values of vehicle speed and engine speed, in case of urban and extra-urban driving. Graphs from Figs. 1 and 2 show that in case of extra-urban environment the values of vehicle speed and engine speed are higher than in urban environment, which was expected.

For example, from Fig. 1 results that in the case of urban driving all 160000 values are in the range of 0–70 km/h; whereas, from Fig. 1b it can be observed that in an extra-urban environment 86.8% of all values are in a superior interval, meaning a range of 70–130 km/h.

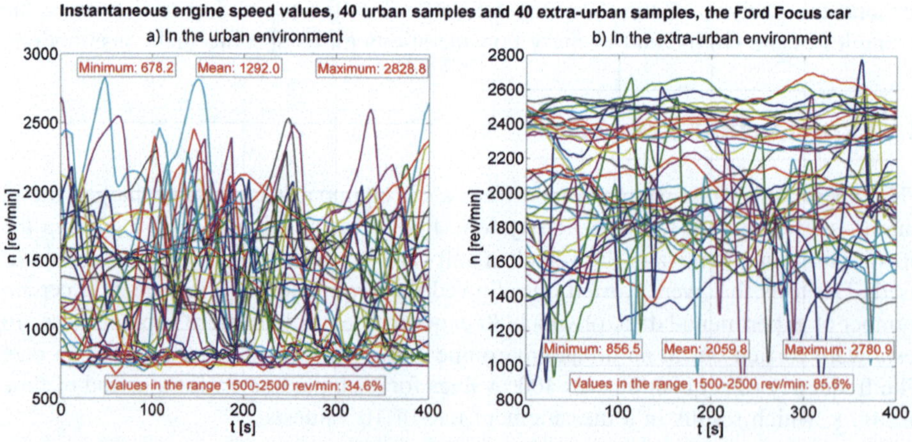

Fig. 2. Instantaneous engine speed values

Similarly, from Fig. 2 results that in an urban environment 34.6% of values are in the range 1500–2500 rev/min, and in case of extra-urban environment 85.6% of the values are within the interval.

Figure 3 shows the instantaneous, minimum, maximum and average values of the throttle position p as a percentage of its maximum opening. As expected, the throttle pedal position has higher values in the extra-urban environment compared to the urban environment.

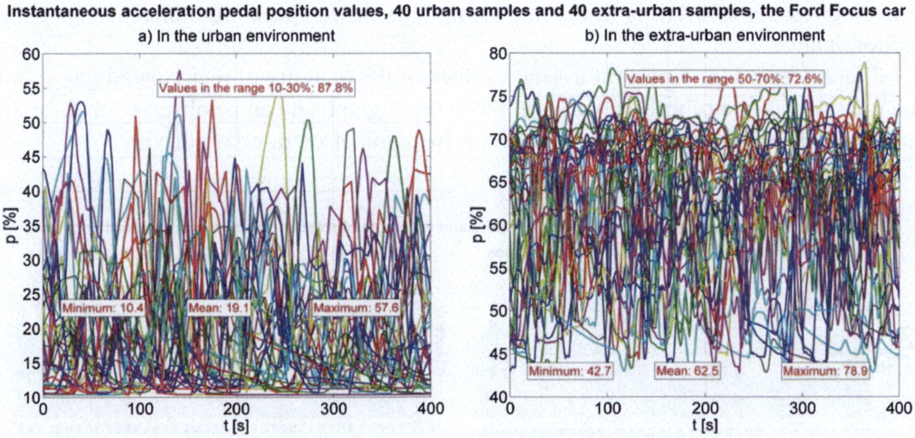

Fig. 3. Instantaneous acceleration pedal position values

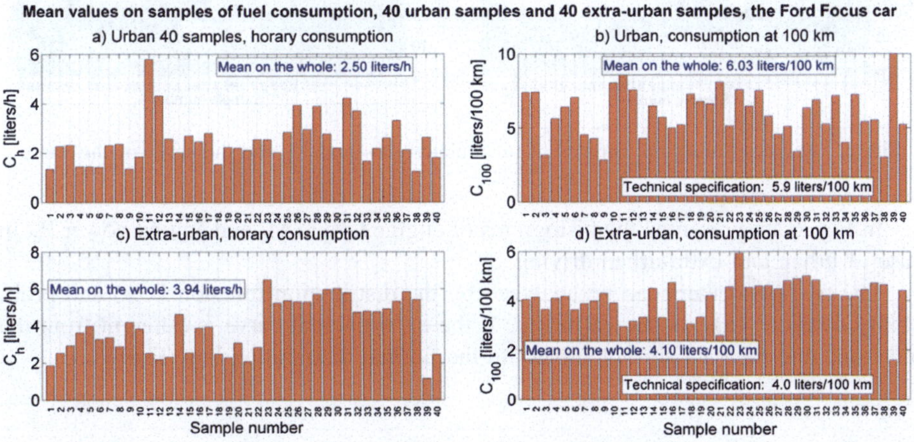

Fig. 4. Mean values on samples of fuel consumption

Figure 4 shows the average values of the hourly fuel consumption and average fuel consumption per 100 km traveled, related to the 40 samples, for the two situations analyzed. From these graphs it results that the two consumptions have higher values in the case of urban driving.

The explanation results from formula (1):

$$C_{100} = 100C_h/V \tag{1}$$

Indeed, even if the hourly consumption is higher for extra-urban driving, the fuel consumption per 100 km is lower due to higher vehicle speeds (the effect of speed is predominant).

Graphs from Fig. 5 present average values of the quantity of recirculated gas γ and the rotational angle of turbine vanes α. From these graphs it can be observed that for all samples, the two parameters are smaller in the case of extra-urban driving.

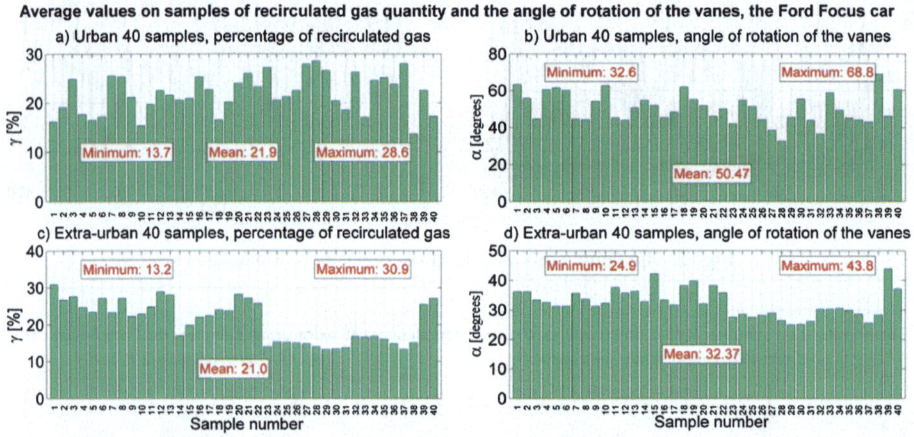

Fig. 5. Average values on samples of recirculated gas quantity and the angle of the vanes

In Fig. 6 are presented the histograms of engine torque M_e and engine power P_e, in case of urban and extra-urban driving.

The two parameters can be analyzed by the distribution function F (x) and probability density f (x). As shown in Fig. 7, the experimental curve is different from the ideal one and the data are not subject to the normal distribution law (Fig. 8).

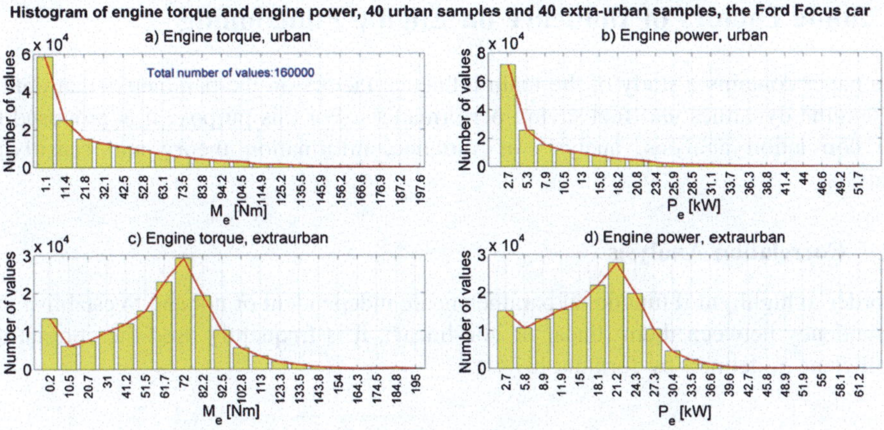

Fig. 6. Histogram of engine torque and engine power

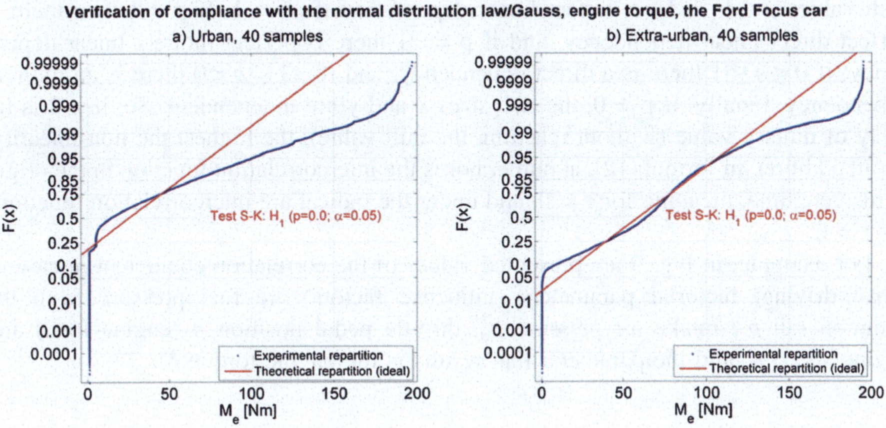

Fig. 7. Verification of compliance with the normal distribution law/Gauss, engine torque

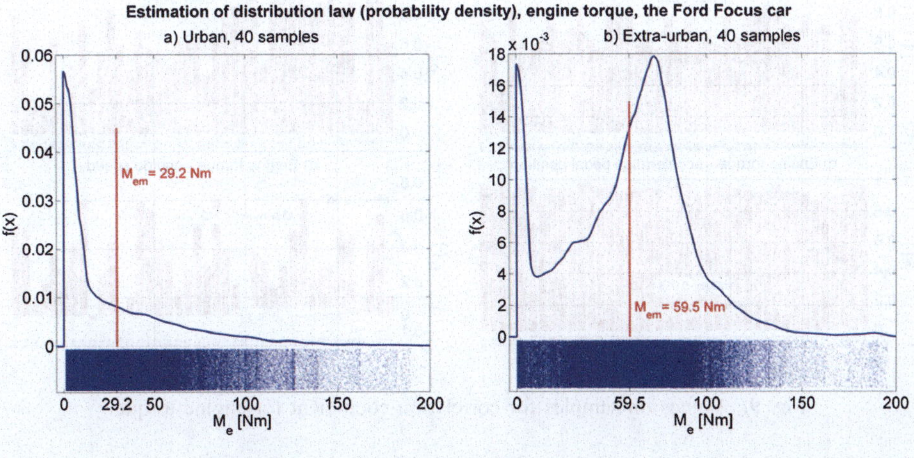

Fig. 8. Estimation of distribution law (probability density), engine torque

4 Some Factors of Influence on Engine Functioning

The paper contains a study of the main influence factors on the parameters that define the engine dynamics and fuel saving performances. For this purpose it is proposed to use correlation analysis, analysis of variance, information theory and sensitivity analysis.

4.1 Correlation Analysis

In order to highlight if functional parameters are independent or not and to establish the dependency between them (linear or non-linear), it is frequently used the correlation coefficient ρ, defined by formula [3, 4]:

$$\rho_{xy} = R_{xy}(0)/\sqrt{R_{xx}(0)R_{yy}(0)} \tag{2}$$

with values $\rho \in [-1; 1]$, a perfect linear dependency if $\rho^2 = 1$. If ρ = 1 then there is perfect direct linear dependency, and if ρ = −1 there is perfect indirect linear dependency; if $0 < \rho \leq 1$ there is a direct dependency, and if $-1 \leq \rho < 0$ there is an indirect dependency. Finally, if ρ = 0, the two sizes x and y are independent. So, if ρ 2 is far away of unitary value (without reaching the null value), the highest the non-linearity.

In addition, in formula (2), at numerator is the intercorrelation function in the origin of discrete time, meaning for τ = 0, and under the radical are intercorrelation functions for τ = 0.

For example, in Fig. 9 are presented values of the correlation coefficient in case of urban driving; factorial parameters (influence factors) are fuel pressure from the common rail p_c, intake air pressure p_a, throttle pedal position p (engine load) and engine speed n, and the parameter that results is the engine torque M_e.

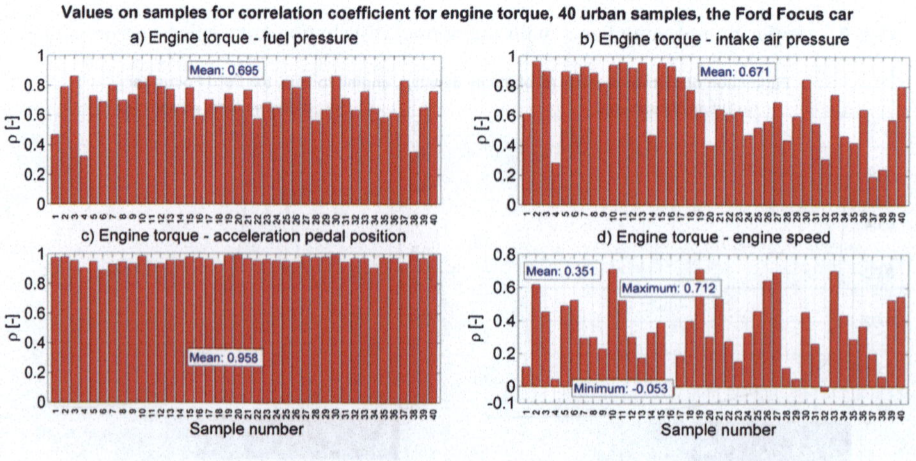

Fig. 9. Values on samples for correlation coefficient for engine torque

As it can be seen from graphs, there is no experimental sample at which the correlation coefficient is nul and because all sizes are not independent, which was functionally expected. Likewise, because for all samples the correlation coefficients are subunitary, it can be concluded that between the parameters there are non-linear dependencies; the most pronounced non-linear dependency is between engine torque and engine speed, the correlation coefficients having the lowest values, including the mean for all samples $\rho = 0.351$.

This final aspect leads to the conclusion that engine functioning must be described by non-linear mathematical models in order to ensure high precision.

4.2 Analysis of Variance

The study of the influence of some factors also uses the analysis of variance (ANOVA - ANalyse Of Variance, MANOVA - Multivariate ANalyse Of VAriance); dispersion plays an important role in analyzing the factors of influence on a dynamic process.

The analysis of the variation consists in comparing two types of differences, factorial and residual. If the factorial variance is higher than the residual one, the respective factor has a sensitive influence on the process. On the other hand, if the factorial variance (individually and in interaction with another factor) is smaller than the residual one, the factor has a negligible influence on the process analyzed.

Figure 10 shows the results after applying the MANOVA algorithm to study the influence of the throttle pedal position p on the engine speed n and the engine torque M_e. In the graph it can be seen that in both situations the engine speed variation is higher than the residual one (89.2% and 79.5% respectively) as well as the variation of the throttle position (9.9% and 14.7% respectively). So in both cases the pedal position has the higher influence on the torque variation.

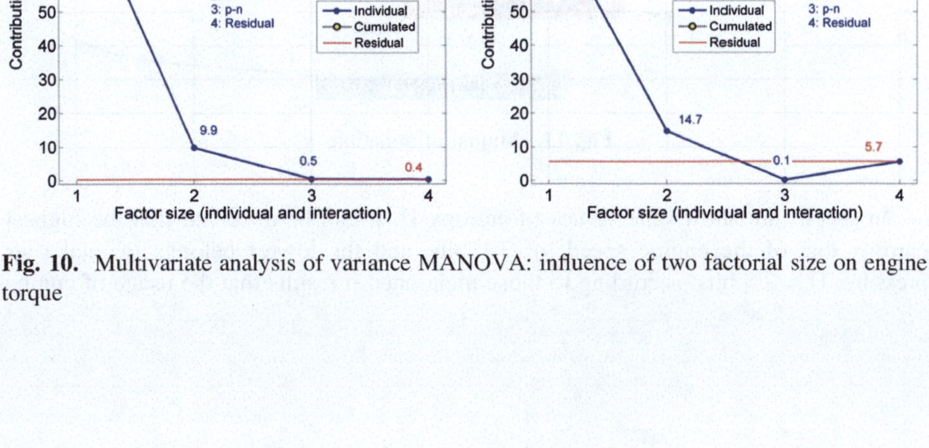

Fig. 10. Multivariate analysis of variance MANOVA: influence of two factorial size on engine torque

4.3 Informational Analysis

The study of factors influence also uses informational theory, which allows establishing significant parameters that define engine functioning. Informational analysis is based on two main concepts from information theory: entropy and mutual information [1, 3, 7–10]. Mutual information represents a concept regarding the quantitative estimation of uncertainty attenuation, meaning the increase in prediction. The higher the values of mutual information, the lower the uncertainties, meaning higher predictions.

Considering this, mutual information represents a basic concept for the study of systems dynamics and represents a measure of interdependencies between variables. Therefore, in establishing mathematical models must be chosen those variables characterized by the highest mutual information, which assures the highest predictions; those are called significant variables, attached to the concept of relevance. Considering all reasons mentioned above, information theory represents a generalization of classical theory, and mutual information represents a measure of relevance.

For example, in Fig. 11 is presented the graph containing results of informational analysis in case of engine torque as a resultant parameter (placed in upper part) and by considering 6 factorial parameters (influence factors): engine load (through acceleration pedal position p), engine speed n, fuel pressure from the common rail p_c, intake air pressure p_a, quantity of recirculated gas γ and rotational angle of turbine vanes α.

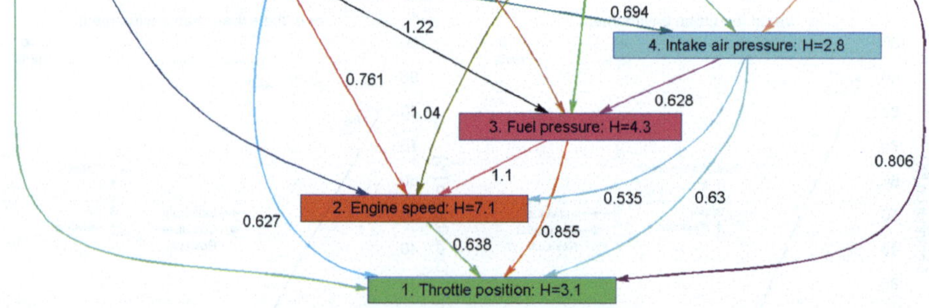

Fig. 11. Mutual informations

In graphs are shown the values of entropy H. It can be observed that the highest entropy that of the engine speed, of 7.1 bits, and the lowest belongs to intake air pressure: H = 2.8 bits; according to those mentioned it results that the usage of engine speed in dynamic calculus has the effect of most reduced prediction. On the graphs are displayed the values of mutual information between the two parameters I. It can be seen that the first two relevant variables (with the highest influence on engine torque) are throttle pedal position (a mutual information with engine torque of 1.165 bits) and fuel pressure (I = 0.485 bits).

In Fig. 11 are also shown mutual information between the 6 factorial parameters; it can be observed that that the highest mutual information is between fuel pressure and rotational angle of vanes, of 2.213 bits, which also exceeds those mentioned above; this aspect confirms the necessity that engine's functional study also target the influence between different factors, not only the influence of factors on resulting parameter.

Similarly, it is made an analysis in case of extra-urban driving, the first two significant sizes being the same as in case of urban driving.

It should also be observed that variance analysis targets only parameters variance, and informational analysis targets the entire parameter, not only the variance.

5 Engine Energy Efficiency

From the graphs presented in Fig. 12 it can be observed that in order to obtain a unit of power and torque, the consumption is higher in urban conditions, so the energy efficiency is higher in the case of extra-urban driving.

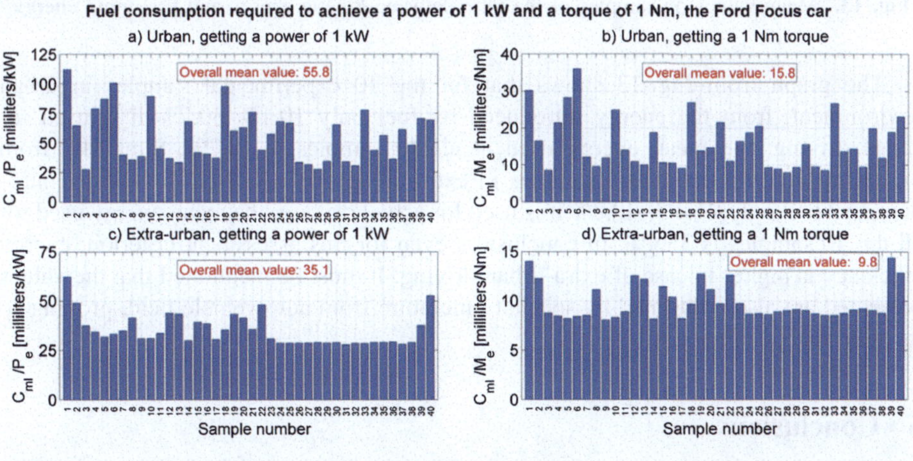

Fig. 12. Fuel consumption required to achieve a power of 1 kW and a torque of 1 Nm

In Fig. 13, the ratio between vehicle kinetic energy W_{cin} and energy introduced by fuel W_i is established with ratio:

$$k_c = W_{cin}/W_i \cdot 100\ [\%] \tag{3}$$

where:

$$W_{cin} = m_a v_a^2/2; \quad W_i = C_h Q_i S/V \tag{4}$$

and in which there are noted: ma and v_a (V) – autovehicle weight and speed; S – distance covered; Q_i – fuel inferior calorific power; C_h – fuel hourly consumption.

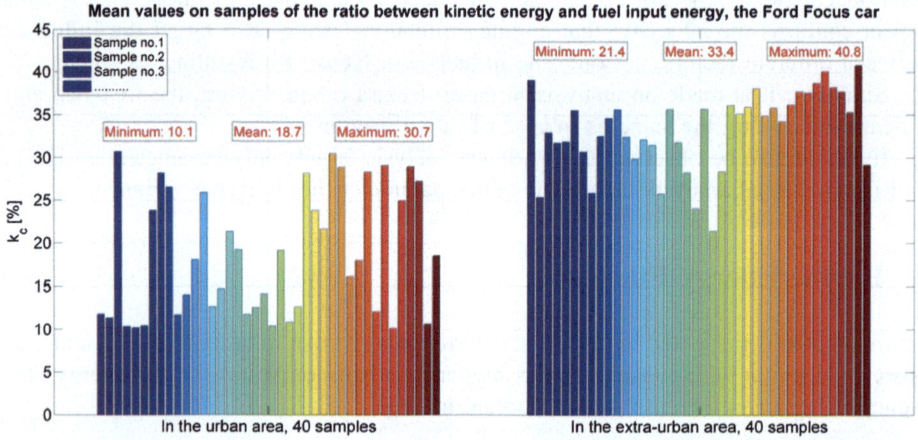

Fig. 13. Mean values on samples of the ratio between kinetic energy and fuel input energy

The graph from Fig. 13 shows that for the 40 experimental samples in urban environment, from the energy introduced by fuel only 10.1%–30.7% it is used for actual driving; the mean energy used at all 40 samples is 18.7%. Similarly, from Fig. 13 results that for all 40 samples in extra-urban driving, from the energy introduced by fuel only 21.4%–40.8% it is used for actual driving; the mean energy used for all the 40 samples is 33.4%. In conclusion, even for this assessment criterion, energy efficiency is higher in case of extra-urban driving. It must be mentioned that the values presented are also confirmed by specialty literature from auto vehicle field, proving so its poor energy efficiency.

6 Conclusions

The equipment of current automobiles with sensors and actuators embedded from fabrication and with an on-board computer, allows conducting complex studies, which go deep in the core of highly rapid dynamic processes that are specific to engine functioning.

Assessment criteria for engine fuel saving, currently used in specialty literature, do not satisfy the requests of a study regarding energy efficiency, that is why there must be established criteria specific to this issue.

The study of factors influence on engine functioning must also take into account interactions between factors, but also the fact that factorial sizes vary simultaneously during driving, these two are the main differences from the classical study presented in technical literature.

References

1. Technical note of Ford Focus car, with 1.6 TDCi engine, 66 kW (90 CP)
2. Diagnostics software FoCOM. http://www.obdtester.com/focom
3. Copae, I., Lespezeanu, I., Cazacu, C.: Dinamica autovehiculelor, ERICOM, Bucureşti (2006)
4. Hyötyniemi, H.: Multivariate Statistical Methods in Systems Engineering. Report 112, Helsinki University of Technology, Helsinki (1998)
5. Carey, G.: Multivariate Analysis of Variance (MANOVA). Colorado State University (1998)
6. Langsrud, Ø.: MANOVA for Matlab, Division of Statistics, Department of Mathematics, University of Oslo, Norway (2003)
7. Gray, R.: Entropy and Information Theory. Stanford University, New York (2007)
8. Hoenselaar, A.: A mutual information as a measure of relevance in neural coding. Thesis. Providence, Rhode Island (2007)
9. MacKay, D.: Information Theory, Inference, and Learning Algorithms. Cambridge University Press, Cambridge (2003)
10. Nemenman, I.: Information Theory, Multivariate Dependence, and Genetic Network Inference, University of California (2002)
11. Norris, M.: Redundancy Through Time. University of Plymouth, Plymouth (2003)

Marine Ship Equipped with Air Turbine – Electric Generator Aggregate

Radu Tarulescu(✉), Stelian Tarulescu, Cristian-Ioan Leahu, and Dana Luca-Motoc

Transilvania University of Brasov, Brasov, Romania
{radu.tarulescu, s.tarulescu, leahu.cristian, danaluca}@unitbv.ro

Abstract. The paper presents the energetic advantageous consequences of equipping a ship with an axial air turbine which capitalizes the relative fluid ship–wind movement. The electric power obtained is calculated, emphasizing the influence factors: random factors (presence, intensity and direction of the wind) and constructive, circumstantial or exploitation factors.

Keywords: Axial turbine · Head resistance · Energy recovery · Wind

1 Introduction

Figure 1 shows the scheme of a ship equipped with an axial aerodynamic turbine, improving the relative fluid movement resulting from the movement.

Fig. 1. Ship equipped with an axial aerodynamic turbine

I. Dumitru et al. (Eds.): SMAT 2019, *The 30th SIAR International Congress of Automotive and Transport Engineering*, pp. 188–194, 2020.
https://doi.org/10.1007/978-3-030-32564-0_22

With v [m/s] is noted the ship speed, with v_∞ [m/s] the wind speed and with $v_{rel,air}$ [m/s] the relative air speed of the turbine [1]. Relative air speed can be expressed in line with the other speeds, according to the relation (Fig. 2):

$$v_{rel,air} = v_\infty + v \cdot \sin\theta = v_\infty \cdot \left(1 + \frac{v}{v_\infty} \cdot \sin\theta\right) [\text{m/s}] \quad (1)$$

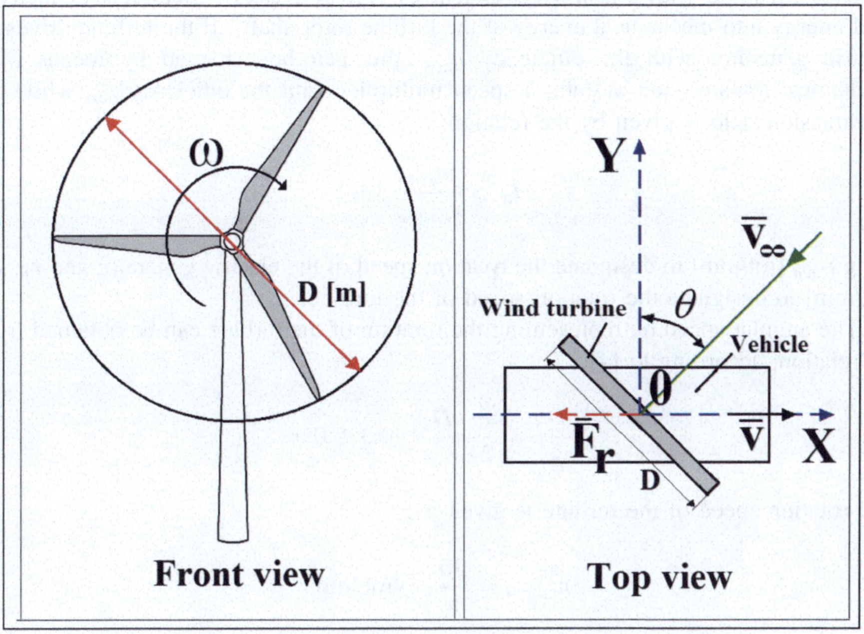

Fig. 2. The scheme of the wind turbine placed on the ship

The turbine and the rotation plane of the rotor shall be perpendicular to the wind direction. The incidence angle of the wind is $\hat{\theta}$ over the 0y axis, perpendicular to the 0x direction of movement of the ship [2].

If the wind blows from the front or from front-side, the incidence angle of the wind is $\hat{\theta} \in (0, \pi)$ [3]. If the wind blows from behind or from behind-side, the relative speed shall be:

$$v_{rel,air} = v_\infty - v \cdot \sin\alpha = v_\infty \cdot \left(1 - \frac{v}{v_\infty} \cdot \sin\theta\right) [\text{m/s}], \quad (2)$$

If $v \cdot sin\alpha > v_\infty$, it is recommended to place the turbine with the plane of the rotor, perpendicular to the direction of movement of the ship. If $v \cdot sin\alpha < v_\infty$, it is recommended to orient the turbine with the plane of the rotor perpendicular to the wind direction.

2 Energy Recovery System with Axial Wind Turbine

In order to recover the wind energy under the conditions described above, the authors suggest the use of a wind turbine with the diameter D [m], the impact section $A = \pi \cdot D^2/4$ [m^2] and with a power coefficient C_p (the conversion coefficient of incident fluid energy into mechanical energy at the turbine rotor shaft). If the turbine drives an electric generator with the efficiency η_{gen}, this can be achieved by means of a mechanical transmission system, a speed multiplier with the efficiency η_{tr}, while the transmission ratio is given by the relation:

$$i_{tr} = \frac{n_{gen}}{n_{turbine}} \text{ [-]}, \tag{3}$$

using n_{gen} [rot/min] to designate the rotation speed of the electric generator and $n_{turbine}$ [rot/min] to designate the rotation speed of the turbine.

The angular speed ω, representing the rotation of the turbine can be obtained from the relation, according to [4]:

$$\frac{\omega \cdot r}{v_\infty} = \frac{\omega \cdot D}{2 \cdot v_\infty} = 0.3 \div 0.4 \tag{4}$$

The rotation speed of the turbine is given by:

$$n_{turbine} = \frac{30}{\pi} \cdot \omega [\text{rot/min}] \tag{5}$$

The wind turbine is a slow machine in which case both, the angular speed and the rotation speed register low values [5]. The fluid power P_∞ [W] is given by the relation:

$$P_\infty = A \cdot \frac{\rho_{air}}{2} \cdot v_\infty^2 [\text{W}], \tag{6}$$

while the electric power P_e [W] is:

$$P_e = C_p \cdot \eta_{tr} \cdot \eta_{gen} \cdot P_\infty [\text{W}] \tag{7}$$

The axial aerodynamic head resistance force $F_r = F_{axial}$ [N], for “pushing” the turbine is:

$$F_{axial} = K \cdot v_\infty^2 [\text{N}], \tag{8}$$

where K [$Pa \cdot s^2$] represents the aerodynamic resistance coefficient.

3 Numerical Application

In terms of applications, the authors suggest the use of a three-blade wind turbine, with the power coefficient of $C_p = 0.24$ and the product $\eta_{tr} \cdot \eta_{gen} = 0.835$. For the wind speed there have been adopted the values 6, 9 and 12 [m/s]. For the ship speed the speed $v = 50$ [km/h] = 13.9 [m/s] has been adopted [6].

For placement on a seagoing ship, a turbine with the diameter $D = 5$ [m] can be inserted. It results an impact section $A = \pi \cdot D^2/4 = 19.6$ [m^2], $A \cdot \rho_{air}/2 = 11.73$ [Pa · s^2] and $K = 0.7038$ [Pa · s^2], where $\rho_{air} = 1.2$ [kg/m^3]. The angular speed resulted is $\omega = 0.14 \cdot v_\infty$, the aerodynamic head resistance force $F_r = 0.7038 \cdot v_\infty^2$ and the fluid power $P_\infty = 11.73 \cdot v_\infty^3$.

For placement on a ship, a turbine with the diameter $D = 2$ [m] can be inserted. It results an impact section $A = \pi \cdot D^2/4 = 3.14$ [m^2], $A \cdot \rho_{air}/2 = 1.884$ [Pa · s^2] and $K = 0.113$ [Pa · s^2], where $\rho_{air} = 1.2$ [kg/m^3]. The angular speed resulted is $\omega = 0.35 \cdot v_\infty$, the aerodynamic head resistance force $F_r = 0.113 \cdot v_\infty^2$ and the fluid power $P_\infty = 1.884 \cdot v_\infty^3$.

Table 1. The computed values for the wind turbine with $D = 5$ [m]

Wind speed	Angular speed	Rotation speed	Transmission ratio	Fluid power	Electric power	Head resistance force
v_∞ [m/s]	ω [s^{-1}]	$n_{turbine}$ [rot/min]	$i_{tr} = 1600/n$ [-]	P_∞ [kW]	P_{el} [kW]	F_r [N]
6	0.84	8.05	199	2.53	0.506	25.3
9	1.26	12	133	8.55	17.1	57
12	1.68	16	100	2.04	40.8	101

Tables 1, 2 and Fig. 3 show the results of the calculations when placing the two turbines on a ship, respectively on a ship.

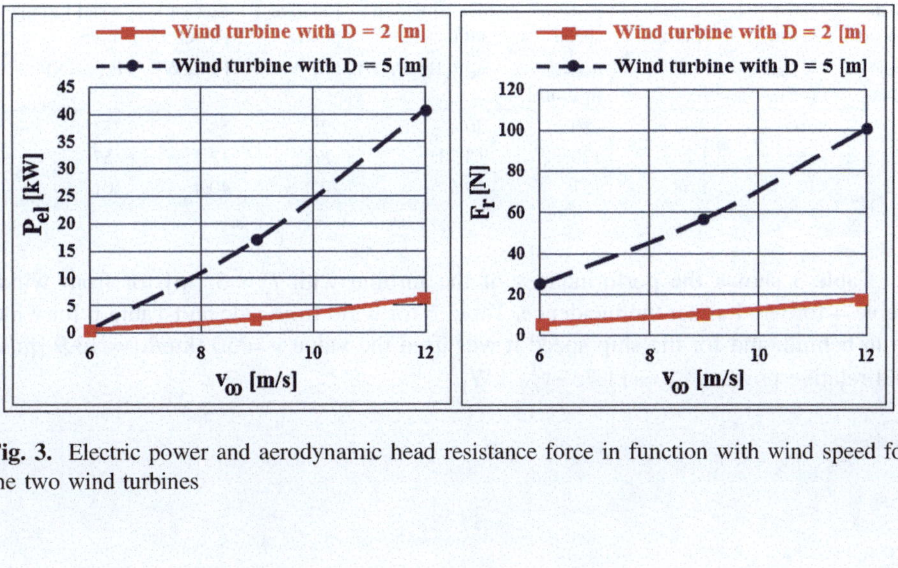

Fig. 3. Electric power and aerodynamic head resistance force in function with wind speed for the two wind turbines

Table 2. The computed values for the wind turbine with $D = 2$ [m]

Wind speed	Angular speed	Rotation speed	Transmission ratio	Fluid power	Electric power	Head resistance force
v_∞ [m/s]	ω [s^{-1}]	$n_{turbine}$ [rot/min]	i_{tr} = 1600/n [-]	P_∞ [kW]	P_{el} [kW]	F_r [N]
6	2.1	20	80	0.405	0.81	4.06
9	3.15	30	53.3	1.38	2.76	9.17
12	4.2	40	40	3.26	6.52	16.2

Table 3. The computed values when wind blows from the front (incidence angle $\theta = \pi/2$)

Wind speed	Relative speed	Angular speed	Rotation speed	Transmission ratio	Relative power	Electric power	Axial turbine force
v_∞ [m/s]	$V_{rel,air}$ [m/s]	ω [s^{-1}]	$n_{turbine}$ [rot/min]	i_{tr} = 1600/n [-]	P_{rel} [kW]	P_{el} [kW]	F_{axial} [N]
6	19.9	2.78	29.2	54.8	93	18.6	278
9	22.9	3.2	33.6	47.6	143	28.6	367
12	25.9	3.62	34.5	46.3	204	40.8	470

Table 4. The computed values when wind blows under the incidence angle $\theta = \pi/4$

Wind speed	Relative speed	Angular speed	Rotation speed	Transmission ratio	Relative power	Electric power	Axial turbine force
v_∞ [m/s]	$V_{rel,air}$ [m/s]	ω [s^{-1}]	$n_{turbine}$ [rot/min]	i_{tr} = 1600/n [-]	P_{rel} [kW]	P_{el} [kW]	F_{axial} [N]
6	15.8	2.21	21.1	76	29.4	5.88	176
9	18.8	2.63	25	64	41.3	8.26	249
12	21.8	3.05	29	55	51.7	10.34	335

Table 5. The computed values when wind blows from the side (incidence angle $\theta = 0$)

Wind speed	Relative speed	Angular speed	Rotation speed	Transmission ratio	Relative power	Electric power	Axial turbine force
v_∞ [m/s]	$V_{rel,air}$ [m/s]	ω [s^{-1}]	$n_{turbine}$ [rot/min]	i_{tr} = 1600/n [-]	P_{rel} [kW]	P_{el} [kW]	F_{axial} [N]
6	6	0.84	80	20	25	5	253
9	9	1.26	120	13.33	86	17.2	57
12	12	1.58	151	10.6	202	40.4	101

Table 3 shows the performances of the turbine with $D = 5$ [m] for front wind, Table 4 for wind under the incidence, Table 5 for wind from side and Table 6 for wind from behind, and for the ship speed it was used the value $v = 50$ [km/h] = 13.9 [m/s] and relative power $P_{rel} = 11.73 \cdot v_{rel}^3$ [kW].

Table 6. The computed values when wind blows from behind (incidence angle $\theta = -\pi/2$)

Wind speed	Relative speed	Angular speed	Rotation speed	Transmission ratio	Relative power	Electric power	Axial turbine force
v_∞ [m/s]	$V_{rel,air}$ [m/s]	ω [s^{-1}]	$n_{turbine}$ [rot/min]	i_{tr} = 1600/n [-]	P_{rel} [kW]	P_{el} [kW]	F_{axial} [N]
6	7.9	1.1	10.5	152	58	11.6	43.8
9	4.9	0.685	6.53	245	13.8	2.76	16.9
12	1.9	0.167	1.59	1000	8	1.6	2.52

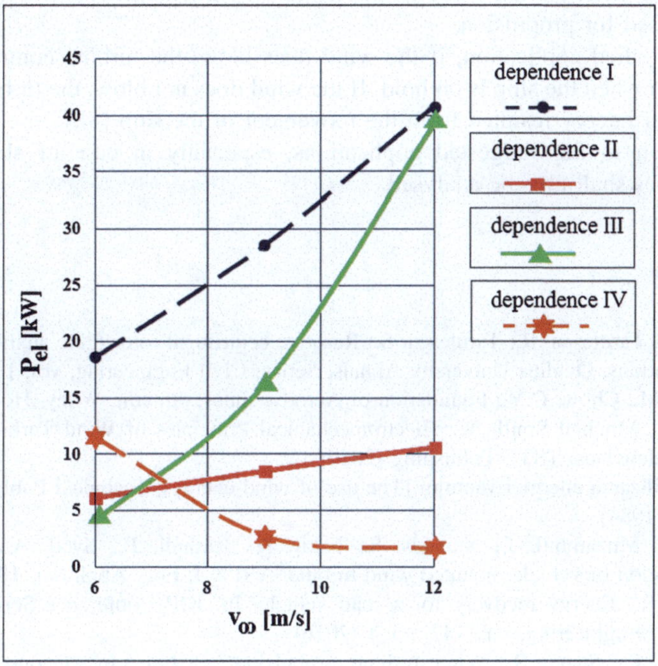

Fig. 4. Electric power and aerodynamic head resistance force in function with wind speed for the two wind turbines

Figure 4 shows the electric power dependences, varying according to the wind direction, for the turbine with the diameter $D = 5$ [m]:

- dependence I – wind blows from the front, $v_{rel,air} = v_\infty + v$ and $F_{axial} = F_r$;
- dependence II – wind under the incidence $\theta = \pi/4$, $v_{rel,air} = v_\infty + v \cdot \frac{\sqrt{2}}{2}$ and $F_{axial} = F_r$;
- dependence III – wind blows from the side, $v_{rel,air} = v_\infty$ and $F_{axial} = F_{side}$;
- dependence IV – wind blows from behind, $v_{rel,air} = v - v_\infty$ and $F_{axial} = F_{rear}$.

As a result of the calculations made by the authors, it can be said that in case the wind blows from the front, the aerodynamic head resistance force F_r opposes to the movement of the ship, in case the wind blows from the side F_r is lateral and in case the wind blows from behind F_r "pushes" the ship.

4 Conclusions

In the last decade, several wind turbines have been implemented on different ships with two specific purposes: one for energy recovery and the other for ship propulsion. In a similar application it was used a Vestas V39/600 wind turbine with the rotor diameter of 39 m, which generate up to 600 kW of power [7]. On smaller vessels wind turbines are usually used for propulsion.

For the studied application, if the wind does blow, the turbine contributes to its recovery even when the ship is on hold. If the wind does not blow, the turbine recovers the relative air energy resulted from the movement of the ship [8].

In addition to the suggested applications, especially in case of ships, various dynamic issues shall also be analysed.

References

1. Benche, V., Tarulescu, R., Tarulescu, S.: Reactive centrifugal rotor – the analytical study of two applications, Ovidius University Annals, Series: Civil Engineering, vol. 1 (2007)
2. Kuethe, A.M., Chow, C.Y.: Foundation of Aerodynamics, 5th edn. Wiley, Hoboken (1998)
3. Plantier, K., Mitchell Smith, K.: Electromechanical Principles of Wind Turbines for Wind Energy Technicians. TSTC Publishing (2009)
4. Ilie, V.: Utilizarea energiei vantului [The use of wind energy], Technical Publishing House, Bucharest (1984)
5. Rabiul, A., Muzammil, J., Nazmus, S., Fakir, H., Rashidi, B., Syed, A.: Design and implementation of vehicle mounted wind turbine. ARPN J. Eng. Appl. Sci. **10** (2015)
6. Tarulescu, R.: Energy recovery for a road vehicle. In: IOP Conference Series: Materials Science and Engineering, vol. 147, no. 1 (2016)
7. Bøckmann, E., Steen, S.: Wind turbine propulsion of ships. In: Second International Symposium on Marine Propulsors, Hamburg, Germany (2011)
8. Tarulescu, R., Tarulescu, S., Leahu, C.: Downforce variation dependence of angle of incidence modification for the rear wing of high speed vehicles. In: IOP Conference Series: Materials Science and Engineering, vol. 252, Conference 1 (2017)

Influence of the Dynamic Vehicle Load on the Power Losses Required to Overcoming the Rolling Resistance

Nicolae Cordoş(✉), Adrian Todoruţ, Călin Iclodean, and István Barabás

Department of Automotive Engineering and Transports, Faculty of Mechanical Engineering, Technical University of Cluj-Napoca, Cluj-Napoca, Romania
{nicolae.cordos, adrian.todorut, calin.iclodean, istvan.barabas}@auto.utcluj.ro

Abstract. The movement of the vehicle is mainly determined by the processes that occur at the interaction between the tread and the tire. Knowledge of the tire behavior under the effect of external loads is very important to study the dynamics of the vehicle, and the operating behavior of the tire in movement affects the reliability, the comfort of the occupants, as well as the economy of the motor vehicles and tread wear. Force and the power losses occurring at the contact patch between the tire and the road is very difficult to detect experimentally. This paper aims to estimate the necessary power need to overcoming the rolling resistance forces of a passenger car for different loads and constant tire pressure using the simulation program IPG CarMaker. The chosen vehicle is one with rear-wheel drive, and the chosen route is a route with cones. The computer simulation was done for the car's curb weight and the for the total weight of the car. The measurement was made on the right rear wheel of the car. The car accelerates uniformly up to a constant speed of 80 km/h at the entrance in the cone zone. The car has performed slaloms through offset cones. The paper follows the influence of the vertical acceleration on the power required to overcoming the rolling resistance force for the two cases of vehicle load and the vehicle speed influence on the power required to overcoming the rolling resistance force.

Keywords: Rolling resistance · Vertical acceleration · Vehicle load

1 Introduction

Rolling resistance is a permanent force that acts in the opposite direction of the vehicle movement. The physical causes of this force are as follows: friction between the tire and the tread, the interpenetration between the tire elements and the treadmill microns, the suction effect produced by the rolling contours on the tread, the inner tire friction and the vehicle's running speed. In the vehicle design calculations, rolling resistance is taken into account by the rolling resistance coefficient. The coefficient of rolling resistance is influenced by certain factors such as: wheel pressure, vehicle load,

I. Dumitru et al. (Eds.): SMAT 2019, *The 30th SIAR International Congress of Automotive and Transport Engineering*, pp. 195–202, 2020.
https://doi.org/10.1007/978-3-030-32564-0_23

condition and nature of the road etc. The rolling resistance coefficient also has an influence on the fuel consumption of the vehicle.

Vehicle weight, aerodynamics, engine, transmission and auxiliary load affects the vehicle fuel economy. Vehicle tires also influence fuel economy by the rolling resistance, resulting in a certain loss of engine power.

In [1], the authors define the rolling resistance as the energy loss per distance traveled by the vehicle due to nonelastic deformations of the tires and losses in the wheel suspension system.

Power lost due to rolling resistance is closely related to fuel economy. The higher it is, the higher is the fuel consumption.

Reduction of the loads on the vehicle axles, increasing the engine efficiency, transmission and other components which generate and transfer of energy on the axles, represents the ways to increase the fuel economy of the vehicle [2]. The tests carried out by the Society of Automotive Engineers (SAE) in accordance with the standards (SAE J1269 and SAE J2452) have fixed values of the tires rolling resistance coefficient for new passenger cars. Measurements under according to such standards have been determined values of the coefficient of rolling resistance between 0.007 and 0.014. For a tire in this range under a load of 450 kg, the rolling resistance is aprox. 3.1 to 6.5 kg, resulting in aprox. 12.7 to 25.4 kg of total force for the four tires on a vehicle weighing 1800 kg, including passenger and cargo load. At 60 mph, a total rolling resistance of 18 kg consumes about 7 horsepower [2]. Also, in this paper, the authors have established that a 10% reduction in rolling resistance corresponds to a 2% reduction in energy demand, or a 2% reduction in fuel consumption.

Following the experimental attempts in [3] at an average speed of 34 km/h in the urban environment, there was a 11% decrease in energy consumption for the vehicles under test, with the rolling resistance coefficient dropping from 0.001 at 0.005.

In [4] the simulations were made regarding the power consumption due to rolling resistance. The simulations were performed on a wheel, with a vertical load of 2500 N. Rolling radius of the wheel was 0.305 m, the inflation pressure of 2.1 bar. As a result of the simulations the values for the power consumption due to the rolling resistance of 758 watts, for a constant speed of 80 km/h have been observed. Also, in this paper, it is noticed that with the increase of the rolling radius for the same tire (inflation pressure higher), the power consumption required to overcome the rolling resistance decreases.

A radial ply tire was used as the tester wheel on clay-loam soil on a single wheel-tester to investigate the influence velocity, inflation pressure, and vertical load on rolling resistance in [5]. The load on the tire was 1, 2, 3, 4, and 5 kN at velocity of 0.7, 1.4 and 2 m/s. The conclusion was that the vertical load applied on wheel has considerable impact on rolling resistance and that increase of vertical load is approximately polynomial with order of two function of rolling resistance. As a result of this experimental research, it has also been established that rolling resistance increases with the reduction in wheel pressure.

In the case of the trucks, the tire pressure has a great effect on rolling resistance, so that a 20% reduction in tire pressure can result in an increase in rolling resistance of 5–8%, respectively an increase in fuel consumption 2–3% [6, 7].

A major factor that contributing to the rolling resistance is the process known as hysteresis. This process is mainly represented by the loss of energy that occurs as a tire rotates on its surface. These losses are directly related to the deformation of the tire, and depend on inflation pressure and wheel load [8]. Energy loss must be compensated by the motor vehicle, which results in wasted fuel.

In [9] the experimental testing was carried out using an *Trailer R2 Mk.2* type, specially built to measure the rolling resistance. As a result of the experimental tests, at a constant speed of 80 km/h, it has been established that at constant tire pressures, with increasing load (from 150 kg to 600 kg), the force required to overcome the rolling resistance increases.

The author from [10] following some computerized simulations observes different results on the rolling resistance for two driving cycles: the New European Driving Cycle and the US FTP-72 driving cycle. In the case of the New European Driving Cycle, the 10% increase in vehicle weight corresponds to a 4% increase in fuel consumption or alternatively in CO_2 emissions. For the US FTP-72 driving cycle the effect of the vehicle's load on rolling resistance is slightly higher.

Approximately 20–30% of the passenger car energy consumption represents the energy consumed to overcome the rolling resistance, depending on the running speed [11]. Therefore, reduction of greenhouse gases and fuel consumption has a direct connection with the power necessary for overcoming the rolling resistance.

This work, following a computer simulation in the IPG CarMaker program, aims primarily to assess the lost power due to the rolling resistance for a rear-wheel drive and correlating these power losses with vertical accelerations on the drive wheel.

2 Materials and Methods

To analyze the influence of the dynamic vehicle load on the power losses required to overcoming the rolling resistance, the simulations were performed using the numerical simulation tool IPG CarMaker. The graphical interface of the software, it can be seen in Fig. 1.

The simulation was performed with a BMW 5 Series rear-wheel drive, engine power of 190 HP, wheelbase of 2.59 m, the curb weight of 1194 kg, total weight of 1569 kg. The tires are 225/60 R16 with a 2.5 bar pressure.

Given the fact that the measurements were performed on the right rear wheel of the vehicle, it was calculated as weight back on the right rear wheel. As a result of the calculations, for the weight of the vehicle, the weight on the right rear wheel is 304 kg, and for the total weight (TW) of the vehicle, the weight of the right rear wheel is 399 kg. Thus, the weight on the right rear wheel for the total weight (TW) of the vehicle is approx. 24% higher than the weight on the right rear wheel in the case of a vehicle curb weight. The calculated rolling radius of the right rear wheel is 0.299 m.

The chosen track has a length of 1100 m. From the 400 m starts the cones area. There are 14 cones, with a distance between them of 36 m (see Fig. 2).

The vehicle was in accelerated motion, with the engine in full load. At the entrance in the cones area, the vehicle's speed is 80 km/h, this speed remaining constant until the computer simulation is complete.

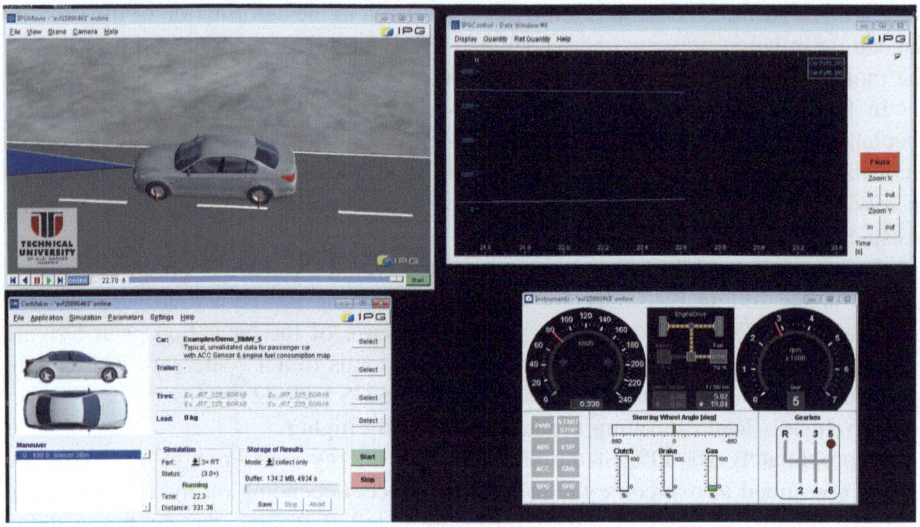

Fig. 1. The graphical interface of the *IPG CarMaker* software with tested vehicle.

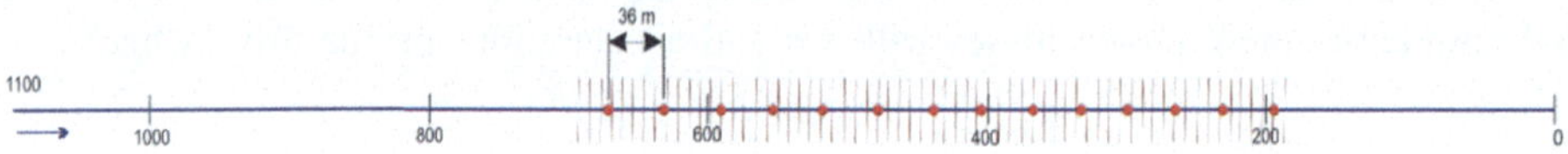

Fig. 2. The track with cones used in the computer simulation

3 Results and Discussions

Figure 3 represented the power required to overcoming the rolling resistance of the right rear wheel of the vehicle based on the speed of the path chosen for both cases: curb weight (CW) of the vehicle and total weight (TW) of the vehicle.

It can be seen that as the speed increase, the power consumption increases due to rolling resistance.

At the total weight of the passenger car it can be seen higher values of the power consumed, due to the rolling resistance. With increasing the speed, the difference between the values of the power needed to overcome the rolling resistance is approaching for the two dynamic loads of the vehicle. This can be explained by the decrease in inertia weight of the vehicle as the speed increases.

The maximum values for the power consumption due to the rolling resistance are observed at 80 km/h, ranging from approx. 600 W and 1500 W, for total weight (TW) of the vehicle, and between 500 W and 1200 W in the case of curb weight (CW) of the vehicle. So the power lost due to the rolling resistance is on average approx. 20% higher in the case of a vehicle with total weight (TW) compared to the curb weight (CW) of vehicle.

In the speed range 35–40 km/h, respectively 60–65 km/h, it can be seen a decrease in power consumption due to the rolling resistance, because of the time needed to

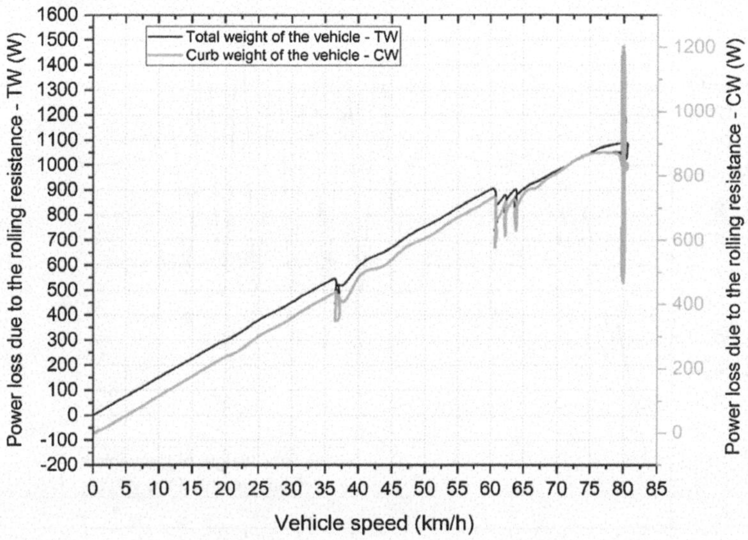

Fig. 3. Variation of the power lost due to the rolling resistance depending on the vehicle speed

change the gear. This can be explained by the fact that during this interval the vehicle's engine is not in the load.

Figure 4 shows the power consumption due to the rolling resistance (see Fig. 4a) and the acceleration on the z-axis tire (vertical acceleration) (see Fig. 4b) to the right rear wheel for the two cases under study, based on time. It can be noticed that with the increase of the vehicle's speed the power needed to overcome the rolling resistance increases.

It is possible to identify the maximum values of the power consumed due to the rolling resistance until to the entrance of the vehicle in the cones area, the difference between them being approx. 10% (about 1100 W for total weight (TW) and about 890 W in the case of the curb weight (CW) of the vehicle) (see Fig. 4a). Maximum values of the power consumed due to the rolling resistance can also be identified in the case of the wheel vertical acceleration values. Higher acceleration values up to the entry of the cone zone are due to the fact that when the vehicle was started, it was loaded with the engine in full load, having to overcome greater inertia due to its weight. As the speed increases, the vertical acceleration values at the rear wheel (the driven wheel) decrease. Higher values for the vertical acceleration can be seen for the total weight (TW) of the vehicle. At the speed of 80 km/h (the constant speed at which the simulation was performed) it can been observe the values of the powers lost due to the rolling resistance.

With the entry of the vehicle in the cones zone, the maximum power required to overcoming the rolling resistance is approx. 1500 W for total weight (TW) of the vehicle and aprox. 1200 W for curb weight (CW) of the vehicle, representing a difference of approx. 20%.

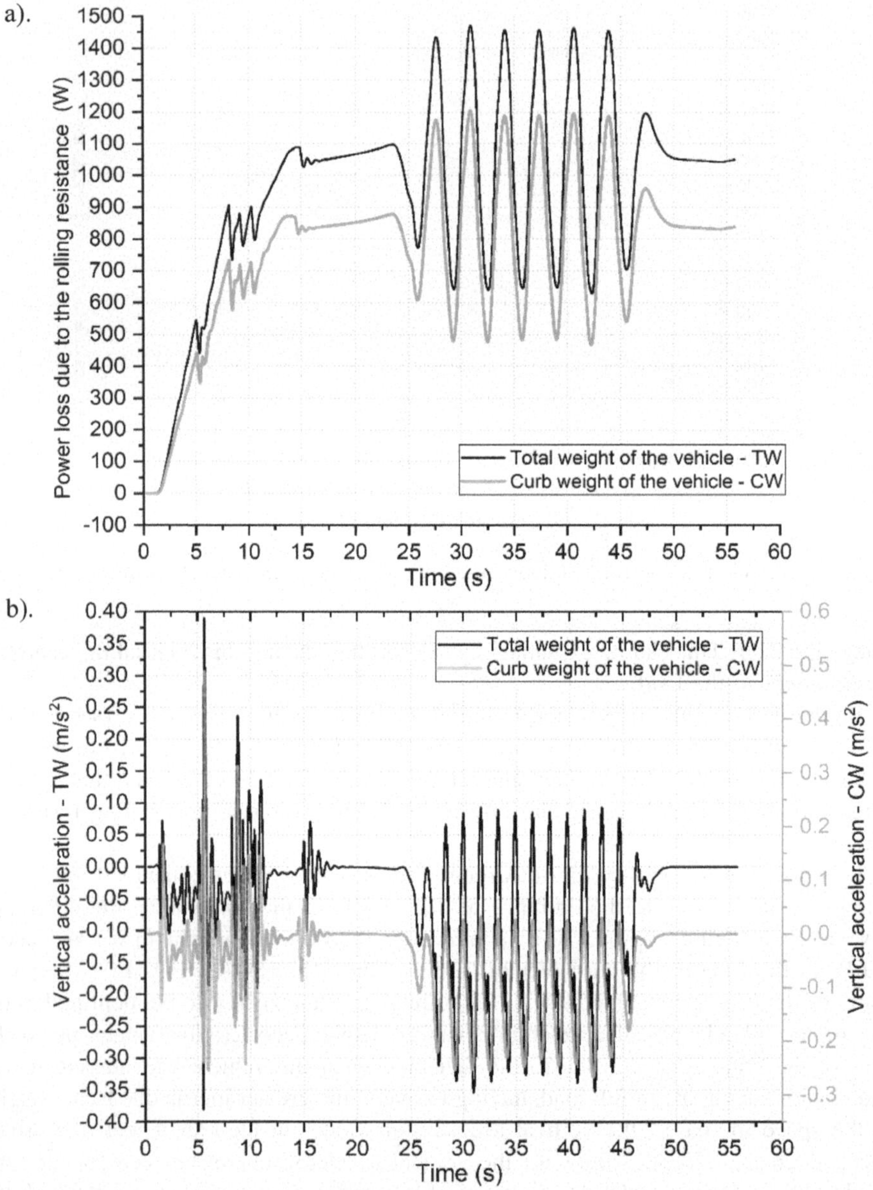

Fig. 4. Variation of the power consumed due to the rolling resistance (a) and vertical acceleration versus time (b)

It can be observed a correlation between the vertical acceleration values and the values of the power necessary to overcome the wheel rolling resistance. Thus, with the increase of the vertical acceleration at the wheel, the power loss due to overcoming the rolling resistance is increasing.

In the cones area, for maximum vertical acceleration values of approx. 0.1 m/s^2, corresponds the values of the power loss due to rolling resistance of approx. 1500 W, and for the vertical accelerations of approx. 0.025 m/s^2, corresponds values of the power loss due to rolling resistance of approx. 1200 W. So, with an increase of approx. 11% of the vertical acceleration values there is an increase in power loss due to the rolling resistance of about 20%.

4 Conclusions

After the simulations performed using IPG CarMaker software it was able to observe the influence of the dynamic loads on the driving wheels of the vehicle on the power consumed due to the rolling resistance. It can be observed an increase in power consumed due to the rolling resistance with increasing the vehicle speed for the curb weight (CW), respectively for the total weight (TW) of the vehicle. It has been noticed that the power lost due to rolling resistance is on average approx. 20% higher in the case of a vehicle with a total weight (TW) compared to a vehicle with the curb weight (CW) of the vehicle. Also, the influence of the vehicle's vertical acceleration on the loss of power due to the rolling resistance has also been observed. Along with the increase in vertical wheel acceleration, increased the power losses due to the rolling resistance were observed.

These results can be developed following the influence of the tire pressure, the state and nature of the road and various types of tires on the power loss due to rolling resistance.

Acknowledgment. The simulations presented in the paper were done using the IPG CarMaker software supported by AVL List GmbH, Austria.

References

1. Lasse, G.A., Jesper, K.L., Elsje, S.F., Bjarne, S., Jeppe, C.D.: Rolling resistance measurement and model development. J. Transp. Eng. (2015). https://doi.org/10.1061/(asce)te.1943-5436.0000673
2. Consumers, Informing: Tires and Passenger Vehicle Fuel Economy. TR News (2006)
3. Świeczko-Żurek, B., Grzegorz, R., Jerzy, E.: Tire rolling resistance and its influence on fuel consumption. Combust. Eng. **56** (2017)
4. VanderWiel, J., Harris, B., Jackson, C., Reese, N.: Exploring the relationship of rolling resistance and misalignment angle in wheelchair rear wheels. RESNA, Arlington, VA, vol. 12 (2016)
5. Taghavifar, H., Aref, M.: Investigating the effect of velocity, inflation pressure, and vertical load on rolling resistance of a radial ply tire. J. Terramech. **50**(2), 99–106 (2013)
6. Leduc, G.: Longer and heavier vehicles – an overview of technical aspects. Office for Official Publications of the European Communities, Joint Research Centre – Institute for Prospective Technological Studies, Luxembourg, JRC 52392EN (2009)

7. NAS 2010: Technologies and approaches to reducing the fuel consumption of medium- and heavy-duty vehicles. The National Academies of Science, Transportation Research Board. The National Academies Press, Washington (2010)
8. Lee, H.W., Cho, J.R., Jeong, W.B., Kim, K.W.: Mesh generation and hysteretic loss prediction of 3-D periodic patterned tire. Int. J. Automot. Technol. **15**(3), 411–417 (2014)
9. Ejsmont, J., Taryma, S., Ronowski, G.: Int. J. Automot. Technol. **17**, 237 (2016). https://doi.org/10.1007/s12239-016-0023-z
10. Juhala, M.: Improving vehicle rolling resistance and aerodynamics. In: Alternative Fuels and Advanced Vehicle Technologies for Improved Environmental Performance, pp. 462–475. Woodhead Publishing (2014)
11. Boere, S., Arteaga, I.L., Kuijpers, A., Nijmeijer, H.: Tyre/road interaction model for the prediction of road texture influence on rolling resistance. Int. J. Veh. Des. **65**(2–3), 202–221 (2014)

Research of Diesel Particle Filter Soot Impact and Regeneration Time Trend-Line

Emilian Vasile Borza, Doru-Laurean Băldean(✉), and Bogdan-Manolin Jurchiş

Technical University of Cluj-Napoca, 08544 Cluj-Napoca, Romania
doru.baldean@auto.utcluj.ro

Abstract. The experimental research paper takes into consideration the possibility of experimental testing and tracking the trend line of the particulate soot impact, particulate filter differential pressure, exhaust gas temperature upstream of particulate filter of using Diesel Standard fuel with compression ignited engines, as well as the influence that each temperature value and the time instances have upon the particle filter regeneration, on the trend line in the testing time. For the practical application i was supplied Diesel Standard fuel and its blends in accordance with operating regime. Specific experimental setting was developed and actual values were collected with accurate digital systems. Practical tests and data collection were realized in order to obtain the significant trend line in the given time of the test in parallel with temperature and pressure. By measuring the latter parameters and voltage on the exhaust gas differential pressure some trend line equations may be traced which sustain the regression squared values for a more complex analyze of the phenomena. The most significant soot impact was reached after thirty minutes of regeneration procedure. Engine speed was above 2000 rpm and coolant's temperature higher than 80 °C.

Keywords: Diesel particulate filter · Exhaust temperature · Pollution · Trend line

1 Introduction

The study of Diesel Particle Filter soot impact and regeneration trend-line in relation to operational characteristics is important due to the fact that it has a great influence upon pollution and environmental impact of the engine.

Fuel type and operational characteristics are clearly influencing engine's exhaust emissions [1–6]. The later ones are controlled and reduced with after treatment systems [7]. Development of the proper control system is based upon the proper design and efficient construction of the whole air/gas flow line, from air intake to the exhaust [8].

2 Materials and Method

In order to study the trend line for particulate filter soot impact and other parameters when supplying the engine with Diesel Standard fuel was considered the complex link between after treatment systems and overall costs and effects, according to Fig. 1.

I. Dumitru et al. (Eds.): SMAT 2019, *The 30th SIAR International Congress of Automotive and Transport Engineering*, pp. 203–208, 2020.
https://doi.org/10.1007/978-3-030-32564-0_24

Method applied for research was experimental testing and measurement upon the power train in operational conditions. Materials used were K9K792 powertrain from Renault with common-rail diesel fuel supply. Injection pressure was located in the 200÷1900 bar interval, as shown in Fig. 2. Compression ignited engine 1, supplied with diesel standard fuel through a common-rail injection system is digitally controlled by the engine control unit 3, and on-board computer 4. Energy for the electrical components is provided by the battery 2. Torque and power output are transferred to the wheels 5.

Fuel blends were realized in volumetric proportions and due to their different chemical compositions there were some challenges with stabilizing procedure. Due to this problem there was used a LAR19 method for realizing the miscibility.

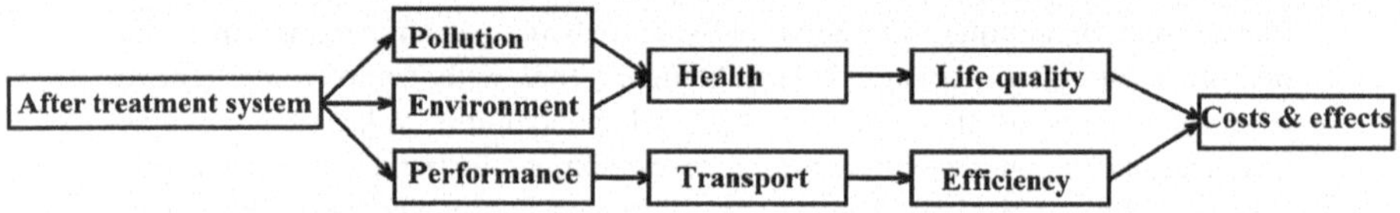

Fig. 1. Link between the after treatment system and other values (health, life quality, and costs).

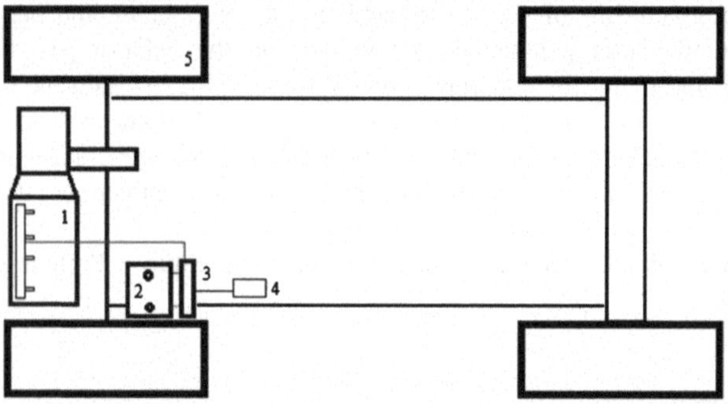

Fig. 2. Structural assembly of the power train used in applied research as a base for experiment.

3 Results of Research

During the practical experimentation and trend line analyze an important set of results was reached and represented concerning particle soot impact (see Fig. 3a), particulate filter differential pressure (see Fig. 3b), exhaust gas temperature upstream of particulate filter (see Fig. 3c), voltage on the exhaust gas differential pressure sensor (see Fig. 3d). It has been outlined the influence of positive impact in diminishing the soot amount when regenerating the particulate filter.

Regenerative chemical mechanisms between some pollutants and the gaseous nitrous oxides take the form as the following equation:

$$PM_{H_mC_n} + G_{NO_x} = G_{N_2} + G_{CO_2} + (L + G)_{H_2O} \tag{1}$$

where: PM_{HmCn} are particulate mater retained in the DPF on the bases of hydrocarbons; G_{NOx} – nitric oxide in the gas state; G_{N2} – nitrogen in the gas state; G_{CO2} – carbon dioxide in the gas state; $L+G_{H2O}$ – water in the liquid and gas phase.

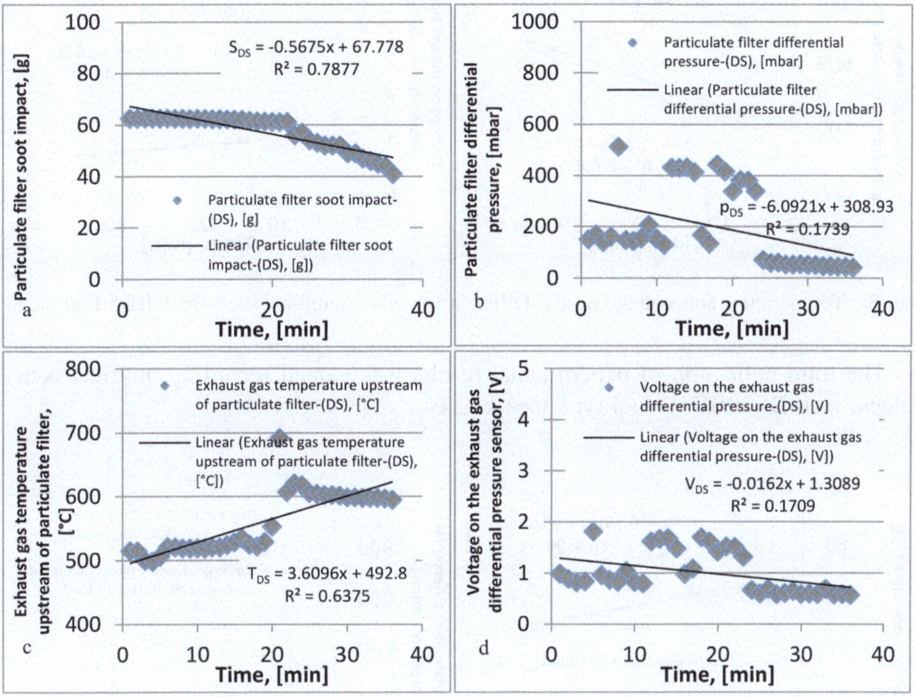

Fig. 3. Actual values measured during experimental DPF testing with Diesel Standard fuel.

The second collection of experimental results was gained regarding physical actual values, with DS95BD5 fuel type (see Fig. 4).

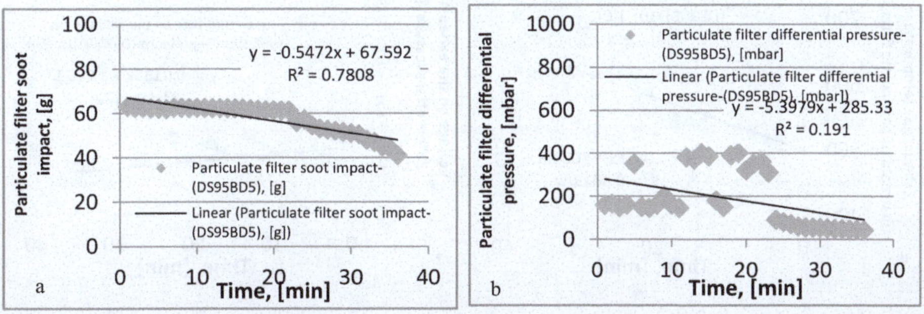

Fig. 4. Real values measured during practical DPF testing with specific DS95BD5 fuel blend.

During testing of the DPF, with DS95BD5 fuel type, were also determined exhaust gas temperature upstream of the particulate filter and the voltage on the exhaust gas differential pressure (see Fig. 5).

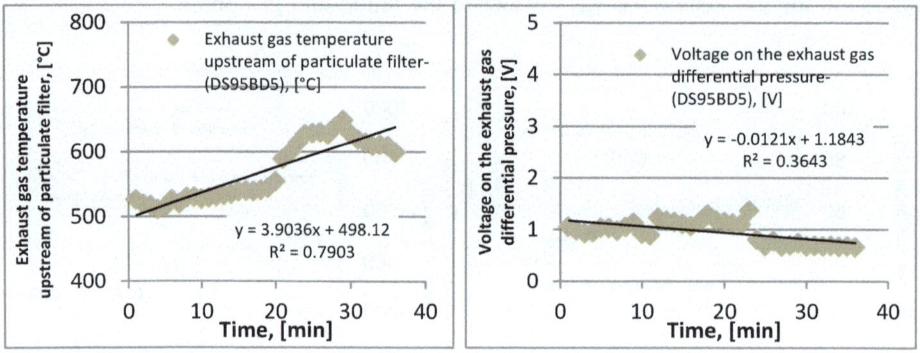

Fig. 5. Temperature and voltage on the DPF sensors when supplied with DS95BD5 fuel blend.

The third collection of experimental results was gained recording physical actual values, with DS90BD10 fuel type (see Fig. 6).

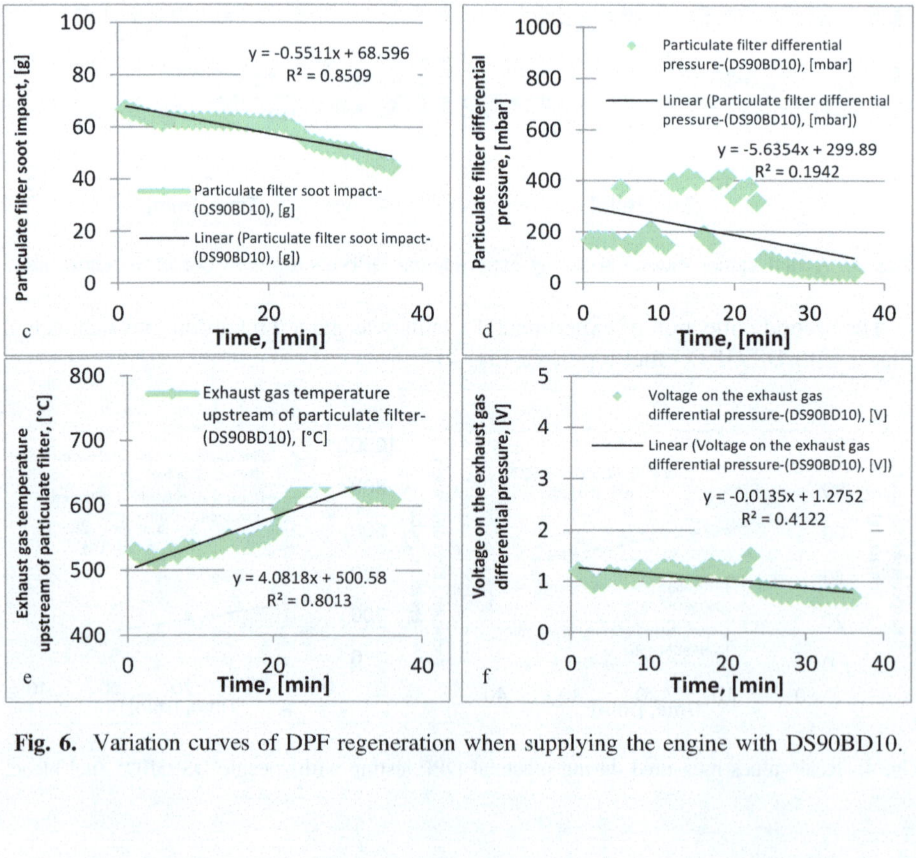

Fig. 6. Variation curves of DPF regeneration when supplying the engine with DS90BD10.

The fourth set of collected results during experimental testing was gained by recording the actual values, with DS85BD10BE5 fuel blend (see Fig. 7).

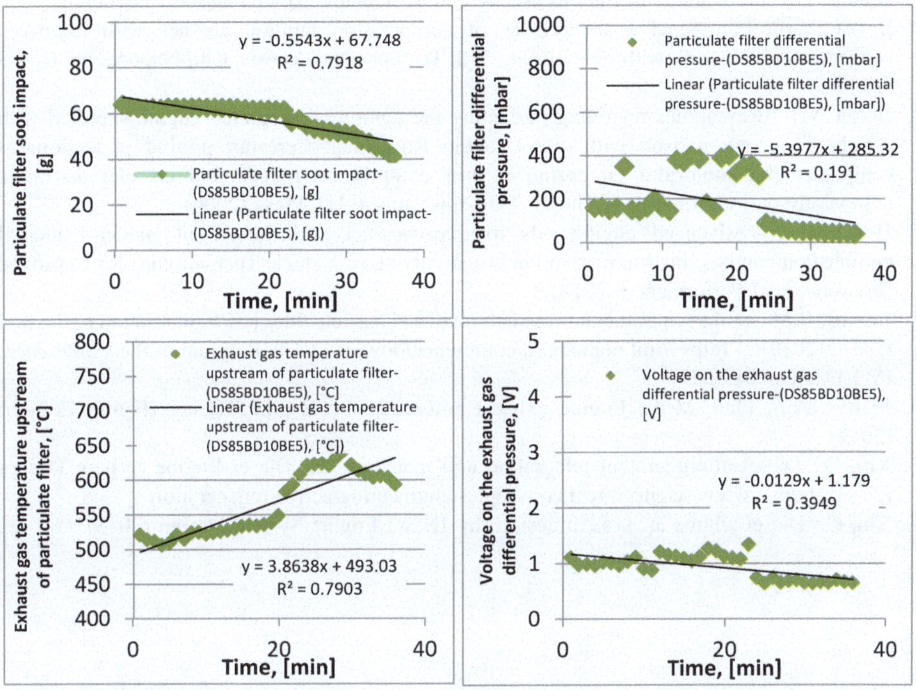

Fig. 7. The fourth set of measurements with DS85BD10BE5 fuel blend on K9K792 engine.

The K9K792 powertrain under testing was adjusted to be operated with four types of Diesel fuels. The first test was conducted on the engine by supplying the common rail injection system with Diesel Standard fuel, and the following three test types were made by changing the reference fuel with three fuel blends, by volumetric mixing with Bio Diesel and Bio Ethanol. Being an exploratory study near the reference case values all the trend lines for particulate filter soot impact were studied as linear equations. For each particular case was recorded the specific equation and also the regression squared value.

4 Conclusion

The overall conclusion of the developed research consists in the idea of beneficial effect of improving the fuel recipe with add on substances and mixing methods in order to create a better combustion reaction and acceptable time for regeneration. These measurements were made on the vehicle in regeneration procedure and soot impact regression.

References

1. Bennett, J.: Application of fuel additives for maintenance of fuel efficiency in modern vehicles. In: Innovations in Fuel Economy and Sustainable Road Transport (2011)
2. Borza, V.E.: Functional characteristics of compression ignition engines with improved equipment which work with rape oil. In: Acta Technica Napocensis, Cluj-Napoca, pp. 67–73 (2009)
3. Borza, V.E.: Researches regarding perfecting the compression ignited engine supplied with canola oil in comparison with diesel fuel/in Romanian: Cercetări privind perfecţionarea echipamentelor motorului cu aprindere prin comprimare funcţionând cu ulei de rapiţă comparativ cu motorină, PhD Thesis, TU Cluj-Napoca, UT Press (2006)
4. Howard, K.: Advanced engine oils to improve the performance of modern internal combustion engines. In: Alternative Fuels and Advanced Vehicle Technologies for Improved Environmental Performance (2014)
5. Jurchis, B.M., et al.: Experimental research on diesel particle filter (DPF) in relation to the fuel type (2019). https://imt.uoradea.ro/conference/download/The.Program.of.the.Conference.IMT.Oradea.2019.pdf
6. Martyr, A.J., Plint, M.A.: Engine exhaust emissions. In: Engine Testing (Fourth Edition) (2012)
7. Xin, Q.: Diesel aftertreatment integration and matching. In: Diesel Engine System Design (2013). https://www.sciencedirect.com/topics/engineering/active-regeneration
8. Xin, Q.: Diesel engine air system design. In: Diesel Engine System Design (2013)

Comparative Analysis of Three Distinct Porsche Models Concerning N.V.H. Aspects

Doru-Laurean Băldean(✉)

Technical University of Cluj-Napoca, 400641 Cluj-Napoca, Romania
doru.baldean@auto.utcluj.ro

Abstract. Operational sounds and frequent displacements are important criteria in checking the noise pollution. Progress of recent efforts in this sector increase vehicles output comfort and operational reliability, also efficiently and economy. To operate all inter-connected systems may become challenging without considering the noise, vibration and harshness (N.V.H.) criteria with specific conditions. N.V.H. criteria influence some distinction between cars in relation with the powertrain specific operation and auxiliary systems contribution. The first objective of this study consists in showing a specific solution using high technology for N.V.H. investigation with three parallel cars test. Regarding three Porsche models (Cayenne, Cayman and Boxter e) at Technical University from Cluj-Napoca with the support of Automotive Engineering and Transportation Department. Particular objectives presuppose a concise theoretical definition of approached subject matter, a description of applied method study and experimental analyze on specific protocol from Automotive Engineering and Transportation Department on all three Porsche models at T.U. Cluj-Napoca.

Keywords: Cabriolet · Noise · Vibration · Harshness

1 Introduction

Noise and vibration are considered when controlling specifically some of the car's physical emissions. Modern acquirements on this level allow the researchers to perform in their quest.

Estimations and statistical analyze [1], investigation of noise, vibrations and comfort performances in applied conditions [2, 3], allowed to further some ideas and to create new opportunities for testing and development of experimental approach.

2 Methodology and Materials

By using photometer and sensitive microphone there are measured the sound levels emitted in low speed range. Also, was determined the range of vibrations on a specialized test bench. Harshness level thus may be appreciated leading to the comfort level assessment. The experimental research outlines a laboratory testing procedure for noise and vibration investigation. Through evaluation of N.V.H. criteria in specific conditions for three Porsche models. The three tested cars were fully registered series

I. Dumitru et al. (Eds.): SMAT 2019, *The 30th SIAR International Congress of Automotive and Transport Engineering*, pp. 209–213, 2020.
https://doi.org/10.1007/978-3-030-32564-0_25

models, which were technically investigated and evaluated in regard to sound level, mass vibrations and assessed in relation to harshness coefficient (considering sound vibration and accessibility). By specialized approach of experimental testing there were spotted out some interesting subjects for future study. The crossing of software applications' analyzing power, on one hand, and N.V.H. aspects, on the other hand, lead to innovative possibilities and creativeness in automotive engineering field.

Experimental specs are researched using applied method and recent technologies. Research efforts and practical symbols/signs for the processing assisted evaluation are: $T_{ambient} = 0° \div 37$ °C; $T = -15° \div 40$ °C; relative humidity (RH) = 5% ÷ 96%; elevation E = 0 ÷ 2500 m. Research paper shows an advanced engineering protocol based on the three vehicles considered (Fig. 2) for investigation in N.V.H. actual values regarding the output of three Porsche cars in operation.

Materials consist in Porsche e-Boxter 2016, Cayman 2016 and Cayenne 2017. They have devices and equipment used for electronic investigation. Table 1 makes a short presentation of specific data for all three cars (see Fig. 1).

Table 1. Materials used in research.

Index	Vehicle 1	Vehicle 2	Vehicle 3
Developer	Porsche	Porsche	Porsche
Body type	Cayman	Cayenne	e-Boxter
Serial	982	3.0V6 250 PS	EV
Energy	Gasoline	Diesel	Electric
Standard	E 6	E 6	ZEV

Fig. 1. Three Porsche models on which were measured NVH parameters.

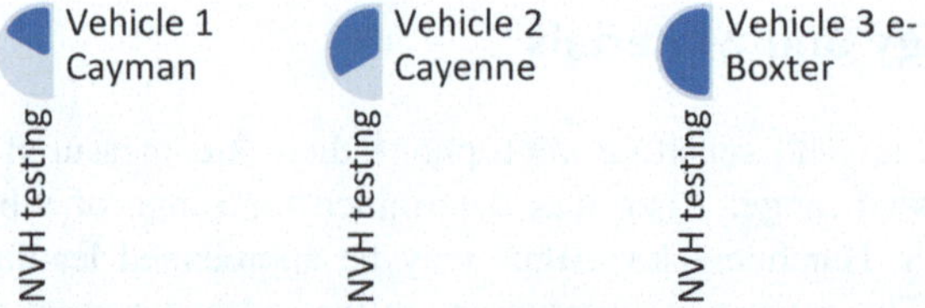

Fig. 2. Study methodology applied in practical tests

Acquisition was made using TS-E 50 mm focal f/2.8L MACRO, with ±8.5° Tilt and ±12 mm and Macro ratio 1:2. Equipment has a frequency that ranges in 1 to 30 Hz, and a displacement of 1 to 10 mm.

Calculating the acoustic pressure and pressure level of noise (P_{LN}), may be done with the following relation [3]:

$$P_{LN} = 20\,\lg\frac{P}{M_{ap}}\,[\mathrm{Np}]; \tag{1}$$

where: M_{ap} is minimum acoustic level of pressure which is corelated with standard value, that has the numerical level of $0.2.10^{-6}$ N/m^2; P - value of actual pressure, expressed in N/m^2.

3 Applied Results

Only some researchers have made a thorough dynamic investigation and analyze of the model, which is useful in sound evaluation in low toward the mid levels of frequency, for multiple vehicles such as considered in this study. All three models and applied devices were monitored through wireless data transfer capability. The vehicles were brought in laboratory N.V.H. test rig in anechoic room, assessing thus components displacements in order to compare their measurements in 1 ÷ 30 Hz. Experimental research results were gained through practical tests in laboratory. Some differences between the three sport vehicles are shown and discussed. After investigating the values and validating some experimental determinations in order to complete the practical procedures the values were analyzed and represented. The noise level was compared. The three Porsche cars have been compared (Fig. 3).

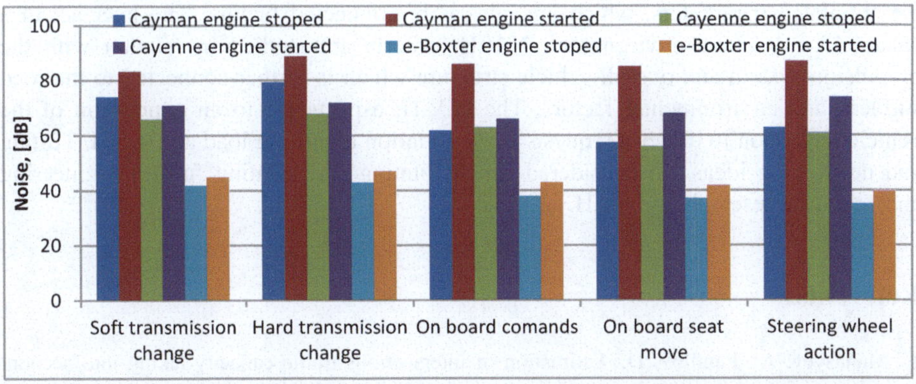

Fig. 3. Noise level measured with all of the three vehicles with electronic assisted noise investigation regarding multiple actions inside the passenger compartment, mutually analyzed according to ISO 17.140 - Sounds and acoustic measurements in vehicle dynamic conditions

There was observed that the Diesel car had lower noise value (Fig. 4) compared with the gasoline supplied vehicle, but not less than the electric powertrain, without A-weighting in order not to gain different values on multiple instruments when some specific ultrasonic source of noise is interfering.

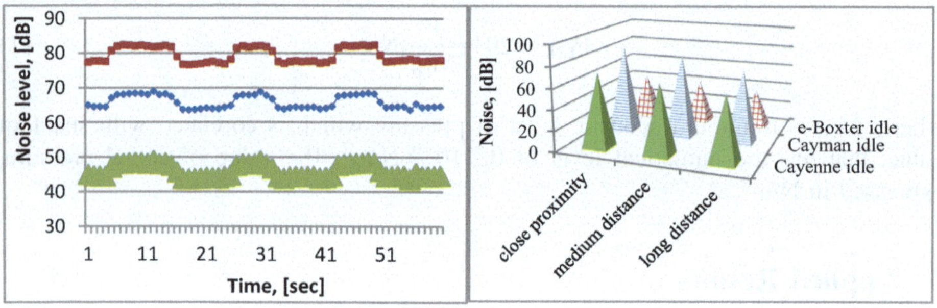

Fig. 4. Noise real values determined on the three Porsche vehicles in various distances

The N.V.H. aspects investigation on the three car models (Cayenne, Cayman and Boxter e) from Porsche, in low speed operating conditions represents a new comparative analyze and an engineering summary regarding capabilities, technologies and expertize.

4 Conclusions

Investigating all three Porsche cars, in regard to the vibration and noise aspects, as well as the appreciation of harshness level, led to the definition of specialized research regarding N.V.H. aspects. The three cars are overall tested and analyzed in regard with the stipulated parameters, which are intricately connected factors. The later act in a cluster like dynamic environment. N.V.H. criteria are studied in relation with the propulsion system and overall vehicle structure which is further connected to the road surfaces and environmental factors. The N.V.H. aspects led to an expression of the vehicle's reaction to driver's requests in exploitation regarding load and speed. Testing and developing ideas are considered for facilitating and creating further engineering and scientific research of N.V.H. aspects.

References

1. Afanasyev, A., Panfilov, D.: Estimation of intersections traffic capacity taking into account changed traffic intensity. Transp. Res. Procedia **20**, 2–7 (2017)
2. Băldean, D.-L., Varga, B.-O., Moldovanu, D., Kocsis, L., Borzan, A.: Contributions in developing advanced engineering methods for researching the N.V.H. aspects in Porsche Cayman operation conditions at Technical University from Cluj-Napoca. In: Proceedings of the 11th Edition of the International Congress of Automotive and Transport engineering

Mobility Engineering and Environment, pp. 13–20. C.A.R., Piteşti, România (2017). https://doi.org/10.26825/bup.ar.2018.002

3. Băldean, D.-L., et al: Contributions in further development of advanced engineering solutions regarding the N.V.H. aspects in Porsche Cayenne vs. Cayman testing conditions at Technical University from Cluj-Napoca. In: Proceedings of the 4th Edition of the International Congress of Automotive and Transport Engineering, AMMA 2018, pp. 13–20. AMMA, Cluj-Napoca, România (2018)

Mathematical Model for Study of Behavior of the Bodywork at Impact

Ana Maria Manea(✉), Daniel Iozsa, Cornelia Stan,
Claudiu Teodorescu, and Gheorghe Frățilă

Politehnica University of Bucharest, Splaiul Independentei nr. 313, Sector 6,
060042 Bucharest, Romania
mannea_annamaria@yahoo.com, daniel_iozsa@yahoo.com,
corraista@yahoo.com, ghe_fratila@yahoo.com

Abstract. With the advances in computer hardware and software, designing programs have been developed, providing a variety of tools for designing modern structures. Each system used in designing structures is based on the fundamental notions of dynamics. Thus, with the help of the programs, many models can be analyzed, ranging from simple to complex models (simple spring-mass models, beam element models and finite element models). To analyze the behavior of the body structure to the frontal impact, a simple mathematical spring-mass model is proposed. The structural elements of the body will be approximated by concentrated mass points, corresponding to the subsystems, connected to each other by elastic elements (springs). The impact energy will be dissipated by the bonds between the elastic members and the mass points. The simple spring-mass model proved to be very useful in developing vehicle structures in the case of an impact. The method provides an easy way to study the vehicle's impact behavior and provide directional guidance on sensitive areas.

Keywords: Bodywork · Spring-mass model · Impact energy

1 Introduction

The fundamental notions of dynamics underlie the impact of two bodies. For theoretical study of the collisions dynamics between two vehicles or between a vehicle and a rigid body reference is made to the fundamental notions of dynamics. Dynamics is part of mechanics that studies the mechanics of bodies when forces act on them [1–3].

Vehicle structures are rated according to how they offer the occupants the best protection during a collision. The structure of the passenger compartment is made up of a closed frame assembly, forming a "safety cell" capable of providing the minimum survival space in the event of an impact. The structures surrounding the passenger compartment are constructed so that they can, by deformation, take up much of the impact energy. Ensures central part integrity and short-time reduction of passenger decelerations (controlled deformation structures) [4, 5].

I. Dumitru et al. (Eds.): SMAT 2019, *The 30th SIAR International Congress of Automotive and Transport Engineering*, pp. 214–220, 2020.
https://doi.org/10.1007/978-3-030-32564-0_26

2 Mathematical Model for Frontal Impact Study

2.1 Presentation of the Simplified One-Dimensional Calculation Model

The deceleration value is proportional to the deformation force generated during the impact. To determine these sizes experimental tests can be performed or numerical models can be developed. Numerical models may be simple or complex. The simple ones provide global information about the structure and kinematic parameters, and by studying a complex model a more precise detailing of the components of the vehicle is made [4, 6].

The one-dimensional computational model is a simplified model, with only one degree of freedom, the vehicle being considered as a unitary mass that strikes a fixed rigid barrier (see Fig. 1).

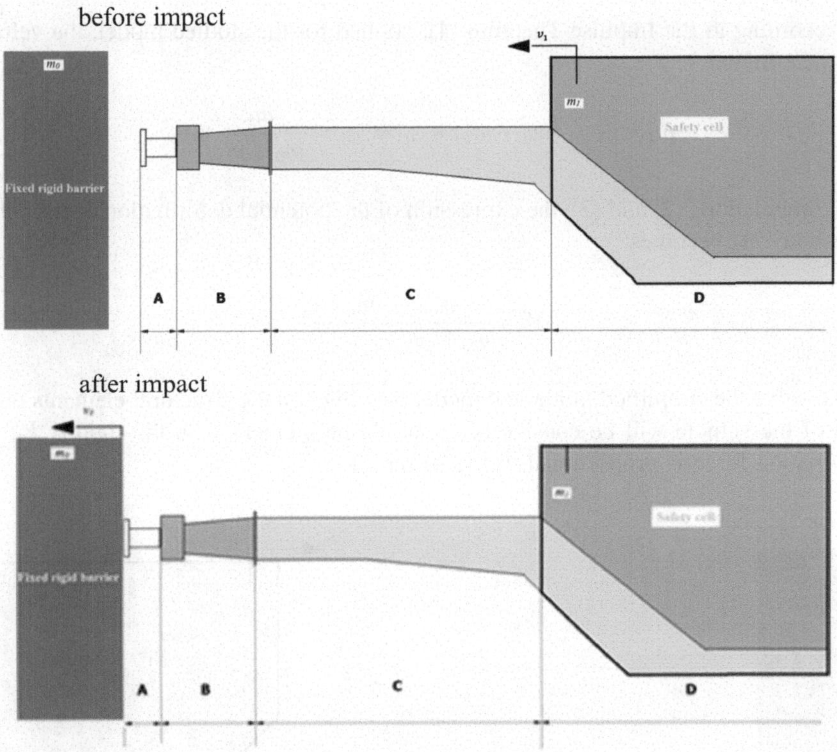

Fig. 1. One-dimensional model for impact study (m_1 - vehicle mass; v_1 - initial speed before impact; v_0 - final speed after impact; m_0 - fixed rigid barrier mass).

During impact, the energies of the bodies involved undergo a series of transformations.

A part of the kinetic energy of the body m_1 in the initial state, before the impact, is lost by deformation, resulting in the potential deformation energy.

The energy balance of the system is:

$$E_{c1i} + E_{c0i} = E_{c1f} + E_{c0f} + V \tag{1}$$

Where: E_{c1i}- is the initial kinetic energy (before impact) of the mass body m_1; $E_{c0i} = 0$ - is the initial kinetic energy (before impact) of the mass body m_0; E_{c1f} - is the ultimate kinetic energy (after impact) of the mass body m_1; E_{c0f} - is the ultimate kinetic energy (after impact) of the mass body m_0; V- is the potential deformation energy of the mass body m_1.

Replacing the expressions of the kinetic energies of the bodies, the relation (1), can determine the energy lost by the deformation of the mass body m_1:

$$V = \frac{m_1 * v_1^2}{2} - \frac{(m_1 + m_0) * v_0^2}{2} \tag{2}$$

According to the Impulse Theorem [4], applied for the studied model, the velocity v_0 is determined:

$$m_1 * v_1 = (m_0 + m_1) * v_0 \Rightarrow v_0 = \frac{m_1}{m_0 + m_1} * v_1 \tag{3}$$

From relations (2) and (3), the expression of the potential deformation energy of the mass body m_1 becomes:

$$V = \frac{m_0}{m_0 + m_1} * \frac{m_1 * v_1^2}{2} \tag{4}$$

To solve the simplified numerical model (see Fig. 2), the structural elements on the front of the vehicle will be considered as an elastic element with the rigidity k_1, (the deformation force is proportional to the stiffness).

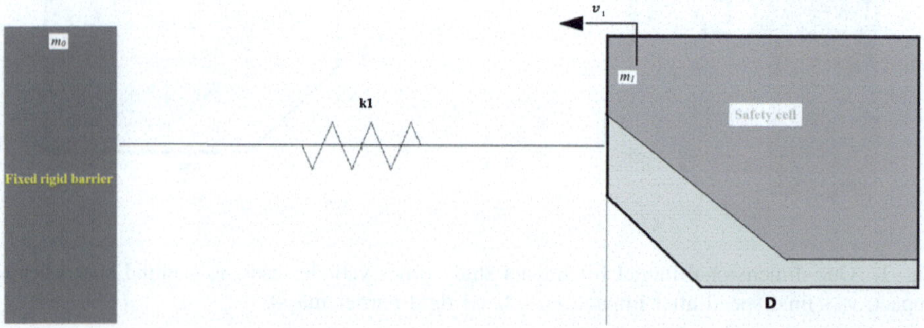

Fig. 2. One-dimensional model with stiffness k_1 for impact study with a fixed rigid barrier

The expression of the deformation energy of the mass body m_1 is:

$$V = \frac{k_1 * x_{1max}^2}{2} \tag{5}$$

Where: k_1 - the elastic constant; x_{1max} - deformation of the body.
By replacing relations (4), (5), it follows:

$$\frac{k_1 * x_{1max}^2}{2} = \frac{m_0}{m_0 + m_1} * \frac{m_1 * v_1^2}{2} \tag{6}$$

From the relation (6) the maximum deformation for the front of the vehicle is determined:

$$x_{1max}^2 = \frac{m_1}{k_1} * v_1^2 * \frac{m_0}{m_0 + m_1} \tag{7}$$

In this test, the fixed rigid barrier will be considered as a concrete wall. The mass m_0 of the fixed rigid barrier is at least 100 times bigger than the mass m_1 of the vehicle, so the ratio $\frac{m_0}{m_0 + m_1} \approx 1$.

Depending on the architecture of the front compartment, during a frontal impact with a barrier, at a speed of approximately 50 km/h, it must record a deformation of 400-700 mm and the deformation of the passenger compartment is limited [7].

2.2 Solving the Simplified Calculation Model

By solving the model, it is desired to determine the total deformation of the front compartment recorded during a frontal impact with a rigid barrier at different speeds from 4 km/h to 50 km/h.

The numerical solving of the model is done by creating block diagram in the Matlab Simulink program (see Fig. 3). To solve the model numerically the following parameters are known: the mass of the vehicle m_1 = 1168 kg; stiffness of the front compartment k_1 = 800000 N/m [4, 6]; vehicle initial speeds before impact $V_{initial}$ = [4, 15, 25, 50] km/h.

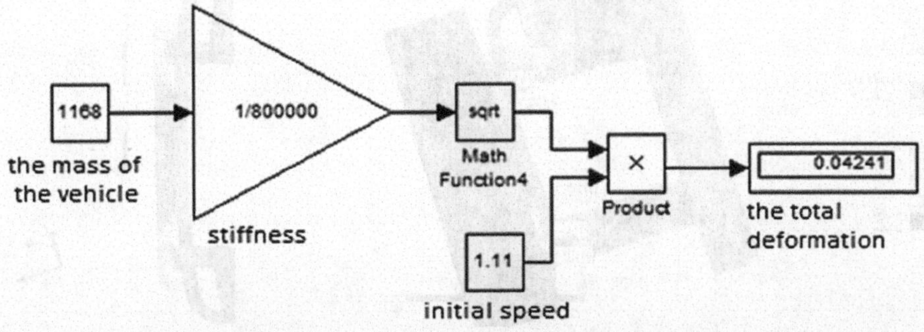

Fig. 3. Block diagram for Mathematical Modeling (Matlab Simulink)

The results obtained are presented in Table 1.

Table 1. The total deformations obtained in the program Matlab Simulink

Test	Speed [km/h]	Total deformation [mm]
1	4	42
2	15	159
3	25	265
4	50	531

Further, to present the behavior and deformation of the compartment structure during a frontal impact, a simplified 3D structure was developed in the Ansys program. The analysis is done by the finite element method of the Explicit Dynamics module.

For the simulation of the impact, the same values are used for the mass of the vehicle m_1 = 1168 kg and vehicle initial speeds before impact $V_{initial}$ = [4, 15, 25, 50]km/h.

Figures 4, 5 and 6 show the results obtained in the Ansys program at the end of the running time. The time end was initially selected to reduce simulation time.

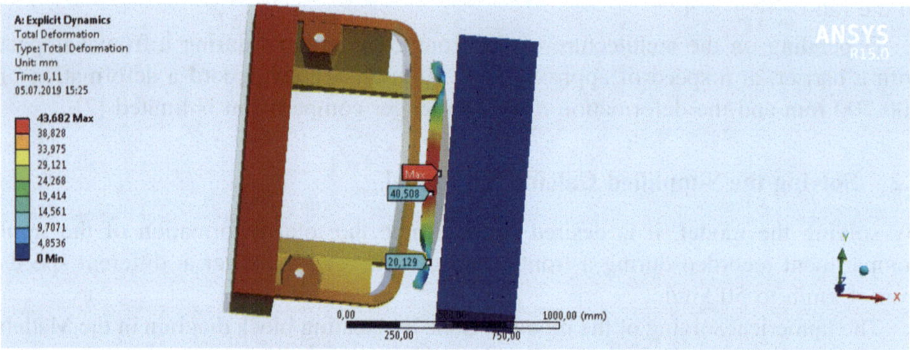

Fig. 4. Total deformation, $V_{initial}$ = 15 km/h (Ansys)

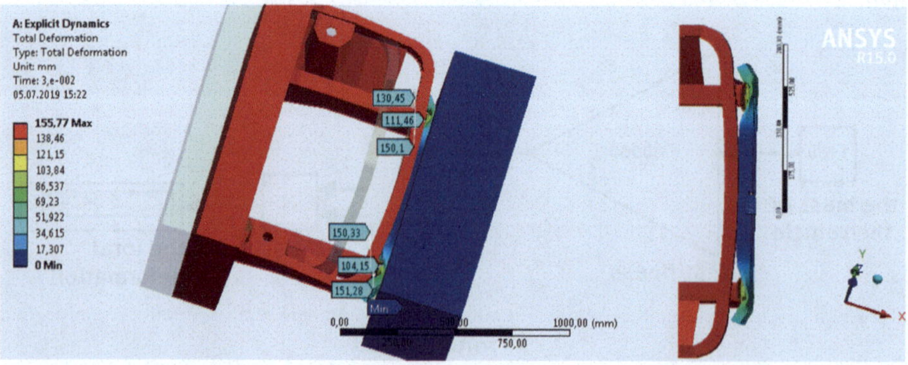

Fig. 5. Total deformation, $V_{initial}$ = 25 km/h (Ansys)

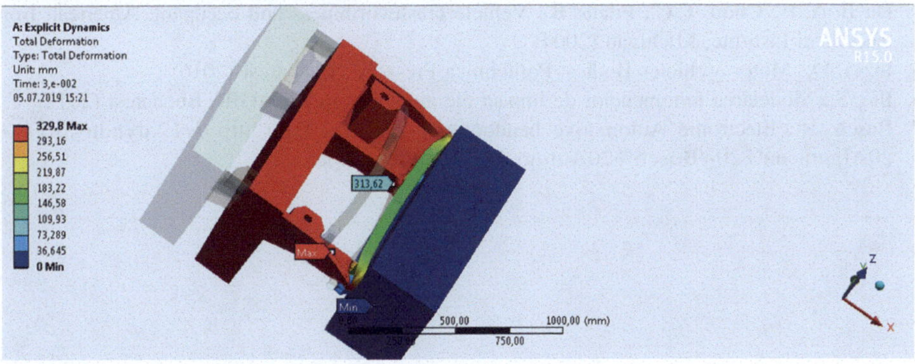

Fig. 6. Total deformation, $V_{initial}$ = 50 km/h (Ansys)

3 Conclusions

The results obtained are presented in parallel in Table 2. It can be noticed that the mathematical computation method obtained higher values of the total deformations compared to those obtained by the impact test of the structure in the Ansys program.

Table 2. The total deformations obtained by the two methods (Matlab Simulink and Ansys)

Test	Speed [km/h]	Total deformation (Matlab Simulink) [mm]	Total deformation (Ansys) [mm]
1	4	42	–
2	15	159	44
3	25	265	156
4	50	531	329

For mathematical models, a constant value was chosen for the rigidity of the front compartment structure.

The controlled and progressive deformation capability of the structure was accomplished with the constructive evolution of the bodywork (by optimizing materials and architecture, layout of deformation initiators).

During the impact, the rigidity of the body structure increases with the evolution of the deformation

In conclusion, the maximum deformation of the structure of the model does not exceed the maximum limit of 700 mm imposed in accordance with [7].

References

1. Radeş, M.: Vibraţii mecanice. Printech, Bucureşti (2008)
2. Atanasiu, M.: Mecanica. Editura Didactică şi Pedagogică, Bucureşti (1973)
3. Alecu, A.: Curs Mecanica. Politehnica din Bucureşti, Bucureşti (2014)

4. Du Bois, P., Chou, C.C., Fileta, B.: Vehicle crashworthiness and occupant. American Iron And Steel Institute, Michigan (2004)
5. Iozsa, D.: Motor Vehicles Bodies. Politehnica Press Ed, Bucharest (2016)
6. Ilie, S.: Modelarea fenomenelor de impact ale autovehiculelor. AGIR, Bucharest (2012)
7. Bosch, R.: Electronic Automotive handbook, p. 1391 (2002). http://v12.dyndns.org/-%20Algemeen%20-/Bosch%20Automotive-Handbook.pdf

Aspects Regarding the Identification of Optimum Driver Comfort Level by Virtual Analysis of the Vertical Oscillations Generated by Road

Marian Florin Mitroi[1] and Anghel Chiru[2(✉)]

[1] Ovidius University, Constanta Bd.Mamaia 124, Constanta, Romania
ejju_marian@yahoo.com
[2] Transilvania University of Brasov, Bd.Eroilor 29, Brasov, Romania
achiru@unitbv.ro

Abstract. The oscillations generated by road are a risk factor for the driver's body, but also discomfort at the same time. Identifying a optimal solutions to reduce them is always a concern for engineers, ergonomists, etc. The main objective of the research is to identify the optimal working level of the driver's seat suspension when driving the vehicle on difficult roads. The research is done by virtually modeling the behavior of the seat suspension system with the AMESim program. The suspension system designed for the seat is analyzed using the technology and behavior of the magnetoreological damping systems. As a result, the optimum working parameters of the suspension are determined so that the driver's comfort factor, when driving the vehicle on the difficult roads, is maintained at a level recommended by international standards. The use of virtual analyzes creates the premises for obtaining performance indices of the suspension system (seat). Through virtual modeling, we can observed the consistency of the theoretical analyzes with the practical applications. This provides an adequate design margin for the final system.

Keywords: Vertical oscillations · Seat suspension · Virtual modeling · Comfort

1 Introduction

The driver's comfort is generally a subjective appreciation of some psychic experiences developed during the vehicle's journey. Permanently optimization of its level of comfort is a major activity for ergonomic engineers, the automotive industry and not only.

The research provides a systematic analysis of the possibilities of lifting the driver's comfort level by virtual modeling of the seat suspension in AMESim. In research, the suspension of the magnetoreological chair and the Skyhook control system are adopted.

I. Dumitru et al. (Eds.): SMAT 2019, *The 30th SIAR International Congress of Automotive and Transport Engineering*, pp. 221–230, 2020.
https://doi.org/10.1007/978-3-030-32564-0_27

2 The Conditions for the Virtual Modeling Road-Vehicle-Driver

Road is the main factor of discomfort while traveling with the vehicle. Its physical characteristics are varied so that the level of bumps and the frequency at which oscillations are produced are specific to each type of road. Depending on the spectral density (PSD) of vertical accelerations and the provisions of BS 8608, for a slope index w = 2, a classification of the road types can be made (Fig. 1). The vertical acceleration, that have random type, has an index of w > 3 for uneven roads.

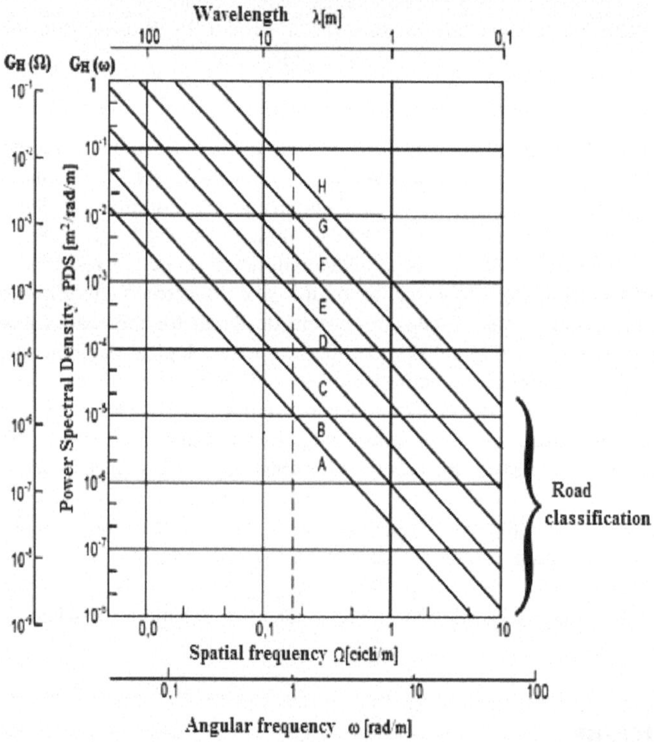

Fig. 1. Road classification and the power spectral density [1]

Under these circumstances, the driver's discomfort is high. VDI 2057 specifies the value of comfort factor (K), depending on the level of vertical accelerations transmitted to the driver and the frequency at which they occur (Table 1).

Table 1. The comfort perception scale, according to VDI 2057 [2]

Level	The threshold of perception K	Subjective appreciation of perception
A	0,1	There is no oscillation
B	0,25	The first sensations appear
C	0,63	Minimum level of sensations
D	1,6	It feels good
E	4	He feels strong
F	10	Very strong level of perception
G	25	Very strong level of perception
H	63	Very strong level of perception

In Fig. 2 is presented the vehicle on a bumpy road and its suspension systems.

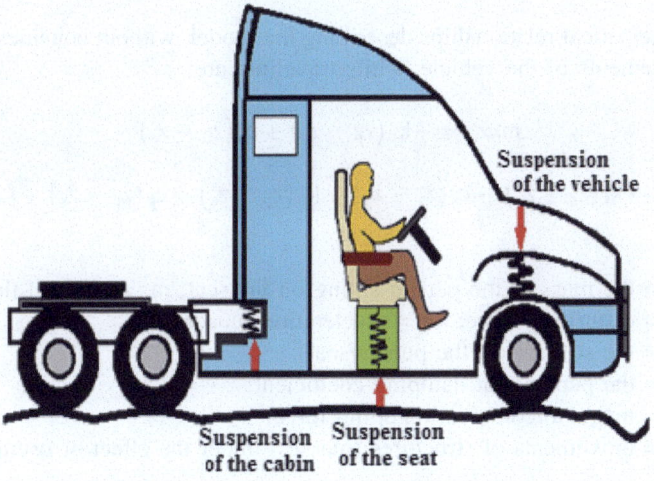

Fig. 2. The suspension systems by which the vibrations induced by the road are transmitted to the driver

In order to identify the optimal level of comfort, it is necessary to observe the behavior of the seat suspension system in relation to the road-induced misalignment. In the research, the influence of vertical random acceleration on the driver was analyzed. The simulation takes into account the driver's mass (80 kg) and the seat (25 kg). The structure of the mathematical model is presented in Fig. 3.

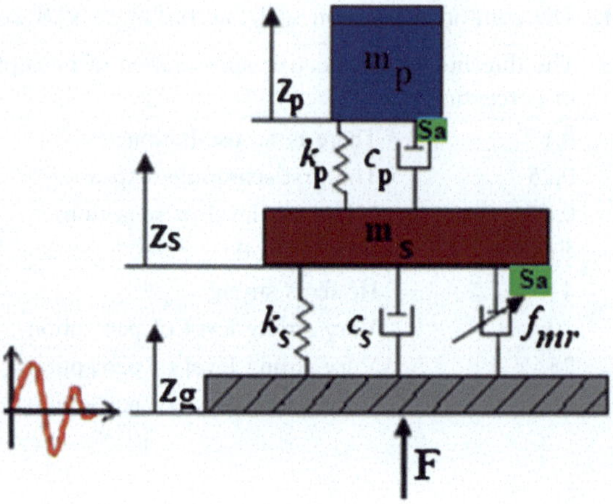

Fig. 3. The structure of the mathematical model for analysis

The mathematical relationships describing the model, without considering the pitch and roll movements of the vehicle, while traveling, are:

$$m_p\ddot{z}_p = -k_p(z_p - z_s) + c_p(\dot{z}_p - \dot{z}_s) \tag{1.1}$$

$$m_s\ddot{z}_s = -k_s(z_s - z_g) - c_s(\dot{z}_s - \dot{z}_g) + k_p(z_p - z_s) + c_p(\dot{z}_p - \dot{z}_s) + f_{mr} \tag{1.2}$$

where:

m_p = the mass of the person sitting on the seat; m_s = mass of the seat;
F = disturbing force; Sa = acceleration sensors;
kp, ks = the stiffness of the person/seat
c_p c_s = the person/seat damping coefficient;
f_{mr} = magnetoreological damping force;
ZpZs, Zg = movements of structures (masses) under the effect of oscillations.

By correlating the working mode of the SkyHook control system and that of the magnetorheological damper, the operating conditions of the controller (I_{max} and I_{min}) can be determined. So:

$$I = \begin{cases} I_{min}, \dot{Z}\cdot(\dot{Z} - \dot{Z}_0) < 0 \\ I_{max}, \dot{Z}\cdot(\dot{Z} - \dot{Z}_0) \geq 0 \end{cases}, \tag{1.3}$$

where:

I_{min} = the supply current of the damper coil;
I_{max} = maximum supply current of the damper coil.

3 Magneto-Reological Suspension for the Driver's Seat

Magnetoreological shock absorbers are performance components used for shock and vibration damping systems. In the research, the classic suspension system was replaced by a magnetoreological one, modeled in AMESim (Fig. 4) [3].

Adoption of this system has been done to obtain an adjustable level of damping factor, depending on the road characteristics. In the Skyhook control strategy a C_{Sky}. med intermediate amortization factor was adopted to provide a mean level of damping (2.1).

$$c_{sky} = \text{med } x > 0 = \; > c_{s.min} < c_{s.med} < c_{s.max} = I_{med}, \tag{2.1}$$

where:

C_{sky} = damping coefficient [Ns/m];

I = the intensity of the supply current of the damper coil [A].

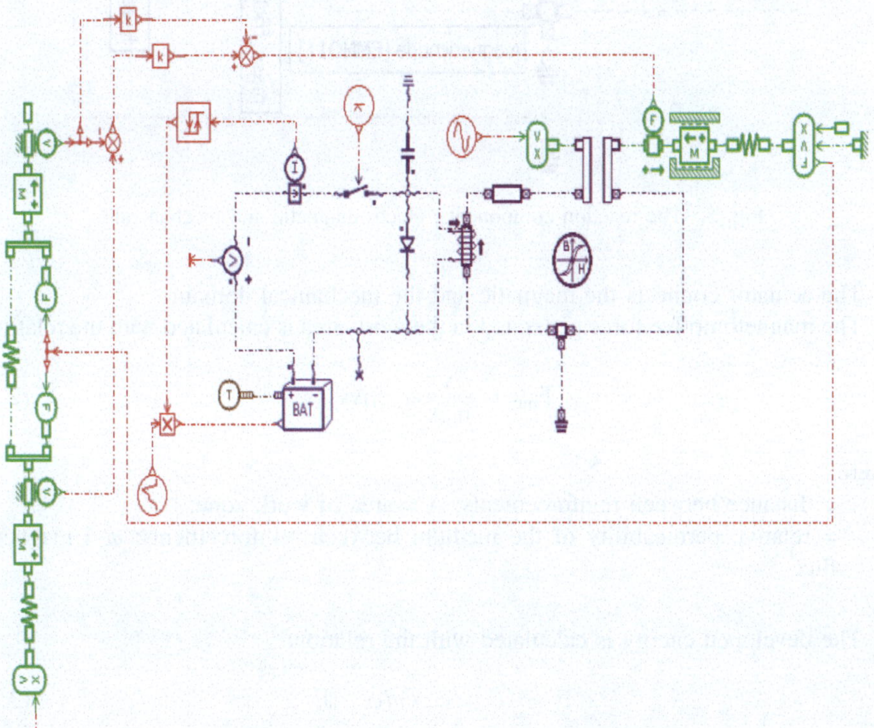

Fig. 4. The virtual model of the magnetoreological shock absorber

The main functional elements of the model are presented in Figs. 5 and 6.

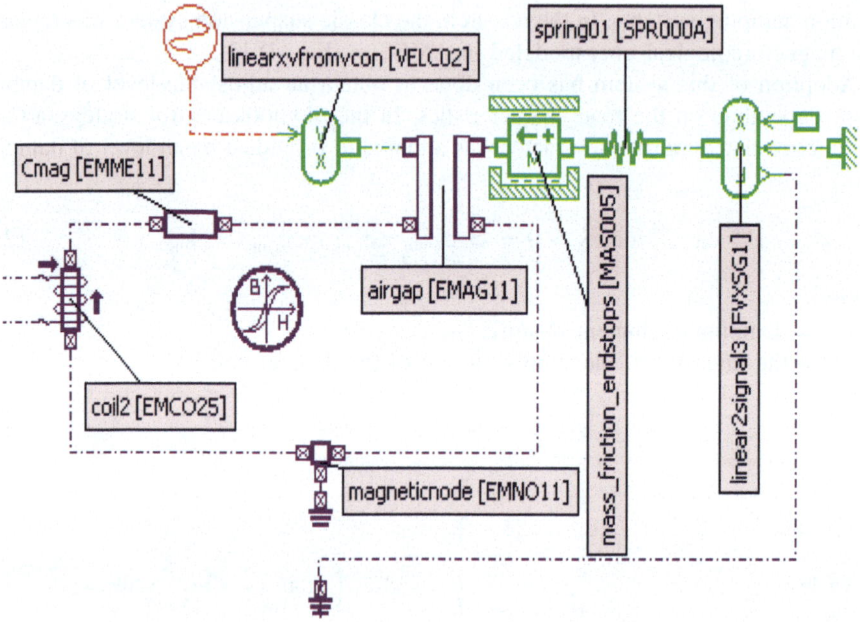

Fig. 5. The reaction components: electromagnetic and mechanical

The actuator connects the magnetic and the mechanical domain.
The magnetomotive force generated in the work area is calculated with the relation:

$$F_{mm} = \frac{x}{\mu_0 A}\varphi \quad [\mathrm{As}], \tag{2.2}$$

where:

x = distance between reinforcements; A = area of work zone;

μ_0 = relative permeability of the medium between reinforcements; φ = magnetic flux.

The developed energy is calculated with the relation:

$$E = \int F_{mm}(\varphi, x)d\varphi \quad [\mathrm{J}], \tag{2.3}$$

$$E = \frac{x\varphi^2}{2\mu_0 A} \quad [\mathrm{J}]. \tag{2.4}$$

The force of the actuator is given by:

$$F = \frac{\varphi^2}{2\mu_0} A \quad [N] \tag{2.5}$$

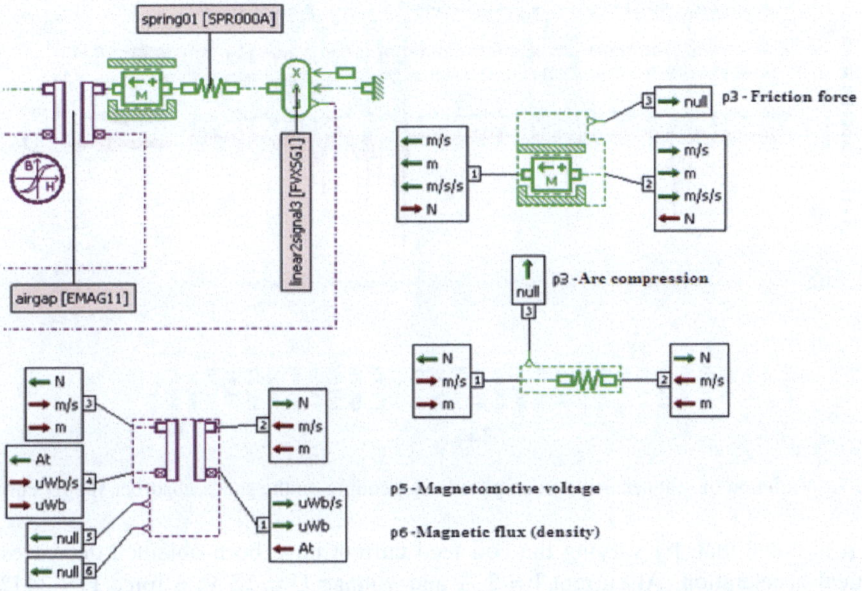

Fig. 6. Reaction components and physical actions induced by ports

The feeding of the coil with different currents allowed the determination of the characteristic measures of work (Fmm, E, F) of the system. In this way, the concordance between the actual operating parameters of the damper and those identified by the simulation was observed.

4 Optimizing the Level of Comfort

The possibilities of adjusting the damping factor of the magnetoreological suspension system create an adaptable level of driver comfort. By modifying the supply current of the shock absorber coil, various resistive forces are obtained for the shock absorber. Analyzes were made by observing the results and comparing them with the classic suspension system.

The signal entered into the system for analysis was random, so that road misalignments could be simulated correctly. The vertical acceleration value was set as $a_z = 0.99\ m/s^2$. The discomfort limit, according to ISO 2631 and BS 6841, occurs at $a_z = 0.8\ m/s^2$.

The simulated magnetoretical system was analyzed by feeding the bobbin with

I = 0.6; 1.2; 1.8; 2 A. Figure 7 present the effect of the supply current of damper on the oscillation of the vertical acceleration.

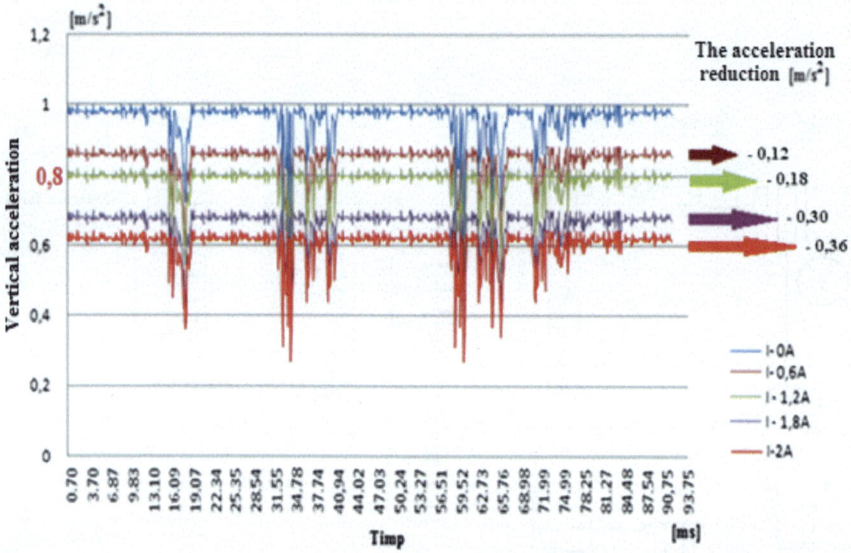

Fig. 7. Variation of seat acceleration amplitude, depending on the shock absorber supply current

It is noted that, by varying the coil feed current have been obtained decreases in vertical acceleration. At current I = 2 A and voltage U = 12 V, a force F = 2442 N was obtained and a vertical acceleration reduction of 0.36 m/s^2. In Table 2 are presented the vertical acceleration values obtained at different currents applied to the damper coil. (short time recorded).

Table 2. The values of vertical accelerations recorded at the different supply currents of shock absorber

Time [ms]	Vertical acceleration [m/s^2] Supply currents [A]			
	I = 0 A	I = 0,6 A	I = 1,2 A	I = 1,8 A
0.70	0,98	0,86	0,8	0,68
0.81	0,97	0,85	0,79	0,67
0.93	0,99	0,87	0,81	0,69
1.04	0,97	0,85	0,79	0,67
1.15	0,99	0,87	0,81	0,69
1.27	0,97	0,85	0,79	0,67
1.38	0,98	0,86	0,8	0,68
1.49	0,99	0,87	0,81	0,69
1.61	0,98	0,86	0,8	0,68
1.73	0,98	0,86	0,8	0,68

As a reference, in order to obtain a proper level of comfort, a supply current of I = 1.97 A was set. From this value, was monitored the comfort level obtained by reducing the vertical acceleration, when the dampers were fed with different currents. In Fig. 8 is observed the vertical acceleration value in comparative mode, for a damping current supply of I = 1,8; 1.9 and 2 A.

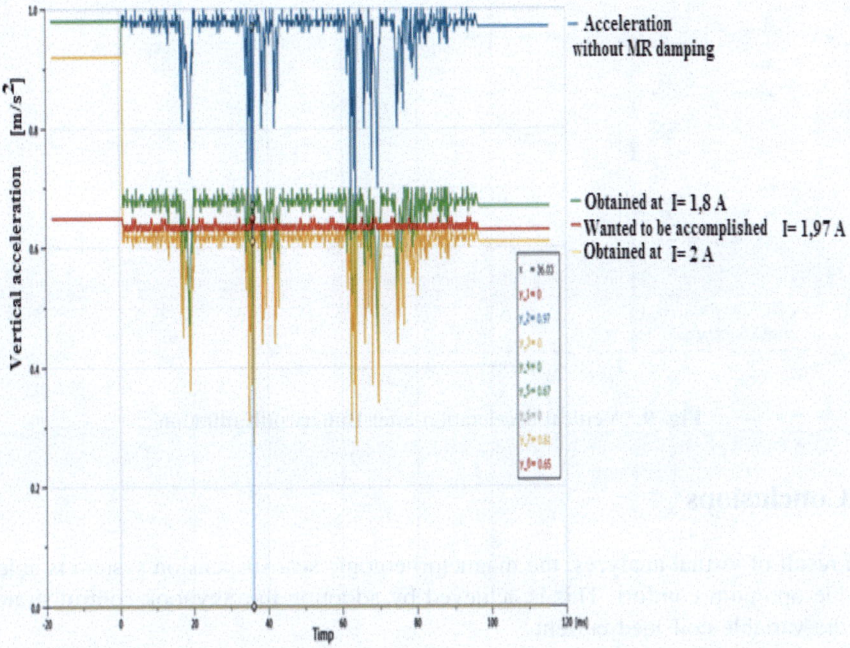

Fig. 8. Vertical acceleration comparison: desired to be achieved/achieved

By applying the Butterworth filter (ord.4), with the cutting frequency set at 5 Hz (human body resonance), the random vertical acceleration is as shown in Fig. 9. Signal processing with this filter turns random acceleration into a sinusoidal type.

The latter is easier to analyze and supporting by the body.

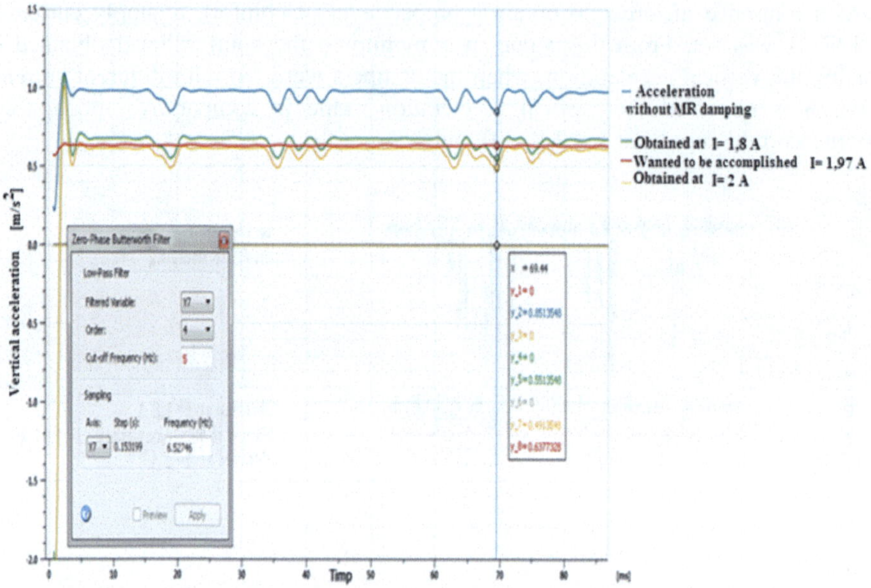

Fig. 9. Vertical acceleration after Butterworth filtration

5 Conclusions

As a result of virtual analyzes, the magnetorheologic seat suspension system is able to provide optimum comfort. This is achieved by adopting the Skyhook control strategy and the variable coil feed current.

For compliance with the recommendations of ISO 2631, BS 6841 and VDI 2057 it is considered that the optimum level of comfort can be ensured as follows:

- For $a_z < 0.6\ m/s^2 \Rightarrow I_{min} = 0.6$ A
- For $a_z < 0.8\ m/s^2 \Rightarrow I_{min} = 1.2$ A
- For $a_z < 1\ m/s^2 \Rightarrow I_{min} = 1.8$ A at the 12 V.

References

1. VDI 2057/2002: Human exposure to mechanical vibrations. Whole-body vibration
2. BS 6841: Guide to measurement and evaluation of human exposure to whole-body mechanical vibration and repeated shock
3. Mitroi, M.F.: Technical solutions for improvement the comfort of driving position of the military vehicles, Ph.d. thesis, Brasov (2019)

Research on the Use of Rheological Systems to Decrease the Vertical Oscillations Perceived by the Driver

Marian Florin Mitroi[1] and Camelia Cerbu[2(✉)]

[1] Ovidius University, Constanta Bd.Mamaia 124, Constanta, Romania
ejju_marian@yahoo.com
[2] Transilvania University of Brasov, Bd.Eroilor 29, Brasov, Romania
cerbucamelia@yahoo.com

Abstract. The Rheological systems prove good damping capabilities in different application areas. The reduction of vertical oscillations generated by rugged roads at the human body through the use of these systems is an important solution in the automotive industry. The main objectives of the research presuppose the determination of the damping forces and energies specific to the operation of the rheological system while driving the vehicle on rough roads. The attempts consist to apply different working frequencies and several supply currents to the shock absorber. The solution adopted for research uses the magnetoreological damper. The performance of the shock absorber is evaluated by cyclical laboratory tests and compared to those observed in the real environment. The expected results are related to the shock absorber's ability to reduce vertical oscillations and shocks at the driver's seat while driving on rugged roads. It is also being watched to operate in accordance with the international standards: ISO 2631-1, VDI 2057, BS 6841. Rheological systems are able to reduce shocks and vertical accelerations when applying to different source energy. In the case of motor vehicles, their use helps to increase the level of comfort of people, especially when traveling on roads with pronounced unevenness.

Keywords: Vertical oscillations · Rheological systems · Damping energies

1 Introduction

Mechanical oscillations felt by the human body while driving whit the cars, are the main factor that affects the level of comfort of personel. Traveling with vehicles on rough roads for long periods of time generates major health problems, for the human body. Reducing its level is a permanent goal for all ergonomists and car manufacturers.

I. Dumitru et al. (Eds.): SMAT 2019, *The 30th SIAR International Congress of Automotive and Transport Engineering*, pp. 231–240, 2020.
https://doi.org/10.1007/978-3-030-32564-0_28

2 Oscillations of Vertical Acceleration and Rheological Systems

2.1 The Effects of Vertical Acceleration Oscillation

Road irregularities cause at human body level, over time, disorders of the spine structure, especially of the lobar area. They also have an effect on the psychic feelings of the passengers in the car. Oscillations of vertical acceleration on the spine produce neuromuscular disorders in the L1-L5 and S1-S2, but also structural changes in the local bone forms (Fig. 1). The frequency of propagation of these oscillations produces some cerebral changes, that generate changes in the behavior of the driver.

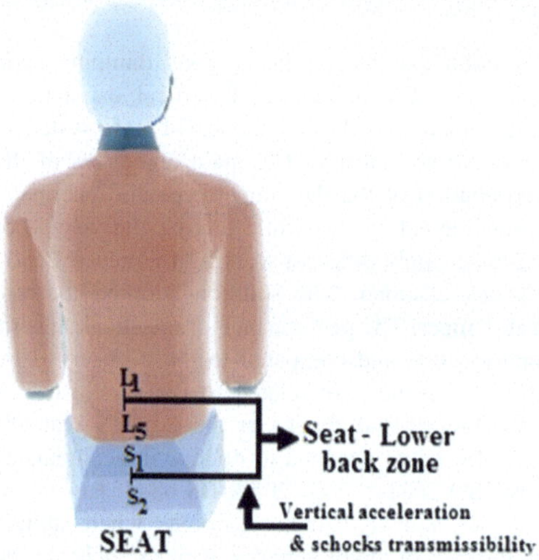

Fig. 1. The area affected by the oscillations of the vertical acceleration and the shocks

The state of fatigue, psycho-physical, can be analyzed by means of brain waves: beta, alpha, tetha and deta. Wikipedia [1] mentions that these brain waves are associated with different frequencies and generate certain states (feelings) for the human body (Table 1):

Table 1. The cerebral waves and associated frequencies

Brain wave	Frequencies [Hz]	The conditions generate on the body
Beta	12,5…30	Normality, vigilance
Alfa	8…12,5	Controlled relaxation
Theta	4…7	Sleepiness
Deta	0,1…3	Deep sleep

Of importance for the driver are theta and deta brain waves, as they are specific to vehicle movements, and frequencies of 2 Hz and 5 Hz produce the most negative effects for human organism. In Fig. 2 we present how the oscillation frequency of vertical acceleration produces changes in neural interactions [2].

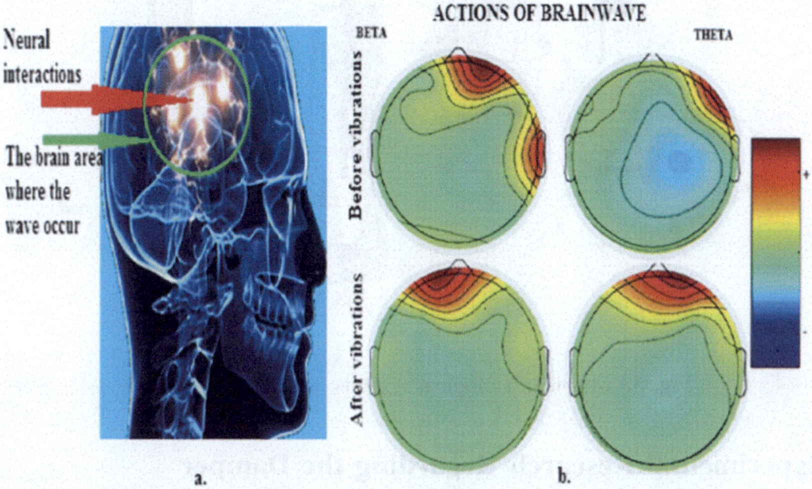

Fig. 2. Neural interactions under the action of random waves: a-area where they occur; b-manifestation. [2]

The effects of mechanical oscillations on the body resulting from traveling on the rough road are, generally:

(a) *Physical effect*: results in decreased performance of some activities, slowdown human processes;
(b) *Mechanical effect*: by the occurrence of lesions, the most common being of the lungs and hearts, ruptures or soft tissue damage;
(c) *Biological effect*: leads to fatigue, decreased attention and work capacity, and in some cases, hormonal changes or changes in blood pressure. This type of effect is also associated with emotional responses, that lead to behavioral changes.

To reduce these negative effects, different damping solutions are used: hydraulic, electrohydraulic, electroreological.

2.2 The Rheological Systems

In the motor vehicle industry, rheological systems are used to reduce shocks and vibrations and can be made of mounts and shock absorbers (Fig. 3). Due to their working properties, these systems have the ability to generate a high and variable force, depending on current and supply voltage.

By using them in the vehicle structure, can be reduced the level of mechanical oscillations, transmitted from the road to the driver by suspended and unsuspended masses. To diminish the oscillation values, researches were carried out using a magnetorheological damper, in the seat suspension.

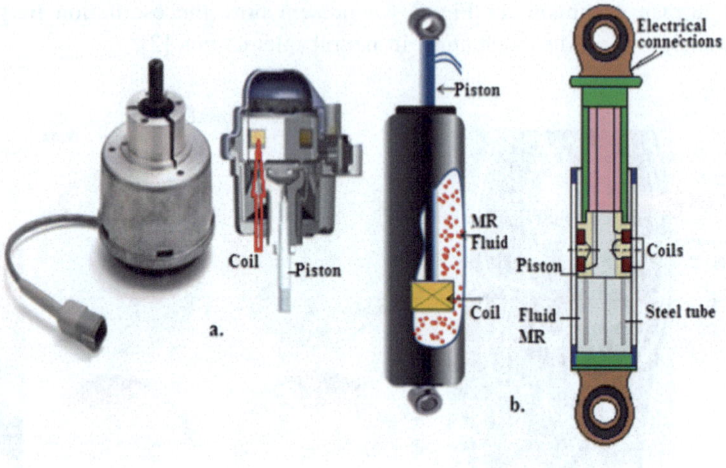

Fig. 3. Magnetorheological systems: a-mount; b-damper.

3 Experimental Research Regarding the Damper Functioning

The dampers' tests have been carried out taking into account the working regime, which must correspond to the driving conditions of the vehicle. To determine these conditions, data on the characteristics of a very rugged road sector were recorded. This road section (Fig. 4) produces a major discomfort to the driver, while traveling with the vehicle.

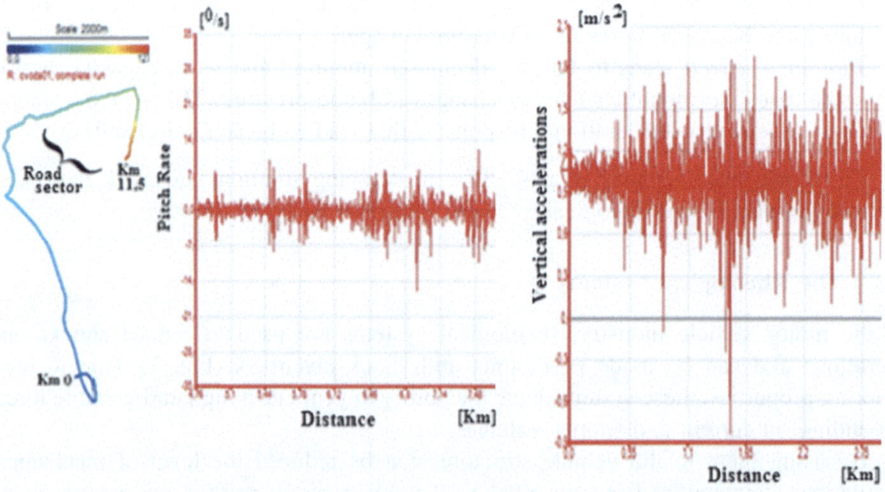

Fig. 4. Road sector: left-the pitch moving; right-vertical acceleration

This sector of the road generated high oscillations of vertical acceleration (a_z max = 1,82 m/s^2) well above the limits of ISO 2631-1/BS 6841.

The recorded values show a level of high discomfort (1,25 m/s^2 = major discomfort). According to VDI 2057 [3], the discomfort perception threshold (K) is calculated by relation (1).

$$K = \left(\frac{18}{\sqrt{1 + \left(\frac{f}{f_0}\right)^2}} \right) \cdot \sigma_{\ddot{x}} \tag{1}$$

where: f = the frequency for which it is calculated (5 Hz);

f_0 = reference frequency (10 Hz); $\sigma\ddot{x}$ = oscillation of the vertical acceleration.

The resulting value of factor K = 32,396 is considered to be a very strong level of perception (step G > 25). Also, the pitch movement indicates high level, with many pits on road, leading to major neural interactions.

To observe the possibilities of reducing oscillation of vertical acceleration and discomfort, the magnetoreological damper was tested on the electro-hydraulic stand Walter-bay Bai 1451-K22305. With electro-hydraulic equipment, different types of cyclical stresses can be defined during work: sinewave, triangular, rectangular loading, etc. For each load cycle, can be defined the working parameters: fatigue by stroke or by force, cycle frequency (max. 35 Hz), number of cycles. In Fig. 5 is observed the damper test system.

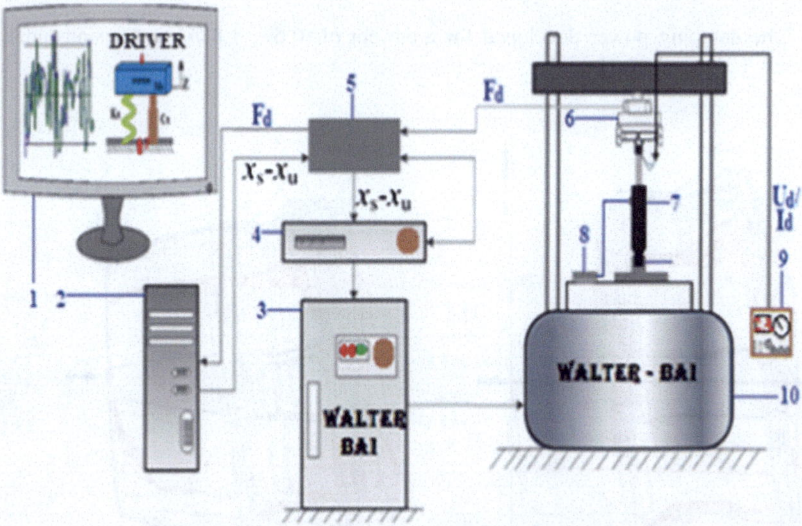

Fig. 5. The system used to test the damper 1-monitor; 2-PC unit; 3-hydraulic control system; 4-control system; 5-conversion unit; 6-clamping mandrel; 7-shock absorber MR; 8-rheological fluid temperature detection equipment; 9-U/I control unit; 10-electrohydraulic equipment.

3.1 Determinations on Damping Force and Energy

Reducing the oscillations generated by the road is determined by the attenuation characteristics developed by the shock absorber, under different working conditions. Maintaining a high force at different applied frequencies is a mandatory operating requirement for the shock absorber. Figures 6, 7 and 8 show the damping energy and the force developed by the shock absorber, for different working conditions, whit 40 mm displacement of piston.

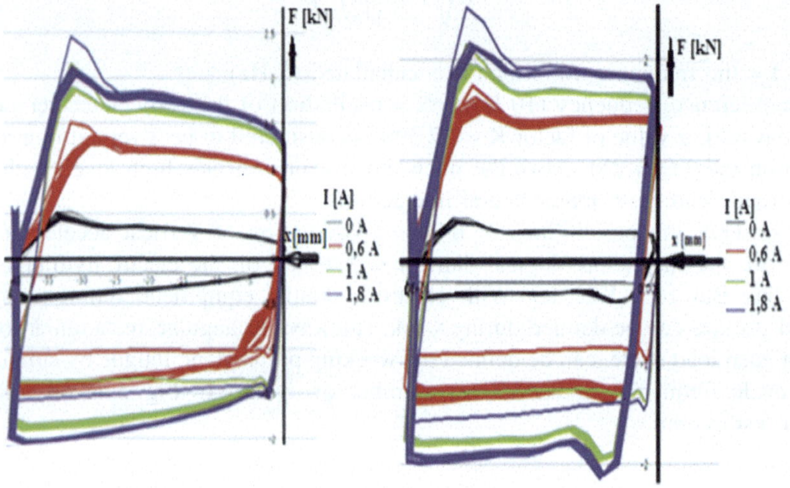

Fig. 6. The damping power developed for a current of (0.6… 1.8 A) and a sinusoidal cyclic charge: left 1 Hz; right 2 Hz.

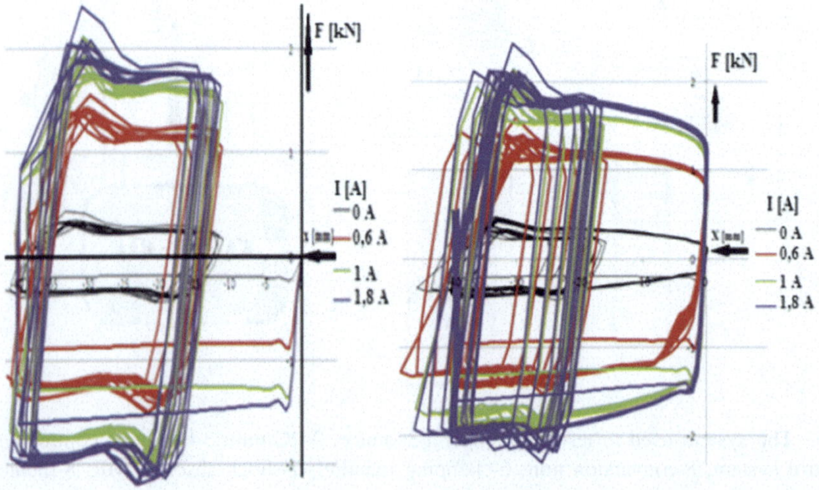

Fig. 7. The damping power developed for a current of (0.6… 1.8 A) and a sinusoidal cyclic charge: left 4 Hz; right 5 Hz.

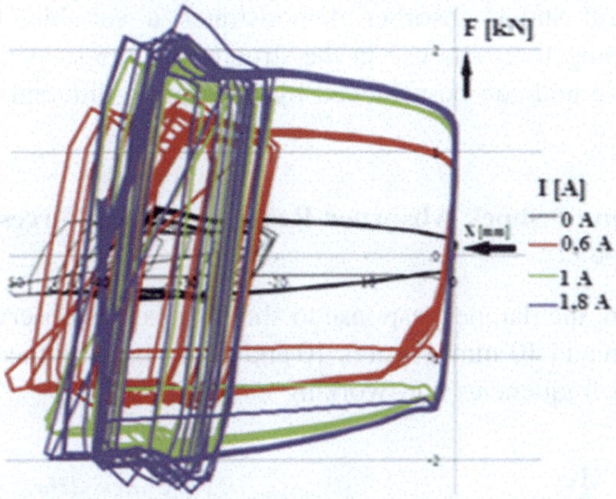

Fig. 8. The damping power developed for a current of (0.6… 1.8 A) and a sinusoidal cyclic charge to 6 Hz frequency.

The tests showed a force between F = 1547 N at a current of I = 0.6 A and 2385 N at a current of I = 1.8 A. The damping energy had values between E = 61.88 and 95.4 J. Without the current supply of the damper, the force values were F = 509 N and the energy E = 20.36 J.

Figure 8 shows the dependence of the damping energy on the supply currents, for frequencies of 4–6 Hz.

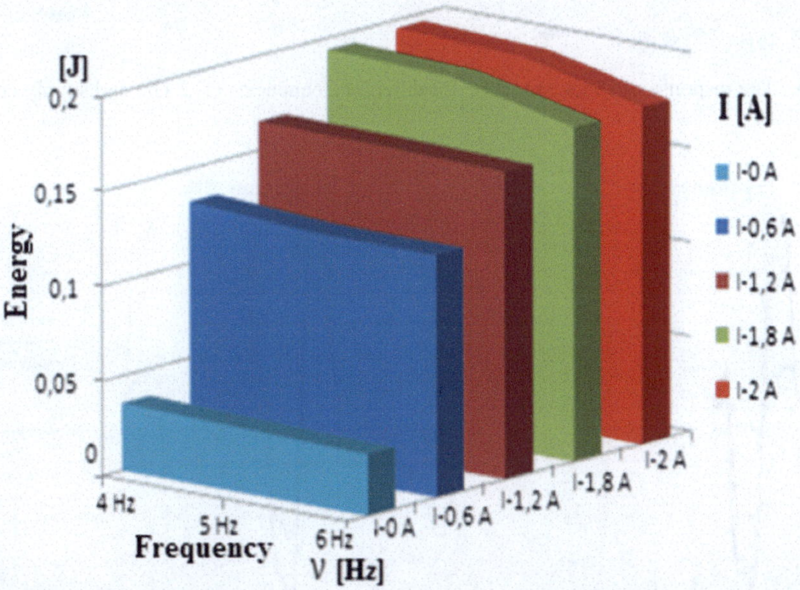

Fig. 9. The spatial charts of damping energy according to the oscillation frequency and the supply voltage of the damping coil

Magnetoreological shock absorber demonstrates a variable, high capability in generating the damping force. Its use in the structure of the suspension system of the driver's seat is viable and can be achieved by integrating different working strategies (Fig. 9).

3.2 Determination of Shock Absorber Response to the Forces of Shock Type

Through comparison, the damper response to shock tests was observed for two strokes of the piston: 20 mm and 40 mm. Figures 10 and 11 is observed the damping response, for different applied frequencies and working currents.

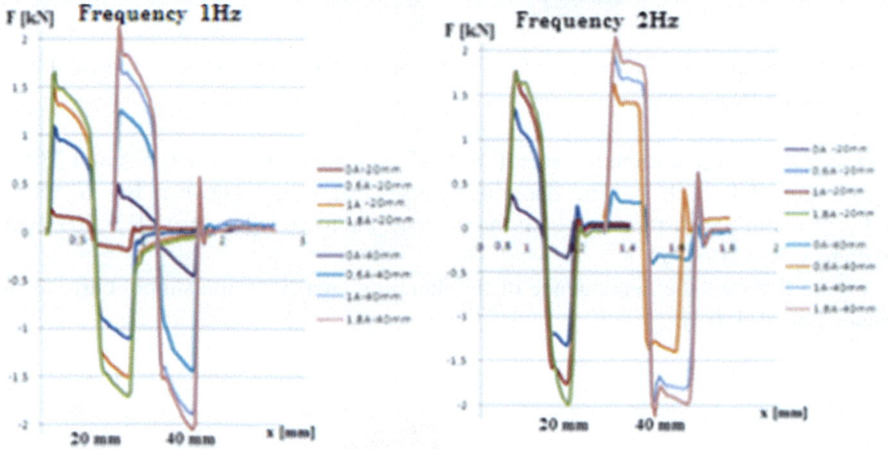

Fig. 10. The response of shock absorber, for different frequencies (1–2 Hz) and supply currents (0.6–1.8 A)

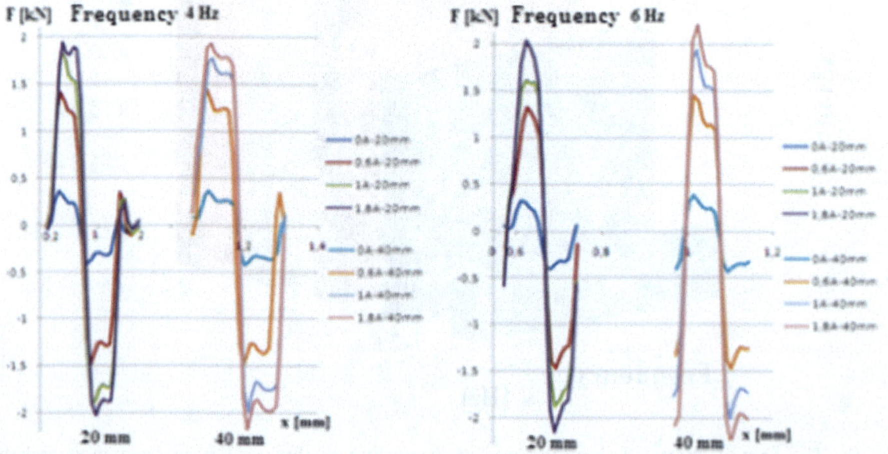

Fig. 11. The response of shock absorber, for different frequencies and supply currents (0.6–1.8 A)

In case of requests made in a short time (shocks), for the piston stroke of 20 mm, the maximum shock absorber force is between 1589 N and 1872 N. The maximum damping power is reached at the 4 Hz frequency load. The 40 mm piston stroke has a maximum force contained of between 2248 and 2381 N. The maximum damping power is developed at a frequency of 2 Hz. The damping energy decreases as the shock frequency increases. So the shock absorber has a low capacity to absorb high frequency shocks. It can be seen that the shock absorber force has higher values for the 40 mm piston stroke. If shock absorbers are used to decrease the shocks, it is recommended to supply the bobbin with a minimum current of I = 1,8 A.

4 Verifying in the Real Environment the Damping Capacity

Laboratory determinations have established different supply currents necessary to coil of the shock absorber, which can provide effective response forces, for the requirements of the seat suspension. In the process of validating the system, the consistency of values obtained through experiments and virtual models, with those obtained in the real environment was pursued. The magnetoreological damper is fixed to the seat structure by means of a cushion support device and the control equipment is caught to the floor of the vehicle (Fig. 12).

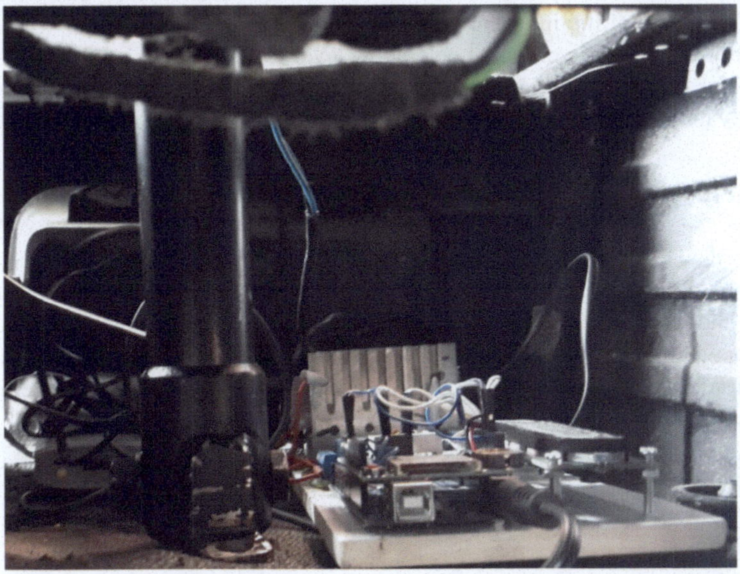

Fig. 12. Position of the damper and of the control and command equipment

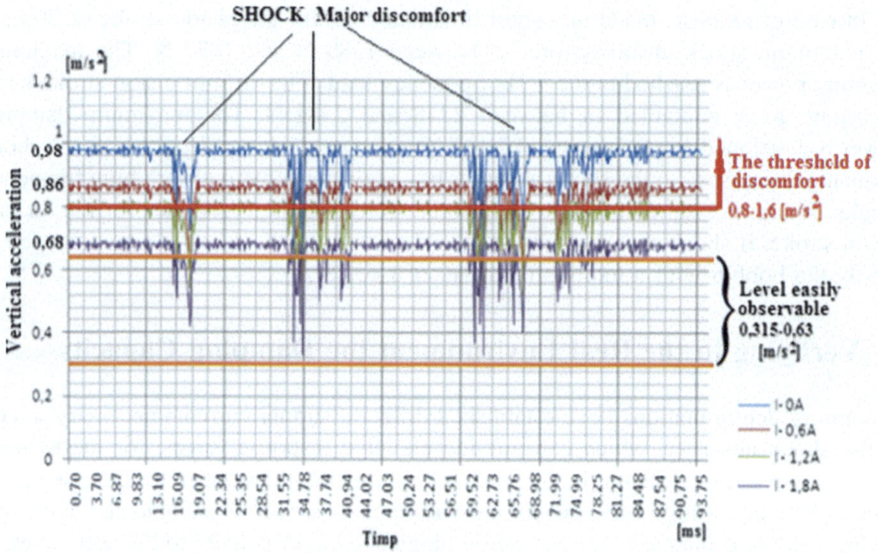

Fig. 13. The evolution of vertical acceleration at the seat level, during on vehicle travel

Figure 13 shows the oscillation of the vertical acceleration recorded without/with the supply current of the magnetoreological damper.

It is noted that, the vertical acceleration level is reduced in accordance with the current value applied to the damper coil. Decreasing this level of vertical acceleration favors elevation of driver comfort.

5 Conclusions

The force developed by the magnetoreological silencer is not significantly influenced by frequency, these it maintaining at a high level. Power supply influences positively the force and damping energy. By using differents current value, a control strategy can be developed, that can provide different levels of damping. Rheological systems, due to the reaction velocity and the generated forces can provide a substantial reduction in mechanical oscillations generated by the road, and improve the comfort of the personnel in vehicle's, during journeys.

References

1. https://en.wikipedia.org/wiki/Alpha_wave
2. Mitroi, M.: The implication of the random vibrations generated by rough route. Lambert Academic Publishing, ISBN 978-620-2-02749-6, Berlin, Germany (2017)
3. VDI 2057 – 1/ 2002: Human exposure to mechanical vibrations. Whole-body vibration

On-Board Measurement of Emissions from Spark Ignition Engine Vehicle for Urban Routes

Stelian Țârulescu(✉), Radu Țârulescu, and Cristian-Ioan Leahu

Transilvania University of Brasov, Eroilor Boulevard No. 29, Brașov, Romania
s.tarulescu@unitbv.ro

Abstract. Currently, testing the ecological performance of light vehicles is made using the New European Drive Cycle. Taking into account the problems of the last years, related to the efficiency of the testing procedures and approval of the motor vehicles on the stand, new testing and approval strategies are required. Our research aims to develop a methodology for measuring on-board emissions from vehicles, for different urban and extra-urban routes. For the present paper, the emission measurement system was tasted on the vehicle equipped with a spark ignition engine. There were 5 routes, of which 4 urban and one combined (urban and extra-urban). The results will help to prepare a methodology for testing on-board vehicles.

Keywords: Vehicle · Engine · Pollution · Emissions · Urban

1 Introduction

In recent years are developed various systems and equipments used for on board vehicle pollutant emissions evaluation. This paper presents experimental methodology for an on-road emissions measurement strategy (using a portable measuring equipment). The collected data from measurements can help the development of an emission inventories methodology.

In order to evaluate and control air pollution produced by the transport sector, it is essential to identify the pollution sources and environmental impact. In present, the assessment of the emission factor is determined using the dynamometer testing. Dynamometer testing is a methodology in which the pollutant emissions of the vehicle are measured in laboratory conditions, using a cycle that simulates road conditions (urban and extra-urban). This cycle is a sequence of stops, starts, rolling at constant speed, accelerations and brakes. Nowadays, different test cycles are used to simulate different running conditions. The problem is that the dynamometer tests are affected from shortcomings associated with non-representativeness of actual driving conditions. On-board measurements represents a more efficient alternative to quantifying vehicle emissions, because the data are collected under real conditions, on any route we want [1].

The light vehicles are tested under controlled laboratory conditions using a standardized cycle. For the pollutant emissions evaluation, the test conditions are well

I. Dumitru et al. (Eds.): SMAT 2019, *The 30th SIAR International Congress of Automotive and Transport Engineering*, pp. 241–248, 2020.
https://doi.org/10.1007/978-3-030-32564-0_29

known and there could be a risk of a configuration of the engine's operating mode for passing the emission tests when operating under test conditions [2].

Most studies at EU level show that the emissions measured in real road conditions were higher than in the case of stand testing with at least 60% and 30% for diesel vehicles and respectively, for almost all pollutants. Also, the carbon dioxide emissions measured on the stand were lower by 10% to 50% compared to the actual emissions measured on board, for different vehicles [3].

In order to resolve these problems, on-board pollution evaluation is needed. Investigation of vehicle emissions with on-board measurement systems is a more efficient alternative in order to understand how these emissions contribute to air pollution [4].

2 Methodology

Our research aims to develop a methodology for measuring on-board emissions from vehicles, for different urban and extra-urban routes. For the present paper, the emission measurement system was tasted on the vehicle equipped with a spark ignition engine. There were 5 routes, of which 4 urban and one combined (urban and extra-urban). The results will help to prepare a methodology for testing on-board vehicles.

The measurements ware made using a portable analyzer, GA-21 plus, an equipment produced by Madur Austria. The GA-21 plus is a flue gas analyzer which can measure the emissions of: Oxygen - O_2, Carbon monoxide - CO, Nitric oxide - NO, Sulphur dioxide - SO_2, Carbon dioxide - CO_2, Nitrogen oxides - NO_x [5].

Experimental research has been made using a light vehicle, Dacia Sandero, equipped with a 1390 cm^3 spark ignition engine. Engine vehicle Power is 55 kW at 5500 rpm [6].

The pollutant emission measurement equipment has been fixed in the vehicle. Its probe was mounted on the vehicle exhaust as instructed in the operating manual of the equipment, as shown in Fig. 1.

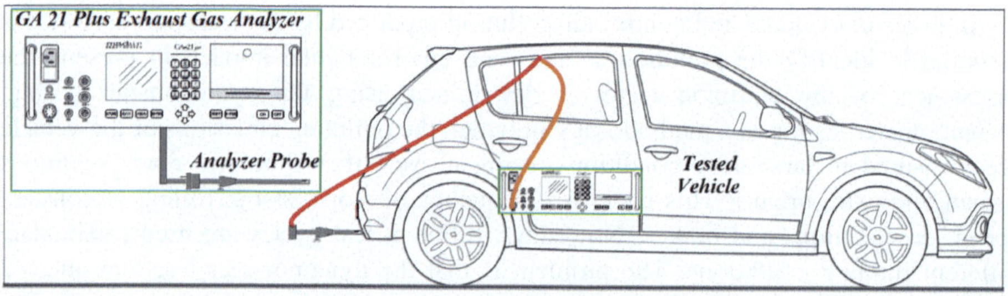

Fig. 1. GA-21 plus Analyzer placement inside tested vehicle.

The 5 routes that were run were chosen as representative for road transportation daily journeys in urban areas in Brasov City: the connection between residential areas and the historic centre, the connection between residential areas and industrial areas, connections between the historical centre and the civic centre or commercial areas, connections between the mall area and new residential neighbourhoods, etc.

The routes were numbered from 1 to 5, as follows:

Route 1 – Transilvania University Institute (Institutului Str.) + Făgăraşului Street + Cristianului Street + Lungă Street + Livada Poştei (see Fig. 2);

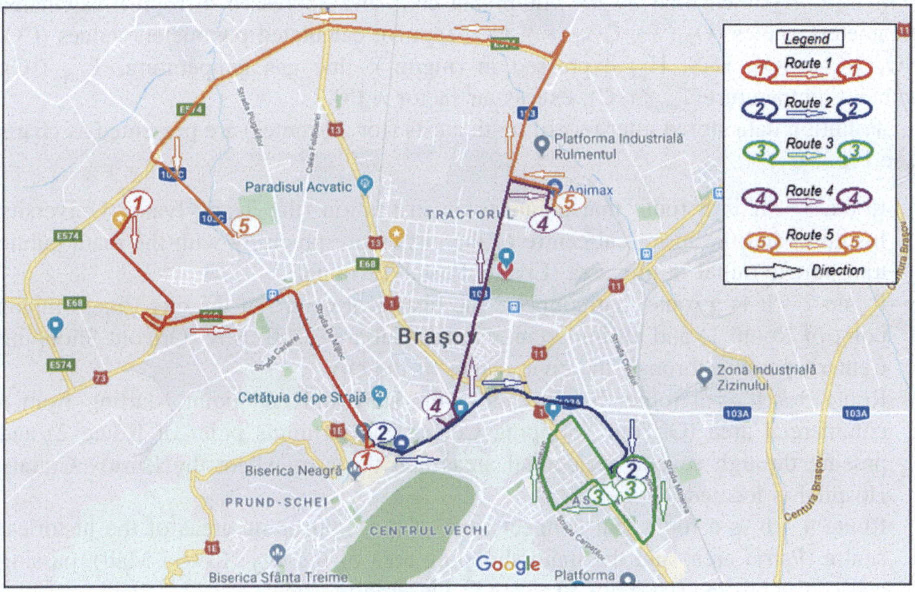

Fig. 2. The 5 analyzed routes from Brasov City.

Route 2 – Livada Poştei + Eroilor Boulevard + 15 Noiembrie Str. + Civic Centre + Zizinului Str. + Gemenii Intersection + Saturn Boulevard + Orizont Shopping Center (see Fig. 2);

Route 3 – Orizont Shopping Center + Saturn Boulevard + Calea Bucureşti Str. + Zorilor Str. + Carpaţilor Str. + County Hospital + Traian Str. + Orizont Shopping Center (see Fig. 2);

Route 4 – Grădinarilor Str. + 15 Noiembrie Boulevard + Castanilor Str. +- 13 Decembrie Str. + Henri Coandă Str. + Coresi Mall (see Fig. 2);

Route 5 – Coresi Mall + Henri Coandă Str. + 13 Decembrie Str. + Brasov Bypass Road - Bartolomeu Residence + Lânii Str. (see Fig. 2).

All five routes were configured to be traveled within 10 min, depending on street categories, intersections, priorities, and road traffic. Every 10 s values for emissions (CO, CO_2, NO, NO_2, NO_x) was stored.

3 Research Results

The results resulted after the five tests were downloaded and it was created a database in order to analyze them [7].

For every report following parameters are saved: parameters values (CO, NO, NO_2, SO_2, H_2S, H_2) measured by the equipment cells and expressed in (ppm); calculated parameters values (O_2 (%), CO_2 (%), NO_x (ppm)); calculated parameters values (CO, NO, NO_2, SO_2, H_2S, H_2) expressed in (mg/m^3); flue gas temperature, T_{gas} (°C); ambient temperature T_{amb} (°C); excess air factor λ [5].

Pollution data stored after recording the tests (for all routes) are presented as charts (see Fig. 3).

Route 1 - It is a route that connects an institution (the Transylvania University Institute) and the historical centre of the city, more precisely with the main public transport terminal in this area (Livada Postei Terminal).

Route 2 - It is a route that connects the historical centre of the city (the terminus point of Route 1) and an important commercial area of Brasov (Orizont Shopping Center), passing through the civic centre of the city.

Route 3 - It is a route that has the same start and end point, starting from a commercial area (Orizont Shopping Center, the terminus point of Route 2) and passing through several residential areas and the area where the Brasov County Hospital is located.

Route 4 - It is a route that connects one of the limitrophic areas of the historical centre (Patria area) and the main shopping area of the city (Coresi Mall), passing one of the busiest streets of Brasov (13 Decembrie Street).

Route 5 - It is a route that connects the city's main shopping area (Coresi Mall) and an expanding residential district (Bartolomeu) through the bypass road of the city.

Each route has peculiarities, but it can be assimilated by the majority of the metropolitan urban routes.

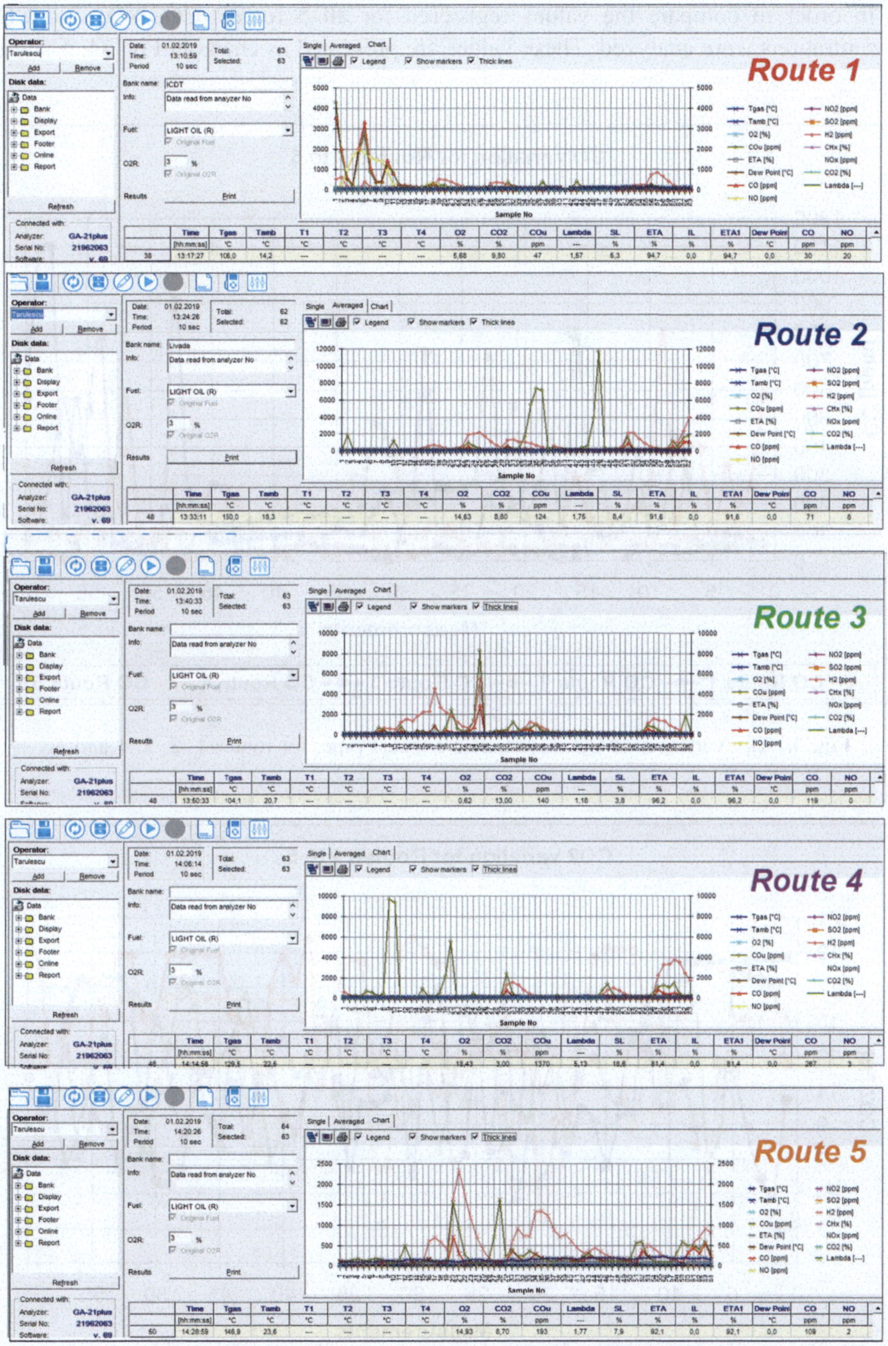

Fig. 3. Pollution data stored from GA21Plus analyzer, for all routes.

In order to compare the values registered for all 5 routes, CO, NO_x and CO_2 concentrations ware analyzed. These values are presented as charts in Figs. 4, 5 and 6.

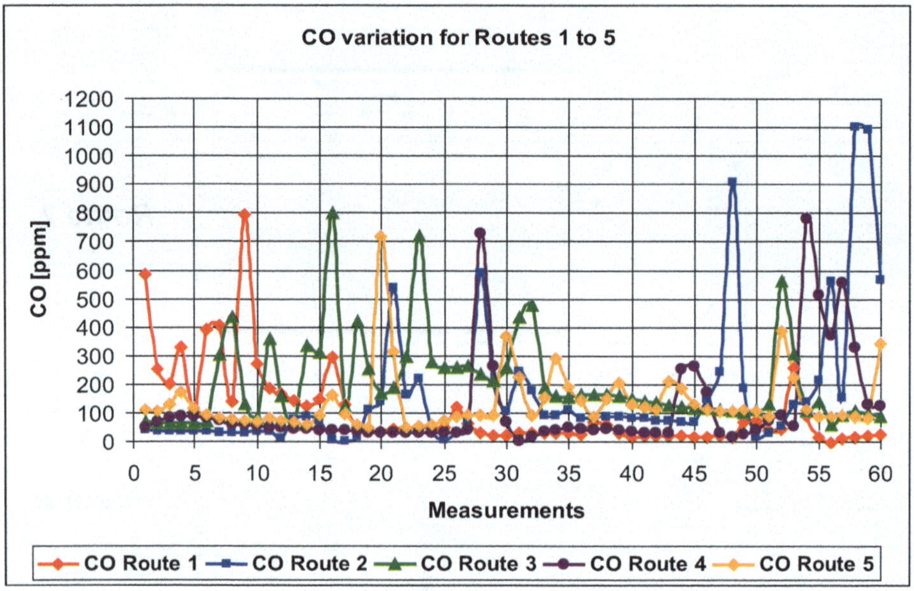

Fig. 4. The variation of CO at vehicle's exhaust pipe, for routes 1, 2, 3, 4 and 5.

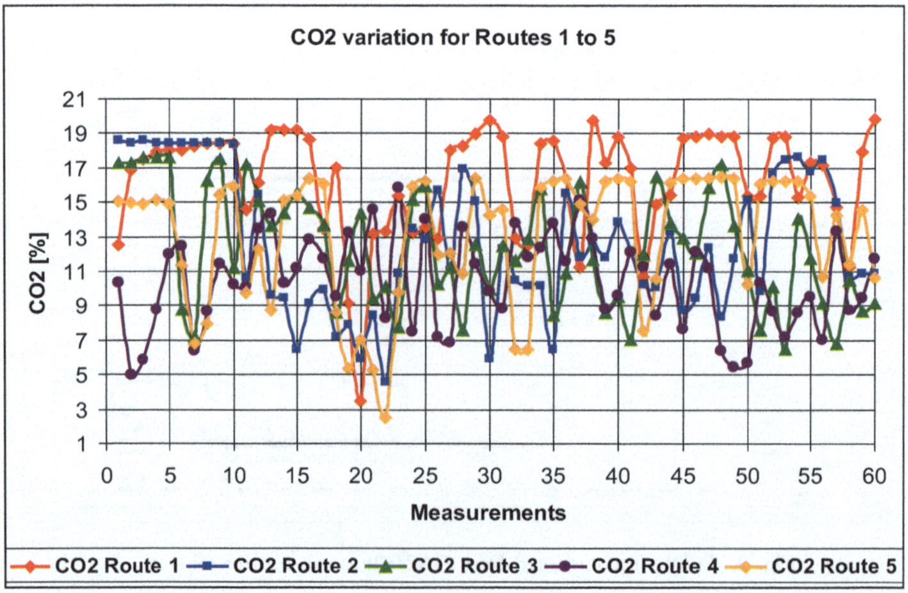

Fig. 5. The variation of CO_2 at vehicle's exhaust pipe, for routes 1, 2, 3, 4 and 5.

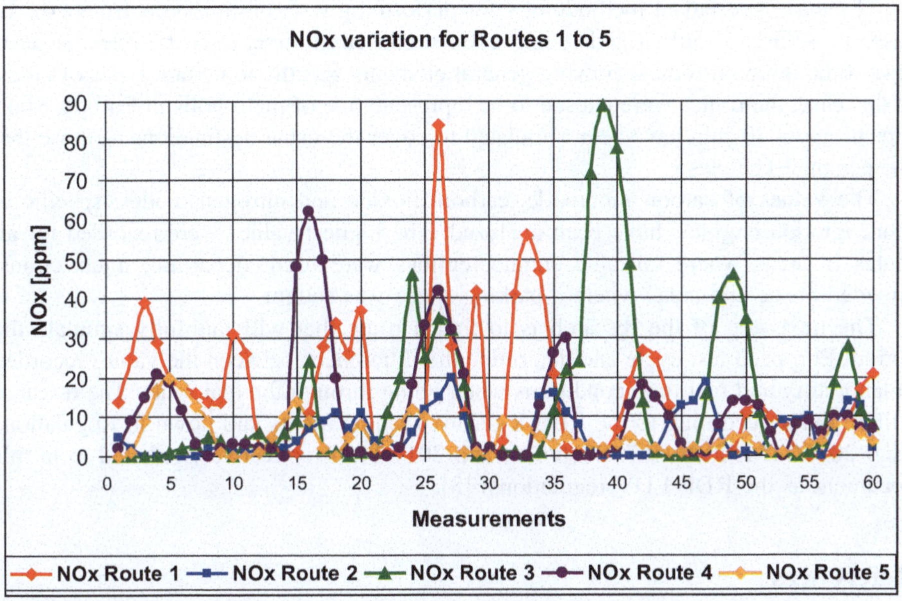

Fig. 6. The variation of NO_x at vehicle's exhaust pipe, for routes 1, 2, 3, 4 and 5.

For carbon monoxide the highest values were recorded for all routes in areas where unstable engine regimes were used: departure, intersections, crowded areas, and areas where a parking space was sought. By comparison, the highest values (peaks or mean values) were recorded for Route 2, Route 3 and Route 4.

In the case of carbon dioxide, the values varied substantially for all routes. By comparison, the highest values (peaks or mean values) were recorded for Route 1 and Route 5, on speed corridors. In the case of Route 1, the highest values were recorded in the starting area, on the Institutului Street and Fagarasului Street, but also in the case of the main arteries with one sense, Lunga Street. For Route 5, the maximum values were recorded when we were on the bypass road of the city.

In the case of nitrogen oxides, the values varied in areas of use of the engine to unstable modes: intersections, turns, pedestrian crossings, crowded areas. The highest values were recorded for Route 1 and Route 3. These two routes are also most fragmented by intersections, pedestrian crossings and priority losses.

4 Conclusions

The on board measurement for vehicles emission is an optimal strategy compared to the strategies based on static measurements, under laboratory conditions. Measurement of emissions on board provides data at any time and location, resulting in a more representative database than can be obtained classically.

The paper presents a methodology for performing tests of emissions from exhaust gases of vehicle in urban and urban/ extra urban environment. To be relevant, tests have to be done on routes showing general elements specific to certain types of cities. In this case, the routes were chosen to be representative of metropolis in Europe. Short trips of about 10 min have been simulated to cover the usual destinations of those that use personal vehicles.

The values of carbon monoxide, carbon dioxide and nitrogen oxides, specific to spark ignition engines, have been analyzed. The highest values were recorded for all routes in areas where unstable engine regimes were used: departure, intersections, crowded areas, and areas where a parking space was sought.

The next step of the research is to find a route that will faithfully simulate the current European test cycle and determine the differences between the values recorded in laboratory and real-time conditions when going through the real route. The research will be made according to the RDE legislative requirements laid down in Regulations (EU) 2017/1151, 2016/467, 2016/626, and 2017/1154 (collectively referred to in this document as the RDE LDV regulations) [8].

References

1. Frey, H.C., Unal, A.: Use of on-board tailpipe emissions measurements for development of mobile source emission factors. Department of Civil Engineering North Carolina State University, Raleigh, NC 27695-7908
2. Rubino, L., Bonnel, P., Hummel, R., Krasenbrink, A., Manfredi, U.: Mobile measurement of pollutant emissions and fuel consumption of road vehicles in real-world driving situations using portable emission measurement systems (PEMS), JRC REF N°28555-2005-08 A1CO ISP
3. Saâdane, C., Boughedaoui, M., Kerbachi, R., Joumard, R.: On-board measurement of emissions from liquefied petroleum gas, gasoline and diesel powered passenger cars in Algeria. HAL Id: hal-01250461. https://hal.archives-ouvertes.fr/hal-01250461
4. North, R.J., et al.: Development of a vehicle emissions monitoring system. Transp.: Proc. Inst. Civ. Eng. **158**(3), 167–177 (2005)
5. Flue Gas Analyser GA-21 plus: Operating manual, Madur Electronics, Vienna - Austria (2013)
6. Ţârulescu, S., Şoica, A.: Emissions level approximation at cold start for spark ignition engine vehicles. Appl. Mech. Mater. **555**(2014), 375–384 (2014)
7. Cofaru, C.: Pollution legislation in road transportation. Transilvania University of Brasov (2002)
8. Valverde Morales, V., Bonnel, P.: JRC technical report, on-road testing with portable emissions measurement systems (PEMS) (2017). https://ec.europa.eu/jrc

Modern Transport Systems and Road Traffic

Relationships Between Accident Severity and Weather and Roadway Adherence Factors in Crashes Occurred in Different Type of Collisions

Alin Drosu(✉), Corneliu Cofaru, and Mihaela Virginia Popescu

Transilvania University of Brasov, 500036 Brasov, Romania
drosu.alin.gabriel@unitbv.ro

Abstract. This paper aims at identifying and analyzing the relationships between road accidents severity and various weather and adherence factors in the crashes occurred in different type of collisions. A three-year injury accident database is analyzed and several types of road crash risks are taken into consideration for accidents' severity assessment. A Damage Severity Index (DSI) is also defined in order to assess the magnitude of the crash damages. Multiple linear regressions have been applied and tested on their global significance. Fatal Risk (FR), Serious Injury Risk (MGIR), Slight Injury Risk (MSIR) and Damage Severity Index (DSI) have been used as dependent variables in the following types of collisions: front – rear, sideswipe (hooking), rollover, and pedestrian hit, head-on, run–off and side collisions. Seven weather factors and eight road adherence conditions have been taken into consideration as explanatory variables and tested for their individual significance. In front-rear collisions, fog significantly explains the MGIR in a proportion of 54.90%. When the number of accidents that occur on fog conditions increase by two units, the seriously injury risk almost doubles its value. In sideswipe collisions, DSI increases on normal road and the FR is positively correlated with the icy roadway and negatively with the roads covered by black ice. The fatality risk FR from rollover collisions is the only variable that is explained by both weather and road adherence conditions. The fog and normal weather conditions explain its variation in a proportion of 61.80%. The number of monthly accidents that occurred on slippery roads and covered by black ice, explain the variation of FR in a proportion of 65.30%. The paper is only focused on studying the monthly accidents and shows which factors better explain the amplitude of the damages and several risks in different type of collisions. The results of this paper may help the insurance system and the Romanian road safety authorities to implement new driving training methods and counter measures in order to reduce the risks of injury and damages' level.

Keywords: Accident severity · Weather · Regression analysis · Road collisions

I. Dumitru et al. (Eds.): SMAT 2019, *The 30th SIAR International Congress of Automotive and Transport Engineering*, pp. 251–264, 2020.
https://doi.org/10.1007/978-3-030-32564-0_30

1 Introduction

At the European level, 25.047 people lost their lives in road accidents in 2018. Romania is the EU state with the higher fatality rate (96 fatalities/1 million inhabitants), the average EU figure showing 49 fatalities/1 million [1] (see Fig. 1).

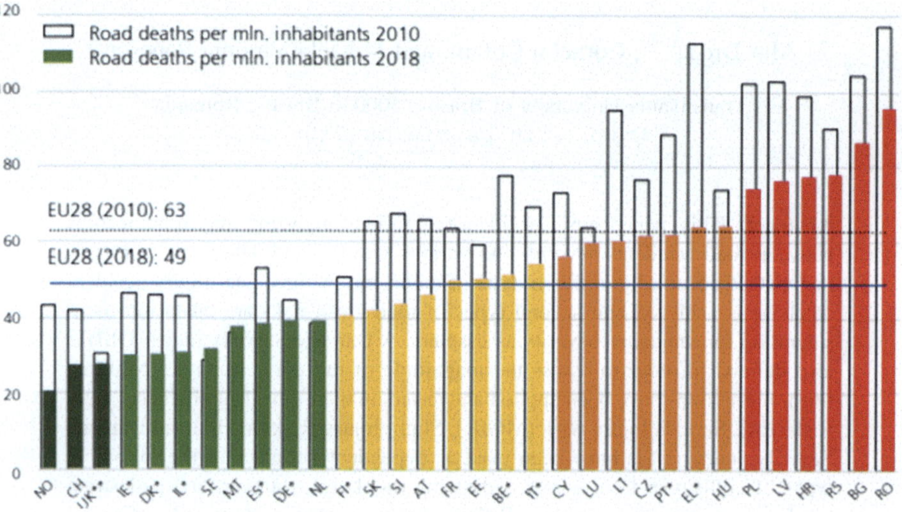

Fig. 1. Fatalities rates (road deaths per million of inhabitants) in the EU states and the average figures in 2010 and 2018.

Weather's influence on road traffic crashes and their severity due to the different type of collisions have been of interest for many researches. The severity of the injuries that resulted from crashes occurred on fog conditions are higher than on other conditions and involve more cars. Under this conditions, head-on and rear-end collisions have the highest crash risk and severity [2]. Other research [3] confirmed that weather is a great risk of accident, especially the rainy and foggy conditions.

Driving on fog or wet roads has a direct relationship with the crash risk [4]. Also, the visibility and road adherence are major risk factors for accidents. Therefore, analyzing the influence of the weather factors on road traffic accidents and the relationships between accident severity and risks and other weather and adherence factors, could lead to a decreasing of the crashes costs.

2 Data Description

To carry out this research, a database containing all road accidents causing bodily injuries recorded over the period 01.01.2014–31.12.2016, produced both inside and outside the localities, was used. To ensure good data consistency, accidents have been grouped monthly, as do their existing attributes in the database. 15411 records were analyzed and information on the number of monthly casualties, resulting victims (deaths, serious injuries and light injuries), as well as meteorological and road adherence conditions were filtered by collision types.

3 Methodology

The degree of severity of an accident is determined not only by the status of the victims, but also by the magnitude of the resulting damage, mainly those caused by the vehicles involved in road collisions. In order to quantify the magnitude of damages resulting from a road accident, one will consider the number of vehicles involved and define the Damage Severity Index (DSI) as follows:

$$\mathrm{DSI} = \frac{NVM}{NM} \tag{1}$$

Where,

NVM = number of vehicles involved in monthly accidents
NM = number of monthly accidents

Road safety at national level is quantified by so-called “road transport and road safety indicators” [5]. Among these, they include those who provide an overview of the risk of injury: lethal risk and risk of injury.

The monthly lethal risk (FR) is defined according to the following relationship:

$$\mathrm{FR} = \frac{\mathrm{F}}{\mathrm{NM}} \tag{2}$$

Where: F = the number of monthly deaths produced in traffic accidents.

The Monthly Injury Risk (MIR) is given by:

$$\mathrm{MIR} = \frac{\mathrm{MI}}{\mathrm{NM}} \tag{3}$$

Where MI is the number of people injured per month in injury-related injuries (NM).

In road safety statistics, there are three degrees of injury to people involved in road traffic accidents: deceased persons (within 30 days of the occurrence of the traffic accident), seriously injured persons and slightly injured persons. Therefore, according to formula (3), one can define the Monthly Serious Injury Risk (MGIR) and the Risk of Slight Injury (MSIR):

$$\mathrm{MGIR} = \frac{\mathrm{MGI}}{\mathrm{NM}} \tag{4}$$

$$\mathrm{MSIR} = \frac{\mathrm{MSI}}{\mathrm{NM}} \tag{5}$$

Where: MGI = number of seriously injured persons and MSI = number of people slight injured in NM traffic accidents.

In this paper, in order to explain what set of factors affect the Damage Severity Index (DSI), Monthly Lethal Risk (FR), Monthly Serious Injury Risk (MGIR), and Monthly Slight Injury Risk (MSIR) in traffic accidents produced by various types' collisions, we will use the multi linear regression method for explanatory purposes and the SPSS statistical software.

Being a multi linear regression used for explanatory purposes, the concepts of dependent and independent variables can be used. The Damage Severity Index (DSI), Monthly Fatality Risk (FR), Monthly Serious Injury Risk (MGIR) and Monthly Slight Injury Risk (MSIR) will be declared as dependent variables, and meteorological factors and those influencing the condition of the road will be used as independent variables (Table 1).

Table 1. Dependent and independent variables to be used in multi linear regressions in the explanatory note.

Dependent variables	Weather and road adherence factors	Independent variables
DSI	Road adherence	Slippery (Slip)
FR		Ice on road (Icy)
MGIR		Slime (Slm)
		Black ice (Bice)
MSIR		Wet (Wet)
		Dry (Dry)
		Snow (Snow)
		Others road adherence (Ora)
	Weather factors	Fog (Fog)
		Sleet (Slt)
		Snowing (Snw)
		Normal weather (Norm)
		Rain (Rain)
		Strong wind (Swind)
		Snow storm (Storm)

The influence of meteorological factors affecting road adherence will be analyzed separately for each independent variable in the 7 types of collisions: front - rear (FR), sideswipe/hooking (HK), rollover (RL), pedestrian hit (PH), head-on (HO), run-off (RO) and side collisions (SC). The weight of each type of collision is shown in Fig. 2.

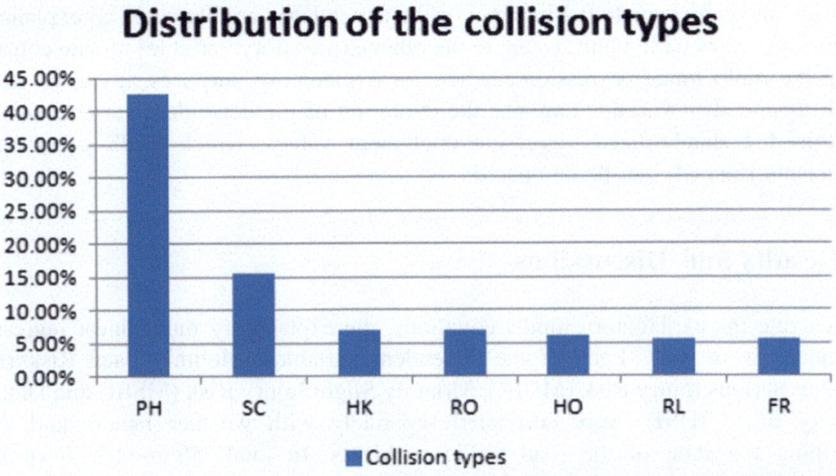

Fig. 2. The type of collisions analyzed in this paper

In order to determine to what extent a certain factor (independent variable) influences the variance of the dependent variables, one will use the multilevel regression method [6] as we will statistically control the influence of the other dependent variables included in the model. The mathematical model of the multiple linear regression equation used for explanatory purposes is the following [7]:

$$\widehat{y}_t = \widehat{a}_0 + \widehat{a}_1 x_{1t} + \widehat{a}_2 x_{2t} + \ldots \widehat{a}_k x_{kt} \tag{6}$$

Where: $\hat{y}_t$ is the dependent variable at time t, x_{kt} is the explanatory k variable at time t, $\hat{a}_0$, $\hat{a}_1$, …, $\hat{a}_k$ are the model's parameters and t = 1, 2,…, n, with n = number of observations.

In order to decide on the best model by which the variables are best explained by the predictors, one will proceed to: testing the global significance of the regression equation, the individual testing of the significance of the predictors and their interpretation.

To test the global significance of a regression equation, the values of two parameters are used: the F (Fisher test) test and the determination coefficient R^2. If there are statistically significant estimates between the values resulting from the regression equation and those based on the mean values, then the F value will be statistically significant. In the ANOVA tables generated by SPSS it is observed whether the generated patterns are statistically significant or not, comparing the Sig value with the significance threshold being less than 0.005 ($p < 0.005$). Moreover, if $p < 0.001$, then that model is considered to have a great explanatory power. Once a model is statistically and globally significant, R^2 shows how % of the value of the dependent variable's dispersion is explained by the common action of all explanatory variables.

In terms of individual testing of the significance of predictors, this test is done using the regression coefficients that SPSS generates (non-standard coefficients b). The non-standardized coefficients b is equivalent to the coefficients $\hat{a}_k$ in the relation [6]. Based on these, one can estimate the increase of the dependent variable when an explanatory variable increases with a unit, when all the other explanatory variables remain constant. Using the multi linear regression method for explanatory purposes, it can be shown which independent variable explains the evolution of the dependent variable at most. For this, the standardized regression coefficient values, which SPSS generates as coefficients Beta (β) can be compared.

4 Results and Discussions

To generate the explanatory model equations, the explanatory multi linear regression techniques were used. Each of the dependent variables, Monthly Fatal Risk (FR), Monthly Serious Injury Risk (MGIR), Monthly Slight Injury Risk (MSIR) and Damage Severity Index (DSI), were correlated separately with weather factors and those describing the state of the road traffic accidents. In total, 56 models have been developed. The determination coefficient R^2, the value of the F test and the Sig values are presented in Table 2.

Of these, after applying the global significance tests, only 9 of them proved to be globally significant.

Table 2. R^2, F and Sig values resulting from running the 56 multi – linear regressions in SPSS, as well as the decision to accept as a result of verifying the significance threshold.

Type of collision	Dependent variable	Weather factors			Roadway adherence factors		
		R^2	F	Sig.	R^2	F	Sig.
PH	DSI	0.18	0.87	0.99	0.19	0.82	0.58
	MFR	0.18	1.11	0.37	0.16	0.64	0.73
	MGIR	0.07	0.41	0.86	0.09	0.34	0.94
	MSIR	0.13	0.72	0.63	0.20	0.86	0.55
SC	DSI	0.34	2.48	0.46	0.28	1.32	0.27
	FR	0.24	1.53	0.20	0.41	2.41	0.04
	MGIR	0.15	0.90	0.50	0.33	1.71	0.14
	MSIR	0.51	5.10	0.00	0.47	3.04	0.01
HK	DSI	0.39	4.94	0.00	0.47	3.00	0.01
	FR	0.24	2.55	0.05	0.52	3.73	0.00
	MGIR	0.17	1.66	0.18	0.26	1.21	0.32
	MSIR	0.15	1.43	0.24	0.43	2.57	0.03
RO	DSI	0.15	1.13	0.36	0.15	0.61	0.75
	FR	0.16	1.15	0.35	0.24	1.11	0.38
	MGIR	0.12	0.86	0.51	0.12	0.46	0.86
	MSIR	0.31	2.70	0.03	0.24	1.08	0.40
HO	DSI	0.12	0.59	0.75	0.38	2.15	0.06
	FR	0.15	0.73	0.64	0.14	0.56	0.79
	MGIR	0.26	1.43	0.23	0.35	1.83	0.11
	MSIR	0.21	1.09	0.39	0.14	0.57	0.78

(continued)

Table 2. (*continued*)

Type of collision	Dependent variable	Weather factors			Roadway adherence factors		
		R^2	F	Sig.	R^2	F	Sig.
RL	DSI	0.10	0.57	0.74	0.18	0.77	0.62
	FR	0.61	7.82	0.00	0.65	6.34	0.00
	MGIR	0.25	1.69	0.15	0.32	1.60	0.17
	MSIR	0.21	1.30	0.28	0.17	0729	0.66
FR	DSI	0.38	3.08	0.01	0.29	1.67	0.15
	FR	0.26	1.78	0.13	0.19	0.96	0.47
	MGIR	0.54	5.87	0.00	0.23	1.21	0.32
	MSIR	0.27	1.78	0.13	0.07	0.33	0.93

Analyzing Table 2, one finds that in the case of PH, RO and HO collisions, no variable describing weather conditions or road adherence contributes to the explanation of DSI, FR, MGIR or MSIR, since the global significance thresholds of the models are higher than 0.005, and these will be removed from the analysis in this paper.

4.1 Side Collisions

In the case of side collisions SC, MSIR is explained by weather factors in a proportion of 51.30%, as the determination coefficient $R^2 = 0.513$. Moreover, this model has a high explanatory power because $p < 0.001$. Following the multiple linear regression technique using SPSS, the data from Fig. 3 are obtained:

Coefficients

Model		Non-standardized Coefficients		Standardized Coefficients	t	Sig.	95.0% Confidence Interval for B	
		B	Std. Error	Beta			Lower Bound	Upper Bound
	(Constant)	1.686	.098		17.270	.000	1.486	1.886
	Fog	.048	.042	.177	1.122	.271	-.039	.134
	Sleet	.253	.103	.414	2.450	.021	.042	.464
1	Snow	.016	.042	.064	.385	.703	-.070	.102
	Normal	-.001	.001	-.146	-.979	.336	-.004	.001
	Rain	-.005	.012	-.062	-.419	.678	-.031	.020
	Strong wind	-.305	.116	-.400	-2.639	.013	-.542	-.069

a. Dependent Variable: MSIR

Fig. 3. Regression coefficients corresponding to MSIR explained by meteorological factors in the side collisions.

As it can be seen from Fig. 3, variables that are predicting the correct MSIR evolution are sleet ($t = 2.450$, $p < 0.05$) and strong wind ($t = -2.639$, $p < 0.05$).

Therefore, in the case of side collisions produced under various weather conditions, the variance of MSIR is defined by the relationship:

$$\text{MSIR} = 1.686 + 0.253 * \text{ Sleet} - 0.305 * \text{ Strong wind} \tag{7}$$

As the two predictors are statistically significant, the regression coefficient corresponding to sleet (b = 0.253) is a positive one, indicating a direct relationship between MSIR and the presence of the sleet at the time of road traffic accidents. The regression coefficient associated with the strong wind variable (b = −0.305) shows an inverse relationship between MSIR and this predictor. The non-standardized coefficients b show how the evolution of the dependent variable changes when the predictor value increases with a unit, as the other predictors remain constant. In the case of side collisions, the risk of slight injuries increases by 0.253 when the number of accidents that occurred during sleet increases with one unit, as the number of accidents produced in strong wind will remain constant.

Standardized coefficients of predictors (b) in Fig. 3 determine the importance of predictors in the evolution of the explanatory variable, which are analyzed in descending order. As can be seen from the Fig. 3, the variable that best explains the evolution of MISR is sleet (β = 0,414), followed by strong wind (β = −0,400).

4.2 Sideswipe/Hooking Collisions

In the case of sideswipe/hooking collisions ($p < 0.005$), one observes that the DSI Damage Severity Index is explained by the meteorological factors in a ratio of 39.00% ($R^2 = 0.390$). Since the monthly traffic frequencies produced during sleet and strong winds are null, they will be removed from the multi linear regression analysis. The resulting regression coefficients are shown in Fig. 4.

Coefficients

Model		Non-standardized Coefficients		Standardized Coefficients	t	Sig.	95.0% Confidence Interval for B	
		B	Std. Error	Beta			Lower Bound	Upper Bound
1	Constant	1.254	.079		15.839	.000	1.093	1.416
	Fog	.256	.156	.251	1.638	.112	-.063	.574
	Snow	.001	.049	.004	.026	.979	-.099	.102
	Normal	.010	.002	.666	4.186	.000	.005	.015
	Rain	.011	.025	.065	.438	.665	-.040	.062

a. Dependent Variable: DSI

Fig. 4. Coefficient values of the regression equation corresponding to DSI explained by the meteorological factors, in the case of collisions by shedding.

It is noticed that the only variable contributing to the IDS explanation from monthly traffic accidents produced in different meteorological conditions (statistically significant) is the normal weather ($t = 4.186$, $p < 0.05$). Therefore, the resulting regression equation is written as:

$$IDS = 1.254 + 0.10 * Normal \tag{8}$$

Concerning the variation of the FR death rate risk of the monthly traffic accidents produced under various conditions of road adherence in this type of collisions, the regression coefficients generated by SPPS are presented in Fig. 5.

Coefficients

Model		Non standardized Coefficients		Standardized Coefficients	t	Sig.	95.0% Confidence Interval for B	
		B	Std. Error	Beta			Lower Bound	Upper Bound
1	(Constant)	.013	.020		.630	.534	-.028	.054
	Others	.009	.027	.052	.352	.727	-.045	.064
	Slippery	-.013	.019	-.133	-.651	.520	-.052	.027
	Ice	.157	.037	1.215	4.197	.000	.080	.234
	Slime	-.013	.018	-.121	-.718	.479	-.050	.024
	Glaze	-.070	.021	-.771	-3.377	.002	-.112	-.027
	Wet	-.002	.003	-.116	-.471	.641	-.008	.005
	Dry	.000	.000	.145	.686	.499	-.001	.001
	Snow	-.004	.022	-.040	-.196	.846	-.050	.041

a. Dependent Variable: FR

Fig. 5. Regression coefficient coefficients of the FR corresponding to the roadway adherence in the case of sideswipe/hooking collisions.

Only two variables are statistically significant in this case: ice-covered road surface ($t = 4.197$, $p < 0.05$) and glaze covered roadway ($t = -3.377$, $p < 0.05$), these two explaining the FR variation in a ratio of 52.50% ($R^2 = 0.525$). In this case, the regression equation can be written as:

$$FR = 0.13 + 0.157 * Ice - 0.070 * Glaze \tag{9}$$

The Ice variable has the highest explanation weight for this model, $\beta = 1.215$. The non-standardized regression coefficient, with a positive sign corresponding to this variable, shows that there is a direct relationship between FR and Ice covered roadway. In other words, if the number of accidents on ice-covered roads increases by one unit, then the risk of death increases by 0.157.

4.3 Collisions with Rollover

In this type of collisions, the variation of FR is explained by both weather conditions and adherence factors of the road.

In the case of traffic accidents produced by rollover collisions in various meteorological conditions, the regression coefficients generated by SPSS are shown in Fig. 6.

Coefficients

Model		Non-standardized Coefficients		Standardized Coefficients	t	Sig.	95.0% Confidence Interval for B	
		B	Std. Error	Beta			Lower Bound	Upper Bound
1	(Constant)	-.054	.056		-.977	.337	-.168	.059
	Fog	.194	.032	.749	6.169	.000	.130	.259
	Sleet	-.017	.052	-.052	-.316	.754	-.124	.091
	Snow	-.010	.037	-.042	-.270	.789	-.086	.066
	Normal	.006	.002	.349	2.743	.010	.001	.010
	Rain	-.012	.012	-.119	-1.022	.315	-.037	.012
	Strong wind	.093	.075	.149	1.239	.225	-.060	.245

a. Dependent Variable: FR

Fig. 6. Coefficients of the regression equation corresponding to FR explained by the weather conditions in the case of rollover collisions.

Fog ($t = 6,169$, $p < 0.05$) and normal weather ($t = 2.743$, $p < 0.05$) are the two predictors that explain the FR variation in a ratio of 61.80% ($R^2 = 0.618$).

In this case, the regression equation is:

$$FR = -0.054 + 0.194 * Fog + 0.006 * Normal \qquad (10)$$

The calculated values and estimated values of FR of the traffic accidents produced in different meteorological conditions by “rollover” collisions are shown in Fig. 7.

The differences between the observed values and the calculated values represent the dispersion of the dependent variable and is described by determination coefficient, in our case $R^2 = 0.618$.

Both non-standardized regression coefficients are positive, with FR and each of the two variables being a direct relationship. For the fog variable ($b = 0.194$), if the number of accidents occurred on fog increases by 5 units, the FR varies by about one unit. The largest share in explaining the variance of FR has the fog variable ($\beta = 0.749$), more than twice the normalized regression coefficient value corresponding to the “Normal” variable ($\beta = 0.349$).

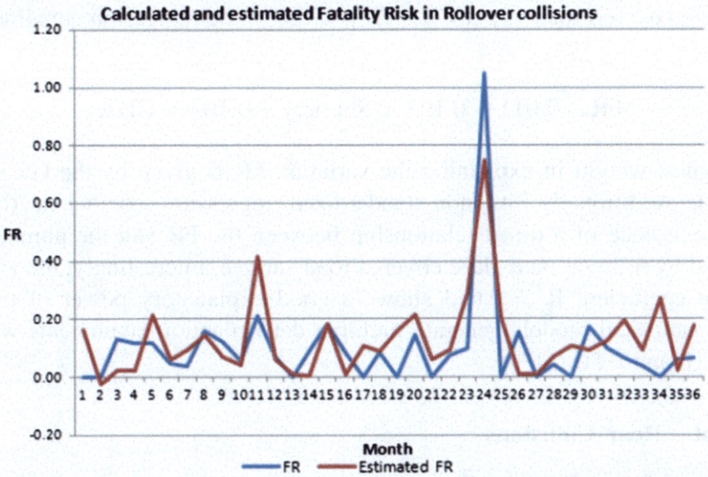

Fig. 7. The calculated and estimated Fatality Risk (FR) values for rollover collisions produced in different weather conditions.

In terms of FR variation in the case of traffic accidents produced under different conditions of road adherence, the regression coefficients resulted from SPSS are presented in the Fig. 8.

Coefficients

Model		Non-standardized Coefficients		Standardized Coefficients	t	Sig.	95.0% Confidence Interval for B	
		B	Std. Error	Beta			Lower Bound	Upper Bound
1	(Constant)	.011	.059		.193	.849	-.111	.133
	Other	-.037	.074	-.060	-.496	.624	-.189	.115
	Slippery	-.193	.047	-.803	-4.080	.000	-.290	-.096
	Ice	.170	.068	.497	2.488	.019	.030	.311
	Slime	-.014	.052	-.053	-.261	.796	-.120	.093
	Glaze	.103	.047	.571	2.189	.037	.006	.200
	Wet	.017	.010	.301	1.744	.093	-.003	.038
	Dry	.001	.002	.047	.345	.733	-.004	.006
	Snow	-.053	.041	-.216	-1.314	.200	-.137	.030

a. Dependent Variable: FR

Fig. 8. Coefficient values of the corresponding regression equation FR explained by the adherence conditions of the road in the case of rollover collisions.

The only variables that are statistically significant and which explain the variation of FR in a ratio of 65.30% ($R2 = 0.653$), is the slippery road ($t = -4.080$, $p < 0.05$)

and the glaze covered roadway ($t = 2.189$, $p < 0.05$), the regression equation being of the form:

$$FR = 0.011 - 0.193 * \text{Slippery} + 0.103 * \text{Glaze} \quad (11)$$

The highest weight in explaining the variation FR is given by the Glaze variable ($\beta = 0.571$). Additionally, its non-standardized regression coefficient ($b = 0.103$) shows the existence of a direct relationship between the FR and the number of accidents caused by rollover on a glaze covered road surface. Interestingly, the value of the determinant coefficient $R^2 = 0.653$ shows a good explanatory power of this model, since most statistical models generate multiple determination coefficients with values between 0.10 and 0.50 [6].

4.4 Front – Rear Collisions

In the case of front-to-back collisions produced in various weather conditions, the regression table generated by SPSS is shown in Fig. 9.

Coefficients

Model		Non-standardized Coefficients		Standardized Coefficients	t	Sig.	95.0% Confidence Interval for B	
		B	Std. Error	Beta			Lower Bound	Upper Bound
1	(Constant)	.176	.085		2.068	.048	.002	.351
	Fog	.485	.142	.613	3.420	.002	.195	.774
	Sleet	.001	.105	.001	.011	.991	-.213	.216
	Snow	-.170	.116	-.215	-1.468	.153	-.407	.067
	Normal	.001	.003	.057	.435	.666	-.005	.007
	Rain	.006	.021	.044	.307	.761	-.037	.050
	Strong wind	.256	.216	.193	1.189	.244	-.185	.698

a. Dependent Variable: MGIR

Fig. 9. MGIR corresponding coefficients of the regression equation explained by weather conditions in the case of front-to-rear collisions.

Figure 9 shows that the only variable that is statistically significant is "Fog" ($t = 3.420$, $p < 0.05$), and the regression equation for this model is:

$$MGIR = 0.176 + 0.485 * \text{Fog} \quad (12)$$

Since the multiple determination coefficient for this model is $R^2 = 0.549$, one can say that the variation of the MGIR is explained by the number of monthly road accidents occurring in fog by front and back collisions, at a rate of 54.90%, the remainder being other factors that have not been taken into account in this paper. The non-standardized regression coefficient for the Fog variable is $b = 0.485$. If the number of monthly crashes caused by front rear collisions increases by two units, the risk of serious injury in this type of accidents increases by 0.97 units. Practice, doubling the number of such crashes leads to an increase in the risk of serious injury by 97%.

5 Conclusions

The most interesting finding of this paper shows that in front – rear collisions, fog significantly explain the seriously injury risk in a proportion of 54.90%. When the number of accidents that occur on fog conditions increase with two units, the seriously injury risk almost doubles its value. As the fog affects the visibility of the drivers, the authorities should implement driver education programs in order to make them aware about the safety distance and driving in bad visibility conditions.

In sideswipe/hooking collisions, the amplitude of the damages increases on normal weather when the traffic volumes are higher comparing with bad weather conditions. In the same type of collisions, the fatality risk is positively correlated with the icy roadway and negatively with the roads covered by black ice, both variables explaining FR in a proportion of 52.50%.

In rollover collisions, the fatality risk FR is the only variable that is explained by both weather and road adherence conditions. The fog and normal weather conditions explain the variation of FR in a proportion of 61.80%. The number of monthly accidents occurring on slippery roads and covered by black ice, explain the variation of FR in a proportion of 65.30%.

The slight injury risk is positively correlated with the sleet and negatively with the strong wind, its variation being explained in a proportion of 51.30% by these two variables.

The paper is focused only on studying the monthly accidents. A deeper investigation of daily and weekly accidents is necessary in order to see what weather and roadway adherence factors best explain the injury risks and amplitude of the damages.

The paper analyses which factors better explain the amplitude of the damages and several risks in different type of collisions. As the road safety research is not very well developed in Romania, the results of this paper may help the insurance system and the Romanian road safety authorities to implement new driving training methods and counter measures in order to reduce the risks of injury and damages' level.

References

1. https://etsc.eu/wp-content/uploads/ETSC-PIN-AR_2019-Final-EMBARGO.pdf. Accessed 01 July 2019
2. Abdel-Aty, M., Ekram, A.-A., Huang, H., Choi, K.: A study on crashes related to visibility obstruction due to fog and smoke. Accid. Anal. Prev. **43**, 1730–1737 (2011). https://doi.org/10.1016/j.aap.2011.04.003
3. Wang, C., Quddus, M.A., Ison, S.G.: Predicting accident frequency at their severity levels and its application in site ranking using a two-stage mixed multivariate model. Accid. Anal. Prev. **43**(6), 1979–1990 (2011)

4. Cohen, S., et al.: Injury prevention: meeting the challenge. In: Book Injury Prevention: Meeting the Challenge. Higher Education Centre for Alcohol and Other Drug Prevention (1989)
5. Todoruţ, A.: Dinamica Accidentelor de Circulaţie. U.T. Press, Cluj Napoca (2008)
6. Sava, F.: Analiza datelor în cercetarea psihologică [Analysis of Data in Psychological Research]. ASCR Publishing House, Cluj Napoca (2011)
7. Duguleană, L.: Statistica în cercetare – Suport curs [Statistics in Research – Course Support]. Transilvania University of Brasov – Doctoral School, Brasov (2012)

Reconstruction of Rollover Collision Using PC Crash 11.1 and 3D Scanner

Laura Diana Grigorie[1(✉)], Dragos Sorin Dima[2], Alina Elena Romanescu[1], and Mihnea Marin[1]

[1] University of Craiova, 200512 Craiova, Romania
lauradianagrigorie@gmail.com
[2] Transilvania University of Brasov, Politehnicii Str. 1, 500024 Brasov, Romania

Abstract. The fundamentals of accident reconstruction are applied to rollover collisions. Typical deceleration rates, roll distances and vehicle deformations are discussed.

The reconstruction of rollover accidents is complex due to the infinite variety of important parameters such as ground geometry, vehicle characteristics, speeds and objects encountered. Although each rollover collision is unique, there are some common features.

This paper presents, starting for the real case, the reconstruction of a rollover accident using the PC Crash 11.1 program. For a faithful reconstruction of road accident, a 3D scanner was used in order to establish the geometry of the place of the road accident. The impacts the vehicle has to the ground may be described by number of rolls, number of ground impacts, ground impact severity and vehicle damage.

Keywords: Accident reconstruction · Evaluation · 3D scanning · Evidence · Trace evidence

1 Introduction

The reconstruction of a road accident should take into account the complex movements and collisions of the vehicles in time and space. To do this, the exact location of the traffic accident must be known, all existing traces must be identified, as well as the end positions of the vehicles.

In most situations, crash sites are analyzed in two spatial dimensions. In the case of overturning and rolling accidents, obtaining information in three dimensions is very important for the reconstruction of the road accident. The traffic accident took place at around 05:30 in the morning, during a starry night, on a dry road, in which a Mazda vehicle was involved. The car was heading of town when, in an attempt to avoid a dog, he used the breaks and suddenly turned left, losing control of the direction of travelling. Subsequently, the Mazda had several sequences of motion, and then it overturned to the left of the car's driving direction (Fig. 1).

Consequences of the accident: The front passenger has a fatal injuries and the Mazda car was damaged.

I. Dumitru et al. (Eds.): SMAT 2019, *The 30th SIAR International Congress of Automotive and Transport Engineering*, pp. 265–272, 2020.
https://doi.org/10.1007/978-3-030-32564-0_31

Fig. 1. The location where the traffic accident took place and the final position of Mazda vehicle

At the site of the road accident, photographs and measurements were made, in which the position of the Mazda was identified in the field on the left of the direction of travelling, overhead, with the front facing the road axle.

Figure 2 shows vehicle after it's back on wheels; the direction impact is also indicated. The largest amplitude was located on the right edge of the car dome.

Fig. 2. Direction impact

Fig. 3. Passenger side window missing, no turret

Following contact with the ground the car rotated around its longitudinal axis, deforming the body to the right and breaking the side windows and turret (Fig. 3), then rotated around its vertical axis, on the dome until it reached the final position.

The movements of the Mazda vehicle are shown further on using the measurements from the crash site and the deformations on the car.

2 Reconstruction of the Movements of the Mazda Vehicle

Existing tracks, the driver's statement, the technical characteristics of the vehicle involved, and the evidences at the scene of the accident were taken into account. Reconstruction of the accident by accident simulation methods was carried out using the specialized program PCCrash 11.1.

The accident occurred on a straight road sector, outside the town, at 05:23 a.m., under conditions of absolute darkness.

In the program, we introduced a car with the same technical characteristics as the car involved in the road accident, a 2D scale sketch was made and the level difference on the left side of the road was also modelled.

After several simulations, which took into account the final position of the vehicle, the speed of the car was obtained, when the skid began, of approx. 55 km/h.

Figure 4 shows the dynamic trajectory of the Mazda, as soon as it entered the skidding sequence (Fig. 5).

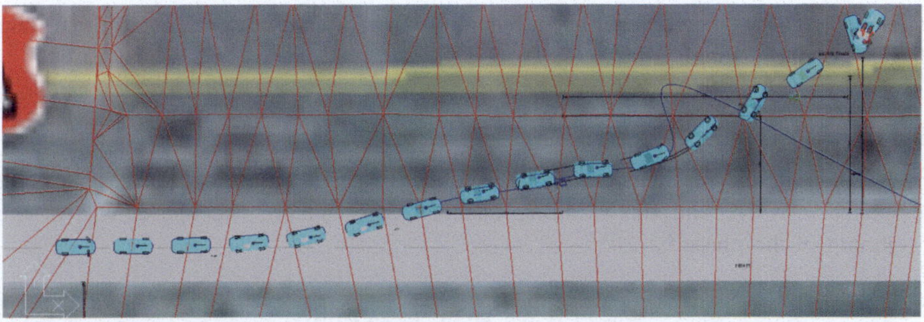

Fig. 4. The dynamic trajectory of the Mazda vehicle

Fig. 5. Leave the road with rotations around its two axes

The Mazda car travels in its direction when driver notices a dog in front of him and initiates the bypass maneuver. As a result of the sudden usage of the braking and steering system, the car is unbalanced because the centrifugal force has become larger than the lateral reaction of the road. At this point, the car begins to perform a lateral translation movement towards the outside of the turn, starting to derail, then leaving the road. At that moment, the back right-hand wheel starts to depressurize, entering the terrain to the left of the road where it collides with the root a tree, then rolls to the right, overturns the dome and stops after a rotation around the mass center due to the irregularities on the terrain.

3 Relations

After determining the dynamics of the road accident, the speed of the Mazda is calculated before the conflictual situation occurs. Speed is calculated, based on the reconstruction of the accident phases [6]. To determine the speed of the car, we start with the reconstruction of the last phase, that of rolling.

A creep coefficient of $c_r = 0,4\ldots0.5$ is adopted and the mass of the car's center of gravity at the end of the skid phase is determined with the relation:

$$W_{r0} = \sqrt{2 \cdot g \cdot c_r \cdot S_{r0}}[m/s] \tag{1}$$

where: $c_r = 0,45$ – a creep coefficient when moving forwards, $S_{r0} = 22$ m – the rolling distance of the vehicle.

$$W_{r0} = 13{,}93[m/s]$$

Reconstruction of the skid phase is done considering a coefficient of rolling resistance $c_d = 1{,}2$ and the skidding distance $S_{d0} = 11$ m. The speed W_{d0} at the beginning of the phase is:

$$W_{d0} = \sqrt{W_{r0}^2 + 2g \cdot c_d \cdot S_{d0}}[m/s] \tag{2}$$

$$W_{d0} = 21, 28[m/s]$$

In the pre-run phase, the car was braked while making a left turn generated by the inequality of the braking forces on the side wheels. The speed of the vehicle W_0 at the time of initial braking is:

$$W_0 = \sqrt{W_{d0}^2 + 2 \cdot g \cdot S_1 \cdot \varphi}[m/s] \tag{3}$$

where: φ – adhesion coefficient to slippage characteristic for dry asphalt slip - 0,7, $S_1 = 27$ m – the distance on which traces of brake have been identified.

Mazda's drive-in speed was: $w_0 = 28{,}61$ m/s, respectively 103 km/h at the moments before breaking.

4 Obtained Results

After the speed setting in the PC Crash program, two simulations were made: in the first simulation it was considered that the two occupants of the car, the driver and the front passenger, did not wear a seat belt and in the second simulation the two occupants had the seat belt.

From the two simulations, it is noted that if the two occupants were wearing a seat belt, their positions during skidding and rolling were stable and did not change in relation to the center of mass as it would have happened if they were not wearing the seat belt.

In the study [3] the influence of the deformation of the car dome and the belt on the probable lesions of the spine in the case of rolling was studied. It has been noticed in the case of the passenger on the right, bound to the seat belt, there is a likelihood of cervical spinal cord injuries in the case of a sprain. In [2] a study was carried out on killed and survivors in 30 collision crashes in Germany, with the result that if passengers wore a seat belt, they survived in 75% of cases and if they did not wear seat belt survived only in 40% of cases.

Fig. 6. Successive positions of the occupants without seat belt

Figure 6 shows the first case: the seat belt is missing and Fig. 7 shows the second case: the two occupants have the seat belt.

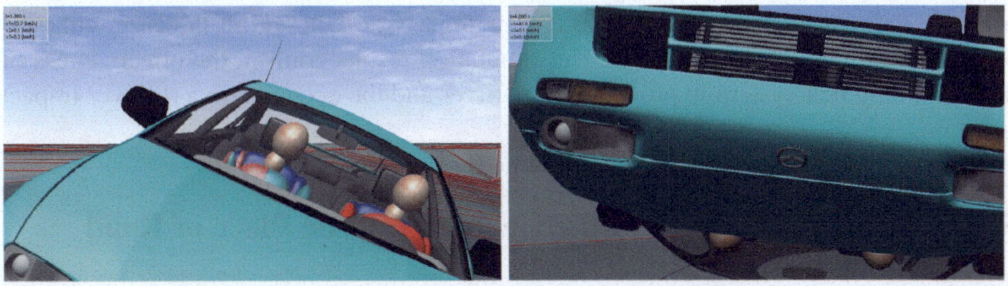

Fig. 7. Successive positions of the occupants with the seat belt

The acceleration curves of the head, neck and torso areas as determined by PC CRASH 11.1 simulations.

Figures 8 shows the diagram of acceleration for the head, torso and neck of the driver and front passenger, simulated in two situations: without the seat belt and with the seat belt.

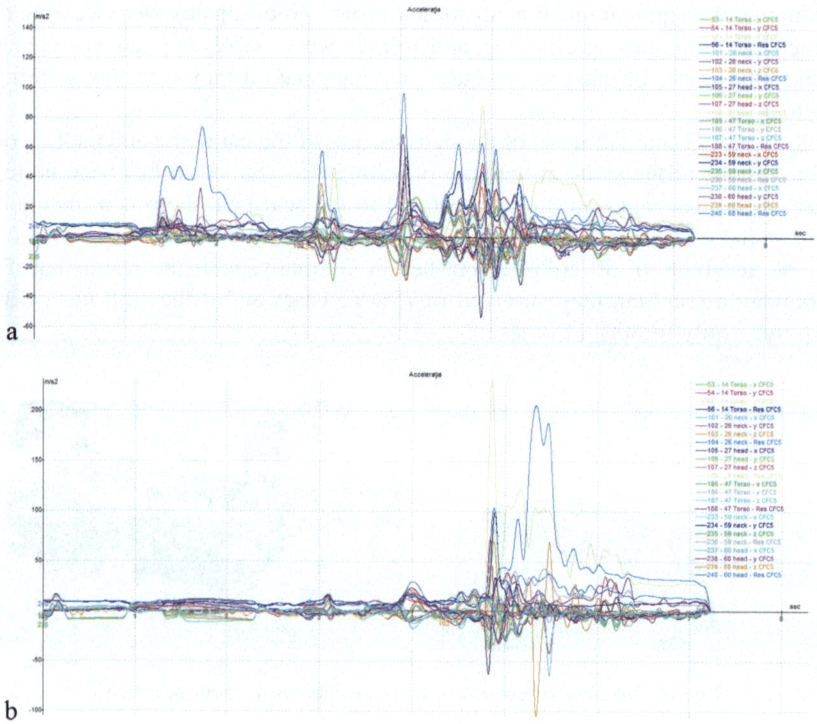

Fig. 8. Diagram of acceleration (a) without seat belt, (b) with seat belt

The results of the accelerations, at the level of the torso, neck and head have higher values in the case without seat belt. The values of the accelerations start to increase from the moment the car skids, in the case without seat belt, and in the case with seat belt increase in the moment of roll. The simulation results confirm the effectiveness of the restraining belt system in protecting the driver and the occupant in a rollover impact situation.

5 3D Scanner to Establish the Geometry of the Place of Road Accident

In the simulation presented above, the existing field on the left side of the tread was reconstructed, taking into account the existing slope, by building polygons, followed by their triangulation.

Instead, using a 3D scanner is useful for determining the real geometry of the road accident site as it scans the surface as a point cloud. The basic technique of 3D scanner is that of optical triangulation. The exact shape of the object is captured every time the laser pulse meets the surface. Once the scanning area and the appropriate resolution are established, you can start the scan operation. This process is carried out completely automatically driven by the software component. After initiating the scanning process the scanner automatically heads to the starting point and start point acquisition.

The point cloud thus obtained was imported into the PC Crash 11.1 Road Accident Rebuilding Program.

After importing, it was crafted as follows: the points that were not of interest were deleted, the runway was selected as the zero point of the point cloud and it was triangulated so that the car would use the points as a road (Fig. 9).

Fig. 9. Point cloud

After performing the simulation under the same conditions as in the first case, it is found that the car has a similar displacement as the original, but there is difference of four meters between the final positions. The difference may occur because the 3D scanning was not done at the time of the crash, and the resulting vegetation which appeared in the meantime generated points that caused errors during the circulation of the vehicle (Fig. 10).

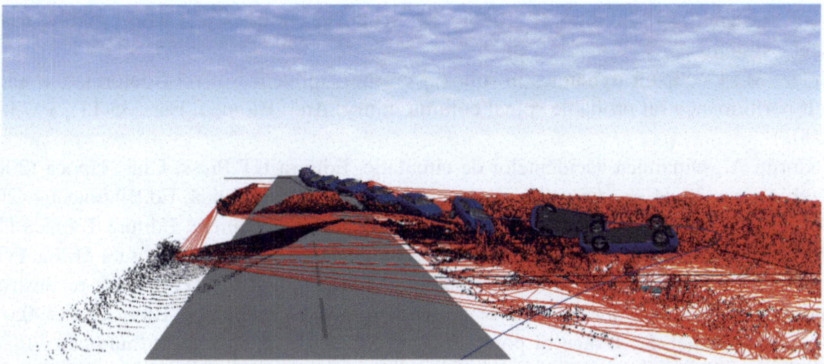

Fig. 10. The dynamic trajectory of the Mazda vehicle on the point cloud

6 Conclusions

From the analysis of the traces of the road crash and the PC Crash 11.1 simulations, it turned out that the Mazda vehicle circulated in its direction outside the town when the driver noticed a dog in front of her and tried to bypass him. Following the operation of the brake and steering system, the car is unbalanced, at this moment the back right-hand wheel is depressurized and the car begins to perform a lateral translation movement towards the outside of the turn, starting to derail, then leaving the road, entering the terrain on the left side of the road, where it collides with the root of a tree, then rolls to the right and rolls over the dome and after a rotation around the mass center, due to the irregularities on the terrain, it stops.

The speed of the car at the beginning of the conflict situation, based on the reconstruction of the crash stages [6], was 103 km/h, and when it's skiding it was 55 km/h.

The cause of the road accident was the sudden action of the steering system and the car started to skid, then left the road, entered the field to the left of the direction of travel, collided with the root of a tree, then rolled and stopped on its dome.

From the two simulations, it can be seen that if the two occupants were wearing a safety belt, their positions during skidding and rolling would be stable and not change in relation to the center of mass, as it would have happened if they were not wearing the seat belt.

Using the 3D scanner is useful for rollover accidents to accurately determine the geometry of the impact site. In this case, 3D scanning was done long after the road accident occurred, so there were errors between the final position of the vehicle identified in the spot research and the final position in the 3D simulation, but nevertheless the dynamic trajectory of the car was the same in both simulations.

References

1. Boboş, B.: Cercetări privind reconstrucţia coliziunii autovehiculelor, Ed Risoprint Cluj-Napoca (2008)
2. Berg, A., Krehl, M., Behling, R., Helbig, M.: Rollover crashes – real world studies, tests and safety systems
3. Bidez, M.: Occupant dynamics in rollover crashes: influence of roof deformation and seat belt performance on probable spinal column injury. Ann. Biomed. Eng. **35**(11), 1973–1988 (2007)
4. Todoruţ, A.: Dinamica accidentelor de circulaţie. Editura UT Press, Cluj Napoca (2008)
5. Fierbinţeanu, M.: Expertiza tehnică auto judiciară şi criminalistică, Ed.Bibliotecha (2006)
6. Gaiginschi, R., Filip, I.: Expertiza Tehnică a Accidentelor Rutiere, Editura Tehnică (2009)
7. Nistor, N., Stoleru, M.: Expertiza tehnică a accidentului de circulaţie, Editura Militară (1987)
8. Stanley, B., Mark, R., David, A.: Combining PC-Crash and MADYMO for reconstruction and occupant motion during rollover crashes. In: EVU Conference, Bratislava (2005)
9. Vertal, P., Stehel, S.: Usage of point cloud in PC crash for insurance fraud case. In: EVU Conference, Dubrovnik (2018)
10. William, C., Moser, A.: Reconstruction of twenty staged collisions with PC-Crash's optimizer. SAE 2001-01-0507 (2001)
11. Wach, W.: Simulation of Vehicle Accidents Using PC Crash. Institute of Forensic Research Publishers, Krakow (2011)
12. Programul PC CRASH 11.1, licenţă nr. 73/2013, Grigorie Laura Diana

The Influence of Vehicle Low Impact Velocity over the Helmet Airbag Deployment and Cyclist Injuries

Ovidiu Andrei Condrea(✉), Anghel Chiru, George Radu Togănel, and Daniel Dragos Trusca

Transilvania University of Brasov, Politehnicii Str. 1, Brasov, Romania
ovidiu.condrea@unitbv.ro

Abstract. The aim of this paper is to assess the influence of the vehicle impact velocity over the deployment of cyclist helmet airbags, as well as the resulting head and neck injury severity. Two crash-tests were carried out in a longitudinal impact configuration (bicycle hit from the rear) using the same vehicle-bicycle-dummy assembly: Test 1 (6.86 m/s impact velocity, 0.4 m offset) and Test 2 (11.1 m/s impact velocity, 0.4 m offset). The head and thorax of the dummy cyclist were instrumented with tri-axial accelerometers. An activated helmet airbag was positioned onto the neck of the dummy for both tests, which did not deploy during the Test 1, but deployed during Test 2 due to the higher head acceleration. A discussion about the causes of the airbag non-deployment during Test 1 is provided. HIC and NIC values were calculated using the recorded head and thorax accelerations and are presented in this paper. Although skull fracture risks were neglectable for Test 1, the NIC value obtained was 56% higher (61 m^2/s^2) than the NIC value obtained for Test 2 (41 m^2/s^2), which shows that helmet airbags do have a potential in mitigating cyclist neck injuries.

Keywords: Cyclist impact · Helmet airbags · Airbag deployment

1 Introduction

Cyclists are traffic participants which have to share the road with motorized vehicles, being characterized by a high degree of vulnerability due to this fact. Although in the EU there is an ongoing initiative which aims to separate cyclist and vehicle trajectories in traffic through cycle tracks, currently there is no notable difference in cyclist fatality reduction, with minimum 2000 deaths each year for the 2010–2015 period [1].

Vehicle-cyclist accidents are characterized by a dominant tendency to result in head injuries for all impact configurations [2]. Maki showed that 72% of cyclist fatalities and 21% of cyclist severe injuries are caused by head trauma, while overall 56% of cyclists were subjected to head impacts with the vehicle's profile, proportion which varies depending on the impact configuration [3]. According to Fredriksson [4], most common severe and fatal cyclist injuries are generated through a head-windshield impact mechanism, 27% respectively. A study carried out by Otte [5] on 876 cyclist accidents

I. Dumitru et al. (Eds.): SMAT 2019, *The 30th SIAR International Congress of Automotive and Transport Engineering*, pp. 273–280, 2020.
https://doi.org/10.1007/978-3-030-32564-0_32

determined that 47.6% of cyclists suffered head injuries and 26.7% suffered AIS2+head injuries.

These statistical studies strongly indicate that the main cause for cyclist severe and fatal injuries are head traumas, while also signaling the dominant tendency for head impacts to occur in vehicle-bicycle accidents. In these circumstances, head protection plays a very important role in mitigating severe and potentially fatal injuries for cyclists, along with the continuous societal effort required to educate cyclists in wearing head safety equipment.

At this point, passive safety equipment for cyclists can be categorized in two distinct types which differ in terms of structure and working principle: traditional helmets and helmet airbags. Traditional helmets represent the classical approach in protecting cyclists and have been extensively used and studied until now. They are molded and produced using expanded polystyrene or polyurethane. Recent design variations (MIPS technology) allow for a better protection against injuries generated through rotational motion, by articulating the inner and outer surface of the helmet. Helmet airbags represent the most recently developed non-conventional approach for cyclist protection, while also constituting to this point the sole alternative to traditional helmets.

The undeployed helmet airbag has the shape of a scarf, and is designed to be worn on the neck, being comprised of two distinct airbags. Each airbag is composed of two layers, the outer layer defining the exterior helmet-shape of the airbag, while the inner layer composed of elongated elastic elements surrounds and generates a protective device for the neck and the entirety of the head except the face. In order to prevent head injuries at the time of deployment, a cold gas inflator is used for the inflation of the airbag [6]. The trigger device used for inflation comprises of a micro-sensor and auxiliary electronic components, being required to be manually activated at the start of each trip in order to function properly. The trigger device prevents airbag deployment for abnormal motions detected on the vertical axis, such that non-dangerous situations would not waste the protective effect of the single-use airbag [7].

Recent studies have shown that helmet airbags perform better than traditional helmets in terms of injury reduction, reducing both head acceleration (linear and angular) and HIC values [8, 9]. Helmet airbags had a better performance than MIPS helmets also, including rotational acceleration reduction, therefore showing a high potential in protecting the human head and neck [8].

Although helmet airbags consistently performed better than traditional helmets, there are still uncertainties regarding the de-facto cyclist head protection offered by these systems, since there is insufficient information in literature about helmet airbag performance in real-scale crash-tests. Furthermore, there is no information regarding the airbag deployment threshold and the criteria which define it. There is also the question of airbag non-deployments caused by low impact speeds -as can be seen in this study- and the helmet airbag's inability to cover all types and directions of injury sustainable by the cyclist at head-level (i.e. frontal head-to-asphalt).

The aim of this paper is to determine the influence of the vehicle impact velocity over the deployment of cyclist helmet airbags and their afferent protective capabilities.

2 Methods

Two crash-tests were performed for this study at different vehicle impact velocities in a longitudinal impact configuration. The bicycles used were hit in the rear wheel by an Opel Corsa vehicle, for a lateral offset of 0.4 m. An anthropometric dummy was positioned on a stationary bicycle during each crash-test. The bicycle-dummy assembly was maintained into position through an electromagnetic "gibbet" mechanism (see Fig. 1), which was deactivated through the means of a laser detection system at the moment the vehicle passed through the laser beam, immediately prior to the collision. The dummy was equipped with an undeployed helmet airbag, which was activated manually before each test in accordance with the indications provided by the manufacturer.

Fig. 1. The gibbet mechanism used for maintaining the bicycle-dummy assembly into position

Cyclist kinematic phases afferent to each crash-test are presented in Fig. 2, using the time of the initial vehicle-bicycle impact as the reference moment ($t_0 = 0$). The other defining moments of the cyclist's kinematic phases are:

- The pelvis-bonnet impact (t_1);
- The head-windscreen impact (t_2);
- The launch of the cyclist off the vehicle (t_3).

Both crash-tests were video recorded with a high-speed camera Fastec HiSpec 5 set at a 600 fps acquisition frequency. The dummy was equipped with two tri-axial accelerometers for data acquisition, mounted inside the head and thorax. The measured head and thorax accelerations were processed in order to determine the resultant accelerations, which were further used to calculate head and neck injury criteria values.

The Head Injury Criterion (HIC) was initially introduced by NHTSA in the 1970's and it is currently the most extensively used of the head injury criteria. HIC is

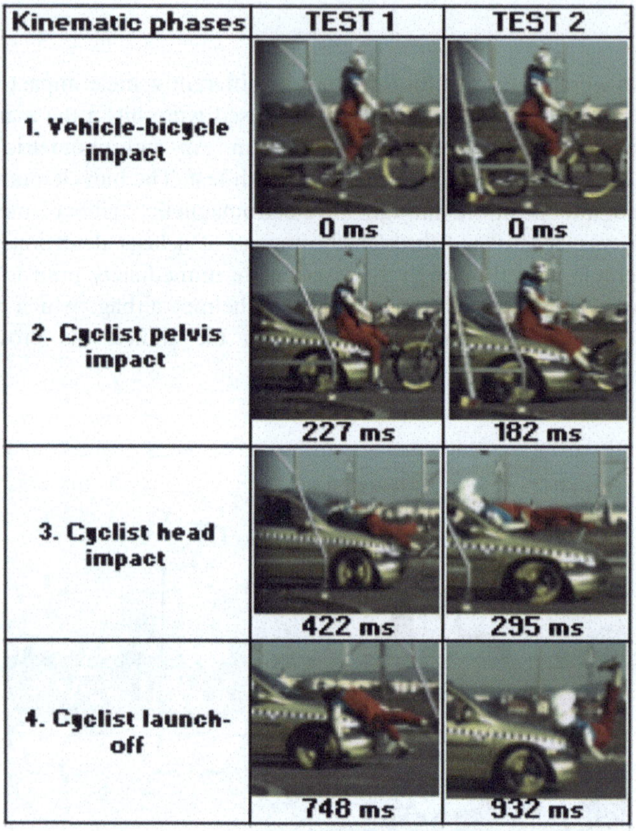

Fig. 2. Cyclist motion for each of the carried-out crash-tests

calculated using the resultant head acceleration as input, which is processed in 15 or 36 ms time intervals, in order to yield HIC 15 or HIC 36 values respectively.

$$HIC = \left\{ \left[\frac{1}{t_2 - t_1} \cdot \int_{t_1}^{t_2} a(t)dt \right]^{2.5} \cdot (t_2 - t_1) \right\} \tag{1}$$

The Neck Injury Criterion (HIC) was proposed in 1996 as a tool to evaluate probabilities of neck injuries. The input parameters used for NIC calculations are the accelerations measured at the level of the head and the first thoracic vertebra (T1), which are processed into 150 ms time intervals.

$$NIC = 0.2 \cdot a_{rel} + v_{rel}^2 \tag{2}$$

$$a_{rel} = a_x^{T1} - a_x^H \tag{3}$$

$$v_{rel} = \int a_{rel} dt \qquad (4)$$

For this study, in order to determine NIC values, the T1 vertebra acceleration was assimilated with the acceleration measured in the center of gravity of the dummy's thorax.

3 Results and Discussion

Test 1 took place at a 6.86 m/s vehicle impact velocity and the inflation of the helmet airbag did not occur during this experiment. The helmet airbag was deployed during Test 2, which was performed at an 11.1 m/s impact velocity. The wrap-around distance which represents the measured distance onto the vehicle's profile, from the ground to the head impact location, was 1670 mm for Test 1 and 2220 mm for Test 2. The head impact location was positioned in the lower part of the windscreen for Test 1 and at the upper edge of the windscreen for Test 2 (see Fig. 3).

Fig. 3. Cyclist head impact location for both crash-tests

Cyclist head and thorax accelerations acquired during the tests (see Figs. 4 and 5 respectively) were used to determine injury head and neck injury criteria values, In order to calculate NIC values, the thorax accelerations measured were assimilated with the acceleration measured at the T1 vertebra level.

The characteristics of each crash-test are shown in Table 1, as they were determined through video analysis and data processing procedure steps, along with the corresponding HIC and NIC values.

The maximum head resultant acceleration obtained in Test 2 (86.9 m/s^2) was nearly three times higher than in Test 1 (30 m/s^2), for a variation in impact velocity of 4.24 m/s.

Test 1 yielded reduced HIC values which consequently determined a null skull fracture probability, furthermore, the obtained HIC values were inferior to the minimum HIC value afferent to the 0% skull fracture probability, as specified in literature [10]. It is clear that for Test 1, the impact was not dangerous in terms of skull fractures or other serious injuries caused by linear motion, however this does not exclude potential injuries caused by rotational motions which cannot be evaluated through HIC. The NIC value obtained for Test 1 was 61 m^2/s^2, which is over four times higher than

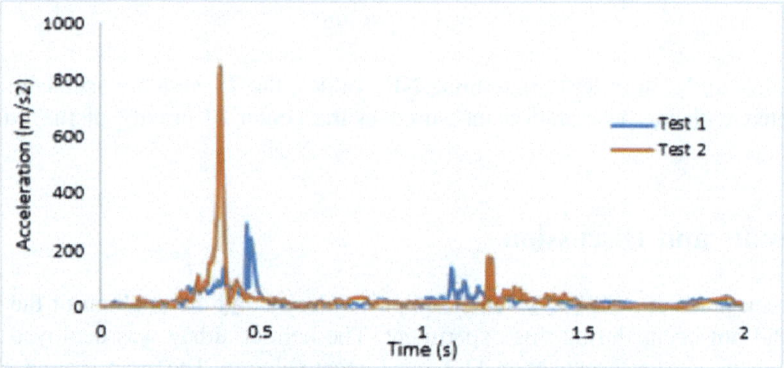

Fig. 4. Dummy head resultant accelerations

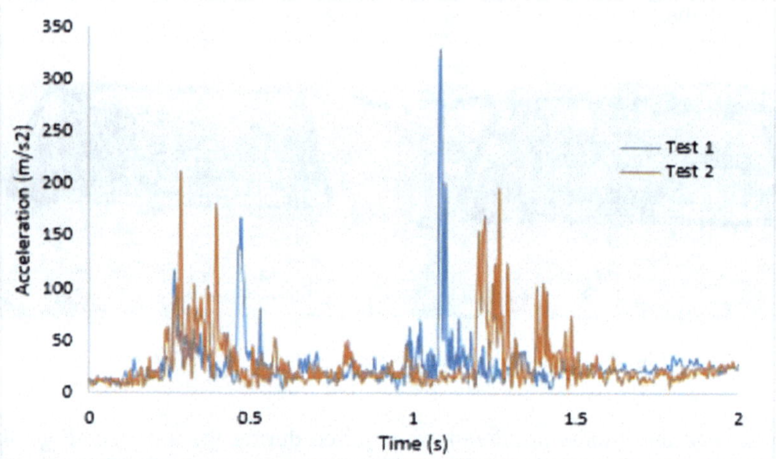

Fig. 5. Dummy thorax resultant accelerations

Table 1. Crash-test characteristics and corresponding injury criteria values

	Impact velocity (m/s)	WAD (mm)	Pelvis impact (ms)	Head impact (ms)	Maximum head acceleration (m/s^2)	HIC 15	HIC 36	NIC (m^2/s^2)
Test 1	6.86	1670	227	422	30	24	16	61
Test 2	11.1	2220	172	295	86.9	268	137	41

the proposed 15 m^2/s^2 threshold specified in literature [11], indicating that the kinematic and dynamic parameters of the collision were more than sufficient to produce severe neck injuries.

HIC values calculated for Test 2 were significantly higher than in Test 1, leading to a 5% skull fracture probability. The NIC value obtained was 41 m^2/s^2, exceeding the 15 m^2/s^2 threshold as well.

A notable result obtained through the comparison of NIC values obtained for each test is represented by the fact that the helmet airbag produced a 33% NIC reduction. Consequently, the airbag significantly diminished the probability for neck injuries, although the impact velocity was superior with 4.24 m/s, which would have clearly generated a higher probability for injury if the helmet airbag would have not been used.

Regarding helmet airbag deployment for Test 2, it can be stated that the 5% skull fracture probability determined implies that a significantly higher skull fracture probability would have been generated if no safety system would have been used. Therefore, the deployment of the airbag was indeed useful and necessary in order to mitigate head injuries.

In Test 1, the relatively low head resultant acceleration of 30 m/s^2 was not sufficient to deploy the airbag, which accordingly to the significantly low HIC values obtained was indeed a dangerless scenario in terms of HIC's solve of severe head injuries. This fact corroborates with the stated characteristic of cyclist accidents according to which cyclist mix with vehicles in relative safety for speeds below 30 km/h.

The cause of the airbag's non-deployment during Test 1 is represented by the low head accelerations on the X and Y axis, yet the system proved to be effective in predicting the neglectable risk of head injuries in this case. However, the results of this study show that even for low impact velocities which generate low head accelerations, the airbag can still have a protective effect by mitigating neck injuries.

4 Conclusion

Traditional cyclist helmets constituted until recently the sole option for protecting against head injuries, yet the appearance of helmet airbags and their effective results show a different viable method, with better results than the conventional approach. The tests carried-out in this study expose another important aspect of passive safety for cyclists, neck protective capabilities respectively. Skull fracture risks were neglectable for Test 1, yet the NIC value obtained was 56% higher (61 m^2/s^2) than the NIC value obtained for Test 2 (41 m^2/s^2), indicating a significant potential of helmet airbags in mitigating cyclist neck injuries.

To this point, there are no passive safety systems for cyclist available related or dedicated to the field of neck injury mitigation. Further improvements of helmet airbags could target the field of neck injuries, for example by implementing a predefined secondary threshold for airbag deployment, lower in value and user-commutable, which would confer the airbag a wider range of protective capabilities.

References

1. European Commission: Traffic safety basic facts on cyclists. European Commission, Directorate General for Transport (2017)
2. Aarts, L., Commandeur, J.J.F., Welsh, R.: Study on serious road traffic injuries in the EU. Publications Office of the European Union, Luxembourg (2016)
3. Maki, T.: The behavior of bicyclists in accidents with cars. JSAE Rev. **21**(3), 357–363 (2000)
4. Fredriksson R.: Priorities for bicyclist protection in car impacts – a real life study of severe injuries and car sources. In: Proceedings of IRCOBI Conference, IRCOBI, Dublin, Ireland, pp. 779–786 (2012)
5. Otte, D.: Possibilities and limitation for protective measures for injury reduction of vulnerable road users. Int. J. Crashworthiness **7**(4), 441–462 (2002)
6. Olsson, D., Sellergren, U.: Airbag suitable for head protection. Hovding Sverige AB. U.S. Patent Application 13/823,986 (2013)
7. Alstin, T., Haupt, A.: System and method for protecting a bodypart. Hovding Sverige AB. U. S. Patent 8,402,568 (2013)
8. Stigson, H.: Consumer testing of bicycle helmets. In: Proceedings of IRCOBI Conference, IRCOBI, Antwerp, Belgium, pp. 173–181 (2017)
9. Kurt, M.: Modeling and optimization of airbag helmets for preventing head injuries in bicycling. Ann. Biomed. Eng. **45**(4), 1148–1160 (2017)
10. Hertz, E.: A note on the head injury criterion (HIC) as a predictor of the risk of skull fracture. In: Proceedings of the 37th Annual Conference of Association for the Advancement of Automotive Medicine, AAAM, San Antonio, Texas, pp. 303–312 (1993)
11. Wheeler, J.B.: Validation of the neck injury criterion (NIC) using kinematic and clinical results from human subjects in rear-end collisions. In: Proceedings of IRCOBI Conference, IRCOBI, Gothenburg, Sweden, pp. 335–348 (1998)

Effects of Vehicle Impact Velocity on Cyclist Injuries and Kinematics for Rear Impacts

Ovidiu Andrei Condrea(✉), Anghel Chiru, George Radu Togănel, and Daniel Dragos Trusca

Transilvania University of Brasov, Brasov, Romania
ovidiu.condrea@unitbv.ro

Abstract. The objective of this study is to assess the influence of the impact configuration and vehicle impact velocity over cyclist injuries. Impact configurations for which the cyclist is collided from the rear were chosen to be studied since this typology is indicated in literature as being the most severe in terms of injuries. In order to complete the objective, a cyclist multibody model was generated, which was further used to carry out simulations of the selected vehicle-cyclist impact typology at different values of the vehicle impact velocity (15 to 100 km/h). The cyclist multibody model was validated through comparison with a real-scale staged test, by comparing kinematic motions, head linear acceleration values and rest positions. Head and neck injury criteria values were assessed by processing accelerations obtained through simulations, and were further used to determine injury probability variance with vehicle impact velocity. The results of this study show not only that head injury probabilities present a polynomial variance with impact velocity, while the neck injury probabilities do not, but also that both are being significantly influenced by cyclist kinematics.

Keywords: Cyclist impact · Injury severity · Multibody simulations

1 Introduction

The bicycle has become throughout the years an increasingly popular mode of transport, both in urban or extra-urban areas. Bicycle use has been promoted globally citing health, environmental and economic advantages, but never for safety reasons. Because cyclists have to share the road with motorized vehicles which are significantly faster, heavier and stiffer, cyclists are very vulnerable in traffic and consequently prone to serious or potentially lethal injuries in case of accidents. In the European Union, cyclists account for 8% of the total accident-induced fatalities [1]. Statistical data show no improvement in cyclist fatality reduction for recent years, with over 2000 deaths each year in Europe only [2].

In order to improve passive safety for cyclists, their generic injury patterns must be well determined and comprehended. Cyclist kinematics can be characterized according to the type of post impact motion: forward projection, wrap projection, fender vault or roof vault [3], the most common being wrap projection at low to medium impact velocities and roof vault at medium to high impact velocities. The former occurs when

I. Dumitru et al. (Eds.): SMAT 2019, *The 30th SIAR International Congress of Automotive and Transport Engineering*, pp. 281–288, 2020.
https://doi.org/10.1007/978-3-030-32564-0_33

the cyclist's body is wrapped upon the vehicle's profile and projected longitudinally in front of the vehicle, while the latter occurs when the vehicle passes under the cyclist after the impact.

One method for assessing cyclist injury patterns consists of processing statistical data regarding accidents (generally collected by police, forensics or medical staff). A study carried out by Maki [4] showed that for 1383 cyclist fatalities, 72% deaths were caused by head injuries, 6% were caused by neck injuries and 8% by thorax injuries. The same study determined that for 17930 cyclist serious injuries, 43% were leg injuries, 21% were head injuries, followed by 13% and 10% for arm and chest injuries respectively. Other study carried out by Otte [5] on 876 cyclist injury cases determined that 26.7% of the seriously injured cyclists (AIS2+) suffered head traumas, 6.7% suffered neck injuries and 22.9% thorax injuries. The injury patterns for vulnerable road users, as highlighted by Otte and Maki, are significantly different. Generically, there is a tendency for cyclists to register less frequently serious leg injuries than pedestrians and motorcyclists, while regarding serious head, neck and thorax injuries, cyclists are more frequently injured than pedestrians and less frequently than motorcyclists. However, statistical data analysis does not offer a complete view over the way cyclist injuries are influenced by impact velocity and configuration. Theoretically, as the vehicle's impact velocity increases, the severity of the cyclist's injuries should increase proportionally, yet as it is shown in this paper, this is not always the case because the cyclist's motion type changes as the impact velocity increases, from wrap projection to roof vault.

The objective of this paper is to determine the influence of the impact configuration and vehicle impact velocity over cyclist injuries, specifically the laws of variance for head and neck injury probability determined through established injury criteria.

2 Methodology

In order to achieve the proposed objective, a real-scale crash-test was performed using an Opel Corsa vehicle and a dummy cyclist/bicycle assembly. The assembly was stationary and maintained into vertical position, while the impact velocity for the vehicle was 5.81 m/s. The test was carried out in a longitudinal rear impact configuration since this impact scenario is indicated in a series of studies as having the highest chance of resulting in lethal injuries for the cyclist [6, 7]. Cyclist motion for the carried-out crash-test was recorded with a high-speed camera Fastec HiSpec 5 and it is shown in Fig. 1.

In order to assess the influence of the vehicle impact velocity over cyclist resulting injuries, it was proceeded to simulate the carried-out crash-test in the PC Crash software, using the same initial conditions. The test was validated by means of achieving the same rest position for the vehicle, cyclist and bicycle (see Fig. 2), similar cyclist and bicycle kinematics and similar peak values for the cyclist's head resultant acceleration.

Using the validated multibody model of the cyclist/bicycle assembly, the simulations were performed for the same impact typology by variating the velocity of the vehicle with 5 km/h periods within the [15; 100] km/h interval.

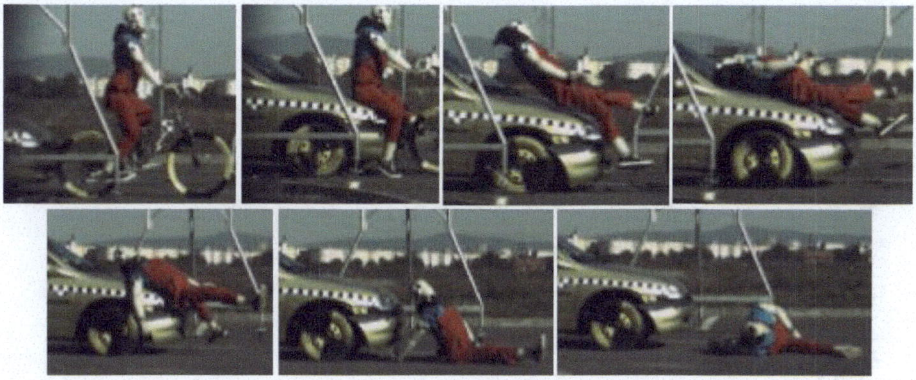

Fig. 1. Cyclist kinematics for the carried-out crash-test

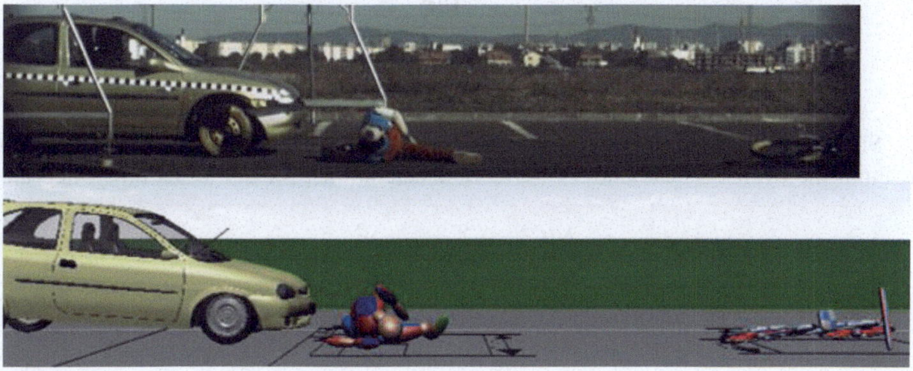

Fig. 2. Vehicle, cyclist and bicycle rest positions for the crash-test and the PC-Crash simulation

Cyclist head and thorax accelerations were extracted from the simulations and were further used as input data to assess head and neck injury criteria values.

The Head Injury Criterion (HIC) was calculated for each simulation by processing the resultant head acceleration of the cyclist, for $(t_2 - t_1)$ time intervals of 15 ms and 36 ms respectively. HIC values were further used to determine skull fracture probabilities.

$$HIC = \left\{ \left[\frac{1}{t_2 - t_1} \cdot \int_{t_1}^{t_2} a(t)dt \right]^{2.5} \cdot (t_2 - t_1) \right\} \quad (1)$$

HIC 15 values were used to assess skull fracture probabilities by using the dependence relation existing in literature [8], depicted in Fig. 3.

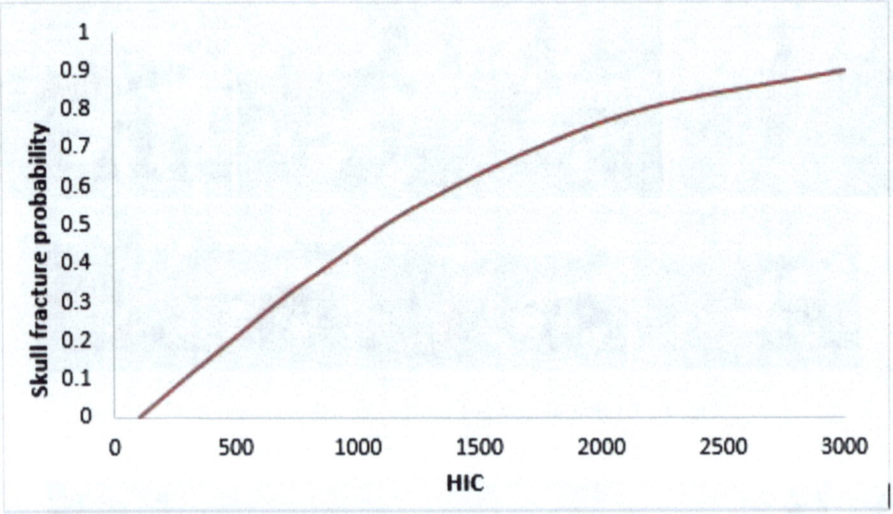

Fig. 3. Relationship between skull fracture probability and HIC values [8]

Similarly, the Neck Injury Criterion (NIC) was calculated for the simulations, using as input the cyclist's head and T1 vertebra accelerations, for a 150 ms time interval.

$$NIC = 0.2 \cdot a_{rel} + v_{rel}^2 \tag{2}$$

$$a_{rel} = a_x^{T1} - a_x^H \tag{3}$$

$$v_{rel} = \int a_{rel} dt \tag{4}$$

3 Results and Discussion

The variation in vehicle impact velocity determines different types of cyclist motion and these changes in cyclist kinematics occur progressively with significant differences in terms of cyclist impact particularities. It has been determined that the area in which the cyclist hits the vehicle is influenced by a series of factors such as the lateral offset, impact angle and configuration, as well as the geometry of the vehicle's frontal profile.

For impact velocities from 15 km/h to 35 km/h cyclist wrap projection occurs and the cyclist is thrown in the vehicle's travel direction. For this interval, the head impact location varies and the wrap-around distance progressively with the impact velocity: for 15 km/h it is positioned on the vehicle's bonnet, for 20 km/h at the windshield's base, for 25 km/h on the center of the windshield and for 30–35 km/h on the roof edge. As the impact velocity is increased, the cyclist will start sliding on the bonnet which will generate a diminished rotational movement, consequently leading to a thorax to windshield impact and a head to roof impact subsequently. The thorax to windshield impact occurred in our study for impact velocities equal or higher than 30 km/h.

In the [40; 100] km/h impact velocity interval, the cyclist is thrown on a longitudinal trajectory, but with significant vertical and transversal velocity components, in a roof vault motion. In this type of motion, the vehicle passes under the cyclist after the thorax-windshield impact, the cyclist hitting the roof of the vehicle with his head during this process.

For impact velocities equal or higher than 75 km/h, the thorax impact location is positioned in the upper right corner of the windshield and the cyclist also impacts the A-pillar and the roof's edge, which induces an additional rotational movement for the cyclist.

Table 1. Injury criteria values, motion types and impact particularities for each simulation

Vehicle impact velocity (km/h)	Cydist motion type	Impact particularities	a max (m/s2)	HIC 15	HIC 36	NIC (m^2/s^2)	Skull fracture probability
15	Wrap projection	Head-bonnet impact	62.71	342	300	75	0.13
20		Head-windshield impact	64.24	455	616	25	0.19
25			45.68	147	110	28	0.02
30		Thorax-windshield and head-roof edge impacts	42.6	125	120	29	0.01
35			42.01	130	201	57	0.01
40	Roof vault	Thorax-windshield and head-roof impacts	53.3	181	277	38	0.04
45			41.45	118	177	46	0.01
50			55.43	310	413	40	0.11
55			56.4	299	605	80	0.11
60			53.36	257	541	49	0.09
65			63.59	443	851	59	0.19
70			75.47	639	1159	130	0.28
75			96.58	823	1761	100	0.37
80			98.91	1222	1752	95	0.54
85			102.03	1342	2100	76	0.58
90			116.23	1775	2643	101	0.72
95			111.67	1661	2635	90	0.69
100			115.44	1872	3011	96	0.75

The results of the simulations in terms of injury criteria values, as well as the corresponding cyclist motion type and impact particularities are presented in Table 1.

The variances of HIC and NIC values with vehicle impact velocity are shown in Figs. 4 and 5 respectively.

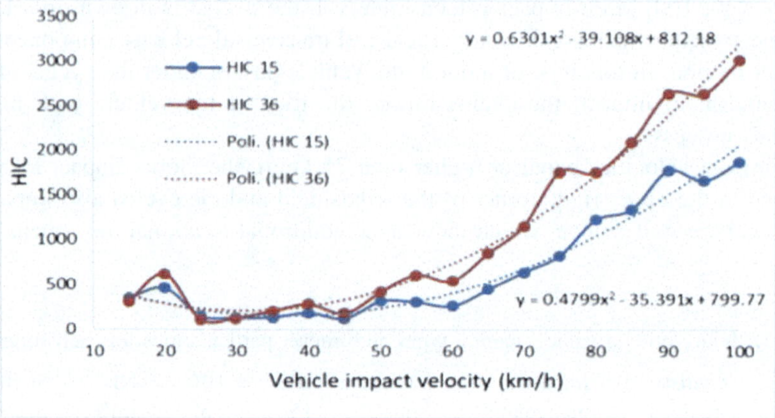

Fig. 4. HIC variance with vehicle impact velocity

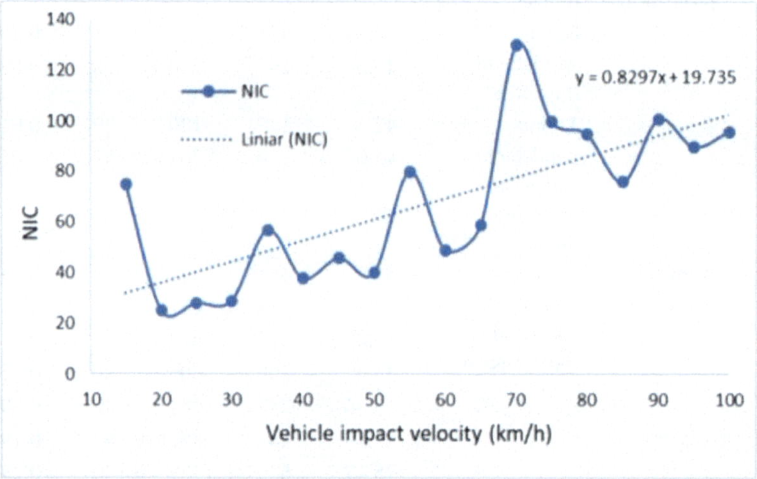

Fig. 5. NIC variance with vehicle impact velocity

The carried-out simulations reflect that injury criteria values vary quasi-progressively with the vehicle's impact velocity, yet a series of exceptions have been observed:

– For the 15 km/h impact velocity, NIC values are superior to the values afferent to the [20; 50] km/h simulations, while HIC values are superior to the [25; 50] km/h simulation values. The skull fracture probability was determined at 13%. The increased injury criteria values were generated as a consequence of the impact between the cyclist's head and the vehicle's bonnet;

- For the 20 km/h impact velocity, HIC values are superior comparing to the rest of the [15; 50] km/h simulations with a skull fracture probability of 19%, due to the impact between the cyclist's head and the base of the windshield;
- For the [25; 100] km/h interval, HIC values and skull fracture probabilities increase progressively, excepting the 45 km/h, 60 km/h, 80 km/h and 95 km/h impact velocities for which HIC values are minorly decreased compared to the previous simulations. Regarding NIC values for this interval, the trend is not similar, with major variances being observed at 35 km/h, 55 km/h, 70 km/h and 85 km/h;
- The most reduced HIC values were obtained for the simulation during which the cyclist's head impacts the center of the windshield, for 25 km/h respectively;
- Skull fracture probabilities are higher than the 50% mark for the [80; 100] km/h interval, the maximum of 75% being determined at 100 km/h.
- The highest NIC value was determined at 70 km/h impact velocity, while the maximum HIC values were determined at 100 km/h;
- It has been determined that the NIC values calculated for the entire set of simulations were superior to the 15 m^2/s^2 threshold indicated in literature [9], implying that neck injuries may occur at any vehicle impact velocity for the rear impact typology. The main cause of this significant potential in neck injury generation is given by the neck's wide extension movement which follows the vehicle-bicycle impact even at low impact velocities, yet another contributing factor is the neck's flexion movement which follows the head impact.

However, it is important to be mentioned that not the entire set of carried-out simulations was validated through comparison with experimental data. Only the 20 km/h simulation was validated with the crash-test presented in Fig. 1, therefore it is possible that the simulations derived from the validated simulation do not reflect a real cyclist kinematic behavior, especially at high vehicle impact velocities.

4 Conclusion

The results of this study show that head and neck injury severity increase quasi-progressively with the vehicle impact velocity, but also that they are significantly influenced by cyclist kinematics:

- For the 15 km/h impact velocity, injury criteria values (HIC15 = 342; HIC36 = 300; NIC = 75 m^2/s^2) are higher than for the [20; 45] km/h interval due to the impact between the cyclist's head and the vehicle's bonnet;
- Head to windshield impacts generate different injury probabilities depending on the impact location: the head impact at the base of the windshield at 20 km/h generated a higher skull fracture probability of 19% than for the head impact positioned in the center of the windshield at 25 km/h (2% skull fracture probability);
- At low vehicle impact velocities when the cyclist hits the central part of the windshield (25 km/h), skull fracture risks are minimal;
- The most reduced HIC values (HIC15 = 147; HIC36 = 110) were obtained for the simulation during which the cyclist's head impacts the center of the windshield, for 25 km/h respectively;

- The highest NIC value was obtained at 70 km/h impact velocity and the highest HIC values were obtained at the 100 km/h impact velocity;
- NIC measured variance with impact velocity is not progressive, yet the NIC trendline is quasi-linear. HIC values determined for both 15 and 36 ms intervals show polynomial variances.

The obtained NIC values were superior to the 15 m^2/s^2 threshold for each carried-out simulation, reflecting high probabilities for cyclists to sustain severe neck injuries at any impact velocity.

References

1. European Commission: Road Safety in the European Union – Trends, Statistics and Main Challenges, European Commission (2015)
2. European Commission: Traffic Safety Basic Facts on Cyclists, European Commission, Directorate General for Transport (2017)
3. Simms, C., Wood, D.: Pedestrian and Cyclist Impact: A Biomechanical Perspective, vol. 166. Springer, Dublin (2009)
4. Maki, T.: Comparative analysis of vehicle–bicyclist and vehicle–pedestrian accidents in Japan. Accid. Anal. Prev. **35**(6), 927–940 (2003)
5. Otte, D.: Possibilities and limitation for protective measures for injury reduction of vulnerable road users. Int. J. Crashworthiness **7**(4), 441–462 (2002)
6. European Commission: ASPECSS Assessment methodologies for forward looking integrated pedestrian and further extension to cyclist safety. European Commission (2011)
7. Maki, T.: The behavior of bicyclists in accidents with cars. JSAE review **21**(3), 357–363 (2000)
8. Hertz, E.: A note on the head injury criterion (HIC) as a predictor of the risk of skull fracture. In: Proceedings of the 37th Annual Conference of Association for the Advancement of Automotive Medicine, pp. 303–312. AAAM, San Antonio (1993)
9. Wheeler, J.B.: Validation of the neck injury criterion (NIC) using kinematic and clinical results from human subjects in rear-end collisions. In: Proceedings of IRCOBI Conference, pp. 335–348. IRCOBI, Gothenburg (1998)

Case Study Regarding the Reconstruction of a Real Vehicle-Bicycle Accident

Ovidiu Andrei Condrea(✉), Anghel Chiru, George Radu Togănel, Daniel Dragos Trusca, and Rares Lucian Chiriac

Transilvania University of Brasov, Brasov, Romania
ovidiu.condrea@unitbv.ro

Abstract. In this paper, a real vehicle-bicycle accident, which was recorded in an urban area in Romania, was reconstructed in PC Crash. This paper describes the prerequisites and the ensemble of procedures needed for the reconstruction of similar accidents, as well as some input parameters which can be extracted from the accident scene or through video analysis. The video presented in this paper was recorded with a moving camera and was further processed in a video analysis software. Vehicle and cyclist kinematic parameters were assessed with the object tracking function and were subsequently corrected in order to compensate the effect of the moving camera. The reconstruction of the accident was carried out by using the kinematic parameters obtained through video analysis as input data for the simulation. Subsequently, the input data and the rest of the undetermined parameters were methodically iterated in order to reconstruct the accident. The validated simulation shows similar cyclist and vehicle kinematics, as well as similar wrap-around distances and rest positions with the video presented as case study.

Keywords: Video analysis · Multibody simulation · Accident reconstruction

1 Introduction

Road traffic became an increasingly dangerous environment for cyclists, as statistical data shows out. Motor vehicles represent the most common collision partner for cyclists. A study carried out by Otte showed that 63.5% of the cyclist accidents recorded in GIDAS database were generated due to collisions with vehicles [1]. A vehicle-cyclist accident can generate profound and long lasting medical effects for the cyclist, fatalities, as well as material losses [2]. Shortly after an accident is produced and signaled through the emergency line, specialist teams comprised of police personnel and medical staff are sent to the accident scene to gather material evidence and conduct rescue operations. This way, a criminal investigation is opened and forensics will subsequently analyze and corroborate the evidence collected by the police with the purpose of determining the causes of the accident and who is legally responsible for it.

The reconstruction of an accident implies the assessment of all definitory kinematic and dynamic parameters of an accident, with the purpose of elucidating the causes which led to the event and studying the common patterns and particularities of accidents in general [3]. Due to the complex nature of accidents, the reconstruction process

I. Dumitru et al. (Eds.): SMAT 2019, *The 30th SIAR International Congress of Automotive and Transport Engineering*, pp. 289–296, 2020.
https://doi.org/10.1007/978-3-030-32564-0_34

requires a significant amount of information which generically is not entirely available based on the material evidence. Simulation software based on multibody approaches such as PC Crash [4] complement the lack of information obtained from the scene of the accident, by allowing unlimited iterations of all kinematic and dynamic parameters. The role of the reconstructionist is to model the accident and bring the simulation to conformance with the real case in terms of throw distances, impact locations and material traces. Numerous studies regarding vehicle-bicycle accidents were carried-out using the multibody approach, with good correlations between simulation and reality [5–8].

In the case of video-recorded accidents, some simulation input data can be extracted through video-analysis, specifically kinematic parameters in the view plane can be assessed and mildly corrected, which limits the number of iterations required. Since the accident kinematics are also recorded, another criteria for verification of the reconstruction is given by the resemblance of the kinematics depicted in the simulation and the video.

This paper presents the methodological process required for reconstructing vehicle-cyclist accidents, described as a logical step procedure, along with a real video-recorded reconstructed accident presented as a case study.

2 Reconstruction Methodology

Accident are characterized by material traces left behind by the vehicle, bicycle or cyclist. The vehicle may l brake marks on the road and the damages produced during the accident will indicate the impact location with the bicycle and the cyclist respectively. The bicycle's damages indicate the vehicle's impacting direction and the impact angle. Regarding the material traces generated by the cyclist, these are comprised of biological stains which form at the impact location on the vehicle (usually hair marks) and in the area in which the cyclist stops in his rest position (blood stains). If the accident is video-recorded, the reconstruction process is facilitated since the definitory moments and kinematic phases of the accident are caught on camera. Furthermore, the distances and velocities of the objects depicted in the video can be measured in the view plane through a video analysis software. The software used in this study is Tracker Video Analysis and Modelling Tool 5.1.2. freeware [9], built upon the Open Source Physics Java code library, which allows for video scale calibrations and object tracking with assessment of kinematic parameters. This can be done by selecting a reference distance within a frame and setting the referential system, followed by marking the interest points during each frame. The marking can be done manually or automatically if the respective point is positioned in a high contrast area. The precision with which kinematic parameters are determined through a video-analysis software is not always enough and it is directly influenced by the acquisition frequency of the camera, more accurate results being obtained for higher frequencies respectively. Multibody simulations can serve as a tool to verify the data obtained from the video recording, while also offering information about the kinematic and dynamic parameters of the accident in a tridimensional manner. The reconstruction methodology (see Fig. 1) is comprised of three phases:

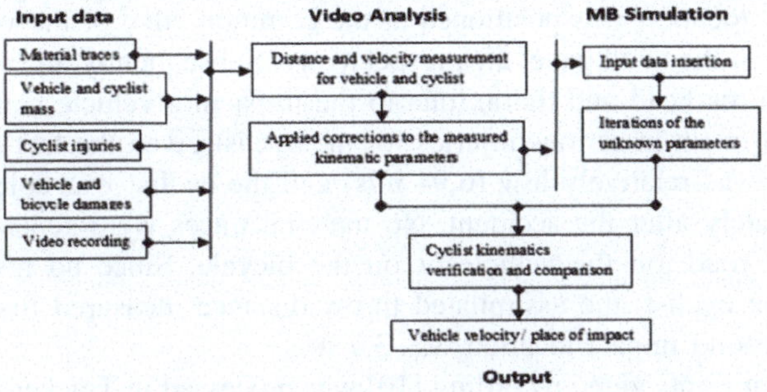

Fig. 1. Summary of accident reconstruction data and procedures

- Input data extraction from material traces and injury patterns;
- Input data extraction through video analysis, if the accident was video recorded;
- Multibody simulation phase during which input data are iterated until all validation criteria are met.

3 Case Study: Reconstruction of a Video-Recorded Vehicle-Bicycle Accident

The vehicle-bicycle accident presented in this paper took place in a roundabout intersection located in the city of Piatra Neamt, Romania. The accident was recorded at a 90 degree angle by a dash-board camera mounted on a vehicle (referred to as R.V. – recording vehicle) which was braked prior to the entrance of the roundabout, (see Fig. 2).

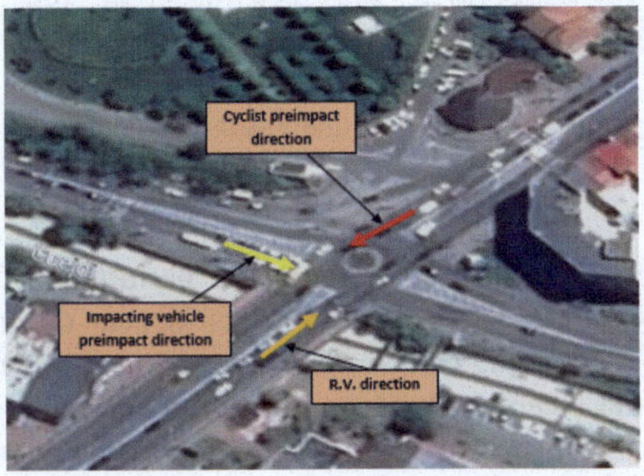

Fig. 2. Vehicle, cyclist and R.V. movement directions

The driver of the vehicle didn't observe the cyclist's presence inside the intersection, therefore the cyclist-bicycle assembly was hit on the right side by the front of the vehicle. After the vehicle-bicycle impact, the cyclist rotated on the vehicle's bonnet and the head impact location was positioned at the common edge of the vehicle's bonnet and windshield, in the cowl panel area respectively. Subsequently, the cyclist impacted the ground with his head and thorax due to the impacting vehicle being braked. The accident did not result in serious injuries for the cyclist given the fact that the vehicle impact velocity was relatively low (6.94 m/s) and the cyclist was able to rise off the ground immediately after the accident. No material traces were generated during the accident on the road, on the vehicle or on the bicycle. Since no rest position was identified for the cyclist, the assimilated throw distance measured from the place of impact to the ground impact location was 3.7 m.

The free from rights video recording [10] was processed in Tracker Video Analysis and Modelling Tool in order to assess the kinematic parameters of the vehicle and of the cyclist's head and thorax. Given the fact that the R.V. was not stationary prior and during the accident, the objects appearing in the video recording were scaled differently for each frame. Dimensional variations, as well as vertical and horizontal shifting of the objects depicted in the video frames were observed due to braked movement of the R. V. and its approach to the roundabout intersection in which the accident took place, as shown in Fig. 3.

Fig. 3. Kinematic sequences afferent to the video-recording presented as case-study

In order to assess the kinematic parameters of the accident properly through video analysis, vertical and horizontal corrections were performed. These corrections are necessary in order to compensate the errors which appear during the video analysis procedure step due to the moving camera. The adjustment was carried out by using the width of the roundabout's center (7 m) as a fixed landmark and assessing the movement of its extremities in the video. Correction diagrams were established in order to assess the deviations of the landmark's extremities positions at each moment on the vertical and horizontal axes. Consequently, using the determined vertical and horizontal deviations, the proportionality scale between the values determined through video analysis and the real values was established globally. The vertical, horizontal and global correction diagrams which were used are presented in Figs. 4, 5 and 6.

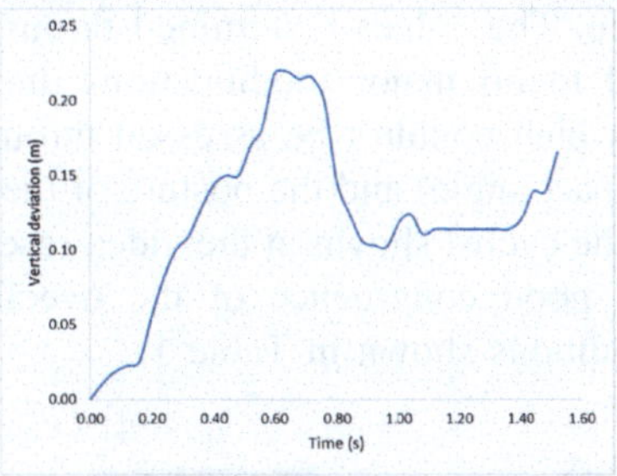

Fig. 4. Correction diagram for the vertical axis

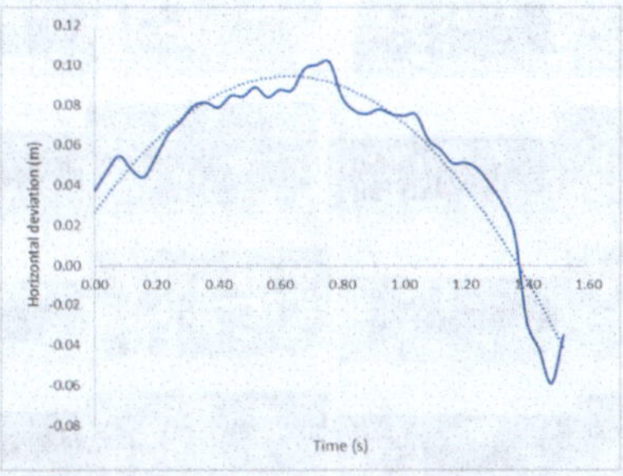

Fig. 5. Correction diagram for the horizontal axis

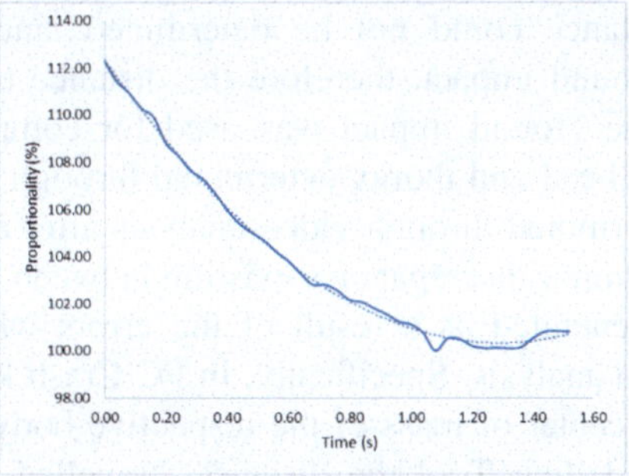

Fig. 6. Global correction diagram

The assessment of the distances and velocities was done only for those positioned in the view plane, for an acquisition frequency of the dash-board camera determined at 25 fps. Vehicle kinematic parameters determined through video analysis were introduced as input data in PC Crash for simulation, while impact and cyclist kinematic parameters were iterated until the cyclist kinematics afferent to the simulation resembled the kinematics generated in reality (see Fig. 7). The initial conditions of the simulation were methodically iterated. Firstly, vehicle impact velocity and braking deceleration were iterated until the braking distance of the vehicle was identical with the one shown in the video. The values determined through video analysis of these parameters were subjected to no major modifications during the iterative process. Secondly, the parameters which couldn't be assessed through video analysis (cyclist impact velocity and the impact angle) and the posture of the cyclist were iterated until the kinematic behavior of the cyclist shown in the video resembled the dummy motion obtained in PC Crash. A good congruence of the overall cyclist kinematics was obtained for the initial conditions shown in Table 1.

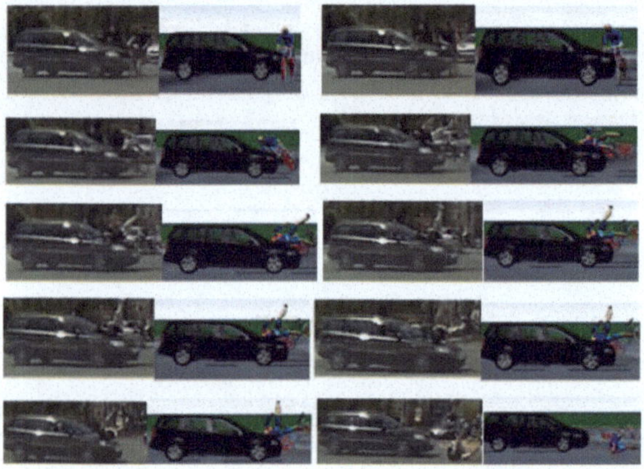

Fig. 7. Cyclist kinematic phases of the accident and simulation

The total throw distance could not be determined since the cyclist stood up immediately after the ground impact, therefore the distance travelled by the cyclist's head and thorax until the ground impact was used for comparison. The trajectories followed by the cyclist's head and thorax determined through simulation show a good match with the ones determined through video-analysis after applying the corrections, as shown in Fig. 8. The minor discrepancies existing between both pairs of trajectories (head and thorax) are generated as a result of the errors which appear during data acquisition through video analysis. Specifically, in PC Crash kinematic parameters are measured directly at the center of mass of the respective body zone, while in Tracker Video Analysis and Modelling Tool the distances travelled by each body zone are determined by following the movement of a point positioned in a specific high contrast area for each frame. However, the object tracking function does not allow for the precise determination of one specific point movement, instead it will acquire the coordinates for a similar high contrast point in the defined area, therefore minorly alternating values. Furthermore, given the rough quality of the video caused by the relatively low acquisition frequency of the dash-board camera, the tracking of the thorax movement was carried-out by defining an area situated in the upper half, above the sternum area respectively, positioned at a superior height than the thorax's center of mass. This explains the initial thorax height variation of 0.1 m between the simulation and the video at the time of the vehicle-bicycle impact.

Table 1. The initial conditions used for simulating the accident

Vehicle impact velocity (m/s)	Cyclist impact velocity (m/s)	Impact angle (deg)	Vehicle braking deceleration (m/s^2)
6.94	2.22	51	−7.8

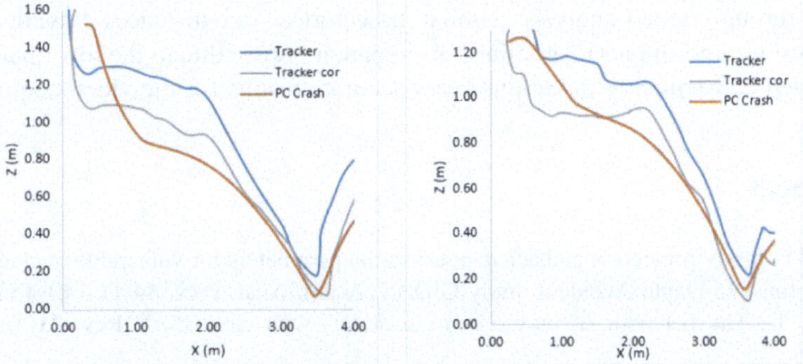

Fig. 8. Cyclist head and thorax trajectories obtained through video analysis and PC Crash

As the video depicts, the cyclist rose off the ground immediately indicating minor injuries sustained. Using the head and thorax accelerations obtained through simulation, head and neck injury criteria values (HIC and NIC) were determined, as shown in Table 2. The comparison with the threshold value afferent to each injury criteria shows that the accident was not particularly dangerous for the cyclist. The HIC 36 value was determined at 69 and the HIC 15 value was determined at 61, the latter representing nearly 9% of the threshold of 700 afferent to the 50% skull fracture probability [11]. The NIC values obtained was 1.72 m^2/s^2, making for 11% of the specified threshold in literature [12].

Table 2. Head and neck injury criteria values

HIC 36	HIC 15	HIC 15 threshold	NIC	NIC threshold
69	61	700	1.72	15

4 Conclusion

This paper presents a reconstruction methodology for cyclist accidents, as well as the reconstruction of a video recorded accident. For recorded accidents, data acquisition through video analysis offers a solid grasp over the evolution of the kinematic parameters afferent to each object involved in the collision (vehicle, cyclist or bicycle) and the phenomena which influence it, although restricted to the ones positioned in the visible plane of the recording. However, for low-quality videos such as the one presented in this study, the video-analysis can not serve as a stand-alone procedure for acquiring data about the accident since it is heavily influenced by the acquisition frequency and luminosity conditions (a high contrast area is required for the acquisition to be carried-out optimally). Kinematically validated multibody simulations are effective for complementing and verifying data extracted from video recordings, while simultaneously being suitable for reconstructing unrecorded accidents. The simulation of the case presented in this study showed a good congruence of cyclist kinematics with the ones depicted in the video. Simulation data show a good correlation with the values acquired through video-analysis (similar trajectories and distances travelled by the cyclist until ground impact), yet minor discrepancies arose due to the low quality of the video which consequently determined less accurate results for the video-acquired data.

References

1. Otte, D.: Injury protection and accident causation parameters for vulnerable road users based on German In-Depth Accident Study GIDAS. Accid. Anal. Prev. **44**(1), 149–153 (2012)
2. Maki, T.: The behavior of bicyclists in accidents with cars. JSAE Rev. **21**(3), 357–363 (2000)
3. Carter, E.L.: MADYMO reconstruction of a real-world collision between a vehicle and cyclist. Int. J. Crashworthiness **14**(4), 379–390 (2009)
4. Moser, A.: The new multibody model in PC-Crash – simulation and validation of single track vehicles. In: Proceedings of the 26th Annual Congress of the European Association for Accident Research and Analysis, pp. 181–187. EVU Conference, Copenhagen (2014)
5. Raslavičius, L.: Accident reconstruction and assessment of cyclist's injuries sustained in car-to-bicycle collision. Procedia Eng. **187**, 562–569 (2017)
6. Van Schijndel, M.: Cyclist kinematics in car impacts reconstructed in simulations and full scale testing with Polar dummy. In: Proceedings of IRCOBI Conference, pp. 800–812. IRCOBI, Dublin (2012)
7. Sun, J.: Identification of pre-impact conditions of a cyclist involved in a vehicle-bicycle accident using an optimized MADYMO reconstruction combined with motion capture. J. Forensic Leg. Med. **56**, 99–107 (2018)
8. Bourdet, N.: Methodology for a global bicycle real world accidents reconstruction. In: Proceedings of ICrash 2012-International Crashworthiness Conference, pp. 335–348. HAL, Milan (2012)
9. Tracker Video Analysis and Modelling Tool Homepage. https://physlets.org/tracker. Accessed 21 Aug 2019
10. https://www.youtube.com/watch?v=_zAh4qELjfs. Accessed 21 Aug 2019
11. Hertz, E.: A note on the head injury criterion (HIC) as a predictor of the risk of skull fracture. In: Proceedings of the 37th Annual Conference of Association for the Advancement of Automotive Medicine, pp. 303–312. AAAM, San Antonio (1993)
12. Wheeler, J.B.: Validation of the neck injury criterion (NIC) using kinematic and clinical results from human subjects in rear-end collisions. In: Proceedings of IRCOBI Conference, pp. 335–348. IRCOBI, Gothenburg (1998)

Image Processing Applied in Road Traffic Using a Quadcopter

Carmen Gheorghe(✉) and Nicolae Filip

Technical University of Cluj-Napoca, Muncii Boulevard 103-105, Cluj-Napoca, Romania
Carmen.Gheorghe@auto.utcluj.ro

Abstract. Video detection of the vehicles in road traffic in order to identify the macroscopic parameters is a technique used very often nowadays, due to the real advantages that it offers. In this present paper the main purpose is to observe road traffic parameters using images obtained from a quadcopter having an attached camera. As part of processing images obtained using the aircraft, in computer science, digital image processing involves the use of computer algorithms to perform digital image processing. By extracting and processing two images, representing the control frame and vehicle frame, can be observed, comparatively, how the pixel intensity changes based on the passing thru of a vehicle within the control frame. Thus, we have developed a case study to analyze an extra-urban area near the city of Cluj-Napoca. In the case study we used the original, RGB (Red, Green, Blue) image and applied a code in MATLAB software in order to obtain a BW (Black and White) image, so we could generate a graph for comparison. As further applications of the developed study, we propose the use of a quadcopter as an alternative method of monitoring traffic parameters, beside the traffic radars and infrared video cameras.

Keywords: Image · Processing · Quadcopter

1 Introduction

Nowadays, the quadcopters or also known as small-scale unmanned aerial vehicles or drones are used for many applications and researches because of its good maneuverability and hovering ability. With the attached camera are obtained images which are used in different fields as flight control, real time navigation systems, exploring the environment but also for surveillance in the military field, recording different kind of sports and festivals or in journalism.

In the logistic field, it was developed a generic algorithm so the transportation of materials inside a factory can be done in short time and with minimal energy consumption. The heavy-duty quadcopters can load up to 20 kg of materials but the real issue is the electric storage in batteries that allow only a few minutes of operations (Olivares et al. 2015).

In the road traffic monitoring field previous studies and tests have shown that the quadcopters can be successfully used as a survey tool for analyzing the traffic dynamics (Salvo et al. 2014). Using recordings from the quadcopter can also be observed single

I. Dumitru et al. (Eds.): SMAT 2019, *The 30th SIAR International Congress of Automotive and Transport Engineering*, pp. 297–303, 2020.
https://doi.org/10.1007/978-3-030-32564-0_35

vehicle road traffic parameters using the spatiotemporal characteristics of the moving synchronized flow patterns at the signal in city traffic (Kaufmann et al. 2018).

Also, the quadcopter camera can be used to determine the vehicle speeds and trajectories through video processing techniques by comparison with images obtained from a high accuracy Global Positioning System (GPS) (Guido et al. 2016).

In this paper the quadcopter was used in order to observe a road path with and without a vehicle on it. The quadcopter's camera recorded a video and then we applied the methodology in two phases: image acquisition using a drone system and the second phases – image processing using a MATLAB algorithm.

2 Methodology and Materials

As methodology, we observed a road path located nearby Cluj-Napoca city. The quadcopter was used in order to obtain, with its camera, relevant and several images which were processed using a MATLAB algorithm from Image Processing Toolbox. In Fig. 1 is presented the large, analyzed area.

Fig. 1. The aerial view of the analyzed area

The materials we used include a quadcopter, a metric measuring wheel and MATLAB software. *HUBSAN H501M X4* is a quadcopter which has many features, including: *Return to Home, Follow Me* or *Waypoint*. It has superior motors so the quadcopter is powerful enough to stay in place, above the road up to 20 min. The Global Positioning System (GPS) allows the quadcopter to enter in the *failsafe mode*, that means it automatically returns to the starting point if they lose the signal. Usually, the tests are made nearby the road so the quadcopter don't lose signal or connection with the remote. As part of the quadcopter system, the remote, or joystick, has a smartphone grip system. In order to be able to use the smartphone, we installed the

X-Hubsan free app. The media recording was downloaded directly on the drone, using a microSD card (Fig. 2).

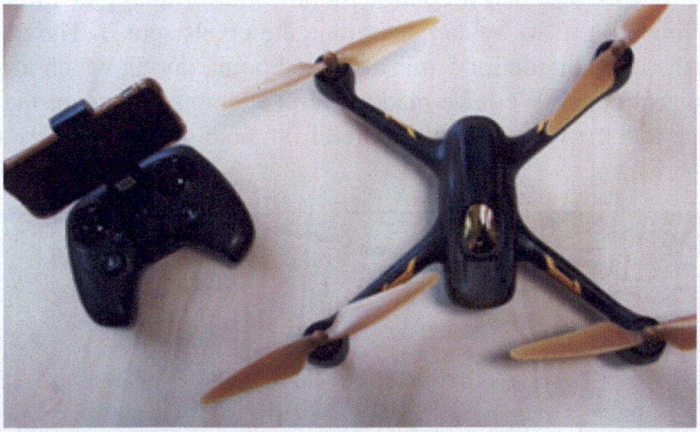

Fig. 2. The quadcopter *HUBSAN H501M X4*

The technical specifications of the quadcopter are as follows:

- Weight (with battery and propellers): 0,41 kg
- Maximum Flight Time: 21 min (0 m/s wind)
- Maximum Hovering Time: 18 min (0 m/s wind)
- Overall flight time: 20 min
- Maximum Flight Distance: 0.29 km (0 m/s wind)
- Operating Temperature: 0°–40 °C
- GPS Mode: Active

The camera specifications are as follows:

- Sensor: Effective pixels: 2.1 MP
- Photography Modes: Single shot
- Video Recording Mode: 720 p
- Maximum Video Bit Rate: 60 Mbps
- Photo extension: JPEG
- Video format: MP4 (MPEG-4 AVC/H.264)

The remote controller specifications are as follows:

- Frequency: 2.4 GHz
- Channels: 11
- Maximum Transmission Distance: 300 m

The Wi-fi Control specifications are the following:

- Frequency: Wi-fi
- Maximum Transmission Distance (Without Relay): 50 m
- Max Transmission Distance (With Relay): 300 m

Using the video camera of the quadcopter, presented above, we have obtained several images of the area. After a brief analysis we have chosen for the processing part, the most relevant image to describe the road.

First of all, it was obtained a control framework, an image of the analyzed road without any vehicle on it, so we can determine the pixels spread. The quadcopter was positioned and held at altitude for 5 min at a fixed point, during which a vehicle passed through the control panel. We have made several attempts to capture the frame image so that it is clear and the vehicles are captured from both directions of the road (Fig. 3).

Fig. 3. The control framework

Starting from an original image captured by the quadcopter's camera, we applied MATLAB codes to determine the intensity of the image. Each image element of an image or each pixel is a specific number in the coordinate system. In the second image we captured a vehicle so we can determine how the pixel's intensity is changing in the area where the vehicle appeared (Fig. 4).

2.1 Image Processing Algorithm

Beginning with an original image, captured by the quadcopter's camera, we have applied MATLAB codes in order to determine the intensity of the image. Each element of an image, or more specifically, each pixel can be represented by a number in the three dimensional coordinate system. We mention that the images were processed in the laboratory, after we downloaded them from the quadcopter's internal memory.

Pixel is defined as the smallest element in a digital image. Bit depth of the digital image indicates the number of bits allocated to a pixel. If an image is made only of black and white pixels is an image on bit.

Each pixel is assigned a bit which may be 0, which means black or may be 1, which means white. The Red Green Blue color image is actually a combination of three 8-bit

Fig. 4. The vehicle framework

images, one for Red, one for Green and one for Blue. The Red Green Blue image is a 8-bit image on a 24-bit channel or 24-bit image capable of rendering $(2^8) \cdot (2^8) \cdot (2^8)$ colors, that would be about 16.7 mil. colors.

Starting from the original image, no matter if it is the control framework or car framework, the algorithm applies the same way, we obtained a matrix in background of the RGB image.

In the first place we considered the *control framework*, so we could determine the intensities of pixels for the road path with no vehicle on it. After Matlab software calculated the matrix in background it was able to plot a surface for visible, understandable results.

Secondly we compared it to another image, with a vehicle on the road path, named *car framework*. The background steps Matlab followed were exactly the same for both frameworks.

In order to achieve real, credible results, the algorithm we followed with the necessary steps helped us to obtain interpretable surface plots, generated by the MATLAB software. The algorithm is presented in Fig. 5 as follows:

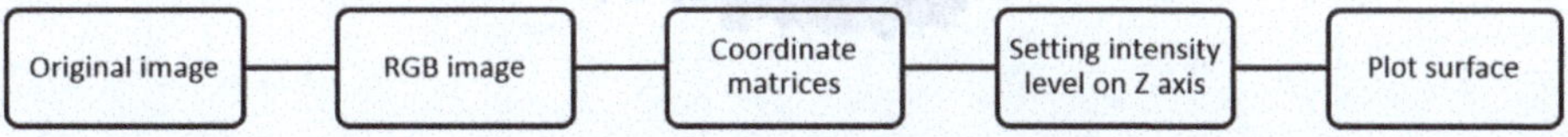

Fig. 5. Image processing algorithm

3 Results

Applying the algorithm in MATLAB Software, we obtained a surface plot with three axes, all of them showing the number of pixels on length, width and height.

In Fig. 6 it can be seen the graphical representation of the number of pixels in the control framework, converted RGB image.

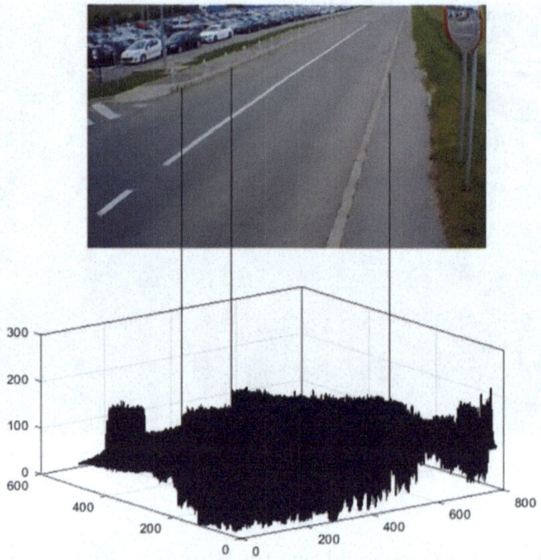

Fig. 6. The pixel's intensity in the control framework, RGB image

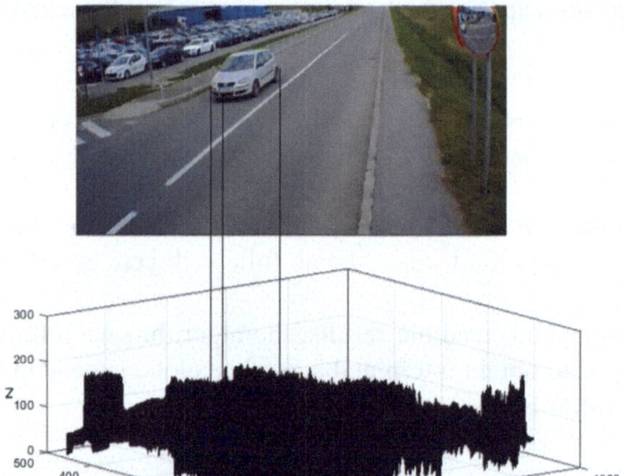

Fig. 7. The pixel's intensity in the vehicle framework, RGB image

It can be observed that the number of pixels on height axis in higher as the colour in the image is brighter, so for the middle of the road path the average number of pixels is 100. The pixel values for the entire image vary from 0 to 210. Minimum values are located in the darkest corners of the image.

In Fig. 7 the number of pixels is slightly increasing in the part of the image where the vehicle appears. So, it is visible on the surface plot that, on the height, the average number of pixels is 120, in contrast to the control framework image where is on average 100. The phenomenon is explainable by the fact that the car is seen as a brighter spot on the road path.

4 Conclusions

The usefulness of the quadcopter in road traffic engineering is obvious; because of its small dimensions, the device can fly over any road path, in a discrete way, without disturbing the traffic in any way.

In this paper the main purpose was to observe a vehicle presence on the road by using images obtained from a quadcopter having an attached camera. In engineering, computer's algorithms are often used with the purpose to process digital images.

As it could be seen, by extracting and processing two images, representing the control framework and vehicle framework, we could observe, comparatively, how the pixel intensity changes on height, in the moment when a vehicle passes thru the control framework.

The research in this paper has demonstrated that the possibility of detecting vehicles on a road path using a quadcopter is realistic. At the point where we arrived with the algorithm, only the vehicle as an entity can be detected on the road path. In the future we will expand the algorithm in order to be able to detect vehicles by shape or color, not only in images, but in videos too.

References

Crisan, H.G., Filip, N.: Traffic modeling aspects using Visum software and effects on the traffic optimization. In: Proceedings of the European Automotive congress EAEC-ESFA 2015, pp. 495–506 (2016)

Chow, J.: Dynamic UAV-based traffic monitoring under uncertainty as a stochastic arc-inventory routing policy. Int. J. Transp. Sci. Technol. **5**, 167–185 (2016)

Filip, N.: Ingineria Traficului Rutier, Cluj-Napoca (2010). ISBN 978-973-713-268-0

Guido, G., Gallelli, V., Rogano, D., Vitale, A.: Evaluating the accuracy of vehicle tracking data obtained from Unmanned Aerial Vehicles. Int. J. Transp. Sci. Technol. **5**, 136–151 (2016)

Kaufmann, S., Kerner, B.S., Rehborn, H., Koller, M., Klenov, S.L.: Aerial observations of moving synchronized flow patterns in over-saturated city traffic. Transp. Res. Part C **86** (2018), 393–406 (2018)

Olivares, V., Cordova, F., Sepulveda, J.M., Derpich, I.: Modeling internal logistics by using drones on the stage of assembly of products. Procedia Comput. Sci. **55**, 1240–1249 (2015). Information Technology and Quantitative Management (ITQM 2015)

Salvo, G., Caruso, L., Scordo, A.: Urban traffic analysis through an UAV. Procedia - Soc. Behav. Sci. **111**, 1083–1091 (2014)

Britannica Homepage. https://www.britannica.com/biography/Louis-Charles-Breguet. Accessed 12 Apr 2019

Wikipedia Homepage. https://en.wikipedia.org/wiki/Quadcopter#Early_attempts. Accessed 12 Apr 2019

The guardian Homepage. https://www.theguardian.com/media/media-blog/2014/feb/12/journalism-drone-capture-floods-protests-wars. Accessed 12 Apr 2019

USA Today Homepage. https://ftw.usatoday.com/2017/02/yes-those-were-drones-at-lady-gagas-super-bowl-51-halftime-show. Accessed 12 Apr 2019

Road Traffic Studies for Urban Development

Georgiana Silvia Muresan, Carmen Gheorghe(✉), and Nicolae Filip

Technical University of Cluj-Napoca, Muncii Boulevard 103-105, Cluj-Napoca, Romania
Carmen.Gheorghe@auto.utcluj.ro

Abstract. The purpose of this paper is to analyze road traffic studies as part of the necessary conditions for the development of urban real estate assemblies. Road traffic studies include its structure and requirements, and traffic analysis is done for three types of scenarios. These scenarios are divided into: the current stage, the stage with the inclusion of the building and the perspective view for a period of 5 years, including the effect of the building. As part of the study, calculations are included to determine traffic flow capacities of road arteries, simulations in Synchro software, and the estimation of effects produced by greenhouse effect gases for all three scenarios. The last part of the road traffic study includes the accesses, to and from the building, so that overall road traffic is less affected. These road traffic studies are an important preliminary part of starting a building construction, as a large number of parking spaces allocated to the building bring a stream of additional vehicles to the urban area.

Keywords: Traffic · Building · Parking

1 Introduction

Density of living and spatial distribution of urban interest points, such as jobs, shopping areas, recreation areas, are key elements in urban development planning. Road traffic diurnal pattern develops mistakes caused by ignoring these issues, escalating urban problems through crowding, accidents, and air pollution. Urban developments have taken a long-lasting impact, which affects road traffic conditions.

Road traffic studies are a necessity at the local government level because the contribution of vehicles that the buildings bring is significant and can disturb the traffic at a fairly high level.

Regarding the accessibility there are many benefits coming along with the private transport and different can be reached by developments owners. Also, the public transport has an impact on property prices and location, aspect which is directly related with the number of building in a city (Cordera et al. 2019).

There were made empirical examinations of the urban development patterns which illustrated that the accessibility and also the availability of the developable pieces of land can be used as a base of a real estate land type of use model (Hansen 1959).

In this paper is presented a pattern of road traffic study which is usually used as a part documentation for the approval in the local administration commission of a building construction in Cluj-Napoca city. First of all, the project must have a purpose

I. Dumitru et al. (Eds.): SMAT 2019, *The 30th SIAR International Congress of Automotive and Transport Engineering*, pp. 304–311, 2020.
https://doi.org/10.1007/978-3-030-32564-0_36

and, based on this, we collect traffic data, then we process these data and calculate the forecast indicators for two scenarios. The two scenarios are the one including the parking lots of the building and the one for the perspective for 5 years ahead. Secondly, we model and simulate in Synchro software the current state of the traffic in the analyzed area, then we do the same thing but including the other two scenarios. Thirdly we calculate the percentage of pollutant emissions brought by the building effect and also the percentage for 5 years ahead, including the building parking lots. Lastly we model the accesses, to and from the building, so that overall road traffic is less affected.

2 Materials and Methodology

In order to develop the traffic study we needed to know the traffic speed and flow, parameters that are measured automatically, using high performance devices as SDR Radar, non-intrusive vehicle detection equipment.

2.1 Traffic Data Collection and Processing

As methodology, the primary data collection is graphically represented, as can be observed in Figs. 1 and 2.

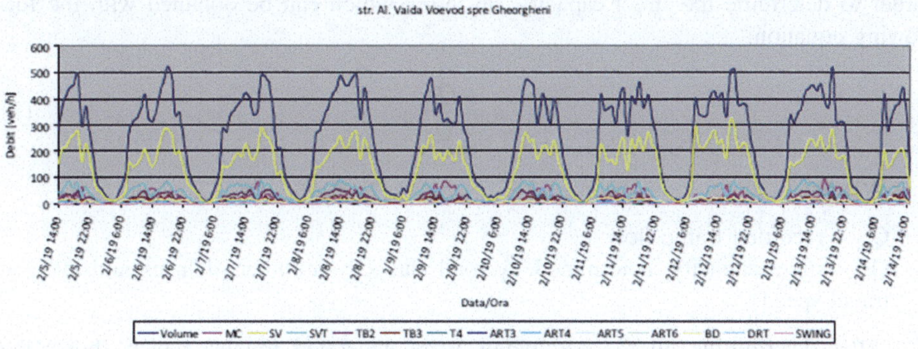

Fig. 1. Horary traffic flow graph

In order to process the traffic data we followed a few steps:

- determining the equivalent vehicle flows for the whole period observation;
- Trafficking statistics for interest categories: tourism, bicycles and trucks;
- Calculating the usage index of the streets monitored;
- Calculating the hourly debits under the terms of the functionality of the objectives;
- Hourly flow forecast for 5 years, including the building.

For the calculation of the equivalent vehicle flow we used the following equation:

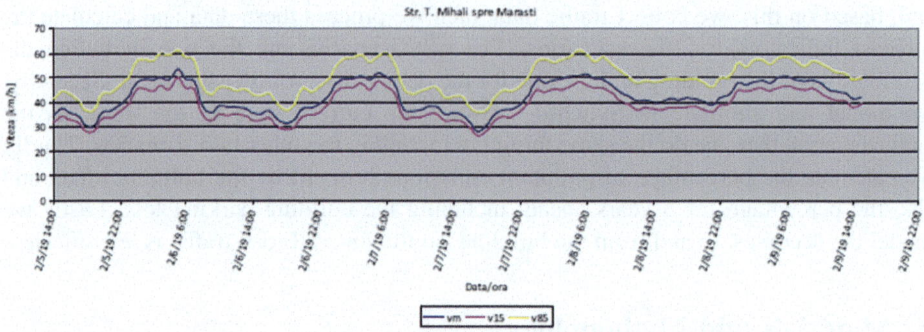

Fig. 2. Horary traffic speed graph

$$Q_{ech} = \sum\nolimits_i Q_i \cdot k_i \tag{1}$$

Where:

Q_i – hourly traffic flow for *i* category
k_i – the equivalence coefficient corresponding to the *i* category of traffic participants

For the calculation of the level of service for the monitoring streets we used, in order to determine the street capacity, an index which can be obtained with the following equation:

$$q = \frac{Q_{ef}}{Q_n} \tag{2}$$

Where:

Q_{ef} – recorded traffic flow
Q_n – street capacity, determined by road category, number of lanes and average travel speed

After determining all the parameters were generated graphs, which shows the variation of these parameters. In Fig. 3 is shown the variation of the actual traffic flow, reported to the street capacity.

In Fig. 4 can be observed the variation of the actual traffic flow, the traffic flow including the parking lots of the building, reported to the street capacity.

The traffic forecast is an indicative assessment tool for the road traffic trend, being related to:

- Motorization index of the municipality;
- Gross Domestic Product growth rate;
- Demographic trend.

Under the terms of Cluj - Napoca, the use of demographically, technically and economically documented information in the Sustainable Urban Mobility Plan (drawn up in 2015) was considered eloquent. According to this document also assumed by the

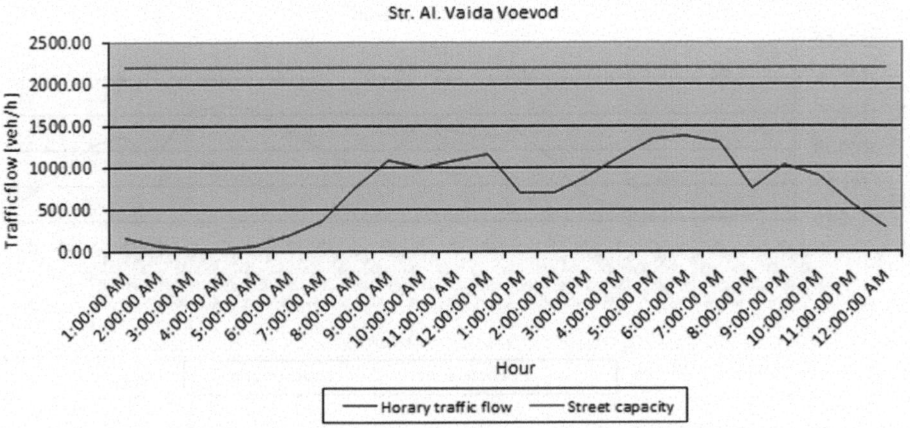

Fig. 3. Hourly traffic flow and street capacity graph

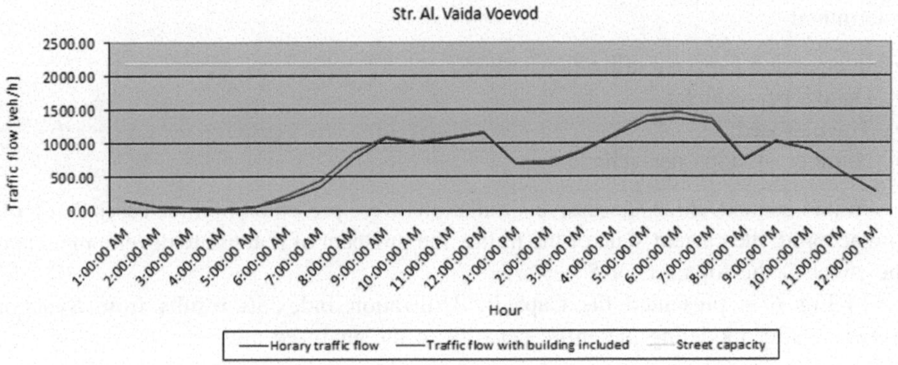

Fig. 4. Hourly traffic flow and street capacity, including the building, graph

local administration Cluj-Napoca, there is a growth rate of demand for transport in 5 years of 17.5%.

It has to be mentioned that the developed prognosis has an orientative character and is valid only in the conditions of preserving the current way of organizing traffic in the area targeted by the study. In Fig. 5 can be observed the variation of the traffic flow forecasted reported to the street capacity.

2.2 Evaluation by Simulating the Building Effect on Road Traffic in the Analyzed Area

In order to observe the effect of the proposed real estate development in the area of influence, it was considered necessary to broaden the scope of analysis to the limit of the main axes of traffic division.

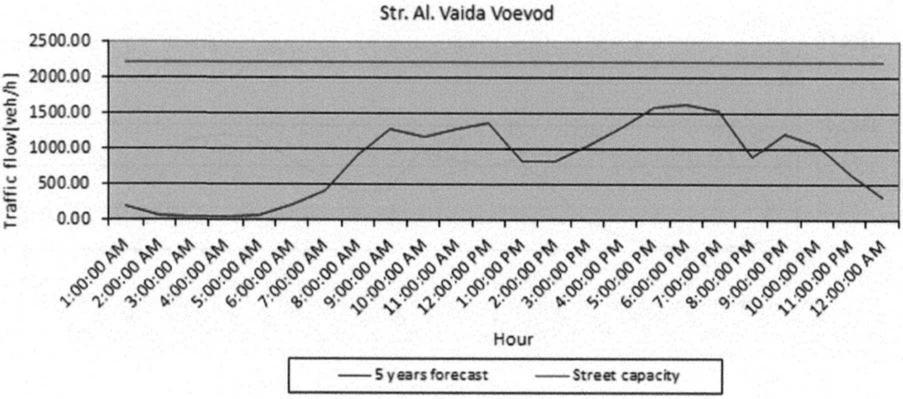

Fig. 5. Hourly traffic flow from the forecast and street capacity graph

The evaluation was performed by simulation and the following indicators were determined:

- Intersection capacity utilization index;
- Delays per vehicle;
- Travel speed;
- Number of stops per vehicle.

In this context, three successive simulations were presented in three cases that took into account: the current state of the traffic, the prospect of real estate development and the effect of the forecast for 5 years.

In Fig. 6 is presented the Capacity Utilization Index as results from Synchro software, after modeling the entire network of the analyzed area.

2.3 Pollutant Emissions Evaluation

For this project, the disaggregated calculation method was used due to the fact that the collected traffic data allowed the traffic participants to be detailed in the number of categories required by this method.

This model allows for the definition of individual speeds, lengths and data on transport flows at the road section and with a higher resolution.

In order to develop the greenhouse gases calculations, the collected traffic data was processed taking into account the restrictions imposed by the data input matrix specific to *Jasper*'s calculation tool. In Fig. 7 can be observed *Jasper* calculation tool.

Years of evaluation were defined, targeting the following reference years:

- Base year, without the effect of the project;
- Base year with the effect of the project implementation;
- Traffic 5 year forecast with project effect.

The calculations were developed for the reference years mentioned in the variants:

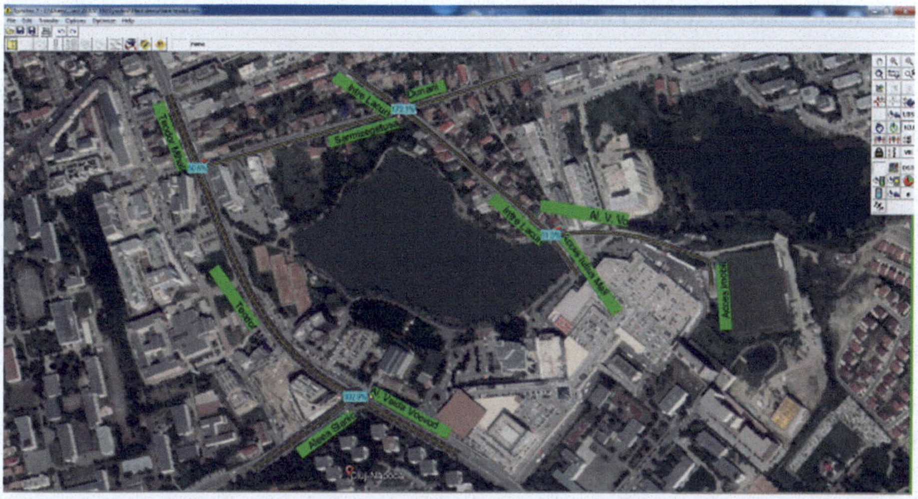

Fig. 6. Example of the Utilization Capacity Index result in Synchro software

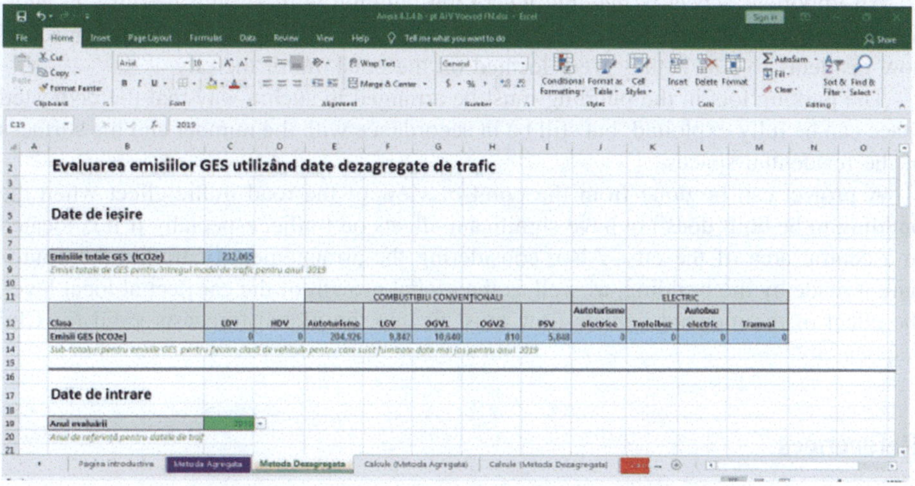

Fig. 7. Example of the spreadsheet used by Jasper calculation tool

- Continuation of current practices;
- The effect of the proposed project implementation.

It is worth to mention that the input data used in matrices are the result of mathematical modeling.

2.4 Road Safety Access Planning

The access to the building is arranged according to the existing traffic conditions, and if necessary, proposals will be made for the rearrangement of the traffic so that the whole area is in harmony from the point of view of traffic.

A concrete example makes it forbidden to turn left when entering or leaving the building's parking area in order not to disturb traffic on an intensely circulated artery, especially at peak hours.

This method can also be used if the building is positioned close to a curve, and visibility at the entrance or exit of the building is prevented from doing so.

The parking lots are designed and arranged by architects according to the regulations in force and consult with the team that elaborates the traffic study.

3 Conclusions

By developing traffic studies when a building includes parking lots is a must. Regardless of the number of parking lots, the contribution of vehicles brought to an area of the city is important.

An important aspect is the area where the building will be built because an additional number of vehicles brought to the centre of a city do not have the same impact on how residents live as if they were located on the outskirts of the city.

The parking lots in the building must be arranged in such a way that the dedicated space can be fully exploited, but still be in accordance with the number and the surfaces of the residential spaces.

A project can be done from the point of view of the road traffic effect when the building to be built does not have significant effects on traffic, especially if it is located in a central area of the city. Also, considering the arrangement of traffic, the contribution made by the building, as well as the global growth of the car fleet at local level, the effect of the induced vehicle flow must fit into the global forecasts valid for Cluj Napoca city.

References

AND 584/2002. Normativ pentru determinarea traficului de calcul pentru proiectarea drumurilor din punct de vedere al capacitatii portante si al capacitatii de circulatie (2002)

AND 600/2015. Normativ pentru amenajarea intersectiilor la nivel pe drumuri publice (2015)

Cordera, R., Coppola, P., Olio, L., Ibeas, A.: The impact of accessibility by public transport on real estate values: a comparison between the cities of Rome and Santander. Transp. Res. Part A: Policy Pract. **125**, 308–319 (2019)

Filip, N., Moldovan, M., Golgot, C.: Access optimization in an urban area: a case study. WIT Trans. Built Environ. **128**, 347–358 (2012)

Gheorghe, C.: Comparative analysis of the performance of one-way and two-way urban road networks. IOP Conf. Ser.: Mater. Sci. Eng. **252**, 012056 (2017). https://doi.org/10.1088/1757-899x/252/1/012056. ISSN 1757-8981. CAR 2017 - International Congress of Automotive and Transport Engineering

Gheorghe, C., Filip, N.: Research regarding on how to assess traffic congestion in relation to vehicle speed and flow rate. In: The 4th International Congress of Automotive and Transport Engineering: Selection of Papers, pp. 199–204. Editura U.T. Press (2018). ISBN 978-606-737-314-1

Gheorghe, C., Filip, N.: Research regarding on choosing optimal traffic organization using ranking criteria. In: Proceedings of the 4th International Congress of Automotive and Transport Engineering (AMMA 2018), pp. 775–783. Springer, Cham (2018). https://doi.org/10.1007/978-3-319-94409-8_90

Hansen, W.: How accessibility shapes land use. J. Am. Inst. Plan. **25**(2), 73–76 (1959). https://doi.org/10.1080/01944365908978307(1959)

Muresan, G., Filip, N.: The relevance of public questionnaires on the development of measures to increase sustainable urban mobility. In: International Congress of Automotive and Transport Engineering, pp. 758–765 (2018)

Rose, J., Martinez, F.: Towards a land-use and transport interaction framework. In: Handbook of Transport Modelling, pp. 181–201 (2007)

STAS 10144/5-89. Calculul capacitatii de circulatie a strazilor (1989)

Urban Mobility Plan for Cluj-Napoca city (2015)

Analysis of Occupant Injury Level in the Case of High-Speed Vehicle Slide-Spinning Collision

Laurentiu Claudiu Manea[1], Adriana Manea[1],
Alexandru Ionut Radu[2(✉)], and Dragos Sorin Dima[2]

[1] Ovidius University, 905900 Constanta, Romania
[2] Transilvania University of Brasov, 500036 Brasov, Romania
alexandru.radu@unitbv.ro

Abstract. The main objective of the paper was to analyze the degree of injury of the vehicle occupants by using virtual accident reconstruction software and calculating the HIC value. The injury level was calculated for the back-seat passengers in the case where the seatbelts were not used. One case study was analyzed where the drivers were racing and caused a collision with a turning vehicle at 130 km/h. The results expected are represented by the injury level probability and to observe if the survival limit was passed. The HIC value was correlated with the AIS injury scale and the acceleration level of the occupants head are compared to the tolerance limit of the human body. The novelty of the paper is that this is a new perspective to study real accidents and correlate them with the injury criterion (HIC) by the means of using virtual reconstruction software.

Keywords: Accident · Reconstruction · Occupant · Injury level · HIC

1 Introduction

Studies show that traffic accidents represent a high risk of death for both pedestrians and vehicle occupants alike in the European Union. [1]. Despite some positive trends, road safety represents still a big challenge for Romania. Together with Leetonia, Bulgaria and Lithuania, Romania is one of the four European Union countries with the biggest road safety accidents rate where victims are involved [2].

The most exposed categories of road users to accidents are pedestrians, cyclists and motorcyclists [3]. The most frequently investigated part of the human body following traffic accidents is the thorax area [4]. The identification and the analysis of the factors that concur to the occurrence of traffic accidents remains still a highly complex issue, given the large number of disturbing factors, the various traffic situations, and, also, the different behavior and reactions manifested by each driver [5].

The head represent the most vulnerable body region [6]. In order to assess the head injury severity was introduced a scale named HIC (Head Injury Criterion). The pedestrian's head is subjected to high loads due to the high decelerations caused by the collision, while HIC is defined as the integration of this acceleration in a specific time interval [7].

I. Dumitru et al. (Eds.): SMAT 2019, *The 30th SIAR International Congress of Automotive and Transport Engineering*, pp. 312–321, 2020.
https://doi.org/10.1007/978-3-030-32564-0_37

The main objective of the paper was to analyze the degree of injury of the vehicle occupants by using virtual accident reconstruction software and calculating the HIC value.

2 Case Analysis

For this study a case (real accident) was analyzed where a race was going on and the ongoing vehicle hit the side of a turning vehicle where its rear passengers did not survive. This case was chosen because the vehicle that got hit started to spin out of control, generating a critical situation for the occupants. In order to assess the head injury probability, the head injury criterion was calculated.

The head injury criterion can be judged by a parameter called HIC, defined by a mathematical formula. The HIC criterion is a way to assess the risk of a cranial trauma injury as a result of an accident. It can be used to assess the consequences of traffic accidents, testing individual protection equipment or safety sports equipment.

The HIC criterion is the maximum standard value of the integral of the head acceleration. Depending on the interval for which it is calculated, the HIC criterion is:

- unlimited - HIC;
- maximum 36 ms - HIC36;
- maximum 15 ms - HIC15.

The mathematical formula for determining the HIC criterion is [8]:

$$HIC = \left\{ \left[\frac{1}{t_2 - t_1} \int_{t_1}^{t_2} a \cdot dt \right]^{2,5} (t_2 - t_1) \right\}_{max}$$

$$a = \sqrt{a_x^2 + a_y^2 + a_z^2} \, [\mathrm{m/s^2}]$$

Where a is the resulting acceleration of the center of gravity of the head, measured in m/s2, and t1 and t2 are the time interval (s) in which HIC has the maximum value. An important observation is that HIC is dimensionless.

At HIC 1000 there is a probability of 18% for severe cranial trauma, a 55% probability of serious cranial trauma, and a 90% probability of a moderate cranial trauma.

The Abbreviated Injury Scale (AIS) is a system for assessing the degree of injury of occupants involved in traffic accidents by expressing injuries in the form of the number of categories from 1 to 6, depending on their severity, from minor injuries, notes with 1, up to serious injuries that can cause death, notes 6 [9] (Table 1).

Table 1. AIS scale

AIS	Category	Probability of death [%]
1	Minor	0–10
2	Moderate	10–30
3	Serious	30–70
4	Severe	70–90
5	Critical	95
6	Maximum	100

Based on a set of experiments using corpses, a correlation was made between the value of the HIC indicator and the AIS scale for the head of the occupant. Note that this correlation is based only on impact tests where head injuries have been analyzed. In the next figure the head injury severity is presented in graphical form [10].

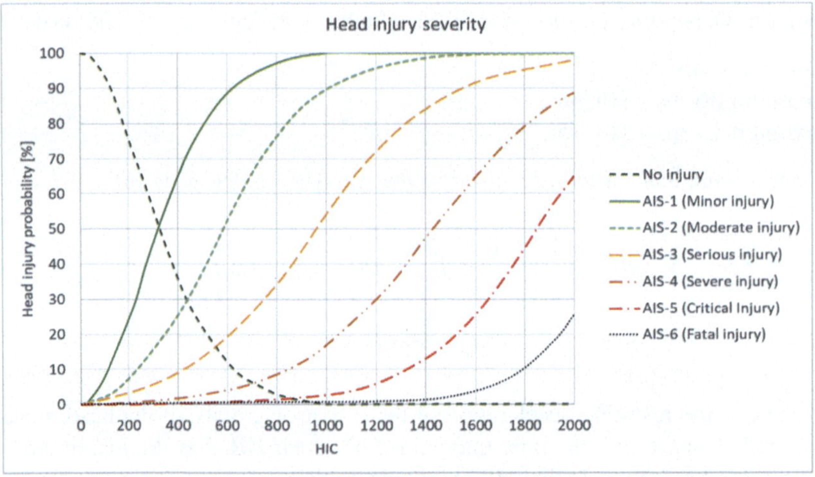

Fig. 1. Correlation between the HIC value and AIS scale

It can be seen that the tolerance value of the HIC = 1000 is the point where AIS-1 is at 100% and injuries occur. Also we can see that a high degree of injury (over 80%) occurs at a HIC value of 2000 where AIS-4 (severe injury) is at 89%, AIS-5 (critical injury) is at 65% and AIS-6 (fatal injury) is at 25%. Using this graphical form, the injury potential is evaluated for the case study (Fig. 2).

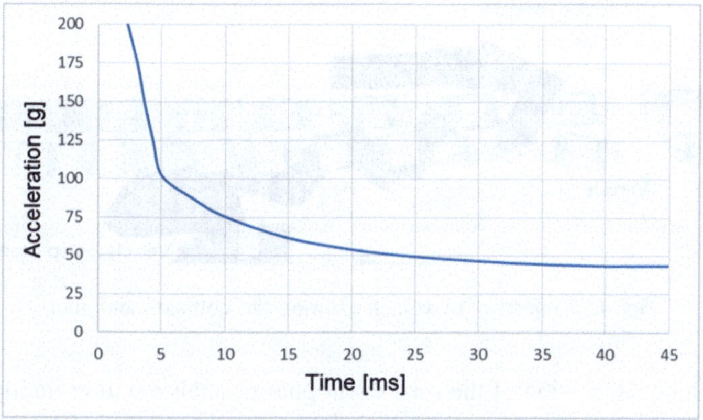

Fig. 2. Head tolerance limit

This graph shows the acceleration value based upon the pulse duration. Thus, it can be seen that the limit of maximum acceleration value on the occupant's head is around 40 g. In the case of impacted pedestrians, the situation is worse. At impact with the ground, much higher headline accelerations 120–200 g) may occur, which can cause severe injuries [11].

In order to determine the cause of death for the backseat occupants, simulations were conducted in PC-Crash and multibody models were added in order to analyze the kinematic and dynamics of the occupants. The key parameter that was studied was the head acceleration levels for the occupants.

The traveling vehicle was driving in a straight line and the second vehicle was turning, making a U turn. The point of impact was at the front. In this accident, a third vehicle that was coming up from behind the first one and hit vehicle number 2 while it was spinning out of control.

The vehicle have been named Vehicle 1 for the first racing vehicle, Vehicle 2 for the turning car and Vehicle 3 for the second racing vehicle. The accident scene schematic is presented in Fig. 3.

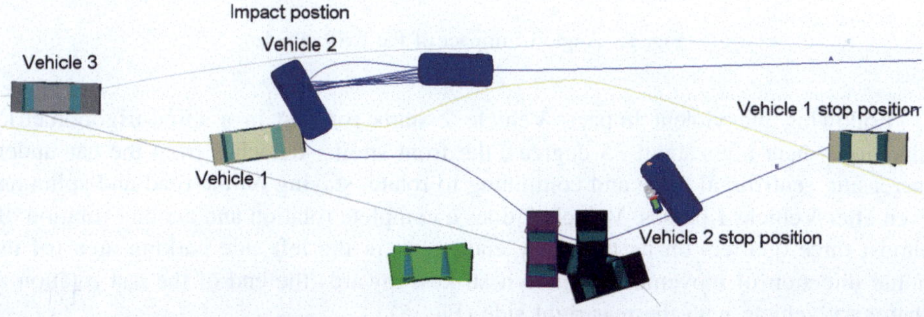

Fig. 3. Impact and stop positions of the vehicles

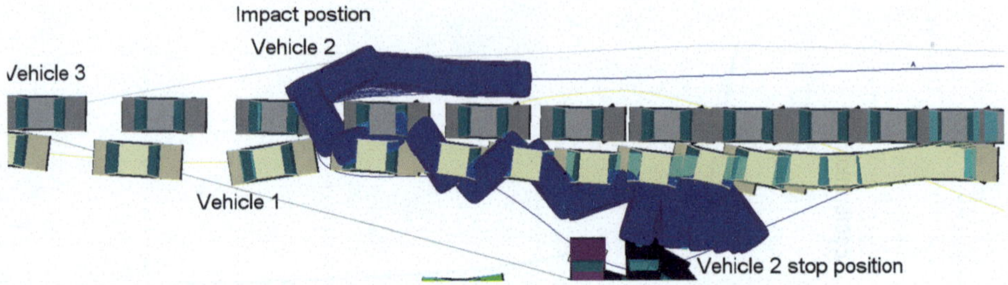

Fig. 4. Trajectory of vehicles during the collision and after

The dynamic succession of the road event phases, analyzed after multiple attempts in the virtual environment software for accident reconstruction PC Crash 10.2. Of the impact variants, with the unconditional preservation of the position of the cars involved in the "post impact" phase, as recorded in the documents drawn up by the bodies empowered with the "on the spot" investigation are, in the opinion of the designated judicial expert, with the assumed probability, represented in Fig. 3. In Fig. 4 the accident dynamic of each of the 3 vehicles is shown with step time of 0.2 s. Vehicles 1 and 3 go in an almost straight line while vehicle 2 was spinning after the impact.

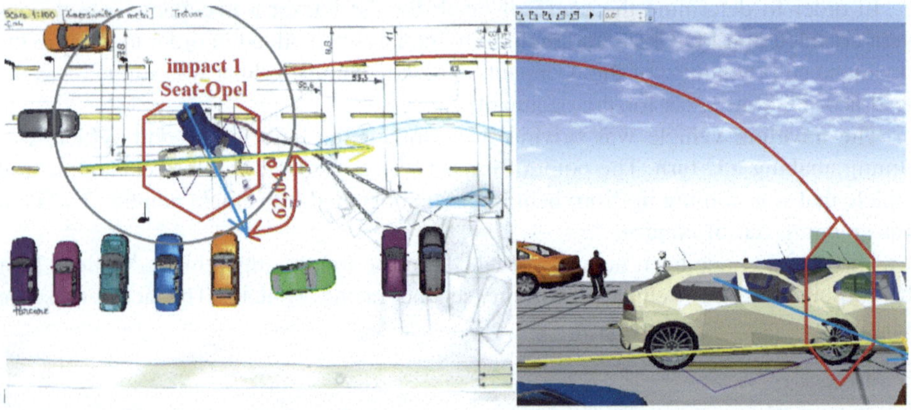

Fig. 5. Angle of impact of the two vehicles

Following the violent impact, Vehicle 2 starts rotating in a spiral-trigonometric direction where after about 35 degrees, the front spoiler detaches from the car under increasing centrifugal force and continuing to rotate, staying on the road and spinning even after Vehicle 1 passes. Vehicle 2 does a complete rotation and another rotation of almost three quarters on its trajectory ending up in the left side parking area (of its initial direction of movement), where it strikes, towards the end of the last rotation a stationary vehicle with the rear right side (Fig. 5).

The accident expertise show that the deceleration and the contact between the occupants and hard parts of the interior of the vehicle were visible and some parts were deformed (the backrest and the headrest of the driver's seat (-deformed (broken) by the impact shock and oriented towards the rear), occupant seat on the right hand side of the driver, right rear door deformed inwards from front to back - which caused the fracture of the lower right limb, the frame of the right pillar B).

The occupants undergo a violent compression of the head and thoracic-abdominal area followed by massive internal hemorrhage due to the complex visceral lesions and although initially the medical crews arriving at the accident place stabilize them, they do not survive the accident, the medical verdict being of "violent death" due to anemia. Severe acute by massive internal hemorrhage as a result of complex traumatic injuries in the thoraco-abdominal viscera within a complex polytrauma suffered by road accident (Fig. 6).

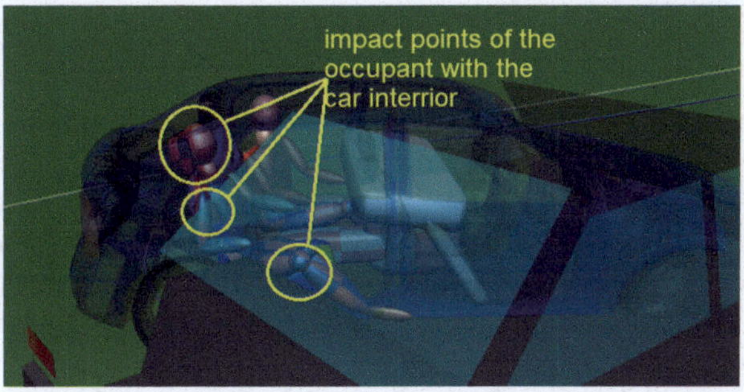

Fig. 6. Rear occupant impact with the interior of the vehicle (headrest and door)

The multibody simulation in PC-Crash show that the occupant hit the internal parts of the vehicle just as described in the medical file of the expert report. The multibody models follow the rotational trajectory of the vehicle under the centrifugal force.

3 Results

After the simulation was conducted, the first results presented in the Fig. 7 below are the vehicle velocities. This accident is composed of 2 impacts, the first was the initial impact between vehicle 1 and vehicle 2 and the second occurred during the spin of the second vehicle between vehicles 2 and 3. The impact moments are indicated in the graph.

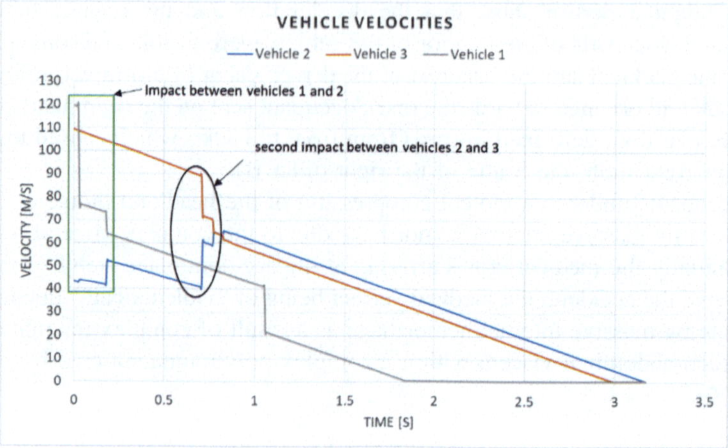

Fig. 7. Vehicle velocities for case 1 (indicating the impact moments)

As shown in the graph, the initial velocity of the striking vehicle, vehicle 1, was 120 km/h, and drops down to 76 km/h during the collision. The struck vehicle, vehicle 2, had an initial velocity of 45 km/h, that increases during the first impact and started to spin, increasing its velocity up to 55 km/h, then gradually drops down to 40 km/h until the second impact, which increases it up to 62 km/h. The third vehicle had a velocity of 110 km/h and slows down to 95 km/h when it strikes vehicle 2. During the second impact we can see a rapid decrease of the third vehicle from 95 to 75 km/h and for the second vehicle a rapid increase from 45 to 62 km/h. After the second impact, the velocity of all vehicles gradually drops to 0 km/h in 2.3 s.

From the multibody model analysis, the head acceleration was analyzed for the backseat occupants. There were 2 occupants, seated to the left and right, with no seatbelts connected. The head acceleration is presented in the next figure for both occupants (Fig. 8).

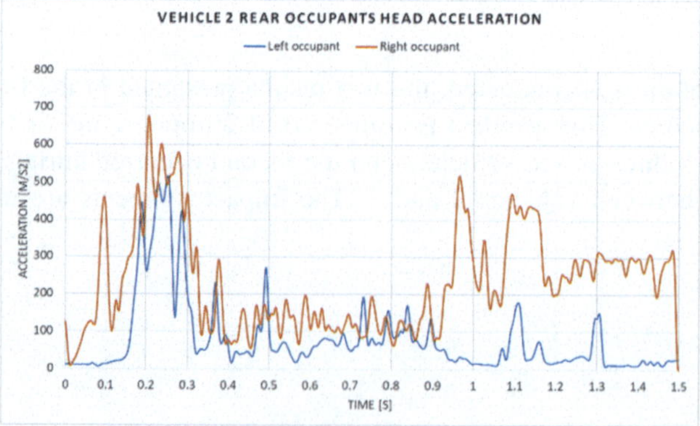

Fig. 8. Backseat occupant head resultant acceleration for case 1

The results show that the acceleration value for the right occupant was higher than the left. Also, during the collision and spin, the acceleration had a variation, meaning higher values over a longer period of time. The right occupant had a maximum of 700 m/s^2 (almost 70 g's) acceleration value and the left had a maximum of 510 m/s^2. (Almost 51 g's).

Table 2. Acceleration tolerance limit

Occupant	Acceleration limit	Time limit	Measured acceleration	Measured time
Left	40 g	Over 35 ms	**42 g**	**150 ms**
Right			**55 g and 45 g**	**170 ms, 200 ms**

In the Table 2, the tolerance level for the head is presented and also the measured acceleration level for the 2 occupants. For the left occupant the measured value was 42 g for about 150 ms during the first impact which is on par with the limit. The right occupant had 2 critical values on the graph. The first was between 0.48 and 0.65 s, with an average acceleration of 55 g for 170 ms. The second was between 1.35 s and 1.55 s, with an acceleration value of 45 g on average for 200 ms. These results suggest that the head acceleration limit was surpassed and thus resulting in serious head trauma with death as a result. The next step was to calculate the HIC value for both left and right occupants and to present the results in graphical form.

As show in the graph, the maximum value of the HIC value was 559 for the left occupant and 833 for the right occupant, values that are not fatal, but critical. Also the HIC value calculated shows high values for a long period of time in the case of the right occupant, HIC max = 750 was present for 200 ms. In the next figure the HIC graph is presented for both occupants (Fig. 9).

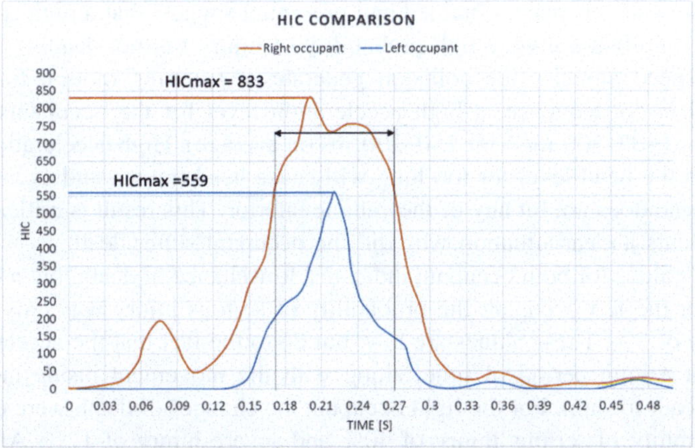

Fig. 9. HIC 36 value for the right occupant

To better evaluate the HIC value, it was correlated with the AIS scale using Fig. 1. This is illustrated in the next table.

Table 3. Probability of head injury based on the HIC maximum value

Abbreviated injury scale	Left occupant	Right occupant
	HIC max = 559	HIC max = 833
AIS-1 (Minor injury)	80%	99%
AIS-2 (Moderate injury)	40%	82%
AIS-3 (Serious injury)	17%	40%
AIS-4 (Severe injury)	5%	12%
AIS-5 (Critical injury)	2%	3%
AIS-6 (Fatal injury)	0%	2%

As we can see, by correlating the HIC value with the AIS scale, we get the results presented above in the Table 3. For the left occupant, we obtained a critical injury potential of 2%, a severe injury potential of 5% and a serious injury potential of 17%. Moderate probability of injury is only 40%.

In the case of the right occupant, the values are a lot higher, probability of moderate injury is 82% and serious injury 40% while severe is 12% and critical 3%. These results show a theoretical probability of injury and are only based on the deceleration values of the head movement. These results show a probability of death based only on the rapid movement of the head and not on the contact forces between the occupant and the interior of the vehicle.

4 Conclusions

The final results of this paper obtained and presented suggest that a high speed vehicle frontal to side collision give a high probability of injury but not deadly.

A high speed vehicle side collision generates a spinning motion for the struck vehicles and in consequence, a high acceleration level for the occupants for a high period of time (400–500 m/s^2 for 150–200 ms on average). High acceleration levels for occupants on the head level for too long will cause head trauma and eventually death even if the head does not hit any of the vehicle interior. This result is critical due to the fact it represents a clear situation why did the occupant suffer death.

The HIC values for both occupant indicate a low chance of death, but a high chance of injury. For the left occupant the probability of serious injury was only 17% and a severe injury of 5%. These values are low, but do to the fact that the acceleration level was high for a long period of time, along with the movement inside the vehicle it resulted in death for him. For the right occupant the changes of death were even higher, with a probability of serious injury of 40% and severe injury of 12%. Also it would mean that if the occupant does not die, it would remain in a serious coma for a long period of time.

Also using this method of analysis can result in accurate results regarding evaluating the occupant injury level by analyzing both the acceleration variation during the impact and also calculating the HIC value.

References

1. Tolea, B., Trusca, D., Antonya, C., Beles, H.: The influence of the frontal profile design of a vehicle upon the pedestrian safety at low velocity. In: 26th Annals of DAAAM and Proceedings (2015)
2. Jovanov, D., Vollpracht, H.J., Beles, H., Popa, V., Tolea, B.A.: Most common road safety engineering deficiencies in South Eastern Europe as a part of safe system approach. In: IOP Conference Series: Materials Science and Engineering, vol. 252, no. 1, p. 012021. IOP Publishing (2017)
3. Otat, O., Matei, L., Victor, O.T., Dumitru, I., Dragovic, M.: The possibilities of using an outline crash sketch in modeling and simulation of accidents. J. Ind. Des. Eng. Graph. **14**(1), 149–154 (2019)
4. Oana, O., Nicolae, D., Ilie, D.: Dynamic models to analyse the influence of the seat belt in a frontal collision. In: IOP Conference Series: Materials Science and Engineering, vol. 252, no. 1, p. 012017. IOP Publishing (2017)
5. Oţăt, O., Dumitru, N., Oţăt, V.: Vehicles' passive safety systems influence on driver's thorax injuries. In: Applied Mechanics and Materials 2016, vol. 823, pp. 187–192. Trans Tech Publications (2016)
6. Longhitano, D., Henary, B., Bhalla, K., Ivarsson, J., Crandall, J.: Influence of vehicle body type on pedestrian injury distribution. No. 2005-01-1876. SAE Technical Paper (2005)
7. Deck, C., Willinger, R.: Improved head injury criterion based on head FE model. Int. J. Crashworthiness **13**(6), 667–678 (2008)
8. McHenry, B.G.: Head injury criterion and the ATB. ATB Users' Group **29**, 5–8 (2004)
9. Gennarelli, T.A., Wodzin, E.: AIS 2005: a contemporary injury scale. Injury **37**(12), 1083–1091 (2006)
10. Prasad, P., Mertz, H.J.: The position of the United States delegation to the ISO Working Group 6 on the use of HIC in the automotive environment. SAE Trans. **1**, 106–116 (1985)
11. Kramer, M.: Sichereitseigenschaften von Verbundglasscheiben bei, pp. 390–397. Germania: ATZ, 33, nr. 10 (1972)

Dynamic Considerations Regarding the Enrolling in the Curve with Small Radius of Railways, of the Locomotives

Iulian Turcanu[1(✉)] and Marian Florin Mitroi[2(✉)]

[1] Polytechnic University Bucharest, Spl. Independentei nr. 313, Bucharest, Romania
oldiulian@yahoo.co.uk
[2] Ovidius University, Constanta Bd. Mamaia 124, Constanta, Romania
ejju_marian@yahoo.com

Abstract. The movement of the locomotives with high-speed on the railway is influenced by the possibilities of approaching the curves with small radius. The traction force of the locomotive is dependent on actual axle loads and adhesion coefficient. The objective of the research is to determine the dynamic characteristics of six-axle locomotives in approaching the sinuous railway. For this, is being watched to adopt different coefficients of adhesion. In the research, the model and technical characteristics of the Siemens ES locomotive class 20 CF were used. The approach of the small curves of the rolling path must be made without the locomotive base touching the rail. This avoids the phenomenon of derailment. Following the calculation of locomotive dynamics, for the established route conditions, the importance and the necessity of using axle guidance systems, forced, when enrolling in the curve, was observed. By using such solutions the transverse and longitudinal sliding of the locomotive (train) is eliminated. The bogies with oriented axle are viable solutions, which can ensure the fluency of the vehicle movement on the tread.

Keywords: Traction running dynamics · Enrollment in curves · Overweight · Gauge

1 Introduction

In the research, for the determination of the traction force of the locomotive, at the limit of skating and the actual loads on the mounted axles, were made according to the constructive parameters of the railway vehicle and the coefficient of adhesion of the motor axles. Traction force is a factor that depends on the available power of the vehicle's engine and has an influence on the speed characteristic of the vehicle. The wheel-rail adhesion coefficient (μ) in the calculations was established to $\mu = 0{,}7$.

I. Dumitru et al. (Eds.): SMAT 2019, *The 30th SIAR International Congress of Automotive and Transport Engineering*, pp. 322–331, 2020.
https://doi.org/10.1007/978-3-030-32564-0_38

2 Determining the Dynamics of the Locomotive in Curves

By analyzing the kinematic chain of the main transmission, the traction characteristics of the vehicle can be determined [2]. The variation in the traveling speed of the vehicle, in time v(t) and space v(s), and the time from start to exit to the engine-limited $F_{lm}(v)$ external traction feature, helped to identify the start-up law and start parameters. The research was conducted on the Siemens ES 20 CF locomotive. This has six axles in the Co-Co formula (Fig. 1).

Fig. 1. Siemens Locomotive Class ES 20 CF (Co-Co)

The technical characteristics of the locomotive used in the measurements are: the length between buffers L = 24 m; minimum enrollment radius R = 250 m; the widened track gauge in the curve is l1 = 1465 mm; track gauge between the outer faces of the lip bandages d = 1410 mm; the bogie wheelbase p = 2.5 m; buffer diameter 2r = 450 mm; the distance between the buffer axes u = 1750 mm; the buffer length l = 620 mm; q + w is the play between the axle spindle and the cumulative bearing, with the play of the traverse; the width of the rolling stock gauge (GMR) is lg = 3150 mm. The principle of the movement of the railway vehicle in curves (Fig. 2) can be described by the following mathematical relations:

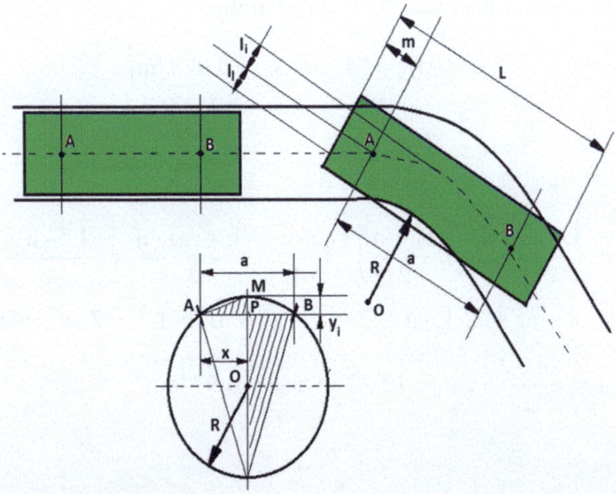

Fig. 2. Schematic of the enrollment in the curve of the locomotive on six axles

Note the inner movement yi = MP. Then:

$$\Delta APM \approx \Delta PBM' \Rightarrow \frac{y_i}{x} = \frac{a - x}{2R - y_i} \Rightarrow y_i(2R - y_i) = x(a - x) \tag{1.1}$$

$$\left.\begin{array}{r} 2Ry_i - y_i^2 = x(a - x) \\ \text{but}: \ 2Ry_i \triangleright y_i^2 \\ \text{where}: \ y^2 \to 0 \end{array}\right\} \Rightarrow y_i = x(a - x) \tag{1.2}$$

The game between the lips of the bandages, in the railway, is:

$$s = l_s - E \Rightarrow l_s = 1435 + s \text{ (the widening of the tread in the curve)} \tag{1.3}$$

where: E = normal track gauge [4]

$$E \Rightarrow j + s = \frac{l_s - d}{2} \tag{1.4}$$

d = gauge of the axle between the outer faces of the bandages

$$\Rightarrow j + s = \frac{1465 - 1410d}{2} \Rightarrow j + s = 27,5.$$

l_i = represents the total play of the rail vehicle in the running track:

$$\left.\begin{array}{l} l_i = y_i + (j + s) + q + w + \frac{p^2}{8R} - k \\ \text{but}: \begin{cases} y_i = \frac{x \cdot (a - x)}{2R} \\ j + s = \frac{l_s - d}{2} \end{cases} \end{array}\right\} \Rightarrow l_i = \frac{x \cdot (a - x)}{2R} + \frac{l_s - d}{2} + q + w + \frac{p^2}{8R} - k \tag{1.5}$$

where: p = is the bogie wheelbase; $l_g = 3150\,[\text{mm}]$,

$$\Rightarrow GMR \triangleright 43 \Rightarrow k = 0.075\,[\text{m}] \tag{1.6}$$

If: $x = \left(a - \frac{a}{2}\right) \Rightarrow l_{i_{max}}$
Then:

$$\frac{a}{2} \cdot \left(a - \frac{a}{2}\right) \Rightarrow \frac{L - a}{2} \cdot \left(a + \frac{L - a}{2}\right) \Rightarrow \frac{a^2}{4} = \frac{(L - a) \cdot a}{2} + \frac{(L - a)^2}{4} \Rightarrow$$
$$\Rightarrow L^2 - 2 \cdot L \cdot a + a^2 + 2 \cdot L \cdot a - 2 \cdot a^2 - a^2 = 0 \Rightarrow L^2 = 2 \cdot a^2 \Rightarrow L = a\sqrt{2} \Rightarrow$$
$$a = \frac{L\sqrt{2}}{2} \Rightarrow a = \frac{24\sqrt{2}}{2} \Rightarrow a = 12\sqrt{2} \Rightarrow a = 16.97\,[\text{m}]. \tag{1.7}$$

$$x = \frac{a}{2} \Rightarrow x = 8.485\,[\mathrm{m}] \Rightarrow y_i = \frac{8.485 \cdot (16.97 - 8.485)}{2 \cdot 250} \Rightarrow y_i = 0.144\,[\mathrm{m}]. \quad (1.8)$$

$$l_i = 144 + 27.5 + 35 + \frac{2500^2}{8 \cdot 250} - 0.0075 \Rightarrow l_i = 3331.425\,[\mathrm{mm}]. \quad (1.9)$$

In the Fig. 3 shows the verification of circularity deviation of the railway radius.

Fig. 3. Verification of the radius run to the railway

3 Analysis of the Movement of the Locomotive in Alignment and Orizontal Way

The internal displacement (y_i) is calculated with the relation 2.1:

$MP = y_i$ is the inner movement (inner retraction).

$$\Rightarrow \frac{4000 - 3150}{2} = 425\,[\mathrm{mm}] \quad (2.1)$$

In Fig. 4 shows the simplified scheme of the railway vehicle, in the case of alignment and orizontal displacement.

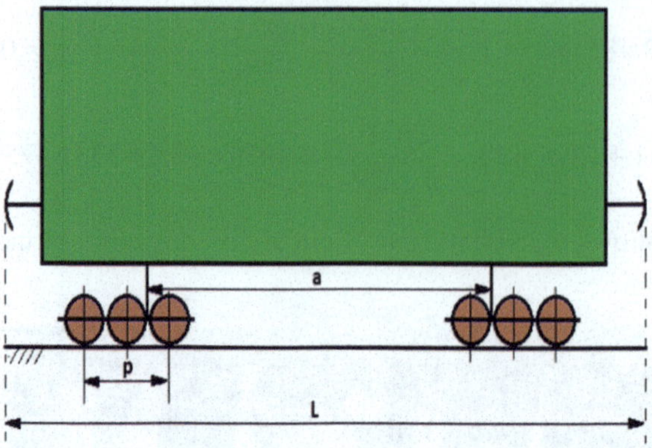

Fig. 4. The railway vehicle in the case analyzed

Then:

$$\Delta APM \approx \Delta GM' \Rightarrow \left.\begin{array}{l} \frac{y_i}{x} = \frac{a-x}{2R-y_i} \\ \text{but}: \ 2 \cdot Ry_i \rhd y_i^2 \end{array}\right\} \Rightarrow y_i(2R - y_i) = x(a - x) \Rightarrow y_i = \frac{x(a - x)}{2R}. \tag{2.2}$$

The total track deviation, to the railway of the vehicle is calculated as follows:

$$e_i = y_i + (j + s) + q + w + \frac{p^2}{2R} - k = \frac{x(a - x)}{2R} + \frac{l_s - d}{2} + q + w + \frac{p^2}{2R} - k. \tag{2.3}$$

where: q = the deviation from spindle to bearing; p = bogie wheelbase.
w = the deviation of the traverse [4];
The external retraction (Fig. 5) is calculated with the relation:

$$\begin{gathered} \Delta APM \approx \Delta M'BP \\ \Rightarrow \frac{y_e}{a + x_1} = \frac{x_1}{2 \cdot R + y_e} \Rightarrow y_e \cdot (2 \cdot R + y_e) = x_1 \cdot (a + x_1). \end{gathered} \tag{2.4}$$

$$\text{But}: \ 2R \cdot y_e \rhd y_e^2 \Rightarrow y_e = \frac{x_1 \cdot (a + x_1)}{2 \cdot R}. \tag{2.5}$$

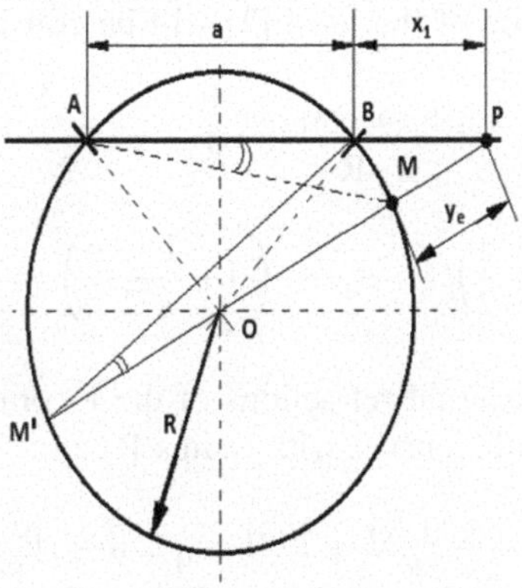

Fig. 5. The external retraction of locomotive

During movement, due to total deviation (l_s), it is assumed that the point (B) of the locomotive becomes (B′) (Fig. 6).

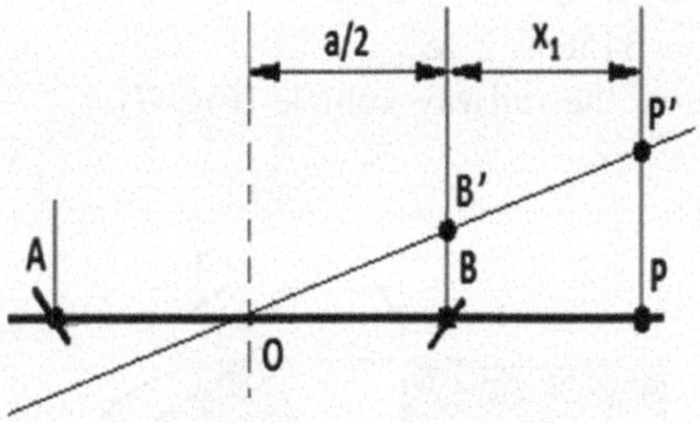

Fig. 6. The schematization of the total deviation, when the point B of locomotive becomes B′

$$\Delta OBB' \approx \Delta OPP' \Rightarrow \frac{OB}{BB'} = \frac{OP}{PP'} \Rightarrow PP' = BB' \cdot \frac{OP}{OB} \Rightarrow PP' = \left(\frac{l_s - d}{2} + q + w\right) \cdot \frac{\frac{a}{2} + x_1}{\frac{a}{2}}$$
$$= \left(\frac{l_s - d}{2} + q + w\right) \cdot \frac{a + 2 \cdot x_1}{a} \tag{2.6}$$

$$e_e = \frac{x_1 \cdot (a + x_1)}{2 \cdot R} + \left(\frac{l_s - d}{2} + q + w\right) \cdot \frac{a + 2 \cdot x_1}{a} - \frac{p^2}{2 \cdot R} - k \tag{2.7}$$

The maximum retraction is:

$$x_1 = n = \frac{L - a}{2} \Rightarrow e_{max}. \tag{2.8}$$

Also, the maximum value of the term [3] will be considered:

$$\frac{x_1 \cdot (a + x_1)}{2 \cdot R}, \text{ so : } x = \frac{a}{2} \quad (2.9)$$

$$\Rightarrow \frac{a}{2} \cdot \left(a - \frac{a}{2}\right) = \frac{L - a}{2} \cdot \left(a + \frac{L - a}{2}\right) \Rightarrow L = a\sqrt{2} \quad (2.10)$$

Thus, the internal and external retractions of the locomotive, which circulate on a railway segment located in the curve, with radius R, are:

$$\begin{cases} e_i = \frac{x \cdot (a - x)}{2 \cdot R} + \frac{l_s - d}{2} + q + w + \frac{p^2}{2 \cdot R} - k \\ e_e = \frac{x_1 \cdot (a + x)}{2 \cdot R} + \left(\frac{l_s - d}{2} + q + w\right) \cdot \frac{a + 2 \cdot x_1}{a} - \frac{p^2}{2 \cdot R} - k \end{cases}$$
$$\Rightarrow \begin{cases} E_i \leq 100 \, \text{pentru} : \alpha = 0 \\ E_e \leq 120 \, \text{pentru} : \beta = 0 \end{cases} \quad (2.11)$$

So,
If: $e_{imax} \triangleright e_{emax} \Rightarrow b = 3150 - 2 \cdot e_{imax}$
If: $e_{imax} \triangleleft e_{emax} \Rightarrow b = 3150 - 2 \cdot e_{imax}$
where: b = the width of the railway vehicle (Fig. 7).

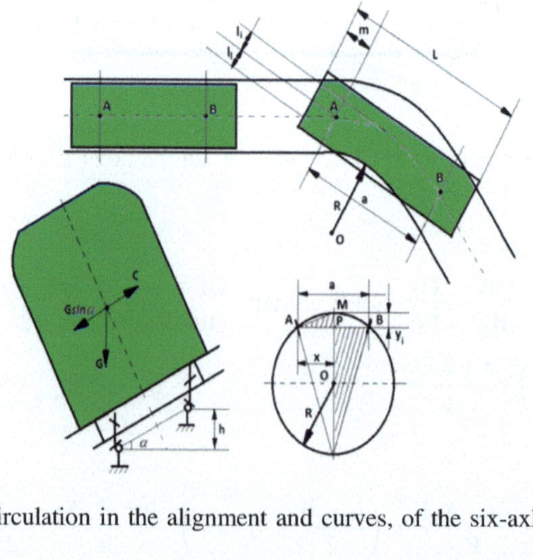

Fig. 7. The circulation in the alignment and curves, of the six-axle rail vehicle.

The Fig. 8 shows how to check the travel of the locomotive in the curve, by using the S20 Berger equipment. This is very important because of the deformations of the tread, when driving at high ambient temperatures. The runway generates disturbances at relatively high frequencies.

Fig. 8. Checking the movement in the curve of locomotive

Thus, depending on its anomalies and the runway elastic system constants, is influenced the excitement of the suspension system, of the suspended masses of the train. The deformation of the railway lines implies a serpentine movement of the locomotive, which is why analyzes of the oscillations of this motion and coupled with the roking-rolling motion are required.

4 The Verifying the Buffers for the Locomotive Gauge

The dynamics of the rail vehicle during journeys require some additional checks. One of these checks is also the exit level of locomotive buffers, out of its width, at the ends. In the Fig. 9 shows how to determine the width of the locomotive.

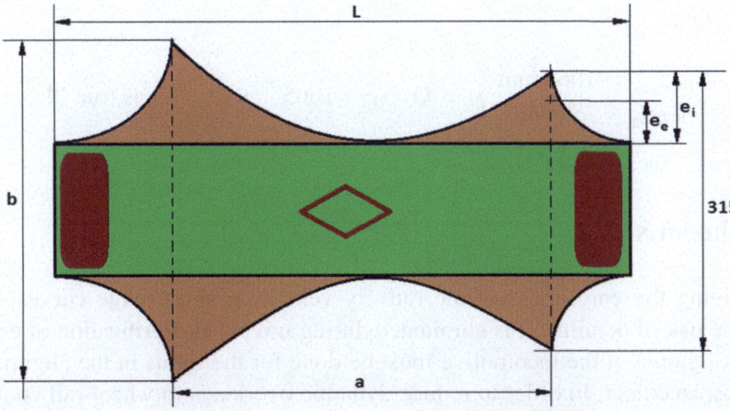

Fig. 9. Determining the width of the locomotive

For checks, account shall be taken of the "International Regulation for Passenger Wagons" [3]. The Fig. 10 presents the analysis scheme for checking the locomotive buffers.

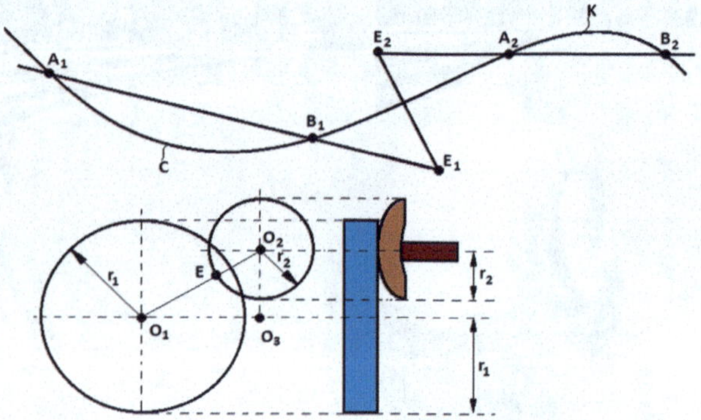

Fig. 10. Scheme for checking the output buffers of locomotive [3]

If: $t = r_1 + r_2 - O_1O_2 \lhd 35$ mm, it results that the buffers is touching at low level:

$$O_1O_2^2 = O_1O_3^2 + O_3O_2^2 \Rightarrow O_1O_3 = e_e^I + e_e^{II} \tag{3.1}$$

where: O_3O_2 = displacement of the two buffers.

Reconsidering the relationships, it result:

If:

$$\begin{cases} E_i = x \cdot (a - x) + \frac{p^2}{4} \rhd 100 \Rightarrow e_i = e_i + \alpha \text{ where } \alpha = \frac{E_i - 100}{750} \\ E_e = x_1 \cdot (a - x) - \frac{p^2}{4} \rhd 120 \Rightarrow e_e = e_i + \beta \text{ where } \beta = \frac{E_e - 120}{750} \end{cases} \tag{3.2}$$

$$\Rightarrow b_{cons.} = 3150 - 2 \cdot e_{emax} \tag{3.3}$$

It result that:

$$\left.\begin{array}{l} N_{max_{tamp}} = 1065\,\text{mm} \\ N_{min_{tamp}} = 940\,\text{mm} \end{array}\right\} \Rightarrow O_3O_2 = 1065 - 940. \text{ This is true !!!} \tag{3.4}$$

5 Conclusions

By determining the enrollment of the railway vehicle in short-range curves, without slippage, the risk of derailment is eliminated during travel [1]. Verification of entry into the clearance gauge of the locomotive must be done for the limits in the alignment and orizontal displacement. In order to reduce dynamic overloads to wheel-rail contact and protect on the bogie components from vibration, it is recommended to use resilient wheels. Longitudinal and transverse landslides can be eliminated, if it is possible for the axle to radial sit and run smoothly across. The gain of real gauge values, which is recorded when entering curves, are the one inside and outside.

References

1. Badea, C.N.: Driving Stability and Safety Guiding on Railway for Freight Wagons, Ph.D.
2. Bruni, S., Vinola, J., et al.: Modelling of suspension components in a rail vehicle dynamics context. Int. J. Veh. Mech. Mobil. (2011). https://doi.org/10.1080/00423114.2011.586430
3. Prudhomme, A.: Les problems que pase, pour la voie, la circulation des rames a grande vitesse, Revue Generale des Chemins de Fer, November 1976
4. Sebesan, I., Mazilu, T.: The Vibration of railway vehicles, Matrix Rom, Bucureşti (2010)

Methods of Analysis and Reconstruction of Traffic Events

Mihai Clinciu[1], Stefanita Ciunel[2](✉), and Bebe Tica[2]

[1] Transilvania University of Brasov, Politehnicii Street, no. 1, Brasov, Romania
[2] University of Craiova, Calea Bucuresti Street, no. 107, Craiova, Romania
ciunel_stefanita@yahoo.com

Abstract. The paper addresses an important issue in the field of road safety because the modeling, simulation and reconstruction of road events is a permanent concern of the specialists in the field. In the contemporary era, dominated by countless and remarkable technical achievements, the diversification of transport means and the significant increase in the speed of movement, the role of technical experts has increased everywhere, and Romania is one of the countries that have taken the safety issues seriously. Road safety is an issue of concern to all citizens, and the creation of road safety conditions for all categories of participants, reducing the number of events and their serious consequences are preconditions that contribute to creating a sense of security in society.

Keywords: Reconstruction · Vehicle · Collision · Restitution coefficient · Speed

1 Introduction

Vehicles in general and cars in particular have experienced a spectacular evolution in a relatively short time, being continuously upgraded, resulting in particularly ingenious solutions that provide high reliability and good road traffic safety. Increasing the number of road events over time is due to the increase in the national motor vehicle fleet and their performance in terms of speed, road infrastructure development and traffic intensification [1, 2]. Therefore, road traffic research is a topical field, which must constantly adapt to technological developments in the field of road transport.

In the automotive field, collision is a process of mutual conditioning between two or more vehicles or between a vehicle and an obstacle in a finite time, so that this interaction does not occur before or after the collision. Mutual contact time is limited and is called collision time. This is much less than the observation time of other states of motion that vary within very wide limits. As a result of the contact, the bodies decelerate each other, suddenly, and thus deform [8].

I. Dumitru et al. (Eds.): SMAT 2019, *The 30th SIAR International Congress of Automotive and Transport Engineering*, pp. 332–340, 2020.
https://doi.org/10.1007/978-3-030-32564-0_39

2 Frontal Collision

The application of the pulse conservation method in the frontal collision study is limited because the impact velocity of one of the vehicles must be known. Knowing the amplitude and shape of the deformation, it is possible to determine the direction of its component vectors by the impulse-strain method [6].

The frontal collision of two vehicles may be centric or non-centered. At centric collisions, the direction of the resulting collision force passes through the mass centers of the vehicles. Both cars, during the collision, describe a common motion with speed, no kickback or giraffe effect. If the velocity vectors of mass centers are not collinear but parallel, then the direction of the resulting force does not pass through the center of mass, and the cars will describe different movements at different speeds in the post-collision phase.

2.1 Centric Frontal Collision

Figure 1 shows the central frontal collision between two cars in which the contact areas will deform at a common speed. Impact is achieved without rebound, and in the post-collision phase cars move in tandem, in the same direction, behaving as a single body, Fig. 1c.

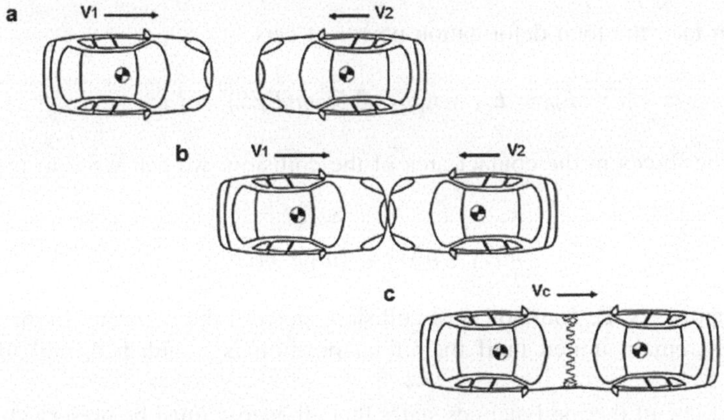

Fig. 1. Phases of centric frontal collision

Starting from the preservation of impulse and kinetic energy, each vehicle exerts the same force during the collision, Fig. 2, in which:

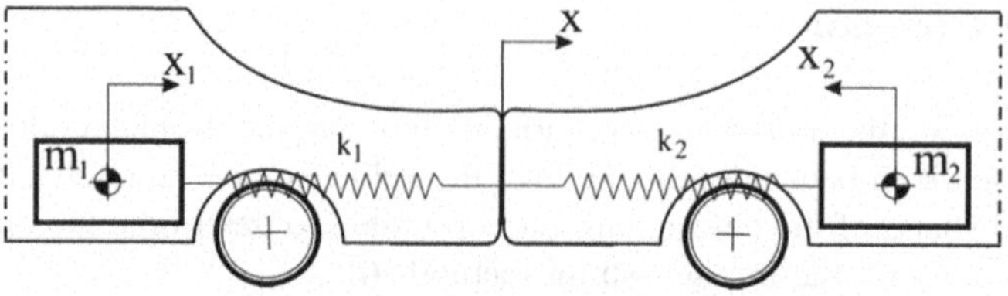

Fig. 2. Centric frontal collision, mathematical model

- m_1 and m_2 are the masses of motor vehicles $[kg]$;- k_1, k_2 elastic constants $[N/m]$;
- x_1, x_2 displacements of mass centers related to motor vehicles during collision $[m]$;
- x the total displacement of the two-vehicle system during collision $[m]$.

The force exerted by each of the vehicles is determined by Hooke's law:

$$F = kx \tag{1}$$

If the front of the vehicles deforms with the size Δx_1 respectively Δx_2, then the total consumed mechanical work will be:

$$L_T = E_{c1} + E_{c2} = \frac{1}{2}k_1(\Delta x_1)^2 + \frac{1}{2}k_2(\Delta x_2)^2 \tag{2}$$

which is, in fact, the total deformation power of cars.

$$E_{td} = E_{c1} + E_{c2} = 0.5 \cdot m(EES)^2 = L_T \tag{3}$$

Slight the forces in the contact area of the collision, we can write to preserve the impulse:

$$m_1 v_1 + m_2 v_2 = (m_1 + m_2) v_c \tag{4}$$

where v_c represents the common post-collision speed of the two cars. In this situation, the vehicles remain united until their final position is established, and there is no refund.

The first law of thermodynamics states that all energy must be preserved. It defines the relationship between mechanical work and kinetic energy, so that the initial energy is equal to the final energy plus the mechanical work performed to produce the deformations:

$$E_T = E_{c1} + E_{c2} = E_{cf} + L_T(E_{td}) \tag{5}$$

$$E_T = 0.5 m_1 v_1^2 + 0.5 m_2 v_2^2 = 0.5(m_1 + m_2) v_c^2 + L_T(E_{td}) \tag{6}$$

from (4) we have:

$$v_c = \frac{m_1 v_1 + m_2 v_2}{m_1 + m_2} \tag{7}$$

using relation (7) we will obtain:

$$E_T = 0.5 m_1 v_1^2 + 0.5 m_2 v_2^2 = 0.5(m_1 + m_2)\left(\frac{m_1^2 v_1^2 + 2 m_1 v_1 m_2 v_2 + m_2^2 v_2^2}{(m_1 + m_2)^2}\right)^2 + L_T \tag{8}$$

$$m_1 v_1^2 + m_2 v_2^2 = \frac{m_1^2 v_1^2 + 2 m_1 v_1 m_2 v_2 + m_2^2 v_2^2}{(m_1 + m_2)} + 2L_T \tag{9}$$

$$(v_1 - v_2)^2 = \frac{2(m_1 + m_2)L_T}{m_1 m_2} \tag{10}$$

$$v_1 - v_2 = \pm\sqrt{\frac{2(m_1 + m_2)L_T}{m_1 m_2}} \tag{11}$$

The final velocity in relation (11) is its size, not its direction, which is given by the positive or negative sign of the result of the difference. If only the negative value is taken into account we amplify each member by -1, and we obtain:

$$v_2 - v_1 = \sqrt{\frac{2(m_1 + m_2)L_T}{m_1 m_2}} \tag{12}$$

The relationship (12) shows that the energy consumed in a collision is a function of the final relative speed of the vehicles and is not an absolute function of the vehicle's speeds. Thus, it can be determined for each vehicle:

$$v_c = v_1 + \Delta v_1 \tag{13}$$

$$\Delta v_1 = v_c - v_1 \tag{14}$$

If we replace v_c of (6) will obtain:

$$\Delta v_1 = \frac{m_1 v_1 + m_2 v_2}{m_1 + m_2} - v_1 \tag{15}$$

$$\Delta v_1 = \frac{m_1 v_1 + m_2 v_2}{m_1 + m_2} - \left(\frac{m_1 + m_2}{m_1 + m_2}\right) v_1 = \frac{m_1 v_1 + m_2 v_2 - (m_1 + m_2) v_1}{m_1 + m_2}$$
$$= \frac{m_1 v_1 + m_2 v_2 - m_1 v_1 - m_2 v_1}{m_1 + m_2} = \frac{m_2 v_2 - m_2 v_1}{m_1 + m_2} = \frac{m_2 (v_2 - v_1)}{m_1 + m_2} \quad (16)$$

$$\text{As well :} \qquad \Delta v_1 = \sqrt{\frac{2 L_T}{m_1 \left(1 + \frac{m_1}{m_2}\right)}} \quad (17)$$

Knowing that the variation of the impulse of a vehicle must be equal and of the opposite direction to the variation of the impulse of the other vehicle, and if the total energy dissipated in the collision is equal to $E_{c1} + E_{c2}$, than:

$$m_1 \Delta v_1 + m_2 \Delta v_2 = 0 \quad (18)$$

also

$$\Delta v_2 = -\frac{m_1}{m_2} \Delta v_1 = -\frac{m_1}{m_2} \sqrt{\frac{2 L_T}{m_1 \left(1 + \frac{m_1}{m_2}\right)}} \quad (19)$$

3 Restitution Coefficient

For an analytical modeling of vehicle speed variation, a parameter is introduced which characterizes the energy consumed in the collision, called the restitution coefficient.

This is defined as the ratio between the relative speed, reduced to the mass centers of the vehicles in the post-impact stage, and the relative velocity at impact initiation, or may be written for the impact with total coverage, and the square root of the ratio between the kinetic energy of the vehicle after collision and the kinetic energy before the collision, or the ratio between the pulse variation corresponding to the separation period and the pulse variation corresponding to the compression period [3, 5, 7]:

$$c_R = \frac{v'}{v} = \frac{v_2' - v_1'}{v_1 - v_2} = \sqrt{\frac{E_c'}{E_c}} = \frac{P_s}{P_d} \quad (20)$$

Figure 3 is a rear-center collision between two cars, where deformed areas describe a common speed v_c. In the collision phase, the cars move in tandem, at the common speed, behaving like a single body, and in the same direction until separation.

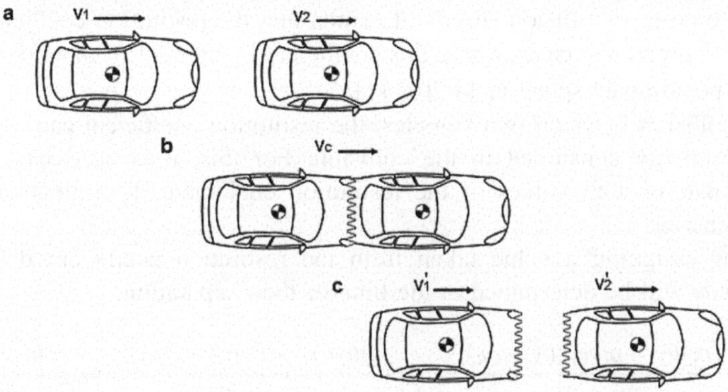

Fig. 3. Phases of rear collision

The common speed of the two vehicles, v_c, can be written as:

$$v_c = v_1 - v_2 \tag{21}$$

from relation (20) and (21) we obtain on $v_2^{'}$ as:

$$v_2^{'} = v_c c_R + v_1^{'} \tag{22}$$

Taking into account the law of energy conservation, in the case of a rigid fixed obstacle collision, one obtains:

$$E_c = E_c^{'} + E_d \tag{23}$$

$$E_d = 0.5mv_1^2 - 0.5mv_1^{'2} = 0.5m\left(v_1^2 - v_1^{'2}\right) \tag{24}$$

If $v_1^{'2}$ is different from zero, then the difference $v_1^2 - v_1^{'2}$, which do not represent Δv^2, must be calculated, which means that the determination of energy with Δv is more difficult, and the procedure is not recommended. This is usually the case for low-speed collisions where the refund rate can reach up to 0.60.

When crashing with a non-deformable fixed barrier $v_1^{'}$ is determined by v_1 and the refund rate:

$$v_1^{'} = v_1 c_R \tag{25}$$

By substituting (24) into (25), we obtain:

$$E_d = 0.5mv_1^2(1 - c_R^2) = E_c(1 - c_R^2) \tag{26}$$

In the collision situation of the vehicle - fixed rigid obstacle, $v_1^{'}$ respective $v_2^{'}$ from (20), become both zero.

There are no post-collision speeds, it results that the restitution coefficient will be zero. However, there are cases where this coefficient is different from zero, the vehicle describes a post-impact speed v_1' [4, 10, 11].

At the collision between two vehicles, the restitution coefficient can overall characterize the energy consumed in the collision. For this, it is necessary to have a statistical database with values of the restitution coefficient determined in different collision scenarios.

Thus, by assigning a value taken from the restitution coefficient database, the vehicle speeds will be determined at the time of their separation:

$$v_1' = v_1 - \frac{(m_2v_2 - m_2v_1)(1 + c_R)}{m_1 + m_2} = v_1 + \frac{m_2(v_1 - v_2)(1 + c_R)}{m_1 + m_2} = v_1 + \frac{m_2v_c(1 + c_R)}{m_1 + m_2} \tag{27}$$

$$v_2' = v_2 + \frac{(m_1v_2 - m_1v_1)(1 + c_R)}{m_1 + m_2} = v_2 - \frac{m_1(v_1 - v_2)(1 + c_R)}{m_1 + m_2} = v_2 - \frac{m_1v_c(1 + c_R)}{m_1 + m_2} \tag{28}$$

Known as these speeds, it is possible to determine the deformation energy consumed during the collision:

$$\Delta E_c = E_d = 0.5\left(m_1v_1^2 + m_2v_2^2 - m_1v_1'^2 - m_2v_2'^2\right) \tag{29}$$

after replacing (35) and (36) into (37), we obtain:

$$E_d = 0.5\frac{m_1m_2(v_1 - v_2)^2(1 - c_R^2)}{m_1 + m_2} = 0.5\frac{m_1m_2v_c^2(1 - c_R^2)}{m_1 + m_2} \tag{30}$$

For the determination of the overall restitution coefficient, which takes into account the stiffness of the cars involved in the collision, it is possible to write the total energy consumed as the sum of the deformation energies consumed by each vehicle:

$$(1 - c_R^2)E_{td} = (1 - c_{R1}^2)E_{d1} + (1 - c_{R2}^2)E_{d2} \tag{31}$$

$$E_{td} = 0.5\frac{m_1m_2(v_1 - v_2)^2}{m_1 + m_2} = 0.5\frac{m_1m_2v_c^2}{m_1 + m_2} \tag{32}$$

Where: E_{d1} and E_{d2} are the deformation energies consumed during the collision of the vehicle sunder going the experimental tests, characterized by there situation coefficient c_{R1} and c_{R2}.

Considering the structures of the two vehicles as elastic (like spring), rigidities k_1 and k_2, also x_1 and x_2 their maximum deformations, we can write:

$$k_1 \cdot x_1 = k_2 \cdot x_2 \tag{33}$$

the energy of the elastic element is:

$$E_c = 0.5 \cdot k \cdot x^2 \tag{34}$$

$$k_1 \cdot E_{c1} = k_2 \cdot E_{c2} = k \cdot E_c \tag{35}$$

$$k = \frac{k_1 \cdot k_2}{k_1 + k_2} \tag{36}$$

k - rigidity of the system.

If the speeds corresponding to the deformation step are $v_d - v_1$ respectively $v_2 - v_d$, we can write the energy balance:

$$\begin{aligned} &0.5m_1\left(1 - c_{R1}^2\right)(v_d - v_1)^2 + 0.5m_2\left(1 - c_{R1}^2\right)(v_1 - v_d)^2 \\ &= 0.5\frac{m_1 m_2 (v_1 - v_2)^2\left(1 - c_{Rm}^2\right)}{m_1 + m_2} = 0.5\frac{m_1 m_2 v_c^2\left(1 - c_{Rm}^2\right)}{m_1 + m_2} \end{aligned} \tag{37}$$

from which result the expression of the weighting coefficient of rebound:

$$c_{Rm} = \sqrt{\frac{c_{R1}^2 m_2 + c_{R2}^2 m_1}{m_1 + m_2}} \tag{38}$$

Restitution coefficients c_{Rk} and c_{Rm} are generally different. The global restitution coefficient is a parameter that characterizes the total energy consumed in the collision, not the energy consumed for deforming each individual vehicle.

It depends on the relative speed of impact and rigidities of the vehicle structure involved in the collision. In the case of the collision study between vehicles, it is advisable to use the restitution coefficient according to the rigidity of the body because this and the amplitude of the deformations contribute to the calculation of the deformation energy [10, 11].

4 Conclusions

In a real collision scenario, the problem is not to determine the ultimate speeds of the cars, usually the opposite, that is, to calculate the initial speeds of the vehicles. However, in order to determine the initial speeds, the final speeds determined by the marks left on the asphaltic apparel by the tires of the vehicles must be known. If only the residual deformations of the vehicles are known, only the variations in speed for each vehicle involved in a collision can be calculated. However, if the vehicle's characteristics are known, ie mass, rigidity and maximum speed, then the problem is resolved [9].

The data measured in experimental trials are published in the literature to support those conducting the construction and/or study of car structure behavior. These models must be carefully designed to become useful applications for investigating and reconstituting road events.

The purpose of investigating the road event is to study the behaviors and the facts that led to its production, taking into account the various aspects of the consequences (injuries, initial and final positions, material damage) and reaching technical, medical, psychological or legal causes. Under these conditions, the analysis will be extended to the operation of passive safety systems, setting out measures to optimize them.

In general, the investigation of the road event can be carried out by means of real-world, laboratory and/or polygon experiments, or virtual using reconstruction information technologies, followed by validation of the results by comparison with existing databases in the literature.

References

1. Cipriani, A.L., Bayan, F.P., Woodhouse, M.L., Cornetto, A.D., Dalton, A.P., Tanner, C.B., Timbario, T.A., Deyerl, E.S.: Low speed collinear impact severity. A comparison between full scale testing and analytical prediction tools with restitution analysis. Accident Reconstruction SP-1666, no. 2002-01-0540, Society at Automotive Engineers, Warrendale, PA (2002)
2. Franch, H., Franch, D.: Mathematical Methods for Accident Reconstruction. CRC Press, Boca Raton (2010)
3. Davis, G.A.: Crash reconstruction and crash modification factors. Accid. Anal. Prev. **62**, 294–302 (2014)
4. Monson, K.L., Germane, G.J.: Determination and mechanisms of motor vehicle structural restitution from crash test data. Accident Reconstruction Technology and Animation IX SP-1407, no. 1999-01-0097, Society of Automotive Engineers, Warrendale, PA (1999)
5. Jones, I.S., Wilson, L.A.: Techniques for the reconstruction of rollover accidents involving sport utility vehicles, light trucks and minivans. Accident Reconstruction: Analysis, Simulation, and Visualization SP-1491, no. 2000-01-0851, Society of Automotive Engineers, Warrendale, PA (2000)
6. Jennings, P.W.: Methods for assessing vehicle speed from impact damage. In: Accident Reconstruction Seminar (1999)
7. Buciu, G., Popa, D.L., Grecu, D., Niculescu, D., Nemes, R.: Comparative Analysis of the Three New Designs of Tibial Nails Which Eliminate the Use of Orthopedic Screws. Lux Libris Publishing House, Braşov (2012)
8. Johnson, N., Hampton, C., Gabler, H.C.: Evaluation of the accuracy of side impact crash test reconstructions. Biomed. Sci. Instrum. **45**, 250–255 (2009)
9. Vatu, M., Craitoiu, M.M., Vintila, D., Mercut, V., Popescu, M.S., Scrieciu, M., Popa, D.L.: Determination of resistance for cesfrom mandibular movements through dynamic simulation using kinematic analysis and finite elements method. Rom. J. Oral Rehabil. **10**(1), 20–28 (2018)
10. Rose, N.A., Fenton, S.J., Beauchamp, G.: Restitution Modeling for Crush Analysis: Theory and Validation, 2006-01-0908 (2006)
11. BrachR, M., Brach, R.M.: Vehicle Accident Analysis and Reconstruction Methods. SAE International, Warrendale (2005)

Behavior of the Cranio-Cervical Biofidelity Complex Compared to the Cranio-Cervical Area of Anthropometric Test Devices

Mihai Clinciu[1] and Stefanita Ciunel[2(✉)]

[1] Transilvania University of Brasov, Politehnicii Street, no. 1, Brasov, Romania
[2] University of Craiova, Calea Bucuresti Street, no. 107, Craiova, Romania
ciunel_stefanita@yahoo.com

Abstract. The article refers to testing of a biofidelic neck of anthropometric test devices, which can also be used in any field of medical, technical interest. The design of this technical system has started from the study of existing technical systems, and compared to the values obtained by the same area of the test dummy, under the same request conditions. From the study of patents, the following observations were made regarding their disadvantages - heavy construction. Many spine constructions have a multitude of mechanical devices; we cannot simulate all the movements of a human neck, but only a part of them. The biofidelic neck type technical system of a test dummy, which by its construction (both vertebrae and intervertebral discs) closely follows the technical, mechanical and constructive characteristics of the human man, and the use of elastic elements actuation to allow simulation of cervical muscles during collision tests of motor vehicles.

Keywords: System · Dummy · Biofidelity · Cervical complex

1 Introduction

1.1 Experimental Realization of Dummy Head

The practical execution of the mannequin model biomechanical model of we studied North America patent inventions, but also of the Humanetics Atd. Corporation, documents showing the mounting/removal process of a anthropometric test device. Upper area being researched area, which includes the head of the dummy, neck vertebrae, intervertebral areas, as well as the devices used in connection with the torso of the manikin (upper chest) [1–3].

In Fig. 1 is a classic system assembly/disassembly of the head of a dummy, as well as parts from which it is made. In this way were carried out several models of head dummy, trying to get a high biofidelity.

I. Dumitru et al. (Eds.): SMAT 2019, *The 30th SIAR International Congress of Automotive and Transport Engineering*, pp. 341–348, 2020.
https://doi.org/10.1007/978-3-030-32564-0_40

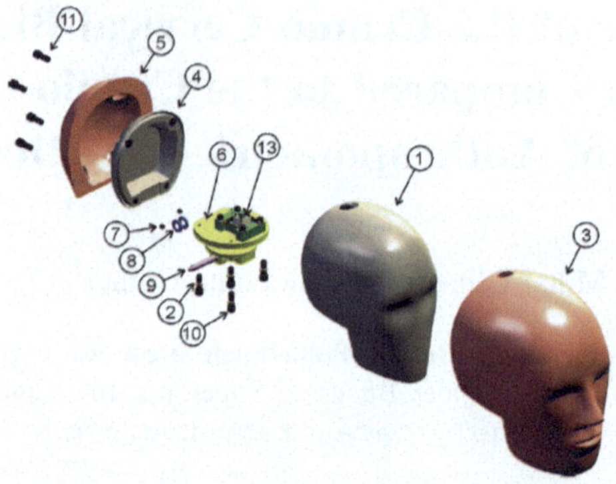

Fig. 1. Dummy head. 1. the front cap, aluminum; flange fastening screws 2. 10-upper neck; 3. face mask; 4. dorsal head dummy side; 5. dorsal head mask dummy; 6. upper fixing flange neck-head dummy; 7, 8, 9. lock flange fixing system; 11. screws fixing the dorsal and dorsal mask dummy head; 13. accelerometer. (Manual Hybrid III 95%, FTSS)

Dummy head assembly mass without mounting bolts fastening flange, and the last vertebra of the cervical area is 3800 g. The totality of the dummy head Assembly must reach the value of 4500 g (the head of a adult man, of average height). Dummy head model was designed to simplify fixing and calibrating the sensor upper flange triax's attachment of the cervical area [4–6].

The dummy's head also executed two spherical caps, providing for a system of assembly/disassembly with screws to the midline. It has executed a dummy head based on the circumference of the head of a mature man, obtaining a value of Ø180 [mm].

For the upper and lower callote we use steel sheet with a thickness of 2 [mm]. For the connection of the two flanges in the median area is fixed, by continuous welding, autogenous two flanges, steel all, thickness 6 [mm], and provided for the attachment of the lower frame (2) with 6-hole M8, equidistant. Lower flange in turn is executed from OL37 steel plate, thickness 6 mm, provided similarly with 6 holes Ø6 [mm]. The two flanges of the polar head are fastened through the 6 screws M8 from the bottom up.

1.2 Experimental Realization of Neck Dummy

Experimental realization of the neck dummy had studied as part a documentary American patent which develops system running, testing, materials, and processes of assembly/disassembly of the cranio-cervical complex of a dummy test on road accidents [1, 3, 12, 19].

Anthropometric test devices have by construction parts of the body similar to human ones, differentiated classes of sex, age, types of road accidents (where they are used in lateral, front-back, pedestrians crash tests), but also depending on the year of manufacture.

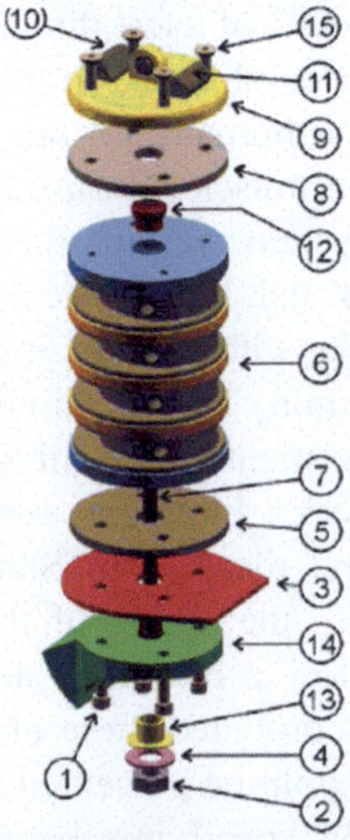

Fig. 2. Cervical complex. 1. Fixing screws 10, 15; 2–4 13, tightening the fastening parts cable; 3, 5, 8 spacers; 6 neck; 7. Cable neck tension system; 9 upper Flange neck; 11 rubber pads; 12 rubber bushing; 14. Lower flange neck. (Manual use dummy Hybrid III male 95%, FTSS)

In the Fig. 2 is presented a cervical complex Hybrid III 95% male, produced by company FTSS. The neck consists of five vertebrae, from Dural, among which there are 4 intervertebral areas, from an elastic material (butyl rubber). Lower flange connects to the trunk of the dummy. Upper flange will link with the dummy's head. Through the median area whole neck is crossed by a cord, which steeled tensed with the help of a whole system ensures rigidity of the cervical spine. For concentricitate in areas where the cable would be able to have contact with the neck are two screw nuts isolators (elimination of vibrations) [7–9].

The upper part of the cable has a fixed bushing attached to the cable, and that it will support a payment/grower washer. Opposite, in the lower flange area the cable ends with a bushing screw system rigidity, thus by tightening with the help of a metal.

For the study of neck biofidelity of a mannequin we made a model that correspond to the characteristics of the neck dimensional type Anthropometric Test Devices used in road accidents [10, 11].

Experimental technical system refers to the realization of a biofidelic neck for experimental crash tests, which can be used and any area of medical interest.

The technical design of this system was started from the study of existing technical systems – US patents for inventions, in which are described the cervical, each of them presenting a particular form, bringing a constructive improvement of the shape as well as the study of neck movements under different workloads.

From the study of these patents have been able to get the following observations regarding the drawbacks:

- technical constructions quite cumbersome:
- many of the spine have constructed a variety of mechanical devices;
- can not get all the movements of a human neck, but only some of them.

Technical system "biofidelic neck" of a test dummy, which through its construction (both of the vertebrae and intervertebral discs) in a heavily mechanical characteristics technical and constructive with the human and the use of elastic elements of action allowing the simulation of cervical muscles during tests of car crash collisions.

Biofidelic cervical technical system presents the following advantages: intervertebral discs are made from a material that has physical and chemical properties of visco-elastic and mechanical compatible with human intervertebral discs; simulates both flexion and extension of the test dummy's head; simulates the movement of the rotating neck biofidelic; simulates the movements of learning of the neck [13–15].

Cervical system consists of seven biofidelic vertebrae, each with a custom construction, four elastic intervertebral discs, with Shape Memory Alloy, fixed on the cervical vertebrae, which simulates the action of the cervical muscles, as well as a mechanism of action that may allow a biofidelic neck rotating around its own axis, enabling to carry out movements combined these elements used [16–18].

Specification for the system should be presented in the cervical biofidelic continue practice neck biofidelic, see Fig. 3a, specifying that all C1–C7 cervical vertebrae have different structural shapes, allowing the attachment of springs thrust with the Shape Memory Alloy, see Fig. 3b.

Biofidelic neck system from Anthropometric Test Device consists of 7 vertebrae denoted as in literature with the letter C (cervical) followed by a numerical index that represents the position of the vertebra, from 1 to 7, as: C1–C7.

(a) (b)

Fig. 3. Experimentally biofidelic neck; (a) overview of cervical biofidelic system; (b) partial to the Sub-Assembly consisting of the elastic elements of the propulsion system biofidelic cervical cervical

Biofidelic cervical system consists of 3 parts, referred to three subassembly into the cervical vertebrae are composed of noted C1–C7, each of them presenting some form design; between the vertebrae C3–C4, C4–C5, C5–C6, C6–C7 disc there is an intervertebral disc, each of which is made of a poly vinyl syloxan type material, with a fluidity and high plasticity, given the physical-mechanical characteristics close to the disks human intervertebral discs. Between the C1 and C2 vertebrae it does not exists intervertebral disk; for the purposes of flexion-extension movements of the head of the manikin used two elastic elements, 12.

In Fig. 4 is shown the cervical system classic: (a) C3–C7 cervical system assembled with bracing system; (b) C1–C7 cervical classic partly assembled).

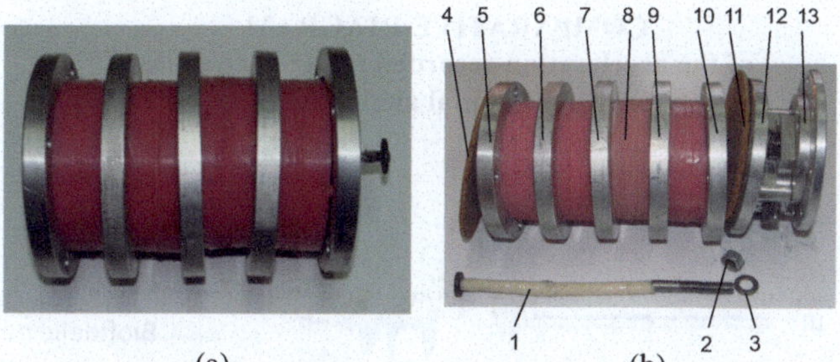

Fig. 4. Classic cervical system with elements of bracing, damping and grip: (a) C3–C7 cervical system assembled with bracing system; (b) C1–C7 cervical classic partially assembled: 1-steel wire for rigidisation: 2 cable-clamping nut, washer, 3–4 - damping disc, 5–6 vertebra vertebra C7–C6, C5 vertebra, 7-8-9-the intervertebral disc, vertebrae C4 vertebra C3, 10-11-disc, 12-13-C2 vertebra vertebra, C1)

2 Discussion. Comparative Analyses

Have compared the data obtained for the dummy's head using the two systems (classical and biofidelic) for each of the initial roll angles type pendulum.

Testing has been carried out under the same conditions, at the same angles in the pendulum. Thus, the initial angle of the pendulum 10° was obtained by comparative chart in Fig. 5.

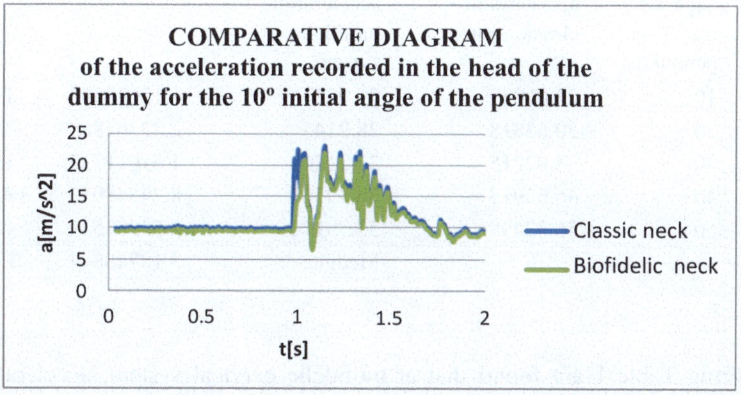

Fig. 5. Comparative chart of the absolute acceleration from dummy's head (ACC3) for the initial angle 10° pendulum

The initial angle of the pendulum of 20° was obtained by comparative diagram of Fig. 6.

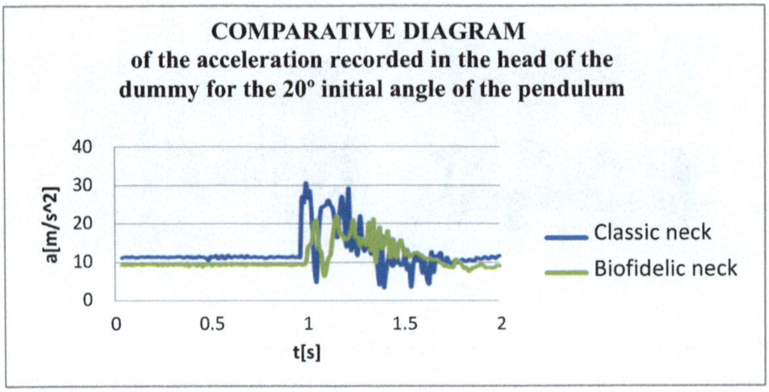

Fig. 6. Comparative chart of the absolute acceleration from dummy's head (ACC3) for the initial angle of pendulum

Similar measurements have been performed for experimental angles of 30°, 40° and 50° to determine maximum accelerations.

After comparing the maximum recorded by accelerometer ACC3 (dummy head) depending on the angle of the pendulum was originally obtained the data in Table 1.

Table 1. Maximum acceleration recorded in the dummy's head to the cranio-cervical systems in relation to the initial angle of the pendulum

No. crt.	Initial angle of the pendulum	Maximum head acceleration - classic system	Maximum head acceleration - biofidelic system	Difference	%
1.	10	23.07839	21.95216	1.12622542	4.88
2.	20	30.63818	28.2145	2.42367888	7.910649
3.	30	38.92215	37.31204	1.61011335	4.136753
4.	40	46.63913	38.39564	8.24348913	17.67505
5.	50	56.32498	53.79114	2.53383541	4.4986
			Mediate	3.18746844	7.82021

Analyzing Table 1 are found that at biofidelic cervical system, acceleration was approximately 7.8% lower in relation to the classic system, under the same conditions.

3 Conclusions

Concerned, in particular, the evolution of temporal acceleration for cranio-cervical system. Data were obtained for angles of the pendulum between 10° and 50°. Biofidelic cranio-cervical system was tested for the same angular values after learning movement left obtained by artificial muscle operation made up of arches with shape memory. The results were transformed into diagrams. For the two devices tested were obtained comparative charts. Starting from the values and results were determined, for both systems, the following parameters for the injury evaluation: gravity index GSI, head injuries and criteria HIC_{15} and HIC_{36} and probability p of injuries, their limit values (minimum and maximum) being determined previously and can be found in the specialized technical literature.

References

1. Abedrabbo, G.: Simulation of a frontal crash test using multibody dynamics. Universite catholique de Louvain, Center for Research in Mechatronics, Belgium (2009)
2. Aldman, B.: Biodynamic studies of impact protection. Acta Physiol. Scand. **56**, 1–80 (1962)
3. Foster, J.K., Kortge, J.O., Wolanin, M.J.: Hybrid III - a biomechanically based crash test dummy. In: 21st Stapp Car Crash Conference, SAE 770938 (1977)
4. Klinich, K.D., Beebe, M.S., Backaitis, S.H.: Evaluation of a proposed Hybrid III hip modification. In: 39th Stapp Car Crash Conference, SAE 952730 (1995)
5. Krug, E.: Injury: a leading cause of the global burden of disease WHO report. Department of Violence & Injury Prevention, WHO, Geneva (2000)
6. Lau, I.V., Viano, D.C.: The viscous criterion - bases and applications of an injury severity index for soft tissues. SAE 861882 (1986)
7. van der Horst, M.I.: Human head-neck response in frontal, lateral and rear end impact loading modelling and validation. Technische Universitet, Eindhoven (2002)
8. Melvin, J.W., et al.: Human head and knee tolerance to localized impacts. SAE 690477 (1969)
9. Mertz, H.J.: Anthropomorphic Models, The Biomechanics of Trauma. Appleton-Century-Crofts, Norwalk, Connecticut (1985)
10. Mertz, H.J., Patrick, L.M.: Strength and response of the human neck, SAE 710855
11. Mertz, J.H., Prasad, P., Nusholtz, G.: Head injury risks assessment for forehead impacts. SAE 960099 (1996)
12. Myers, B.S., McElhaney, J.H., Richardson, W.J., Nightingale, R.W., Doherty, B.J.: The influence of end condition on human cervical spine injury. In: Proceedings from the 35th Stapp Car Crash Conference, SAE 912915, pp. 391–399 (1991)
13. Netter, F.H.: Interactive Atlas of Human Anatomy v.3.0. Medicală Callisto, Bucureşti (2003)
14. Vatu, M., Craitoiu, M.M., Vintila, D., Mercut, V., Popescu, M.S., Scrieciu, M., Popa, D.L.: Determination of resistance forces from mandibular movements through dynamic simulation using kinematic analysis and finite elements method. Rom. J. Oral Rehabil. **10**(1), 20–28 (2018)

15. Lap-shoulder belt restraint. J. Biomech. Eng. (1980)
16. http://www.humaneticsatd.com. Accessed June 2019
17. http://internationaltransportforum.org. Accessed 21 June 2019
18. http://www.spineuniverse.com/anatomy/intervertebral-discs. Accessed 11 June 2016
19. http://standardizare.wordpress.com/tag/antropometrie-si-biomecanisme. Accessed 12 June 2019

Study of the Possibilities to Improve the Service Level of Traffic Light Intersections by Road Traffic Micro-simulation

Andrei Alexandru Boroiu, Elena Neagu(✉), Alexandru Boroiu, and Sebastian Parlac

University of Pitesti, Targul din Vale 1, 110040 Pitesti, Romania
elena.neagu@upit.ro

Abstract. In the last few years, one of the traffic issues that has been constantly increasing in Pitesti is the traffic congestion that occurs at peak hours in the Maior Sontu – Targul de Vale area and at the Targul din Vale roundabout, situated in the proximity. Research to solve this problem was developed with the help of the micro-simulation software for road traffic PTV Vissim. The results obtained revealed that by the micro-simulation of road traffic at traffic light intersections it is possible to demonstrate the limits of improvement of the level of service by redesigning the traffic light plan and the need to look for other ways to solve the traffic problems. Thus, the traffic analysis by micro-simulation has demonstrated in the present case that the most effective way to solve the problem is to increase the capacity of the traffic light intersection by creating additional lanes either to increase the capacity of a groups of lanes or to eliminate overlapping the traffic flows adjacent with those that enter the traffic area of the intersection. The putting into operation by the City Hall of Pitesti Municipality of the proposals advanced based on these studies confirmed the disappearance of road congestion in the area under analysis, thus confirming the usefulness of the analysis by micro-simulation of the road traffic for solving the problems at traffic light intersections.

Keywords: Traffic light intersection · Service level · Traffic micro-simulation

1 Road Traffic Problems Identified in the Central Area of Pitesti

The observations made in the last few years on road traffic in the central area of Pitesti [4] have revealed the phenomenon of traffic congestion that occurs at peak hours in the Maior Sontu - Targul din Vale area and at the Targul din Vale roundabout, situated in the proximity (see Fig. 1).

Thus, it was found that the length of the waiting string in the North-South direction (to which a single traffic lane is allotted) reaches high values during peak hours, the traffic light intersection not being transited during the first phase of Green light, but after 2–3 cycles of traffic lights, a fact that substantially increases the waiting time of the vehicles in this stream.

I. Dumitru et al. (Eds.): SMAT 2019, *The 30th SIAR International Congress of Automotive and Transport Engineering*, pp. 349–358, 2020.
https://doi.org/10.1007/978-3-030-32564-0_41

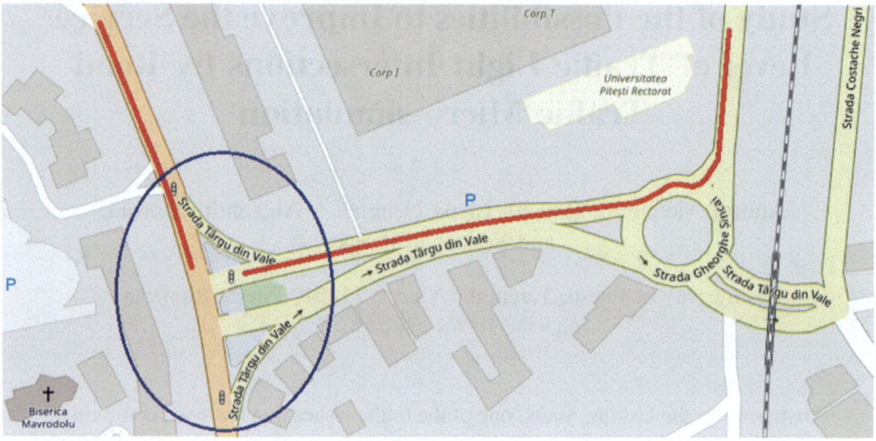

Fig. 1. The Maior Sonțu – Targul din Vale traffic-light intersection (blue colour) and the road sections where road congestion occurs (red colour).

Also, although access to the traffic light intersections for the streams following the East-South direction is achieved on two lanes, the waiting time for vehicles on the right lane increases significantly when the waiting string length exceeds about 15 m because this lane is also used by the vehicles that turn right onto the special brace that bypasses the traffic light area of the intersection, there being no area for separating the two road streams before the intersection.

In peak hours the length of the waiting string reaches values greater than the distance to the intersection in the Targul din Vale roundabout, upstream, the traffic congestion propagating at this intersection and even upstream, on the North Arm especially so as can be seen in the 3D simulation made with the Vissim software [10] – see Fig. 2.

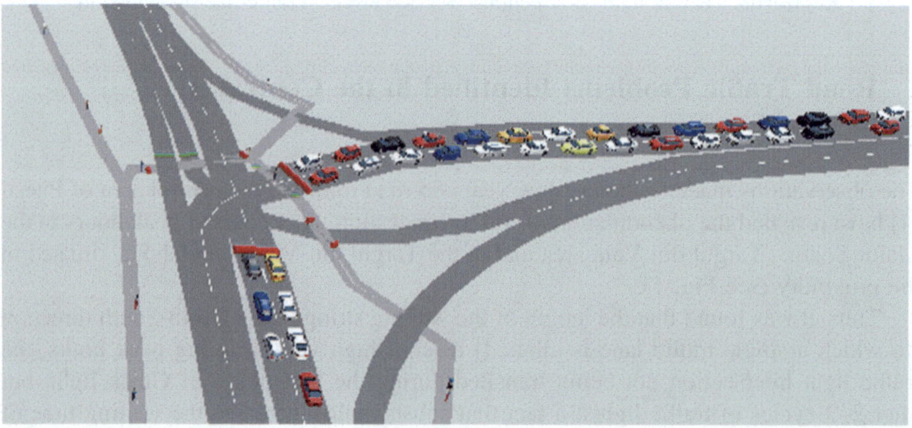

Fig. 2. Traffic congestion at the Maior Sontu – Targul din Vale intersection on the E-S arm.

2 Evaluating the Capacity of the Traffic Light Intersection

By observations made on the ground in March 2019, the geometric elements of the Maior Sontu-Targul din Vale traffic light intersection - and the road facilities (see Figs. 3 and 4) - were identified.

Fig. 3. The Maior Sontu-Targul din Vale traffic light intersection.

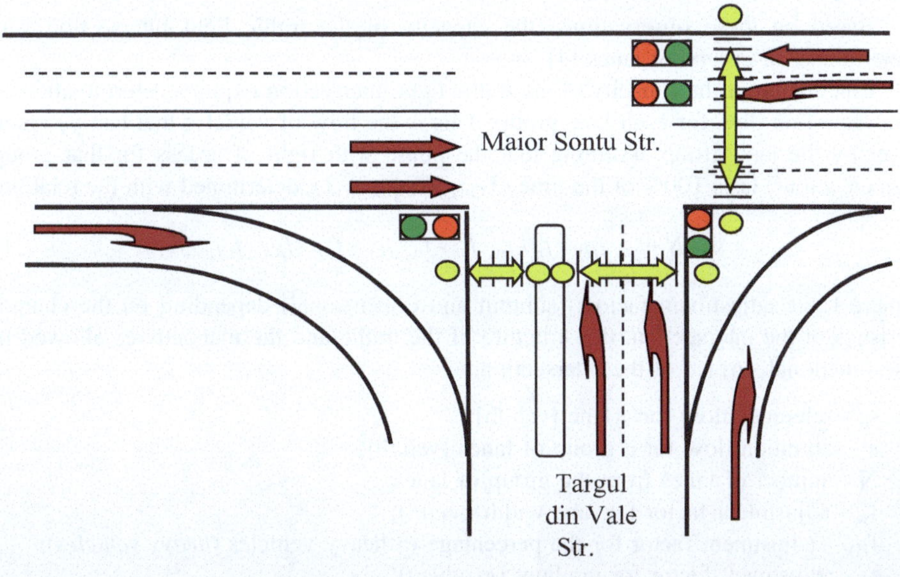

Fig. 4. Organising road traffic in the Maior Sontu – Targul din Vale traffic light intersection.

At the same time, road traffic regulation was analysed at the intersection, and it was found to be performed with a total time for the traffic light cycle TC = 94 s and 4 Green Phases for the vehicles, organized as shown in the Fig. 5 with the Vissim PTV traffic microsimulation software [5].

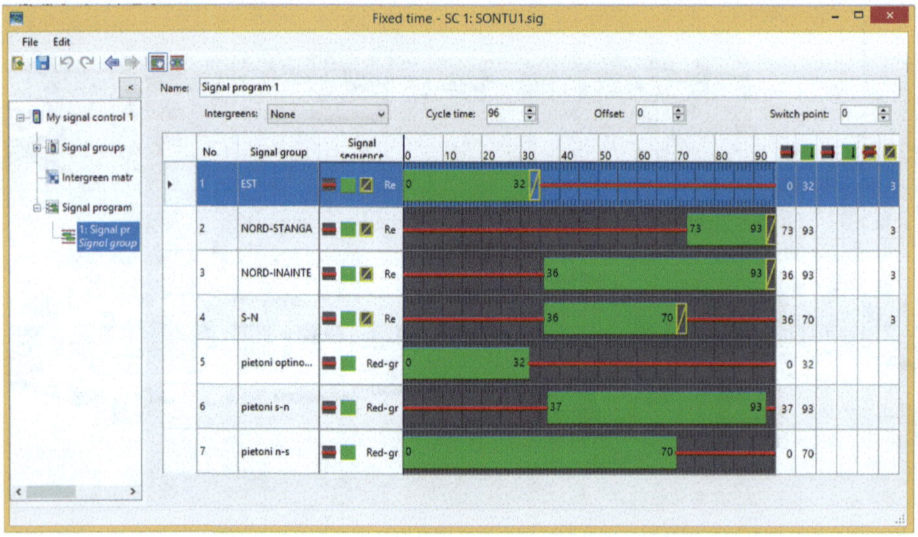

Fig. 5. The Maior Sontu – Targul din Vale traffic-light intersection (blue colour) and the road sections where road congestion occurs (red colour).

Based on these observations, the capacity of the traffic light intersection was determined on groups of lanes [4].

Calculation of the capacity of the traffic light intersection requires determination of the saturation flow for each lane group. This is the flow of vehicles that can be taken over by the lane group assuming that the phase with right of access for that group (green colour) lasts 100% of the time ($T_{Green}/T_{Cycle} = 1$), determined with the relation:

$$s = s_0 \cdot N \cdot f_w \cdot f_{HV} \cdot f_g \cdot f_p \cdot f_{bb} \cdot f_a \cdot f_{LV} \cdot f_{LT} \cdot f_{RT} \cdot f_{LTp} \cdot f_{RTp}, \quad (1)$$

where f_i are adjustment factors (subunit and dimensional) depending on the characteristics of the intersection, the structure of the traffic and the manoeuvres allowed in the functional surface of the intersection:

- s_0 - reference flow for a lane [veh./h];
- s - saturation flow for a group of lanes [veh./h];
- N - number of lanes from the group of lanes;
- f_w - adjustment factor for lane width (*width*);
- f_{HV} - adjustment factor for the percentage of heavy vehicles (*heavy vehicles*);
- f_g - adjustment factor for gradient (*gradient*);

- f_p - adjustment factor for the number of parking manoeuvres;
- f_b - adjustment factor for the number of bus stops (*bus*);
- f_a - adjustment factor for the area where the intersection is located (*Area*);
- f_{LU} - adjustment factor for lane use (*Lane Use*);
- f_{LT} - adjustment factor for left turns from the group of lanes (*Left Turn*);
- f_{RT} - adjustment factor for right turns from the group of lanes (*Right Turn*);
- f_{LTp} - adjustment factor for left turn due to pedestrians (*Left Turn due to pedestrians*);
- f_{RTp} - adjustment factor for right turn due to pedestrians (*Right Turn due to pedestrians*).

The final results of the calculations are presented in the Table 1.

Table 1. The capacities of the lane group of the traffic light intersection.

	Group 1	***Group 2***	***Group 3***	***Group 4***
	N-S	***S-N***	***N-E***	***E-S***
Date		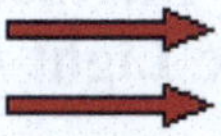		
s_i [veh./h]	1665	3145	1582	3140
$T_{Green,i}$ [s]	58	35	21	30
$T_{C,ef}$ [s]	86			
$c_i = s_i \cdot (T_{Green,i}/T_{C,ef})$	**1123**	**2560**	**386**	**1096**

The saturation flow for each group of lanes is the flow of vehicles that can be taken over by the lane group assuming that the phase with right of access for that group (green colour) lasts 100% of the green time of the entire traffic light cycle (ratio $T_{Green,i}/T_{C,ef} = 1$). It is found that the traffic light intersection has (according to the intersection geometry, the regulations regarding the access of the vehicle categories to the intersection and the traffic light plan [9]) the total capacity:

$$C = \sum_{1}^{4} c_i = 5165\,[\text{veh-et/h}] \quad (2)$$

The capacity on lane groups during a traffic light cycle is also of interest here [1, 3], because these theoretically determined values can be compared to the measured peak traffic hour values (if the waiting strings are higher than these values, on the respective group of lanes the vehicles will not cross the intersection at the first Green signal). Considering that one hour comprises

$$n_c = \frac{3600}{T_C} = \frac{3600}{94} = 38.3 \; \textit{traffic light cycles} \tag{3}$$

And that these values will be equal to the ratio:

$$c_{i\,cycle} = \frac{c_i}{n_c} \; [\text{veh-et/cycle}] \tag{4}$$

the following values can be extracted:

- Group 1 N-S: 29.3 veh./cycle;
- Group 2 S-N: 66.8 veh./cycle;
- Group 3 N-E: 10.1 veh./cycle;
- Group 4 E-S: 28.6 veh./cycle.

3 Research to Increase the Intersection Service Level by Optimizing the Traffic Light Cycle

There are various methods for drawing the traffic lights programme [7]:

- equalization of saturation degrees of critical movements;
- minimizing the average waiting times of vehicles crossing the intersection;
- ensuring the principle of "average equal waiting time between users (equity)";
- minimizing the amount of saturation degrees.

The working algorithm for determining the service level in the Norms for setting up level intersections on public roads [9], which is adapted after the model presented in the reference work Highway Capacity Manual [8], is based on the criterion of minimizing the average waiting times of vehicles crossing the intersection.

The working algorithm operates with two components that are determined for each lane group:

- flow/capacity ratio (X_i);
- average waiting time of a vehicle of d_i;

In the case identified via observation (the capacity is exceeded for one of the traffic movements), the first indicator: the flow-capacity ratio (X_i) is interesting.

The flow-capacity ratio X_i is defined as the ratio between actual traffic flow and capacity. It is calculated independently for each group of lanes with the relation:

$$X_i = \frac{v_i}{c_i} \tag{5}$$

Based on the manual recordings carried out with traffic observers (see Fig. 6), the maximum traffic flow values for each group of lanes were determined (for group 4,

Fig. 6. Measurement of traffic on the group of lanes of the intersection.

which includes heavy vehicles, the equation in standard vehicles was performed). The proportions of vehicles arriving at the traffic light on each group of lanes during the "Green" signal were visually estimated.

The values for the *flow-capacity ratios* X_i were calculated with the relation (5). We then worked with the Microsoft Excel program in the same table, taking data the known data that were required and calculating the delays with the respective relations in the table.

The aggregation of the delays at the level of the intersection arms (one arm may have more groups of lanes) was performed only for the North arm, which has two lane groups, group 1 and group 3.

By aggregating the intersection delays for the peak traffic hours, the control delay at the intersection level [sec/veh.] was obtained, and it is found to correspond to service level E (difficult traffic, permanent waiting strings, reduced speed).

The service level of the intersection is not satisfactory and a first study direction for improving the traffic conditions by intersection and classifying it at a higher service level is the optimization of the traffic light cycle [2, 6].

This study was carried out, but the reduction in control delay was insignificant, maintaining the same level of service.

Considering also that the result for the E-S group is not the one expected because one of the traffic lanes also overlaps with the right turn during the peak traffic hours, a search has been carried out with the purpose of identifying a solution involving works of larger dimensions, but one that could take action from the origin of the problems: on the basis of the results obtained by traffic microsimulation in new conditions of intersection development, proposals were made to the City Hall of Pitesti for the creation of additional intersection lanes, as presented below.

4 Increase the Intersection Capacity by Increasing the Number of Traffic Lanes

On the basis of the proposals made, the following additional arrangements were made at the traffic light intersection;

1 - an additional traffic lane for the North-South direction was created (see Fig. 7), so that during the same traffic light cycle the capacity for this group of lanes doubled, reaching the value CN-S = 2 × 1123 = 2246 veh./h, respectively, a double capacity on the traffic light cycle: Ci, cycle, N-S = 2 × 29.3 = 58.6 veh./cycle

Fig. 7. Setting up of a second lane for the North-South direction.

2 - an extra lane for the East-North direction, separated from the East-South lane group (Fig. 8), has been set up, so that the two streams do not overlap at the intersection, the capacity of the two left turns (1096 veh./h) being now fully deployed, eliminating the phenomenon of congestion propagation towards the Targul din Vale roundabout intersection.

Fig. 8. The third traffic lane on the East arm, intended only for East-North road traffic.

After this rearrangement of the intersection, it was found that, although the traffic light cycle did not have to be modified (because it was demonstrated by microsimulation that there is not a sensible increase in the capacity of the intersection), the reported road traffic problems disappeared.

5 Conclusions

The results obtained throughout the research presented revealed that by road traffic microsimulation at traffic light intersections it is possible to demonstrate the limits of improvement of the level of service by redesigning the traffic light plan and the need to look for other ways to solve the traffic problems.

In the present case redirecting road traffic can be an effective solution, but projects to build the road passages to take over some of the road traffic flows [5] have very long deadlines.

Thus, it was proved that the most effective way to solve the problem is to increase the capacity of the traffic light intersection by creating additional lanes either to increase the capacity of a groups of lanes or to eliminate overlapping the traffic flows adjacent with those that enter the traffic area of the intersection.

The putting into operation by the City Hall of Pitesti Municipality of the proposals advanced based on these studies confirmed the disappearance of road congestion in the area under analysis.

Thus, the usefulness of the analysis by road traffic microsimulation for solving the problems at traffic light intersections was confirmed.

References

1. Boroiu, A.: Circulaţie rutieră (Road Traffic). University of Pitesti Publishing House, Pitesti (2003)
2. Boroiu, A.: Transporturi de persoane (Public Transport). University of Pitesti Publishing House, Pitesti (2009)
3. Boroiu, A.A., Neagu, E.: Trafic rutier şi siguranţa circulaţiei. Aplicaţii (Traffic and Road Safety. Applications). University of Pitesti Publishing House, Pitesti (2015)
4. Boroiu, A.A.: Poluarea fonică rutieră (Road Noise Pollution). University of Pitesti Publishing House, Pitesti (2018)
5. Boroiu, A.A., Neagu, E., Boroiu, A., Vieru, I.: Proposals for the reorganization of road traffic in the central area of Pitesti municipality based on microsimulation-performed traffic analyses. In: 8-th International Conference on Advanced Concepts in Mechanical Engineering-ACME 2018, WOS: 000467443600128, Iaşi (2018). https://doi.org/10.1088/1757-899x/444/7/072021
6. Neagu, E.: Trafic rutier şi siguranţa circulaţiei (Traffic and Road Safety). University of Pitesti Publishing House, Pitesti (2003)
7. Roşca, M.A., Dinu, O.: Aspecte privind optimizarea semaforizării în intersecţiile izolate (Aspects concerning the optimization of the traffic lights at isolated intersections). AGIR Bulletin, Supplement no. 1 (2012)
8. Transportation Research Board (TRB) of National Academies of Science: Highway Capacity Manual. National Academies of Science, Washington (2010)
9. CNADNR: Normativ pentru amenajarea intersecţiilor la nivel pe drumuri publice (Standard for the Development of Intersections at Public Roads). Search Corporation (2009)
10. PTV AG Visum 11,5: User Manual. PTV (Planung Transport Verkehr) AG Karlsruhe, epubli GmbH, Berlin (2012)

The Influence of the Color of the Reflective Vest on the Visibility of the Cyclist at Night

Bogdan Cornel Benea(✉), Daniel Dragos Trusca,
George Radu Togănel, and Alexandru Ionut Radu

Transilvania University of Brasov, Politehnicii Str. 1, Brasov, Romania
b.benea@unitbv.ro

Abstract. Bicycle fatalities make up 8% of the total number of road accident fatalities in 2016 in the EU countries. In these countries, 2.015 people riding bicycles were killed in road accidents in 2016. During the night, the number of accidents involving cyclists is 26% higher than the number of accidents during the day. This work shows the influence of the reflective elements worn by cyclists on their visibility during the night. The tests were carried out on a public road, with cyclists wearing reflective vests of different colors (orange, yellow and white). The measurements were made using the low beam and the high beams for vehicles equipped with different types of bulbs.

Keywords: Night visibility · Cyclist · Reflective vest

1 Introduction

A road accident is a casual and unexpected event that causes a breakdown and brings injuries, mutilation or the death of a person.

Two major factors are generally involved in the road traffic accident: the vehicle and the man in his multiple qualities (driver, passenger, pedestrian, bicyclist, biker, etc.) [1].

In 2016, about 25.600 people were killed in road accidents throughout the EU, a reduction of 40% from the 2007 total of 43.150 [2]. This fact sheet explores the known characteristics of cyclist fatalities. Cyclists, while relatively small in proportion with respect to motorized vehicles, have a high level of vulnerability, creating a significant need to better understand the characteristics specific to this user group. A good insight into the problem provides an opportunity to improve the safety of this cheap, convenient and environmentally safe mode of transport [3].

Bicycle fatalities make up 8% of the total number of road accident fatalities in 2016 in the EU countries. In these countries, 2.015 people riding bicycles were killed in road accidents in 2016 (excluding Lithuania and Slovakia) [2].

Figure 1 shows the number of cyclist fatalities in the EU between 2007 and 2016. In this period the decrease of bicycle fatalities was 24%.

Night-time cycling has been shown to be more dangerous than cycling in daylight, with 40% of cyclist fatalities occurring at night despite much lower exposure rates than

I. Dumitru et al. (Eds.): SMAT 2019, *The 30th SIAR International Congress of Automotive and Transport Engineering*, pp. 359–365, 2020.
https://doi.org/10.1007/978-3-030-32564-0_42

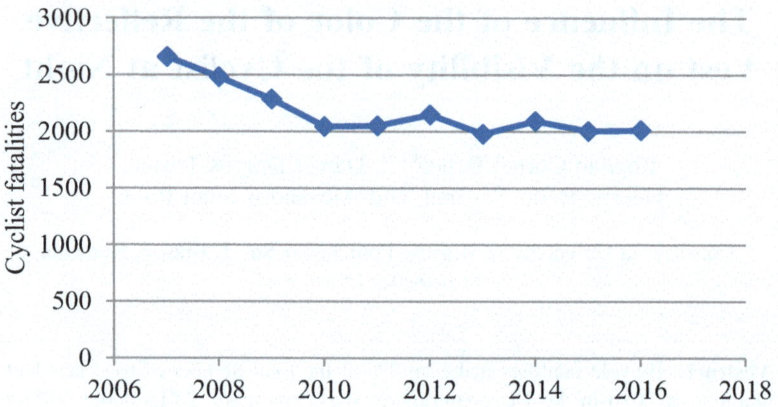

Fig. 1. Number of cyclist fatalities EU 2007–2016 [2].

in the daytime [4]. The drivers involved in night-time collisions with cyclists report that they did not see the cyclist until it was too late to avoid a collision [5].

Herslund and Jørgensen suggested that increasing the visibility of cyclists could reduce the number of 'looked but failed to see'-accidents between motorized vehicles and cyclists [6].

To improve the visibility of cyclists, they can wear fluorescent and/or reflective clothing. Also, an important element in nighttime visibility is the lighting system of the vehicle [7].

The objectives of the paper are the determination of the influence of the bicycle's vest and the lighting system of the vehicle on the visibility of the bicyclist during the night.

2 Method

The experimental study aimed to determine the visibility characteristics and the distances of perceiving cyclists by the driver during the night time using the low beam and the high beam, depending on the characteristics of the reflective vest.

To perform this experiment, the measurements were made on the test track, with a length of 150 m, without street lighting.

To achieve the objectives were used three cyclists wearing different reflective vests and two cars, one equipped with halogen headlight (low and high beam) and other with LED headlight (low beam) and halogen headlight (high beam). The cyclist was located at a distance of 0.5 m from the right merge of the road, and the car was positioned at a distance of 10 to 100 m from the cyclist (Figs. 2 and 3).

A Fujifilm X-T1 camera was used to take pictures. This camera allowed taking photos at night without changing the focus. The picture were taken without using the flash.

For image processing was used the PC-RECT 4.2 software, developed by DSD Austria, specializing in photogrammetry. This software determines the brightness on

Fig. 2. The test track.

Fig. 3. Cyclist's vest.

different areas of the image and can calculate the contrast between the two analysed areas, so the software gives the visibility of the pedestrian by comparing the two luminance values and the contrast determination. The program uses Adrian Werner's method, which is the method of stability of the visibility of the object under different light conditions (Fig. 4).

The analyzed area of the cyclist was the chest area, the wearing area of the reflective vest.

The program determines the value of the luminance of the object in the area of the cyclist's chest (1) and compares it with the luminance of the environment (2), thus obtaining the contrast value.

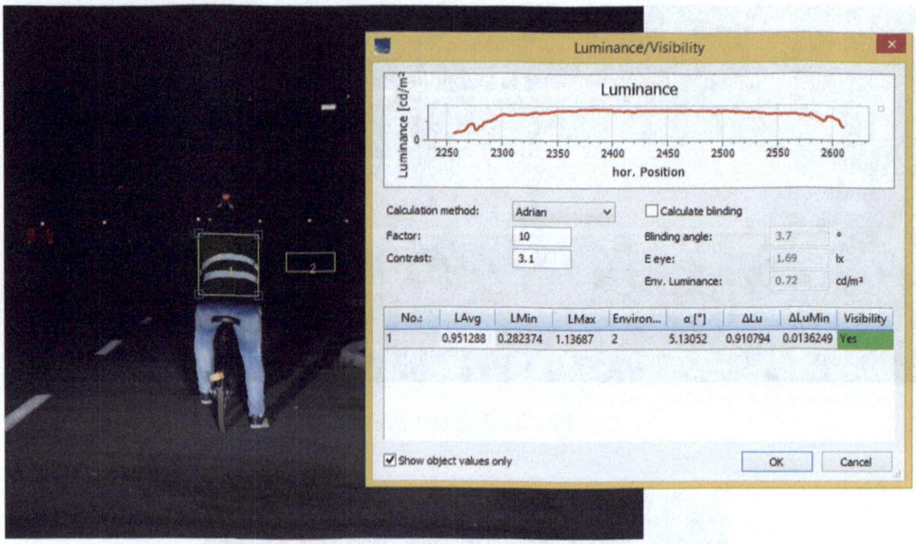

Fig. 4. Photo processing to determine the visibility of the cyclist.

The software extracts all the data in a table showing the object's visibility based on the ΔLu parameter, which represents the difference between the two selected areas.

Lavg – average luminance; LMin, LMax – shows the minimum and maximum luminance of the respective area; α – object size; Environment – assigned background area; ΔLu – luminance difference, it is the difference between LAvg of the defined area and LAvg of the background area; ΔL – luminance difference threshold.

3 Results

The cyclists wore an orange, yellow and white reflective vest.

Figure 5 shows the contrast obtained for the car equipped with halogen headlight. The most visible was the cyclist who wears white reflective vest. The lowest contrast was obtained for the orange vest. In this case, from the distance of 60 m, the cyclist is almost invisible.

Figure 6 present the contrast when the road was illuminated with low beam of a led headlight. The contrast is higher for all three vests than when using a halogen headlight. And in this case, the cyclist that wears the orange vest becomes hardly visible from a distance of 70 m.

In Fig. 7 is presented the contrast obtained when was use the high beam from the cars equipped with halogen headlights. In the case of white vest the cyclist is clear visible from 100 m, while the cyclist who wears orange vest is hardly visible.

In Fig. 8 is presented the contrast of cyclist when it was used high beam of the second car. In this case, the light was produced by LED headlight (low beam) and halogen headlight (high beam). In this case was obtained the greater contrast for cyclist.

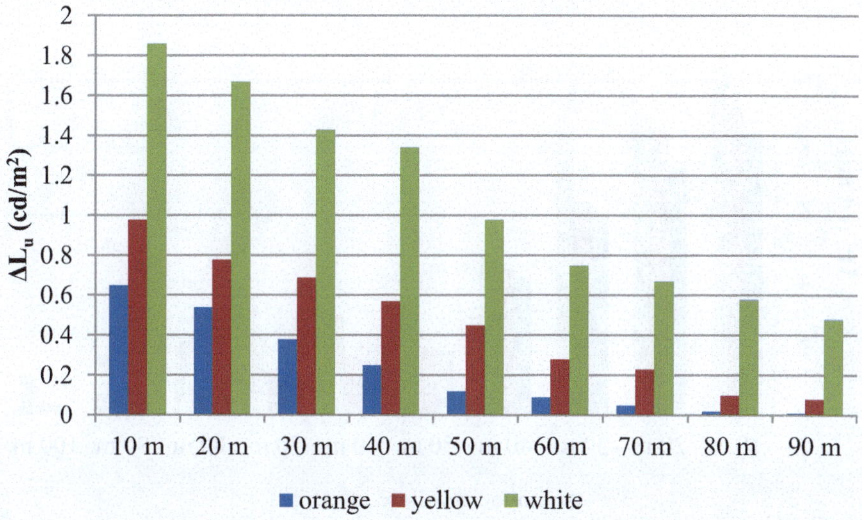

Fig. 5. Contrast for low beam from halogen headlight (car 1).

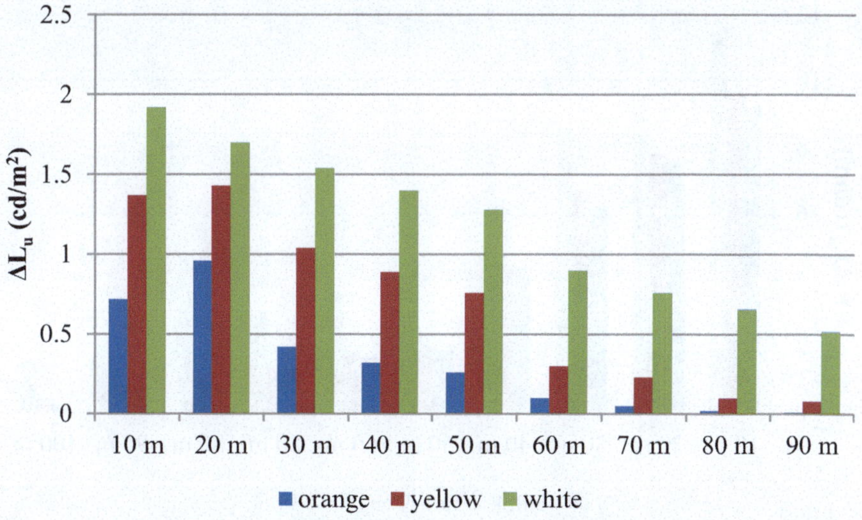

Fig. 6. Contrast for low beam from LED headlight (car 2).

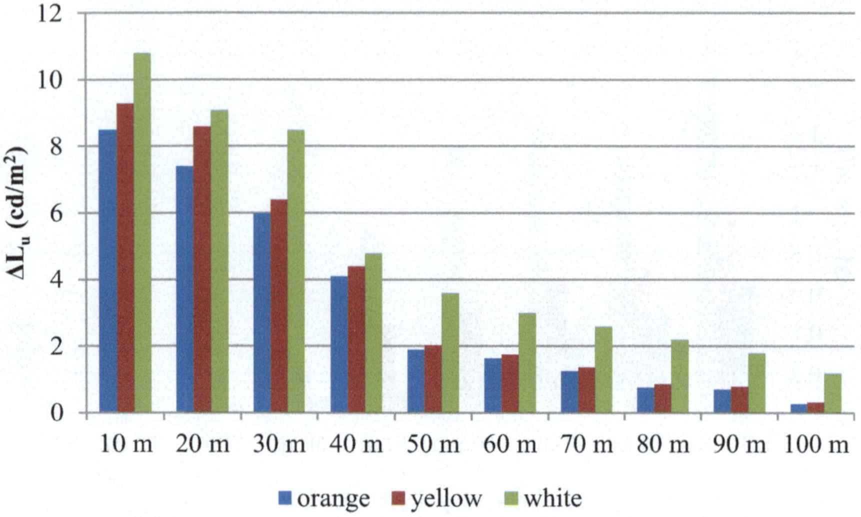

Fig. 7. Contrast for high beam from halogen headlights (car 1).

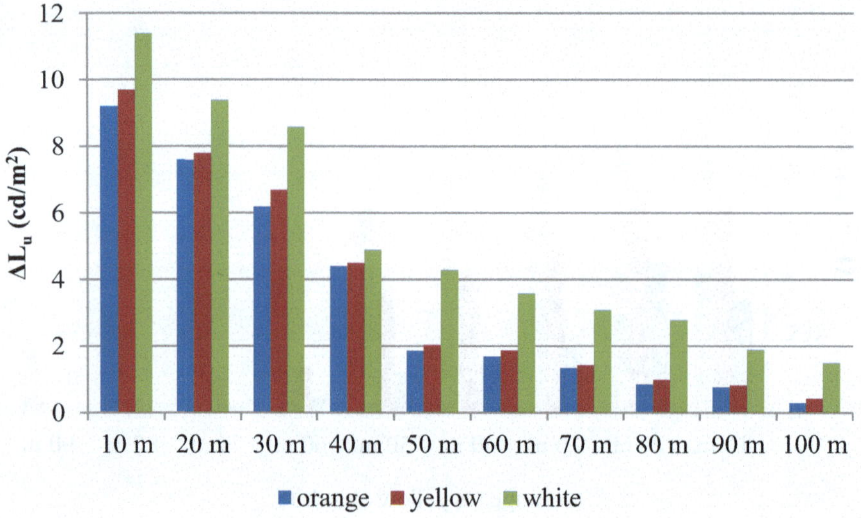

Fig. 8. Contrast for high beam from LED and halogen headlights (car 2).

4 Conclusions

In this study we sought to determinate how the color of the cyclist vest and the type of headlights of the car influenced the cyclist visibility during the night. The data demonstrate that cyclist's clothing and the type of headlight affect the ability of drivers to recognize cyclist in night-time driving conditions.

The visibility of cyclists during the night depends on the used clothing and the surface of the reflective elements. The best visibility was obtained for the white reflective vest (large reflecting surface). Fluorescent materials have little influence on the visibility of cyclists, as they are activated by UV radiation, which is not present in the headlights or streetlights.

The headlight type has an important role. The cyclist's visibility was greater when it was used the led headlight than when was used halogen headlight. The led headlights provide much focused rays and can be also played with to create different shape.

References

1. Chen, P., Shen, Q.: Built environment effects on cyclist injury severity on automobile-involved bicycle crashes. Accid. Anal. Prev. **86**, 239–246 (2016)
2. European Road Safety Observatory. https://erso.eu. Accessed 25 Feb 2019
3. Räsänen, M., Summala, H.: Attention and expectation problems in bicycle-car collisions: an in-depth study. Accid. Anal. Prev. **30**, 657–666 (1998)
4. Rodgers, G.B.: Bicyclist deaths and fatality risk patterns. Accid. Anal. Prev. **27**(2), 215–223 (1995)
5. Wood, J.M., Lacherez, P.F., Marszalek, R.P., King, M.J.: Drivers' and cyclists' experiences of sharing the road: incidents, attitudes and perceptions of visibility. Accid. Anal. Prev. **41**(4), 772–776 (2009)
6. Herslund, M., Jørgensen, N.O.: Looked-but-failed-to-see-errors in traffic. Accid. Anal. Prev. **35**(6), 885–891 (2003)
7. Sonkin, B., Edwards, P., Roberts, I., Green, J.: Walking, cycling and transport safety: an analysis of child road deaths. J. R. Soc. Med. **99**, 402–405 (2006)

Cross-Validation of Two Collision Modeling Software Based on the Reconstruction of a Real Collision - A Possible New Approach to Collision Reconstruction

Attila Iuliu Gönczi(✉) and Adrian Cipleu

Politehnica University Timisoara, Timisoara, Romania
{attila.gonczi,adrian.cipleu}@upt.ro

Abstract. Road accident reconstruction is a complex task in almost every case. The analysis of a real collision is often based on the usage of a collision modeling software. Usually a collision is reconstructed using only one collision simulation software. In order to improve the level of certainty of the reconstruction, the authors used two commercial collision reconstruction software, namely PC-Crash 11.1 and Virtual Crash 4.0 for the reconstruction of a real collision as a possible new approach. The objective of this possible new approach was to improve the level of certainty and the precision of the results, comparing the two simulations, which were realized using somehow different modeling. In the same time, this way we were able to treat at a certain level the problems caused by the simplifications used in the modeling as we used two different approaches for the two programs.

Keywords: Accident reconstruction · Cross-validation · PC-Crash · Virtual Crash

1 Introduction

Road accident reconstruction is a complex work, which today is normally based on computer-aided modeling of the collision(s) and simulation of the pre-crash and post-crash events, in order to help the judicial system.

The authors have the licenses and are experienced users of two different road accident reconstruction software that have the capability to model and simulate collisions and pre- and/or post-crash movements, namely PC-Crash and Virtual Crash. The scope of this paper is neither to present their models in details, nor to compare their general capabilities. The authors simply use both of them in certain ways (not completely independently) in solving concrete cases and make a comparison between the results of the two software and, very importantly, of two different approaches to the reconstruction of an accident.

In the case of PC-Crash, the authors have used the 11.1 version, while for the simulations in Virtual Crash the used version was 4.0, both of them having the latest updates at the moment of the modeling. We have to emphasize, that for the simulation of the collision and of the post-collision movements, both programs use complex

I. Dumitru et al. (Eds.): SMAT 2019, *The 30th SIAR International Congress of Automotive and Transport Engineering*, pp. 366–374, 2020.
https://doi.org/10.1007/978-3-030-32564-0_43

mathematical modeling, based mainly on momentum conservation (the Kudlich – Slibar model as the basic method in the case of a collision between two passenger cars).

A very important characteristic of the model is, that the momentum change is modeled as taking place in an infinitely short time, at a given moment after the first contact, which is defined as the "depth of penetration". The value of the depth of penetration is variable from 0 s to 0.075 s in PC-Crash [3] and from 0 s to 0.06 s in Virtual Crash [2].

As a methodological approach, there is a common opinion among experts, that the most convenient (and the most simple) way to reconstruct an accident is to make a kinetic simulation starting with the crash, i.e. when the two colliding objects are in the position when the impact change takes place [1, 3]. After finding the best solution for the crash and the post-crash, one could model the pre-crash movements using kinematic backward simulation. In the opinion of the authors, this approach has a critical point, i.e. if at least one of the two colliding vehicles has an angular velocity that is significantly different from zero at the collision, it is difficult to estimate the right value of that angular velocity, which in this case is an input. At the same time, leaving the angular velocity at zero, could lead to biased results. Instead, it is very difficult to make an all-kinetic simulation starting somewhere before the crash within at least 1 or 2 s. Another issue is the fact that the usage of the collision optimizer in PC-Crash is possible **only** if the simulation is started at the moment of the momentum change during the impact.

2 A Short Description of the Collision

On a two lane main rural road, in a curve, two passenger cars, an AUDI A3 1.8i and a SMART ForFour 1.3 collided frontally at relatively high speeds. As both cars suffered heavy damages (both engines included), and the Smart caught fire (it was almost immediately extinguished), the surface of the road was covered with oil and the particles eliminated by the extinguisher. Therefore, even if there were tire marks on the road, showing the pre– and post–crash movements, they were not documented as evidence.

The photographs in Fig. 1 show the position of the two cars after the collision.

Fig. 1. The positions of the two cars after the collision.

Both drivers sustained that the car driven by the other driver crossed the axis of the road and entered into the lane he uses legally.

Obviously, both versions could not be true at the same time. One or both of the drivers did not present the truth.

On the one hand, based on the experience of the authors, it was somewhat obvious from the beginning that the POI (Point of Impact) of the collision was located in the lane in which the Audi ran legally (but probably very close to the longitudinal axis of the road) and the SMART was clearly on the wrong side (i.e. not in the lane in which legally could be driven). On the other hand, it was very important to check the exact location of the POI and the positions of the two cars at the collision moment, because a partial guilt of the driver of the Audi could arise based on that analysis.

3 The Analysis of the Collision

Based on the qualitative analysis of the damages (a quantitative analysis was done too, but we will not present it in this paper), it was clear that the impact was an almost 100% overlap head-on collision, but the depth of the damage was significantly deeper on the right front corner than on the left front corner of the Audi (see Fig. 2).

Fig. 2. The image of the damages of the Audi on the right front corner (left) and the left front corner (right).

This means that the crash was somehow oblique and the Smart approached the Audi not from Audi's left (from the point of view of the Smart it was a right curve) but from its right. This means that it is certain that the Smart in the pre-crash movement entered deeply into the lane of the other car and before the moment of the crash, it was in a panic maneuver, trying to drive back into his lane, causing yawing of the car in a clockwise rotation. It was clear that both cars had angular velocities significantly different from zero.

As a first step in the reconstruction, we made a kinetic simulation of the crash, using the collision optimizer function of the PC-Crash [3].

3.1 The First Phase: Kinetic Simulation of the Crash and Post-crash Movements Using the Collision Optimizer of PC-Crash

As main inputs, the stopping positions were clear and documented, but we needed to estimate the approximate positions (linear and angular) of the two cars at the collision, their pre-crash speeds and their angular velocities. The optimisation included all the possible parameters, i.e. impact velocities, the point of impact (POI), the height (z coordinate) of POI, the contact plane angular position, the pre-impact directions and the positions of the two cars, the value of the coefficient of restitution and the contact friction coefficient (see Fig. 3).

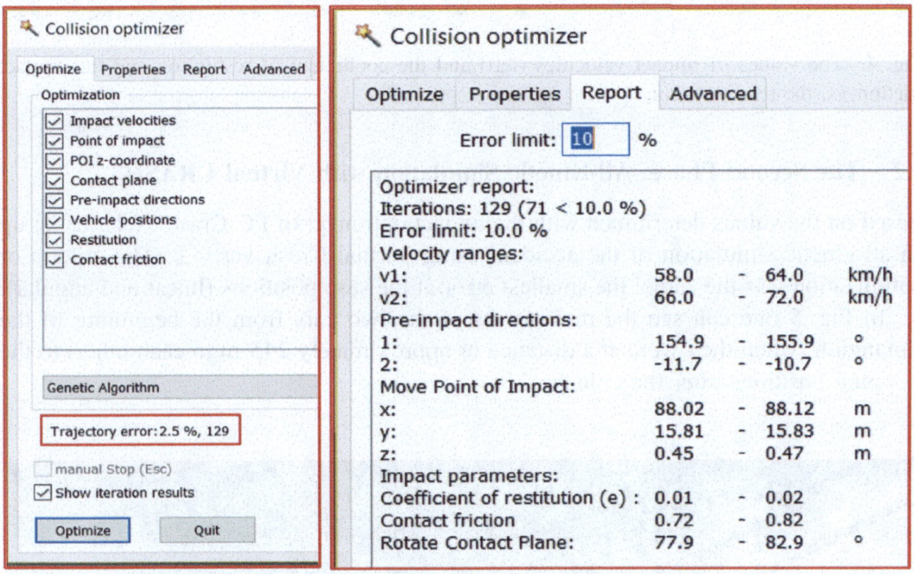

Fig. 3. The main optimized parameters and the interval in which their values were optimized.

It is important to emphasize that the angular velocities cannot be optimized using the collision optimizer function and, therefore, the authors optimized somehow the value in a traditional way, using "trial and error".

Due to the length limit of such a paper, we cannot present all the results, but in Fig. 4 one can see the optimization results for the impact velocities (left) and the value of the friction coefficient at the contact surface and the restitution coefficient (right) vs. the corresponding relative trajectory error, which was very low for the optimal values, being 2.5%.

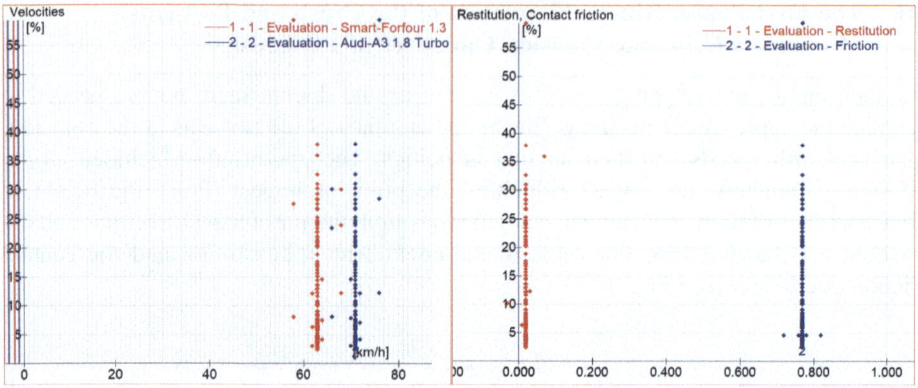

Fig. 4. The values of impact velocities (left) and the coefficient of restitution and the contact friction vs. the relative error.

3.2 The Second Phase: All-Kinetic Simulation with Virtual CRASH

Based on the values determined with the impact optimizer of PC-Crash 11.1, we set up an all-kinetic simulation of the accident using Virtual Crash ver. 4.0. The criteria of optimization was the same: the smallest error at the stop positions (linear and angular).

In Fig. 5 one can see the movements of the two cars from the beginning of the simulation (when they were at a distance of approximately 245 m to each other) to the stopping positions after the collision.

Fig. 5. The graphic representation of the result of the all-kinetic simulation in Virtual Crash 4.0.

In Fig. 6, the graphic representation of the collision is presented, based on the two different modeling. On the left, the positions are based on the automated optimization using the collision optimizer of PC-Crash. It is important to repeat, that in this case all the parameters of the movements of the two cars are input parameters, including the angular velocity caused by the curve trajectory in the case of the Audi (the red one) and by the yawing in the case of the Smart (the yellow car). In the image on the right, the

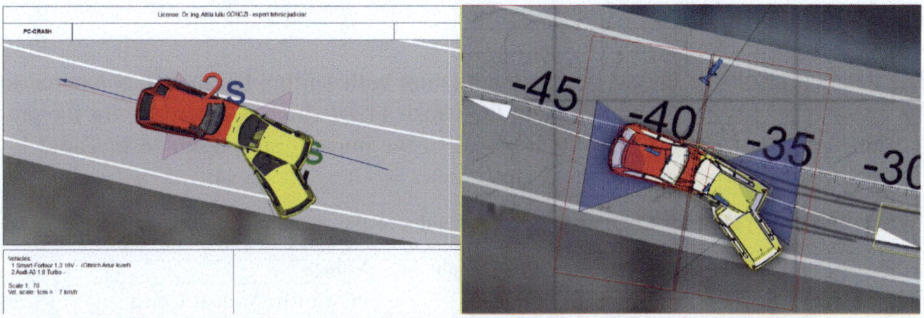

Fig. 6. The positions at the collision and at stop after the collision. Left: based on the collision optimizer in PC-Crash 11.1. Right: based on the all-kinetic simulation in Virtual Crash 4.0

positions are the result of the all-kinetic simulation in Virtual Crash. As a result, one can see the yawing marks of the tires of the Smart, before the collision.

In Fig. 7 one can see the variation of the velocities of the two cars as a function of time, based on the simulation done in PC-Crash (left) and Virtual Crash (right).

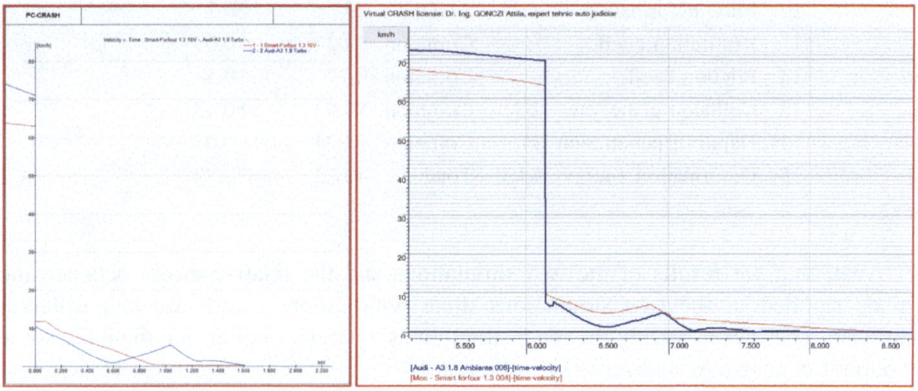

Fig. 7. Diagrams of velocities as a function of time. Left: PC-Crash. Right: Virtual Crash.

It is important to emphasize that the angular velocity in the case of the Smart, used in PC-Crash was estimated to −0.22 rad/s, while the value resulted in the kinetic simulation in Virtual Crash was −0.656 rad/s. The relative difference is quite big and could affect the processes (even from a judicial point of view, even if in this file, this was not the case), so it suggests that a kinetic simulation even in the pre-crash phase it is important to be done.

4 Conclusions

The detailed results of the all-kinetics simulation with Virtual Crash 4.0, compared to the results of the simulation made with PC-Crash 11.1 (which is kinematic for the pre-impact period and kinetic from the moment of the impact) are presented in Table 1.

Table 1. Comparison of the results of the two simulations.

Nr.	Parameter	Unit	Vehicle	Values	
				PC-Crash	Virtual Crash
1	Pre-impact velocity	km/h	SMART	63	63.21
2	Pre-impact velocity	km/h	AUDI	71	69.6
3	Post-impact vel.	km/h	SMART	11.94	10.496
4	Post-impact vel.	km/h	AUDI	10.5	7.536
5	EES	km/h	SMART	65.96	69.247
6	EES	km/h	AUDI	65.1	58.837
7	Delta V	km/h	SMART	71.44	73.6
8	Delta V	km/h	AUDI	60.7	64.14
9	Max. deformation	cm	SMART	25	68.4
10	Max. deformation	cm	AUDI	29	56.7
11	Restitution coeff.	–	Common	0.02	0.09
12	Friction coeff.	–	Common	0.77	0.54
13	Collision plane yaw	[°]	Common	80.42	80.42
14	Depth of penetration	s	Common	0.025	0.03
15	Deformation energy	kJ	Total	407.2	378.82

Analyzing the results of the two simulations and the relative errors between the values resulted in the two simulations done with different software and different approaches besides assessing the reconstruction as a whole, implies that there are some important conclusions and remarks to be made.

The positions of the two vehicles at the moment of the impact, determined using the two different software and approaches, are very close to each other, the differences being around 0.05…0.08 m, which is much smaller than the distances to the axis of the road. Therefore, the errors are not relevant from a judicial point of view. In both cases the Smart was entirely in the left lane (from its own point of view), while the Audi in both simulations was totally in its legal lane, but very close to the axes. Nevertheless, even if we are to take into account the possible errors, the Audi, in the worst case, was with its left wheels on the marking. However, it is important to emphasize that, in the Romanian legislation it is compulsory to drive as close as possible to the right side of the road when another vehicle is coming from the other way. Therefore, this is a starting point of a possible partial guilt of the driver of the Audi.

For the values of the pre-impact velocities the relative errors are −0.33% (for the Smart) and 1.97% (for the Audi). These values are much smaller than the measurement errors at the accident scene (which are very important as they are the main inputs for the reconstruction) and their values are of great importance from a judicial point of view. Naturally, we did not know what the real velocities at the impact were. But the fact that, two different approaches, using two different (even if, from the point of view of the mathematical models used in this case, almost similar) software offer solutions which are almost identical, demonstrates that the result(s) are probably very close to the real values.

The errors of the post-impact velocities are higher, 12.1% (for the Smart) and 28.22%, but their values are not relevant from a judicial point of view. The cause of these higher errors is the fact that the movements after the collision are done by the damaged vehicles on a road surface which is contaminated by the debris and the oil and water from the vehicles, facts which cannot be modeled. Another source of the errors is the value of the real height of the COG of the two vehicles and of the point of impact (POI) in the modeling.

The values of the energy equivalent speed (EES), the velocity change during the impact (Delta V or ΔV) and the total deformation energy are very close, the relative errors being under 10%, which implies a very high level of confidence for the reconstruction of a road collision without data from an event data recorder (EDR).

In the case of the restitution coefficient, the differences are very high. The value given by the impact optimizer in PC-Crash is 0.02 and seems too small, even if both cars were relatively old (i.e. their frontal stiffness could be lower than the normal value). The optimal value given by the simulation in Virtual Crash is 0.09, which is in the normal interval for the impact speeds.

The friction coefficient, even if the difference between the values is relatively high, is not important, as the impact was an almost full frontal crash in which relative movements appears only because of the deformations of both cars and are mainly rotations rather than linear movements.

Another important conclusion is that it could be important and useful in some cases, to make a kinetic simulation of the movements of the vehicles before the impact, in order to refine the values of the input parameters (not only the angular velocities, but even the angular positions of the two cars) for the kinetic simulation of the crash and the post-crash movements. In these cases, the proposed methodology is to simulate the movement after a first, *grosso modo* correct simulation of the crash was found, using the approximate correct values of the pre-crash velocities.

The example of the usage of two different vehicle collision simulation software (PC-Crash and Virtual Crash, in this case) demonstrates that this approach could lead to reconstruction solutions that have an unusually high level of confidence.

Even if from a scientific point of view it is not relevant, the driver of the Audi, whose partial guilt was demonstrated by the reconstruction, in an informal phone call accepted that he was talking on his mobile phone, did not realize the close and present danger caused by the other driver and he really drove very close the axis of the road and his reaction time was unusually long. The authors have demonstrated that, normally he could have been able to avoid the collision if he had abided by the speed limit of 50 km/h and he had been attentive at a normal level. The fact that he recognized his guilt (even if only in an informal way) demonstrates that he is telling the truth.

Therefore, one could be sure that the results of the reconstruction were correct qualitatively, even if quantitatively the results of the reconstruction obviously have some errors.

References

1. Burg, H., Moser, A. (eds.): Handbook of Accident Reconstruction. Accident Investigation – Vehicle Dynamics – Simulation, 1st edn. Amazon – E-book (2013)
2. Virtual Crash 3. Virtual Crash 4. Users Guide. vCRASH, Americas, Inc. (2018). www.vcrashusa.com/guide
3. Wach, W.: Simulation of Vehicle Accidents Using PC-Crash. Institute of Forensic Research Publishers, Cracow (2011)

Study of a Car-to-Pedestrian Collision in Case of Vehicle's Post-impact Braking

Bogdan Adrian Tolea[1(✉)], Alexandru Ionut Radu[2], Horia Beles[1], George Dragomir[1], and Sorin Moca[1]

[1] University of Oradea, Universitatii Str. 1, Oradea, Romania
bogdan.tzolea@gmail.com
[2] Transilvania University of Brasov, Politehnicii Str. 1, 500024 Brasov, Romania

Abstract. The objective of the paper was to assess the pedestrian's head injury risk, in case of vehicle's post-impact braking. In order to reach the established objective, two crash-tests were carried, using a crash-test dummy, equipped with tri-axial accelerometers, mounted in its head. In the first test, the vehicle was braked before the collision, while in the case of the second experiment, the braking occurred in the post-impact phase. Using the data collected from the crash-test, we were able to calculate HIC values in both cases, and assess the head injury risk. At the same time, we were able to determine the influence of the vehicle's post-impact braking with respect to head injury severity risk.

Keywords: Accidents · Crash-test · Pedestrian · HIC · Injury assessment

1 Introduction

Road accident is an incident involving the collision of at least two vehicles, or a vehicle with a pedestrian, bicyclist or other stationary obstruction (e.g. tree, pole, etc.) resulting in injury or the fatality of the persons involved, respectively, result of material damage [1]. After a car-to-pedestrian accident, often results the pedestrian injury. This type of collision, is classified by the researchers in 5 main categories [2–4]: (I) wrap collision (45% of overall car-to-pedestrian accidents); (II) wrap collision when the pedestrian is thrown in front of the vehicle (35% of overall car-to-pedestrian accidents); (III) collision in which the pedestrian fall on the side of the vehicle (2% of overall car-to-pedestrian accidents); (IV) collision where the pedestrian impacts the vehicle's roof, and arrives in the rest position, behind the car (5% of overall car-to-pedestrian accidents); (V) collision in which the pedestrian has its rest position behind the vehicle, without entering in contact with the vehicle's roof (13% of overall car-to-pedestrian accidents).

When a car impacts a pedestrian, the vehicle can be braked before the collision, or the braking occurs after the impact takes place. The second type of accident often occurs during night-time, when the driver observes the pedestrian very late, and the impact takes place within its reaction time.

In Fig. 1 is presented a graph that indicates the visibility conditions in fatal car-to-pedestrian accidents in Romania, in the last years.

I. Dumitru et al. (Eds.): SMAT 2019, *The 30th SIAR International Congress of Automotive and Transport Engineering*, pp. 375–382, 2020.
https://doi.org/10.1007/978-3-030-32564-0_44

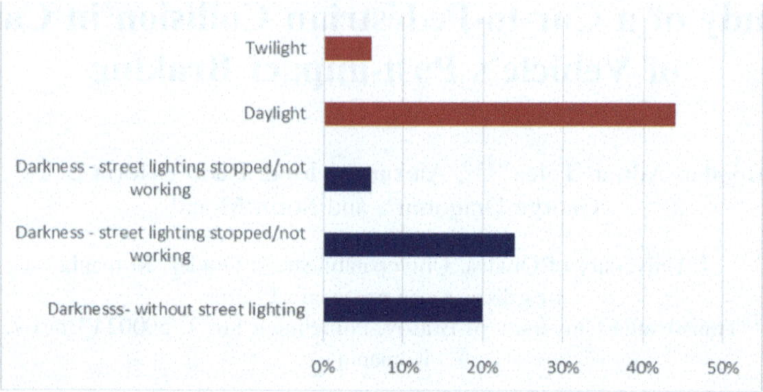

Fig. 1. The visibility conditions in fatal car-to-pedestrian accidents in Romania, in the last years [7]

As can be observed, 50% of the collisions between cars and pedestrians occurred in night conditions, out of which 26% of the accidents street lighting was stopped or did not work. Therefore, in these cases, there is a great chance that the braking occurred after the collision.

2 Methodology

Two crash-tests at similar velocities were carried, using a dummy and two different vehicles. A tri-axis accelerometer was mounted in the dummy's head, such that, we were able to measure head accelerations on the three axis, during the collision.

In order to record the data of interest, the following devices were used: PIC DAQ 5 produced by DSD Austria was used to determine head acceleration; GPS DS5, was used to determine vehicle's velocity; Hi-Spec 5 camera, able to record at 1400 fps, was used to extract the impact phases; DJI Phantom 3 drone, was used to determine the pedestrian kinematics seen from above.

For the first crash-test, the configuration was the following: the collision velocity was 30 km/h, the vehicle was braked before and after the impact, with an average deceleration of 5 m/s^2, the impact angle was 157°, while the overlap was 100%.

The crash configuration in case of the first test is presented in Fig. 2.

From the high speed camera recordings, the detailed impact phases were extracted (Fig. 3):

- Phase 1: First contact that takes place between the vehicle's bumper and the pedestrian's lower limb.
- Phase 2: Wrapping the pedestrian on the bonnet – Due to the impact beneath the pedestrian center of gravity, a rotational moment is applied, raising the dummy on the vehicle's bonnet.
- Phase 3: Impact between the pedestrian's head and the base of the windshield.
- Phase 4: Phase of flight – pedestrian is thrown away from the vehicle.
- Phase 5: Landing phase – pedestrian hits the ground and moves to the rest position.

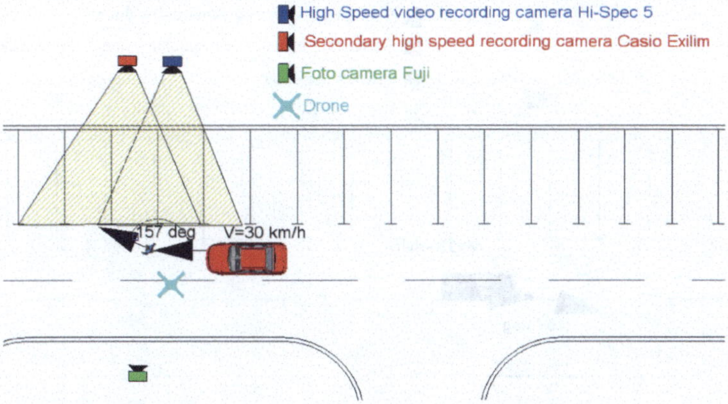

Fig. 2. Configuration of the first experiment

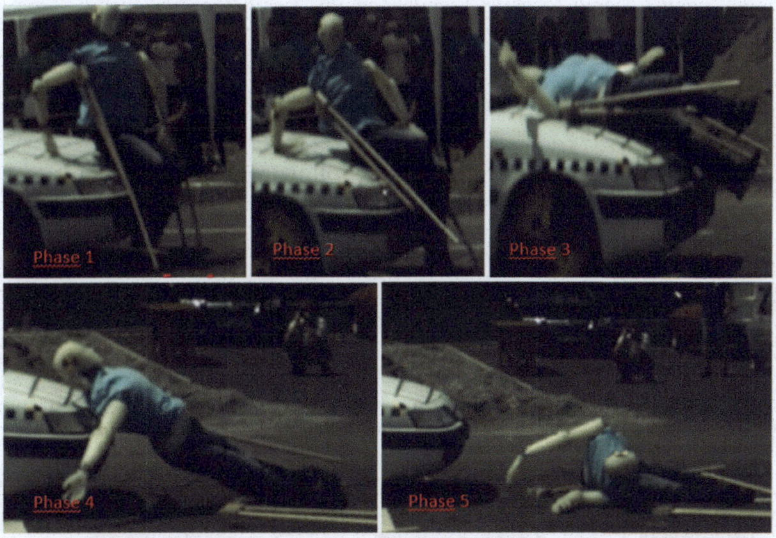

Fig. 3. Impact phases extracted from the high speed video

For the second crash-test, the impact configuration was the following: the collision velocity was 33 km/h, the vehicle was braked only after the collision, during the pre-impact, respectively impact phase the braked was not applied, the impact angle was 157°, while the overlap was 50%.

This type of collision can be classified as wrap-collision, the most common type of vehicle-to-pedestrian accident.

The configuration for the second experiment is presented in Fig. 4.

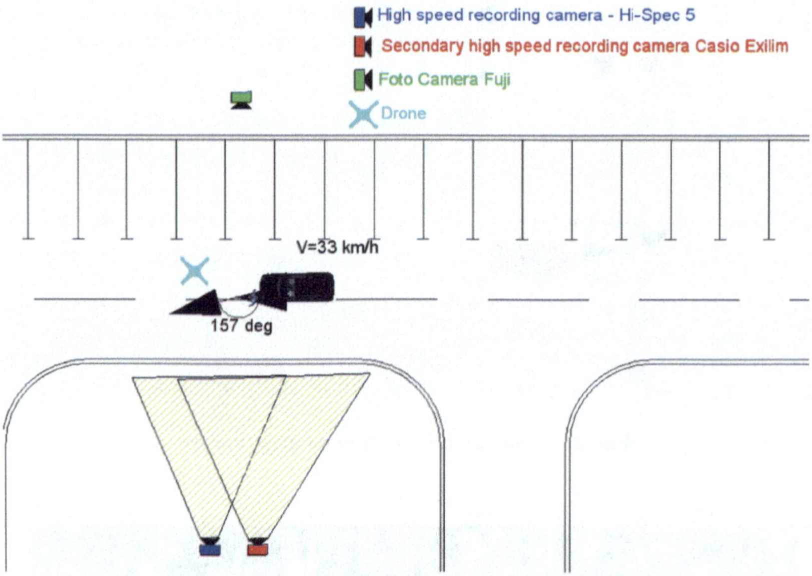

Fig. 4. Configuration of the second experiment

From the high speed camera recordings, the following impact phases were extracted (Fig. 5):

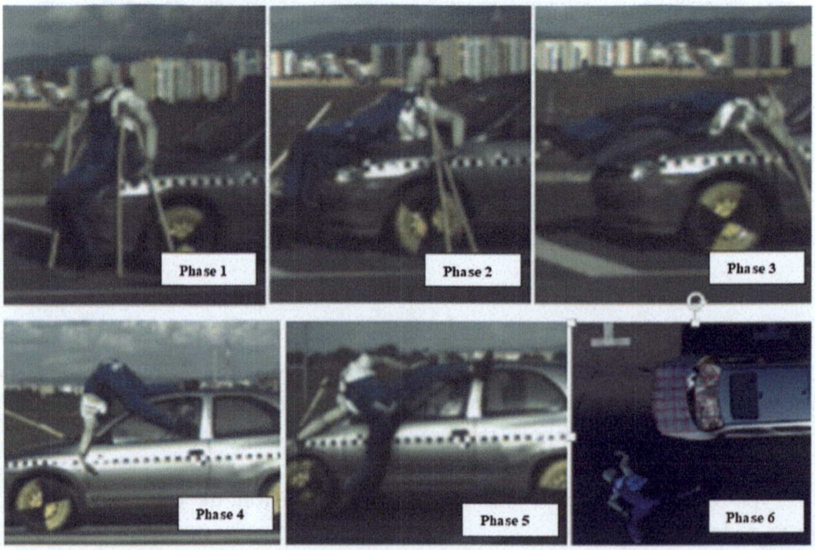

Fig. 5. Impact phases for the second crash-test

- Phase 1: Contact that takes place between the vehicle's bumper and the pedestrian's lower limb.
- Phase 2: Wrapping the pedestrian on the vehicle's bonnet.
- Phase 3: Contact of the pedestrian's thorax and head with the vehicle's hood.
- Phase 4: Due to the lack of deceleration, the pedestrian rolls through an accelerated motion towards the vehicle's roof, coming into contact with it.
- Phase 5: Pedestrian slides from the vehicle roof to the ground.
- Phase 6: The pedestrian landing phase - the phase in which it lands on the ground and travel to its rest position.

Unlike the case discussed above, this type of collision is difficult to classify as a combination of wrap collision in the first two phases, and when the pedestrian enters in contact with the vehicle's roof, the collision can be placed in the fourth category presented in the first chapter (pedestrian enters in contact with the vehicle's roof), while in the final phase, the impact classification can be placed in the third category, where the pedestrian is falling on the left side of the car. It can be noticed that in this test, the flight phase is missing and the pedestrian slips from the vehicle to the ground, reaching its rest position.

3 Results and Discussion

With the collected data, we were able to calculate the Head Injury Criteria (HIC), for both crash-tests. Thus, the obtained values were compared with the values available on Mertz injury prediction curve [5]. The HIC value is determined using the following formula [6, 8, 9]:

$$HIC15 = \max t1, t2([\frac{1}{t_2 - t_1} \cdot \int_{t1}^{t2} a_{res}(t)dt]^{2.5} \cdot (t_2 - t_1)) \qquad (1)$$

where:

HIC15 – Head Injury Criteria, for 15 ms time interval,
t1 and t2 – time interval,
a_{res} – resultant head acceleration, measured in g (gravitational acceleration).

The obtained pedestrian's resultant head acceleration for the first experiment is presented in Fig. 6.

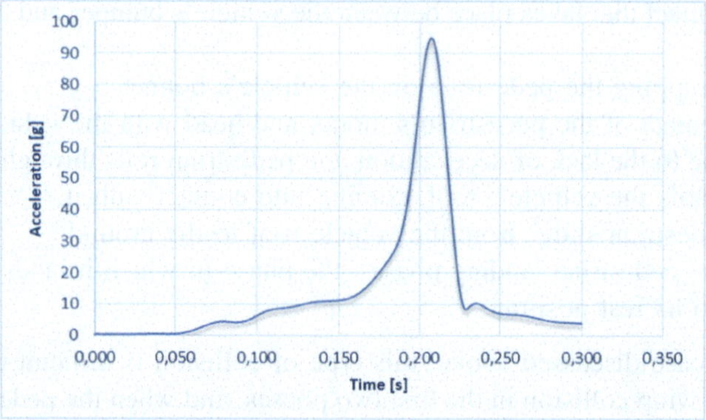

Fig. 6. Resultant head acceleration in case of the first test

Therefore, the value of HIC in case of the first experiment was calculated:

$$HIC15_1 = \max t1, t2([\frac{1}{t_2 - t_1} \cdot \int_{t1}^{t2} a_{res}(t)dt]^{2.5} \cdot (t_2 - t_1)) = 1050 \qquad (2)$$

The resultant head acceleration for the second experiment is shown in Fig. 7.

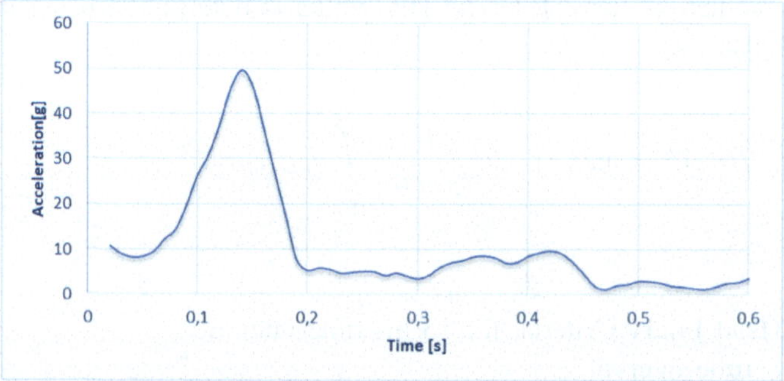

Fig. 7. Resultant head acceleration for the second test

Using HIC formula, we determined the result for the second test:

$$HIC15_2 = \max t1, t2([\frac{1}{t_2 - t_1} \cdot \int_{t1}^{t2} a_{res}(t)dt]^{2.5} \cdot (t_2 - t_1)) = 280 \qquad (3)$$

Using Mertz' curve, we are able to predict pedestrian's brain injury risk. Therefore, the pedestrian's brain injury risk in case of the two tests was summarized in the following diagram (Fig. 8).

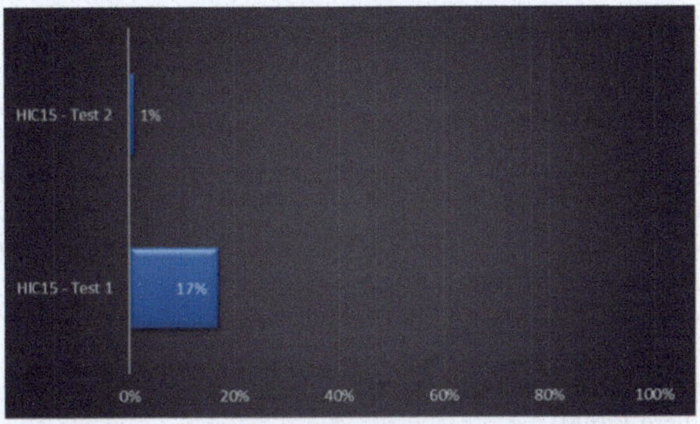

Fig. 8. Brain injury risk for the two crash-tests

The brain injury risk was higher in the first crash-test, obtaining a value of 17% head injury risk, while in the second crash-test, the brain injury risk was significantly smaller, obtaining a 1% risk.

Therefore, in case of vehicle's delayed braking (post-impact braking), the head injury risk at the contact with the vehicle is reduced, in comparison with the case where the vehicle is braked before the collision. The main reason is the pedestrian's head impact location: in case of the first test, the pedestrian impacts the base of the windshield, while in the second crash-test, the pedestrian head impacts the windshield, which is a softer area that is not generating high acceleration. The head impact location is mainly influenced by the vehicle design, but also of the braking sequence (before or after collision). If in the moment of impact the velocity remains constant, or it increases, the head contact point may be modified.

However, in the second case, the pedestrian kinematics is atypical, and during the pedestrian's landing phase, the brain injury risk may increase, due to violent contact with the ground. In this particular case, the pedestrian landed on its feet, and then it felt to the ground, resulting smaller head acceleration during the landing phase, and therefore, HIC value was not calculated at the contact with the ground.

4 Conclusions

Based on the obtained values, the difference of brain injury risk between the two crash-tests was 16%. The injury risk was higher in the first crash-test, mainly due to the head impact location (base of the windshield), which was a stiffer area, in comparison with the second crash-test, where the head impact location was in the center of the windshield. The head impact location is influenced by the vehicle's design, but also from the braking sequence. In the second crash-test, where the vehicle braked after the collision, the head impact location was higher on the vehicle, and impacted the vehicle in a softer area. However, in the second crash-test, the pedestrian kinematics was atypical, because during the pedestrian's landing phase, the brain injury risk may increase, due to violent contact with the ground. In this case, the dummy landed on its feet, reducing the head acceleration caused by the landing phase.

References

1. National Safety Council (eds.): American national standard manual on classification of motor vehicle traffic accidents
2. Brooks, D., Wiechel, J., Sens, M., Guenther, D.A.: A comprehensive review of pedestrian impact reconstruction. SAE Technical Paper, No. 870605 (1987)
3. Ravani, B., Brougham, D., Mason, R.T.: Pedestrian post-impact kinematics and injury patterns. SAE Technical Paper, No. 811024 (1981)
4. Simms, C., Wood, D.: Pedestrian and Cyclist Impact: A Biomechanical Perspective, vol. 166. Springer, Heidelberg (2009)
5. Mertz, H.J.: Injury risk assessments based on dummy responses. In: Accidental Injury, pp. 89–102. Springer (2002)
6. Burg, H., Moser, A.: Handbook of Accident Reconstruction (2013)
7. https://ec.europa.eu/transport/road_safety/specialist/statistics_en. Accessed 10 Mar 2019
8. Oţăt, O.V., Dumitru, I., Oţăt, V., Matei, L.: Modeling the vehicle-bicycle collision-the analysis of the projection distance and acceleration at the cyclist's head. In: Applied Mechanics and Materials, vol. 880, pp. 177–182. Trans Tech Publications (2018)
9. Oţăt, O.V., Castravete, Ş.C., Oţăt, V.: Dummy kinematic behaviour and head injuries analysis in frontal collisions. In: Proceedings of the European Automotive Congress EAEC-ESFA 2015, pp. 723–731. Springer, Cham (2016)

Study Regarding the Impact Between a Vehicle and a Two-Wheel Vehicle

Oana Oțăt(✉), Ilie Dumitru, Victor Oțăt, Lucian Matei, and Dragoș Tutunea

Faculty of Mechanics, University of Craiova, Craiova, Romania
otatoana@yahoo.com

Abstract. One of the consequences of the ever-increasing demand of transport needs is reflected in traffic accident statistics and the recorded number of road accident victims. In case of motorcycle accidents, the highest and most serious injuries are caused following a car crash.

The effects following a collision between a motorcycle and a car depend on a number of factors, among which, the most important are directly connected with the type and the mass of the car, the place and impact direction, the characteristics of the road and the environment.

For a more accurate analysis that would involve lower costs compared to experimental tests, several software platforms have been developed. Such virtual platforms are designed to model and simulate various traffic accident scenarios, taking into account the pre-and post-impact vehicle kinematics, but also can determine the dynamic parameters of the collision.

The main purpose of this paper is to perform a comparative analysis of the kinematic and dynamic parameters of the vehicle, the motorcycle and the motorcycle driver who were involved in a collision. The simulation and analysis of vehicle - motorcycle type collision is carried out via the Virtual Crash software platform.

The analyzed situations are: • side vehicle - front motorcycle; • front vehicle - front motorcycle; • front vehicle - rear motorcycle.

Keywords: Traffic accident · Vehicle-motorcycle collision · Virtual Crash software

1 Traffic Safety from the Perspective of Vulnerable Road Users

1.1 Statistical Data

In the European Union, motorcyclists account for 15% of all deaths. For each 100,000 registered motorcycles, 11 motorcycle deaths are reported, compared with 5 vehicle occupant deaths for every 100,000 registered vehicles [1], this proving that motorcyclists are more exposed in the event of an accident. Most, 56%, lose their lives on interurban roads. In this area, Estonia, Romania and Bulgaria have the lowest deaths, motorcyclists accounting for less than 5% of all deaths, Fig. 1. This percentage may

I. Dumitru et al. (Eds.): SMAT 2019, *The 30th SIAR International Congress of Automotive and Transport Engineering*, pp. 383–391, 2020.
https://doi.org/10.1007/978-3-030-32564-0_45

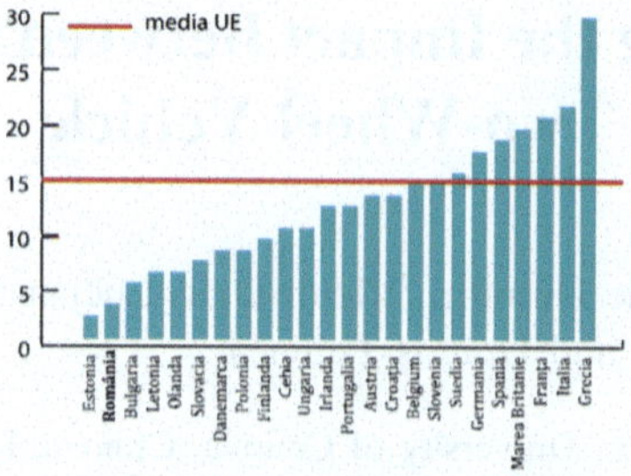

EE	2%	AT	13%
RO	3%	HR	13%
BG	5%	BE	14%
LV	6%	SI	14%
NL	6%	SE	15%
SK	7%	DE	17%
DK	8%	ES	18%
PL	8%	UK	19%
FI	9%	FR	20%
CZ	10%	IT	21%
HU	10%	EL	29%
IE	12%		
PT	12%	EU	15%

Fig. 1. Rate of death among motorcyclists

also be due to the fact that the total number of motorcyclists in these countries is much lower than the number of auto vehicles.

1.2 The Evolution of Road Accidents According to the Type of Vehicle Involved

Most vehicles involved in serious road accidents over the last 5 years are cars. During 2010–2015 there was a decrease in both the number and their share (from 68% in 2010 to 56.5% in 2014); a significant decrease was also observed for moped drivers (57% compared to 2010), according to [2]. Also, there have been significant increases in the number of motorcycles (+68.9%) involved from the driver's fault in serious road accidents.

1.3 Category of Traffic Participants

Starting from the hypothesis that age affects differently the victimization frequency for each category of traffic participants, Fig. 2 presents a distribution analysis for each of these categories, depending on the age group.

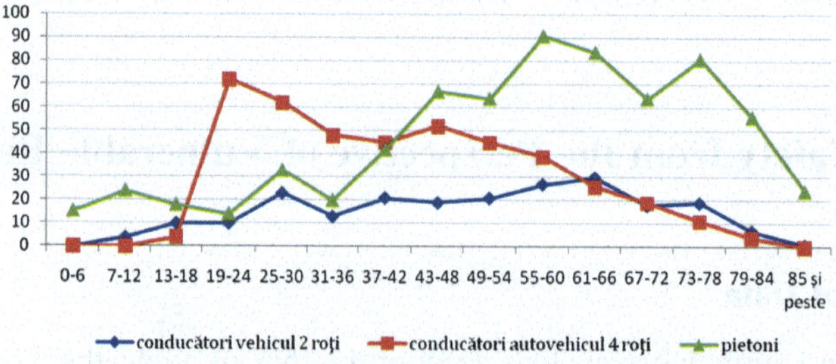

Fig. 2. Distribution of people killed in road accidents, by age group and category of road users

The highest share of pedestrians in serious road accidents is among the elderly. Also, according to [3], geometry of the vehicle's frontal profile is influencing the pedestrian's head accelerations (linear and angular) in car-to-pedestrian accidents. For the drivers of a four-wheeled vehicle, the most vulnerable age segment is the young population (aged 19 to 30). Another important aspect concerns the way the collision is produced and the place where it occurred. Thereby, in [4], is presented a method to develop computational programs by which all traffic flows from the roundabouts are calculated.

2 Typical Situations of Vehicle Collision - Motorcycle

The consequences of a collision between a motorcycle and a car depend on a number of factors, among the most important being the type and mass of the car, the place and direction of impact, the characteristics of the road and the environment.

Of the motorcycle accidents, the highest and most serious weight is the car collisions. A graphical representation of the typical situations of a motorcycle collision is shown in Fig. 3 [5]:

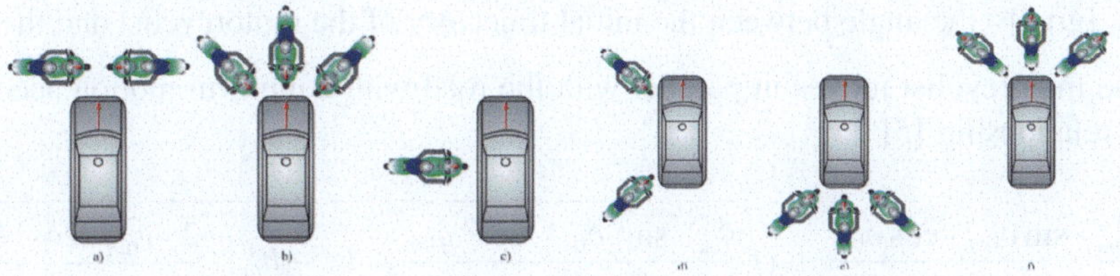

Fig. 3. Typical situations of vehicle collision - motorcycle

(a) Type 1 collision (90° ± 20°): frontal car - lateral motorcycle; (b) Type 2 collision (180° ± 70°): frontal car - frontal motorcycle; (c) Type 3 collision (90° ± 20°): lateral car - frontal motorcycle; (d) Type 4 collision: car in the corner - frontal motorcycle; (e) Type 5 collision: rear car - frontal motorcycle; (f) Type 6 collision: frontal car - rear motorcycle.

3 Determining the Parameters that Characterize the Impact of Motorcycle-Car

The vertical displacement of the motorcyclist is characterized by relationship [5] (Fig. 4):

$$y = h_m + v_{0my} \cdot t - \frac{g \cdot t^2}{2} = h_m + (v_{om} \cdot \sin \delta_{0m}) \cdot t - \frac{g \cdot t^2}{2} \,[\mathrm{m}] \tag{1}$$

where: t - random time

h_m [m] - the height of the motorcyclist center of gravity at the throwing moment

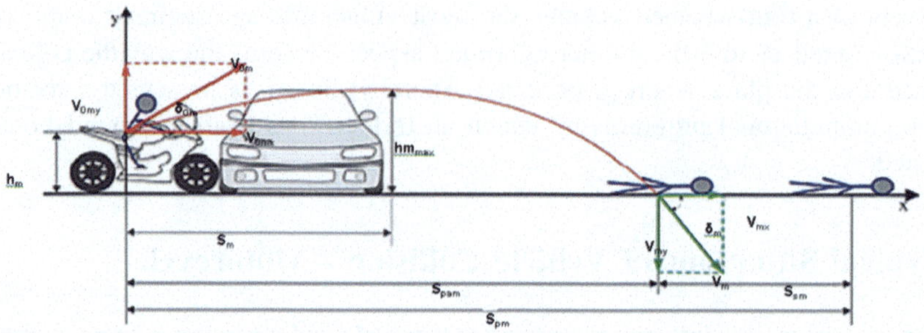

Fig. 4. Sketch of a motorcycle and car collision

v_{Omy} [m/s] - the component in the vertical direction of the motorcyclist's initial speed
v_{Om} [m/s] - the initial speed of the biker
g [m/s2] - gravitational acceleration
δ_{Om} [grad] - the angle between the initial trajectory of the motorcyclist and the road.

The motorcyclist moves in parallel with the road with a uniform motion according to the relationship [5]:

$$x = \frac{v_{0m}^2 \cdot \sin\delta_{0m} \cdot \cos\delta_{0m}}{g} + \sqrt{\frac{v_{0m}^4 \cdot \sin^2\delta_{0m} \cdot \cos^2\delta_{0m}}{g^2} - (y - h_m) \cdot \frac{2 \cdot v_{0m}^2 \cdot \cos^2\delta_{0m}}{g}} [\mathrm{m}] \tag{2}$$

If the ground on which the motorcyclist is thrown is at the same level as the road, then y = 0 and x = S_{pam}, so [5]:

$$S_{pam} = \frac{v_{0m}^2 \cdot \sin\delta_{0m} \cdot \cos\delta_{0m}}{g} + \sqrt{\frac{v_{0m}^4 \cdot \sin^2\delta_{0m} \cdot \cos^2\delta_{0m}}{g^2} + \frac{2 \cdot h_m \cdot v_{0m}^2 \cdot \cos^2\delta_{0m}}{g}} [\mathrm{m}] \tag{3}$$

where: S_{pam}[m] - the distance traveled by the motorcyclist from the time of throwing to landing (flight distance through air).

The slip distance on the ground results from the conditions of evenly decelerated motion of the motorcyclist [6]:

$$S_{sm} = \frac{v_m^2}{2 \cdot g \cdot f_m} [\mathrm{m}] \tag{4}$$

where: S_{sm} [m] - the slip distance of the motorcyclist on the ground by a slow deceleration
v_m [m/s] - the resulting speed with which the motorcyclist's body starts to move after touching the ground

f_m [-] - the overall friction coefficient of the motorcyclist's body on the ground.

4 Simulation and Analysis of Vehicle - Motorcycle Collision Type

The development of software platforms in the field of road traffic accident reconstruction makes it possible to accurately simulate collision conditions as well as to analyze the consequences of these collisions.

Simulation and analysis of the vehicle - motorcycle collision type is done through the Virtual Crash software platform [7].

In this application, a comparative analysis will be carried out on the kinematic and dynamic parameters of vehicles and motorcycle driver who were involved in a collision. The analyzed situations are presented below:

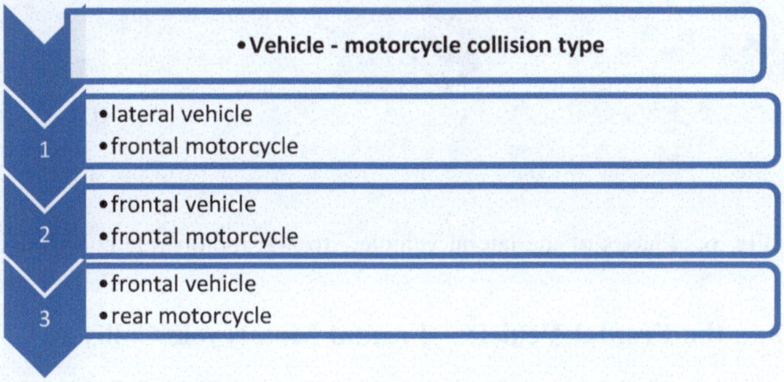

4.1 Modeling the Lateral Vehicle – Frontal Motorcycle Collision

To simulate a lateral vehicle-frontal motorcycle-type collision, the motorcycle was positioned at an angle of 90° to the vehicle. The vehicle's contact area is located on the right side.

The pre-impact speed was 50 km/h for both the vehicle and motorcycle, as shown in Fig. 5:

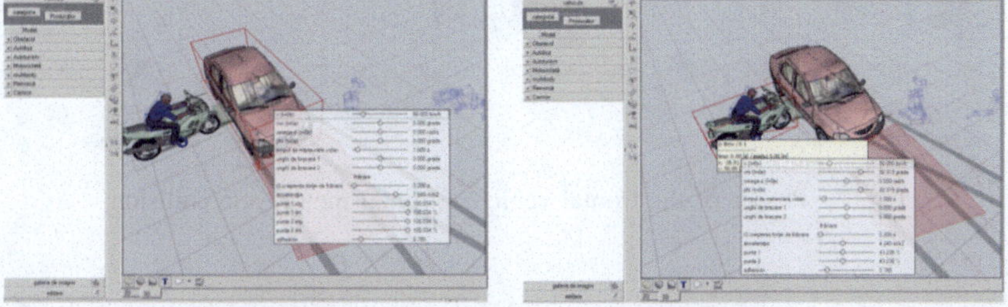

Fig. 5. Vehicle and motorcycle initial speed

Regarding the dynamic state of the two vehicles, at the moment of impact they were in braking, with a vehicle deceleration of 7.64 m/s^2, respectively 4.24 m/s^2 for the motorcycle.

The main phases of the collision dynamics of the lateral car - frontal motorcycle are presented and described below:

- Pre-collision phase: represents the moment before collision
- Collision phase: Fig. 6a presents the collision plane, represented in red, between the two vehicles
- Motorcyclist's detachment: following the collision between the vehicle and the motorcycle, the moment of detachment of the motorcyclist's body begins, Fig. 6b
- Projecting the motorcyclist on the roadway
- Final position of the motorcyclist's body - Fig. 6c.

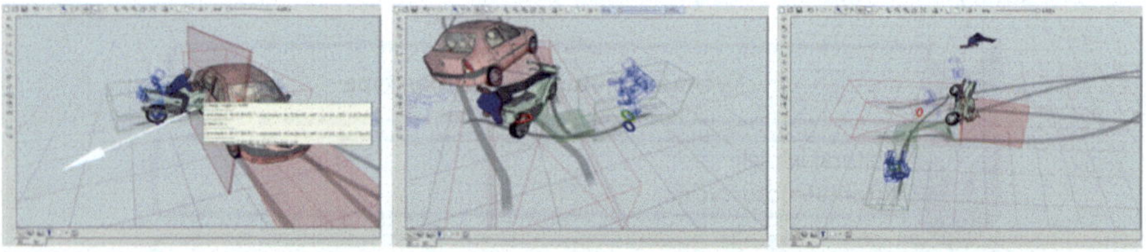

Fig. 6. Phases of the lateral vehicle - frontal motorcycle collision

4.2 Modeling the Frontal Vehicle – Frontal Motorcycle Collision

To simulate a frontal vehicle – frontal motorcycle collision, the motorcycle was positioned in the front of the vehicle, being oriented in the opposite direction to the vehicle's direction of travel.

The pre-impact speed was 50 km/h for both the vehicle and motorcycle. Regarding the dynamic state of the two vehicles, at the moment of impact they were in braking, with a vehicle deceleration of 7.64 m/s^2, respectively 3.9 m/s^2 for the motorcycle. Figure 7 shows the collision phases:

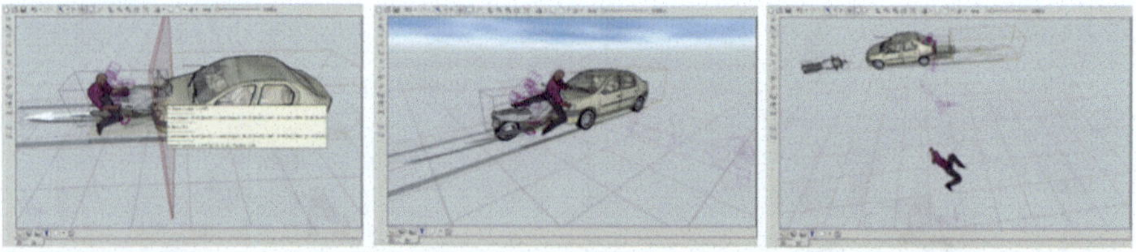

Fig. 7. Phases of the frontal vehicle - frontal motorcycle collision

4.3 Modeling the Frontal Vehicle – Rear Motorcycle Collision

To simulate a frontal vehicle - rear motorcycle collision, the motorcycle was positioned in the front of the vehicle, being oriented in the same direction as the vehicle's direction of travel.

The pre-impact speed was 50 km/h for the vehicle and for the motorcycle and motorcyclist was assigned a zero speed value, considering that the motorcycle was stopped at the moment of impact, as shown in Fig. 8:

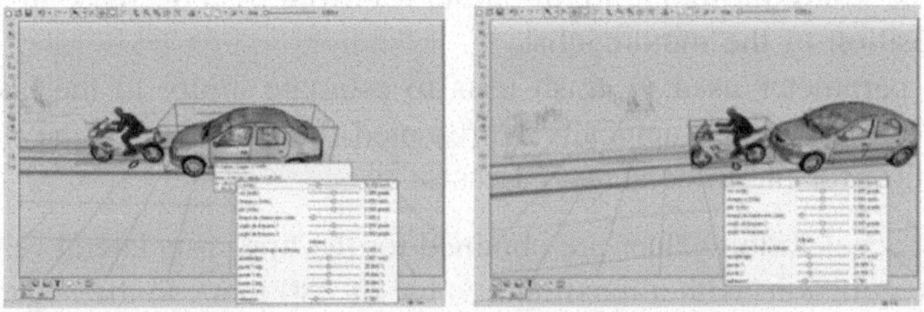

Fig. 8. Vehicle and motorcycle initial speed

Regarding the dynamic state of the two vehicles, at the moment of impact they were in braking, with a vehicle deceleration of 3.90 m/s^2, respectively 2.41 m/s^2 for the motorcycle.

5 Comparative Analysis of the Obtained Results

To assess the impact of the collision on the motorcyclist, a comparative analysis of the obtained results for the three analyzed situations was carried out.

The speed values at the motorcyclist head level were plotted in Fig. 9:

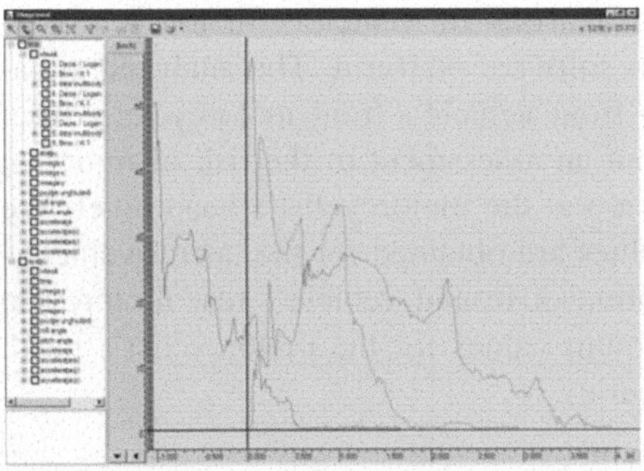

Fig. 9. Speed variation according to time

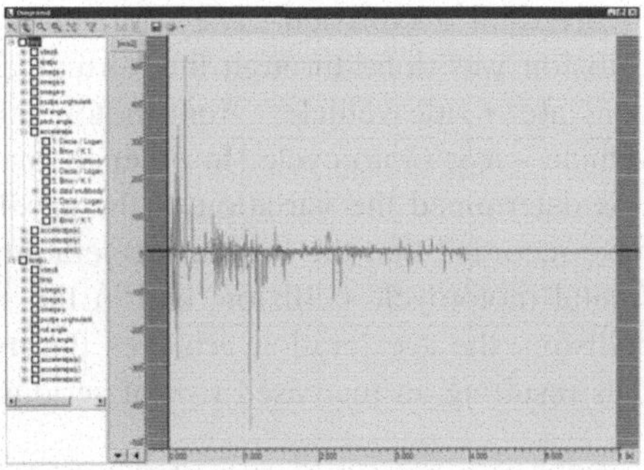

Fig. 10. Acceleration variation according to time

- case I – lateral vehicle - frontal motorcycle collision (represented by blue)
- case II – frontal vehicle - frontal motorcycle collision (represented by purple)
- case III – frontal vehicle - rear motorcycle collision (represented by red).

The maximum value recorded for the motorcyclist's head is 57 km/h for case II.

In case I, the speed decreases from an initial speed of 50 km/h to 20 km/h, followed by a slight increase of up to 30 km/h in the flight phase and projection of the motorcyclist's body on the ground.

In the third case, the initial speed of the motorcyclist's head is zero, reaches a maximum of 28 km/h during the flight phase, followed by a decrease in speed to zero in the final position of the motorcyclist.

The main parameter used in crash tests to estimate injury to the biker's head is acceleration. In Fig. 10 is graphically represented the variation of the motorcyclist's head acceleration for the three analyzed cases.

- The lowest acceleration values are obtained for the first case, lateral vehicle - frontal motorcycle collision, with a maximum value of 280 m/s^2, in the negative sense, as it corresponds to the motorcycle's projection phase on the ground.
- Similar variation is also obtained for case II, but with a maximum value of 480 m/s^2, in the negative sense.
- The third type of collision produces serious head injuries, reaching acceleration values above 500 m/s^2 during the take-off and flight phase of the motorcyclist.

6 Conclusion

The consequences of a collision between a motorcycle and a car depend on a number of factors, among the most important being the type and mass of the car, the place and impact direction, the characteristics of the road and the environment.

The main purpose of this paper was to perform a comparative analysis of the kinematic parameters of vehicle, motorcycle and the motorcycle driver who were involved in a collision. The simulation and analysis of vehicle - motorcycle type collision was done through the Virtual Crash software platform. The analyzed situations are: • side vehicle - front motorcycle; • front vehicle - front motorcycle; • front vehicle - rear motorcycle. In order to carry out an assessment of the risk of injury, it was determined the variation of the acceleration at the motorcyclist's head level. We have recorded that the lowest acceleration values are obtained for the lateral vehicle - frontal motorcycle collision, and in the situation of frontal vehicle - rear motorcycle collision, the acceleration achieves the maximum values for the motorcyclist's head, thus resulting an increased risk of serious injuries.

References

1. Toadere Roxana, F.: Contribuţii la realizarea unor drumuri mai sigure. Teză de doctorat. Universitatea Tehnică din Cluj, Cluj Napoca (2015)
2. Strategia Naţională de Siguranţă Rutieră pentru Perioada 2016–2020 – Anexa I
3. Tolea, B., Radu, A.I., Beles, H., Antonya, C.: Influence of the geometric parameters of the vehicle frontal profile on the pedestrian's head accelerations in case of accidents. Int. J. Autom. Technol. **19**(1), 85–98 (2018). https://doi.org/10.1007/s12239-018-0009-0. ISSN 1229-9138
4. Boroiu, A.A., Tabacu, I., Ene, A., Neagu, E., Boroiu, A.: Computing programs for determining traffic flows from roundabouts. In: International Congress of Automotive and Transport Engineering - Mobility Engineering and Environment (CAR 2017), vol. 252 (2017). ISSN 1757-8981
5. Todoruţ, A., Cordoş, N., Barabás, I., Bălcău, M.: Aspects regarding the numerical modeling of traffic incidents between motorcycles and passenger cars. Acta Technica Napocensis Ser.: Appl. Math. Mech. Eng. **59**(II) (2016). Technical University of Cluj-Napoca
6. Todoruţ, A., Barabás, I., Branzas, P., Gavrilas, C.A.: Reconstrucţia Accidentelor Rutiere Motocicletă – Automobil. A X-a Conferinţă Naţională multidisciplinară – cu participare internaţională. "Profesorul Dorin PAVEL - fondatorul hodroenergeticii româneşti", Sebeş (2010)
7. Virtual Crash User's Manual. http://www.vcrash.com/download/manual/english_short1.pdf

Practical Use of Accelerometers in Road Traffic Studies

Dinu Covaciu(✉), Daniela Florea, and Janos Timar

Transilvania University of Brasov, Politehnicii Str. 1, 500024 Brasov, Romania
dinu.covaciu@unitbv.ro

Abstract. Vehicles acceleration is a parameter with many practical uses, from estimation of road roughness to modeling of fuel consumption. Models of acceleration or speed profiles of vehicles in road traffic are used in microscopic traffic simulation. Longitudinal and lateral accelerations can be used for this purpose. From many years, the acceleration noise, as a traffic parameter, is used to measure the disutility associated with acceleration variations in road traffic. It was found that there is a relationship between the acceleration noise and the characteristics of the traffic flow. In this paper is introduced a dedicated device developed for measuring and recording the vehicle accelerations in real traffic conditions, the method for processing the data acquired and some experimental results.

Keywords: Acceleration measurement · Road traffic · Acceleration noise · Acceleration profile

1 Introduction

The road traffic can be analyzed using two groups of parameters: microscopic and macroscopic. The microscopic parameters describe the behavior of the individual vehicle, as an entity moving on road and analyzed through the driving particularities [1, 2]. This group of parameters is connected to vehicle dynamics, and as consequence a series of terms are explained through formulae derived from dynamic analysis. Such parameters are [2, 3]: *time headway*, *space headway*, *speed* and *acceleration*. *Time* and *space headway* consist of the difference in time and position (in seconds or meters) between the fronts of two successive vehicles. The speed is characterized by three values: the *instantaneous speed*, resulted from direct measurements, the *average speed*, resulted by processing data collected on certain road, and the *velocity field* (velocity statistics), which is the area of the domain of variation of an individual vehicle velocity, on a road segment (velocity is a function of both position and time). *Acceleration* is the microscopic indicator that allows the rigorous determination of the vehicle dynamics. The ascertainment of acceleration can be done analytical, by processing the speed records, or directly, using an accelerometer.

I. Dumitru et al. (Eds.): SMAT 2019, *The 30th SIAR International Congress of Automotive and Transport Engineering*, pp. 392–399, 2020.
https://doi.org/10.1007/978-3-030-32564-0_46

2 Methodology

2.1 Acceleration as a Traffic Parameter

The variables used to describe the individual vehicle motion are: speed and acceleration. Considering x(t) is the distance traveled as function of time (measured in meters), the speed and acceleration are defined simple, as follows:

$$v(t) = \frac{dx}{dt} \tag{1}$$

$$a(t) = \frac{dv}{dt} = \frac{d^2x}{dt^2} \tag{2}$$

$$b(t) = \frac{da}{dt} = \frac{d^2v}{dt^2} = \frac{d^3x}{dt^3} \tag{3}$$

where *v(t)* is the travel speed as function of time [m/s] (instantaneous speed), *a(t)* is the acceleration as function of time [m/s^2] and *b(t)* is the acceleration variation as a function of time [m/s^3].

Considering an initial time t_0, at which the kinematic variables have the values x_0, v_0 and a_0, the following equations of motion can be written:

$$x(t) = x_0 + \int_{t_0}^{t} v(t)\,dt \tag{4}$$

$$v(t) = x_0 + \int_{t_0}^{t} a(t)\,dt \tag{5}$$

$$a(t) = x_0 + \int_{t_0}^{t} b(t)\,dt \tag{6}$$

and, consequently:

$$x(t) = x_0 + \int_{t_0}^{t} v_0\,dt + \int\int_{t_0}^{t} a_0\,dt\,dt + \int\int\int_{t_0}^{t} b(t)\,dt\,dt\,dt \tag{7}$$

As function of the travelled distance, the equations of motion can be obtained by applying the transformation:

$$v(x) = \frac{1}{dt/dx} \tag{8}$$

resulting:

$$t(x) = t_0 + \int_{x_0}^{x} \frac{dx}{v(x)} \tag{9}$$

The acceleration by space is:

$$a(x) = \frac{d\left[\frac{1}{2}v^2(x)\right]}{dx} \tag{10}$$

and further:

$$v(x) = \sqrt{v^2 + 2\int_{x_0}^{x} a(x)\,dx} \tag{11}$$

The time-mean speed is the parameter that highlights the vehicle behaviour on a road segment. This traffic indicator is one of the basic parameters for develop and compare the vehicle test cycles, for emission certification. The time-mean speed for a travelled route can be ascertained using the formula below:

$$v_{med} = \frac{\int_{t_0}^{t} v(t)\,dt}{t} \tag{12}$$

Acceleration is also an indicator that allows a precise evaluation of the vehicle dynamics, especially in an urban environment. The measurements in urban area [2] have shown a variation of acceleration in an interval of ±2.5 m/s^2. This corresponds to random traffic conditions. The mean acceleration in these conditions does not overtake a value of 1 m/s^2, most of the measurements giving mean values of 0.3–0.5 m/s^2.

The mean acceleration for a duration of $(t - t_0)$ is given by the formula:

$$a_{med} = \frac{\int_{t_0}^{t} a(t)\,dt}{t} \tag{13}$$

The acceleration indicator is useful for defining the traffic quality, through two synthetic indicators: dispersion (or the acceleration noise) and relative dispersion of acceleration.

The *acceleration noise* is resulting from the statistic evaluation of the dispersion of acceleration, and is given by the formula below.

$$a_\sigma = \sigma^2 = \frac{\int_{t_0}^{t} \left[a(t) - a_{med}\right]^2 dt}{t} \tag{14}$$

This indicator allows the evaluation of the traffic instability in urban area and is also named *acceleration noise*. The relative dispersion of acceleration is given by the formula:

$$D_{rel} = \frac{\sigma}{v_{med}} \tag{15}$$

where σ is the standard deviation.

The dispersion of acceleration is considered to be an useful traffic parameters for many years [4]. It is a rapid and convenient method of traffic evaluation.

From the value of the relative dispersion of acceleration it can be ascertained the traffic fluency. Some example values are given in Table 1 [2, 3].

Table 1. Example values of the relative dispersion of acceleration

Relative dispersion D_{rel}	Degree of fluency	Comments
$0 < D_{rel} < 0{,}02$	Fluent traffic	Arterials with non-intensive circulation, without risk of saturation
$0{,}02 < D_{rel} < 0{,}1$	High density traffic	Arterials used between the limits of their circulation capacity
$0{,}1 < D_{rel} < 0{,}2$	Difficult traffic	Congestion points and over-saturation of arterials for short periods of time
$0{,}2 < D_{rel}$	Very difficult traffic	Traffic congestions during the entire day

2.2 Acceleration Measurement

The vehicle acceleration is easily measured and recorded using an instrumented vehicle. In case of acceleration calculated from measured speed values, the data acquisition equipment is, in most cases, based on a GPS receiver [5]. The speed, in this case, is determined based on the Doppler effect, and the acceleration is calculated by deriving the speed over time. The direction of the velocity, and of acceleration, is the direction of movement. The longitudinal and lateral components of the acceleration vector can be calculated using the vehicle course (heading). Another way to obtain the vehicle speed is by reading data from the vehicle CAN bus (through the OBD II interface), but the resolution is lower and the direction of movement is difficult to determine.

Direct measurement of acceleration is possible using accelerometers - devices that measure the acceleration in a particular direction. The acceleration sensors most used today are manufactured in MEMS (Micro-Electro-Mechanical Systems) technology. A MEMS accelerometer integrates an accelerometer and the electronics into a single silicon chip.

The most common MEMS accelerometers are capacitive devices consisting of a suspended proof mass/plate assembly. As acceleration acts on the proof mass, micro-machined capacitive plates sense a change in acceleration from deflection of the plates [6]. The automotive applications for MEMS include [7]: airbags, anti-lock braking systems, active suspension, navigation control systems, vibration monitoring, fuel sensors, noise reduction, rollover detection, seatbelt restraint and tensioning.

Accelerometers are widely used in Inertial Navigation Systems (INS), together with gyro units, or combined INS-GPS systems.

3 Solutions for Vehicle Acceleration Measurement and Processing

3.1 Data Acquisition Device Based on MEMS Accelerometer

There are acceleration measurement devices available on the market, but many of them are of low performance (like those included in smartphones), or are too expensive. In order to collect many data, using multiple devices installed on different vehicles, a dedicated, customized devices was designed and developed. The block diagram of the data acquisition system is shown in Fig. 1. The system consists in two main sections: the data acquisition device (as hardware component) and the post-processing software.

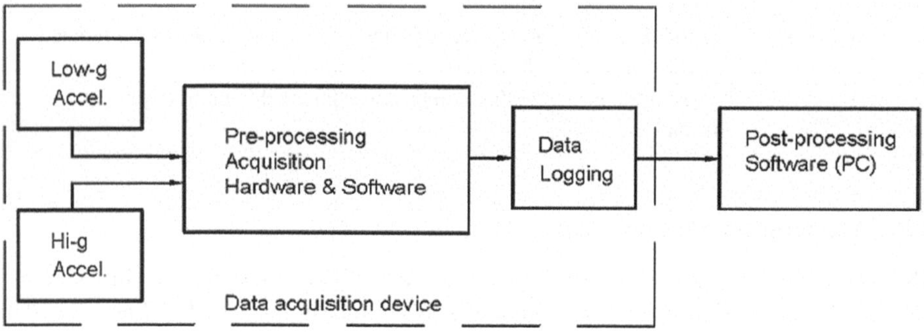

Fig. 1. Block diagram of the data acquisition system

The components of the data acquisition device are: the sensors (accelerometers), the pre-processing unit, based on microcontroller, and the data logging unit. The range of accelerations measured in normal traffic conditions is under ±1 g. This interval is important when choosing the most appropriate sensor to be used. However, the device was equipped with two independent accelerometers, with different measuring range: one for low accelerations, to be used in normal traffic conditions, and another one for high accelerations, to be used in case of some special events, like traffic collisions.

The Accelerometer. Low-g means accelerations acceleration values in the range of ±3 g (where g is the gravity acceleration: about 9.8 m/s^2). The accelerometers use g as a reference value, since the ±1 g value is easily used for calibration. The chosen sensor, for low-g range, was ADXL 337, from Analog Devices [8]. This is an analog accelerometer, with a measuring range of ±3 g, on three axes (x, y, z).

The Control Unit. The core of the system is the microcontroller, included in a development board. The board is Feather M0 Basic Proto (Feather is a family of development boards produced by Adafruit [9]), with the microcontroller ATSAMD21

Cortex M0, that features enough memory for the program (256K of flash memory, 32K of RAM) and 6 analog inputs (12 bits), covering the need for reading data from two tri-axial accelerometers at the same time.

Data Logging. Over the development board is installed a shield, containing the micro-SD card unit and a RTC unit, allowing to control the time of recording. The file format is text, and each data sample includes time and raw accelerations read on the 6 analog inputs of the microcontroller.

The software embedded in the control unit is written in C++ language and allow the control of the acquisition system and also communication with an external computer, through USB interface. The system can be configured using this communication, using a dedicated program (on computer side) or any virtual terminal.

The sampling rate can be set up to 500 Hz, without the risk of data loss, and for traffic and vehicle dynamics analysis a sampling rate of 50–100 Hz is recommended.

3.2 Data Processing Software

Since data is stored as text files, any spreadsheet calculation software (like Excel) can be used to process. However, the text files contains raw data, which are difficult to understand by the final user. The dedicated software, developed for this kind of data, allows to calibrate (Fig. 2), filter, display and export the data [10].

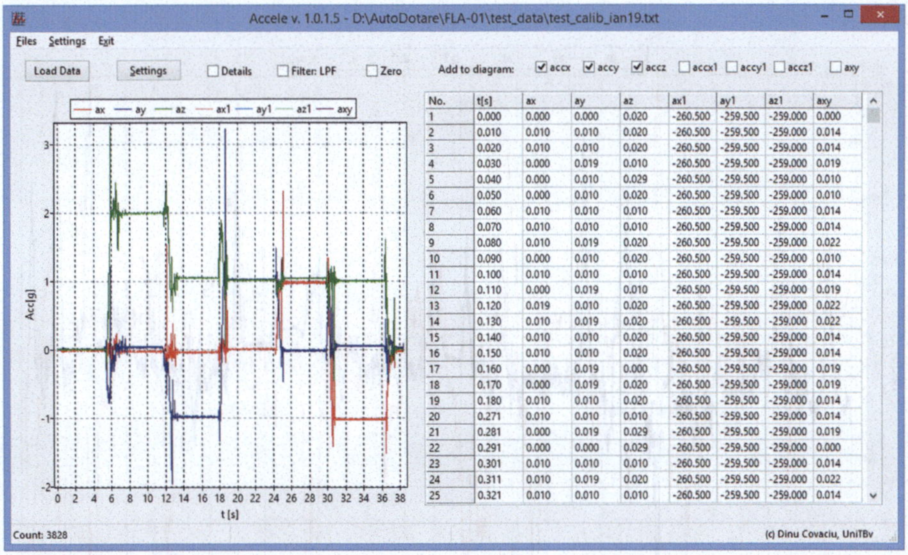

Fig. 2. Calibration of the acceleration logger using the dedicated software

The digital filter is one of the most important parts of the data processing software. Recorded values are affected by white noise and vehicle's vibrations, and all of these should be eliminated. The filters used are low-pass filters, of different types (including Butterworth and Kalman filters), and can be chosen by the user.

The Eqs. (1–15) are also included in the software algorithm. The calculated values of accelerations are shown in the table, and the chosen ones can be represented as curves on the diagram. The graphic part of the interface (the diagram) can be zoomed in/out, and includes also some interactive functions, allowing to show in the table the value of a clicked point from the graph. The values from the displayed part of the diagram can be exported in another file, and this is useful to select the data needed for analysis, without affecting the raw data recorded on field.

4 Experimental Data

The data acquisition device (acceleration logger) was installed on a vehicle, in order to collect real traffic data. When installing the device, it is important to align the accelerometer axes with the vehicle axes, and to avoid any uncontrolled movement of the device relative to the vehicle.

An example of data collected is shown in Fig. 3. The red curve represents the evolution of acceleration on x direction (longitudinal acceleration), and the blue curve is the acceleration on y direction (lateral acceleration). The sampling rate was in this case 50 Hz. The limit values obtained, from multiple data collecting sessions, all in

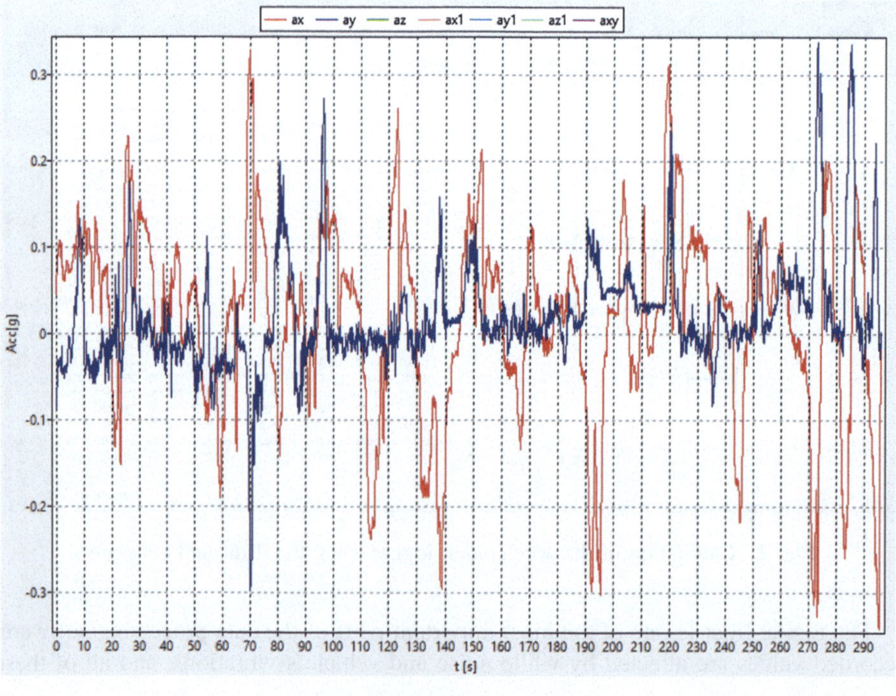

Fig. 3. Example of data collected

urban traffic conditions, were about 3.5 m/s^2 for longitudinal acceleration and 3.3 m/s^2 for lateral acceleration.

The *acceleration noise*, calculated with Eq. 14, resulted from the collected data, was 0.556, and the relative dispersion is 0.075, which means, according to Table 1, a high density traffic.

5 Conclusion

The main objective of the paper was to present a solution for collecting and processing the vehicle accelerations, in real traffic conditions, in order to use the results for traffic evaluation. The solution consists in a data acquisition device, developed in the own laboratory, used to record raw acceleration values, and a computer program for processing the recorded values.

The experimental results presented are just examples of using the developed system, and are particular for the routes traveled and traffic conditions at that time. It was not the intention of the authors to characterize the road traffic in the city or on certain routes.

A known issue when only acceleration values are recorded is to determine the speed of the vehicle. Even after filtering and applying other corrections to compensate the accelerometer bias, the calculated speed is still subjected to errors. It is preferable to use also another system for speed measurement, like a GPS receiver, which allow also to record the waypoints.

References

1. Florea, D.: Road Traffic Management, 2nd edn. Transilvania University Publishing, Braşov (2000). (in Romanian: Managementul traficului rutier)
2. Filip, N.: Road Traffic Engineering. Editura Mediamira, Cluj-Napoca (2010). (in Romanian: Ingineria Traficului Rutier)
3. Hobbs, F.D.: Traffic Planning & Engineering, 2nd edn. Pergamon International Library: Civil Engineering Division, Cox and Wyman Ltd., London (1979)
4. Jones, T.R., Potts, R.B.: The measurement of acceleration noise - a traffic parameter. Oper. Res. **10**(6), 745–763 (1962)
5. Covaciu, D., Florea, D., Preda, I., Timar, J.: Using GPS devices for collecting traffic data. In: Proceedings of SMAT 2008 Congress. University of Craiova, Craiova (2008)
6. Dima, D.S., Covaciu, D.: Use of MEMS sensors for data acquisition in crash tests. In: Proceedings of EVU 2016 Congress, Bratislava (2016)
7. Fernandes, R.: Five decades of innovations in MEMS technology. Int. J. Recent Dev. Eng. Technol. **3**(1), 131–137 (2014)
8. Analog Devices, Datasheet ADXL337, Small, Low Power, 3-Axis ±3 g Accelerometer. http://www.analog.com/en/products. Accessed 01 July 2019
9. Adafruit Homepage. https://www.adafruit.com/product/2772. Accessed 01 July 2019
10. Dima, D.S.: Research regarding development of methods used in analysis and reconstruction of traffic accidents. Ph.D. thesis. Transilvania University of Brasov (2018)

A New Method to Improve Computing Time for TSP Little Algorithm

Adrian Cipleu(✉), Werner Ştefănescu, Ionel Vandici, and Attila Iuliu Gönczi

Politehnica University Timisoara, Timisoara, Romania
{adrian.cipleu,werner.stefanescu,ionel.vandici, attila.gonczi}@upt.ro

Abstract. The paper describes a method to speed up the finding of the solution of the NP-complete Travelling Salesman Problem, using the classic Little algorithm. The Little algorithm is belonging to the class of branch-and-bound methods for solving the TSP problems or the determination of Hamiltonian optimum circuits in directed or undirected graphs. We propose changes to be made in the steps of the algorithm to speed up solving and eliminate the possibility of error. In goods distribution, in warehouse order – picking etc. the real time decision regarding the optimal route choice influence the cost of operations. The proposed method reduces the solving time of a TSP problem and so decreases the overall costs and diminish the environmental impact of transport.

Keywords: Travelling Salesman Problem · Little algorithm · Operational logistics

1 Introduction

We can say that the traveling salesman problem (TSP) emerged from Euler's studies on the "Knight's Tour Problem" in 1766, but some may say that the first encounter of a travelling salesman type problem is in Homer's poem, the Odyssey, in the route chosen by the hero Ulysses/Odysseus in his Mediterranean adventure. The mathematicians Sir W. R. Hamilton and T. P. Kirkman carried out the first studies and discussions in the 18th century on the problem. Their work was about finding paths and circuits on the dodecahedral graph, satisfying certain conditions. Hamilton is regarded as the forerunner of the TSP that is why TSP is also known as "Optimal Hamiltonian circuits". Quoted as the origin of TSP are the work of Carl Menger done in Vienna and at Harvard University in the 1920–1930s, the problem was later further developed by H. Whitney and M. M. Flood. It was the year 1948 when Flood presented the travelling salesman problem at the RAND Corporation. The Linear Programming was investigated as a vital force in computing solutions to combinatorial optimization problems in the 1950s. In 1954 G. Dantzig, R. Fulkerson and S. Johnson working for RAND found a method for solving the TSP. They have proved the effectiveness of their method by solving a 49-city instance. However, it became evident, as early as the mid 1960's, that the general instance of the TSP could not be solved in polynomial time using Linear Programming techniques. Finally, these categories of problem became known as

I. Dumitru et al. (Eds.): SMAT 2019, *The 30th SIAR International Congress of Automotive and Transport Engineering*, pp. 400–405, 2020.
https://doi.org/10.1007/978-3-030-32564-0_47

NP-hard (nondeterministic polynomial time hard). Great progress appeared in the following years by developing new techniques and methods for TSP, namely Heuristics algorithms, Cutting plane methodology, Branch-and-Bound technique, Ant Colony Optimization Algorithm, Particle Swarm Optimization Algorithm, Genetic Algorithm etc. [1–3].

In transport and logistics, numerous possibilities and practical examples of application of the TSP model exist. Some typical examples are distribution of products from distributor to customers/shops, orbital inspection walks or checkups from a certain point and leading through individual checking points, warehouse order – picking, public transport circuits etc.

The paper presents a method to improve Little's branch and bound algorithm by increasing the calculus speed of solving TSP. The Little TSP algorithm [4], was developed in 1963 by John Little and his collaborators KattaMurty, Dura Sweney and Caroline Karel.

2 TSP Formulation

The simplest formulation is: a salesman wishes to find the shortest route through a number of locations and back to initial start location [5].

Given a finite set of location to visit, N, and a cost matrix $C = [c_{ij}]_{i,\, j\, \epsilon\, N}$, that can express distance, travel time, actual cost travel etc., determine:

$$min_{\pi} \sum\nolimits_{i \in N} c_{i\pi(i)}, \quad (1)$$

where π runs over all cyclic permutations of N. If N = {1, 2, … n}, then a similar formulation is:

$$min_{\sigma}\left(\sum\nolimits_{i=1}^{i=n-1} c_{\sigma(i)\sigma(i+1)} + c_{\sigma(n)\sigma(1)}\right), \quad (2)$$

where σ runs over all permutations of N and the $\sigma(k)$ is the k location in the tour. If G represents the complete directed graph on the vertexes (nodes) set N with the cost matrix C for each arc (i, j), then the optimal tour is the Hamiltonian circuit on G with the minimum cost. Where the Hamiltonian circuit is the circuit that pass through each vertex once and return to starting node.

If $c_{ij} = c_{ji}$ the problem is called symmetric otherwise is called asymmetric and if for all (i, j, k) from N the $c_{ij} \leq c_{ik} + c_{kj}$ the problem is called Euclidean.

3 The Little Algorithm and the Improvement Proposal

The algorithm is a heuristic, "branch and bound" type that is used for discreet optimization problems such as the TSP in which the variables are finite, they have discreet values, satisfying a restriction system and it is called to optimize a function defined on the restriction set.

We will simultaneously present the classical Little algorithm with an example and in parallel the proposed improvement method.

Let us consider a travel cost matrix C of N = {1, 2, 3, 4, 5, 6} vertexes/locations. The c(i, j) represents the cost of travel (distance, time or effective cost) from location i to location j.

$$C = \begin{array}{c} \\ 1 \\ 2 \\ 3 \\ 4 \\ 5 \\ 6 \end{array} \begin{array}{c} \begin{array}{cccccc} 1 & 2 & 3 & 4 & 5 & 6 \end{array} \\ \begin{bmatrix} \infty & 8 & 5 & 3 & 10 & 8 \\ 5 & \infty & 10 & 8 & 3 & 2 \\ 13 & 11 & \infty & 8 & 4 & 11 \\ 2 & 3 & 6 & \infty & 8 & 5 \\ 6 & 4 & 7 & 9 & \infty & 13 \\ 9 & 7 & 3 & 11 & 5 & \infty \end{bmatrix} \end{array}$$

Step 1. Determine the reduced matrix, save the sum of reduction as the lower bound of the cost of the tour. The reduced matrix is obtained by subtracting the smallest element of the row from each value of corresponding row and subsequently subtracting the smallest element of the column from each value of corresponding column.

$$C^r = \begin{array}{c} \\ 1 \\ 2 \\ 3 \\ 4 \\ 5 \\ 6 \end{array} \begin{array}{c} \begin{array}{cccccc} 1 & 2 & 3 & 4 & 5 & 6 \end{array} \\ \begin{bmatrix} \infty & 5 & 2 & 0 & 7 & 5 \\ 3 & \infty & 8 & 6 & 1 & 0 \\ 9 & 7 & \infty & 4 & 0 & 7 \\ 0 & 1 & 4 & \infty & 6 & 3 \\ 2 & 0 & 3 & 5 & \infty & 9 \\ 6 & 4 & 0 & 8 & 2 & \infty \end{bmatrix} \end{array}$$

Where the bound value is: h = 3 + 2 + 4 + 2 + 4 + 3 = 18;

Step 2. For every arc (k, l) of zero value in the reduced matrix, we calculate a penalty defined as:

$$\theta(k,l) = \min\{c^r_{ks} : s \in N - \{l\}\} + \min\{c^r_{hl} : h \in N - \{k\}\} \tag{3}$$

Step 3. We choose the zero, which has the highest penalty. In this example, it turns to be arc (1, 4). We branch this arc into the TSP solution. Establish the location pair (p, r) which would join the ends of the connected path that includes the newly chosen arc (1, 4). In the first case is simple: it is the arc (4, 1). We will delete row "1" and column "4" from the reduced cost matrix. Set the cost of arc (p, r) infinite (c (4, 1) = ∞). This phase involves a procedure that must branch all chosen arcs in order to establish the arc that closes partial loops.

We propose to eliminate this branching and so to reduce the computing time with the following procedure: If the chosen arc is (k, l) we will replace the row index "l" with number "k" (in our case 1 replace the 4, 1 → 4). In addition, doing so in the cost matrix one will emphasize the arcs (i, i) which cost becomes infinite. The procedure will eliminate the need for branching for partial loop determination.

$$C^{r1} = \begin{array}{c} \\ 1 \\ 2 \\ 3 \\ 4 \\ 5 \\ 6 \end{array} \begin{array}{c} \begin{array}{cccccc} 1 & 2 & 3 & 4 & 5 & 6 \end{array} \\ \begin{bmatrix} \infty & 5 & 2 & 0^{2+4} & 7 & 5 \\ 3 & \infty & 8 & 6 & 1 & 0^{1+3} \\ 9 & 7 & \infty & 4 & 0^{4+1} & 7 \\ 0^{1+2} & 1 & 4 & \infty & 6 & 3 \\ 2 & 0^{2+1} & 3 & 5 & \infty & 9 \\ 6 & 4 & 0^{2+2} & 8 & 2 & \infty \end{bmatrix} \end{array}$$

$$C^{r1} = \begin{array}{c} \\ 1 \\ 2 \\ 3 \\ 4 \\ 5 \\ 6 \end{array} \begin{array}{c} \begin{array}{cccccc} 1 & 2 & 3 & 4 & 5 & 6 \end{array} \\ \begin{bmatrix} \infty & 5 & 2 & 0^{6} & 7 & 5 \\ 3 & \infty & 8 & 6 & 1 & 0^{4} \\ 9 & 7 & \infty & 4 & 0^{5} & 7 \\ 0^{3} & 1 & 4 & \infty & 6 & 3 \\ 2 & 0^{3} & 3 & 5 & \infty & 9 \\ 6 & 4 & 0^{4} & 8 & 2 & \infty \end{bmatrix} \end{array}$$

Step 4. Reduce rows and columns if possible and add the amount of reduction to the lower bound "h" and return to Step 2 until the cost matrix reduce to 1 × 1 dimension matrix.

Iteration 1

Little's algorithm

$$C^{r1} = \begin{array}{c} \\ 1 \\ 2 \\ 3 \\ 4 \\ 5 \\ 6 \end{array} \begin{array}{c} \begin{array}{cccccc} 1 & 2 & 3 & 4 & 5 & 6 \end{array} \\ \begin{bmatrix} \infty & 5 & 2 & \mathbf{0^{6}} & 7 & 5 \\ 3 & \infty & 8 & 6 & 1 & 0^{4} \\ 9 & 7 & \infty & 4 & 0^{5} & 7 \\ 0^{3} & 1 & 4 & \infty & 6 & 3 \\ 2 & 0^{3} & 3 & 5 & \infty & 9 \\ 6 & 4 & 0^{4} & 8 & 2 & \infty \end{bmatrix} \end{array}$$

(1, 4) first chosen arc

The loop 1 → **4** → **1**

$c_{(4,1)}=\infty$

h1 = h + 1+2=21

Proposed method

$$C^{r1} = \begin{array}{c} \\ 1 \\ 2 \\ 3 \\ 1 \rightarrow 4 \\ 5 \\ 6 \end{array} \begin{array}{c} \begin{array}{cccccc} 1 & 2 & 3 & 4 & 5 & 6 \end{array} \\ \begin{bmatrix} \infty & 5 & 2 & \mathbf{0^{6}} & 7 & 5 \\ 3 & \infty & 8 & 6 & 1 & 0^{4} \\ 9 & 7 & \infty & 4 & 0^{5} & 7 \\ 0^{3} & 1 & 4 & \infty & 6 & 3 \\ 2 & 0^{3} & 3 & 5 & \infty & 9 \\ 6 & 4 & 0^{4} & 8 & 2 & \infty \end{bmatrix} \end{array}$$

(1, 4) first chosen arc

Row index replacement 1 → 4

$c_{(1,1)}=\infty$

h1 = h + 1+2=21

Iteration 2

Little's algorithm

$$C^{r2} = \begin{array}{c} \\ 2 \\ 3 \\ 4 \\ 5 \\ 6 \end{array} \begin{array}{c} \begin{array}{ccccc} 1 & 2 & 3 & 5 & 6 \end{array} \\ \begin{bmatrix} 1 & \infty & 8 & 1 & 0^{3} \\ 7 & 7 & \infty & \mathbf{0^{8}} & 7 \\ \infty & 0^{2} & 3 & 5 & 2 \\ 0^{1} & 0^{0} & 3 & \infty & 9 \\ 4 & 4 & 0^{5} & 2 & \infty \end{bmatrix} \end{array}$$

(3, 5) second arc

The loop 3 → **5** → **3**

$c_{(5,3)}=\infty$

$h^2 = h^1 + 0 = 21$

Proposed method

$$C^{r2} = \begin{array}{c} \\ 2 \\ 3 \\ 1 \\ 3 \rightarrow 5 \\ 6 \end{array} \begin{array}{c} \begin{array}{ccccc} 1 & 2 & 3 & 5 & 6 \end{array} \\ \begin{bmatrix} 1 & \infty & 8 & 1 & 0^{3} \\ 7 & 7 & \infty & \mathbf{0^{8}} & 7 \\ \infty & 0^{2} & 3 & 5 & 2 \\ 0^{1} & 0^{0} & 3 & \infty & 9 \\ 4 & 4 & 0^{5} & 2 & \infty \end{bmatrix} \end{array}$$

(3, 5) second arc

Row index replacement 3 → 5

$c_{(3,3)}=\infty$

$h^2 = h^1 + 0 = 21$

Iteration 3

Little's algorithm

$$C^{r3} = \begin{array}{c} \\ 2 \\ 4 \\ 5 \\ 6 \end{array} \begin{array}{c} \begin{array}{cccc} 1 & 2 & 3 & 6 \end{array} \\ \begin{bmatrix} 1 & \infty & 8 & 0^3 \\ \infty & 0^2 & 3 & 2 \\ 0^1 & 0^0 & \infty & 9 \\ 4 & 4 & \mathbf{0^7} & \infty \end{bmatrix} \end{array}$$

(6, 3) third arc

The loop 6 → 3 → **5 → 6**

$c_{(5,6)}=\infty$

$h^3 = h^2 + 0 = 21$

Proposed method

$$C^{r3} = \begin{array}{c} \\ 2 \\ 1 \\ 6 \to 3 \\ 6 \end{array} \begin{array}{c} \begin{array}{cccc} 1 & 2 & 3 & 6 \end{array} \\ \begin{bmatrix} 1 & \infty & 8 & 0^3 \\ \infty & 0^2 & 3 & 2 \\ 0^1 & 0^0 & \infty & 9 \\ 4 & 4 & \mathbf{0^7} & \infty \end{bmatrix} \end{array}$$

(6, 3) third arc

Row index replacement 6 → 3

$c_{(6,6)}=\infty$

$h^3 = h^2 + 0 = 21$

Iteration 4

Little's algorithm

$$C^{r4} = \begin{array}{c} \\ 2 \\ 4 \\ 5 \end{array} \begin{array}{c} \begin{array}{ccc} 1 & 2 & 6 \end{array} \\ \begin{bmatrix} 1 & \infty & \mathbf{0^3} \\ \infty & 0^2 & 2 \\ 0^1 & 0^0 & \infty \end{bmatrix} \end{array}$$

(2, 6) fourth arc

The loop 2 → 6 → 3 → **5 → 2**

$c_{(5,2)}=\infty$

$h^4 = h^3 + 0 = 21$

Proposed method

$$C^{r4} = \begin{array}{c} \\ 2 \\ 1 \\ 2 \to 6 \end{array} \begin{array}{c} \begin{array}{ccc} 1 & 2 & 6 \end{array} \\ \begin{bmatrix} 1 & \infty & \mathbf{0^3} \\ \infty & 0^2 & 2 \\ 0^1 & 0^0 & \infty \end{bmatrix} \end{array}$$

(2, 6) fourth arc

Row index replacement 2 → 6

$c_{(2,2)}=\infty$

$h^4 = h^3 + 0 = 21$

Iteration 5

Little's algorithm

$$C^{r5} = \begin{array}{c} \\ 4 \\ 5 \end{array} \begin{array}{c} \begin{array}{cc} 1 & 2 \end{array} \\ \begin{bmatrix} \infty & \mathbf{0^\infty} \\ \mathbf{0^\infty} & \infty \end{bmatrix} \end{array}$$

(4, 2) fifth arc

(5, 1) closing arc

$h^5 = h^4 + 0 = 21$

Proposed method

$$C^{r5} = \begin{array}{c} \\ 1 \\ 1 \to 2 \end{array} \begin{array}{c} \begin{array}{cc} 1 & 2 \end{array} \\ \begin{bmatrix} \infty & \mathbf{0^\infty} \\ \mathbf{0^\infty} & \infty \end{bmatrix} \end{array}$$

(1, 2) fifth arc

Row index replacement 1 → 2

(1,1) closing arc

$h^5 = h^4 + 0 = 21$

The TSP solutions is given by the arches chosen by the Little algorithm and total cost is given by the value of the lower bound:

TSP: 1 → 4 → 2 → 6 → 3 → 5 → 1

In the proposed method the solution is reconstructed by following in reverse and unidirectional the chosen arches:

1 → 1
1 → 2 →1
1 → **2 → 6** → 1
1 → 2 → **6 → 3** → 1
1 → 2 → 6 → **3 → 5** → 1
1 → 4 → 2 → 6 → 3 → 5 → 1

4 Conclusions

The proposed method eliminates the construction in parallel with the algorithm the tree of alternative branches in order to eliminate the arc's that close the partial loops. The method also induces a linearity to the Little algorithm and as a consequence, it increases the speed of solving the TSP and eliminates the possibility of error.

References

1. Demez, H.: Combinatorial optimization: solution methods of traveling salesman problem. MSS thesis. Eastern Mediterranean University, Gazimağusa, North Cyprus, January 2013
2. Dantzig, G., Fulkerson, R., Jonson, S.: Solution of a large scale traveling salesman problem. J. Oper. Res. Soc. Am. **2**, 393–410 (1954). Paper P – 510, the RAND Corporation
3. Filip, E., Otakar, M.: The travelling salesman problem and its application in logistic practice. WSEAS Trans. Bus. Econ. **8**(4), 163–173 (2011)
4. Little, J., Murty, K., Sweney, D., Karel, C.: An algorithm for the travelling salesman problem. Massachusetts Institute of Technology, 1 March 1963
5. Lenstra, J.K., Rinnooy Kan, A.H.G.: Some simple applications of the traveling salesman problem. Oper. Res. Soc. **26**(4), 717–733 (1975)

Experimental Research on Brake Behavior for Different Types of Commercial Vehicles

Mădălin Florin Popa(✉), Marius S. D. Capăță, and Nicolae Burnete

Technical University of Cluj Napoca, B-dul Muncii, 103-105, 400461 Cluj-Napoca, Romania
madalin.popa@live.com

Abstract. The dynamics of the society's development in recent years has underlined and strengthened the position of cars in everyday life. The current vehicle dependence of the society, leads to the continuous development and improvement of the means of transport and to the improvement of the operating conditions. In this respect, traffic safety becomes an essential aspect and as the braking system is one of the most important safety systems on a vehicle, its development becomes a central element. On the other hand, the development of the braking system and the achievement of higher braking performance are topical issues in the context of increasing driving speeds. This study proposes a practical comparative analysis of several brake systems, both constructively and in terms of performance. The study proposes to analyze dynamic braking behavior by measuring decelerations for different types of trucks under certain established conditions and routes. It also analyzes the behavior on the brake stand for each type of truck used.

Keywords: Brake system · Brake performance · Commercial vehicles braking

1 Introduction

Increasing the number of vehicles, the need to transport large masses of goods and a growing number of people are key issues on which the continuous and accelerated development of motor vehicles is based. At the same time, in order to increase the quality of life, a series of measures are needed to increase traffic safety. The development of elements in the braking system to improve braking performance or increase reliability contributes significantly to increased traffic safety.

This paper proposes a series of experimental measurements on the dynamic parameters of braking in commercial vehicles. Commercial vehicles are most commonly used on long journeys and with loads close to the maximum load capacity. These aspects and how they are used bring additional braking loads. Moreover, light commercial and distribution vehicles have to bear different urban paths in addition to the load, which are usually more demanding for the braking system.

I. Dumitru et al. (Eds.): SMAT 2019, *The 30th SIAR International Congress of Automotive and Transport Engineering*, pp. 406–413, 2020.
https://doi.org/10.1007/978-3-030-32564-0_48

2 Basics of the Study. Elements of Analysis

Most current studies address braking problems by analyzing the processes and materials used in friction couplings. A smaller number of studies focus on the study of commercial vehicles and the problems caused by heavy loads and the factors that cause brake fades are also encountered in very few works.

Improvement of braking performance can be achieved by analyzing the processes during operation. Thus, the study of real-time braking parameters can be used, on the one hand, to analyze the intensity and frequency of braking to increase the reliability of the braking system and, on the other hand, to identify different situations and conditions specific to the type of operation. As shown in study [1], the driving style combined with road conditions frequently causes braking overloads, which in turn lead to lower brake capacity or even premature removal of brake system components.

Hill et al., in [2], analyzes the driving style and based on the intensity and frequency of braking (especially careful analysis of sudden braking) they attempts to identify an algorithm for the reduction of traffic accidents. In this case, driving style has an essential role and no emphasis is placed on improving the technical properties of the vehicle. Thus, experimental analysis based on braking, braking frequency and dynamic parameter analysis can be used to create algorithms used to identify the probability of producing a crash by the driver. Also based on the analysis of the braking behavior, algorithms can be created to avoid front and rear collisions [3].

It is known that the braking system is the most important safety system on a vehicle. Also very important is the problem of adapting the braking system to autonomous vehicles, in which case no operating errors or unexpected situations are accepted. Hill et al., in [2] notes that experienced drivers have a lower brake frequency, but the study also highlights other factors that interfere with braking, such as: driver's age, route type, road surface or driving time (fatigue).

Last generation trucks feature rolling and braking assistance systems that meet a wide range of requirements but are not yet fully prepared for autonomous driving. This could be corrected by collecting and analyzing operating data under real road conditions and transforming them into algorithms so as to cover a wide range of situations that may become pernicious. In study [3], emergency braking is divided into two phases: Phase 1 - the driver identifies the obstacle and brakes normally (1–2 m/s^2 deceleration level); phase 2 - the driver brakes even more (3–4 m/s^2); while the maximum braking capacity of the vehicle studied is 10 m/s^2. For braking to be achieved with maximum efficiency, it is necessary to interfere with an electronic system, based on different algorithms, to complement the actions of the driver in the boundary situations.

Another important aspect of braking, also in the field of traffic safety, is the braking stability and the easy keeping of the direction of travel during braking. In this idea, Mirzaeinejad et al. in [4] seeks solutions for reducing the speed of the cornering, while maintaining the desired trajectory under different road conditions. In order to increase the stability of the vehicle during braking, the study finds that it is not enough to know only the braking force but it is also necessary knowledge of the force distribution, the wheel-running interaction and lateral forces.

The current paper also contains a set of experimental tests in which the accelerations on the 3 axes have been measured so that the data sets can then be processed in such a way as to obtain details about the braking behavior in a slope or ramp, moment of inertia, frequency and intensity of braking. The main idea is to compare the data obtained with the average brake number of 136.5 braking per 100 km (maxim 228 brakes and minim 76 brakes) obtained for 8 trucks on certain routes [5].

To increase the performance of braking systems, modern trucks also use auxiliary braking systems, which help reduce speeds and reduce the braking of the main braking system, but come with the main disadvantage of mass and size. It is also known that, in most cases, braking is achieved by transforming kinetic energy into heat. In the current context of energy efficiency, it is more and more common to recover energy from braking [6] and Pugi et al. in studies like [7], propose an auxiliary braking system that uses the heat taken in braking to heat the passenger compartment of a bus. Auxiliary braking systems are found on most commercial vehicles, but the heat released from braking is not reused.

3 Experimental Procedures

The analysis and processing of data to improve braking performance is done in most cases by computerized braking process simulations or recreation of friction torque conditions. Most of the times, the patterns are much too simplified or include a smaller array of elements. In order to be able to simulate the real conditions as accurately as possible so that the results are as accurate as possible, it is necessary to take over and analyze the data during the vehicle's operation. This paper proposes such an analysis of the data taken during vehicle operation or braking capacity measurements on the stand. The data obtained is stored in a database created for this purpose.

The data were taken in two ways: direct take-off from the vehicle, on an established route, using an accelerometer with a data storage application, and the second mode was concretized by analyzing the data processed after the braking capacity measurement on a roller stand. For even better analysis, the vehicles were grouped according to the mode of operation, the maximum authorized mass, the type of braking system.

Thus, it has been found that in vehicles with a maximum authorized mass of less than 3500 kg the braking system does not differ greatly from that of small cars (larger dimensions). In the category of cars with a mass less than 3500 kg, small freight transport vehicles were analyzed and in the 3500 kg category, cargo and passenger minibuses were included. The braking system used on these vehicles is with discs and brake pads hydraulically operated by a servo-brake mechanism (vacuum brake servo). In the analyzed trucks, the most widely used braking system is with brake discs and brake pads, driven by air pressure, due to the need for much higher braking forces. The proper braking is done by actuating the brake piston mechanism (by means of levers) from the caliper, which in turn presses the plate from the inside towards the disc (see Fig. 1). At the same time, the brake caliper slides on its guide pins and allows the other brake pad to be depressed from the outside towards the brake disc.

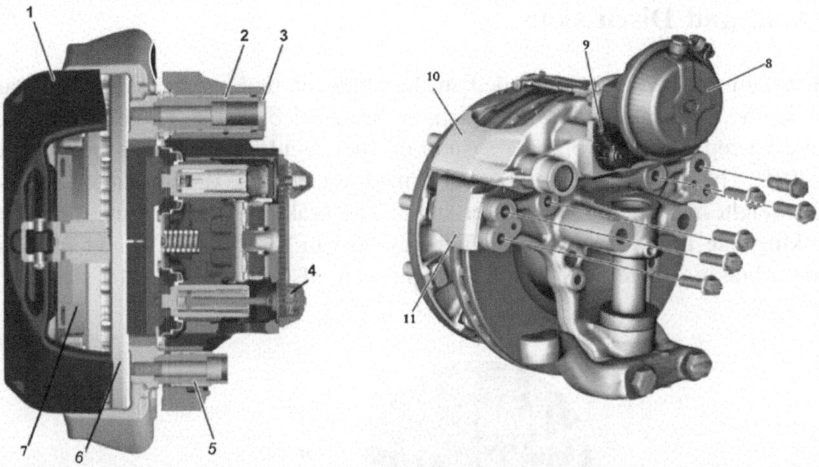

Fig. 1. Brake and disc brake assembly: 1-brake calipers; 2-pilot bushing (fixed bearing); 3-bolt cover; 4-piston actuation mechanism; 5-pilot bushing (loose bearing); 5-retaining ring; 6-brake disc; 7-brake pads; 8-brake cylinder (diaphragm brake cylinder); 9-brake wear sensor; 10-brake calipers; 11-brake carrier [8]

Although the performances of the disc brakes and the brake pads are superior, the study also analyzed trucks equipped with a drum brake system (Figs. 2 and 3). These types of swing and drum brakes are commonly used on trucks that work under construction sites or on mud or stone roads. There are no significant differences in the braking force on the stand, but the main advantage of using the drum brakes lies in the closed environment in which the friction coupler is located. This ensures efficient braking without excessive wear in site environments with mud and dust surfaces, conditions that accentuate certain types of wear, as is shown in study [1].

Fig. 2. Drive mechanism for brake with drum and drum (Z-Cam Brake System)

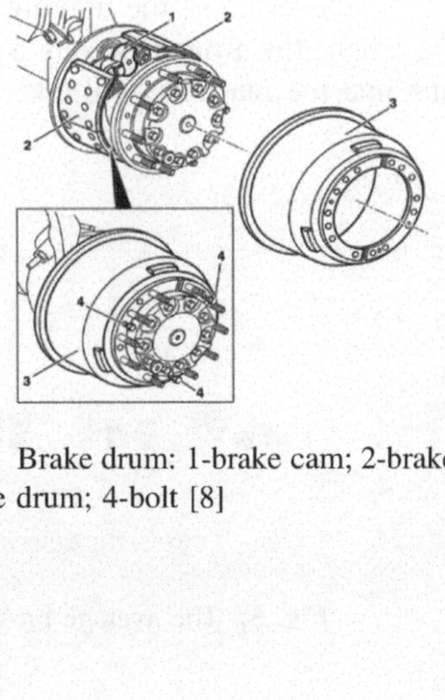

Fig. 3. Brake drum: 1-brake cam; 2-brake shoe; 3-brake drum; 4-bolt [8]

4 Result and Discussions

The distribution of vehicles included in the study of braking forces on the stand is shown in Fig. 4. Brake stand reports have been selected for 70 vehicles tested in service after repairs to the braking system or after regular technical inspection during 2017–2019. The main grouping was performed after the maximum authorized weight and the weight measured on the roller stand. The braking force was measured in kN, the braking rate in %, the braking efficiency also in %, the axle weight in tons, the maximum braking force and the total weight were also recorded on the stand.

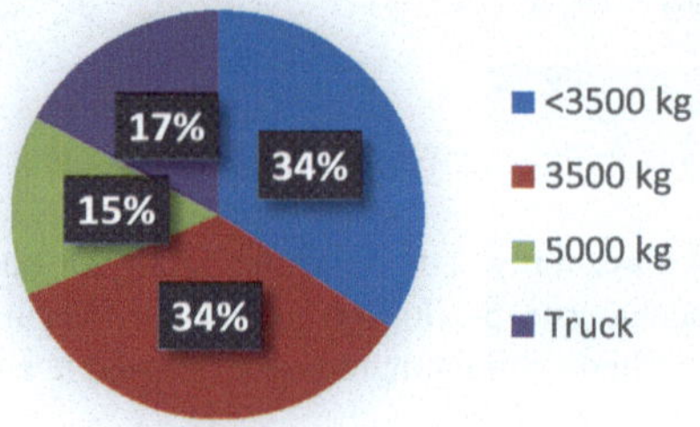

Fig. 4. Vehicle distribution.

The first observation of the measurements refers to the dependence of the braking force on the weight. Thus it was found that on the stand the measured braking force is higher for vehicles equipped with an auxiliary cooling system (a van for transporting food at lower temperatures) than for the same vehicle without the respective equipment. This is due to the extra mass of the auxiliary equipment. The same is true of passenger minibuses.

As a result of the measurements, an average brake force was obtained for each category of vehicle. It should be noted that for vehicles with a mass of less than 3500 kg due to their compact dimensions, the weight distribution on the decks is more balanced, so the mean of the braking forces measured is almost equal to the 3500 kg vehicles, where the front axle has a higher load due to the presence of the engine, following that the rear axle will take over the carried cargo (Fig. 5).

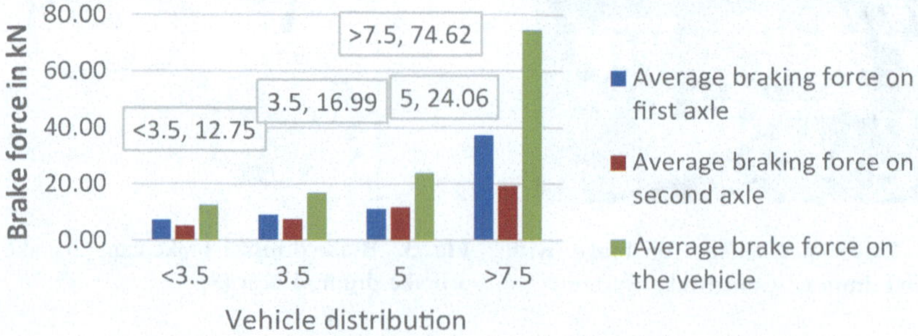

Fig. 5. The average braking force obtained for each category

When measuring the braking force on the wheels of the same axle, there is a slightly higher braking force on the right (both on the front axle and on the rear axle), this being the case where there is no greater weight on the right side. Figure 6, shows the graph relative to the left/right measurements, relative to the weight measured on each wheel.

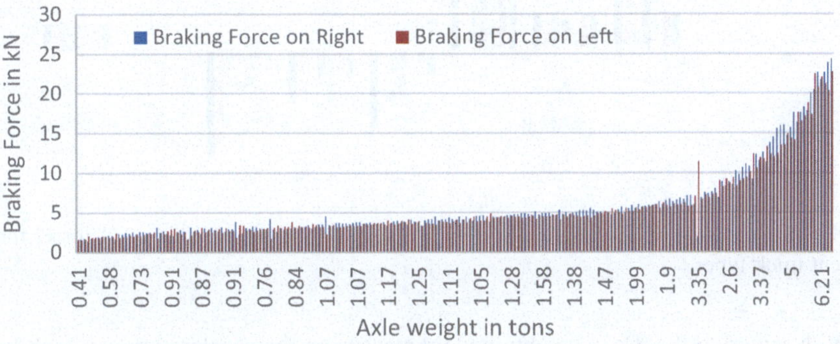

Fig. 6. Brake force dependence on axle weight

In order to check the braking, dynamic behavior, traffic influence and driving style, the second measurement mode, based on the acceleration record on the x, y and z axes, was introduced with the aid of an accelerometer and of an application. The data sets obtained by this method have the form shown in Fig. 7 and reveal aspects related to the maximum axle and deceleration on each axle or the frequency and intensity of the brakes. At the same time, these data sets can be used to describe the behavior of the vehicle in different displacement situations (climb, descent, corners, slips, etc.) or to calculate the yaw rate.

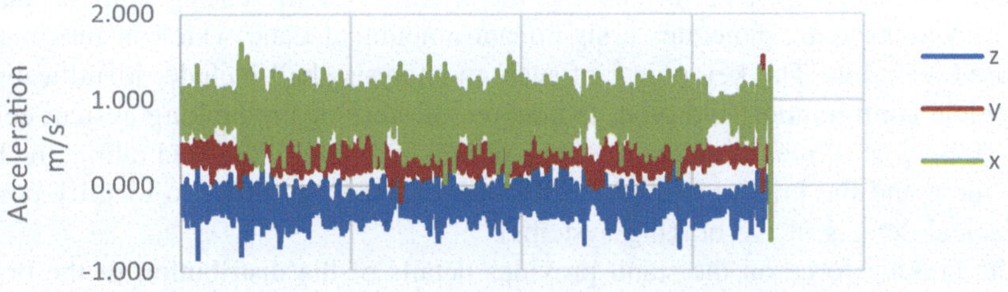

Fig. 7. Data received from acceleration sensor

Following the measurement of the three-axle accelerations for four different trucks, the data were grouped as shown in Fig. 8, where *truck 1* and *truck 2* were tractor units (Mercedes and Ford) without semi-trailer; *truck 3* and *4* are trucks of 12 tons with different configurations to the first two. *Truck 3* is a special truck with two axles (garbage truck) and *truck 4* is a 3 axle truck (Ford tipper) with Z-Cam Brake System.

The track used for the measurements was the same, having a length of 9.5 km and including both urban traffic and sloping traffic.

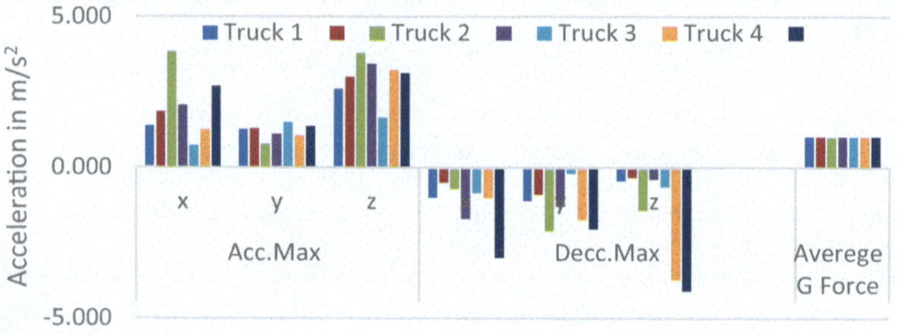

Fig. 8. Measurement of accelerations on the 3 axes and the mean value of G force, for four different truck types

With regard to the data sets obtained by measuring the accelerations on each axis, it has been found that in some cases the accelerations have lower or higher values than expected. It was later found that this was due to the inappropriate positioning of the measuring device, which led to the measurement of additional accelerations. In the current study, the measuring device was positioned on the floor of the cabin and not in the center of gravity of the truck. In the case of small cars, acceleration and deceleration errors in a straight line do not have the same deviation as in the case of trucks, where the damping movements of the truck suspension system also add the moves from the soft suspension of the cab.

5 Conclusions

The study of the dynamic performance of the braking system, starting from the analysis of the operating data, generates a significant amount of data, which in turn requires increased attention. The key points of such an analysis shall include: identification of the vehicle configuration and type, the degree of wear of the braking system components at the time of measurement, the type of braking system, the load rating, the brake pedal force and the brake pressure the braking chamber (in order to identify possible mechanical defects of the braking systems).

The braking force on the stand provides details of the distribution of the braking force on each wheel and on the wheels of the same axle, being proportional to the weight of the respective wheel or axle. In this way, the comparative analysis of different types of commercial vehicles was performed and the two analyzed braking systems analyzed: disc brakes or drums, but no significant differences were identified on the stand.

As regards the measurement of accelerations, it should be emphasized that the values obtained are significantly influenced by the position of the measuring apparatus. Also, the measurement of decelerations, in the context of increased traffic safety, must be related to the speed of travel, and the tests must be performed at certain speeds under different specified conditions. If it is desired to analyze the accelerations on a wider route, it is also necessary to consider the very large data volume, which makes it difficult to process and analyze the data.

Acknowledgement. The practical part of the paper and the data acquisition took place at RMB Inter-Auto, which is Mercedes-Benz and Ford Trucks partner.

References

1. Popa, M.F., et al.: Aspects of wear braking process on brake discs and brake pads in the construction of commercial vehicles. In: Proceedings of the Annual Session of Scientific Papers, IMT Oradea, vol. XVIII, no. XXVIII (2019)
2. Hill, A., et al.: Computer-based hazard perception test scores are associated with the frequency of heavy braking in everyday driving. In: Accident Analysis and Prevention, vol. 122 (2019)
3. Xiong, X., et al.: A forward collision avoidance algorithm based on driver braking behavior. Accid. Anal. Prev. **129**, 30–43 (2019)
4. Mirzaeinejad, H., et al.: A novel technique for optimal integration of active steering and differential braking with estimation to improve vehicle directional stability. ISA Trans. **80**, 513–527 (2018)
5. Dynafleet online homepage. https://www.dynafleetonline.com/AS. Accessed 18 June 2019
6. Pugi, L., et al.: Design of a hydraulic servo-actuation fed by a regenerative braking system. Appl. Energy **187**, 96–115 (2017)
7. Ye, L., et al.: Performance analysis and test of a novel eddy-current braking & heating system for electric bus. In: Energy Conversion and Management, vol. 183, pp. 440–449 (2019)
8. Daimler media website and Xentry portal. https://www.media.daimler.com. Accessed 07 July 2019

Considerations Concerning the Study of Dynamic Equilibrium for Engine Axle and Bogies for Locomotives

Iulian Turcanu(✉) and Ioan Sebesan(✉)

Politehnica University of Bucharest, Splaiul Independentei nr. 313, Bucharest, Romania
oldiulian@yahoo.co.uk, ioan_sebesan@yahoo.com

Abstract. The study of the conditions of transport of railway vehicles are of major importance, in the context of the constant increase of the performance (speed of movement) of locomotives. The railway and traffic conditions are important factors, which together with technical characteristics generate performance for railway vehicles. The research is a comparative analysis of the engine power transmission systems, existing in different types of locomotives. Also, motor axle drive solutions are presented and analyzed. The value of research has the effect in the developing of performance of transmission components of railway vehicles.

Keywords: Engine axle · Load symmetry · Axle attack · Torsion torque · Dynamic balance

1 Introduction

The research focuses on identifying complex demands for engine axles, at the different types of locomotives. The analysis of the different types of dynamic axle loads, determines the optimal solutions for the drive and use, in exploitation, of the dynamic drive systems [1]. Bogies are designed after the arrangement of the B'-type axles. Each bogie has two drive axles that are grouped, and the wheels are single block construction, of the ER8 type. The motion of each axle is done with attacks both simple axle and double axle. Depending on the results obtained (moments, forces), can choose the work components for locomotives.

2 Determination of the Dynamic Equilibrium of Wheel Set Axles and Engine Bogies for Locomotives

The research takes into account the different situations of transmitting power, through the attack pinions of the drive axles. The analyzes carried out comprise three types of locomotives, similar in terms of technical characteristics: Astride, Novocherkassk and Taurus.

Case 1. The "Astride SLM 36000" locomotive.
In Fig. 1 one can see the model of such a locomotive, which has the Bo-Bo type axle. The diagram of the distribution of bending moments and the resulting cutting effort is

I. Dumitru et al. (Eds.): SMAT 2019, *The 30th SIAR International Congress of Automotive and Transport Engineering*, pp. 414–425, 2020.
https://doi.org/10.1007/978-3-030-32564-0_49

shown in Fig. 2. Thus, the power P [kW] of attack is transmitted through the toothed crown ①. The brake discs are represented by ② and ③. The axle of the locomotive rotates whit speed (n).

Fig. 1. Electrical locomotive "Astride", serie 36000 SLM

$$P = Mt \cdot \omega, \tag{1.1}$$

where: w = angular speed.

But,

$$\omega = \frac{\pi \cdot n}{30} \Rightarrow P = Mt \cdot \frac{\pi \cdot n}{30} \Rightarrow Mt = \frac{30}{\pi} \cdot \frac{P}{n}, \tag{1.2}$$

$$M_{int} = M_{ext}$$

The connection equation between powers is:

$$P_1 = P_2 + P_3 \tag{1.3}$$

where: P_1 is the given yield power and $P_2 + P_3$ represents the received power.

$$\triangleleft \mathrm{Mt} \triangleright_{SI} = 1\,[KN \cdot m] \Rightarrow Mt_{2-1} = \frac{30}{\pi} \cdot \left(-\frac{P_2}{n}\right) \Rightarrow Mt_{1-3} = \frac{30}{\pi} \cdot \frac{P_3}{n} \tag{1.4}$$

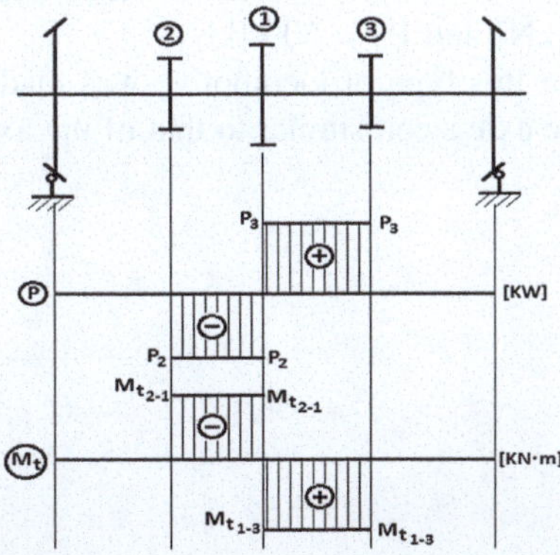

Fig. 2. The distribution diagram of moments of the "Astride" electrical locomotive axle

Determination of engine axle moments, shall be calculated with the relations:

Condition 1 for sizing:

$$W_{Pnec} = \frac{Mt_{max}}{\tau_a} \cdot \frac{\pi \cdot d^3}{16} \text{(if the circular section of the axle is full)} \tag{1.5}$$

$$\Rightarrow d = \sqrt[3]{\frac{16 \cdot Mt_{max}}{\pi \cdot \tau_a}} \tag{1.6}$$

Condition 2 for deformation:

$$\theta_{max} = \theta_a$$

where:

$$\theta_a = \left(\frac{1}{4}\right) {}^{\circ}\!/\mathrm{m} \tag{1.7}$$

$$\Rightarrow \theta_a = \frac{1}{4} \cdot \frac{\pi}{4 \cdot 180 \cdot 10^3} \, [\mathrm{rad/mm}] \Rightarrow \frac{Mt}{G \cdot Ip} = \theta_{a_{lim}} \Rightarrow Ip_{nec} = \frac{Mt}{G \cdot \theta_a} \tag{1.8}$$

but: $Ip = \frac{\pi \cdot d^4}{32} \Rightarrow \frac{\pi \cdot d^4}{32} = \frac{Mt_{max}}{G \cdot \theta_a}$

$$\Rightarrow d = \sqrt[4]{\frac{32 \cdot Mt_{max}}{\pi \cdot G \cdot \theta_a}} \tag{1.9}$$

From the two conditions required for sizing, its choose the diameter (d) with the highest value.

Case 2. Locomotive "Novocherkassk" (Fig. 3).
In Fig. 4 we notice the sizing of the attack shaft, of the Novocherkassk EP 1 electric locomotive, with the axle formula Bo-Bo-Bo. This has three attacking pinions, which induce the effort explained in the diagram (Fig. 4). For this type of locomotive, the installed forces, which interact with the level of mounted axle, of the locomotive, are: P1 = 80 [kN], P2 = 300 [kN] and P3 = 500 [kN].

The axle dimension for this type of locomotive was made taking into account the value of the tensions in the axle steel, similar to that of the axles of the previous Swiss locomotive.

Fig. 3. Novocherkassk EP 1 locomotive

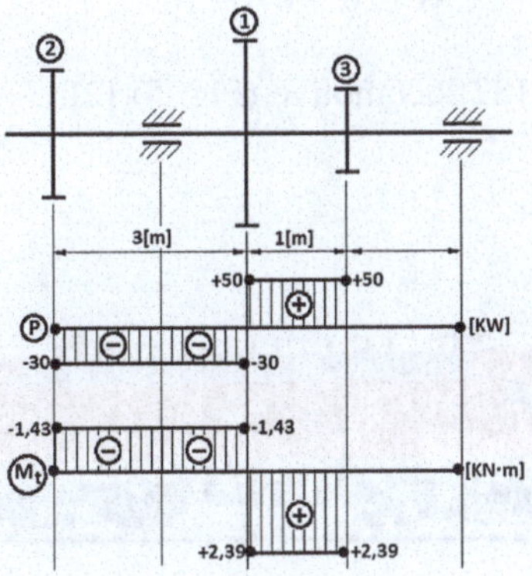

Fig. 4. The simplified scheme of the locomotive attack tree, type "Novocherkassk" EP1

So: $\tau_a = 120\,[\text{MPa}]$, $\theta_a = \frac{1}{4}\,\left[{}^{\circ}\!/\!_{\text{m}}\right]$ and $G_{steel} = 81 \cdot 10^3\,[\text{MPa}]$

$$Mt_{2-1} = \frac{30}{\pi} \cdot \left(-\frac{30}{200}\right) \Rightarrow Mt_{2-1} = -1.43, \tag{1.10}$$

$$Mt_{1-3} = \frac{30}{\pi} \cdot \frac{50}{200} \Rightarrow Mt_{1-3} = +2.39 \tag{1.11}$$

Condition 1:

$$W_{Pnec} = \frac{Mt_{max}}{\tau_a} \Rightarrow W_{Pnec} = \frac{2.39 \cdot 10^6}{120} \Rightarrow W_{Pnec} = 12.92 \cdot 10^3\,\left[\text{mm}^3\right], \tag{1.12}$$

$$\frac{\pi \cdot d^3}{16} = 19.93 \cdot 10^3 \Rightarrow d = 46.6\,[\text{mm}]. \tag{1.13}$$

Condition 2

$$Ip_{nec} = \frac{Mt_{max}}{G \cdot \theta_a} \Rightarrow Ip_{nec} = \frac{2.39 \cdot 10^6 \cdot 4 \cdot 180 \cdot 10^3}{81 \cdot 10^3 \cdot \pi} \Rightarrow Ip_{nec} = 6.67 \cdot 10^4 \left[\text{mm}^3\right], \tag{1.14}$$

$$\text{If: } Ip = \frac{\pi \cdot d^4}{32} \Rightarrow d = 10 \cdot \sqrt[4]{\frac{32.676}{\pi}} \Rightarrow d = 91\,[\text{mm}]. \tag{1.15}$$

Always its choose the highest value of the diameter, and the spin from the point 1, will be:

$$\varphi = \sum\nolimits_{i=1}^{n} \frac{Mt_i \cdot l_i}{G_i \cdot Ip_i} \Rightarrow \varphi_{3-1} = \frac{10^6 \cdot 10^3}{81 \cdot 10^3 \cdot 676 \cdot 10^4} \Rightarrow \varphi_{3-1} = -0.00347\,[\text{rad}]. \tag{1.16}$$

Case 3. The “Taurus” BR 182 locomotive (Fig. 5) [2].

Fig. 5. Taurus BR 182 locomotive

The traction system of the “Taurus” electric locomotive class BR182, is loaded by the drive bar, which is supported in connection nodes A and B (Fig. 6). On the drive bar are fixed the three toothed crowns, for which are known: the moment of torsion (Mt), the moment bending (Mi), rotation [rpm], gearing power $P_1 = 90$ [kW], $P_2 = 60$ [kW] and $P_3 = 30$ [kW], respectively the tension $\sigma_a = 60$ [MPa], lengths $l_1 = 1$ [m], $l_2 = 1$ [m], $l_3 = 3$ [m] and $l_4 = 2$ [m] and reactions $V_1 = V_2 = V_3 = 0{,}5$ [kN].

The identification of the dangerous section in the drive bar of one of the locomotive motor axles can be done after the dimensioning calculation. To do this, the effort diagrams are drawn and determine the optimum diameter of the drive bar.

$$\sigma_{ech1} = \frac{\sigma}{2} + \frac{1}{2}\sqrt{\sigma^2 + 4\varsigma^2}, \tag{1.17}$$

$$\sigma_{ech2} = \frac{1-\nu}{2} \cdot \sigma + \frac{1-\nu}{2} \cdot \sqrt{\sigma^2 + 4\varsigma^2}, \tag{1.18}$$

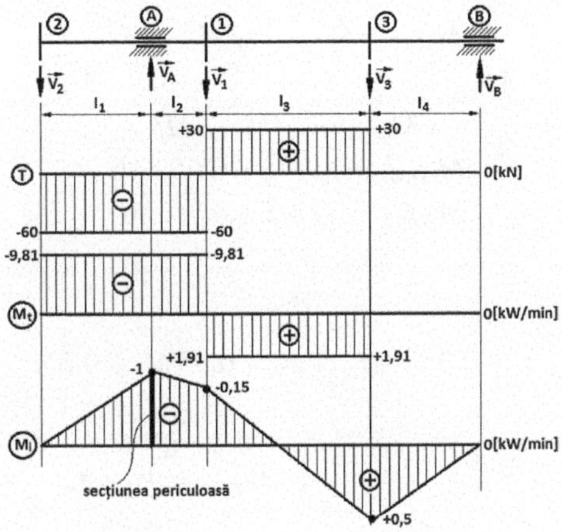

Fig. 6. Diagram of cutting forces and a bending moments, for the drive axle

$$\sigma_{ech3} = \sqrt{\sigma^2 + 4\varsigma^2}, \quad (1.19)$$

$$\sigma_{ech4a} = \sqrt{\sigma^2 + 2(1+v)\cdot\varsigma^2}, \quad (1.20)$$

$$\sigma_{ech4b} = \sqrt{\sigma^2 + 3\varsigma^2}. \quad (1.21)$$

In the case of twisting bend, the relationships will be:

$$\begin{cases} \sigma_x = \frac{M_y}{W_y} \\ \sigma_y = \frac{M_z}{W_z} \\ \varsigma_{xz} = \frac{M_x}{W_p} \end{cases} \quad (1.22)$$

In the case of circular or annular sections, sizing is done in a similar manner to the bending request and is considered an equivalent moment (M_{ech}), which will replace in the calculations, the moments of bending:

$$M_y \text{ and } M_z \Rightarrow W_y = \frac{M_{ech}}{\sigma_a} \quad (1.23)$$

For circular or ring sections, the relationships will be:

$$\varsigma_{xz} = \frac{M_z}{2W_y} = \frac{M_t}{W_p} \Rightarrow \frac{M_{ech}}{W_y} = \frac{M_i}{2W_y} + \frac{1}{2}\sqrt{\frac{M_i^2}{W_y^2} + y\cdot\frac{M_t^2}{4W_y^2}} \Rightarrow \quad (1.24)$$

$$\begin{aligned} M_{ech1} &= \tfrac{M_i}{2} + \tfrac{1}{2}\sqrt{M_i^2 + M_t^2} \\ M_{ech2} &= \tfrac{1-\nu}{2} \cdot M_i + \tfrac{1+\nu}{2}\sqrt{M_i^2 + M_t^2} \\ M_{ech3} &= \sqrt{M_i^2 + M_t^2} \\ M_{ech4a} &= \sqrt{M_i^2 + 0.65 \cdot M_t^2} \\ M_{ech4b} &= \sqrt{M_i^2 + 0.75 \cdot M_t^2} \end{aligned} \tag{1.25}$$

If:

$$\nu = 0.3 \Rightarrow M_{ech2} = +0.35\,M_i + 0.65, \tag{1.26}$$

$$\sqrt{M_i^2 + M_t^2} \Rightarrow \begin{cases} \sigma_{ech} = \frac{M_{ech}}{W_y} \leq \sigma_a \\ M_{ech_{cap}} = W_y \cdot \sigma_a \end{cases}, \tag{1.27}$$

where:

$$\sigma_{ech} = \frac{M_{ech}}{W_y} \leq \sigma_a \tag{1.28}$$

represents the condition of strength to the breaking

$$M_t = \frac{30}{\pi} \cdot \frac{P}{n} \Rightarrow M_{t_{2-1}} = \frac{30}{\pi} \cdot \left(-\frac{60}{50}\right) \Rightarrow M_{t_{2-1}} = -3.52\,[\mathrm{kNm}], \tag{1.29}$$

$$M_{t_{2-3}} = \frac{30}{\pi} \cdot \frac{30}{150} \Rightarrow M_{t_{2-3}} = +1.91\,[\mathrm{kNm}], \tag{1.30}$$

$$V_A = \frac{0.5(8+5+2)}{6} \Rightarrow V_A = 1.25\,[\mathrm{kN}], \tag{1.31}$$

$$V_B = \frac{0.5(4+1-2)}{6} \Rightarrow V_B = 0.25\,[\mathrm{kN}], \tag{1.32}$$

$$M_{ech} = \sqrt{1 + 0.75 \cdot (3.82)^2} \Rightarrow M_{ech} = +3.456\,[\mathrm{kNm}], \tag{1.33}$$

$$W_{y_{nec}} = \frac{3.456 \cdot 10^6}{60} \Rightarrow W_{y_{nec}} = 57.6 \cdot 10^3\,\left[\mathrm{mm}^3\right]. \tag{1.34}$$

If the circular section of the drive shaft is full, then:

$$W_{y_{nec}} = \frac{\pi \cdot d^3}{32} \Rightarrow d = \sqrt[3]{\frac{32}{\pi} \cdot 57.6 \cdot 10^3} \Rightarrow d = 179\,[\mathrm{mm}]. \tag{1.35}$$

3 Determination Regarding the Actuation of the Drive Shaft, in the Group

For development of the different solutions for power transmission to the engine axles are required dynamic analyzes. In the research, the axial and radial forces acting on the locomotive drive axle pinion are analyzed. In case of operating the drive axle in group by attack whit vertical reaction arm, (Fig. 7), the equilibrium shaft equations are performed, of the gearbox housing and the traction motors of the electric motor. Figure 7 shows the drive scheme of the drive axle with a vertical reaction arm.

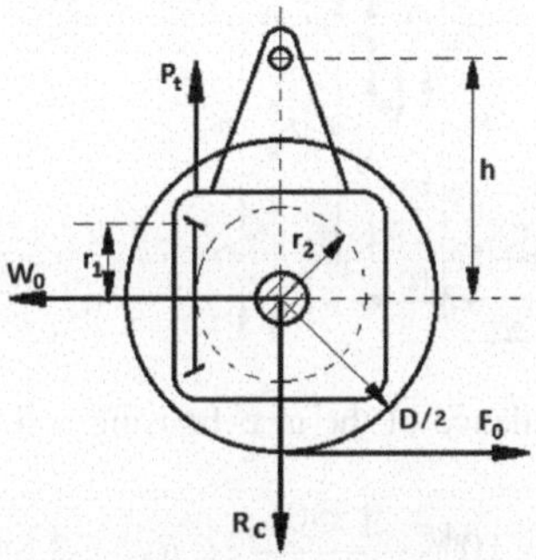

Fig. 7. Axle action in group by attack with vertical reaction arm

The reaction is calculated by:

$$P_t \cdot r_2 = F_0 \cdot \frac{D}{2} \Rightarrow P = F_0 \cdot \frac{D}{2_t \cdot r_2} \Rightarrow F_0 = 8 \cdot 10^4 \cdot \frac{1250}{2 \cdot 350} \Rightarrow F_0 = 14.28\ [\mathrm{kN}], \tag{2.1}$$

$$R_c = P_t \Rightarrow R_c = F_0 \cdot \frac{D}{2 \cdot r_2} \Rightarrow R_c = 14.28 \cdot 10^4\ [\mathrm{N}], \tag{2.2}$$

$$F_0 = W_0.$$

where:

F_0 = represents the traction force at the locomotive wheel;
W_0 = represents the resistance to advancing of the locomotive.

Determination of the equilibrium of the gear housing and of the electric traction motor (Fig. 8) shall be calculated as follows:

$$R_A \cdot h = P_t \cdot r_2 \Rightarrow R_A = P_t \cdot \frac{r_2}{h}, \tag{2.3}$$

But: $\mathrm{P_t} = \mathrm{F_0} \cdot \frac{\mathrm{D}}{2 \cdot \mathrm{r_2}} \Rightarrow \mathrm{R_A} = \mathrm{F_0} \cdot \frac{\mathrm{D}}{2 \cdot \mathrm{r_2}} \cdot \frac{\mathrm{r_2}}{\mathrm{h}} \Rightarrow \mathrm{R_A} = \mathrm{F_0} \cdot \frac{\mathrm{D}}{2 \cdot \mathrm{h}}$

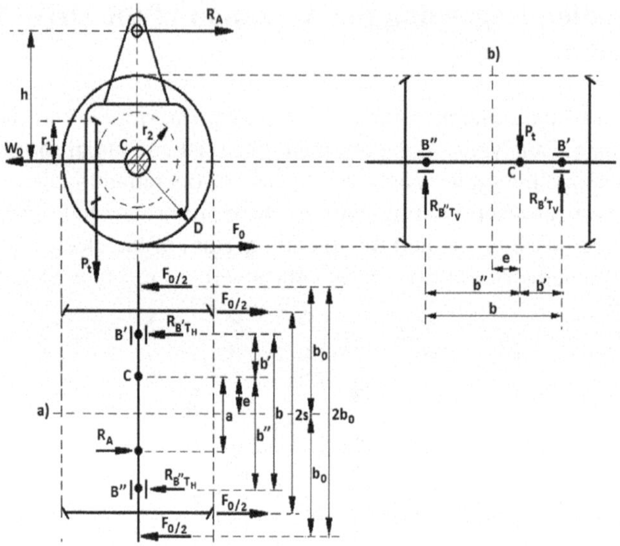

Fig. 8. Determining the balance of the gear housing and the electric traction motor

$$\Rightarrow \mathrm{R_A} = 8 \cdot 10^4 \cdot \frac{1250}{2 \cdot 1300} \Rightarrow R_A = 3.84 \cdot 10^4\,[\mathrm{N}]. \tag{2.4}$$

Determinations will be made for two situations: in horizontal plane (a) and vertical plane (b).

(a) Horizontal:

$$\left(\sum M_{B''} = 0\right) \Rightarrow R_{B'T_H} \cdot b - R_A \cdot (b'' - a) = 0 \Rightarrow R_{B'T_H} = R_A \cdot \frac{b'' - a}{b} \Rightarrow$$
$$R_{B'T_H} = 3.84 \cdot 10^4 \cdot \frac{600 - 300}{1000} \Rightarrow R_{B'T_H} = 1.15 \cdot 10^4\,[\mathrm{N}], \tag{2.5}$$

$$\left(\sum M_{B'} = 0\right) \Rightarrow R_{B''T_H} \cdot b - R_A \cdot (b' + a) = 0 \Rightarrow R_{B''T_H} = R_A \cdot \frac{b' + a}{b} \Rightarrow$$
$$R_{B''T_H} = 3.84 \cdot 10^4 \cdot \frac{400 + 300}{1000} \Rightarrow R_{B''T_H} = 2.68 \cdot 10^4\,[\mathrm{N}]. \tag{2.6}$$

(b) Vertically:

$$\left(\sum M_{B''} = 0\right) \Rightarrow R_{B'T_V} \cdot b - P_t \cdot b'' = 0 \Rightarrow R_{B'T_V} = P_t \cdot \frac{b''}{b}$$
$$\Rightarrow R_{B'T_V} = 14.28 \cdot 10^4 \cdot \frac{600}{1000} \Rightarrow R_{B'T_V} = 8.56 \cdot 10^4\,[\mathrm{N}], \tag{2.7}$$

$$\left(\sum M_{B'} = 0\right) \Rightarrow R_{B''T_V} \cdot b - P_t \cdot b' = 0 \Rightarrow R_{B''T_V} = P_t \cdot \frac{b'}{b} \Rightarrow$$
$$R_{B''T_V} = 14.28 \cdot 10^4 \cdot \frac{400}{1000} \Rightarrow R_{B''T_V} = 5.71 \cdot 10^4\ [\mathrm{N}]. \quad (2.8)$$

The horizontal forces due to the radial and axial forces from the locomotive drive axle sprocket are introduced by the axle attack (Fig. 9).

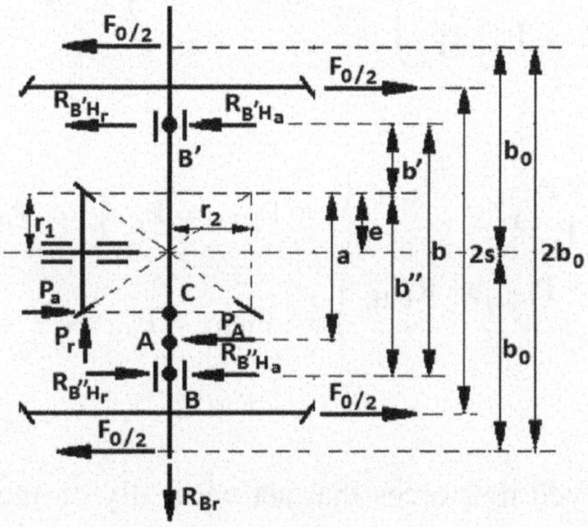

Fig. 9. Horizontal forces due to the radial (P_r) and axial (P_a) forces from the attack sprocket of the locomotive drive axle

R_{Br} = the force that acting to the axle of the axle.

$$\left.\begin{aligned}&\begin{cases} P_r = \lambda P_t \\ P_a = \nu P_t \end{cases}\\ &\text{But: } P_t = F_0 \cdot \tfrac{D}{2r_2}\end{aligned}\right\} \Rightarrow \begin{cases} P_r = \lambda F_0 \cdot \frac{D}{2r_2} \\ P_a = \nu F_0 \cdot \frac{D}{2r_2} \end{cases} \Rightarrow \begin{cases} P_r = 0.4 \cdot 14.28 \cdot 10^4 \\ P_a = 0.3 \cdot 14.28 \cdot 10^4 \end{cases}$$
$$\Rightarrow \begin{cases} P_r = 5.71 \cdot 10^4 \\ P_a = 4.28 \cdot 10^4 \end{cases}, \quad (2.9)$$

$$\left(\sum M_{B''} = 0\right) \Rightarrow \begin{cases} R_{B'H_r} \cdot b - P_r \cdot r_2 = 0 \\ R_{B'H_a} \cdot b - P_a \cdot r_1 = 0 \end{cases} \Rightarrow \begin{cases} R_{B'H_r} = P_r \cdot \frac{r_2}{b} \\ R_{B'H_a} = P_a \cdot \frac{r_1}{b} \end{cases}$$
$$\Rightarrow \begin{cases} R_{B'H_r} = 5.71 \cdot 10^4 \cdot \frac{350}{1000} \\ R_{B'H_a} = 4.28 \cdot 10^4 \cdot \frac{600-150}{1000} \end{cases} \Rightarrow \begin{cases} R_{B'H_r} = 1.99 \cdot 10^4\ [\mathrm{N}] \\ R_{B'H_a} = 1.92 \cdot 10^4\ [\mathrm{N}] \end{cases}. \quad (2.10)$$

$$\left(\sum M_{B'} = 0\right) \Rightarrow \begin{cases} R_{B''H_r} \cdot b - P_r \cdot r_2 = 0 \\ R_{B''H_a} \cdot b - P_a \cdot (b' + r_1) = 0 \end{cases} \Rightarrow \begin{cases} R_{B''H_r} = P_r \cdot \frac{r_2}{b} \\ R_{B''H_a} = P_a \cdot \frac{b' + r_1}{b} \end{cases} \tag{2.11}$$

$$\Rightarrow \begin{cases} R_{B''H_r} = 5.71 \cdot 10^4 \cdot \frac{350}{1000} \\ R_{B''H_a} = 4.28 \cdot 10^4 \cdot \frac{400+150}{1000} \end{cases} \Rightarrow \begin{cases} R_{B''H_r} = 1.99 \cdot 10^4\,[N] \\ R_{B''H_a} = 2.35 \cdot 10^4\,[N] \end{cases}.$$

$$\left.\begin{array}{l} \left.\begin{array}{l} R_{B_r} = P_r \\ P_r = \lambda P_t \end{array}\right\} \Rightarrow R_{B_r} = \lambda P_t \\ P_t \cdot r_2 = F_0 \cdot \frac{D}{2} \Rightarrow P_t = F_0 \cdot \frac{D}{2 \cdot r_2} \end{array}\right\} \Rightarrow R_{B_r} = \lambda F_0 \cdot \frac{D}{2 \cdot r_2} \Rightarrow R_{B_r} = 5.71 \cdot 10^4\,[N]. \tag{2.12}$$

$$\begin{cases} \left.\begin{array}{l} P_{B'H_r} = -(-R_{B'H_r}) \Rightarrow P_{B'H_r} = R_{B'H_r} \\ \text{But: } P_{B'H_r} = P_r \cdot \frac{r_2}{b} \end{array}\right\} \Rightarrow P_{B'H_r} = P_r \cdot \frac{r_2}{b} \Rightarrow P_{B'H_r} = 1.99 \cdot 10^4\,[N] \\ \left.\begin{array}{l} P_{B'H_a} = -(-R_{B'H_a}) \Rightarrow P_{B'H_a} = R_{B'H_a} \\ \text{But: } P_{B'H_a} = P_a \cdot \frac{b'' - r_1}{b} \end{array}\right\} \Rightarrow P_{B'H_a} = P_a \cdot \frac{b'' - r_1}{b} \Rightarrow P_{B'H_a} = 1.92 \cdot 10^4\,[N] \end{cases} \tag{2.13}$$

In Fig. 10 are observed the forces that act vertically on the drive axle.

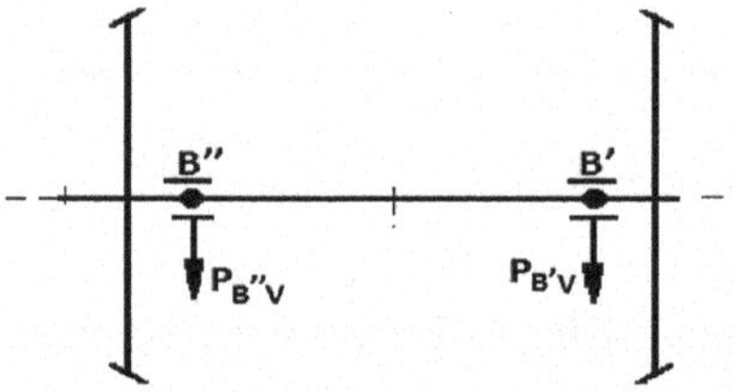

Fig. 10. Horizontal forces due to the radial and axial forces from the drive axle sprocket of locomotive

$$\left.\begin{array}{l} P_{Br} = -R_{Br} \\ \text{But: } R_{Br} = P_r \end{array}\right\} \Rightarrow \left.\begin{array}{l} P_{Br} = -P_r \\ \text{But: } P_r = \lambda P_t \end{array}\right\} \Rightarrow \left.\begin{array}{l} P_{Br} = -\lambda P_t \\ \text{But: } P_t = F_0 \cdot \frac{D}{2r_2} \end{array}\right\} \Rightarrow P_{Br} = -\lambda F_0 \cdot \frac{D}{2r_2}$$
$$\Rightarrow P_{Br} = 5.71 \cdot 10^4\,[N] \tag{2.14}$$

$$\begin{cases} P_{B'H} = P_{B'H_r} + P_{B'H_a} \\ P_{B''H} = -P_{B''H_r} + P_{B''H_a} \end{cases} \Rightarrow \begin{cases} P_{B'H} = P_r \cdot \frac{r_2}{b} + P_a \cdot \frac{b'' - r_1}{b} \\ P_{B''H} = -P_r \cdot \frac{r_2}{b} + P_a \cdot \frac{b' + r_1}{b} \end{cases} \Rightarrow$$

$$\begin{cases} P_{B'H} = \frac{P_t}{b}[\lambda r_2 - (b'' - r_1)] \\ P_{B''H} = \frac{P_t}{b}[-\lambda r_2 - (b' + r_1)] \end{cases} : \begin{cases} P_{B'H} = (1.99 + 1.92) \cdot 10^4 \\ P_{B''H} = (-1.99 + 2.35) \cdot 10^4 \end{cases} \Rightarrow \begin{cases} P_{B'H} = 3.91 \cdot 10^4 \\ P_{B''H} = 0.36 \cdot 10^4 \end{cases}. \tag{2.15}$$

$$\begin{cases} \left.\begin{matrix} P_{B'V} = R_{B'V} \\ \text{But: } R_{B'V} = P_t \cdot \frac{b''}{b} \end{matrix}\right\} \Rightarrow P_{B'V} = P_t \cdot \frac{b''}{b} \Rightarrow P_{B'V} = 8.56 \cdot 10^4\ [\mathrm{N}] \\ \left.\begin{matrix} P_{B''V} = R_{B''V} \\ \text{But: } R_{B''V} = P_t \cdot \frac{b'}{b} \end{matrix}\right\} \Rightarrow P_{B''V} = P_t \cdot \frac{b'}{b} \Rightarrow P_{B''V} = 5.71 \cdot 10^4\ [\mathrm{N}] \end{cases} \tag{2.16}$$

4 Conclusions

The kinematic analysis of the engine axle is an important part in the study of the dynamics of rail vehicles. Determining precisely the forces and working moments existing at the pinion of the drive axles are important factors for observing the performance of locomotives. During the movement, the friction, centering and suspension forces on each axle of the rail vehicle must be in dynamic balance. Also, the forces in the two floors of the vehicle suspension, acting on the bogie chassis, must be in balance with the force applied to the locomotive pivot. The balancing calculation of locomotive engine axles can be done with precision. An ideal example is that of the bogie model with infinite longitudinal flexibility.

References

1. Badea, C.N.: Driving stability and safety guiding on railway for freight wagons. Ph.D., thesis, University ‘Politehnica’ of Bucharest, Faculty of Transportation, Bucharest, Romania, 17th of July 2015
2. Ioan, S.: About the synthesis of the rolling surfaces profiles of the railway vehicles wheels. In: International Conference on Railway Engineering-Challenges for Railway Transportation (2008)
3. Sebesan, I., Mazilu, T.: The vibration of railway vehicles, matrix Romania, Bucureşti (2010)

Automotive Black Box and Development Platform Used for Traffic Risks Evaluation and Mitigation

Silviu Zamfir[1(✉)] and Radu Drosescu[2(✉)]

[1] Allianz Tiriac, Branch Iasi, Iasi, Romania
Silviu.Zamfir@allianztiriac.ro
[2] Mechanical Faculty, Technical University of Iasi, Iasi, Romania
radu.drosescu@tuiasi.ro

Abstract. This paper presents a "Black-Box" device designed and accomplished by authors to monitor, acquire and record some functional and dynamic parameters for motor vehicles in order to detect and report traffic risk situations, violation of legislation as well as the possibility of reconstructing road accidents or special traffic events. The materialization of this equipment resulted from the need to know and to access useful data and information which can help forensic investigators in the case of an accident, insurers in the event of damage, specialists in road safety and traffic monitoring, to medical staff dealing with victims or medical emergencies resulting from road accidents, or even drivers looking to monitor the evolution of an event in which they were directly or indirectly involved as witnesses.

All these data can also be saved to an internal eMMC flash memory or to an external micro SDXC card making from this device a data logger. Stored information in black-box memory can be encrypted to ensure the data confidentiality and protect the individual property rights. Stored data can be accessed via a CAN interface using a key and a password. The Black-Box device can be simple integrated in vehicle's electronic control architecture, using provided CAN interface.

Keywords: Black-Box · OBD · CAN

1 Motivation and Introduction

Each airbag control module (ACM) has a small memory that continuously records certain vehicle dynamic data (typically change in velocity, frontal and lateral accelerations, seat belt wearing state, or pedals angle) for a limited duration, typically 5 s. In the event of a crash resulting in triggering SRS, ACM can permanently record this data for the purpose of understanding and checking whether the deployment of restrain devices was correct or not. Data recorded in ACM EEPROM memory are not the same and changes depending upon the year, make, or vehicle model. As the stored information has diversified and their number has increasing, the name of ACM, Airbag ECU or RCM (Restrain Control Module) has changed to EDR (Event Data Recorder).

I. Dumitru et al. (Eds.): SMAT 2019, *The 30th SIAR International Congress of Automotive and Transport Engineering*, pp. 426–438, 2020.
https://doi.org/10.1007/978-3-030-32564-0_50

Although there is no standard to regulate data stored in EDR devices, or the format in which they are stored, EDR data has been accepted in criminal prosecutions and civil litigation by courts of law.

Even though the storage of these data had a whole different purpose, more and more forensic and investigators use this information in accident reconstruction, so that the name of these devices has erroneously changed from EDR in Black-Boxe's, probably in analogy with the aviation field.

In the case of cars, EDR devices have the primary goal to detect the crash and trigger the restrain systems and as an auxiliary function, the recording of control data. On the contrary, airplanes Black-Box has as primary purpose the full recording of flight parameters for the reconstruction of an airplane crash. The second major difference concerns the mode and amount of data stored by the two devices. In the black box of an airplane, the data is continuously memorized as the flight progresses, while in vehicle case there is a limited memory area in which new data continuously replace the old ones with the time evolution. Upon detecting a collision strong enough to trigger the SRS passive protection devices, the last recorded 5 s. of data will be permanently stored and can only be read.

Usually, EDR are not designed for the purpose of assisting judicial experts in accident reconstruction expertise. Experts should consider some possible limitations.

The presented Black-Box device is not designed to capture data and information from a collision which are useful in airbags deployment and seat belt pretensioners, but less useful in accidents reconstruction. This device is intended to be used in the reconstruction of a traffic event due to the storage of most used information such as vehicle speed, accelerations, yaw and roll angular speed, brake and acceleration pedal angles, odometer value, current time, GPS coordinates of the route traveled, video images from inside and outside of the cabin. To reach these objectives, a series of sensors not found as standard equipment like inside/outside cameras, 9-DOF Inertial Measuring Unit (IMU), GSM/GPRS, GPS and LIDAR were integrated in a sensor fusion manner.

A more complex version of the Black-Box was built around a new released Raspberry PI 4 platform. Upgraded version of the device was supplemented with TI mm Wave radar sensors and an accelerator pedal force transducer. The storage space and speed were also enhanced with a solid-state drive (SSD) connected with the new USB 3.1 available interface. Consequently, black-box system can be provided with new features to predict the driver behavior and vital sign, to detect the occupant's presence, traffic monitoring, detecting fail safe rear-end inter-vehicle distance for collision avoidance, or make automated emergency calling (e-Call).

2 Black-Box Design

A Black-Box performant design for vehicle industry must solve some specific technical challenges both related to hardware implementation as well as to the software program. Specifically, the device must reach computational power and speeds at least similar with the latest smartphones, extended possibilities of remote high speed communication, extended space for data storage, interfacing possibilities with a varied range of sensors for the measurement of distance, pedal travel, automobile speed or for recording sounds and images in a Sensor Fusion conception. From the point of view of automobile's network integration and to achieve diagnostic capabilities, the Black-Box device must have one or two CAN interfaces, access to the OBD network but also to operate without issues in a hostile environment specific to automobiles (temperature $-10° \div +50$ °C, vibrations, shocks, vapors, dust). The dimensions and clearances as reduced as possible are also very important as well as the possibility of supply from the automobile electric network, a simple and fast installation, an optimal location being the passenger cabin below the board. Connection to the automobile OBD socket represents the optimal solution to satisfy most of the above-mentioned requirements.

The access to the OBD socket offers advantages such as additional functions of the Black-Box device with the capability of local and remote diagnostics via the transmissions of some OBD specific data to the central server. Another major advantage is the simultaneous access to the supply network, OBD diagnostics network and CAN communication network of the automobile by using the same connector.

Since the start of designing, the operating system Ubuntu (Linux) was chosen due to the extended execution possibilities in real time of the basic program, because of software drivers availability and so many "free of charge" or "open source" programs and examples for serial communication, WI-FI, Bluetooth, GSM, GPS or GNSS (e.g. GitHub portal). Based on the same operating system a series of graphical and processing libraries operate natively for images processing such as OpenCV (Open source Computer Vision Library), OpenCPN (open Chart Plotter Navigation) or QT5 (a library of widget type control graphical elements – e.g., odometer, tachometer, speedometer, board clocks etc. – used in designing a graphical interface for multi-platform applications).

From the point of view of the hardware design, the development solution was chosen on existing embedded SoC type platform, in order to significantly reduce the costs and execution times. In this context, we may select a SoC based system (System on Chip) of ARM type or a FPGA (Field-Programmable Gate Array) structured platform. The first category has the most extended range of available devices at the lowest prices and does not need special integrated programming and developing interfaces and media which are quite expensive.

The decision to carry out the design on an 32 bit or 64/32 bit ARM controller, respectively on a single-core or multi-core SoC took into account the necessary calculation speed, the handling, processing and transfer requirements of a highest number as possible of data, therefore the access requirements to high capacity fast memories but also the future development possibilities, incorporation of new functions without hardware modifications or the extension possibilities of the number of sensors or devices that could be coupled in the future (e.g., mm Wave radars, thermographic camera, 3D lidar), their specific interfaces and drivers. We also considered that most controllers not addressing to the Windows system are designed from one or more ARM cores on 32 or 64-bit, using the Linux/Python or C developing and programming medium, respectively Ubuntu as an operating system.

Initially, the design and programs and also the first Black-Box prototype were made on a Raspberry Pi 3 B+ board, but finally, in the perspective of product technological upgrading as well as the new functions to be added, and looking to the latest released devices on the marked, three platforms were taken into account:

- **Raspberry Pi 4 B** [1] which, depending on the LPDDR4 SDRAM memory, costs 35$ (1 GB), 45$ (2 GB), respectively 55$(4 GB), redesigned based on a SoC Broadcom 2711 Quad-core 64 biţi Cotrex A72 and VideoCore VI GPU graphical nucleus which supports OpenGL ES 3.0. Compared with the previous Raspberry Pi 3 B+ version, it was found out that the performances were triple to ten as better especially for LAN, USB and WI-FI communications. The platform is equipped with 2 USB 3.0 interfaces, 2 USB 2.0, 2 microHDMI 4k graphical ports (dual display), MIPI CSI video camera interface, Bluetooth 5.0, Ethernet Gigabit 2.4/5 GHz, WiFI5 (802.11b/g/n/ac). The 40 GPIO inputs/outputs remain unmodified on 3.3 voltage level. 5 V, and max. 4 A power supply when both 4k graphical interfaces are active. As a weak feature, the increased heat dissipation mainly in the case of 4k graphics is not supported by a heater and/or integrated fan and thus an adequate cooling is the responsibility of the buyer;
- **Odroid N2 Platform** [2] with 4 GB DDR4RAM-1320 MHz that can be acquired at 79$ designed on a SoC Amlogic S922X (ARM Cortex 73 4-cores operating at 1.8 GHz, ARM Cortex A53 dual cores and a GPU Mali-G52 integrate graphical processor). Available interfaces include 4 USB 3.0, 1 HDMI 2.0, Ethernet RJ45 10/100/1000 Mbps, port for eMMC memory, microSDX card slot, Port GPIO (3,3 V)- 40 pins. Power supply 12 V, 1.5A. GPIO 40 pin header with signals of 3.3 V voltage except the analog inputs on the ADC converter which have 1.8 V. As a remark: the price may increase by procuring an eMMC program memory (with a transfer speed higher than the SDX card) for 24€ (16 GB), 36€ (32 GB), 60€ (64 GB) and 84€ (128 GB);
- **Nvidia Jetson Nano Platform** [3] 4 GB LPDDR4, SoC Tegra X1 (Model T210) 1.43 GHz have one of the most developed and powerful range of image processing tools, image recognition, object detection, "self-learning" based on TENSOR FLOW neuronal networks, robot control software programs for robots, autonomous vehicles or drones control, the best known being ROS (Robot Operating System) [4] JetPack and ISAAC SDK. GPIO header connector includes 40 multifunctional signals (3-UART, 2-SPI, 4-I2C, PWM) all at 3.3 V level. The purchase price from the supplier is 99 USD.

Although the simplest solution would be the hardware upgrading from Raspberry 3 to Raspberry 4, all three models were procured in order to make an exhaustive evaluation of the performances and behavior of each platform. The recent launch data showed in the Fig. 1, the waiting time from placing the orders to the delivery but also the large list of buyers did not allow these tests to be performed simultaneously in the resting time, however, their procurement was carried out and the benchmark tests will be achieved in the immediate future.

From the perspective of their use as Black-Box devices, we may observe that the weak points of all three systems are the absence of a built-in CAN controller, lack of A/D and D/A convertors except the ODROID N2 platform, however, in this case, there are only two analog inputs with maximum level of the 1.8 V signal and 10-bit resolution as well as the absence of an inertial device (accelerometer/gyroscope/ magnetometer – built-in IMU). From the perspective of the extended storage space, none of the three platforms offer the possibility of direct connection of a SSD on M2 MVMe-2280 or SATA interface. The available USB3.0 and USB 3.1 interfaces compensates to some extent for this deficiency allowing the coupling of external SSD.

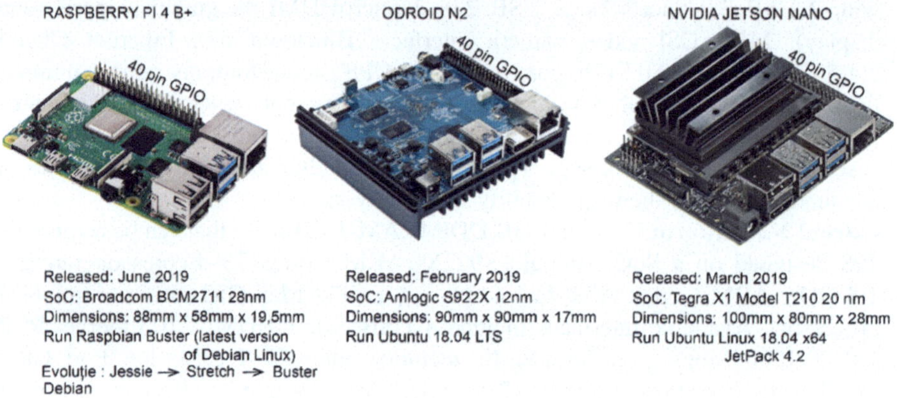

Fig. 1. Hardware platforms selected for Black-Box design

Why run from USB SSD instead local microSD card? Obviously due to the speed increasing from 50 MBps for SD card to theoretically 500 MBps for Samsung SSD T5, practically to 300–350 MBps. Unfortunately, Raspberry Pi 4 cannot boot from an external USB SSD or network like 3B+ , at least for now because bootloader is not on card but in an 4 MB serial EEPROM. But is possible after booting from SD Card to store and run the Raspbian OS entirely from SSD as shown in Fig. 2. Correspondingly, all data and installed programs can be resident on the SSD, including movies and pictures from video cameras. This is a great advantage for a black-box device, significantly increasing the video recorded length and resolution, or to the parameters number taken from the vehicle via CAN bus.

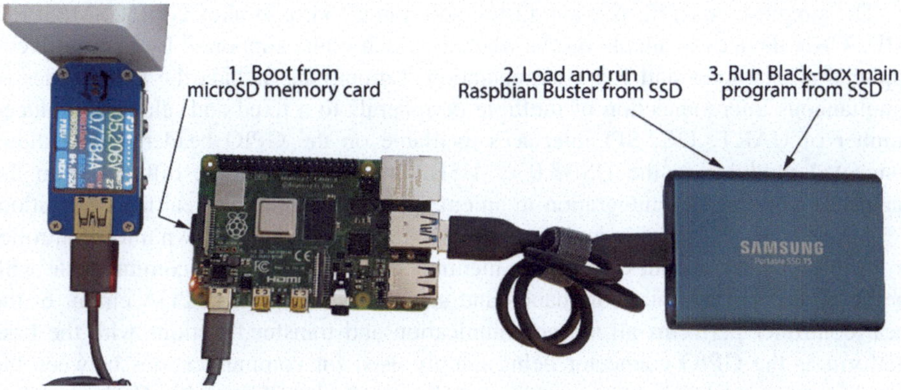

Fig. 2. Running Black-Box program sequence on Raspberry Pi 4

3 Black-Box Interfacing

As shown above on any of the platforms, there are several components necessary for the operation of a black-box device (at least one CAN, GPS, accelerometers, gyros, A/D converters, a true Real Time Clock, GPRS 4G). All these modules are coupled to the platform through the interfaces on the 40-pin GPIO connector. More interfaces included on the connector pins means several modules that are connected to the platform.

All the systems include SP, I2C, UART, PWM interfaces and digital inputs/outputs on a 40-pin header and all signals have the same voltage level of 3.3 V- Table 1.

Table 1. SPI, I2C, UART, PWM interfaces on GPIO 40 pins expansion header

	UART	I2C	SPI	HRW. PWM	Level
Raspberry Pi 4	1 (#8, #10)	1	1	2(#32, #33)	3,3 V
Odroid N2	1* (#8, #10)	2	1	4	3,3 V**
Jetson Nano	1 (#8, #10)	2	2	2*** (#32, #33)	3,3 V

*Alternative conf. on pins (#8, #10) UART_EE_A or (#15, #35) UART_EE_B

**Analog input signals AIN2 (#40), AIN3 (#37) are at 1.8 V level

***There's no PWM support for Jetson Nano (for ex. servo control)

The number and layout of these interfaces on the GPIO 40 pin header are important since on these interfaces the peripherals are connected which are not found on the mother platform but are necessary for the operation of a Black-Box or other embedded type systems (GSM/GPRS, GSM/GNSS, IMU, CAN, A/D and D/A converters). All these subsystems will be programmed and controlled by the mother platform via UART, I2C, SPI interfaces. All three versions are compatible with the voltage, signals and even their location on the 40 pins GPIO header (see Fig. 3).

The simplest, cost effective and fastest solution is to use turnkey GPS, GSM/GPRS, IMU, CAN devices available on the market, which come supported by many drivers, application examples and full documentation. Limitations include the possibilities of simultaneous interconnection of multiple peripherals to a fixed and relatively reduced number of UART, I2C, SPI interfaces available on the GPIO header. Using direct connected modules to the USB3.0 or USB2.0 interface is more difficult from the perspective of software integration in an embedded program with real time execution.

Another method consists in designing a separate board with its own microcontroller or a FPGA which includes the above-mentioned modules and can communicate with the base platform all data, commands and status. The integrated FPGA circuit or the local controller performs all the communication and transfer functions with the base platform as the GPIO connector being mainly used for communications between the two systems (usually via the pins corresponding to the UART circuit). This solution is more complicated since its design needs a licensed and expensive layout design program EDA (OrCAD, EAGLE, Proteus, Altium), an integrated platform for the microcontroller or FPGA programming, debugging and flashing and last but not least being the costs related to the manufacturing of the designed board which may be considerable, especially for a reduced number of pieces or if designing defects are found out later on.

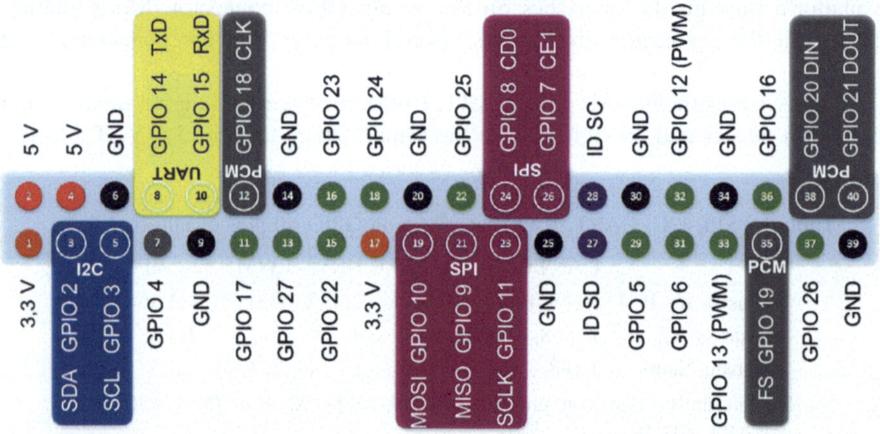

Fig. 3. Raspberry Pi B (3 & 4) 40 pins GPIO header -serial interfaces pins location

We also must consider the solution of using ELM327 adapter/dongle devices that can communicate via Wi-Fi, Bluetooth or USB with an Android, Linux, Windows or IOS stations, transmitting OBD codes between the vehicle and the and a Black-Box device based on Raspberry Pi. An original ELM327 adapter [6] supports the following protocols: ISO 9141-2, ISO 14230-4 (KWP 2000), ISO 15765-4 [7, 10, 11] (CAN 11/29 pin ID, 500 Kbps, 250 Kbps), User CAN (11 bit ID, 125/50 Kbps), SAE J1850 PWM. See all these three variants in Fig. 4.

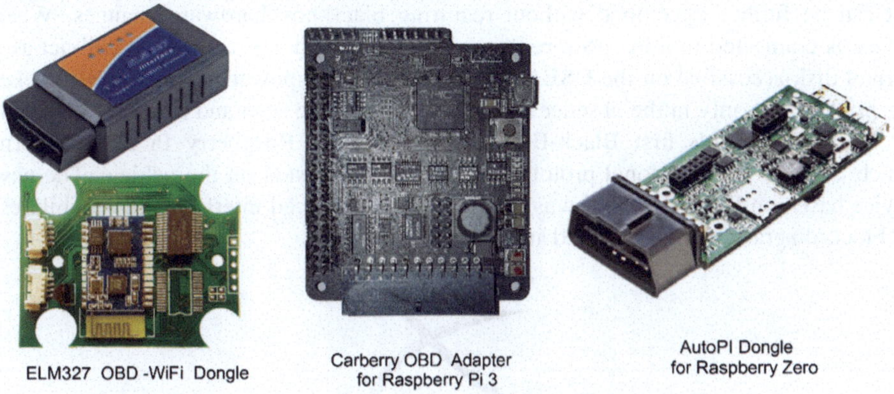

Fig. 4. OBD adapters and dongle for Raspberry boards [4, 5]

Particular attention must be paid to the fact that the market is being invaded by a great number of fake adapters that are not able to process all these stipulated protocols. It should be considered that all these adapters are intended for OBD diagnosis being able to read, interpret and convert error codes from engine and transmit them to a scanner or receiver via multiple protocols, but these adapters are not capable of covering the whole range of functions requested from a Black-Box device.

4 Black-Box Construction and Functions

Black-Box device presented in this paper was designed from the beginning to access, store and process data from three sources:

Signals coming from internal Black-Box sensors like a 9DOF accelerometer, gyro and magnetometer, temperature, pressure or gases (CO_2, alcohol), or device like cameras, microphone - some of them not yet physically implemented;
Data and information from external devices like video cameras, thermal sensor or camera, TI mm Wave Radar sensor, LTE Garmin 2D Lidar;
Automotive CAN and OBD networks allowing access to various parameters from engine, dashboard cluster, central panel, drivetrain.

Considering the USB3.0 interfaces on the Raspberry Pi 4 platform, the device has been redesigned to function as an on-line interface for downloading vehicle parameters on a PC. Connected to the OBD/CAN interface, the device receives power from it will function as a real-time data storage system on CAN. When Black-Box is subsequently connected to a PC, the data can be stored internally and simultaneous sent over USB to the PC where they can be displayed, analyzed, scaled, interpreted by a corresponding program. This PC program is not yet developed because it's not a target of this contract but can be further developed without requiring black-box hardware changes. When device is connected to a PC powered by a wall outlet and 5 V adapter, it will act as a normal disk accessible on the USB interface. Always the power from the OBD socket has priority and only in the absence of it the device can be seen and accessed via USB.

Figure 5 presents first Black-Box device based on Raspberry Pi 3B+ platform which is already a functional prototype calibrated and tested on the vehicle. The new device based on Raspberry Pi 4 was provided with enhanced interface and capabilities, its block diagram being presented in Fig. 6.

Fig. 5. First Black-Box device based on Raspberry PI 3B+

For data transfers with a remote server the black-box device was provided with a high speed 4G GSM/GPRS module which can achieve max. transfer rates of 150 MB/s (1,2 Gbps) and GPS access to all satellites (GPS, GLONASS, BeiDou, Galileo, QZSS).

In addition, the Raspberry Pi 4 B+ has a Bluetooth 5.0, 2.4 GHz and 5 GHz Wi-Fi, and Gigabit Ethernet (LAN) at max. 1000 Mbps. The ETH interface thus achieves transfer speeds of up to 100–125 MB/s. Under these conditions, the transfer is approx. 4 times faster (or transfer capacity 4 times higher) compared to the Raspberry Pi 3 B + predecessor platform.

One of the basic functions of a Black-Box device is to provide a buffer to store as many data as possible when the speed at which they are received from cameras, sensors, and OBD exceeds the download capability to the central server. For this purpose, the black-box device has been provided with extensive storage capabilities consisting of a Samsung EVO Plus Class 10 microSD memory card capable of write speeds of 90 MBps and 100 MBps read (4K video capability) as well as a 500 Gb Samsung T5 portable SSD with USB 3.1 C Gen.2 interface (10 Gbps) which ensure a maximum theoretical transfer speed of 540 MBps. In this case, the results are more spectacular, resulting in a transfer rate of 363 MBps compared to only 33 MBps between the Pi4 and Pi3 B+ versions.

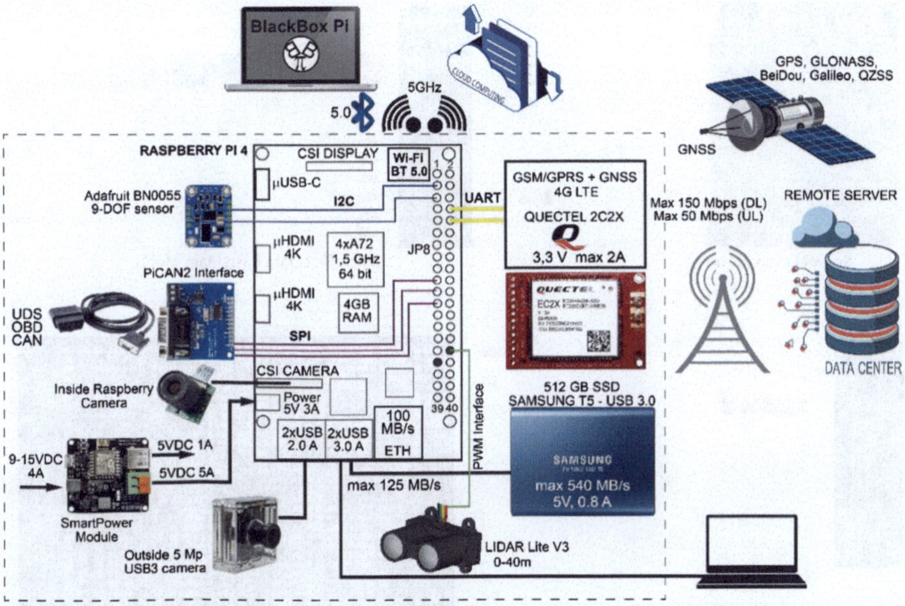

Fig. 6. Block diagram for the new Black-Box based on Raspberry Pi 4

The main program is designed based on two parallel running scripts. The first script collects data from the three available sources mentioned above, with priority being given to those that are transmitted from the OBD socket, followed by internal ones (accelerometer, GPS, and Lidar) and then those from video cameras. The data is saved on the SSD if it exists or on the microSD memory card. Acquisition starts as soon as the line 15 of the vehicle (Positive battery voltage passed through the ignition key) is powered and the operating system has succeeded booting. Main program will run continues into an infinite loop until the voltage on line 15 is stopped. The data retrieval speed from sensors can also be configured between 1–20 Hz.

The second script checks if exist an internet connection and sends the data saved by the previous script to the local database. Internet connection is made via a SIM card or a USB dongle. As soon as the server has successfully responded data is sent every second via an HTTPS POST. There is no proprietary communication protocol.

Screen captures with the Black-Box database running on the Ubuntu server are presented in Fig. 7.

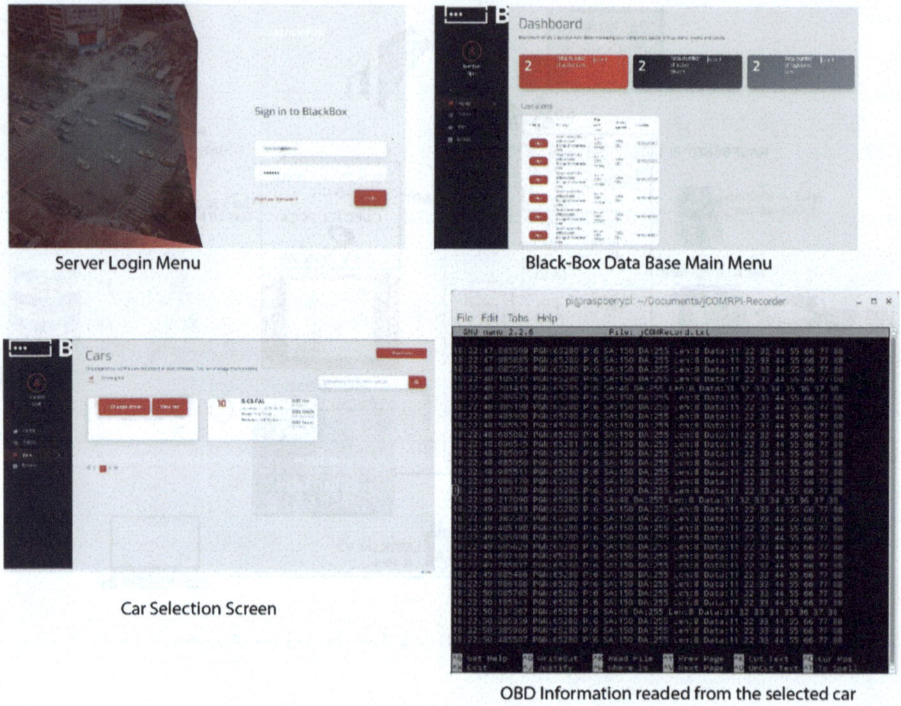

Fig. 7. Screen captures from remote server database

5 Conclusions

EDR and Black-Box are different devices. The first is a passive safety control unit with functions related to optimal seat belt and airbag deployment. So, the information stored in EEPROM memory is very limited and strictly for testing the triggering algorithms, and memorized parameters are not yet standardized. Stored data is also limited by the self-disconnected of the vehicle power in crashes occurrence and to the facts that SRS devices deployment actions are priority to those of data storage. The black-box is a type of driver assist system (ADAS) having extended capabilities to store all kind of data and information useful in accident reconstruction, driver behavior and driving skill estimations, car diagnosis, car parameters analysis like fuel consumption or CO_2 emission, front-end, rear-end and front-side inter-vehicle distance estimation, traffic signs and traffic legislation violation, emergency Call. At present, neither the black box, nor its features are standardized.

What distinguishes the Black-Box from all other car ECUs is how stored data and information can be communicated and accessed to/from outside. This device operates in an open and accessible manner, protected only by passwords and eventually by data encryption which would result in a slower running time and lower storage capabilities. As early as the design phase, methods to protect information and data should be considered, as well as the existence of a legislative framework that would legislate ownership rights over their Black-Box information, as well as the access rights to stored data.

The contradiction between the primary purpose of the black-box device to provide an impressive amount of data and information for purposes of reconstruction, evaluation, research, bonus-malus and prevention and how to grant access to them rests to be solved.

Within this project two functional Black-Box were designed and physically prototypes built, along with a dedicated database implementation and development on a remote server allowing devices to communicate over GSM/GPRS existing networks after a successful login. This allow to remote retrieve car parameters, location and trajectory, images from inside or outside of the vehicle. Car information can also be read off-line by connecting black-box to laptop, tablet, or other devices running Ubuntu OS and disposing at least one USB interface, or on-line (but only those data coming from OBD connector) with black-box acting as an interface between car and PC.

Our Black-Box device, which is the object of a research contract, was designed to accumulate as much as possible from the functions listed above, to have processing power and speed, a large storage space, transmission and communication capabilities surpassing those available onboard in car ECUs, but at a much lower price. Finally, if all tests to be carried out will prove the efficiency and usefulness of this type of equipment, redesigning all modules on a single board and making a more durable carcass, specific to the automotive field, is self-imposed.

The success of a Black-Box device will only be ensured if the manufacturer can convince each driver of the utility to store in a relatively accessible device of a large amount of data and artificial intelligence offering an real support as long he is an occupant of the cabin, making him feel safe in traffic, ensuring him that in case of accident the location coordinates and images from inside and outside of the vehicle will be automatically and immediately sent to the nearest emergency station, offering a clear and detailed indication of a technical failures, providing vehicle protection by functioning similar to that of a surveillance camera accessible from a distance or other facilities. If, first of all, the benefits provided to the technical experts and forensic investigators, insurers, lawyers, researchers and engineers in the traffic safety area are considered, then the success of the Black-Box will only be ensured by legalizing it as standard equipment for any vehicle, as for example to airplanes.

References

1. RaspberryPi homepage. https://www.raspberrypi.org/products/raspberry-pi-4-model-b/
2. Hardkernel homepages. https://www.hardkernel.com/shop/odroid-n2-with-4gbyte-ram/
3. Nvidia Jetson. https://developer.nvidia.com/embedded/jetson-nano-developer-kit
4. Carberry homepage. http://www.carberry.it/
5. AutoPI io homepage. https://www.autopi.io/
6. ELM electronics. https://www.elmelectronics.com/products/
7. SAE International: J1939_201206, serial control and communications heavy duty vehicle network. SAE Standard

8. SAE International: J1939_200612, data link layer. SAE Standard
9. SAE International: J1939-73 200609, application layer –diagnostics. SAE Standard
10. Paret, D.: Multiplexed Networks for Embedded Systems. SAE Books, Willey (2007)
11. Walter, R., Walter, E.: Data acquisition from light-duty vehicles using OBD and CAN. SAE Book, Willey (2019). R-458
12. Wajape, M., Elamana, N.B.: Study of ISO 14229-1 and ISO 15765-3 and implementation in EMS ECU for EEPROM for UDS application. In: IEEE International Conference on Vehicular Electronics and Safety (2014)

Experimental Studies of the Same Car Behavior in Different Pile-up Scenarios

Nicolae Ispas(✉) and Mircea Nastasoiu

Transilvania University of Braşov, 29 Eroilor Blvd., 500036 Braşov, Romania
inicu@unitbv.ro

Abstract. Pile-up accidents are frequent, especially in motorways traffic. The consequences of such accidents depend among other things on the characteristics of the car and the type of collision it is involved. Experimental study of the behavior of the same car (or car with the same characteristics) in different scenarios of multiple impact was the main goal of this paper. The paper is based on experimental research in which the same car (or same car type) was used, which was involved in various pile-up cases. The results obtained, including the damage magnitude and the acceleration values of the cars involved, allow a realistic assessment of the effects of such multiple collisions. The results obtained can be used to best estimate the behavior of a particular type of car involved in a particular accident scenario.

Keywords: Pile-up · Safety features · Multiple collisions · Pile-up scenarios

1 Introduction

Road traffic accidents are an important factor in the mortality rate across the European Union. A large number of these fatal accidents occur in traffic on motorways. Although the overall number of road fatalities has decreased rather steadily, the trend for motorway fatalities has been more variable. However, clear the increase of 10% was recorded between 2014 and 2015 in the EU motorway fatalities [1].

In the Fig. 1 are presented the involution number of fatalities on motorways and all road fatalities in EU, between 2006 and 2015 [1].

The experiments in this paper are typical of those encountered in motorway traffic. Thus, the collision behind a truck equipped with under ride guard system, the rear-end collision, or the rear-end collision followed by a front collision at speeds encountered after braking, are specific to highway traffic [2].

Studies from [3] has revealed that 72% of crashes involve factors related to road user factors and observation, interpretation and planning errors are relatively common. There is remarkably little variation in these errors between the main road user types [3].

In many paper are consider that reducing the severity of injury in crashes is more important than reducing incidence and ranking incidence by broad categories [3].

Vehicle and road modifications are generally more effective than behavioral change efforts with the exception of certain laws such as required use of seat belts, motorcycle helmets and graduated licensing of teenagers [3].

I. Dumitru et al. (Eds.): SMAT 2019, *The 30th SIAR International Congress of Automotive and Transport Engineering*, pp. 439–447, 2020.
https://doi.org/10.1007/978-3-030-32564-0_51

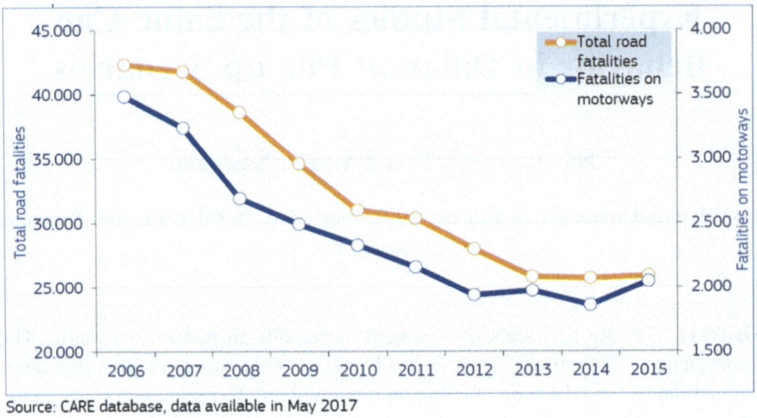

Fig. 1. Number of fatalities on motorways and all road fatalities, EU, 2006–2015 [1].

The experimental results described in this paper were obtained with the entire logistic support and research equipment available in DSD laboratories, Dr. Steffan Datentechnik GmbH - Linz, Austria, during the Easter 2018 PC-Crash Seminar research.

2 Experimental Results and Discussions

The experimental studies presented in the paper have as a specific element that several collision tests were involved, the same vehicle, with the same old age. This guarantees a similar behavior of the vehicle structure involved in different collision scenarios (Fig. 2 and Table 1):

Fig. 2. First collision between P02 Opel Zafira and L13 MAN truck with underride guard system (collision schema in the left and damaged vehicles in the right).

1. First collision between year 2000 P02 Opel Zafira and year 1997 L13 MAN truck with under ride guard system (Fig. 2).
2. Second impact between yar 2002 P04 Opel Zafira and year 2002 P03 Opel Zafira
3. Third collision between P03 and P02
4. Fourth impact between P02 and L12

Table 1. From second to fourth collision scenarios and involves vehicles characteristics.

Label	P04	P03	P02	L13	
Make	Opel	Opel	Opel	MAN	
Model	Zafira-A	Zafira-A	Zafira-A	19.403 FLLC	
Year	2002	2002	2000	1997	
Velocity	90.7	0	0	0	km/h
Weight	1418	1437	1395	10754	kg
Length	4.32	4.32	4.32	9.6	m
Width	1.74	1.74	1.74	2.5	m
Height	1.68	1.68	1.68	3.6	m
Wheelbase	2.69	2.69	2.69	5.5	m
Trackwidth	1.48	1.48	1.48	2.04	m
Axle load front	868	872	811	5400	kg
Axle load rear	550	565	584	5354	kg
Dist. COG to front axle	1.05	1.06	1.13	2.74	m

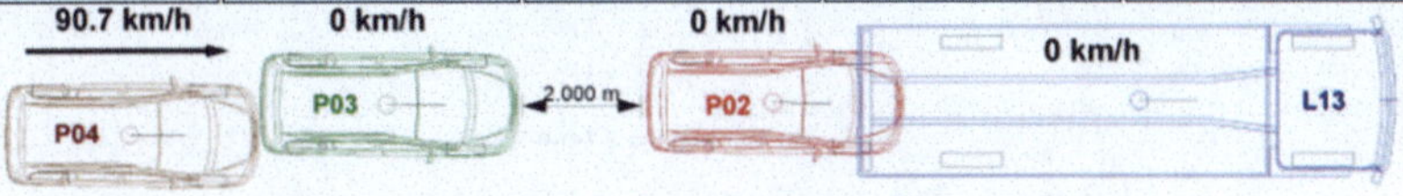

The physical quantities determined during the experiments had as reference the coordinate system shown in Fig. 3.

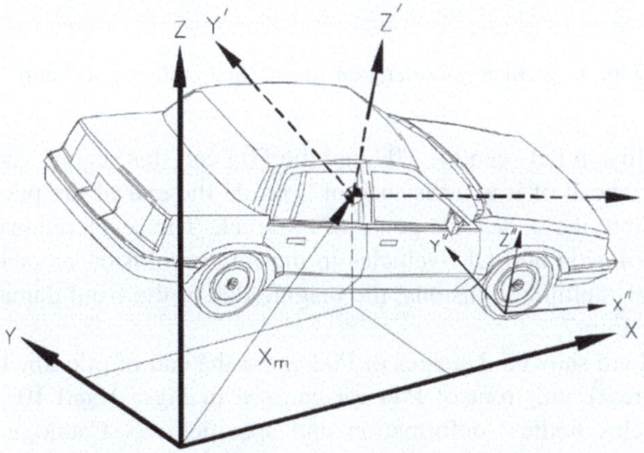

Fig. 3. Coordinate system for cars and truck.

To estimate the magnitude of danger to virtual vehicles occupants in the various impact scenarios, accelerations (measured in vehicle weight centers) were determined in all four cases of impact. Thus, in Fig. 4 the measured accelerations for P02 and L13 are shown (the speed of the collision was 70,8 km/h), and in Fig. 5 there are presented the accelerations of the cars P04 and P03 (the speed of the P04 car at the collision was 90,7 km/h): also here can see the acceleration for third and fourth impact scenario for P004 and P03 cars.

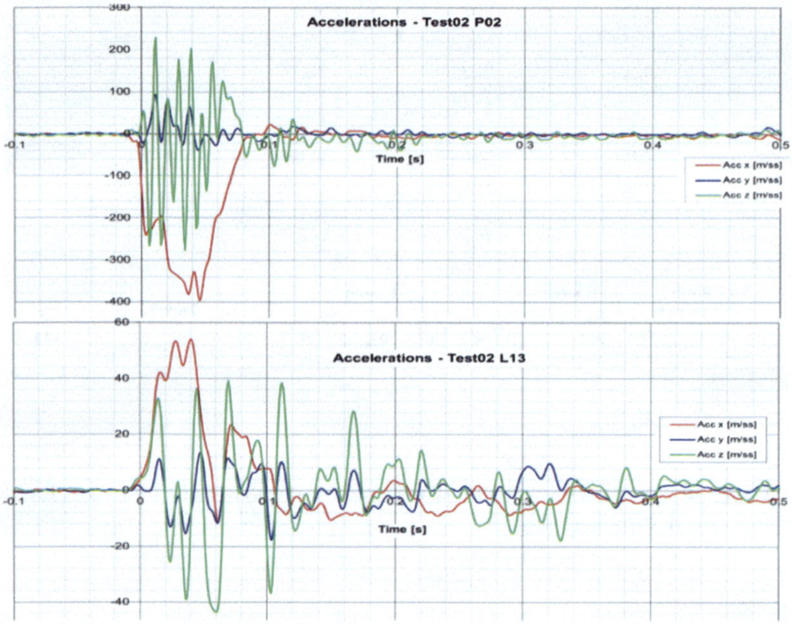

Fig. 4. Vehicles acceleration for the first collision scenario.

After the collision between the P04 and the P03 car, this second car was projected into the P02 car ahead of it at a distance of 2 m. At the end of the pile up, the P02 is again projected into the under ride guard of the truck. The accelerations of the P02 for third and four collision and L13 vehicles in the fourth collision are shown in Fig. 6.

At the end of multiple collisions, the magnitudes of the front damages of P02 are presented in the Fig. 7.

In the Fig. 8 are showed damages of P02 car at the end of pile-up. The damages of P03 (front and rear) and front of P04 car can see in Figs. 9 and 10. Using detailed images of vehicles bodies' deformation and specific EES Catalogue can estimate Energy Equivalent Speed for each car involved in a single collision. For cars involved in multiple collisions, it is almost impossible to estimate the magnitude of bodies deformations determined by each collision, so it is impossible to determine the Energy Equivalent Speed for each variation in the speed of the cars.

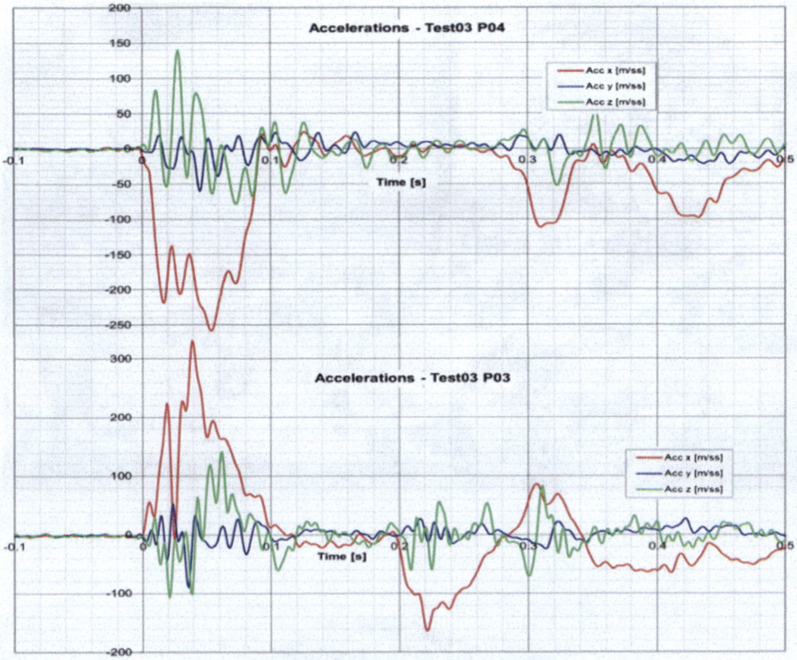

Fig. 5. Cars accelerations in the second, third and fourth scenarios collision test.

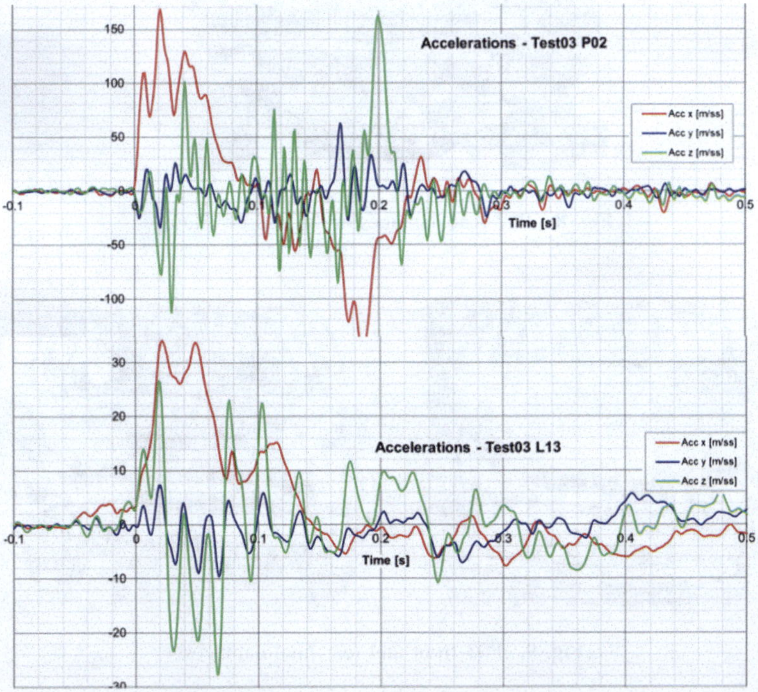

Fig. 6. P02 and L13 vehicles acceleration in the third and fourth collision experiments.

Fig. 7. P02 front damages at the end of multiple collision experiments.

Fig. 8. P02 rear damages at the end of pile up.

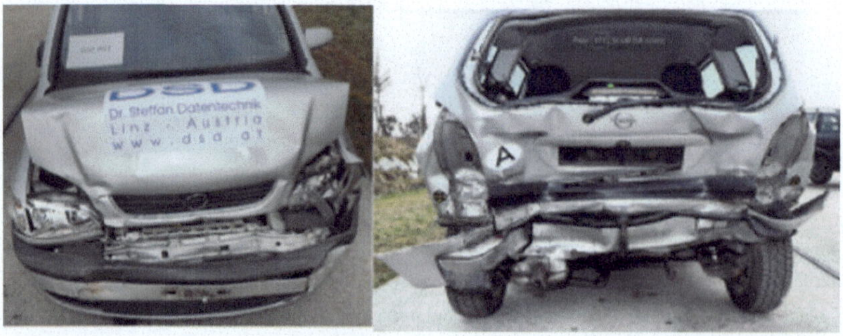

Fig. 9. P03 front and rear final damages.

Fig. 10. Car P04 front final damages.

3 Conclusions

In the paper were experimentally researched the multiple collisions specific to the traffic on the motorway. In order to better highlight the security measures needed to provide the best protection for occupants, the behavior of the same type of car in different collision scenarios were studied experimentally. A synthesis of these tests is depicted in Fig. 11 where the central part can be seen the damage of the vehicles at the end of the pile-up, at the upper part of the deformations of the sides of the cars, and at the bottom of the deformations of the rear parts of the cars.

Fig. 11. Synthesis of multiple collisions with deformations of the front parts of the cars (top views), side views of central faults and deformations of the rear parts at the bottom of the image.

It is demonstrated the direct connection between the size of the absolute acceleration of the car and the severity of the injury of the occupants of that car [4]. To determine exactly the absolute acceleration values, Ox, Oy and Oz were used for the performance devices from DSD Linz Austria [5] and [6].

For each of the four collision scenarios, the authors of this paper can outline the following main conclusions:

A. First collision between year 2000 P02 Opel Zafira and year 1997 L13 MAN truck with under ride guard system:
   1. Car front damages are major. The presence of the under ride guard system prevents damage to the passenger compartment of the car (Fig. 11);
   2. The absolute value of the acceleration of the vehicle in the Ox direction was 400 m/s^2, and in the Oz direction of 280 m/s^2; the absolute value of the truck's acceleration in the Ox direction was 53 m/s^2, and in the Oz direction of 43 m/s^2.
B. Second impact between yar 2002 P04 Opel Zafira and year 2002 P03 Opel Zafira:
   1. The P04 car is seriously damaged at the front of the car while the P03 car is damaged in this collision at the rear (Fig. 11);
   2. The absolute acceleration of the P04 in the Ox direction reaches 260 m/s^2 and in the Oz direction of 110 m/s^2; for P03, in the second collision scenario, the Ox direction recorded an absolute acceleration value of 330 m/s^2 and in the Oz direction a value of 140 m/s^2 of absolute car acceleration.
C. Third collision between P03 and P02:
   1. In this third collision scenario, the rear side of the P02 car is damaged (Fig. 11);
   2. The absolute acceleration of P03 in the Ox direction was 160 m/s^2, and in the Oz direction of 58 m/s^2; for P02 the absolute acceleration on the Ox direction was 170 m/s^2, and on the Oz direction of 112 m/s^2.
D. Fourth impact between P02 and L12:
   1. In the fourth impact case the front and rear parts of the P02 are even more severely damaged; for P02 car the absolute acceleration on the Ox direction was 150 m/s^2, and on the Oz direction of 167 m/s^2.
   2. The absolute value of the truck's acceleration in the Ox direction was 34 m/s^2, and in the Oz direction of 28 m/s^2; for P03 car the absolute acceleration on the Ox direction was 82 m/s^2, and on the Oz direction of 70 m/s^2; also P04 was damages and is recorded a maximum of 110 m/s^2 absolute acceleration in Ox direction.

Acknowledgment. The authors keep on this way to thank Prof. PhD eng. Hermann Steffan from the Technical University of Graz, who provided full theoretical and experimental support in the DSD research, Dr. Steffan Datentechnik GmbH - Linz, Austria during the Easter 2018 PC-Crash Seminar.

References

1. European road safety observatory traffic safety basic facts 2017 – Motorways. https://ec.europa.eu/transport/road_safety/sites/roadsafety/files/pdf/statistics/dacota/bfs2017_motorways.pdf. Accessed 20 May 2019
2. Thomas, P., Morris, A., Talbot, R., Fagerlind, H: Identifying the causes of road crashes in Europe. In: Annals of Advances in Automotive Medicine, vol. 57, pp. 13–22, September 2013

3. National motorway safety administration-NHTSA Report: Analyses of rear-end crashes and near-crashes in the 100-car naturalistic driving study to support rear-signaling countermeasure development, October 2007. https://www.nhtsa.gov/sites/nhtsa.dot.gov
4. Aarts, L.T., Commandeur, J.J.F., Welsh, R., Niesen, S., Lerner, M., Thomas, P., Bos, N., Davidse, R.J.: Study on serious road traffic injuries in the EU, October 2016. https://ec.europa.eu/transport/road_safety/sites/roadsafety/files/injuries_study_2016.pdf. Accessed 20 May 2019
5. Ispas, N., Năstăsoiu, M.: Cars crashes with cars of same or different generation – occupants loads, timings and accelerations effects. In: CONAT 2016 International Congress of Automotive and Transport Engineering, pp. 765–781 (2017). https://doi.org/10.1007/978-3-319-45447-484
6. Ispas, N., Năstăsoiu, M.: Analysis of cars frontal collision against pole. In: International Congress of Automotive and Transport Engineering- Mobility Engineering and Environment (CAR), IOP Conference Series-Materials Science and Engineering, vol. 252. (2017) https://doi.org/10.1088/1757-899x/1/012012

Advanced Engineering Methods

Modeling of a Standardized Testing Track with the Use of Specialized Software

Daniela Voicu(✉), Ramona-Monica Stoica, and Laszlo Barothi

Military Technical Academy, Bd. G. Cosbuc nr. 39-49, District 5, Bucharest, Romania
daniela.voicu@mta.ro

Abstract. Present paper aims to model and validate several simulation tracks used for the movement of transport vehicles in different off-road situations, in order to conduct a time analysis of vibrations and shocks transmittable to equipments. With the use of such a model there can be performed high accuracy simulations of the movement of logistical transport vehicles on off-roads and it has the advantages of low costs and lower periods required to evaluate vibrations and shocks than in case of experimental research.

Keywords: Testing track · Vibration · Modeling-simulation software

1 Introduction

The movement of vehicles destined to cargo transport on different road categories, with uneven surface, as well as on unpaved roads, can lead to high accelerations transmitted to the cargo, which can produce damages.

Also, road quality can induce roll and pitch movement of suspended mass with direct consequence on transported commodities.

In order to avoid these effects, the driver should adapt vehicle speed according to state and road quality.

In comparison with experimental research, modeling tracks represents a more cost-effective method to conduct a real time analysis of vibrations recorded on transported goods, as well as to determine maximum speed.

2 Experimental Research

During road transport result, on short periods, longitudinal, vertical and lateral forces which are transmitted to transported materials. These forces can also cause vibrations, variable in amplitude and frequency, based on suspension type, road category or driving style. Moreover, occasionally, shocks might be registered, which are the most severe stresses that can occur during an impact, for example, in case of materials hitting container walls or floor, materials falling due to improper pile-up, or shock experienced during manipulation (tilt over, hitting, falling etc).

I. Dumitru et al. (Eds.): SMAT 2019, *The 30th SIAR International Congress of Automotive and Transport Engineering*, pp. 451–457, 2020.
https://doi.org/10.1007/978-3-030-32564-0_52

Therefore, most common stresses are caused by structural parameters of transported goods, vehicle's building parameters, disposal manner inside the container, load degree of container, road condition etc.

Taking these aspects into consideration, IMO/ILO/UNECE standard indicates that during road transport, the force acting on transported goods is defined as the product between load mass (m) in kg and gravitational acceleration ($g = 9.81\ m/s^2$), but because during transport tend to appear accelerations due to sudden braking, lane change, uneven track surface etc., this guide offers following equation, able to prevent incidents which can cause damages of transported goods [1]:

$$F = m \cdot c_x \cdot g \tag{1}$$

where c_x is a safety coefficient which takes into account all before-mentioned matters.

These coefficient values consider both the secure manner of goods (which can be on longitudinal and transversal direction), and the direction of input force (Table 1).

Table 1. Values of acceleration coefficients.

Road transport				
The secure direction of goods	Acceleration coefficients			
	Longitudinal (c_x)		Transversal (c_y)	Vertical (c_z)
	Forward	Backward		
Longitudinal direction	0.8	0.5	–	1
Transversal direction	–	–	0.5	1

3 Results

For the analysis of vibrations recorded in case of transported goods, there are standardized tracks (ex. APG track).

APG track consists of multiple obstacles, disposed both on left and right of the track, with a certain shift or on the same transversal line, so that vehicle's wheels from the same axle can come in contact at the same time with the obstacles, or at short periods of time. When the wheels of an axle pass at the same time over obstacles, the truck is submitted to roll movement and lift up. Main dimensions of obstacles and disposal pattern is presented in Fig. 1.

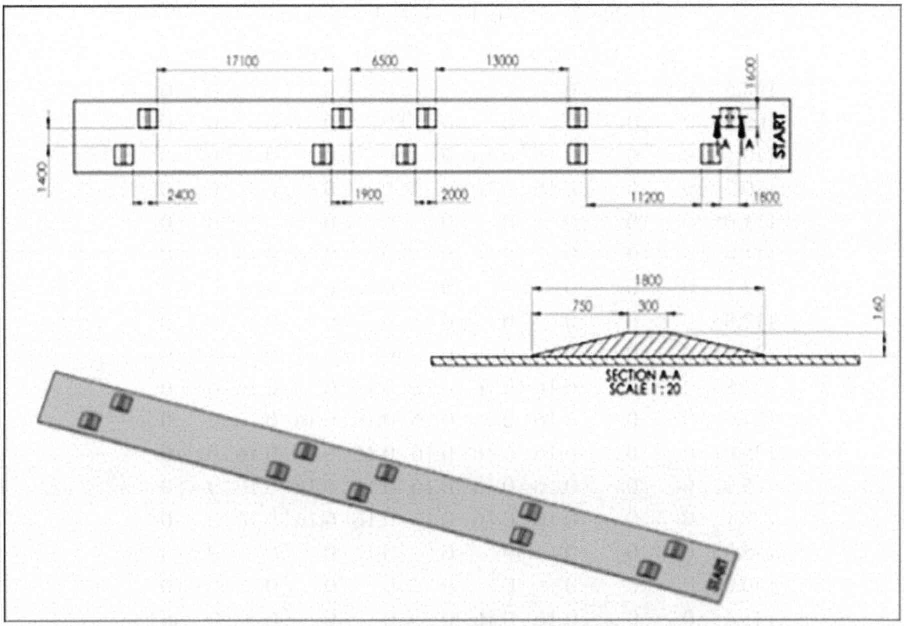

Fig. 1. Obstacles disposal pattern

TruckSim modeling-simulation software can be used to model such a testing track. Educational version of the software contains within the library, different testing tracks for vehicles destined to transport goods and passengers, including a lay-out named left/right bumps and which is similar to the track presented in Fig. 1.

Differences between what is offered by the modeling-simulation software and the track from Fig. 1 consist mainly in geometrical dimensions of obstacles, as well as their disposal pattern. For that reason, the new lay-out of obstacles was set as it is presented in Fig. 1.

However, the modeling-simulation software allows to define obstacles at different dimensions, as well as main characteristics of tested vehicle (suspension type, number and type of axles, type of container mounted on the vehicle chassis, load mass placed within the transport container etc.), but also to set movement speed (which can be variable or constant, by actuating the throttle or the braking system).

Considering previous data specific to geometrical dimensions of obstacles and distances between them, in the modeling-simulation software was created a 25 by 11 matrix (Table 2), in Excel software, with values which describe obstacles height, length and width, as well as the distance between them, both lateral and frontal. The final table was inserted in TruckSim software.

Table 2. Matrix for obstacles disposal.

0	−10	−1.5	−0.9	−0.6	0	0.01	0.6	0.9	1.5	10
108.5	0	0	0	0	0	0	0	0	0	0
109.25	0	0	0	0	0	0	0	0	0	0
110	0	0	0.16	0.16	0	0	0	0	0	0
110.3	0	0	0.16	0.16	0	0	0	0	0	0
111.05	0	0	0	0	0	0	0	0	0	0
111.8	0	0	0	0	0	0	0.16	0.16	0	0
112.1	0	0	0	0	0	0	0.16	0.16	0	0
112.85	0	0	0	0	0	0	0	0	0	0
124.05	0	0	0	0	0	0	0	0	0	0
124.8	0	0	0.16	0.16	0.16	0.16	0.16	0.16	0	0
124.9	0	0	0.16	0.16	0.16	0.16	0.16	0.16	0	0
125.01	0	0	0.16	0.16	0.16	0.16	0.16	0.16	0	0
125.02	0	0	0.16	0.16	0.16	0.16	0.16	0.16	0	0
125.1	0	0	0.16	0.16	0.16	0.16	0.16	0.16	0	0
125.85	0	0	0	0	0	0	0	0	0	0
140.65	0	0	0	0	0	0	0	0	0	0
141.4	0	0	0.16	0.16	0	0	0	0	0	0
141.7	0	0	0.16	0.16	0	0	0	0	0	0
142.45	0	0	0	0	0	0	0	0	0	0
142.65	0	0	0	0	0	0	0	0	0	0
143.4	0	0	0	0	0	0	0.16	0.16	0	0
143.7	0	0	0	0	0	0	0.16	0.16	0	0
144.45	0	0	0	0	0	0	0	0	0	0
148.95	0	0	0	0	0	0	0	0	0	0

Linear 2d interpolation and extrapolation of data from Table 2 are presented graphically in the software, thus:

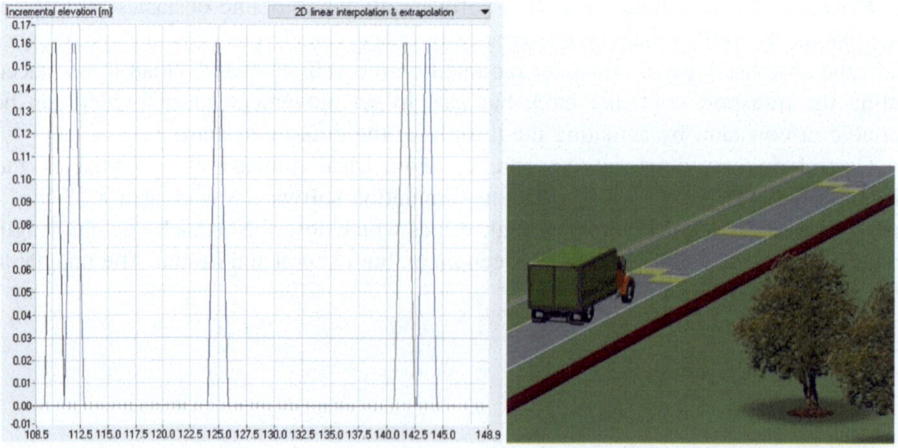

Fig. 2. Linear 2d interpolation and extrapolation of data

Fig. 3. Modeled truck

After modeling the track, simulation research continued with modeling an auto-truck destined to transport goods (Fig. 3), exactly it was modeled the suspension [2]: suspension with leaf springs, specific to logistic transport, with main front/rear characteristics: 5.5t/8.5t, travel range: ±150 mm/ +100 mm, −60 mm). The truck had attached a standardized cargo container of 20 ft (Fig. 3) and total mass was set to 6.2 tone [5], so that to obtain a more real transport situation.

At the same time, present study also involved mounting 6 accelerometers on the vehicle, aiming to measure acceleration values which appear when moving on the simulated track [3].

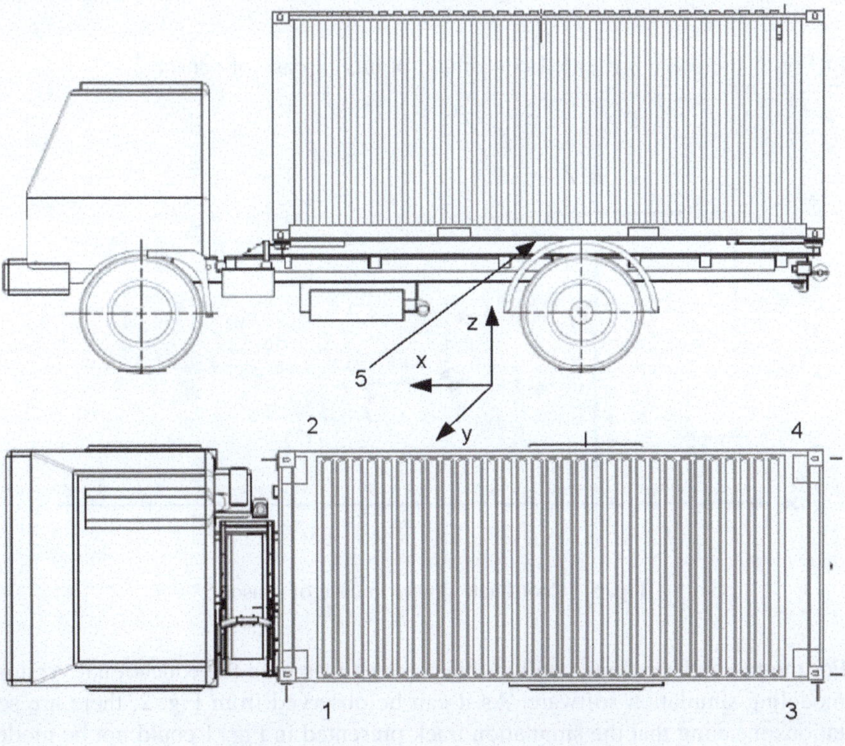

Fig. 4. Layout of all 5 accelerometers

Accelerometers layout on the truck is presented in Fig. 4: there have been placed one accelerometer at each of the four container corners, one accelerometer was mounted on the longitudinal section of the vehicle and the sixth one was placed on the front axle. In order to include in the simulation software the mounting position of all 6 accelerometers at the points of interest, there have to be known dimensional characteristics for both truck chassis and container. The coordinate system represented by x, y, z points was used to define disposal place for all 6 sensors and it was set based on a global coordinate system whose origin was on median plane, at the contact between front wheel and ground. Therefore, by reference to the global coordinate system, in Table 3 are presented disposal coordinates of accelerometers, used within the modeling-simulation software [4].

Table 3. Accelerometers disposal coordinates.

Accelerometer	X	Y	Z
1	−1600 mm	1175 mm	1200 mm
2	−1600 mm	−1175 mm	1200 mm
3	−7579 mm	1175 mm	1200 mm
4	−7579 mm	−1175 mm	1200 mm
5	−1103 mm	1175 mm	1200 mm
6	0 mm	0 mm	510 mm

In Fig. 5 there are presented coordinate points in case of sensor 5:

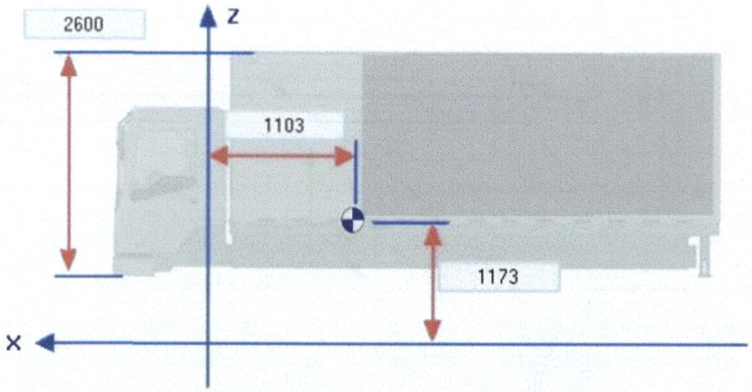

Fig. 5. Coordinate points in case of sensor 5

For the research presented within this paper, it was used the educational version of the modeling-simulation software. As it can be observed from Fig. 2, there are some limitations meaning that the simulation track presented in Fig. 1 could not be modeled entirely because the program did not allow inserting repeated sequences (obstacles) and it allowed modeling only the first three series of obstacles (Fig. 2). Nevertheless, with the use of these three series of modeled obstacles it was able to obtain targeted information.

4 Conclusions

Due to the fact that the simulation presented in this paper is an ideal case, being performed based on simplified hypotheses (movement with constant speed, no usage of throttle or braking system, use of steering to maintain the trajectory of the truck so that it passes over every obstacle, no gear change, lack of lateral/frontal wind), it is considered necessary to conduct experimental tests in order to compare the results from simulation with ones from real experiments.

As a general conclusion, it can be stated that the use of software based on "multibody" approach allows taking into account the constructive solution of the vehicle and offers the possibility to obtain relevant data regarding accelerations on different points. For a more detailed analysis of results, it is necessary to use Matlab and Mathcad applications which can allow a more advanced analysis of frequency, characteristic to movement.

References

1. ***Information. https://wiki.unece.org/display/TransportSustainableCTUCode/Chapter+5.%09General+transport+conditions
2. Pierce, C., Singh, S.P., Burgess, G.: A comparison of leaf spring to air cushion trailer suspensions in the transportation environment. J. Packag. Technol. Sci. **5**, 11–15 (1992)
3. Zhou, R., Yan, L., Li, B., Xie, J.: Measurement of truck transport vibration levels in China as a function of road conditions, truck speed and load level. Packag. Technol. Sci. **28**(11), 949–957 (2015)
4. Vilău, R., Grosu, D., Lespezeanu, I.: Considerations concerning the dynamic performances improvement of the wheeled military vehicles braking systems, using modern modelling and simulation techniques. In: The 9th International Congress on Automotive "CAR 2005", Pitești, 2–4 November 2005. ISBN 973-690-450-4, CAR20051061
5. Vilău, R., Puncioiu, A.: Considerations concerning the dynamic performances analysis of a special vehicle steering servomechanism. In: The 5th International Conference on Advanced Composite Materials Engineering and The 3rd International Conference, "Research Innovation in Engineering" "COMAT 2014", 16–17 October 2014, Brașov, Romania, vol. II, pp. 148–153 (2014). ISBN 978-606-19-0411-2

Fuzzy Decision Algorithm for Driver Drowsiness Detection

Tiberiu Vesselenyi, Alexandru Rus(✉), Tudor Mitran, Sorin Moca, and Csokmai Lehel

University of Oradea, Universitatii Street no. 1, 410087 Oradea, Romania
alexandru.rus.61@gmail.com
https://www.uoradea.ro

Abstract. The paper presents the development of a multi-criterial fuzzy decision algorithm applied to the monitoring and warning of drowsy drivers in order to prevent accidents, based on EEG signals and eye state images. The drowsiness warning system is based on four main components, which learn, analyze and decide whether the driver is or is not in a drowsy or sleepy state. This result can be then used to warn the driver if he/she is in a drowsy state.

Driver sleepiness is a complex phenomenon involving a progressive decrease in attention to road and traffic requirements, having the effect of lowering driving performance, longer reaction time and an increased risk of crash involvement.

The development of a system for detecting tiredness while driving, based on EEG (Electroencephalography), and eye state (closed or opened) image classification has been studied in previous works. For eye state detection we used two methods in order to increase the accuracy of the system.

Detection of driver fatigue based on images acquired while driving, was done by analyzing driver eye condition: open, and closed. For the classification of driver fatigue we used artificial neural networks.

The objective of this study is to develop a multi-criterial fuzzy decision algorithm which has as inputs EEG signals and eye state images acquired by two methods. Biological data that are acquired by the sensors will be stored, processed and evaluated in real time, through a system able to detect early signs of fatigue, because the physiological variables are closely related to this phenomenon.

Keywords: Fuzzy inference systems · Eye image detection · Image recognition · Video signal processing

1 Introduction

In goods and persons transport the driver's drowsiness leads to deaths and important material losses. According to the U.S. National Traffic Safety Administration NHTSA from 2017, in 91,000 from the accidents reported by the Police drowsiness drivers were implied. These accidents lead to an estimated value of 50,000 wounded persons and almost 500 deaths. By studying 700 accidents, the AAA Foundation for the traffic safety has found that 9,5 to 10,8% of all the accidents have implied drowsiness drivers.

I. Dumitru et al. (Eds.): SMAT 2019, *The 30th SIAR International Congress of Automotive and Transport Engineering*, pp. 458–467, 2020.
https://doi.org/10.1007/978-3-030-32564-0_53

The car manufacturers have developed systems for the detections of driver's fatigue. Volvo uses a system for the face recognition in order to detect the driver's drowsiness, and the Bosch Group has developed a system that detects the driver's drowsiness based on an algorithm that starts to register the behavior of the driver in the moment that the travel starts.

Fatigue detection systems have been developed based: on the cardiac monitoring and the analysis of cardiac frequency variability (HRV) [11], on the EEG (Electroencephalography) [6] or EOG (Electrooculography) [12].

The performance of the hybrid detection systems, that combine two of the previous presented methods, increases. So, in [6] is presented the system for the fatigue detection based on EEG, EOG and the on the eye status (closed or opened).

The system for the detection of the driver's fatigue state presented in [7] realizes the analysis of the driver's eye state (opened, opened and closed), based on the images acquired during the driving. For this purpose, two artificial neuronal networks were used: one with a hidden layer and one with self-encoders [1]. The development in this field is supported by the efforts to create the brain-computer interfaces [2] for different applications. The acquisition and the processing of the EEG signals were presented in [3, 9]. In the incipient research [8] are addressed the possibilities to acquire the EMG signals from three sensors, and for the classification of the driver's fatigue states one used artificial neuronal networks [4, 5, 13].

The objective of this research is to analyze the possibility to use soft computing techniques (Artificial Neural Networks, Fuzzy Inference Systems) to for processing and decision making in applications implying driver drowsiness detection systems.

2 Drowsiness Detection System

The drowsiness detection system proposed in this research is using two types of inputs. The first is the EEG signal recorded during driving while the second one is the driver's face image streamed by an onboard camera. The EEG signal processing was analyzed by the authors in a previous work [6, 7]. The image of the driver's face is first processed through a face recognition and tracking algorithm and then is fed to two different processing units: the first one being an Artificial Neural Network (ANN) trained to recognize the state of the driver's eyes (opened or closed) and the other one is the Eye Aspect Ratio (EAR) algorithm which essentially also gives information about the state of the eye (opened or closed) but through a different computational process. The last stage of processing is a unit based on ANN which is trained to recognize the driver's behavior (like for example the frequency and distribution of blinks) and decide if is an alert state behavior or a drowsy state one. The schematic of the system is presented in Fig. 1.

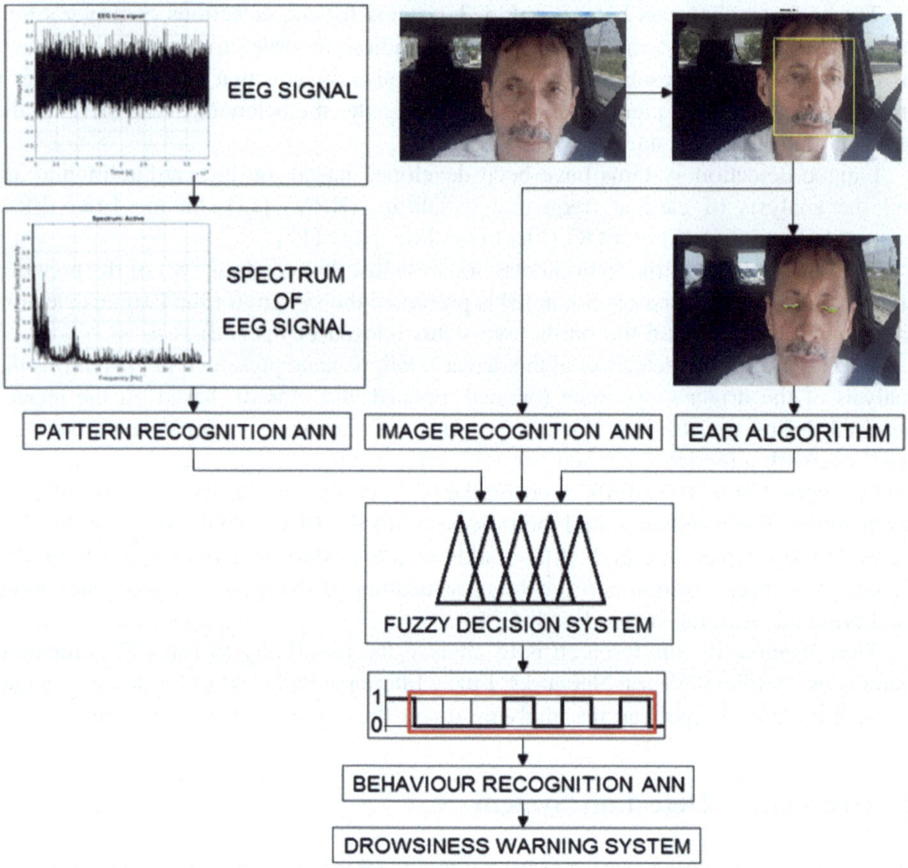

Fig. 1. Schematics of the drowsiness detection system.

In the following we will present the processing units and the results we obtained by training and testing the system. The processing unit for the EEG signal spectrum was described in a previous For training and testing of the system we used a video stream of 60 s (1975 frames) recorded during driving. The Matlab (Mathworks) environment had been used for the ANN image recognition, face detection, and the development of the fuzzy inference system [10].

3 Image Recognition with ANN

In order to train the ANN for image recognition the video stream is read frame by frame and face recognition and tracking functions of the Computer Vision System Toolbox was used to select the region of interest from each frame. These functions are developed applying algorithms described in [14, 15].

The method used to detect and track the face is based on the skin color and is applying a histogram of pixel values of the Hue channel which are extracted from the nose region (Fig. 2). The result of the face detection and tracking is given by an example in Fig. 3.

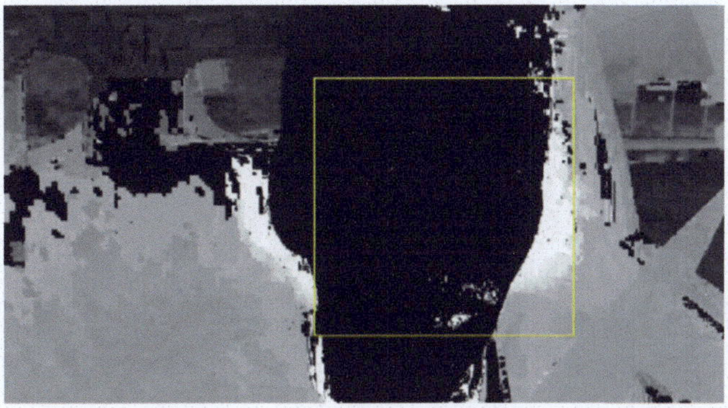

Fig. 2. Hue channel pixels detection.

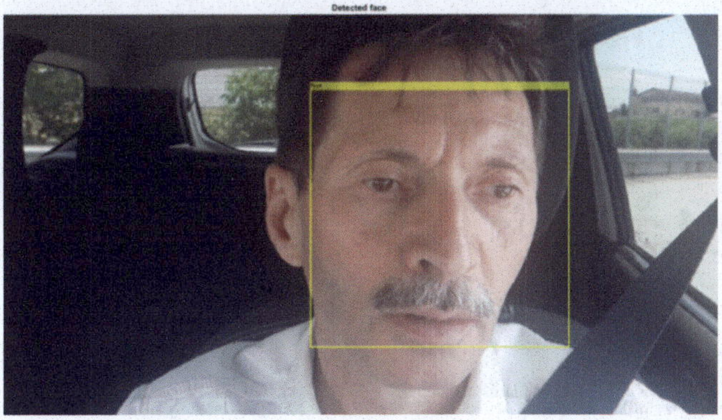

Fig. 3. Face detection and tracking.

The ANN was trained on labeled images obtained by cropping the region of interest detected with the previously described algorithm. For inputs we used 100 images with eyes opened and 100 images with eyes closed. The training of the network was made using the Neural Network Toolbox. As it can be seen from the training performance diagram presented in Fig. 4, which shows a very small error after 33 epochs. Also from the confusion diagrams (Fig. 5) we can conclude that training, validation and testing results were successful as al images were classified correctly. Additionally we tested the developed network on 100 labeled images which were not used in the training process and the results are shown in Fig. 6. Although the network worked properly on the test data there are still uncertainties regarding the network performance. In order to increase the accuracy of drowsiness detection we apply simultaneously with the neural network, another algorithm which is described in the following chapter.

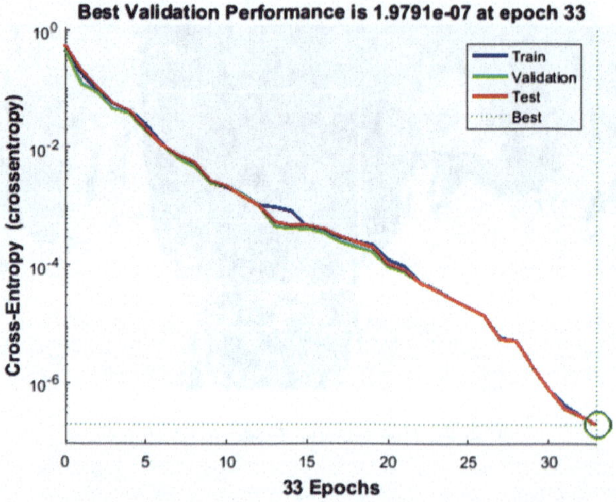

Fig. 4. Performance diagram of the ANN training.

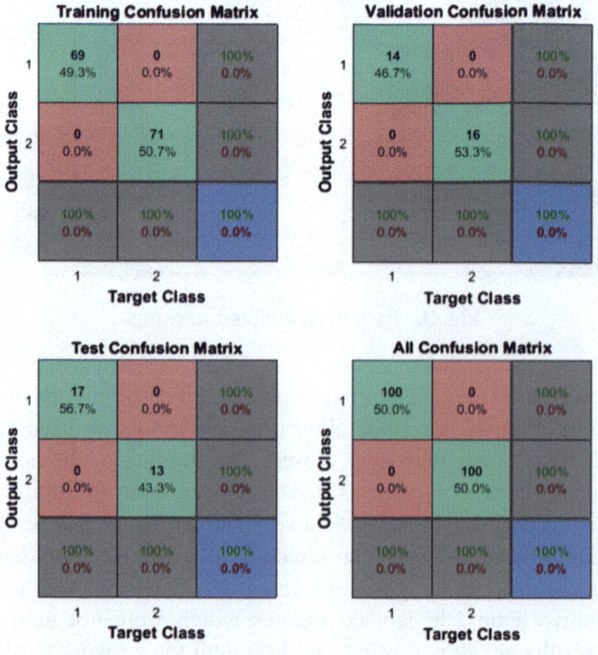

Fig. 5. Confusion diagrams for the ANN training.

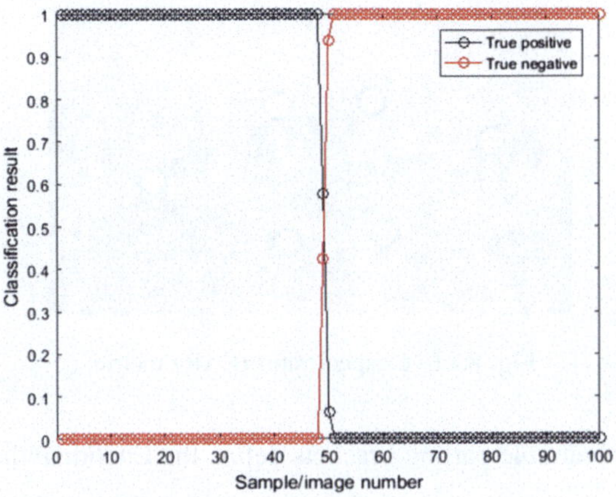

Fig. 6. Classification results for the eye state detection ANN for 100 test frames.

4 Eye Aspect Ratio (EAR) Algorithm

The Eye Aspect Ratio (EAR) algorithm used in our case is described in [16] and is based on researches presented in [17, 18]. This algorithm also uses face detection and tracking with labeled facial landmarks (coordinates of regions surrounding each facial structure) and was trained on the 68 point iBUG 300-W dataset (dlib facial landmark predictor).

Applying the eye detection algorithm we can track the eyes of the driver as it is shown in Fig. 7.

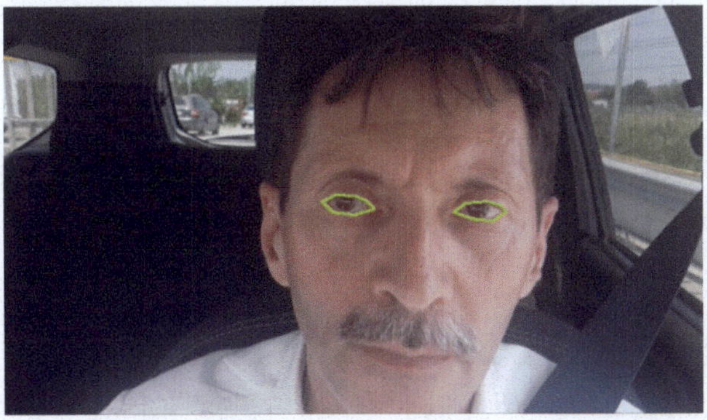

Fig. 7. Eye detection in streamed video.

On the detected contour of the eyes 6 characteristic points are defined as it is presented in Fig. 8. The EAR is then computed using relation (1). High values for EAR are obtained for the opened eye and low values are obtained for the closed eye.

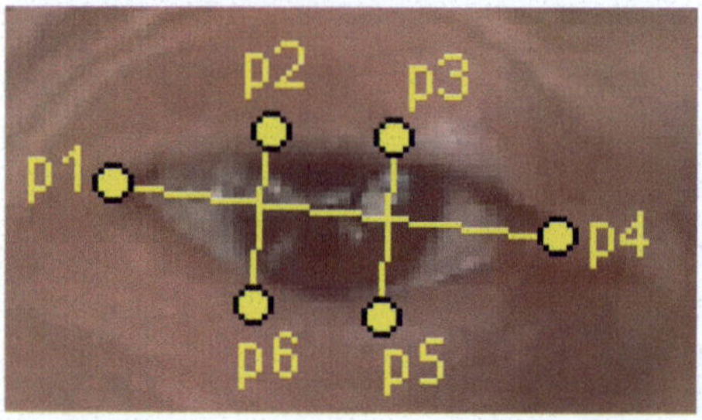

Fig. 8. Eye aspect ratio (EAR) metric.

In relation (1) the double parting brackets define the Euclidian distance between the specified points.

$$EAR = \frac{\|p2 - p6\| + \|p3 - p5\|}{2\|p1 - p4\|} \quad (1)$$

Applying the EAR algorithm to the same video stream used in the ANN analysis we obtain a value of EAR for each frame which had been represented in Fig. 9.

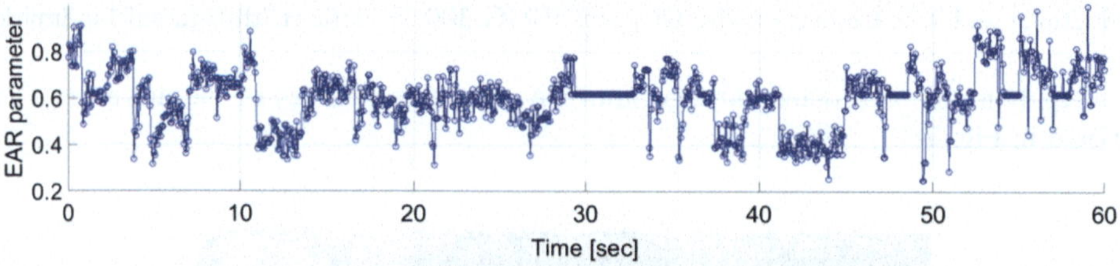

Fig. 9. EAR algorithm results for 60 s of video streaming.

In Fig. 10, we labeled the portions of the graph corresponding to the states of the eyes in the video stream.

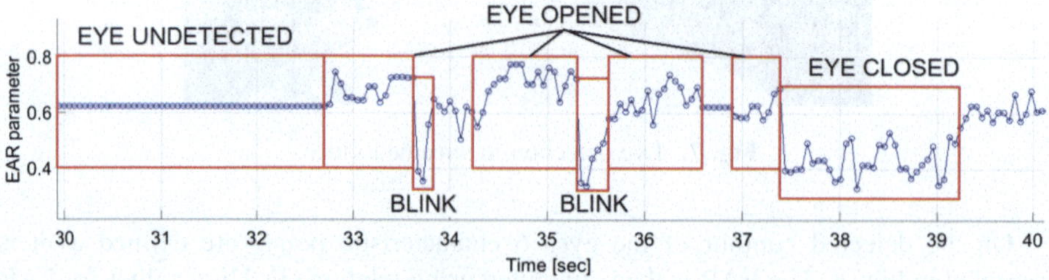

Fig. 10. Labeling of eye states in EAR results.

Analyzing the results of the EAR algorithm we can see that there are portions of the video stream where the EAR parameter could not be computed and also there are values of the EAR parameter which are hard to decide in which category are contained (opened or closed eye).

5 Fuzzy Inference Decision System

As it had been presented in the previous chapters image processing algorithms are prone to errors due to the complexity of the image subject (driver face movements, interior of the car, environment changes seen through windows) and also some issues in the detection accuracy. Also in a previous work for the EEG detection [6] we saw that EEG signals also have issues with noisy signals and sensor accuracy. One way to overcome these inconvenient and to improve accuracy is to use these methods simultaneously and employ a multi-criterial decision system.

A well-known multi-criterial decision method is the use of fuzzy inference systems. In this research we used the Fuzzy Logic Toolbox in order to develop such a system. The system has as inputs the results of EEG processing and the ANN and EAR image processing. The system is defined by the input membership functions, trapezoidal in our case (example for EEG signal is presented in Fig. 11).

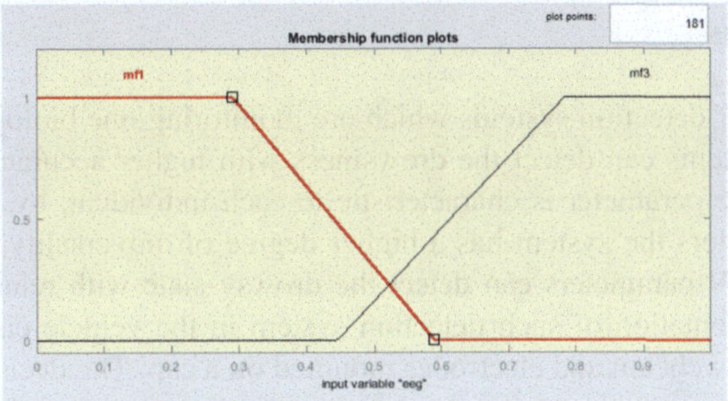

Fig. 11. Input membership function for the EEG signal.

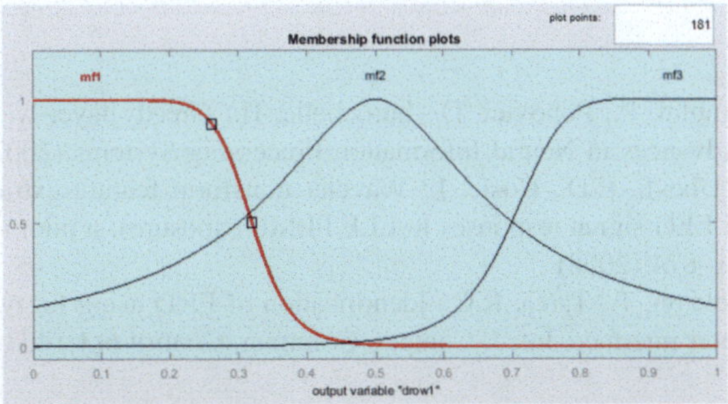

Fig. 12. Output membership function of the fuzzy inference system.

Figure 12 presents the shape of the output membership function. The chosen output method was of Mamdani type. The rules had been defined so that if at least two of the inputs are representing eyes closed states than the system decides the drowsy state.

Example of the decision surface is given in Fig. 13 for EEG and ANN image processing inputs.

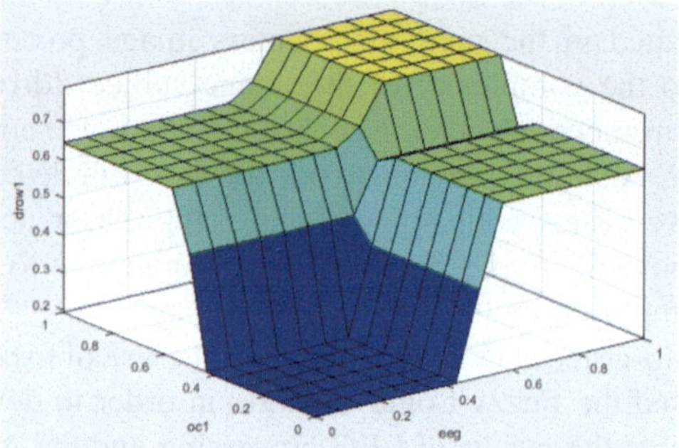

Fig. 13. Decision surface (representing the fuzzy rules) for EEG and ANN image processing inputs.

6 Conclusions

Compared to other detection systems which are monitoring one biologic parameter the multi-criterial systems can detect the drowsiness with higher accuracy.

If one biologic parameter is characteristic to each individual, by monitoring more biological parameters the system has a higher degree of universality.

EEG and ANN parameters can detect the drowsy state with relatively high accuracy. The implementation of such detection system in the vehicle can be easily done with the help of a webcam and electrodes mounted on a cap. The decisional system can be used in a soft manor only to alert the driver as the drowsy state is installed or can be a harder one which can actuate the propulsion system of the car.

References

1. Bengio, Y., Lamblin, P., Popovici, D., Larochelle, H.: Greedy layer-wise training of deep networks. In: Advances in Neural Information Processing Systems (2007)
2. Cvetkovic, D., Übeyli, E.D., Cosic, I.: Wavelet transform feature extraction from human PPG, ECG, and EEG signal responses to ELF PEMF exposures: a pilot study. Digit. Signal Proc. **18**(5), 861–874 (2008)
3. Dzitac, I., Vesselenyi, T., Tarca, R.C.: Identification of ERD using fuzzy inference systems for brain computer interface. Int. J. Comput. Commun. Control **6**(3), 403–417 (2011). ISSN 1841-9836
4. Hinton, G.E., Osindero, S., Teh, Y.-W.: A fast learning algorithm for deep belief nets. Neural Comput. **18**, 1527–1554 (2006)

5. Robert-Bela, N., Popentiu-Vladicescu, F., Vesselenyi, T.: An analysis of electro-oculogram signals processing using an artificial neural network. In: International Scientific Conference - eLearning and Software for Education, vol. 3, pp. 560–567 (2017). https://doi.org/10.12753/2066-026x-17-257
6. Vesselenyi, T., Rus, A., Mitran, T., Tataru, B., Moldovan, O.: Vehicle driver drowsiness monitoring and warning system. In: 12th International Congress of Automotive and Transport Engineering (CONAT), Brasov, pp. 873–880 (2016)
7. Vesselenyi, T., Moca, S., Rus, A., Mitran, T., Tataru, B.: Driver drowsiness detection using ANN image processing. In: International Congress of Automotive and Transport Engineering - Mobility Engineering and Environment (CAR 2017), Pitesti, Romania, 08–10 November 2017. Proceedings ISI (2017)
8. Vesselenyi, T., Dzitac, I., Dzitac, S., Hora, C., Porumb, C.: Preliminary issues on brain-machine contextual communication structure development. In: IEEE Conference, 3rd International Workshop on Soft Computing Applications, Szeged, Hungary, pp. 35–40 (2009)
9. Ting, W., Guo-zheng, Y., Bang-hua, Y., Hong, S.: EEG feature extraction based on wavelet packet decomposition for brain computer interface. Measurement **41**(6), 618–622 (2008)
10. MATLAB – Mathworks, Users Manual, Neural Network Toolbox, Deep Learning, Autoencoders (2016)
11. Purnamasari, P.D., Hazmi, A.Z.: Heart beat based drowsiness detection system for driver. In: Proceedings - 2018 International Seminar on Application for Technology of Information and Communication: Creative Technology for Human Life, iSemantic 2018, pp. 585–590 (2018)
12. Schmidt, J., Laarousi, R., Stolzmann, W., Karrer-Gauss, K.: Eye blink detection for different driver states in conditionally automated driving and manual driving using EOG and a driver camera. Behav. Res. Methods **50**(3), 1088–1101 (2018)
13. Antonya, C., Butnariu, S., Beles, H.: Parameter estimation from motion tracking data. In: 6th International Conference on Digital Human Modeling, DHM 2015 Held as Part of 17th International Conference on Human-Computer Interaction, HCI International 2015, Los Angeles; United States, 2–7 August 2015. Lecture Notes in Computer Science (including subseries Lecture Notes in Artificial Intelligence and Lecture Notes in Bioinformatics), vol. 9185, pp. 113–121, Code 123809. Springer (2015). https://doi.org/10.1007/978-3-319-21070-4_12. ISSN 03029743. ISBN 978-331921069-8
14. Bradski, G.R.: Real time face and object tracking as a component of a perceptual user interface. In: Proceedings of the 4th IEEE Workshop on Applications of Computer Vision (1998)
15. Viola, P.A., Jones, M.J.: Rapid object detection using a boosted cascade of simple features. In: IEEE CVPR (2001)
16. Rosebrock, A.: Drowsiness detection with OpenCV, 8 May 2017. https://www.pyimagesearch.com/2017/05/08/drowsiness-detection-opencv
17. Kazemi, V., Sullivan, J.: One millisecond face alignment with an ensemble of regression trees. In: Computer Vision and Pattern Recognition, Columbus, Ohio, USA, June 2014. https://doi.org/10.13140/2.1.1212.2243
18. Soukupova, T., Cech, J.: Real-time eye blink detection using facial landmarks. In: Čehovin, L., Mandeljc, R., Štruc, V. (eds.) 21st Computer Vision Winter Workshop, Rimske Toplice, Slovenia, 3–5 February 2016 (2016)

Contribution to Research the Applied Engineering Protocol to Implement a Fuzzy Regulator for Autonomous Driving of an Automotive Model Implemented in Virtual Reality

Florin Covaciu and Doru-Laurean Băldean(✉)

Technical University of Cluj-Napoca, 400641 Cluj-Napoca, Romania
florin.covaciu@muri.utcluj.ro,
doru.baldean@auto.utcluj.ro

Abstract. Monitoring and controlling the autonomous driving in automotive sector where high traffic density and risk factors are recorded it's today a common trend in global research and high potential for artificial intelligence implementation. But reaching to the large scale market and the series production in the same time it's a quite difficult problem that has to be approached and solved by engineers. The objective of the present research paper is mainly to innovate the application of a fuzzy regulator and to model a specific vehicle type in virtual reality through multidisciplinary components. It aims to implement some authentic methods for data collection, transfer and whole digital management, in the vehicle control system for automotive sector correlated with the virtual reality protocols, meaning that the algorithms and protocols of the vehicle in different conditions will be modeled and digitally tested in a special environment a priori to their implementation on full scale model. Specific objectives consist in a short technical presentation of the fuzzy regulator operational protocol for system communication method in the case of an road vehicle with autonomous driving capabilities and a practical set-up conducted in the Laboratory of Artificial Intelligence for Automotive Sector at Technical University from Cluj-N with the technical data retrieved from virtual reality tested model. Artificial intelligence applied to the control system is integrated in the model when protocols are defined in virtual reality for optimizing the driving conditions and improving the stability and steering, thus positively impacting the environment, society and autonomous vehicles.

Keywords: Artificial intelligence · Autonomous driving · Fuzzy · Virtual reality

1 Introduction

1.1 Research Objectives

Modeling and defining the logic behind the autonomous driving process in transport sector, where high traffic density and risk factors are often recorded, it is nowadays a

I. Dumitru et al. (Eds.): SMAT 2019, *The 30th SIAR International Congress of Automotive and Transport Engineering*, pp. 468–476, 2020.
https://doi.org/10.1007/978-3-030-32564-0_54

common trend and a demanding task in global research and development creating a high potential for artificial intelligence application [1–9]. On the other hand, achieving the scale for series manufacturing and large marketing will be a difficult problem with many aspects that have to be solved by engineering approach [7–10]. The main objective of the present research paper consists in the innovation of the fuzzy regulator application to a specific vehicle model with the support of virtual reality by using multidisciplinary technics. The target is the implementation of a specific logic and method for data recording, computation, transfer and complete digital management for the vehicle's autonomous driving system. Automobile sector backed up by fuzzy logic is analyzed with the virtual reality protocols. This means that the mathematical algorithms and protocols of the autonomous driving vehicle (in different conditions) will be mathematically modeled and digitally validated, using a virtual environment, a priori to their application in full scale car. Specific objectives are the following: a specific technical description of the fuzzy logic and regulator; defining an operational protocol for system communication method in the case of a small road vehicle with autonomous driving capabilities and also an experimental set-up model prepared in the Laboratory of Artificial Intelligence for Automotive Sector at Technical University from Cluj-Napoca, based on the data retrieved from virtual reality validated car.

1.2 State of the Art

A significant bulk of research deals with fuzzy drive expert systems for autonomous driving systems and auto-cruise cars [1–11]. This system consists of four rule sets such as surroundings recognition rules, driving control rules, learning-evaluation rules, and management meta-rules. The structure of those rules is hierarchical. The fuzzy drive expert system is structured with live units; the distance extraction and image processing unit, the environment recognition unit, the control unit, the learning-evaluation unit, the I/O unit, the knowledge rule base, and the man-machine interface. Each unit drives the fuzzy production rules which are described by sentence and symbols based on the if-then type format. Antecedent parts and consequent parts of those rules include the fuzzy words such as big, positive, wide, short, and so on. On the basis of the recognized result of environment, the control unit manipulates the steering and the throttle valve (or fuel injectors, brake pressure) for direction control and speed control of vehicle. The vehicle drive controls on the straight road and the corner are simulated on the digital computer. The overtaking control, the tracking control, and the avoidance control of obstacles are successful and smooth [11].

In the autonomous driving, lane keeping control and lane changing control are among the first problems to be analyzed and solved properly. One major challenge for lateral controller design is the smoothness and robustness under different vehicle conditions. A multi-model fuzzy controller, which includes four local controllers, is then proposed for both the operations of lane keeping and lane changing. These four local controllers are set up for four vehicle speed regions respectively. A fusion block is then implemented to provide a smooth and accurate transition between the different controllers. The proposed controller inherits the advantages of both the multi-model control and fuzzy control. Simulations show that it could get good performances in the whole range of operation speed, and could also repel the system uncertainties such as changes in vehicle load, movement inertia and wheel cornering stiffness [12–21].

Classic or conventional logic has its sets in which there is a collection of any kind of objects. In such a classical set a certain object belongs or not to a set. There may be applied some kind of operations to these sets, such as: union, intersection and complementary.

Meanwhile fuzzy sets are represented by their member like function. Figure 1 shows a U fuzzy set with its member function. In this example, may be said that 100 doesn't belong to the U set. 200 belongs to 0,2 member degree, 250 belongs to 0,3 member degree and 300 belongs to 0,4 member degree. Fuzzy operators are defined for realizing the union, intersection and complementation. These operators are fitted with linguistic significance AND, OR and respectively NOT. The AND operation may be determined with a function of T-Norm class. An AND operator used in usual mode may be the smallest from the two member functions.

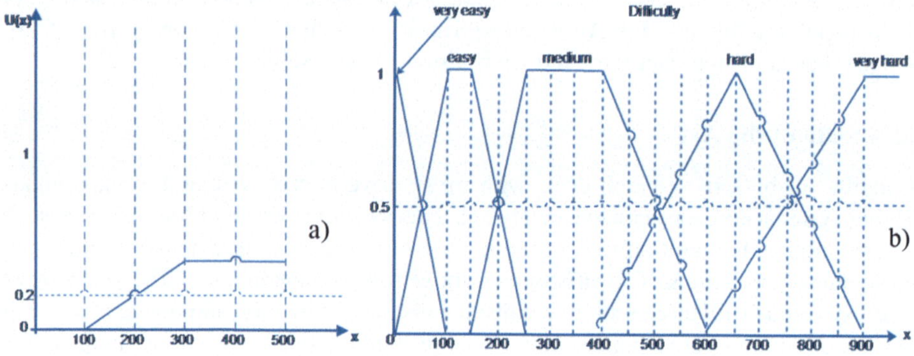

Fig. 1. Graphical representation of fuzzy sets (a) and linguistic variables (b).

Fuzzy approach extends the classical sets by returning to the function F(x) a specific value in domain [0, 1], meaning that the variable x belong in the set (1), doesn't belong to the set (0), but it may belong even 0.6 or 0.6/1.0, as it is shown in Fig. 1. The linguistic variables are data used in fuzzy systems, replacing numerical values.

Fuzzy logic was introduced as an alternative approach for solving problems when the classic theory and discreet mathematics are inappropriate or too complex to be applied. In order to approach a specific set of problems that a human being may solve not with the support of precision and mathematical accuracy, but based on a tolerant perspective toward errors it is imposed the capability to manage partial truths. Fuzzy calculus manages the values qualitatively and not quantitatively. It may define so called linguistic variables replacing classical numeric variables and may develop the calculation with them using fuzzy rules imitating and simulating processes of human reason. Incomplete or uncertain data may be used for calculus, as follow: -small increase of speed; -high increase of speed; -this object is big; -this object is small; -turn a little bit right etc.

In all these cases the significance of linguistic variables such as little, low, high, small, big, easy, hard, very hard, etc. is more relevant (then the precise numerical value) for solving the specific problem. Operating, calculating and representing the problems in this perspective there are used Fuzzy sets, which are quite common in daily human communication.

2 Methodology and Materials

2.1 Methodology

Applying state of the art artificial intelligence and other available engineering protocols and methods from multiple sectors of modern science in order to develop the most advanced automotive applications, is the optimal way to get from present state to a new and improved set of results when vehicles may have available programs and sufficient data in order to drive itself autonomously without direct human interaction in driving process. This is supported by a complex interaction as shown in Fig. 2.

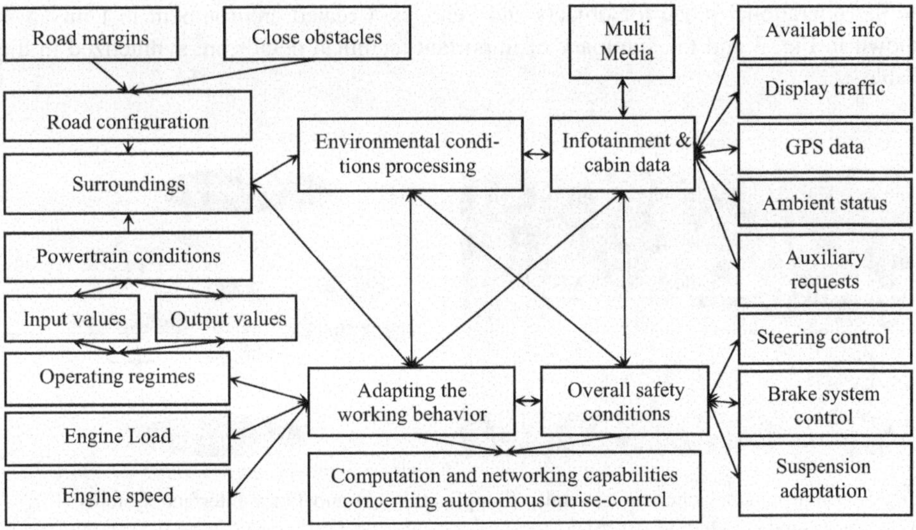

Fig. 2. Protocol summary for aspects that have to be considered in common driving scenarios.

The present research develops a program for virtual reality modeling and testing of artificially intelligent (AI) car that is driving itself in real time. Computer aided data transfer and processing of the significant parameters using virtual reality both for power-train control and for the AI protocol. The cyber engineering proposed methodology is applied in virtual reality modeled vehicles and systems, thus implementing latest achievements in the field of AI management and data networking, as well as autonomous driving available data and some infotainment capabilities that are quite important to define a whole systemic car-to-car communication protocol. Virtual reality for innovative achievements in artificial intelligence and cyber physical systems are combined in the present work for improving automotive sector. Applying virtual reality potential and network-data transferring capability in the automotive engineering development protocol is the main methodological protocol for innovative achievements and new feature in autonomous driving process.

2.2 Materials for Research

Virtual Reality (VR) is one of the state of the art technologies which has recorded an important development and an increased popularity. From its label and denomination, it is understood that the virtual reality is not anything but a specific technology that supports experimentation of a non-existent reality, an environment created by a computer. The concept of virtual reality dates from a past age, but in the same time it is (was) a futuristic aspect detailed and defined in lots of literary documents and movies (i.e. Star Trek, Matrix etc.), increasing the potential of changing the real life.

In order to generate the virtual reality application space it was implemented Unity 5 environment, which is an advanced state of the art program facilitating VR creation.

First, the environment generation has to be taken into consideration because it will be the operational place for objects and vehicles. Created environment in Unity 5 is shown in Fig. 3 and the summary of important technical details are synthetized in the Table 1.

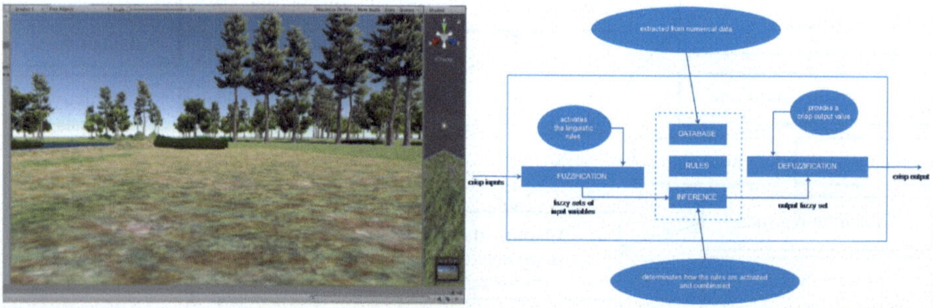

Fig. 3. Environment generation in unity 5 application (a) and fuzzy interface systems (FIS) – based on Mamdani Method (b) [5].

Table 1. Important data regarding the VR creation and environment [21]

Parameter	Actual values
Implemented application	Unity 5
Programing potential	3 codes: Unity Javascript, C# and Boo
Export capabilities	Windows; Linux; Mac; Windows Phone; BlackBerry; iOS; Unity Web Player: PlayStation, Xbox and Wii
Compatibilities	Virtual Reality (VR) and Augmented Reality (AR)
Alternative solutions	Personal Unity and variant Pro Unity

After generating the environment in the Virtual Reality program Unity 5 it has to be generated the driving track.

2.3 Operations and Properties of Fuzzy Sets [15]

Fuzzy sets through generalization allow elements to belong in a given set with a certain degree as it was mentioned previously. Instead of placing the characteristic functions with value in {0, 1} they are considered valued in [0, 1]. Considering the fuzzy subset F of a set X is a function $\mu_F(x)$ assisting to every element x of X the degree of membership of x to F [15]:

$$x \in X \mapsto \mu_F(x) \in [0, 1], \tag{1}$$

The operations that are used in fuzzy sets consist in union, intersection and complement, defined by the models below:

$$\text{Fuzzy op. 1} \left\{ \begin{array}{ll} \chi_{A \cup B}(x) = \max\{\chi_A(x), \chi_B(x)\} & \mu_{F \cup S}(x) = \max\{\mu_F(x), \mu_S(x)\} \text{ union} \\ \chi_{A \cap B}(x) = \min\{\chi_A(x), \chi_B(x)\} & \mu_{F \cap S}(x) = \min\{\mu_F(x), \mu_S(x)\} \text{ inters.} \\ \chi_{A'}(x) = 1 - \chi_A(x) & \mu_{F'}(x) = 1 - \mu_F(x) \text{ complement} \end{array} \right\} \tag{2}$$

There are also available other operations that are applicable to the fuzzy sets as these:

$$\text{Fuzzy op. 2} \left\{ \begin{array}{l} \chi_{A \cap B}(x) = \min\{\chi_A(x), \chi_B(x)\} = \chi_A(x) \times \chi_B(x) \\ \chi_{A \cap B}(x) = \min\{\chi_A(x), \chi_B(x)\} = \max\{0, \chi_A(x) + \chi_B(x) - 1\} \\ \mu_F(x) \cap \mu_S(x) = \mu_S(x) \cap \mu_F(x), \text{commutative intersection} \end{array} \right\} \tag{3}$$

Fuzzy output is the significance of system output, but in many situations, there is imposed a supplementary numerical output equivalent with the fuzzy output. There is applied the method of centroid, where it is calculated mass center of X axis, as follow:

$$y = \frac{\sum_{i=1}^{n} \mu_0(x_i) x_i}{\sum_{i=1}^{n} \mu_0(x_i)} \tag{4}$$

where n is number of steps, because it is an approximation method for area center [5].

3 Applied Part to Virtual Modeling

3.1 The Novelty of the Conducted Research

The artificial intelligence, infotainment developing trend and cyber physical systems in automotive engineering sector today is the most innovative and further high potential vector regarding the future transport and road traffic. Thus, defining, modeling and implementing new features or capabilities in the artificial intelligence and on-board data management module facilitate other opportunities for innovative cyber system development and vehicle automation. The research staff is highly experienced in artificial intelligence, automotive engineering and electro-machines building. In each specific case advanced methods and virtual data management are significant for the automotive sector development. The research of artificial intelligence, digital modeling and virtual reality capabilities on modern autonomous vehicles as mobile drive by itself units in traffic specific conditions represents an original achievement in cyber research and automotive trans-disciplinary approach. The application protocol for specific fuzzy regulator integration and the virtual reality model in operation condition of data transfer based on real infrastructure e and laboratory testing-modeling expertize at Technical University from Cluj-N. gives the innovative feature to the paper.

3.2 Track Generation and Vehicle Setting

Results are consisting in applied program for fuzzy regulator data management and transfer from virtual environment to car power-train management module and a protocol for artificial intelligence capability available on-board. There is a data acquiring protocol which is modeled, analyzed and defined to spot the most important ideas related to automotive artificial intelligence and virtual reality improvement opportunities. There are also centralized specific results gained in the first phases of modeling and implementing the data transfer protocol to the virtual reality environment in the vehicle on-board management module.

Figure 4 represents the testing track and the vehicle generated for VR testing.

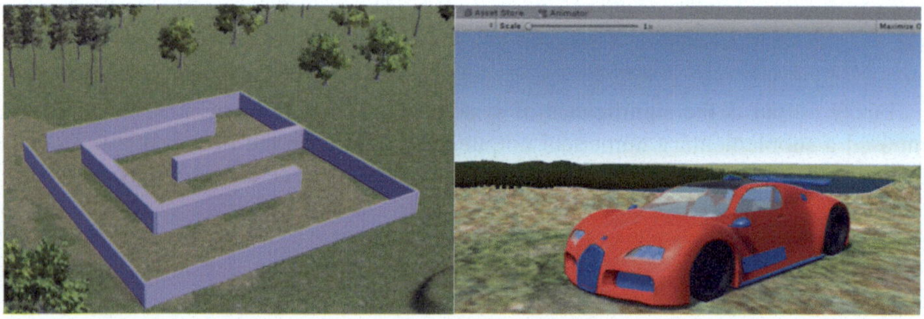

Fig. 4. Placing the track and the vehicle in VR environment. Left-track generation; right-vehicle introduction in the VR environment.

4 Conclusions

Improving the level of control, the data regulatory process by fuzzy characteristics, and management integration capability of the AI available today, with high potential in automotive field creates an added value to the engineering sector. Innovating data management in road vehicles generates an optimal feature for autonomous vehicles of the future. Fuzzy regulators from specific applications are important because they are contributing with other factors (traffic configuration, environmental conditions etc.), making the architecture more complex and more responsive. Data are studied in the virtual reality model for interaction of the vehicle in the multi-factorial traffic environment. AI and data management create a complex info-traffic environment in which robotics and automotive sector lead to the optimizing possibility for studied model. The AI applied to the control system is integrated in the model when protocols are defined in virtual reality for optimizing the driving conditions and improving the stability and steering, thus positively impacting the environment, society and autonomous vehicles. Also, AI, as operational data package, generates the decisions to be made due to predictions regarding the contextual autonomous driving of the vehicles in real traffic conditions, creating opportunities for efficiency improvement and for reducing the time spent in jams. Innovative methods and virtual reality modeling were also applied in this research to support development of autonomous driving solutions.

References

1. AlGhamdi, A.S., et al.: Using FUZZY logic to control vehicles' velocity in traffic system. IFAC Proc. Vol. **46**(10), 54–59 (2013)
2. Atagoziyev, M., et al.: Lane change scheduling for autonomous vehicles. IFAC-PapersOnLine **49**(3), 61–66 (2016). https://doi.org/10.1016/j.ifacol.2016.07.011
3. Attia, R., et al.: Longitudinal control for automated vehicle guidance. IFAC Proc. Vol. **45** (30), 65–71 (2012). https://doi.org/10.3182/20121023-3-FR-4025.00049
4. Bayard, D.S., et al.: Implicit dual control for general stochastic systems, First published July/September 1985. https://doi.org/10.1002/oca.4660060307. Accessed 22 Mar 2018
5. Caversan, F.L.: Fuzzy Computing: Basic Concepts. AForge.NET Framework, Articles, Forums. http://www.aforgenet.com/articles/fuzzy_computing_basics/. Accessed 22 Mar 2018
6. Chebly, A., et al.: Coupled longitudinal and lateral control for an autonomous vehicle dynamics modeled using a robotics formalism. IFAC PapersOnLine **50**(1), 12526–12532 (2017)
7. Fustier, B.: The mathematical framework of fuzzy logic. In: Achille, L., Betti, G. (eds.) Fuzzy Set Approach to Multidimensional Poverty Measurement (2006). ISBN 978-0-387-34251-1
8. Jiménez, F., et al.: Advanced driver assistance system for road environments to improve safety and efficiency. Transp. Res. Procedia **14**, 2245–2254 (2016)
9. Katrakazas, C., et al.: Real-time motion planning methods for autonomous on-road driving: state-of-the-art and future research directions. Transp. Res. Part C: Emerg. Technol. **60**, 416–442 (2015)
10. Krolikowski, A., et al.: Constrained suboptimal dual control algorithms for discrete-time stochastic systems. In: ICINCO 2009 - 6th International Conference on Informatics in Control, Automation and Robotics (2009). ICINCO_2009_41_CR.pdf. Accessed 22 Mar 2018
11. Maeda, M., et al.: Fuzzy drive expert system for an automobile. IFAC Proc. Vol. **26**(2, Part 3), 105–108 (1993). https://doi.org/10.1016/1069-0115(94)00076-E

12. Maslowski, A., et al.: Hybrid simulation system for analysis of autonomous vehicles. IFAC Proc. Vol. **29**(1), 8168–8173 (1996)
13. Riedewald, F.: Comparison of deterministic, stochastic and fuzzy logic uncertainty modelling for capacity extension projects of DI/WFI pharmaceutical plant utilities with variable/dynamic demand. Ph.D. thesis, University College Cork (2011)
14. Singh, A.K., et al.: Fuzzy based controller for lidar sensor of an autonomous vehicle. Energy Procedia **117**, 1160–1164 (2017). https://doi.org/10.1016/j.egypro.2017.05.241
15. Spada, L.: Introduction to fuzzy sets and fuzzy logic, department of mathematics and computer science. University of Salerno, REASONPARK, Foligno, 17–19 September 2009 (2009). http://logica.dipmat.unisa.it/lucaspada/wp-content/uploads/foligno_handout.pdf. Accessed 19 Apr 2018
16. Taha, M.A., et al.: Traffic simulation system based on fuzzy logic. Procedia Comput. Sci. **12**, 356–360 (2012). https://doi.org/10.1016/j.procs.2012.09.084. Accessed 22 Mar 2018
17. Ulivi, G.: Emergent techniques for mobile robots and autonomous vehicles. IFAC Proc. Vol. **29**(4), 31–38 (1996). https://doi.org/10.1016/S1474-6670(17)44681-7. Accessed 22 Mar 2018
18. Yusupbekov, N.R., et al.: Application of soft-computing technologies to the traffic control system design problems. Procedia Comput. Sci. **102**, 540–546 (2016)
19. Zhao, J., et al.: Integrated longitudinal and lateral control system design for autonomous vehicles. IFAC Proc. Vol. **42**(19), 496–501 (2009)
20. Zhao, J., et al.: Lateral control of autonomous vehicles using multi-model and fuzzy approaches. IFAC Proc. Vol, **43**(8), 514–520 (2010)
21. Working in Unity. https://docs.unity3d.com/Manual/UnityOverview.html. Accessed 02 Apr 2018

Study of Vehicle Dynamics Using Tensor Analysis

Radu Vilău, Ion Lespezeanu(✉), Constantin-Iulian Pântece, Alexandru Vasile, and Ion Copae

Military Technical Academy "Ferdinand I", Bucharest, Romania
ion_lespezeanu@yahoo.com,
vasile.alexandru05@gmail.com

Abstract. This paper highlights a vehicle's dynamics study using elements from tensor calculus, considered as a superior step of the matrix calculus. The experimental data used in the paper was collected from vehicle's on-board computer. Based on what the paper presents, we managed to conclude that specific elements of tensor calculus can be used for studying the vehicle's dynamics and efficiency, given the fact that the achieved results are very close to the experimental data.

Keywords: Tensor analysis · Vehicle dynamics · Fuel consumption · Mathematical model

1 Introduction

Frequently we find in technical literature as an appreciation criteria for vehicle dynamics indicators such as, acceleration time until a certain speed is reached, space covered in the same conditions, average speed and average acceleration [1, 2, 4]. Also the technical literature uses as appreciation criteria for vehicle efficiency, the fuel consumption for 100 covered kilometers, hourly fuel consumption, or fuel consumption reported to the vehicle's mileage unit.

If experimental data are available, then the vehicle's dynamic behavior benefits of concrete values of its functional parameters; to approach vehicle dynamic behavior and efficiency or to approach engine functionality, vectors of discrete values are obtained on each test run. For example, Fig. 1 present the average values of all 30 test runs that we had available when experimenting with a Logan Laureate vehicle: Fig. 1a specific for the vehicle's speed V, and Fig. 1b for the vehicle's acceleration a.

As we can see here, in order to study the vehicle's dynamic behavior, value vectors were used for speed and acceleration; so in this case we call on vector analysis for vehicles dynamic behavior. As we can notice from graphs, if we pay attention to the average vehicle speed, from Fig. 1a we can deduct that the most dynamic test run was L19 (reaching 100,5 km/h); otherwise if we pay attention to the average acceleration we can see that the most dynamic test run is L26 (with 0,59 m/s^2).

So if we call on vector analysis, which is frequently used, there are some ambiguities that emerge when appreciating the vehicle's dynamic behavior as do for its

I. Dumitru et al. (Eds.): SMAT 2019, *The 30th SIAR International Congress of Automotive and Transport Engineering*, pp. 477–484, 2020.
https://doi.org/10.1007/978-3-030-32564-0_55

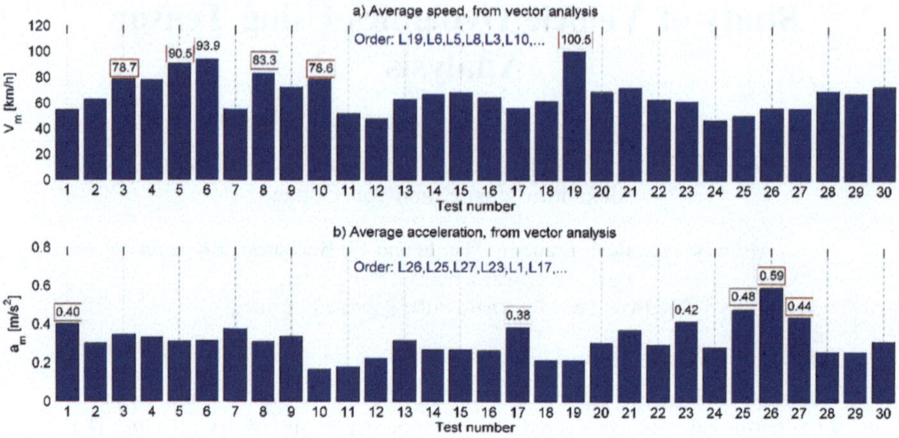

Fig. 1. Dynamic study: average values for vehicle speed and acceleration, Logan vehicle

efficiency; the mentioned ambiguities are accentuated if we look at the other appreciation indices used when judging the vehicle's dynamic behavior (acceleration time or ground covered until reaching a certain speed).

The next step relies on matrix analysis, which uses value matrix, for example the average values for the vehicle's speed that were recorded during all 30 experimental test runs. As we can see from Fig. 1, the ambiguity in appreciating the vehicle's dynamic behavior is maintained, the two criteria offering a different classification of all 30 test runs (the graphs offer a decreasing order for the vehicle's dynamic behavior during all its test runs). So the next step is to call on tensor analysis, meaning to move towards tensors that engulf all the functional parameter data from all available test runs.

2 Tensor Analysis

Tensor analysis is an extension of matrix analysis, which in its turn is an extension for vector analysis; vector analysis calls on value vectors, multivariable analysis calls on matrix, and tensor analysis calls on tensors [3, 5]. Thus experimental data and data achieved through calculus based on it may be processed in three different ways. So if we take into consideration only one parameter (for example the vehicles speed) from only one experimental test run (for example P1), then we have a value vector, so a single-dimensional array z (Fig. 2a). If we take into consideration two or more functional parameters (for example speed V, vehicle acceleration a, hourly fuel consumption Ch. etc.) registered during one experimental test run (for example P1), than we deal with a matrix, thus a two dimensional array Z (as seen in Fig. 2b). Finally, if we target several functional parameters from several experimental data (for example $P1, P2, \ldots, P30$), then we get a tensor, in this case a three dimensional array (Fig. 2c).

As already known, a tensor represents a multidimensional array of data; a tensor's order is equal to the space's dimension, Fig. 2c represents a third degree tensor N = 3 (in the three dimensional space). A scalar (a number/a value) is a zero degree tensor

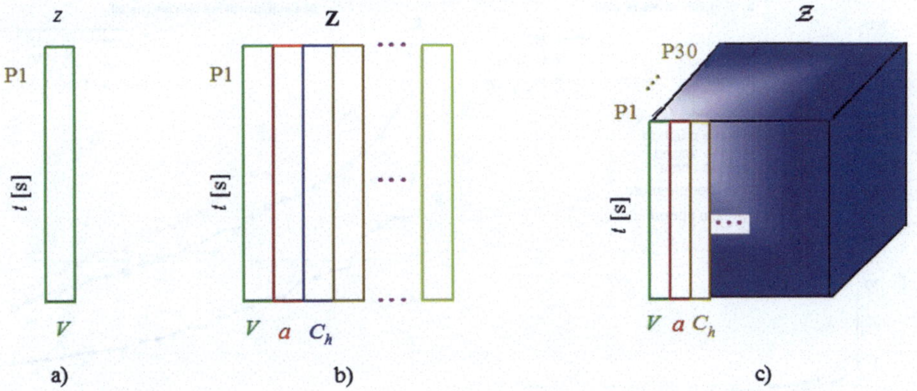

Fig. 2. Principal sketches for applying vector, matrix and tensor analysis

(N = 0), a vector represents a first degree tensor (N = 1), a matrix represents a second degree tensor (N = 2); so we can say that tensor analysis represents a generalization of vector analysis. Figure 2c presents a tensor in three dimensions (third degree tensor N = 3) when considering 30 experimental test runs (noted P1, P2, . . ., P30). As we can see from Fig. 2c, the frontal plane of the tensor contains a matrix which displays the values for the functional parameters recorded on test run P1; the rows contain their values at a certain moment in time t for vehicle's speed, vehicle acceleration etc.

As the case of matrix, spectral analysis relies on tensor decomposition (factorization), in order to obtain its eigenvalues, eigenvectors and it's singular values [3, 5]. Likewise as matrix case data tensor data compression means establishing singular values that are the highest (highest eigenvalues), in order to ensure tensor estimation with imposed precision. For example, Figs. 3 and 4 present the establishment of singular values for a tensor $\mathcal{Z}(175 \times 3 \times 50)$ in the case of three groups of three functional parameters that are mentioned on graphs (each with 175 values) from all 50 experimental test runs data that were available on the Logan Laureate vehicle; we achieve this way $3 \times 50 = 150$ singular values for each group, presented in Figs. 3a and 4a. Figure 3 targets the vehicle dynamic behavior (through the vehicle's speed), and Fig. 4 targets the vehicle's efficiency (through fuel consumption recorded when covering 100 km).

In order to establish the contribution of each, Fig. 3a present the relative singular values σ_r, meaning the current values divided by the maximum one. In order to establish the dominant singular values (that has the highest contribution to the vehicle's dynamics), a precision of 3% is imposed as we can see from Fig. 3b where a detail of Fig. 3a is presented. The graphs from Fig. 3 show different singular values for various groups of functional parameters; thus it is confirmed that various functional parameters have different contribution to the vehicle's dynamic behavior. Other than that Fig. 3 presents a horizontal line representing an imposed error of 3%. As we can see from Fig. 3b, if we adopt an error of 3%, then in the case of group 1 of parameters (ξ, n, V) only 14 singular values may be retained; in other words, in the case of group 1 of parameters it is sufficient to use only 14 relevant singular values when describing the

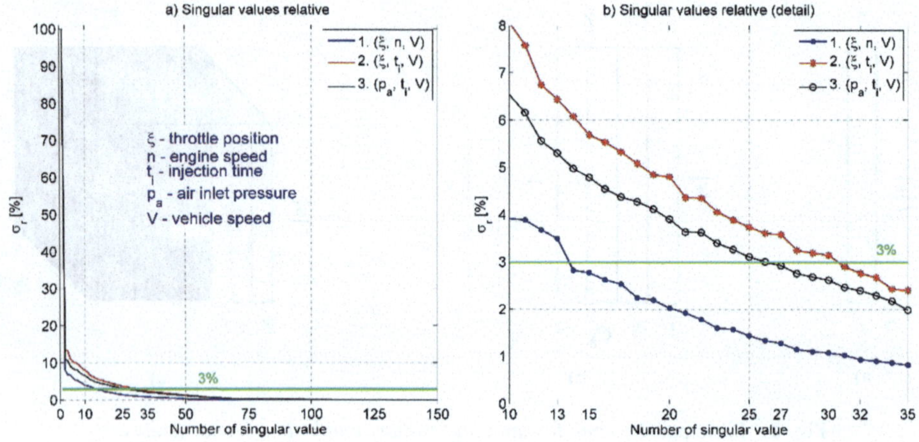

Fig. 3. Dynamic study: establishment of relevant singular values for a three dimensional tensor

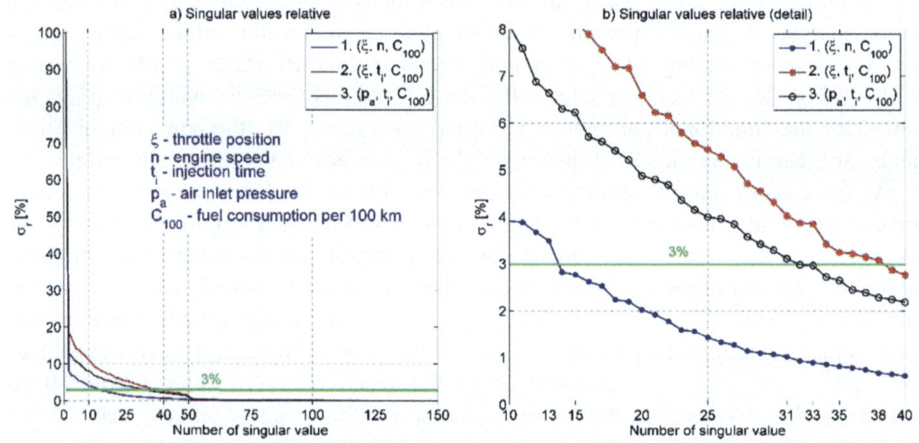

Fig. 4. Efficiency study: establishment of relevant singular values for a three dimensional tensor

vehicle's dynamic behavior, for example when establishing a mathematical model for the vehicle's movement. Otherwise if we target the second group of parameters (ξ, t_i, V) and the third group (p_a, t_i, V) of parameters, we have to hold 27 respectively 31 singular values, thus the number of relevant singular values is increased. In other words, in order to achieve calculus errors of under 3%, more singular values should be kept when the experimental dynamic array is rebuilt (when estimating it through a mathematical model).

Similarly, from Fig. 4b, if we adopt a calculation error of 3%, then in the parameter group 1 (ξ, n, C_{100}) we can keep only 14 singular values, same number from Fig. 3b; in other words, in the case of group 1 it is sufficient to keep only 14 relevant singular values when studying the vehicle's efficiency. If we target group 2 of parameters

(ξ, t_i, C_{100}) and group 3 of parameters (p_a, t_i, C_{100}), we have to keep 33 respectively 39 singular relevant values (dominant).

As we can see from Figs. 3b and 4b, the engine's speed is the most important influence factor onto the vehicle's dynamic behavior, as well as onto the vehicle's efficiency; the number of relevant singular values being the same in both cases (equal to 14), therefor we can deduct that engine's revolution speed influences in the same amount both the dynamic behavior as well as the vehicle's efficiency. In exchange, because the other parameters generate different singular values (27 and 31 compared to 33 and 39), we can deduct that these parameters (throttle's position, injection duration, and intake air pressure) influence the dynamic behavior in a higher degree than the vehicle's efficiency. In addition, from Figs. 3b and 4b we can deduct that intake air pressure has a higher influence than the throttle's position ξ in both dynamic behavior as well as in the vehicle's efficiency.

What has just been presented reveals that the engines speed is the one parameter that best defines vehicle dynamic behavior; thus if it is to establish a mathematical model for vehicle dynamic behavior or if it is wished to study the influence of certain parameters onto the vehicles mode of operation, the engine's revolution speed is a parameter that must be taken into account.

As mentioned, the graphs from Figs. 3 and 4 are obtained by decomposition (factorization) of the targeted tensor, thus obtaining its eigenvectors and eigenvalues. As the case of matrix, for a third degree tensor we get the spectral decomposition:

$$A = S \text{ x } U_{(1)} \text{x } U_{(2)} \text{x} \ldots\ldots \text{x } U_{(N)} \tag{1}$$

Where the inferior index marks the tensors mode (1, 2 and respectively 3). In expression (1) S is called nuclei and represents a tensor of equal dimension with A; if we were to make an analogy with matrices S is to some extent similar with the matrix that contains on its principal diagonal the singular values of the matrix that is targeted for decomposition. Added to that, in relation (1) U(i) matrices are called modal matrices (afferent to mode $1, 2, \ldots, N$ tensor); these are orthogonal and contain the main sub unitary vectors of A(n) matrices which in their turn are achieved by transforming the tensor A into matrices.

3 Obtained Results Regarding the Vehicle Dynamics

Coming back to our efficiency study, the superior graphs from Fig. 5 resume those from Fig. 1; in exchange Fig. 5c presents the results of applied tensor analysis, using the tensor that engulfs both appreciation indices for vehicle dynamic behavior (average speed and average acceleration). The graphs from Fig. 5 contain the decreasing order of the 30 test from dynamic point of view; as it can be noticed, the established order of this tensor analysis (the correct one) coincides with the order established by vector analysis of the average speed.

As mentioned before, tensor analysis has the advantage that allows for taking into consideration at the same time several functional parameters, influence factors or appreciation indexes for vehicle dynamics. For example, Fig. 6 targets another

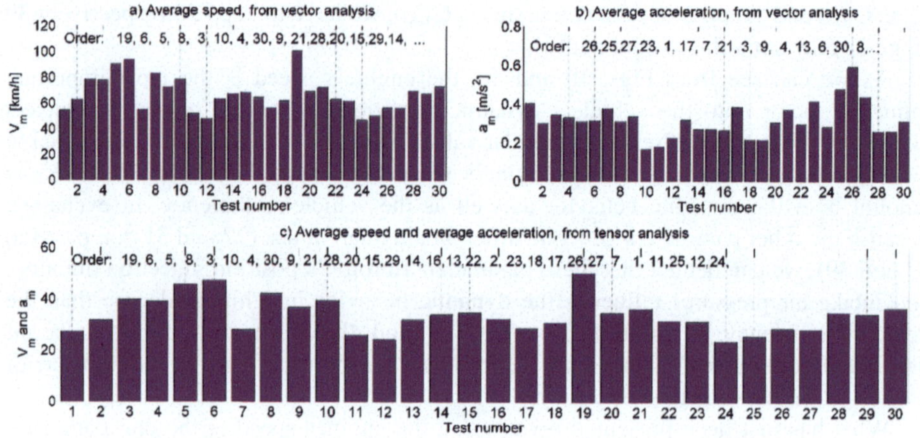

Fig. 5. Dynamic study considering two appreciation criteria

appreciation criterion that of space covered the vehicle when it's accelerating next to the other two that were presented in Fig. 5. As we can see from Figs. 5c and 6d, adding another appreciation criterion leads to a different classification for the test runs applying tensor analysis. We can conclude that applying tensor analysis and by taking into consideration all parameters the right conclusions will be drawn.

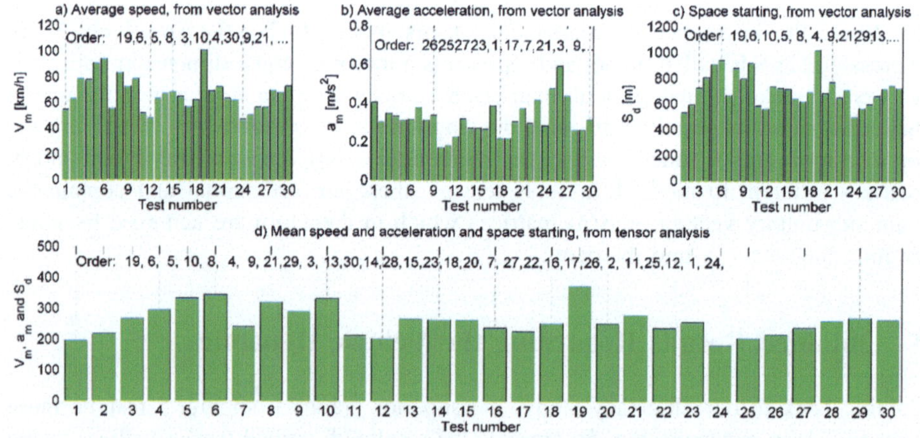

Fig. 6. Dynamic study taking into consideration three appreciation criteria

Similarly, Fig. 7 presents the results when applying vector analysis (upper graphs) and tensor analysis (Fig. 7c), using the tensor that contains both appreciation criteria for vehicle efficiency (average values for fuel consumption when traveling 100 km and hourly fuel consumption). As we can see, tensor analysis leads this time to a different classification of test runs from the original classification achieved by vector analysis; as it can be observed from Fig. 7c, test run L24 is characterized by the most efficient test run, and test run L19 has the lowest efficiency.

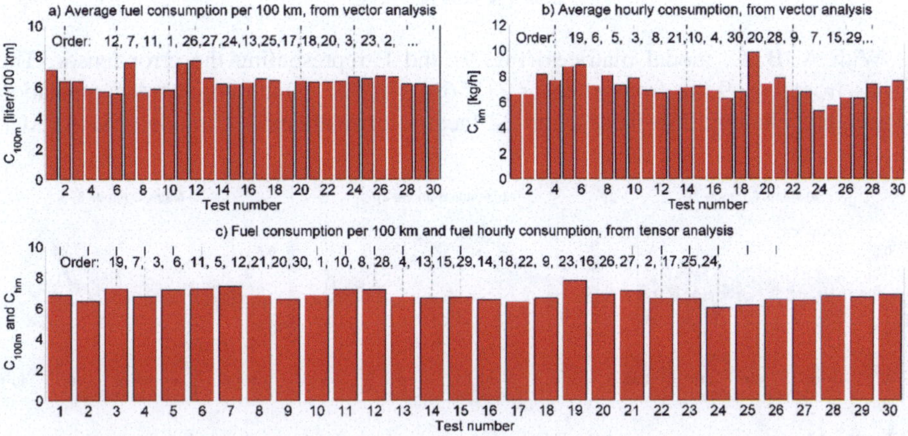

Fig. 7. Vehicle efficiency study considering two appreciation criteria

The presented study may be applied also for performance comparison of different vehicles as we can see from Fig. 8. The graphs reveal that on the overhaul 50 experimental test runs available the dynamic behavior of Logan vehicle is superior to Tacoma by 29.7% obviously in certain test conditions for each vehicle.

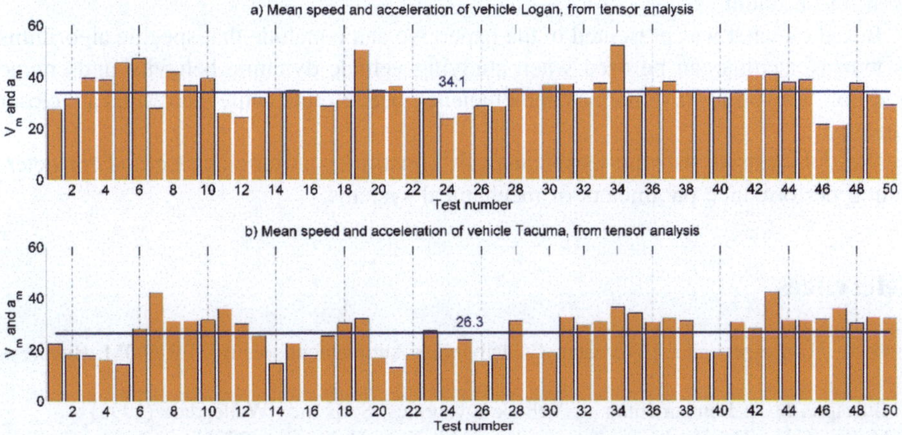

Fig. 8. Comparative study of dynamic behavior for two different vehicles using tensor analysis.

The spectral analysis previously presented based on decomposition that calls on singular values, was extended and improved, leading to multidimensional mathematical models (modal models). One of the most popular one is PARAFAC (PARAllel FACtor), described by equation [5]:

$$Z = (\mathbf{A} \circ \mathbf{B})\mathbf{C}^{T} + E \tag{2}$$

With A, B, C, modal matrices (Fig. 9) and E representing the error tensor. The graphs from Fig. 9 are from a tensor $Z(256 \times 4 \times 32)$, which engulfs 4 appreciation criteria, each having 256 values, and 32 factors for mathematical model PARAFAC.

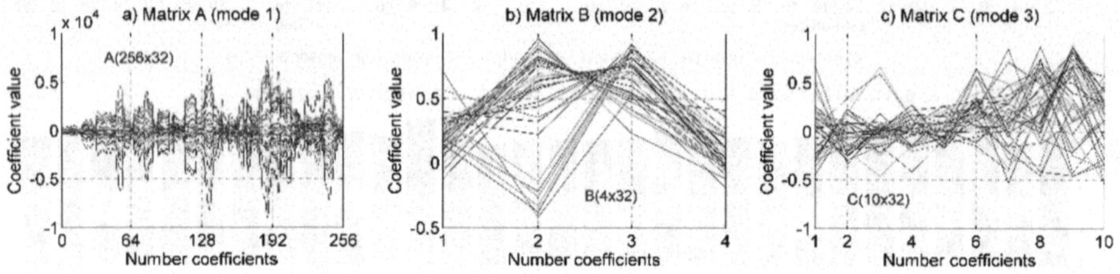

Fig. 9. Modal matrices for PARAFAC mathematical model for a three dimensional tensor.

4 Conclusions

The algorithm presented reveals that the engines speed is the one parameter that best defines vehicle dynamic behavior; thus if it is to establish a mathematical model for vehicle dynamic behavior or if it is wished to study the influence of certain parameters onto the vehicles mode of operation, the engine's speed is a parameter that must be taken into account.

Based on what was presented in the paper, we can conclude that specific algorithms for tensor calculus can be used when studying vehicle dynamic behavior and engine operation, as well as its dynamics and efficiency because the achieved results are closer to reality.

We demonstrate that the tensor analysis constitutes an efficient method for determining performance parameters of mechanical systems.

References

1. Copae, I., Lespezeanu, I., Cazacu, C.: Dinamica Autovehiculelor. Ed. ERICOM, Bucureşti (2006)
2. Gillespie, D.T.: Fundamentals of Vehicle Dynamics. SAE Inc., Wallendale (1992)
3. Martin, C.D.: The Rank of a Tensor. James Madison University (2006)
4. Pereş, Gh., Stoicescu, A.: Dinamica Autovehiculelor. Tipografia Universităţii Transilvania, Braşov (1988)
5. Sharipov, R.: Quick Introduction to Tensor Analysis. Bashkir State University (2004)

Aerodynamic Study of an Electric Vehicle Prototype

Liviu Scurtu[1(✉)], Ancuta Jurco[1], Emilian Vasile Borza[1], Florin Mariasiu[1], Nicolae Vlad[1], and Silvan Morariu[2]

[1] Technical University of Cluj-Napoca, Cluj-Napoca, Romania
liviu.scurtu@auto.utcluj.ro
[2] Inovo Finance SRL, Cluj-Napoca, Romania

Abstract. In this paper is presented the aerodynamic study of a prototype structure for an electric vehicle. In the first part of the paper is presented the actual research of body shape vehicles using the iterative CFD methods and the theoretical aspects of this method. In the second part of this study, using advanced modelling techniques the CAD model of the car body is realized. The three-dimensional model of the car body is modelled in SolidWorks software. In the third part of the work, the CAD model is prepared and imported in the CFD environment, where is simulated for four different car speeds: 14 m/s, 25 m/s, 34 m/s and 55 m/s. The shape of the car body is improved in order to achieve a lower drag coefficient. The improved car body shape is tested at the same conditions and parameters like as the first simulation. A comparative study is presented between the first and second car body shape simulations results. In the last part of the paper are presented the conclusion and the future research directions of this study.

Keywords: Electric vehicle · Car body shape · CFD · Drag coefficient

1 Introduction

Electric vehicles are becoming more and more accepted by consumers in the current automotive market. By the time these vehicles were proposed as sustainable alternatives and less polluting solution in the medium and long term compared to internal combustion engine vehicles, there were a number of market barriers that prevented massive penetration into the auto market of this vehicle type.

Market barriers such as: acquisition price, energetic performance, necessary/dedicated charging infrastructure etc.; have been overcome and/or removed through the following actions: providing tax incentives and subsidizing the purchase price, creating more performant batteries and propulsion systems, increasing the number of public charging stations, parking spaces reserved exclusively for electric vehicles, etc. [1, 2].

However, the main issues related to massive penetration of the car market by electric vehicles are related to the still of small number of models available and the maximum range that these vehicles can achieve at a single full charge of battery.

In this context, the development of new models of energy-efficient electric vehicles becomes a contemporary necessity. The main challenges for designers in the

I. Dumitru et al. (Eds.): SMAT 2019, *The 30th SIAR International Congress of Automotive and Transport Engineering*, pp. 485–493, 2020.
https://doi.org/10.1007/978-3-030-32564-0_56

automotive field remain to define, design and calibrate the electric or hybrid power train to achieve high dynamic performances and a minimum autonomy requested by consumers.

The factors that directly influence the autonomy and operation of an electric vehicle are quite numerous and can be remembered as: gross weight, passenger number, body type, volumetric luggage capacity, rolling train, specific front area, type and shape type, HVAC (air conditioning, opened windows, use of lights, entertainment systems, etc.) Their influence on the overall energy efficiency of an electric vehicle was widely presented in papers that addressed this issue [3–5] (Fig. 1).

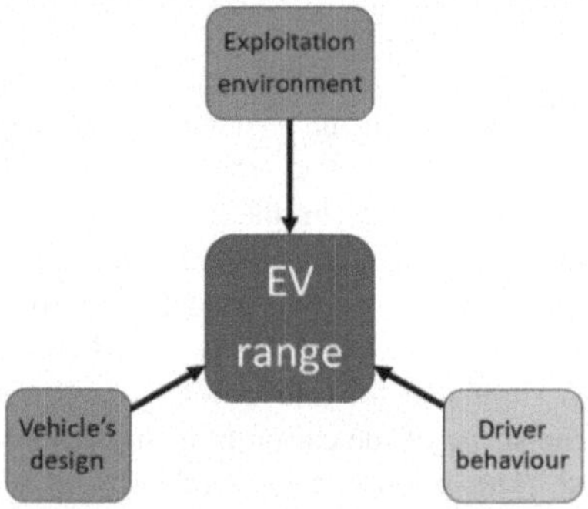

Fig. 1. Classes of influence factors on electric vehicle's range.

By studying and analyzing the results of researches on these issues, the aerodynamic resistance coefficient is an important parameter in defining the amount of energy lost by the vehicle in motion. Although the influence of this parameter in road traffic conditions in the main cities is low (due to the fact that the average traffic speed is between 12 … 25 km/h [6]), the electric vehicle must be designed also for use in the highway regime, where high speeds are achieved which lead to increased energy losses due to the air and rolling resistance.

Thus, it is important to pay attention to reducing the energetic losses when designing an electric vehicle, and the main method by which this analysis (aerodynamic optimization) can be achieved easily is the use of computerized CFD simulation methods.

The present paper aims to analyze by computerized simulation methods the aerodynamic characteristic of an electric vehicle prototype to be developed in the near future. Are presented the initial conditions that were taken into consideration and parameters of variation of geometric shape of the prototype (body), in order to identify the optimal constructive shape from the aerodynamic point of view. Numerical computational simulation is performed by the CFD method to obtain the necessary information about the airflow around the body of the electric vehicle prototype.

2 Method

The methodology of the iterative CFD analysis that is applied in this research is presented in Fig. 2, where are presented the initial steps. In the first step the car model design is modeled in CAD software. The car model is introduced into a computational domain definite by the user. In this domain are applied required boundary conditions and the airflow velocity in the second step.

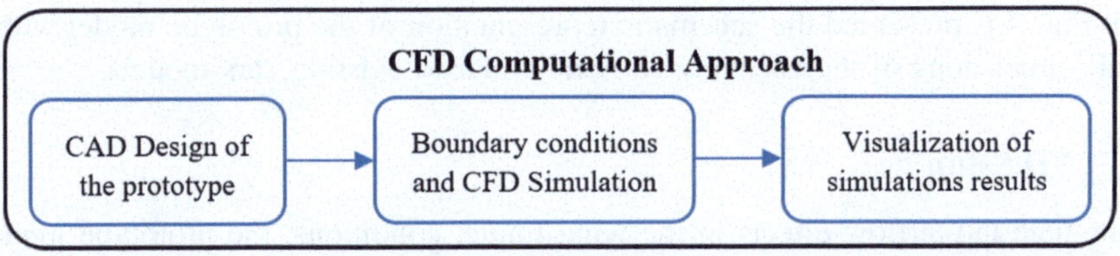

Fig. 2. The steps of the CFD process

The end step of the CFD simulation are given by the results collecting, the airflow streamlines trajectories visualizations, the pressure and velocity distribution.

2.1 CAD Design of Car Prototype

In car prototype design approach, the car is modeled on the SolidWorks, using surface module [7]. In Fig. 3 are presented the front and rear axonometric view of the car model. The hood geometry is designed to generate a greater down force, achieving a better vehicle stability at the high speeds.

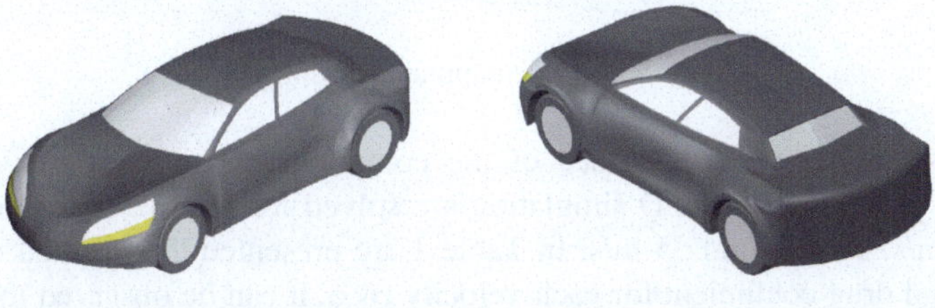

Fig. 3. Axonometric representation of the CAD prototype base model

The rear side of the car is modelled in order to optimize the shape of the model, by changing the rear slant the shape can be optimize to streamlines have a linear trajectory without turbulent flow.

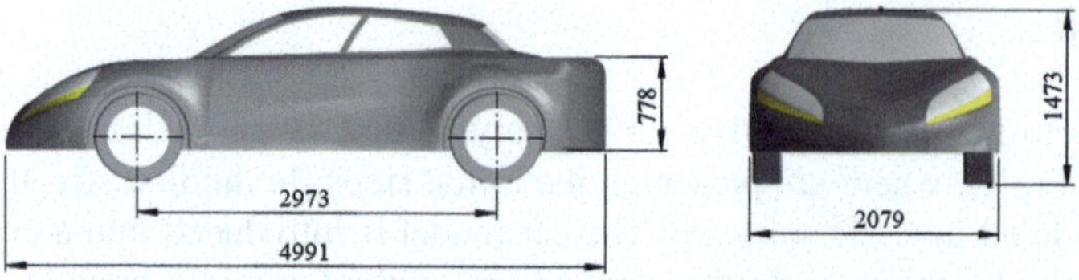

Fig. 4. Schematic representation of the CAD prototype model

In Fig. 4 is presented the schematic representation of the prototype model with the overall dimensions of the car body chosen similar to existing cars models.

2.2 CFD Setup

To simulate the airflow effects in the wind tunnel conditions, the prototype model is introduced into a virtual computational domain, where the airflow conditions are the same like as into the wind tunnel. Dimensions of the computational domain are very important to view the airflow trajectories. In Fig. 5 are presented the virtual wind tunnel and the position of the car model into the computational domain. The width of the computational domain is 5 m, the height is 4 m and the length 25 m. The front side of the car model is situated at 5 m length from the air inlet and computational domain has a length of 15 m from the rear side of the car, this length is recommended to view the streamline behavior on the CFD simulations.

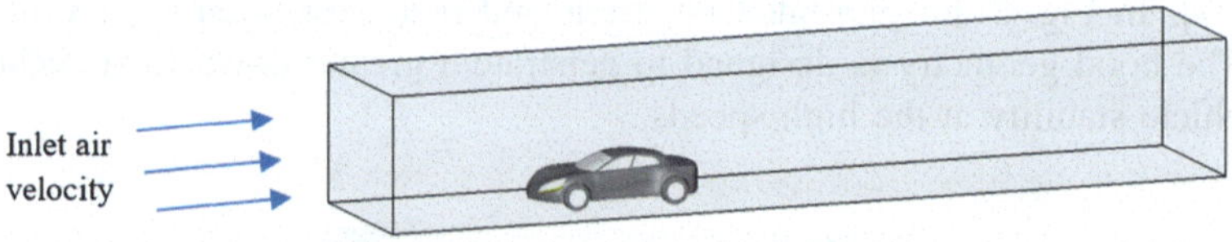

Fig. 5. CFD computational domain

The pressure of external surface of the computational domain are setup at the atmospheric pressure. The CFD simulation are solved for four velocity airflow cases: 14 m/s, 25 m/s, 34 m/s and 55 m/s. In Table 1 are presented the resulted value of the drag force and drag coefficient for each velocity case. It can be observed that the value of the drag force and the drag coefficient increase smoothly in relation to the increase the airflow velocity.

Table 1. Results simulation of the base model.

Velocity [m/s]	Drag force [N] - averaged value	Drag coefficient - averaged value
14	80.038	0.293
25	260.808	0.300
34	485.899	0.302
55	1276.659	0.303

In Fig. 6 are presented the front, right and top views of the streamline's distributions at a velocity of 34 m/s. In the present case is chose this velocity because this is almost the reference car velocity used for the European approval cycle of the aerodynamic coefficient. For a better visualization of a streamline's distributions are represented only the area around the car model. In the unrepresented area of the computational domain there not are the airflow turbulences generated by the car geometry. It can be observed that the airflow is not laminar in the behind of the car, there was a swirl caused by the shape of the car model. The maximum value of the velocity of the air is situated on the front side of the car and on the windscreen and the roof.

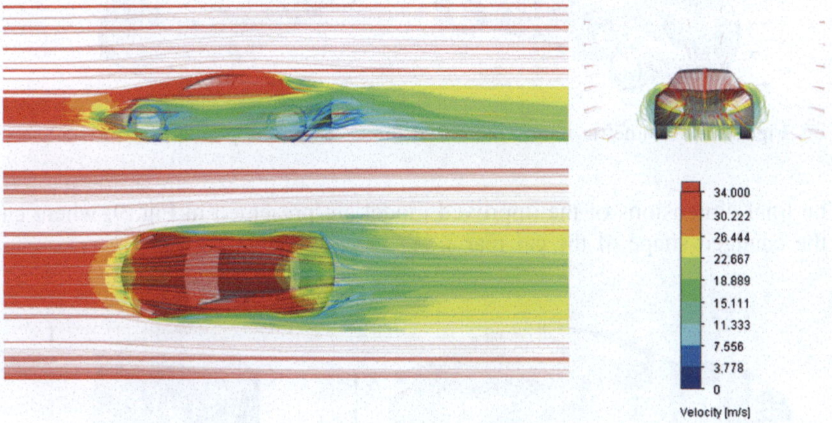

Fig. 6. Distribution of the streamlines of a 34 m/s velocity for the base car model

In Fig. 7 are presented the air pressure distributions when the airflows velocity is 34 m/s. It can be observed that the maximum value of the pressure is located on the front bumper surface and the front side of the hood.

Fig. 7. Pressure distribution of the air at 34 m/s velocity for the base car model

In order to decrease the drag force, the drag coefficient value and to reduce the airflow turbulences resulted in the rear side of the car. The base car model is improved by changing the rear side geometry from a height of 778 mm to 627 mm and adding a rib in the middle of the rear window as can be seen in Fig. 8, where are presented the front and rear side of the improved CAD model. It can be observed in Fig. 8 that it is designed the place for the license plate, the light and the rear bumper.

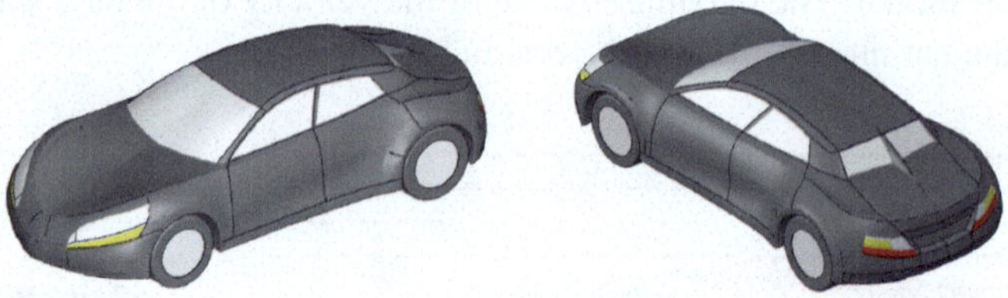

Fig. 8. Axonometric representation of the CAD prototype improved model

The final dimensions of the improved model are presented in Fig. 9, where can be seen the changed shape of the car rear side.

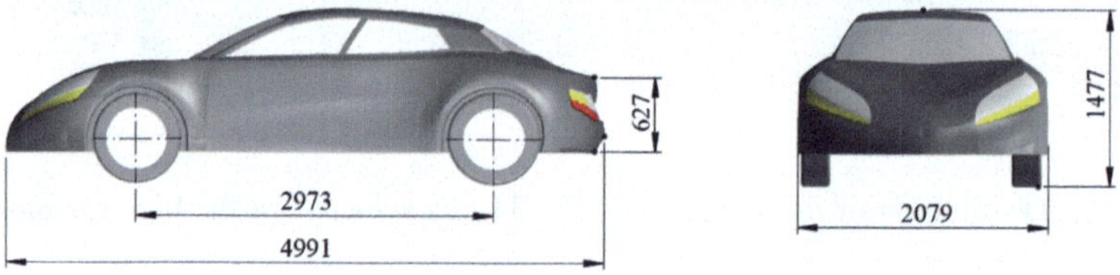

Fig. 9. Schematic representation of the improved CAD prototype model

To view the improved model results the CFD simulation is done at the same parameters used for simulation of the base model. In Table 2 is presented the resulted value of the drag force and the drag coefficient for the improved model.

Table 2. Results simulation of the improved model.

Velocity [m/s]	Drag force [N] - averaged value	Drag coefficient - averaged value
14	80.590	0.295
25	258.183	0.298
34	452.797	0.281
55	1189.960	0.282

In Fig. 10 are presented the front, right and top view of the airflow distributions from the improved model at a 34 m/s air velocity.

The plot of the air pressure distribution is presented in Fig. 11, where can be observed that the maximum value of the pressure is situated on the front bumper surface.

To a better view of air velocity distribution in all computational domain at 34 m/s is presented in Fig. 12 the axonometric view of the improved model.

In Fig. 13 is presented the axonometric view of the pressure distribution for the improved car model at a velocity of 34 m/s.

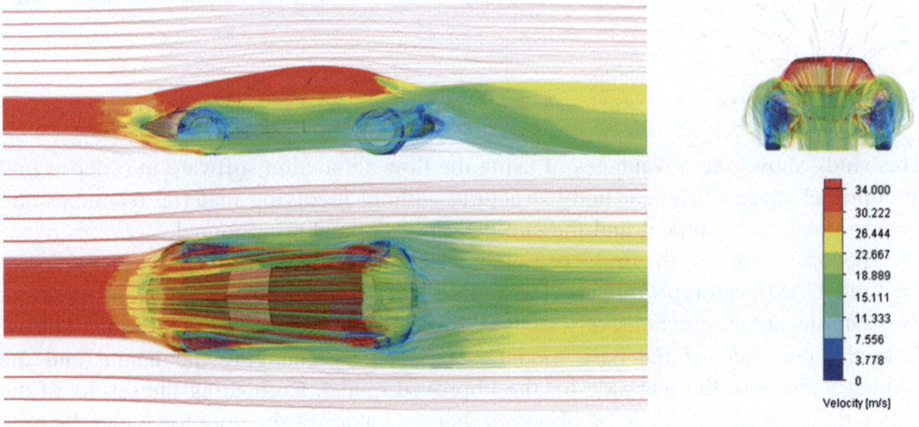

Fig. 10. Distribution of the streamlines of a 34 m/s velocity for the improved car model

Fig. 11. Pressure distribution of the air at 34 m/s velocity for the improved car model

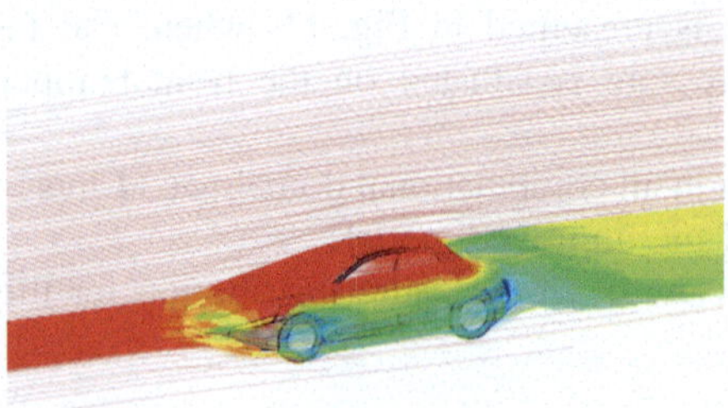

Fig. 12. Velocity distribution of the air at 34 m/s or the improved car model.

Fig. 13. Pressure distribution of the air at 34 m/s velocity for the improved car model.

3 Conclusions

This study shows the advantages of using the flow simulation software in order to find the optimal shape of the car body structure, without involving material resources and time to create a car model and the simulation in the real wind tunnel.

The drag force and the drag coefficient are calculated from the different air velocity for both CAD prototype models. Flow trajectories of the air, velocity and pressure distributions are comprehensively studied. It can be observed that the turbulent airflow from the rear side of the base model is reduced by changing the height and the geometry shape of the rear side for the improved model. Comparing the results of the both simulations cases it can be observed that the values of the drag force and the drag coefficient are lower at the improved model than the base model.

Acknowledgment. This research was funded by the Program "POC-A1-A1.2.3-G-2-15 Parteneriate pentru transfer de cunoştinţe, nr. contract de finanţare: 11/01.09.2016, titlul proiectului: Tehnologii pentru vehicule electrice urbane inteligente URBIVEL, ID: P_40_333, MySMIS: 105565".

References

1. Park, E., Lim, J., Cho, Y.: Understanding the emergence and social acceptance of electric vehicles as next-generation models for the automobile industry. Sustainability **10**, 662–675 (2018)
2. Egbue, O., Long, S.: Barriers to widespread adoption of electric vehicles: an analysis of consumer attitudes and perceptions. Energy Policy **48**, 717–729 (2012)
3. Varga, B.O., Sagoian, A., Mariasiu, F.: Prediction of electric vehicle range: a comprehensive review of current issues and challenges. Energies **12**(5), 946 (2019)
4. Mruzek, M., Gajdáč, I., Kučera, L., Barta, D.: Analysis of parameters influencing electric vehicle range. Proc. Eng. **134**, 165–174 (2016)

5. Iora, P., Tribioli, L.: Effect of ambient temperature on electric vehicles' energy consumption and range: model definition and sensitivity analysis based on nissan leaf data. World Electr. Veh. J. **10**, 2–17 (2019)
6. Trigg, T.: Cities where it's faster to walk than drive (2015). https://blogs.scientificamerican.com/plugged-in/cities-where-it-s-faster-to-walk-than-drive/. Accessed 18 May 2019
7. Bodea, S.M., Prodan, C.V., Scurtu I.L.: The Aerodynamic study of a body truck. In: International Congress of Automotive and Transport Engineering, pp. 73–79. Springer, Cham (2018)

Drag Coefficient Analysis for Side Mirrors of an Electric Vehicle Prototype

Liviu Scurtu[1(✉)], Adela-Ioana Borzan[1], Florin Mariasiu[1], Nicolae Vlad[1], Bogdan Varga[1], and Silvan Morariu[2]

[1] Technical University of Cluj-Napoca, Cluj-Napoca, Romania
liviu.scurtu@auto.utcluj.ro
[2] Inovo Finance SRL, Cluj-Napoca, Romania

Abstract. The geometry of the car side mirrors influences the aerodynamics of the car by increasing the drag coefficient. In the present paper is designed a CAD model of the external side mirrors for a new electric car body. The shape of the mirror is geometrical optimized in order to decrease the drag coefficient. In the first part of the paper are presented the actual studies regarding the electrical cars, aerodynamics and side mirrors shape. The optimization of the shape for side mirrors was determined using the iterative computational methods, CFD for determination the airflow streamlines. In the second part of the paper, using advantages modelling techniques are modeled two CAD model of the side mirror. The CFD simulation is performed for six different car velocity in the third part of the paper. A comparative study is carried out to choose the better side mirror shape. Conclusions regarding of this work are presented in the last part of the paper.

Keywords: Drag coefficient · CFD analysis · Rear view mirror · Optimization

1 Introduction

The electric vehicle is one of the important solutions to reduce pollutant emissions and GHG emissions from road transport systems. Pollutant emissions of an electric vehicle are zero for road traffic conditions, which offers an immediate increase in citizens' quality of life, especially in large urban agglomerations [1]. However, indirectly, and the electric vehicle cause pollution also due to the pollutant emissions caused by the electricity production process (depending on the electricity mix used to charge the battery) [3]. It is well known that in the energy balance's relation, due to the forces acting on a moving vehicle, the losses due to the forwarding resistance are related to the air resistance and the rolling resistance of the tires with the road surface (Fig. 1).

Mathematically, the energetic balance of a moving vehicle can be expressed as [4]:

$$F_t = F_g + F_{ar} + F_r = mg\sin\alpha + 1/2\ \rho_a\ C_d\ A\ (\upsilon + \upsilon_{fw})^2 + mgc_r, \qquad (1)$$

where: F_t is traction force to propel a vehicle forward, F_g – gravitational force, F_r – rolling resistance, F_{ar} – aerodynamic (air) drag, m is the mass of vehicle,

I. Dumitru et al. (Eds.): SMAT 2019, *The 30th SIAR International Congress of Automotive and Transport Engineering*, pp. 494–501, 2020.
https://doi.org/10.1007/978-3-030-32564-0_57

g – gravitational acceleration, α – air density, C_d – drag coefficient, A – vehicle's frontal area, v – speed, υ_{fw} – frontal wind speed and c_r is the rolling resistance coefficient.

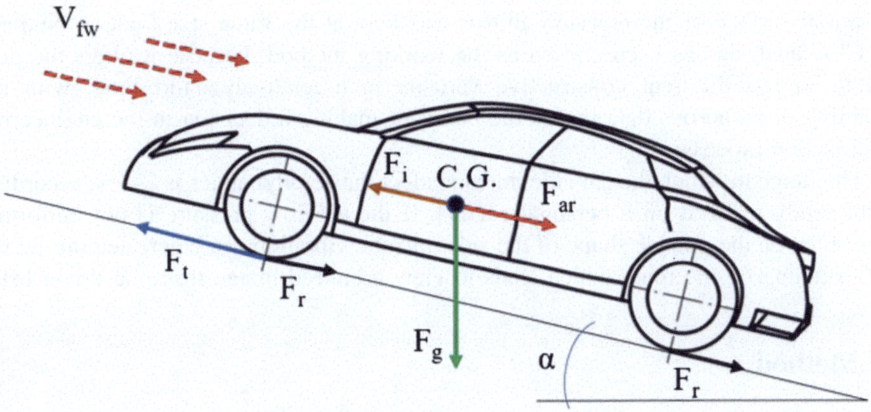

Fig. 1. Forces acting on a moving vehicle.

It's can be noticed that the term defining losses due to air resistance when the electric vehicle is in move, directly depends on the speed of the vehicle and its frontal area.

$$F_{ar} = 1/2\ \rho_a\ C_d\ A(\upsilon + \upsilon_{fw})^2 \tag{2}$$

In this context, it is noticed that the frontal area is a factor of direct influence by its size, and thus, theoretically, the lower the resistance of the electric vehicle is for less frontal area. However, there are some constructive conditioning imposed by the comfort that must be provided to the passengers, conditioning that limits the minimum dimensions of the vehicle's frontal area, namely the minimal space of comfort of the passenger compartment.

For these reasons, one of the directions by which the air resistance can be reduced is also the optimization of the external structure of the body from the aerodynamic point of view, allowing the air to flow in a laminar regime around it [5].

The rear-view mirrors are an element which, due to its construction and its location on the vehicle body, creates specific areas of turbulence that influence laminar airflow around the vehicle. This increases the vehicle's drag resistance and, implicitly, energy loss. To achieve an energy-efficient electric vehicle (with as much autonomy as possible) energy losses should be reduced/limited or even eliminated.

There are solutions to remove these rear-view external mirrors by implementing video cameras that can provide driver necessary information about the vehicle's exterior environment (by displaying them on monitors located on the vehicle's dashboard). However, the major disadvantage of this technology is the associated costs and the need to use complex, electronic and electrical components, equipment and systems with direct effect on the overall vehicle reliability.

The paper aims to analyze the possibility of aerodynamic optimization of a rear-view mirror that will fit an electric vehicle. As a variable parameter of CFD numerical simulations that were performed was considered the curvature of the mirror support, as the initial surface of the rearview mirror was kept at the same size (value). Numerical CFD analysis has been chosen as the working method, because it offers the possibility to test different constructive variants in a relatively short time, with the obtaining of numerous data that is the basis for making a decision in the engineering optimization process.

The range in which the car side mirrors affect the aerodynamics is 2–5%, according to the study realized on a personal car [7]. If the air flow pressure is not uniformly distributed on the outside shape of the side mirrors, vibration can be created during the car's running on the road, which leads to view a blurred image from the driver [8].

2 Method

2.1 CAD Model Design

The CAD model of the side mirror is designed by using the advantage modeling techniques in SolidWorks software [6]. For a better rear view the size of the mirror was chosen in accordance with Fig. 2. Another important role for improving the aerodynamic of the body car and the noise resulted by the air flow is the support that connect the mirror to the car body.

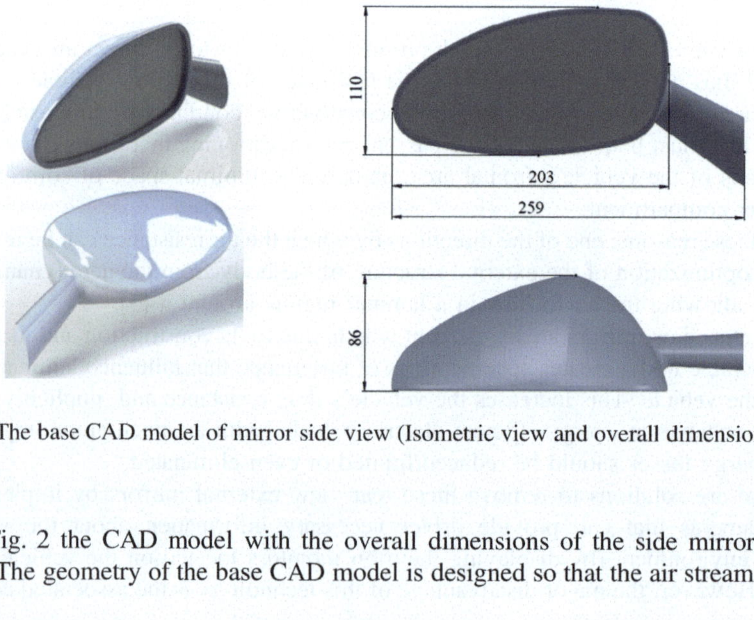

Fig. 2. The base CAD model of mirror side view (Isometric view and overall dimensions of the mirror).

In Fig. 2 the CAD model with the overall dimensions of the side mirror is presented. The geometry of the base CAD model is designed so that the air streamline has a laminar flow without turbulent flow in the front side of the mirror, and to decrease on the rear mirror side the turbulent flow.

2.2 CFD Setup

In the present optimization study an enclosed computational domain was created to simulate the distribution of the air flow on the mirror. The airflow simulation process is performed into the computational domain volume showed in Fig. 3. The width of the computational volume is 0.6 m, height is 0.4 m and the length are 1.2 m.

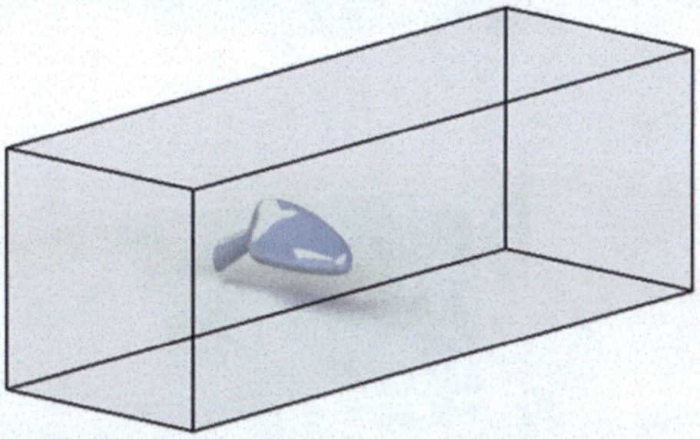

Fig. 3. Enclosed computational domain.

In this optimization analysis it is simulated the airflow on the mirror model, the mirror is stationery and fluid flow. In order to a better visualization of the streamline flow, the mirror model is placed into a computational domain at a distance of 0.3 m from the air inlet. The CFD simulation process is performed for six airflow cases: 10 m/s, 20 m/s, 30 m/s, 40 m/s, 50 m/s and 60 m/s.

3 Results and Discussions

The results of this optimizing analysis, the simulations results are represented in Table 1. It can be seen that the value of the drag coefficient and the drag force increases gradually with the increasing of the airflow velocity.

Table 1. Results of the base model simulations.

Velocity [m/s]	Drag force (Base CAD model) [N]	Drag coefficient (Base CAD model)
10	0.65627	0.4351
20	2.60387	0.4355
30	5.86396	0.4359
40	10.4585	0.4372
50	16.3732	0.4382
60	23.6064	0.4386

The air streamline distribution at the velocity case of 40 m/s is presented in Fig. 4. For a better viewing of the airflow effect, the right view of the mirror and the top view are presented in Fig. 4. It can be observed that in the rear side of the mirror the airflow is turbulent. After the turbulence area, the air streamline tends to have an increase flow over the mirror body.

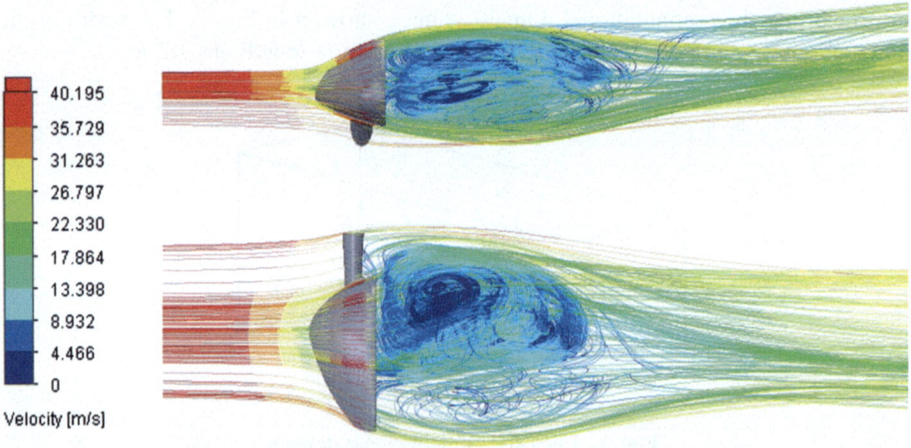

Fig. 4. Streamline distribution at velocity of 40 m/s.

Another important aspect of the mirror aerodynamics is the distribution of the air pressure on the mirror surface. Figure 5 shows the pressure resulting after airflow at 40 m/s.

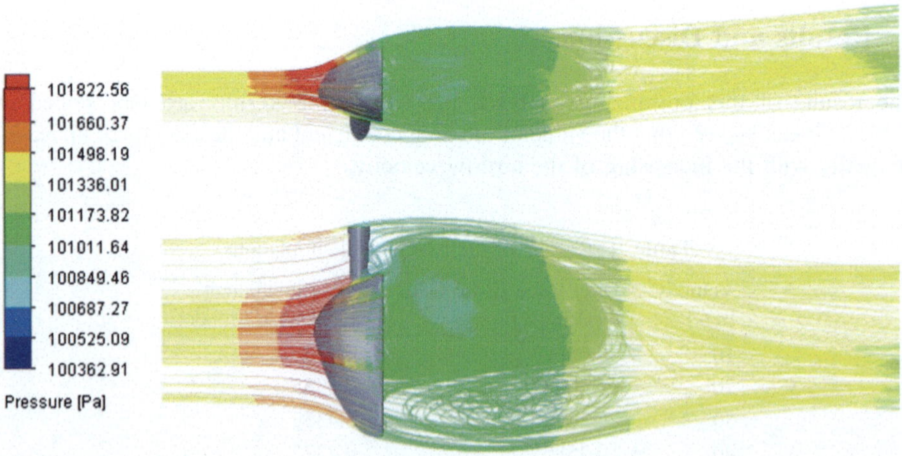

Fig. 5. Airflow pressure distribution at velocity of 40 m/s.

It can be observed that the maximum value of the pressure is on the frontal mirror surface. Analyzing the obtained result, it is desired to decrease the value of the drag coefficient, the drag force and a linear flow of the air in the rear side of the body mirror by optimization of the mirror geometry.

In Fig. 6 is presented a top view and isometric view from front side and rear side of the optimizated model of mirror geometry. The shape of the mirror is changed by the radius geometry value. The geometry of the mirror becoming more sharper, as can be seen in the picture.

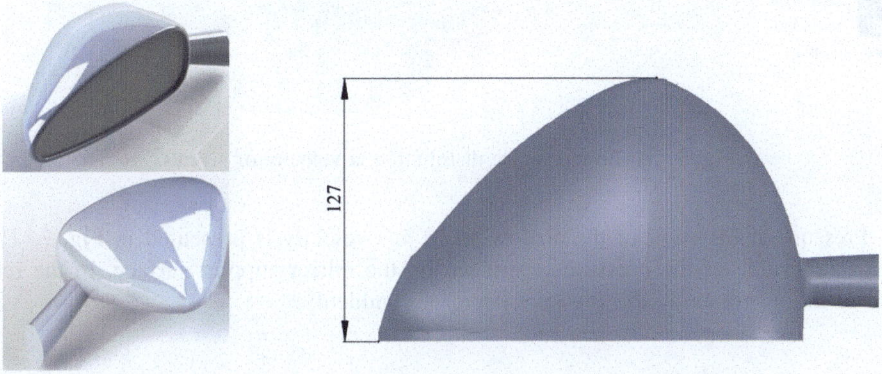

Fig. 6. The optimized CAD model of mirror side view.

The width of the side mirror house is larger, which gives it a half three-dimensional water drop. Due to this shape, the air flow pressure exerted on the mirror housing surface will be distributed to the outside of the mirror, reducing possible vibration that can could blur the image.

Table 2. Results of the optimized model simulations.

Velocity [m/s]	Drag force (Improved CAD model) [N]	Drag coefficient (Improved CAD model)
10	0.60804	0.4067
20	2.41672	0.4041
30	5.40586	0.4018
40	9.56547	0.3999
50	14.9456	0.4000
60	21.5455	0.4003

In Table 2 the resulted values of the drag force and drag coefficient are presented. It can be observed that the values of the drag force and the drag coefficient decrease slowdown in relation to speed up.

Fig. 7. Airflow pressure distribution at velocity of 40 m/s.

Pressure distribution of the airflow at 40 m/s velocity is presented in Fig. 7. The maximum value of the pressure is situated on the mirror support surface. In the rear side of the mirror body the pressure decreases suddenly.

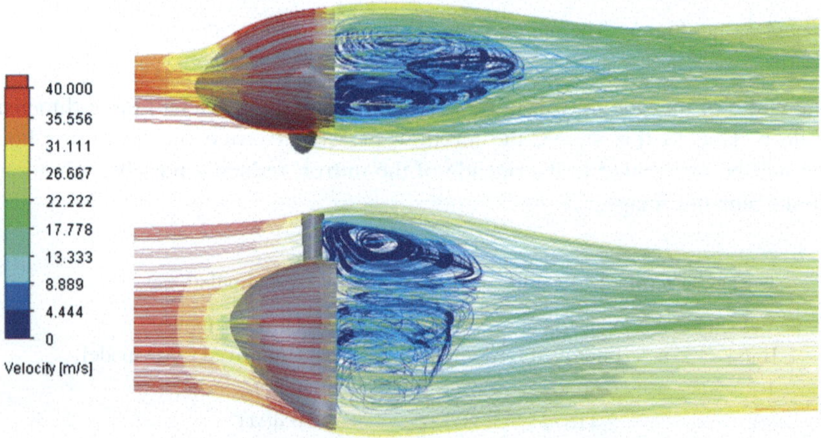

Fig. 8. Streamline distribution at velocity of 40 m/s.

In Fig. 8 are presented the top view and right view of the mirror model simulated at a velocity of 40 m/s. It the rear side of the mirror it can be observed that the streamline velocity drops suddenly, creating a swirl due to the mirror geometry support. It observed that the air streamline flows linear in the rear side of the optimizated mirror body comparted to the base model mirror.

The visual results are presented for the speed of 40 m/s because it is the closest speed compared to the maximum speed, during the short time, which the cars are tested by the European approval cycle. The test of the car speed is done at 120 km/h (33.3 m/s).

4 Conclusion

In this study are simulated two mirror geometry, to view the airflow distribution. The effect of the increase the radius of housing curvature reduces the drag coefficient of the side mirror. The resulted drag coefficient values for the base model are in the range 0.4351 N and 0.4386 N for a velocity variation of 10 to 60 m/s. The drag force increase gradually increases depending on the velocity increase, in the range of 0.665627 N and 23.6064 N. The values of the drag coefficient for improved model is situated in the range of 0.4067 N and 0.4003 N and the value of the drag force is in the range of 0.60804 N and 21.5455 N. Comparing the obtained resulted from this two simulations model it can be viewed that the value of drag coefficient decrease depending on the change of the model geometry, as well as the value of the drag force. The value of the drag force and the drag coefficient are lower for the optimized model than of the base model.

Acknowledgment. This research was funded by the Program "POC-A1-A1.2.3-G-2-15 Parteneriate pentru transfer de cunoştinţe, nr. contract de finanţare: 11/01.09.2016, titlul proiectului: Tehnologii pentru vehicule electrice urbane inteligente URBIVEL, ID: P_40_333, MySMIS: 105565".

References

1. Maroufmashat, A., Fowler, M.: Policy considerations for zero-emission vehicle infrastructure incentives: case study in Canada. World Electr. Veh. J. **9**(3), 38 (2018)
2. Ramachandran, S., Stimming, U.: Well to wheel analysis of low carbon alternatives for road traffic. Energy Environ. Sci. **8**, 3313–3324 (2015)
3. Varga, B.O., Mariasiu, F.: Indirect environment-related effects of electric car vehicles use. Environ. Eng. Manag. J. **17**(7), 1591–1599 (2018)
4. Varga, B.O., Sagoian, A., Mariasiu, F.: Prediction of electric vehicle range: a comprehensive review of current issues and challenges. Energies **12**(5), 946 (2019)
5. Hucho, H.: Aerodynamics of Road Vehicles, vol. 10, 4th edn, p. 398. Warrendale, SAE International (1998). no. 1026
6. Bodea, S.M., Prodan, C.V., Scurtu, I.L.: The aerodynamic study of a body truck. In: International Congress of Automotive and Transport Engineering, pp. 73–79. Springer, Cham (2018)
7. Al-Obaidi, A.S.M., Otten, W.A.: Aerodynamic analysis of personal vehicle side mirror. J. Eng. Sci. Technol. **13**, 52–64 (2018)
8. Alam, F., Jaitlee, R., Watkins, S.: Aerodynamic effects on an automotive rear side view mirror. In: 16th Australasian Fluid Mechanics Conference Crown Plaza, Gold Coast, Australia, pp. 762–765 (2007)

Design and CFD Simulation of an Automotive Air Heat Exchanger

Monica Balcau(✉) and Liviu Scurtu

Technical University of Cluj-Napoca, Cluj-Napoca, Romania
monica.balcau@auto.utcluj.ro

Abstract. In this paper is presented a CAD design and the CFD study of the air heat exchanger. In the first part of the paper are presented the actual studies of CFD simulation with dedicated software. In the second part of the paper the CAD geometry of the air heat exchanger is modelled. The CFD analysis of the intercooler is done by applying the different flow parameters of the fluid the base geometries. After viewing the simulation results, a new model of the heat exchanger is proposed in order to uniformize the airflow distribution and temperature. The improved model is simulated at the same condition like the base model. In the final part of the paper is presented the comparative study between these two heat exchanger models and the conclusions regarding this CFD study.

Keywords: CAD design · CFD simulation · Air heat exchanger · Intercooler

1 Introduction

Heat exchanger are a vital component of a car engine assembly realizing the cooling of thermal engines, the turbocharger air cooling, more recently, the battery cooling of the electric vehicle. The cooling system of automotive with thermal engine contributes significantly to the environmental, having benefits in fuel consumption reducing by maintaining optimum engine temperature and reducing the pollutant emission [2].

In this paper, a CFD study was performed on an air heat exchanger structure used to air cooling from the engine turbocharger. The types of heat exchanger are called intercooler and use air as cooling agent. The intercooler is placed in the front of the vehicle between the turbocharger and the intake manifold, in direct access of the wind, having a better cooling of the heated air from the turbocharger exhaust gases.

The intercooler is an important part because when the air is cooled will have better oxygen, thus increasing the power and better combustion of the fuel mixture resulting a lower fuel consumption. An important aspect of the intercooler design is given by the cooling capacity, which can be influenced by the geometric shape and the material of the heat exchanger. CFD simulation software can be used to determine the geometric shape of the intercooler, which provide numerical and visual information of the air distribution in the internal volume of the intercooler during operation.

An CFD analysis study performed on an intercooler radiator used to air cooling during the operation of a thermal engine is performed in [7]. In this study an intercooler model is designed and CFD simulated at different value of the air mass that flow into model, presenting the distribution and the values of the air temperature passing through

I. Dumitru et al. (Eds.): SMAT 2019, *The 30th SIAR International Congress of Automotive and Transport Engineering*, pp. 502–507, 2020.
https://doi.org/10.1007/978-3-030-32564-0_58

the radiator. Based on filled synergy principle, Huang presented the CFD analysis that shows the temperature and velocity of the air distributions for two different intercooler tank model. The CFD analysis and the heat transfer simulation is performed in Fluent 13 software [4].

A comparative CFD study and performance analysis of copper, aluminum and Cuprobraze radiators is studied in [3]. Following the study results has been noticed that de aluminum structure is better than copper or Cuprobraze. The rate of dissipation is higher for aluminum radiator then the other two radiators.

2 Method

The intercooler is made up of rows of tubes through which the air passes and a secondary part attached outside the string of tubes through which the air flows and keeps the model body cold. The sting of tubes can be arranged vertically or horizontally. If the tank is placed on the top and the bottom the tubes are arranged vertically and the air flows vertically from top to bottom, it is the named the down flow radiator, if the tanks are placed on the side of the radiator, the tubes are arranged horizontally, the air flows across the radiator from inlet tank to outlet tank, it is named cross flow radiator [2].

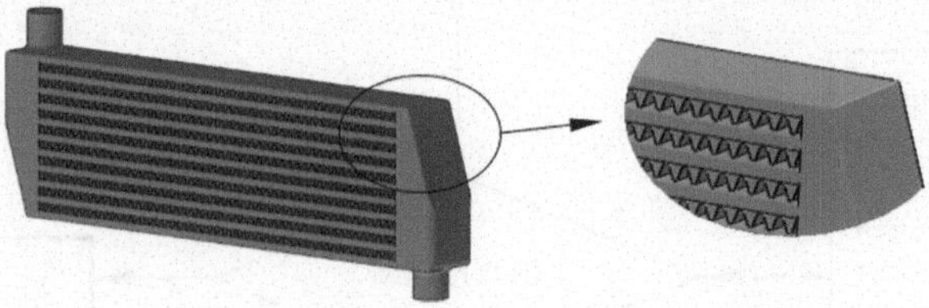

Fig. 1. Base CAD model of the intercooler

At the ends of the tubes is placed the tank that achieve a uniform flow horizontally, of the air. These tanks are slanted at the ends in order to uniformize the air flow through intercooler cells. In Fig. 1 is presented the base proposed intercooler. In this study, an aluminum alloy model is used because it represents advantages over a model made of Cu Brass alloy. Some of the advantages are: lower mass, lower costs and better heat transfer [1].

2.1 CAD Base Model

Using advanced modelling techniques, the CAD geometry of the intercooler is modeled in SolidWorks software. In Fig. 2 is presented the basic geometry of the intercooler, the overall dimensions and the size of the inlet and outlet diameters. The wall section of the radiator is 1 mm.

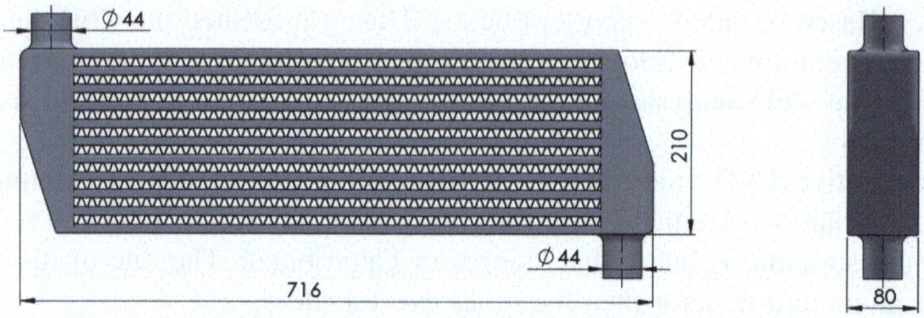

Fig. 2. Front and right view of the CAD model of the intercooler.

2.2 CFD Setup

The CFD method is a branch of fluid mechanics that uses numerical analysis to solve numerical problems that involving fluid flow [5]. In this paper, the CFD simulations is performed in FlowWorks module which is a fluid analysis solver integrated into the SolidWorks software package. To perform the CFD simulation it is necessary to go through a basic step presented in the flowchart from Fig. 3.

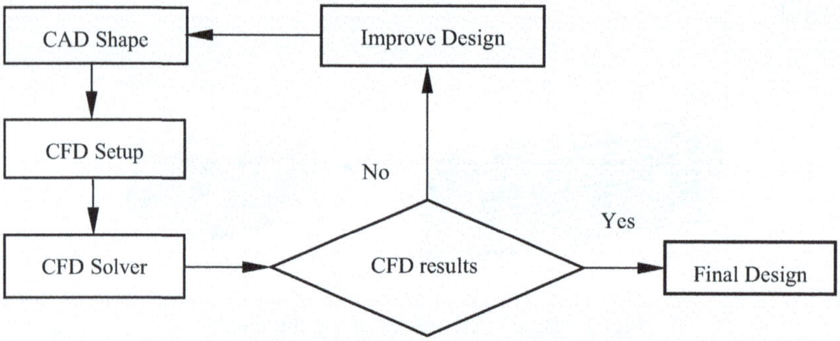

Fig. 3. Flowchart of the iterative CFD process [6].

The first step for preparation of the CFD process is the acquisitions of the CAD geometry, followed in the second step by the definitions of the computational domain in which the study will be performed, the material properties, the air properties and the imposed simulation conditions.

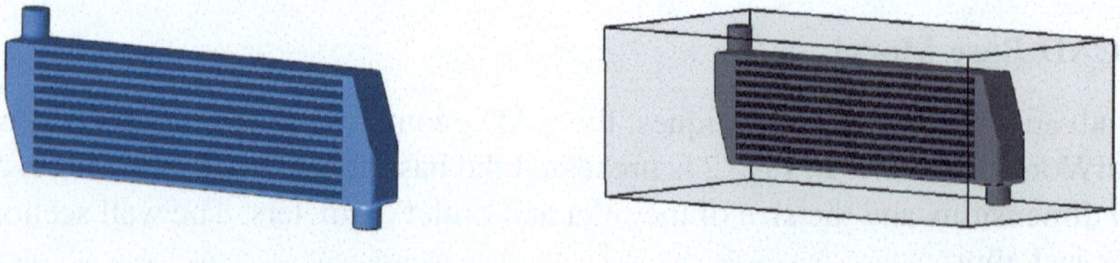

a) Internal computational subdomain **b)** Computational domain

Fig. 4. CFD computational domain.

After running the CFD simulation if the obtained results not satisfy the conditions the process entering an iterative calculation loop until the moment when the obtained results will be fulfilled, the process is stopped resulting the final model. In Fig. 4 is presented the internal computational subdomain and the external computational domain.

The CFD study is performed at a 60 °C at the moment when the air into the intercooler model. The temperature of the intercooler aluminum alloy body is 20 °C, the inlet mass flow is 0.18 kg/s and the wind velocity are 10 m/s and has a temperature of 10 °C.

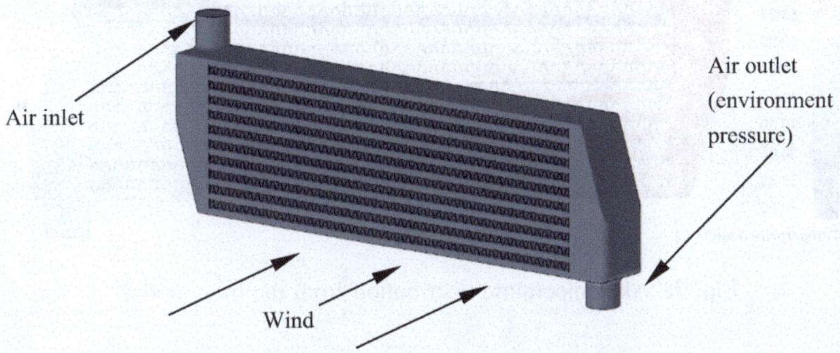

Fig. 5. CFD load case.

The CFD load case is presented in Fig. 5, where can be observed the air inlet, the wind direction and the air outlet. The air outlet is represented by the environment pressure.

2.3 CFD Simulation Results

The distribution of the air velocity inside the intercooler is presented in Fig. 4, it can be seen that the distribution of the air velocity is not uniform in all intercooler cells.

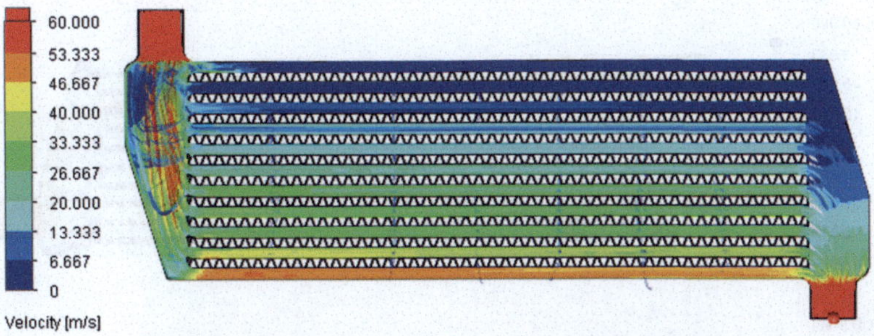

Fig. 6. Air flow distribution from the base model.

In Fig. 6 can be observed that the air velocity is very lower in upper side of the intercooler. Due to the side tank geometry in the upper side is created a vortex that prevent the uniform air flow from the cooling tubes inside of the radiator.

The temperature distribution of the air is presented in Fig. 7, where can be observed that in the second tube of the radiators the temperature is lower compared to other tubes. In that area the air velocity is minimum, the air stay more time there and is cooled better, but not the temperature distribution is not uniform.

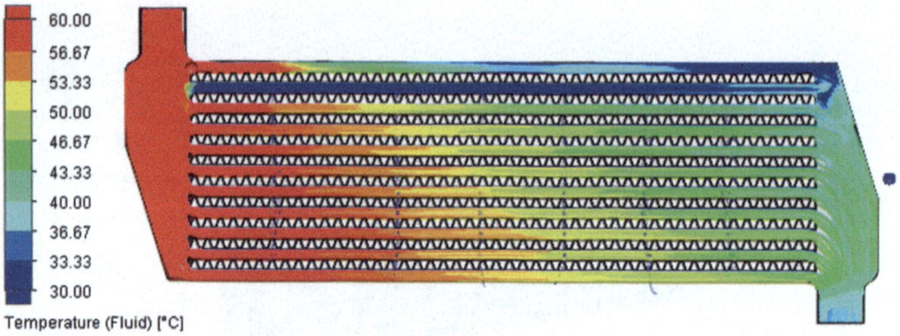

Fig. 7. Air temperature distribution from the base model.

The air temperature range is between 30–60 °C. It is desired to achieve an improved radiator structure, capable to offer a better air flow velocity and temperature.

2.4 Improved Model

In order to improve and uniformize the air flow cooling a new intercooler model is designed. The inlet of the air vent is placed horizontally on the left side of the intercooler and the outlet air vent is placed horizontally on the right side of intercooler and the, at can be seen in Fig. 8, where can be observed the distribution of the air flow velocity. In can be observed that in the second tube the air velocity has greater velocity than the others tubes, because the air inlet is placed on the line of this tube.

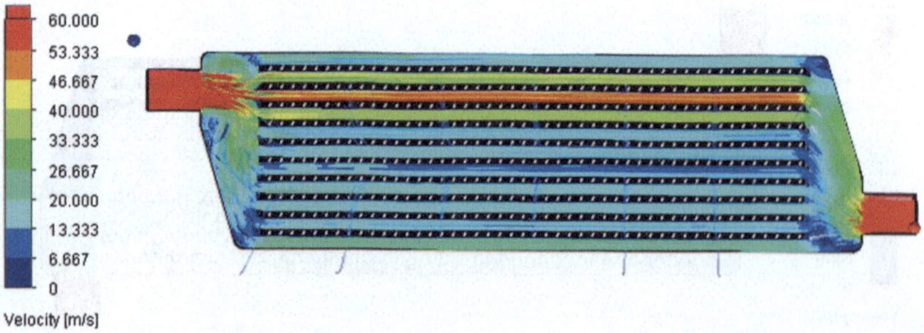

Fig. 8. Air velocity distribution on the improved model.

In the Fig. 9 is presented the air temperature distribution of the improved model, it can be observed that the temperature of the air is almost uniform distributed into internal radiator volume.

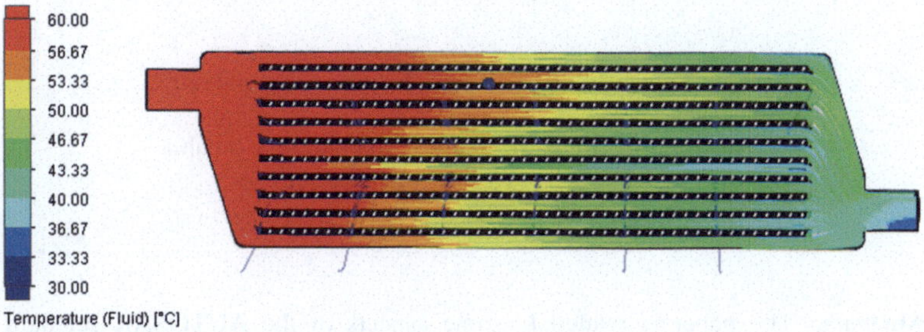

Fig. 9. Air temperature distribution on the improved model.

3 Conclusions

In this paper it was studied an intercooler model that can be used for air cooling in the automotive industry. The CDF simulation result of the base model shows that the airflow and temperature is not uniformly distributed into internal volume of the intercooler. In the improved model the air flow and the temperature distribution are better than the first model. The present CFD study shows a numerical method that can be used before manufacturing the intercooler radiator.

References

1. Patel, S., Deshmukh, S.P.: Analytical design and verification of automotive radiator using 1-D simulation. Int. J. Res. Appl. Sci. Eng. Technol. **5**(6), 2349–2360 (2017)
2. Pathade, V.C., Satpute, S.R., Lajurkar, M.G., Pancheshwar, G.R., Karluke, T.K., Singitvar, N. H.: Design and analysis of car radiator by finite element method. IJARIIE **3**(2), 1375–1382 (2017)
3. Jaya kumar, R.K., Ashokkumar, T.: Comparative study and performance analysis of copper, aluminium and Cuprobraze radiators using flow simulation software. Int. J. Eng. Tech. **3**(5), 78–83 (2017)
4. Huang, Y., Yu, X., Liu, Z., Lu, G.: Simulation, fabrication and optimization of the vehicular intercooler based on field synergy principle. In: 9th International Conference on Heat Transfer, Fluid Mechanics and Thermodynamics (2012)
5. Subba, R.G., Kishore, K.B.: Modeling and CFD analysis of radiator by using nanofluids. Int. J. Prof. Eng. Stud. **9**(2), 358–368 (2017)
6. Scurtu, I-L., Jurco, A-N.: Airflow and thermal comfort of the bus passengers. In: 8-th International Conference on Engineering Graphics and Design, vol. 14, no. 1, pp. 171–174 (2019)
7. Patil, V., Malipatil, A.: CFD analysis of intercooler with sine wave fins. Int. J. Eng. Tech. **3** (6), 764–769 (2017)

Application Model in AUTOSAR Software Development for Control Systems Design Through Fuzzy Methods

Cristian Spirleanu[1(✉)] and Eugen Diaconescu[2]

[1] Institute of Solid Mechanics, Bucharest, Romania
cristi_spirleanu@yahoo.com
[2] University of Pitesti, Pitesti, Romania

Abstract. The paper is related to some aspects of the AUTOSAR standard application in software design and development of CVT type powertrain transmission control system in a vehicle. To overpass the difficulties generated by the powerfull non-linear characteristics of the process and of the involved components, a fuzzy methods approach by using AUTOSAR methodology was proposed. In this context, there are presented the structure and the rule base of the proposed fuzzy regulator model, the software components modelling by using the artext language in ARTOP environment and their connections to virtual functional bus VFB of CVT-ECU. It is also presented the SWC-CVT components interaction with others ECU that are connected to the same Ethernet or CAN communication bus through hardware communication abstractization. This approach leads to control algorithms design simplicity and flexibility of software development process ensured by the Autosar standard.

Keywords: AUTOSAR · Fuzzy control · CVT · ECU · Software component · VFB

1 Introduction

The paper presents a model for implementing a fuzzy control algorithm using the AUTOSAR methodology. The model chosen refers to the control of a chain-type propulsion process of a hybrid vehicle and is developed only to exemplify the design using AUTOSAR, yet it is quite close to the actual case encountered in practice.

The opportunity to implement the model within the AUTOSAR methodology resulted from the finding that the AUTOSAR partnership has become de facto a standard in the world of vehicle manufacturers in recent years [13]. The main purpose of this partnership is to standardize a common methodology for defining and using basic systemic functions and functional interfaces. The result of the AUTOSAR collaboration are the integration, transfer and reuse of the functions of a network with the specific architecture of the vehicles in an efficient way and that will stimulate its development. Starting from the ECU-based paradigm as individual entities, we have switched to a car software model capable of conveniently managing the complexity of the system while respecting the technological and economic requirements [8].

I. Dumitru et al. (Eds.): SMAT 2019, *The 30th SIAR International Congress of Automotive and Transport Engineering*, pp. 508–517, 2020.
https://doi.org/10.1007/978-3-030-32564-0_59

The analytical model of the hybrid vehicle's propulsion system is complex and strongly nonlinear. A classical control solution based on the analytical model is not the best solution in this case, being difficult to achieve. As a result, numerous control approaches based on fuzzy logic have been applied to vehicle propulsion systems that reduce and simplify the difficulties specific to nonlinear models [1–5].

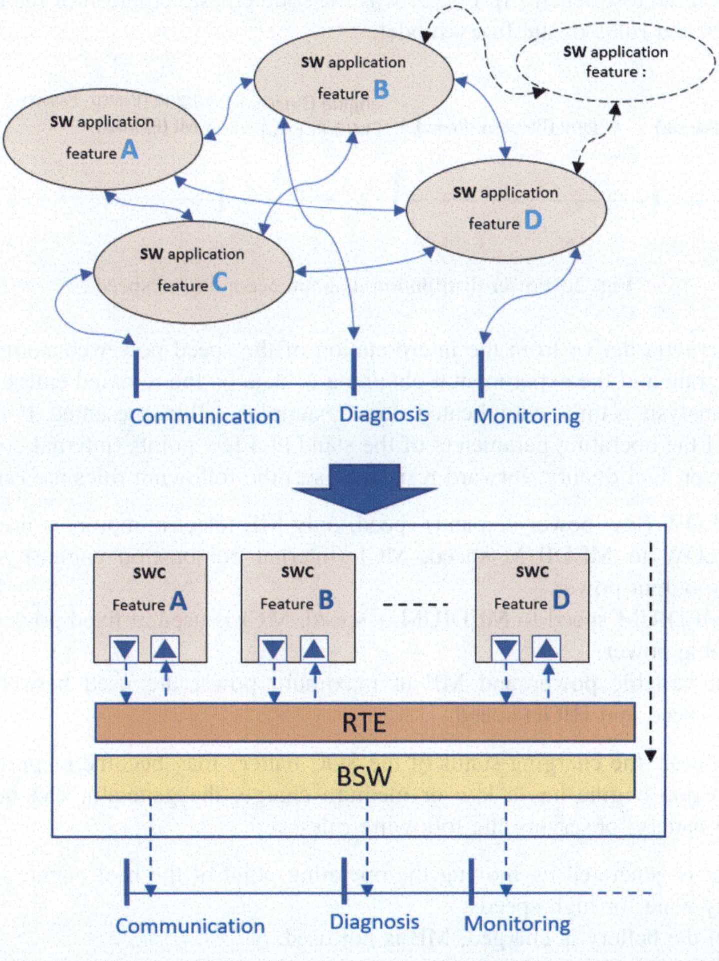

Fig. 1. Embedded software, the common approach in comparison to AUTOSAR

The authors have also been involved in designing fuzzy control applications of the propulsion system using different hardware supports for the fuzzy controller. In the application presented in [6], the controller used in an experimental laboratory stand was an ECOMAT 360 PLC type unit from IFM with CoDeSys IEC 61131-3 development system. In book [7], the description of the fuzzy model (the initial analytical model, the fuzzification of the variables, the membership functions, the deduction of the basic rules and the realization of the rules, the decision making) is repeated and an implementation of the fuzzy controller on a simulator based on a LabVIEW programmable National Instruments equipment.

In this paper we will describe, as far as space allows, the approach of implementing, broadly, the same model of fuzzy controller, but this time using the tools made available by the AUTOSAR standard.

Power distribution diagram with regard to the considered speed of the model simulated on the test bench [6], Fig. 2, was the source/basic criterion of the deduction of variables and rules of the fuzzy model.

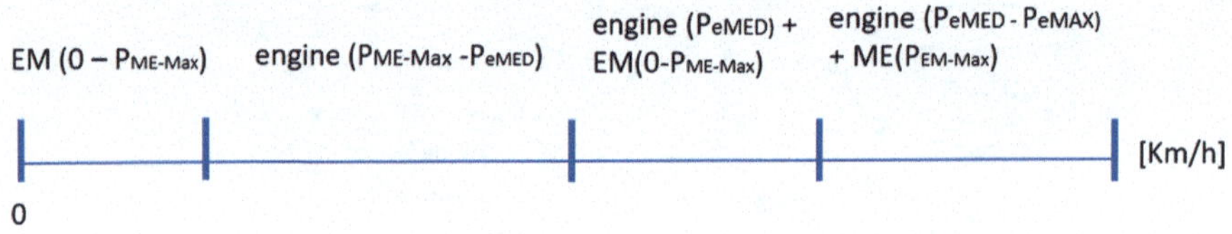

Fig. 2. Power distribution diagram according to speed

Other criteria derive from the interpretation of the speed-power-consumption efficiency diagram and the experimental obtaining of data on the released emissions. As a complete analysis is quite complicated, some examples will be presented. For a certain structure of the operating parameters of the stand at a few points (internal combustion engine power, fuel quality, forward resistance set) the following rules are established:

- Up to LOW (low power demand) speed, only ME (electric motor) is used.
- From LOW to MEDIUM speed, MCI (internal combustion engine) is used at variable output power.
- From MEDIUM speed to MEDIUM + speed, MCI is used at fixed power and ME at variable power.
- MCI at variable power and ME at maximum power are used between MEDIUM + speed and HIGH speed.

In operation, the charging status of the SOC battery may become inappropriate. If the combustion engine has a low or medium charge, the generator can be used to charge the battery, observing the following rules:

(a) Power is generated by moving the operating point of the heat engine to an efficiency zone for high speeds;
(b) When the battery is charged, ME is not used;
(c) The SOC is maintained in the range 0.6–0.8 for efficient charging/discharging and battery protection.

2 Fuzzy Control Model

For the model of the hybrid propulsion system corresponding to the current configuration of the laboratory stand for which tests and experiments were carried out, a distribution of inputs and outputs was obtained as in Fig. 7.

The variables that were merged were chosen: the desired speed, vehicle speed, SOC (charging state) of the battery, the speed of the thermal engine and the speed of the electric motor.

Outputs were set the controls for the thermal engine, the electric motor, the generator (to simulate the vehicle's rolling resistance and recuperative braking) and the control value i (i1 = 1, i2 = 2) of the transmission ratio of the CVT device. The defuzzification method is the "center of the area".

Based on the model in Fig. 3, which establishes the linguistic variables of input and output that are merged, the membership functions are attached to them and a set of rules has been developed that describes the functioning of the fuzzy controller based on the knowledge of an expert. Decision making is done by defusing output variables using a certain appropriate method, for example that of the "center of gravity".

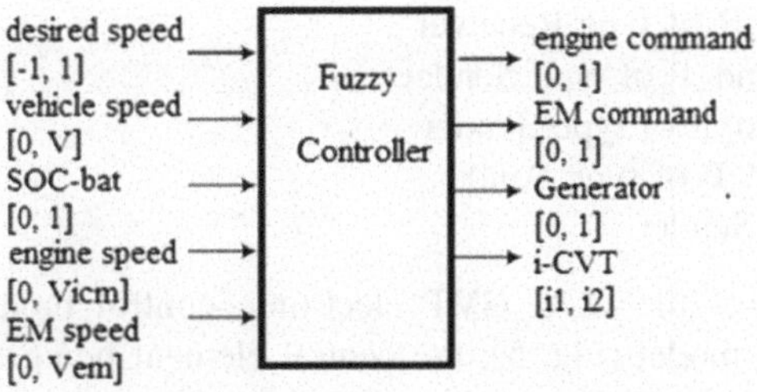

Fig. 3. Fuzzy controller input and output

3 The Work Environment Architecture and the Software Technologies Used

The RTE environment is responsible for the implementation of the Virtual Functional Bus (VFB), through which the communication between ECUs is realized [14]. As an intermediary, RTE uses the Basic Software (BSW) level for inter-ECU communication on CAN, Ethernet or LIN communication buses.

During the design, the SWC components are allocated within an ECU. Communication on the virtual functional bus (VFB) is of an intra-ECU type when the components are connected in a single ECU, and of an inter-ECU type when the components of an ECU are connected with the components of other ECUs (see Fig. 4).

An alternative solution for generating RTE APIs related to CVT_FuzzyControl SWC ports, is the use of ARUnit [12].

Starting from the model shown in Fig. 1, the signals used in the composition of the ECU-CVT electronic control module, which can be transmitted on the CAN or Ethernet communication buses, need to define the following sender or receiver type ports, within the CVT_FuzzyControl component.:

- StateOfCharge_R of type Receiver
- DesiredSpeedSense_R of type Receiver
- VehicleSpeed_R of type Receiver

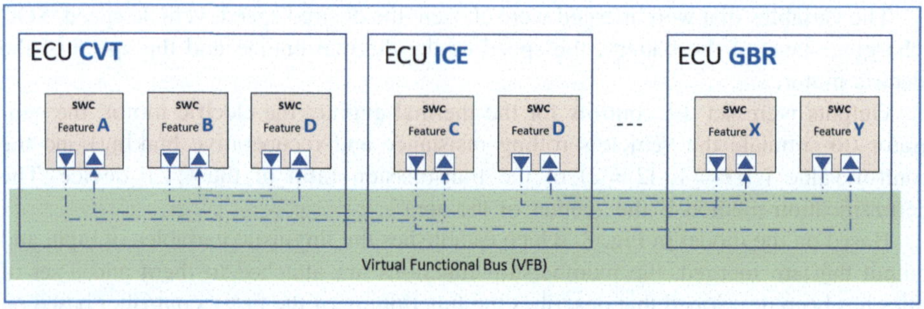

Fig. 4. Communication between SWC components in AUTOSAR

- N_IntCombEngine_R of type Receiver
- N_ElectricEngine_R of type Receiver
- IntCombEngineCmd_P of type Sender
- ElectricEngineCmd_P of type Sender
- GenRecBreakCmd_P of type Sender
- i_CVT_P of type Sender

The composition of the ECU_CVT electronic control module is made for the following application model (Fig. 5), the central element being the CVT_FuzzyControl SWC component, which interacts with other electronic ECU control modules connected to the same CAN bus:

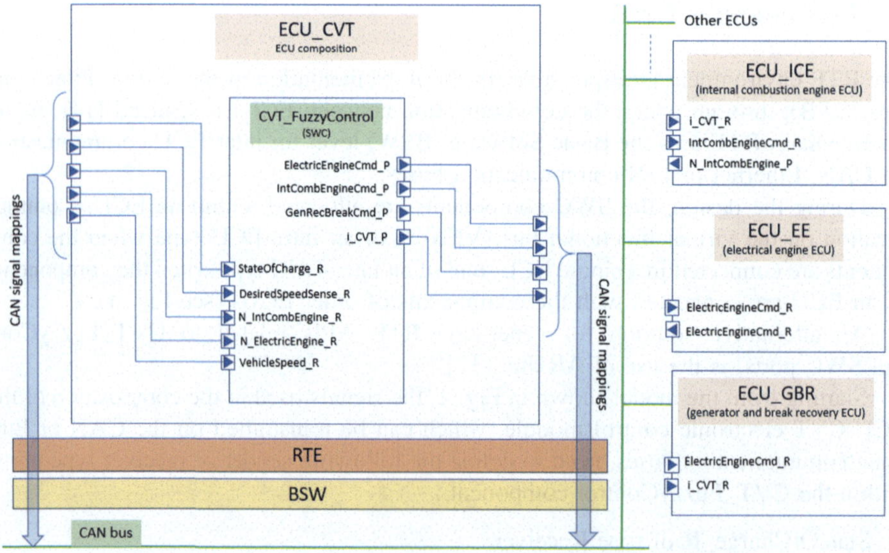

Fig. 5. CVT_FuzzyControl application model inside CVT_ECU composition

- ECU_ICE (internal combustion engine ECU)
- ECU_EE (electrical engine ECU)
- ECU_GBR (generator and break recovery ECU)

4 The AUTOSAR Software Implementation of the Fuzzy Control Model

The interfaces and ports defined for the exchange of information between the CVT control ECUs are contained in:

- CVTPortInterfaces.swcd and
- CVT_FuzzyControl.swcd files, shown in Fig. 6 by the description in artext language [12].

```
package CVT_ECU_composition
import AUTOSAR_Platform.ImplementationDataTypes.uint8
import AUTOSAR_Platform.ImplementationDataTypes.uint16

interface senderReceiver IfStateOfCharge{ data uint8 ChargeValue initValue 0 }
interface senderReceiver IfDesiredSpeedSense{ data uint8 SpeedSenseValue initValue 0 }
interface senderReceiver IfVehicleSpeed{ data uint8 SpeedValue initValue 0 }
interface senderReceiver IfnEngine{ data uint16 nEngineValue initValue 0 }
interface senderReceiver IfEngineCmd{ data uint8 CmdValue initValue 0 }
interface senderReceiver IfGeneratorCmd{ data uint8 CmdValue initValue 0 }
interface senderReceiver IfiCVT{ data uint8 iCVTValue initValue 0 }

component application CVT_FuzzyControl {
        ports {
                receiver StateOfCharge_R requires IfStateOfCharge
                receiver DesiredSpeedSense_R requires IfDesiredSpeedSense
                receiver VehicleSpeed_R requires IfVehicleSpeed
                receiver N_IntCombEngine_R requires IfnEngine
                receiver N_ElectricEngine_R requires IfnEngine
                sender IntCombEngineCmd_P provides IfEngineCmd
                sender ElectricEngineCmd_P provides IfEngineCmd
                sender GenRecBreakCmd_P provides IfGeneratorCmd
                sender i_CVT_P provides IfiCVT
        }
```

Fig. 6. CVT_FuzzyControl ports and interfaces description in artext language

After completing the description of the SWCs used for the ECU_CVT composition, the next step is exporting in *.arxml file format, named CVT_ECUextract. For the implementation code, the mandatory section for reference to the internal behaviour IbCVT_FuzzyControl of the SW-C CVT_FuzzyControl component has been defined also in artext language.

The next step is to run the RTE generator tool, having as input the exported * arxml file. The RTE generator tool will produce the files “Rte_CVT_FuzzyControl_Type.h”, “Rte_CVT_FuzzyControl.h” and “Rte_Type.h”, as presented in Fig. 7.

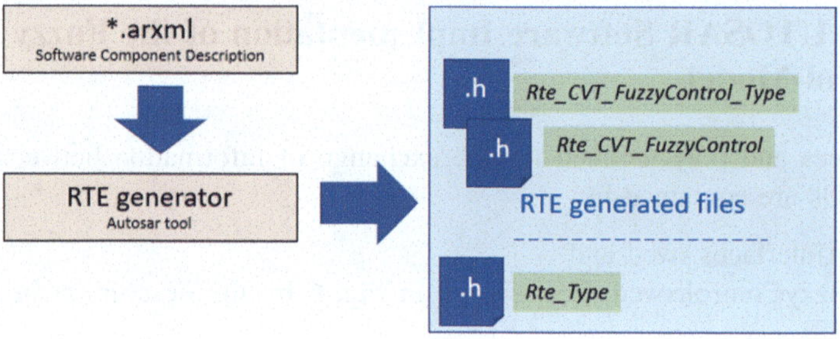

Fig. 7. Running the RTE API generation tool

The following APIs were obtained after running the RTE generator tool:

- Std_ReturnType Rte_Read_StateOfCharge_R_ChargeValue (uint8 *data);
- Std_ReturnType Rte_Read_DesiredSpeedSense_R_SpeedSenseValue (uint8 *data);
- Std_ReturnType Rte_Read_VehicleSpeed_R_SpeedValue (uint8 *data);
- Std_ReturnType Rte_Read_N_IntCombEngine_R_IfnEngine (uint8 *data);
- Std_ReturnType Rte_Read_N_ElectricEngine_R_IfnEngine(uint8 *data);
- Std_ReturnType Rte_Write_IntCombEngineCmd_P_CmdValue (uint8 data);
- Std_ReturnType Rte_Write_ElectricEngineCmd_P_ CmdValue (uint8 data);
- Std_ReturnType Rte_Write_ GenRecBreakCmd_P_ CmdValue (uint8 data);
- Std_ReturnType Rte_Write_ i_CVT_P_ iCVTValue (uint8 data);

For the conversion from the “triangle” type representation to the “trapeze” type representation, the middle points have identical values. Next, we will use for the representation of the linguistic variables the “triangle” type. For simplicity, we will choose the representation on 8 bits, the range of variation being 0..255.

Figure 8 shows the structure used for cvtFuzzy membership functions, with the following implementation in C code:

```
struct membershipFunct_cvtFuzzy_type {
 char mf_name[255];
 int value;
 int p_xL;   // trapezoidal membership function value: X left point
 int p_xR;   // trapezoidal membership function value: X right point
 int s_Left; // trapezoidal membership function value: slope on Left side
 int s_Right; // trapezoidal membership function value: slope on Right side
 struct membershipFunct_cvtFuzzy_type *nxt;
 };
```

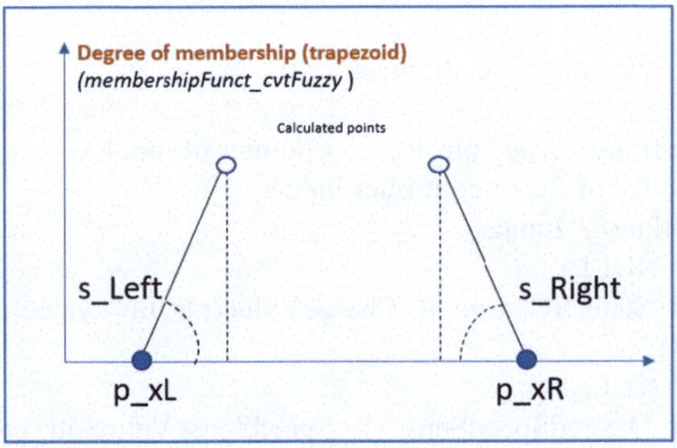

Fig. 8. Definition of the structure for the cvtFuzzy membership function

The runnable function "CvtFuzzyControl_Main" is called every 50 ms, and will include the following sequences:

```
void CvtFuzzyControl_Main (void)
{
//01: Read input signals by calling Rte_read_ functions
 CvtFuzzyControl_readInputs();
//02: Fuzzification of CvtFuzzyControl-SWC inputs
 CvtFuzzyControl_fuzzification();
//03: Rules evaluation
 CvtFuzzyControl_ruleStrength();
//04: Defuzzification
 CvtFuzzyControl_defuzzification();
//05: Outputs update / command signals
 CvtFuzzyControl_writeOutputs();
}
```

The function CvtFuzzyControl_readInputs() has the following implementation code in AUTOSAR:

```
void CvtFuzzyControl_readInputs(void)
{
struct io_cvtFuzzy type *ptr_in;      //pointer of input(s)
//update the list of fuzzy-controller inputs
ptr_in = cvtFuzzy_Inputs;
if(ptr_in != NULL) {
  Rte_Read_StateOfCharge_R_ChargeValue(ptr_in-> value);}
ptr_in ->nxt;
if(ptr_in != NULL) {
  Rte_Read_DesiredSpeedSense_R_SpeedSenseValue(ptr_in->value);}
ptr_in ->nxt;
if(ptr_in != NULL) {
  Rte_Read_VehicleSpeed_R_SpeedValue(ptr_in-> value);}
ptr_in ->nxt;
if(ptr_in != NULL) {
  Rte_Read_N_IntCombEngine_R_IfnEngine(ptr_in-> value);}
ptr_in ->nxt;
if(ptr_in != NULL) {
  Rte_Read_N_ElectricEngine_R_IfnEngine(ptr_in-> value);}
}
```

5 Conclusions

The paper presented a fuzzy controller model for the transmission system of a vehicle, developed within the AUTOSAR standard. It started from the modeling of the SWC component as part of an ECU, continuing with the generation of the communication APIs and finally with the implementation code of the SWC's main runnable function.

The advantages of this way of working using the AUTOSAR standard are the high efficiency of designing the software solution which is decoupled from the hardware aspects and the possibility of simpler /easier introduction of more complex algorithms within the software system, even belonging to artificial intelligence.

As a future work, the next steps will be focused on the integration of the cvtFuzzyController inside a real ECU, continuing with the validation and realtime testing by using CAN communication messages in combination with a log&trace tool.

References

1. Schouten, N.J., Salman, M.A., Kheir, N.A.: Energy management strategies for parallel hibrid vehicles using fuzzy logic. Control Eng. Pract. **11**, 171–177 (2003)
2. Kheir, N.A., Salman, M.A., Schouten, N.J.: Emissions and fuel economy trade-off for hybrid vehicles using fuzzy logic. Math. Comput. Simul. **66**, 155–172 (2004)

3. Poursamad, A.: Design of genetic-fuzzy control strategy for parallel hybrid electric vehicles. Control Eng. Pract. (2007)
4. Ippolito, L., Loia, V., Siano, P.: Extended fuzzy c-means and genetic algorithms to optimize power flow management in hybrid electric vehicles. Fuzzy Optim. Decis. Making **2**, 359–374 (2003)
5. Syed, F.U., Filev, D., Ying, H.: Fuzzy rule-based driver advisory system for fuel economy improvement in a hybrid electric vehicle. Document Ford Motor Company (2007)
6. Diaconescu, E., Lefter, E.: PLC fuzzy control solution for a hybrid powertrain with CVT. In: International Conference on 4th Edition Electronics, Computers and Artificial Intelligence, ECAI 2011, Pitesti, Romania, (2011)
7. Preda, I., Ciolan, Gh., Diaconescu, E., Cristea, D.: Transmisii mecanice cu frictiune pentru autovehicule, Ed. Univ. din Pitesti (2012)
8. AUTOSAR. http://www.autosar.org. Accessed 12 Aug 2019
9. Michels, K., Klawonn, F., Kruse, R., Nurnberger, A.: Fuzzy Control – Fundamentals, Stability and Design of Fuzzy Controllers. Springer, Heidelberg (2006)
10. Kiencke, U., Nielsen, L.: Automotive Control Systems – For Engine, 2nd edn. Driveline and Vehicle. Springer, Heidelberg (2005)
11. Husain, I.: Electric and Hybrid Vehicles Design Fundamentals. CRC Press, Taylor & Francis Group, Boca Raton (2005)
12. ARTOP. https://www.artop.org. Accessed 12 Aug 2019
13. Fürst, S., Dörr, H.: AUTOSAR – an open standardized software architecture for the automotive industry, In: 1st AUTOSAR Open Conference & 8th AUTOSAR Premium Member Conference. BMW. Cobo Center, Detroit, MI, USA (2008)
14. Naumann, N.: AUTOSAR runtime environment and virtual function bus. Technical report. Department for System Analysis and Modeling, Hasso-Plattner-Institute für Softwaresystemtechnik, Potsdam (2009)

Sensors Used for Human Gait Monitoring

Daniela Tarnita(✉) and Alin Teofil Oncescu

Faculty of Mechanics, University of Craiova, 13, A.I. Cuza str., 200318 Craiova, Romania
tarnita.daniela@gmail.com

Abstract. The present paper presents the main wearable sensors used in order to monitor the human gait, to monitor the training and performance in sports, to monitor the normal and pathological human gait, as well as the movements of the driver and the occupants of the car during driving.

Keywords: Human gait · Wearable sensors · Goniometers · EMG sensors · Force platforms

1 Introduction

The importance of normal and pathological human gait analysis in diagnosis, patient monitoring and rehabilitation is based on the use of wearable sensors. Nowadays, the literature presents the methods of human walking analysis based on portable sensors and they are described in the dynamics of human walking, the kinetics of human walking and the electromyography of human walking [1–39]. In the analysis of human gait the most used sensors are: accelerometers, gyroscopes, electrogoniometers, inertial sensors, electromyographic sensors and force platforms. Based on these types of sensors, both individual and combinations of sensor types, they are presented in the literature.

2 Types of Sensors Used in Human Gait Analysis

2.1 Wearable Sensors

The wearable sensors are used together with the optical systems for tracking a human movement, such as the Vicon system, the Simi Motion system or Biometrics, for quantifying, measuring human walking capacity and processing acquired data [1–5]. The advantage of using sensors in human walking analysis is the facility of assembling these sensors on different parts of the body, such as legs, thighs or hips, allowing data to be retrieved for extended periods, thus facilitating the subject's comfort. The most representative wearable sensors in the category are accelerometers, gyroscopes and goniometers, used in human walking analysis by placing them on the body of subjects in order to perform a complex analysis to determine normal or pathological walking parameters [3–6]. Lately, a great deal has been given to the analysis of human walking using releasable targets, and the advances in this new field of research have been

I. Dumitru et al. (Eds.): SMAT 2019, *The 30th SIAR International Congress of Automotive and Transport Engineering*, pp. 518–524, 2020.
https://doi.org/10.1007/978-3-030-32564-0_60

materialized by identifying and establishing three areas of human walking analysis, namely, the kinematics of human walking, the kinetics of human walking and electromyography of human walking (EMG). By defining the three registers for human walking analysis, the types of sensors described above will underpin the methods and techniques used in human walking analysis. The kinematics of human walking describes the movements of the joints of the lower extremities. The kinetics of human walking is based on the study of the moments and forces that occur now of the movements, and the electromyography describes the analysis of muscle activity during walking [1]. In the paper [1], there have been analysed concepts such as the human walking analysis and the impact of using the portable sensors shown in Fig. 1.

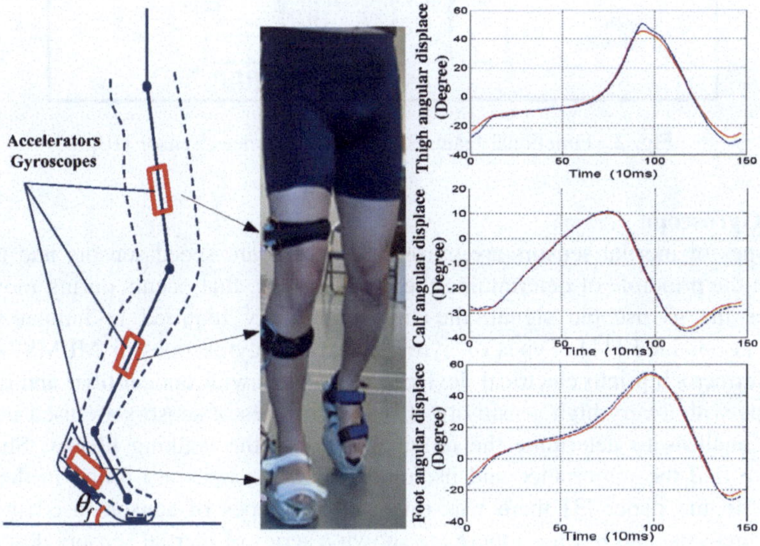

Fig. 1. Kinematic measurement of human walking based on portable sensors, namely, accelerometers and gyroscopes [1].

2.1.1 Accelerometers

Accelerometers are types of inertial sensors designed to monitor human motion and measure acceleration along the axis while walking.

Types of accelerometers: piezoelectric accelerometers - they use the piezoelectric effect of sensitive elements to generate the electrical charge at the output; piezoresistive accelerometers and capacitive accelerometers, the latter presents a much greater stability. The advantage of using accelerometers placed on the legs is that they are easy mounted and they measure the acceleration and the speed of the feet during movement [7]. In the papers [8] and [9] we find studies describing the importance and usefulness of accelerometers. In the paper [7], the chest expansion forces were studied using the Vicon and accelerometers. Figure 2 shows the functional characteristics of an accelerometer.

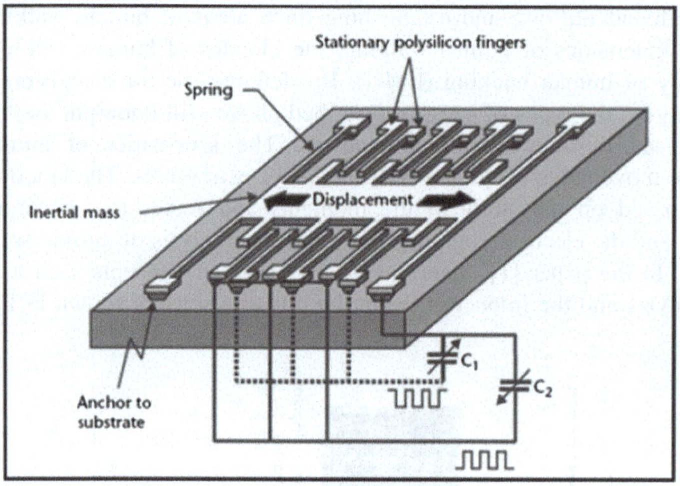

Fig. 2. Functional features of an accelerometer sensor [10]

2.1.2 Gyroscope

Other types of inertial sensors are gyroscopes, these are speed sensors and they are based on the principle of determining the Coriolis force that occurs during movement, and after the gyroscopic signal, the angular velocity required in human walking analysis is obtained [11]. Types of gyros: Electronic gyroscopes or MEMS, it incorporates a microchip into electrical devices; gyroscopes with optical fiber and quantum gyroscope with a very high sensitivity [12]. These types of sensors are used in human walking analysis to determine the reorganization of the walking phases. Studies in which we find the importance and usefulness of gyroscopes are found in the papers [13–15]. In the paper [3] there was reviewed the types of sensors used in human walking analysis, and in Fig. 3 there are shown a series of inertial sensors described in the paper [3].

Fig. 3. Inertial sensors [3]

2.1.3 Goniometers

The goniometers are portable sensors made up of two piezoresistive layers and an electric layer used for monitoring the biomechanical parameters of human joints, especially applied to the knee, hip and ankle joint [3, 6, 7]. The flexible goniometer is used to measure the relative rotation between the two segments of the human body. Flexible goniometers used in human walking analysis are also classified as flexible inductive mechanical goniometers and fiber optic goniometers. A flexible mechanical goniometer is designed to measure the flexion-extension angle and rotation in the knee joint's front plane during walking [6, 7 and 17]. Laskoski and colab. have developed a goniometer with an inductive sensor for measuring human motion [18]. Optical fiber goniometer is used to monitor human joint movement [19, 20]. Electro-goniometers are types of sensors used to evaluate dynamic activities such as human walking based on a three-dimensional optoelectronic system with high precision [21]. Flexible electro-goniometers are types of electrogoniometers, used to give a functional evaluation of several joints, such as the wrist, knee. The advantage of using this type of electrogoniometers is that they are easy to apply and have a high degree of accuracy [22, 23]. These types of sensors have also been studied in literature [24, 25]. In the papers [26, 27] there were carried out studies on how to mount the goniometer and electrogoniometer sensors on human subjects; they are shown in Fig. 4, while in [28, 29] the electrogoniometers were used in order to study the nonlinear dynamics of human gait.

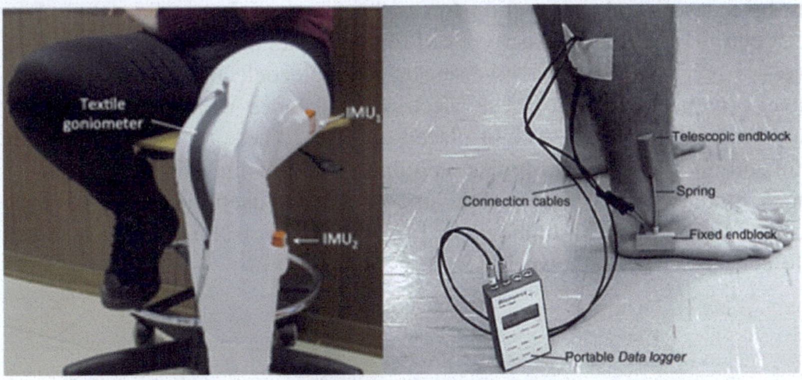

Fig. 4. Textile goniometer and electrogoniometer [26, 27].

2.1.4 Electromyography (EMG) Sensors

There are sensors that measure muscle activity to detect and provide information about the intensity of muscle contractions that occur during walking. Electromyographic sensors are essential in monitoring and assessing the walking performance of subjects or patients with amputated inferior limbs or with various problems [30]. These sensors have also been studied in the paper [31] on the reflexes that occur during the movement in response to the electromyographic sensors' electric stimulus, but also in the papers [32, 33] treating the walking of healthy subjects but also patients who have suffered amputations of lower limbs. In the paper [15] there are presented a series of types of

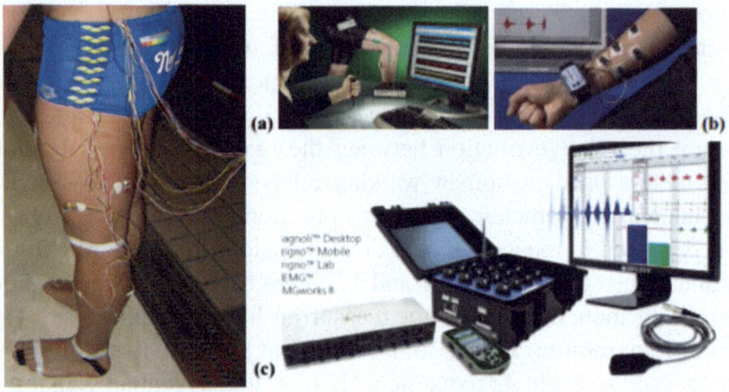

Fig. 5. Fixing system and electromyographic sensors [15, 34].

portable sensors used in the analysis of human walking, and in Fig. 5 there are presented different types of portable sensors composed of electromyographic sensors (EMG).

2.2 Power Sensors

Ground reaction force play an important role in the biomechanical analysis of human walking by measuring these reaction forces that occur during walking but also to the pressure centers [3, 5, 6, 35]. The types of force sensors used in the literature are force resistive, piezoelectric and capacitive sensors. The piezoelectric force sensors are large and have very good linearity. The electrical resistance of the resistive sensors is inversely proportional to the weight placed on them. These types of sensors are used predominantly in literature, in studies on human walking analysis by introducing them into shoes. Studies in which we find the importance and usefulness of force sensors are found in the papers [36–38]. In the paper [37], the human walking analysis was studied using force sensors [39], shown in Fig. 6.

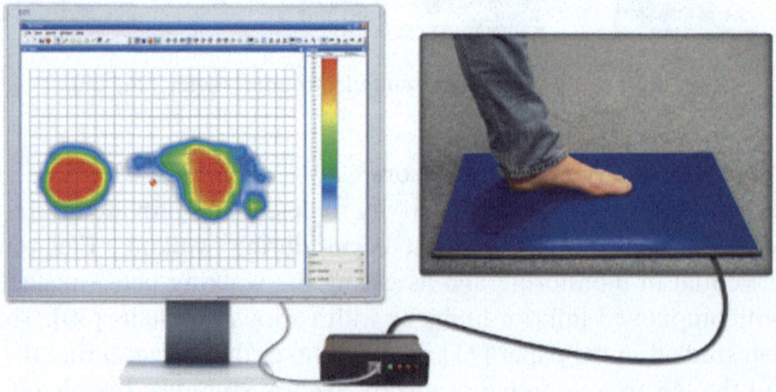

Fig. 6. Forces platforms [39]

3 Conclusions

The present paper overviews the human walking analysis using portable sensors. The main advantages of portable sensors are that they are small, lightweight, some of them having wireless communication. Portable Sensor Systems overcome the boundaries of classical measurement systems, making measurements and doing experiments anywhere, indoors or outdoors. Their usefulness is to monitor the training and performance in sports, to monitor the normal and pathological human gait, as well as to monitor the movements of the driver and the occupants of the car during driving.

References

1. Weijun, T.: Gait analysis using wearable sensors. Sensors **12**, 2255–2283 (2012)
2. Tadano, S., Takeda, R., Sasaki, K., Fujisawa, T., Tohyama, H.: Gait characterization for osteoarthritis patients using wearable gait sensors (H-Gait systems). J. Biomech. **49**, 684–690 (2016)
3. Tarnita, D.: Wearable sensors used for human gait analysis. Rom. J. Morphol. Embryol. **57**, 373–382 (2016)
4. Catana, M., et al.: Modeling, simulation and optimization of a human knee orthotic device. Appl. Mech. Mater. **371**, 549–553 (2013)
5. Tarnita, D., Geonea, I., Petcu, A., Tarnita, D.N.: Numerical simulations and experimental human gait analysis using wearable sensors. In: New Trends in Medical and Service Robots, vol. 48, pp. 289–304. Springer (2018)
6. Tarnita, D., Geonea, I., Petcu, A., Tarnita, D.N.: Experimental characterization of human walking on stairs applied to humanoid dynamics. In: Advances in Robot Design and Intelligent Control, pp. 293–301. Springer (2016)
7. Zeng, H., Zhao, Y.: Sensing movement: Microsensors for body motion measurement. Sensors **11**, 638–660 (2011)
8. Adkin, A.L., Bloem, B.R., Allum, J.H.J.: Trunk sway measurements during stance and gait tasks in Parkinson's disease. Gait Posture **22**(3), 240–249 (2005)
9. Domínguez, G., Cardiel, E., Arias, S., Rogeli, P.: A digital goniometer based on encoders for measuring knee-joint position in an orthosis. In: Proceedings of 2013, USA (2013)
10. Arthittayapiwat, K., Pirompol, P., Samanpiboon, P.: Chest expansion measurement in 3-dimension by using accelerometers. Eng. J. **23**, 71–84 (2019)
11. Ayrulu-Erdem, B., Barshan, B.: Leg motion classification with artificial neural networks using wavelet-based features of gyroscope signals. Sensors **11**, 1721–1743 (2011)
12. Tuncel, O., Altun, K., Barshan, B.: Classifying human leg motions with uniaxial piezoelectric gyroscopes. Sensors **9**, 8508–8546 (2009)
13. Strohrmann, C., Harms, H., Kappeler-Setz, C., Tröster, G.: Monitoring kinematic changes with fatigue in running using body-worn sensors. IEEE Trans. Inf. Technol. Biomed. **16**, 983–990 (2012)
14. Kun, L., Inoue, Y., Shibata, K., Enguo, C.: Ambulatory estimation of knee-joint kinematics in anatomical coordinate system using accelerometers and magnetometers. IEEE Trans. Biomed. Eng. **58**, 435–442 (2011)
15. Guo, Y., Wu, D., Liu, G., Zhao, G., Huang, B., Wang, L.: A low cost body inertial sensing network for practical gait discrimination of hemiplagia patients. Telemedicine e-HEALTH, 748–754 (2012). https://doi.org/10.1089/tmj.2012.0014
16. Tarnita, D., et al.: Design and simulation of an orthotic device for patients with osteoarthritis. In: New Trends in Medical and Service Robots, pp. 61–77. Springer (2016)

17. Roduit, R., Besse, P.A., Micallef, J.P.: Flexible angular sensor. IEEE Trans. Instrum. Meas. **47**, 1020–1022 (1998)
18. Laskoski, G.T., Martins, L.D.L., Pichorim, S.F., Abatti, P.J.: Development of a telemetric goniometer. In: Proceedings of the World Congress, vol. 25, pp. 227–230 (2009)
19. Donno, M., Palange, E., Di Nicola, F., Bucci, G., Ciancetta, F.: A flexible optical fiber goniometer for dynamic angular measurements, pp. 1–5 (2007)
20. Donno, M., et al.: A new flexible optical fiber goniometer for dynamic angular measurements: application to human joint movement monitoring. IEEE Trans. Instrum. Meas. **57**(8), 1614–1620 (2008)
21. Moriguchi, C.S., Sato, T.O., Coury, G.: Ankle movements during normal gait evaluated by flexible electrogoniometer. Rev. Bras. Fisioter. **11**(3), 205–211 (2007)
22. Tesio, L., Monzani, M., Gatti, R., Franghignoni, F.: Flexible electro-goniometers: kinesiological advantages with respect to potentiometric goniometers. Clin. Biomech. **10**, 275–277 (1995)
23. Rowe, P.J., Myles, C.M., Hillmann, S.J., Hazlewood, M.E.: Validation of flexible electrogoniometry as a measure of joint kinematics. Physiotherapy **87**, 479–488 (2001)
24. Van der Linden, M.L., Rowe, P.J., Nutton, R.W.: Between-day repeatability of knee kinematics during functional tasks recorded using flexible electrogoniometry. Gait Posture **28**, 292–296 (2008)
25. Moriguchi, C.S., Sato, T.O., Gil Coury, H.J.C.: Ankle movements during normal gait evaluated by flexible electrogoniometer. Rev. Bras. Fisioter **11**(3), 205–211 (2007)
26. https://jneuroengrehab.biomedcentral.com/articles/10.1186/1743-0003-11-56
27. http://www.scielo.br/scielo.php?script=sci_arttext&pid=S1413-35552007000300006
28. Tarnita, D., Marghitu, D.B.: Nonlinear dynamics of normal and osteoarthritic human knee. Proc. Rom. Acad. **18**, 353–360 (2017)
29. Tarnita, D., Georgescu, M., Tarnita, D.N.: Applications of nonlinear dynamics to gait analysis on plane & inclined treadmill. In: New Trends in Medical and Service Robots, vol. 39, pp. 59–73. Springer, Cham (2016)
30. Rainoldi, A., Melchiorri, G., Caruso, I.: A method for positioning electrodes during surface EMG recordings in lower limb muscles. J. Neurosci. Methods **134**, 37–43 (2004)
31. Zehr, E.P., Komiyama, T., Stein, R.B.: Cutaneous reflexes during human gait: electromyographic and kinematic responses to electrical stimulation (1997)
32. Inman, V.T., Ralston, H.J., Todd, F.: Human walking. Williams & Wilkins, Philadelphia (1981)
33. Ralston, H.J., Todd, F.N., Inman, V.T.: Comparison of electrical activity and duration of tension in the human rectus femoris muscle. Electromyogr. Clin. Neurophysiol. **16**, 271–280 (1976)
34. https://www.intechopen.com/books/electrodiagnosis-in-new-frontiers-of-clinical-research/underwater-electromyogram-for-human-health-exercise
35. Veltink, P.H., Liedtke, C., Droog, E., van der Kooij, H.: Ambulatory measurement of ground reaction forces. IEEE Trans. Neural Syst. Rehabil. Res. **13**, 423–427 (2005)
36. Bae, J., Tomizuka, M.: A tele-monitoring system for gait rehabilitation with an inertial measurement unit and a shoetype ground reaction force sensor. Mechatronics **23**(6), 646 (2013)
37. Wang, W.C., Ledoux, W.R., Sangeorzan, B.J., Reinhall, P.G.: A shear and plantar pressure sensor based on fiber-optic bend loss. J. Rehabil. Res. Dev. **42**, 315–326 (2005)
38. Lackovic, I., Bilas, V., Santic, A.: Measurement of gait parameters from free moving subjects. Measurement **27**, 121–131 (2000)
39. https://www.sensorprod.com/foot-plate-pressure-sensor.php

Conceptual Design of an Urban Electric Microcar

Mehmet Murat Topaç(✉), Merve Karaca, and Levent Bilal

Dokuz Eylül University, 35397 Izmir, Turkey
murat.topac@deu.edu.tr

Abstract. In the scope of this work, design steps of an L6e-class vehicle which is one of the most suitable alternatives for future urban transportation was introduced. Firstly, technical requirements have determined. Design features such as the vehicle mass, motor power and maximum speed were indicated, and the limitations that are determinative on them were pointed out. Effects of these parameters on driving capabilities and driving comfort were investigated. Proper types of suspension and steering linkage were employed based on the axle loads and the driving dynamics, as well as the major dimensions concerning the useful design volume. The suitable electric motor which satisfies the power requirements was chosen. Finally, the driveline scheme was established.

Keywords: Electromobility · Electric vehicle · Microcar · Urban transportation · Vehicle design · Lightweight design

1 Introduction

As a consequence of a large number of the world's population living in cities, city centers are becoming increasingly wearisome and challenging for citizens to live in. According to the statistics, the ratio is expected to reach 80% until 2050 [1]. On account of more people mean more congestion, traffic problems that are already tough will become even worse. As reported by the World Business Council of Sustainable Development (WBCSD) in 2004, light road vehicles are expected to reach 2 billion in the next 30 years [2]. Traffic jams, whose length may currently be over 150 km in crowded cities will become more wearisome soon. The vast majority of drivers travel alone. Even though they provide some advantages such as long driving range and high speed, conventional passenger vehicles are designed to reach more than enough velocity to cities. They also have a notable role in air and noise pollution, as well as the contribution of occupying more than enough space for the number of passengers in jammed city traffic. Therefore, world countries will have to change their transportation habits as soon as possible to the efficient and sufficient ways, which are enough for more people. L-class vehicles, which do not provide more than necessary for city travel, can be one of the many solutions for all these traffic problems caused by overpopulation. Because of their similarities to conventional vehicles such as having four-wheels and safe frame structure rather than many of two-wheeled alternatives, L6 category is the most advantageous vehicle category to the others included in this class.

I. Dumitru et al. (Eds.): SMAT 2019, *The 30th SIAR International Congress of Automotive and Transport Engineering*, pp. 525–532, 2020.
https://doi.org/10.1007/978-3-030-32564-0_61

A design for an intercity utilization brings some additional problems to L6-class vehicles such as short range and limited design envelope.

In the scope of this study, the design process of subsystems of an L6-class vehicle was summarized. Firstly, the technical requirements, such as motor power, vehicle dimensions were appointed based on the restrictions given in the literature for L6-class vehicles. Afterward, essential measurements were specified to compose initial body design, and an approximate vehicle mass was established. The essential properties of body structure and the variation of different body materials were introduced. In the light of axle load values, the suspension and steering system characteristics were determined to satisfy both safety and comfort demands of the vehicle. Depending on the mechanical values, driving resistances were evaluated. Subsequently, the appropriate electric motor was chosen by assigning the most efficient driveline composition. Finally, potential design alternatives were discussed. The design steps of the L6-class electric vehicle were summarized in Fig. 1.

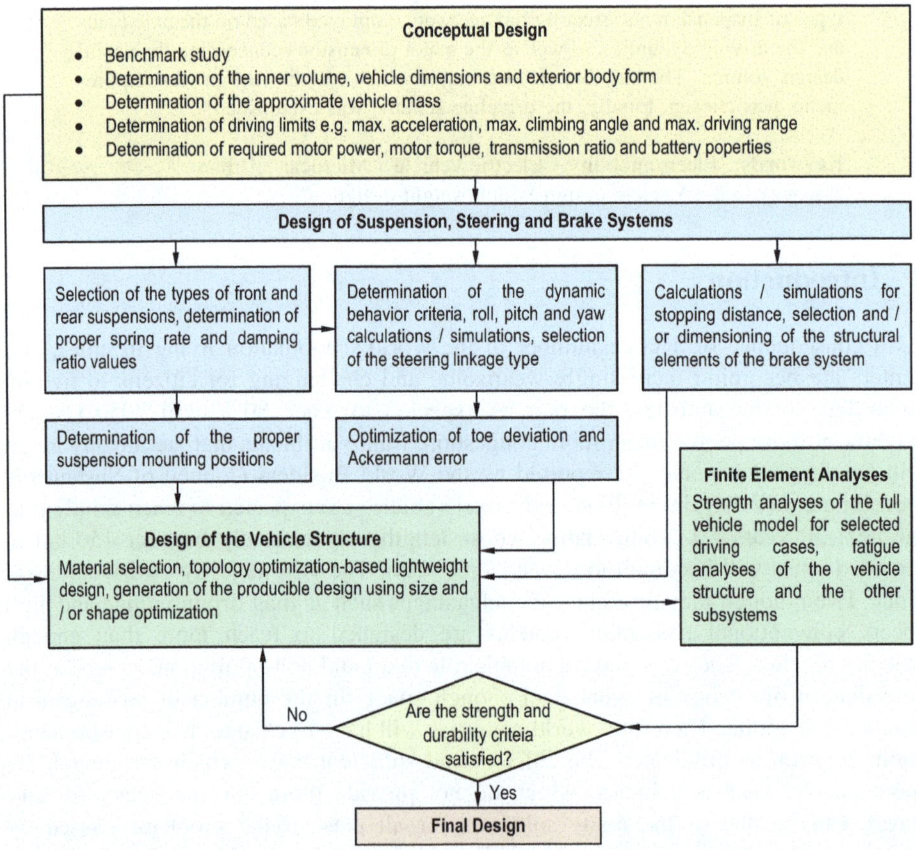

Fig. 1. Design steps of the electric microcar.

2 Design Aspects

2.1 General

In order to offer low-cost transport for a single occupant, L6-class vehicles have some limitations during the design process. So that they service intercity, it is not necessary for the maximum speed value to achieve more than 45 kph. As a subcategory, L6e-B, which is a light quadricycle, is manufactured for passenger transport and has 6 kW of maximum motor power. Its overall dimensions do not exceed 3000 mm of length, 1500 mm of width, and 2500 mm of height depend on ECE R 618 standards. Furthermore, the mass value remains less than 425 kg excluded batteries and passengers [3]. The mass distribution of the vehicle, which consists of the subsystems, components, and the payload of the vehicle is given, and the conceptual design of the vehicle is illustrated in Fig. 2.

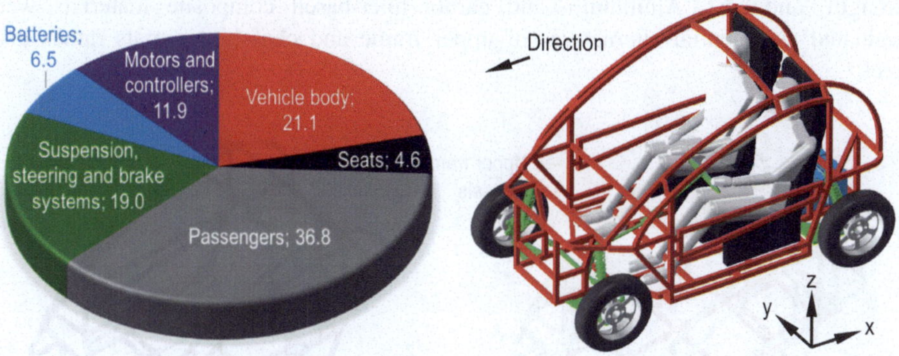

Fig. 2. Mass distribution percentages (left) and design concept (right) of the microcar.

Here, wheelbase and track width are determined as 1800 mm and 1200 mm respectively. The sitting position of the occupants is considered as side by side. The frontal projection area of the vehicle does not exceed 1.7 m^2.

2.2 Vehicle Structure

As long as it takes long hours recharging batteries, range extension is one of the main design challenges for an electric vehicle, which refers to travel longer in one battery cycle. The mass has a significant effect on the total range and the performance of the vehicle for the same power and battery characteristics. Body, which consists of the upper frame structure and chassis forms a considerably large portion of the vehicle weight [4]. Thus, reducing the mass of these subsystems provides a significant advantage in extending range. A chassis must resist variable loads which cause bending, torsion, or shear stresses within an appropriate compliance value since it bears all the components of a vehicle, particularly passengers [5]. Therefore, a chassis must be stiff as well as it is light.

General view of the microcar structure is seen in Fig. 3. In the design phase of the chassis, a three-stage optimization was applied to reduce mass while maintaining comfort and safety standards operated. Firstly, an approximate weight was estimated taking the mass-volume ratio of electric vehicles on the market into consideration [3]. In order to achieve the primary form of the chassis, topology optimization was operated to establish the basic form by using this mass value. Afterward, Design of Experiments-Response Surface Methodology (DOE–RSM) was utilized to specify the final geometry of the chassis under certain loading conditions. The positions of lateral chassis bars determined by finite element (FE) – based topology optimization were adjusted longitudinally by parameterizing the positions of the bars in the x-axis. Vertical loads have a direct relationship with the moment of inertia of the cross-section of a chassis bar. Thus, subsequently, the cross-section dimensions of the chassis components were determined with the same method to obtain a reasonable stress distribution (Fig. 3, right). Finally, the final geometry was verified under the same loading conditions. The effects of different chassis materials on mass were also investigated in terms of mass, strength, and cost. Aluminum, and carbon-fiber-based composite materials were evaluated as potential alternatives of upper frame and chassis materials rather than steel.

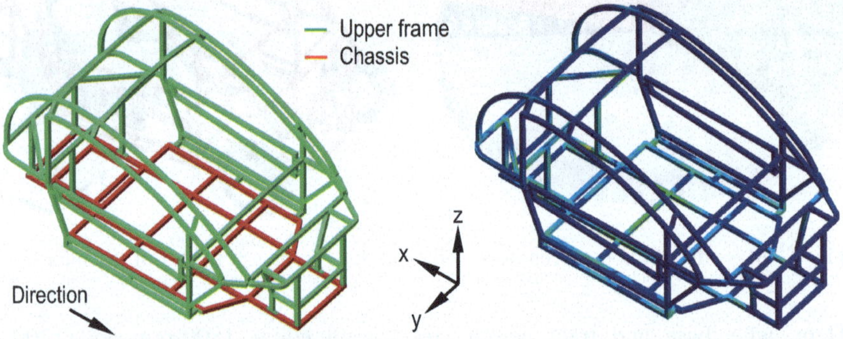

Fig. 3. General view (left) and FE analysis of the microcar structure (right).

2.3 Suspension and Steering Systems

Kinematics of suspension and steering mechanisms are the most critical factors that affect the dynamic behavior of a vehicle. The MacPherson strut-type suspension, which offers reasonable performance compared to its cost, was used in the front axle. Since it eliminates the upper wishbone, it provides lower unsprung mass. Although it offers more space laterally and longitudinally, its vertical space demand is more than double wishbone suspension system for many cases. Nevertheless, having a longer vertical length means that the wheels have more vertical travel. Also, all of its design elements are combined into one assembly which submits design simplicity. On the other hand, the MacPherson strut suspension is more sensitive to imbalances on the road [6]. The trailing arm-type suspension (also known as the "crack-axle suspension") was also preferred in rear axle due to its design simplicity and lower volume requirements [6].

Since it has a longitudinal wishbone, it provides enough space for batteries. In the design stage, firstly, desired spring rates for both rear and front axle wheels were calculated according to the axle loads, and some kinematic variables such as camber angle and roll behavior were determined by using multibody dynamics (MBD) approach. In the light of the data obtained, the positions of suspension joints, and the proper hardpoints of the steering mechanism were established by taking kinematic requirements such as turning radius into account. Shock absorbers which provide proper damping for both front and rear suspensions were also selected. Figure 4 shows the MBD models of the full vehicle and the front suspension.

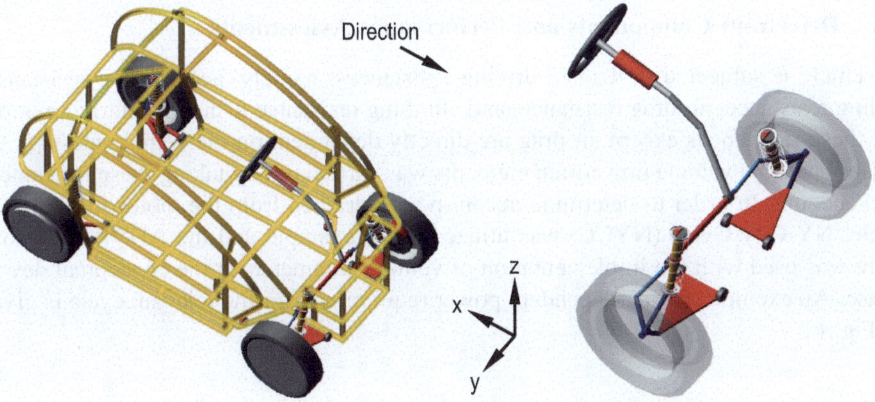

Fig. 4. MBD models of the full vehicle (left) and front suspension (right).

The steering system has a significant role in road stability while suspension system geometry affects some vital kinematic factors such as the roll angle directly. A rack-and-pinion steering mechanism, which is shown in Fig. 5, converts rotational motion on the steering wheel to linear motion on tie rods via rack and pinion gears. It is the simplest steering system which is suitable for independent suspension systems. Besides its lower space requirements, rack-and-pinion steering mechanism has a lower cost. Furthermore, it provides a better road feel. However, the steering angle is directly dependent on the length of the rack. This situation sometimes causes shorter tie rod lengths resulting in kinematic problems and higher force reactions on joints [7, 8].

The main design issue of the steering system in microcars is low-packaging space due to its limited design volume. Therefore, in this study, the rack-and-pinion mechanism, which is commonly used in passenger cars, was employed as the steering system. Basic kinematic behaviors of the vehicle, such as turning motion or toe deviation during bump depend on the position of tie-rod directly. In order to enhance dynamic behavior, DOE-RSM-based design optimization studies were carried out for both suspension and steering systems.

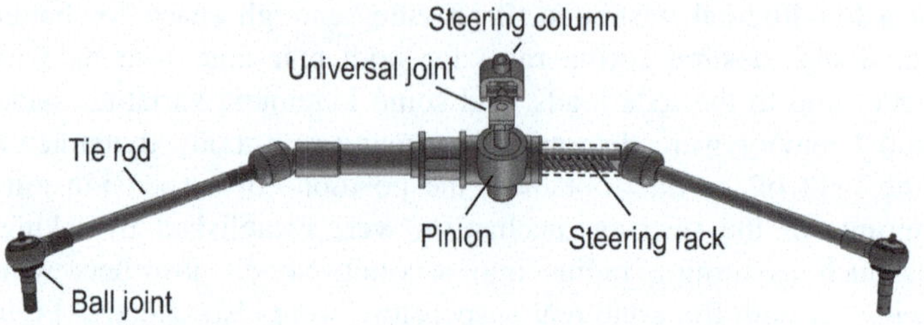

Fig. 5. The rack-and-pinion steering linkage.

2.4 Drivetrain Components and Performance Assessments

A vehicle is subjected to 4 main driving resistances; namely, acceleration resistance, rolling resistance, air drag resistance, and climbing resistance to achieve during motion. All these resistances except air drag are directly dependent on the vehicle mass [5, 9]. Selection of the vehicle powertrain elements was carried out by taking these resistances into account. In order to determine instant power drained from the motor for an urban cycle, NY-City Cycle (NYCC) was utilized as an example and the ADVISOR® software was used with the implementation of vehicle parameters at the conceptual design phase. An exemplary time-dependent power requirement for the relevant cycle is given in Fig. 6.

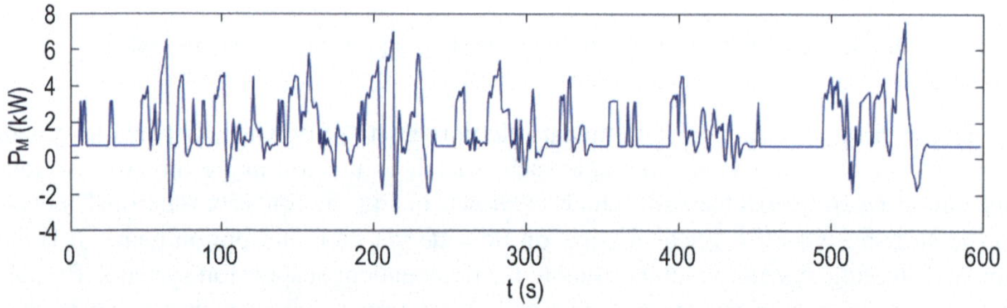

Fig. 6. Time-dependent power requirement for the NY-City Cycle

Drivetrain consists of the batteries, motor and the transmission elements including a constant – ratio gearbox and optionally a final drive which transmits the motor power to the wheels mechanically [9]. The automotive differential provides different torque values for wheels to maintain road stability since inner and outer wheels take different paths during cornering motion. With developing automotive technology, mechanical differentials have started to be replaced by electronic differentials in electric vehicles because of its some significant advantages such as avoiding mechanical friction. It also allows controlling the torque on each wheel separately while the working principle of the mechanical differential is balancing the torque value on each wheel which causes energy loss as a consequence of employing multi-step mechanical elements between the engine and the wheels. Electronic differential utilizes two electric motors whose speed is controlled by speed controllers in each rear wheel. A speed controller uses an algorithm based on the input data of the turning angle of the front wheels, turning radius and, vehicle speed in accordance with Ackermann turning geometry. Furthermore, motorizing each rear wheel is more efficient and stable than using a central motor due to less trans-mission elements which cause mechanical losses through the driveline. An exemplary wheel-motor mounting application is seen in Fig. 7. Moreover, limited space is the most critical problem for a microcar throughout the mass reduction for an electric car. Using two electric motors in each driven wheel not only offers more space for cabin volume and other vehicle subsystems but also provides less weight than mechanical differential since it does not utilize unnecessary mechanical elements such as propeller shafts. It also provides a fast response to the sudden changes in course and road stability [10–12]. In this study, two rear wheels were motorized by 3 kW electric motors.

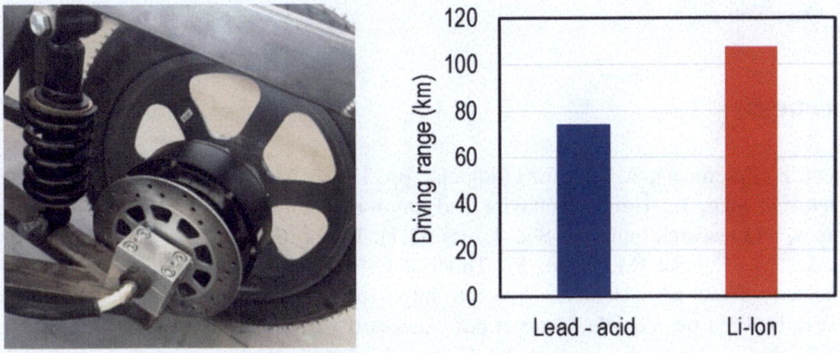

Fig. 7. Rear axle motor (left) and effect of the battery type on driving range (right)

In electric vehicles, different battery types are considered by their advantages and disadvantages. Lead-acid battery, which is the oldest and traditional type used in electric vehicles, can be manufactured in desired sizes and shapes. Moreover, it offers a long life cycle and stands to overcharge. It is also cheaper than other battery types. However, it is heavier compared to the other batteries and offers low specific energy besides its shorter life cycle. Furthermore, its substances are not friendly for the environment [9]. Lithium-ion batteries are preferred more widely with the latest electric vehicle technologies because of its higher energy density compared to its weight. Although it is more expensive than lead-acid batteries, it is preferred more frequently day by day due to its noteworthy advantages, for instance, high life cycle, low discharge values and short recharging time [9, 13].

In this study, these two battery types were considered in terms of their advantages and disadvantages. Since the main design issues are low packaging space and lightweight design, it was shown that Li-Ion batteries are more suitable for an electric microcar according to their low mass value and packaging space requirements within the same voltage V and battery capacity Ah arrangement in comparison to lead-acid batteries. The effects of battery type on the driving range of the microcar were investigated for lead-acid and Li-Ion batteries as a guide for the conceptual design. The simulations, which were carried out by the ADVISOR® software revealed that Li-Ion batteries ensure approximately 45% increase on the driving range of the vehicle. The driving range of the microcar dependent on the battery type is also given in Fig. 7.

3 Conclusions

In the scope of this study, common design features of a microcar were summarized. Thereby, the conceptual design of the microcar was completed. Key design features were introduced. The study is planned to follow by a feasibility study, whose subject consists of the detailed design study of the relevant L6-class urban vehicle.

Acknowledgement. This study was supported by TÜBİTAK, The Scientific and Technological Research Council of Turkey (Grant no: 218M105). The authors are grateful for the contributions of Mr. Onur Solmaz.

References

1. Fleet, B.: Microcars: A Cure for Gridlock?, pp. 17–20. Municipal World, June 2017
2. Vlek, C., Steg, L.: Human behavior and environmental sustainability: problems, driving forces, and research topics. J. Soc. Issues **63**(1), 1–19 (2007)
3. Reske, M., Funcke, M., Date, P., Thonhofer, S., Gauthier, Q.: Feasibility Study on a Radically New L6e Vehicle Concept. http://eu-live.eu/wp-content/uploads/2018/01/EU-LIVE_D_4.3_L6e_Vehicle_Concept.pdf. Accessed 14 May 2019
4. Stodolsky, F., Vyas, A., Cuenca, R., Gaines, L.: Life-cycle energy savings potential from aluminum-intensive vehicles. SAE technical paper 951837 (1995)
5. Reimpell, J., Hoseus, K.: Fahrwektechnik: Fahrzeugmechanik, 2nd edn. Vogel Verlag, Würzburg (1992)
6. Reimpell, J., Stoll, H., Betzler, J.W.: The Automotive Chassis: Engineering Principles. Society of Automotive Engineers Inc, Warrendale (2002)
7. Simionescu, P.A., Smith, M.R.: Initial estimates in the design of rack-and-pinion steering linkages. J. Mech. Des. **122**, 194–200 (2000)
8. Reimpell, J.: Fahrwerktechnik, vol. 3. Vogel-Verlag, Würzburg (1974)
9. Larminie, J., Lowry, J.: Electric Vehicle Technology Explained. Wiley, West Sussex (2003)
10. Magallan, G.A., De Angelo, C.H., Bisheimer, G., Garcia, G.: A neighborhood electric vehicle with electronic differential traction control. In: 34th Annual Conference of IEEE Industrial Electronics, pp. 2757–2763. IEEE (2010)
11. Hartani, K., Miloud, Y., Miloudi, A.: Electric vehicle stability with rear electronic differential traction. In: International Symposium on Environment Friendly Energies in Electrical Applications, Algeria (2010)
12. Haddoun, A., Benbouzid, M.E.H., Diallo, D., Abdessemed, R., Ghouili, J., Srairi, K.: Modeling, analysis, and neural control of a EV electrical differential. IEEE Trans. Ind. Electron. **55**(6), 2286–2294 (2008)
13. Yong, J.Y., Ramachandaramurthy, V.K., Tan, K.M., Mithulananthan, N.: A review on the state-of-the-art technologies of electric vehicle, its impacts and prospects. Renew. Sustain. Energy Rev. **49**, 365–385 (2015)

New Approach for the Integration of CAD/CAPP/CAM Systems for Bearings Manufacturing

Dan George Prutica(✉) and Gheorghe Brabie

"Vasile Alecsandri" University of Bacău, 600115 Bacău, Romania
dan.prutica@yahoo.com

Abstract. Due to the high potential of improving efficiency and productivity in a competitive and challenging industrial environment, the automation and integration of CAD, CAPP and CAM have been under discussion and research since the beginning of their existence. Even so, diverse systems have evolved on distinct levels, CAD and CAM systems have been intensively developed and implemented as dedicated tools for their specific served functions and CAPP systems are still under research for development of applicable solutions in industry. This paper proposes an original approach for the integration of CAD with CAPP and CAM by the development of a complete knowledge-based system for automatic generation of product design data and process planning data of bearings. Through the work presented in this paper, a prototype of a complete knowledge-based system has been realized by developing a new process planning module integrated with an already existing module for product design data generation. Furthermore, the system has been integrated with an existing commercial CAM system for the generation of NC simulations and NC programs. The application of the proposed solution was verified by a case study for turning process of the outer rings for large-size single row cylindrical roller bearings. The time for realization of the process plans using this approach has been radically decreased compared with the manual approach, generating cost savings and shorter time to market. The presented solution can be further extended and developed for all the processes required for bearings manufacturing.

Keywords: CAD/CAPP/CAM integration · Automated process planning · Bearings manufacturing

1 Introduction

The idea of utilizing computer technology for design (CAD), process planning (CAPP) and manufacturing (CAM) activities was initiated more than fifty years ago and has been generating a large amount of research for the development and integration of these systems. Information technology can provide an efficient communication interface between management, design, process planning and production through the integration of CAD, CAM and CAPP systems in a complex and cost-efficient manufacturing system [1–5].

I. Dumitru et al. (Eds.): SMAT 2019, *The 30th SIAR International Congress of Automotive and Transport Engineering*, pp. 533–539, 2020.
https://doi.org/10.1007/978-3-030-32564-0_62

The usage of computers in design, process planning and manufacturing, does not necessarily mean that these activities have been integrated in an automated manufacturing system applicable on a large scale in industry. The development of CAD, CAM and CAPP systems has reached various levels of evolution, on one level is the CAD and CAM systems development as assisting tools for the specialists in design or manufacturing activities and the extended implementation of this systems in industry. On another level is the development of CAPP systems and the integration of CAD with CAPP, which are still requiring additional research to develop solutions that can be implemented in industry [1–5].

Process planning is a complex multi-task for the transformation plan of a product idea, from the design phase, into a physical product, in the manufacturing phase. A knowledge-based CAPP system incorporates the manufacturing knowledge data containing all the constraints and rules necessary for making optimal decisions to generate products process plans. The integration of CAD with CAPP, represented by the part features recognition, is the first essential step in improving the speed and the quality of design information converted into manufacturing instructions in the planning process. The main two approaches for part feature recognition are design by features (DBF) and automated feature recognition (AFR). In the DBF approach, a part is designed using a set of predefined features stored in system libraries and for each predefined design feature can be created a set of rules for the conversion and generation of manufacturing features. The AFR approach consists of different methods, specific for various applications area, utilized for the conversion of geometric entities created in CAD systems into manufacturing features [2].

2 System Concept

This paper proposes an approach for the integration of CAD with CAPP and CAM by the development of a complete knowledge-based system for automatic generation of product design data and process planning data of bearings, connected with an existing commercial CAD/CAM system for 2D/3D design files generation, NC simulations and NC programs generation.

The proposed approach of the system is presented in Fig. 1 containing a system framework, integrated modules for product design and process planning, connected data bases and connected commercial CAD/CAM system. The process planning module has been developed and implemented as an automated computer aided process planning (ACAPP) system, having as a start trigger for the automated generation of the process plan the released design of the product in the integrated design module. The system module for process planning includes all the required functionality for automated generation of process plans for bearings components manufacturing, starting with the part feature recognition for the generation of manufacturing features and continued with the technology selection, machine selection, clamping methods selection, tools selections, process parameters calculation, and machining sequence generation. All these steps are executed according to the specific manufacturing knowledge defined in the system and according to the selected objective function for process optimization. The manufacturing knowledge data is introduced and maintained within

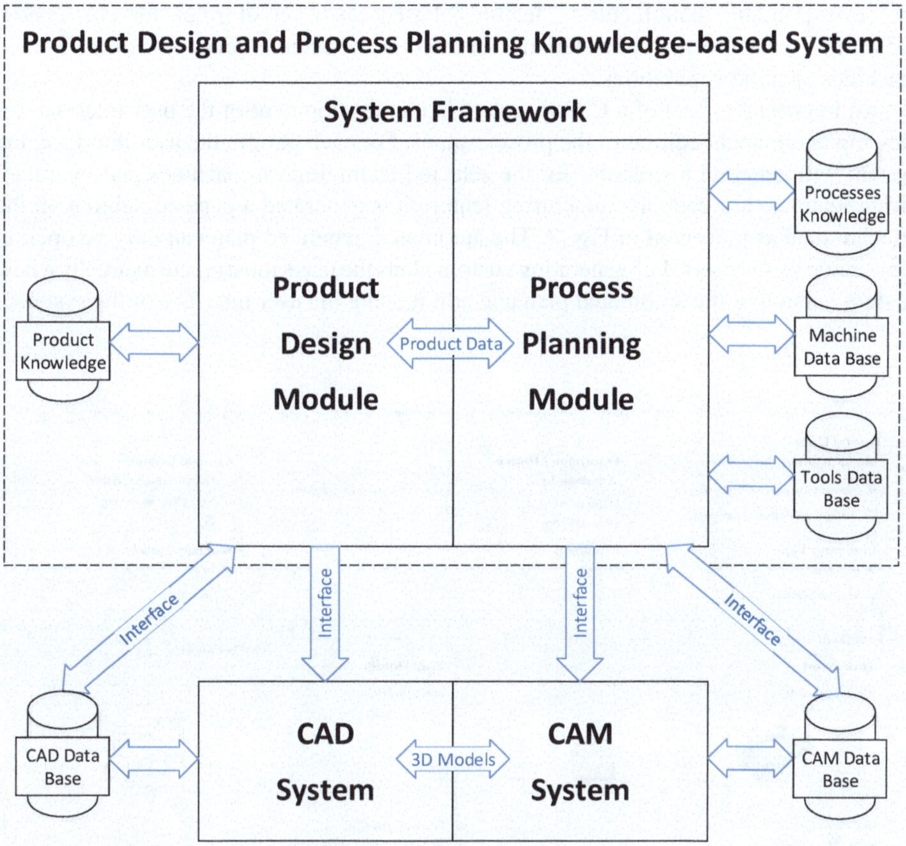

Fig. 1. Product design and process planning knowledge-based system

the system by planning specialists, with the possibility, in the future, of increasing and improving the knowledge base through intelligent learning methods. The raw material data, machine data and tools data are stored in specific databases connected to the system.

Most of bearings components are parts composed from a combination of predefined design features selected and calculated to form the finished bearing component. These types of parts are offering a high potential for applying a form of design by features approach for the features recognition in the integration between CAD and CAPP. One of the key steps in the current work is represented by the developed method for the generation of the manufacturing features corresponding to each design feature without any conversion of actual CAD files. The module for product design of the knowledge-base system has been previously developed to calculate and generate the bearings design configurations. All generated data is available, through the connection interface, for the commercial CAD system to generate the 2D/3D CAD files. The generated part configuration contains the design features of the bearings components which will be used as input data for the process planning. The process planning module is generating the corresponding manufacturing features, based on a set of rules for conversion, according to the specific characteristics of the design features, process technology and machines specific capabilities.

An important aspect of a CAPP system is the development of the user interface for viewing and manual editing of the process plans. For each part, in the user interface, the system will generate a structure for the selected technologies, machines and manufacturing features. For each manufacturing feature it is generated a page containing all the relevant data as presented in Fig. 2. The automated generated plan can only be open in view mode by the users. For generating custom plans the users must create manually a new custom version of the automated plan and edit it using the user interface of the system.

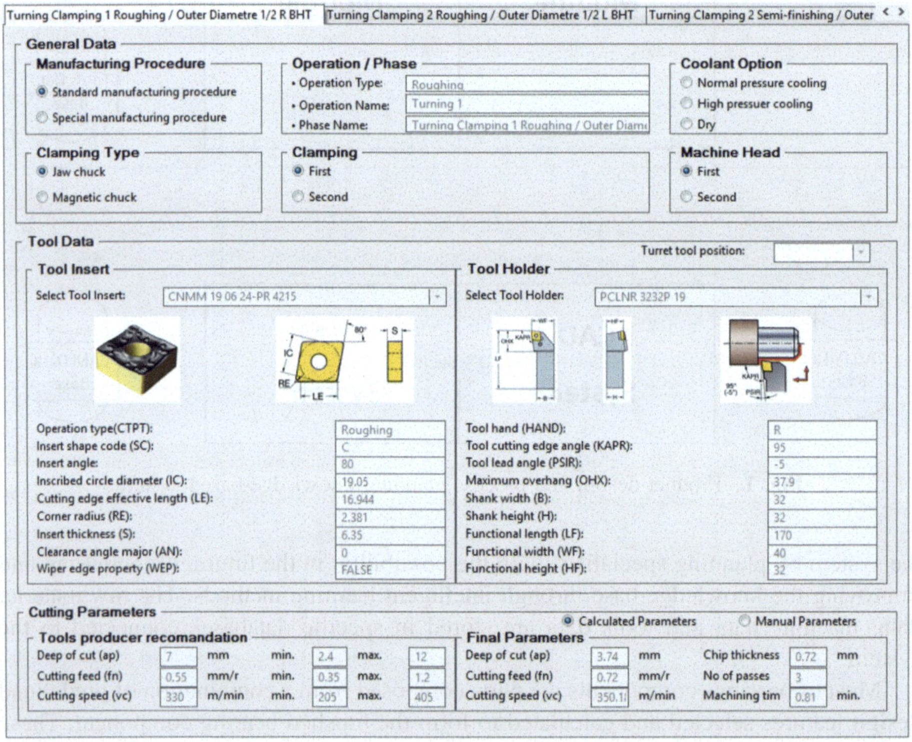

Fig. 2. The user interface model for displaying the data of each manufacturing feature

3 System Prototype

In the manufacturing process of the bearing components, the main manufacturing process and the most time consuming is turning, resulting that the automation of the process planning for turning would bring the greatest cost-benefit ratio, where the cost represents the development effort of the system.

Through the work presented in this paper, a prototype of a complete knowledge-based system has been realized by developing a new process planning module for turning process of the outer rings for large-size single row cylindrical roller bearings, integrated with an already existing module for product design data generation. The single row cylindrical roller bearings are a type of products with a very large variety of dimensions and types, containing different combinations of design features, being representative for bearings products. Creating the prototype for one of the rings of this bearing type demonstrates the applicability of the system for all bearings types.

The knowledge-based system is developed to be independent of the CAD/CAM system, offering flexibility to use different other systems in case of specific requirements. In the current implementation, we have used Creo Parametric, from Parametric Technology Corporation (PTC) as CAD/CAM solution due to its parametric capabilities and due to the possibility of using native CAD files directly in the CAM system.

4 Case Study

In this case study, we have selected five representative types of large-size cylindrical rollers bearings of type NU, NJ, NUP, NF and N. The outer rings of the selected bearings are representing three distinct constructive types within a wide range of dimensions, as it can be observed from the data presented in Table 1.

Table 1. Main dimensions of the selected bearings outer rings, Source: Schaeffler

Number	1	2	3	4	5
Bearing type	NU	NJ	NUP	NF	N
Bearing catalog designation	NU10/560-TB-M1	NJ252-E-TB-M1	NUP2344-EX-TB-M1	NF230-E-M1	N334-E-TB-M1
Outer diameter (D)	820 mm	480 mm	460 mm	270 mm	360 mm
Width (B)	115 mm	80 mm	145 mm	45 mm	72 mm
Raceway diameter (E)	754 mm	429 mm	413 mm	242 mm	318 mm
Board diameter (D1)	728,4 mm	406,6 mm	385,8 mm	233,2 mm	–

The average time required for the manual creation of the process plan and the NC program for a bearing outer ring, is about 4 h. The average time for automatic generation of the process plan and the NC program using the developed system is about 15 min and, optional, the check of the generated data would take another 45 min, in total one hour, resulting a reduction of the time required for the automated process plans generation compared to the manual approach with approx. 75%.

For turning process, the system will generate all the manufacturing features corresponding to the design features for roughing and semi-finishing or finishing. The test has shown a direct connection between the number of design features, their complexity and the number of manufacturing features generated by the system. In the same time has been confirmed the repeatability regarding the generation of the same manufacturing features for the same type of rings with the same configuration of design features, independent of the dimensions (see Fig. 3).

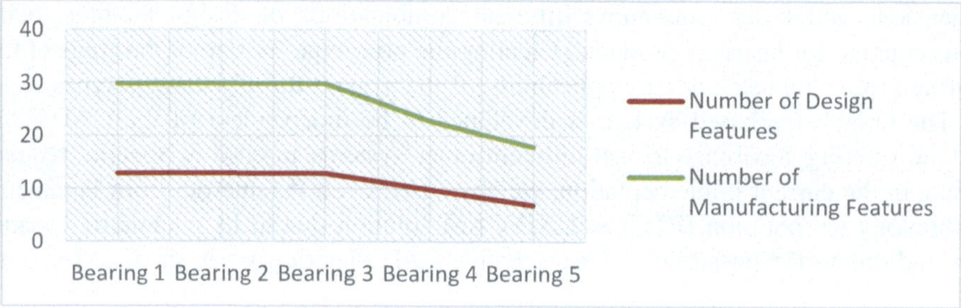

Fig. 3. Number of design features vs. Generated number of manufacturing features

5 Conclusion

Through the work presented in this paper, a prototype of a complete knowledge-based system has been realized by developing a new process planning module for turning process integrated with an already existing module for bearings design data generation. The process planning module has been developed and implemented as an automated computer aided process planning system (ACAPP), capable of generating the process plan without any user intervention, for all the dimensions and types of the selected product range. After the generation of the process plan, all generated data are available, through the connection interface, for Creo NC to generate the NC simulations and NC programs.

The system has been developed in collaboration with one of the world top three bearings producers company and the quality of the generated data has been tested for several bearings produced by this company. The presented solution can be further extended and developed for all the processes required for bearings manufacturing.

References

1. Yusof, Y., Latif, K.: Survey on computer-aided process planning. Int. J. Adv. Manuf. Technol. **75**(1–4), 77–89 (2014)
2. Al-wswasi, M., Ivanov, A., Makatsoris, H.: A survey on smart automated computer-aided process planning (ACAPP) techniques. Int. J. Adv. Manuf. Technol. **96**, 1–4 (2018)
3. Sanfilippo, E.M., Borgo, S.: What are features? An ontology-based review of the literature. Comput. Aided Des. **80**, 9–18 (2016)

4. Lau, H.C.W., Lee, C.K.M., Jiang, B., Hui, I.K., Pun, K.F.: Development of a computer-integrated system to support CAD to CAPP. Int. J. Adv. Manuf. Technol. **26**, 1032–1042 (2005)
5. Yusof, Y., Case, K.: STEP compliant CAD/CAPP/CAM system for turning operations. Lecture Notes in Engineering and Computer Science, vol. 2173 (2008)

Design Optimization of the Towing Hook Used in Passenger Cars for Light Trailers

Mario Trotea(✉), Augustin Constantinescu, and Loreta Simniceanu

University of Craiova, 200585 Craiova, Dolj, Romania
mtrotea@yahoo.com

Abstract. The paper presents the design optimization of the towing hook used in passenger cars for light trailers. For this goal the cross-section and the geometry of the longitudinal axis of the hook were parameterized so the design optimization analysis took into consideration 5 parameters as design variables. The hook's mass and the global structural safety factor were considered as design function and constraint, accordingly. A new design was built using the optimized values of the parameters and the results of the structural analysis of the optimal design were emphasized.

Keywords: Design optimization · Fatigue analysis · Towing hook

1 Introduction

At present the need for a portable towing system has increased so much so that, due to competition, equipment with low design, production, operation and reliability costs is essential. The design of such equipment plays a fundamental role. Thus, the handling and the utilization of the equipment should be made quick and easy.

As they have been initially used for towing trailers and caravans, towing hooks have acquired, along with technical improvements, other utilities such as the transport of bicycles or skiing equipment, and other means of transportation.

A towing device is the primary component that connects a vehicle to the towed object. This device requires specific additional components, such as tow frame and the tow hook for the connection to be completed.

The sphere of the towing hook is the connecting point between the towing vehicle and the object that is being towed. Along with the tow bar, a spherical ball facilitates approaching turns and overtaking road obstacles without losing the connection between the vehicle and the towed object. The tow bar is mounted and connected above the spherical ball, ensuring a swiveling motion and articulation around it when driving.

Some spherical devices are equipped with a ball that is already attached, be it welded directly on the superior bar, or screwed in with a nut on a supporting system.

Different studies for testing the towing devices and numerical structural analyses or fatigue analyses were done and found in technical literature [1, 4].

I. Dumitru et al. (Eds.): SMAT 2019, *The 30th SIAR International Congress of Automotive and Transport Engineering*, pp. 540–549, 2020.
https://doi.org/10.1007/978-3-030-32564-0_63

2 Design, Loading Calculation and Testing of the Towing Hook

The design, loading calculation and the establishment of a testing scheme for the towing hook for vehicles are implemented through the 94/20/CE Direction of the European Parliament and the Council of Europe, regarding mechanical towing devices of vehicles and their trailers, as well as their safe attachment to these vehicles.

2.1 The Design of the Towing Hook

The 94/20/CE Directive applies to mechanical towing devices for vehicles and trailers, as well as the mounting of these devices to road vehicles. Some of the requirements that have to be met in order to allow the use of mechanical towing devices between ensembles of vehicles [5] are as follows:

- to ensure the compatibility of the attachment of vehicles with different types of trailers;
- to ensure the safe attachment of vehicles in conditions of use;
- to ensure that safety procedures have been established.

Tow balls and the tow bar for which the design optimization will take place are part of category A (according to [5]), a category that consists of one-piece tow balls (see Fig. 1).

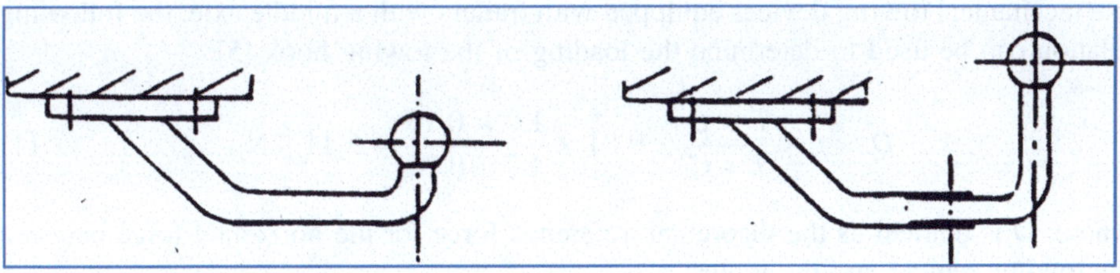

Fig. 1. Tow ball with a one-piece tow bar - constructive variations [5]

Category A tow balls have to correspond to the design displayed in Fig. 2 from the perspective of shape and dimensions in order to be compatible with requirements and to eliminate the possibility of deformation in conditions of towing, as stated in the 94/20/CE Directive.

The range of attachment between the tow ball and the tow bar has to be tangential to the tow bar, and also the inferior horizontal surface of the tow ball. The shape and the dimensions of the tow ball are in concordance with ISO/R 468 and ISO 1302 standards. The N9 stringency refers to an Ra value of 6.3 μm.

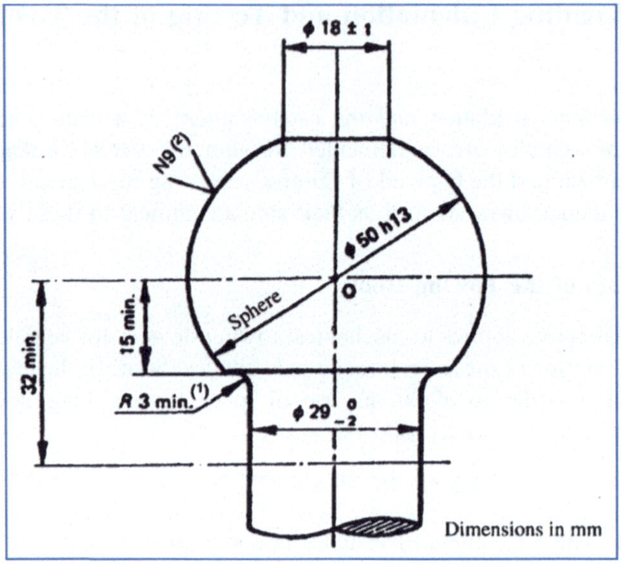

Fig. 2. Tow ball – standardized dimensions [5]

2.2 Loading Calculation of the Towing Hook

In the case of the towing hook for vehicles with a weight smaller than 3.5 tons which use mechanical towing devices equipped with trailers with a middle axle, the following relation can be used to determine the loading of the towing hook [5]:

$$D = g \times \frac{T \times C}{T + C} = 9.81 \times \frac{1.7 \times 0.75}{1.7 + 0.75} = 5.11\ [\mathrm{kN}] \quad (1)$$

where: D is defined as the theoretical reference force for the horizontal force between the towing vehicle and the trailer;

T is the maximum allowed gross vehicle mass, expressed in tons (T = 1.7 tons for a certain case);

C is the total mass of the trailer with a middle axle that carries the maximum allowed mass, expressed in tons (C = 0.75 tons);

g is the gravitational acceleration (g = 9.81 m/s^2).

2.3 Testing of the Towing Hook

Towing devices have to undergo resistance tests, as well as functional tests. These tests cannot produce fissures, fractures or otherwise visible external defects to the devices, or any excessive and permanent deterioration. The plastic deformation after release cannot be larger than 10% of the maximum deformation. The resistance of the towing devices has to be verified through a dynamic test (endurance test), with a number of loading cycles that is appropriate for the material of which the device is made [5].

The towing hook shall be subjected under an alternating force ($F_{res} = D$) and must be applied to the coupling ball at an angle of 15° ± 1°, as shown in Fig. 3.

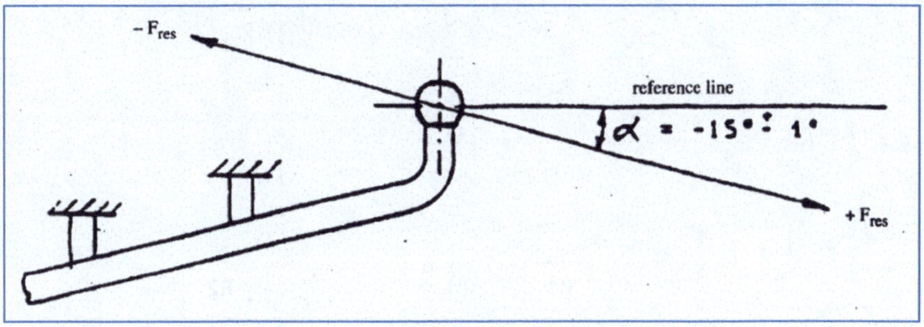

Fig. 3. The angle of applying the testing load [5]

3 Design Optimization of the Towing Hook

Using ANSYS Workbench, a 3D model of a towing hook was built (see Fig. 4) using the sketch defined in Fig. 5. Five design variables were chosen for defining the geometry of the towing hook, along with the minimum and maximum values:

- d, the cross-section's diameter (35 mm $\leq$ d $\leq$ 50 mm)
- L1, the length of the horizontal section (120 mm $\leq$ L1 $\leq$ 180 mm)
- α, the angle of the inclined section (10° $\leq$ d $\leq$ 20°)
- R1, the fillet radius between first two sections (120 mm $\leq$ R1 $\leq$ 150 mm)
- R2, the fillet radius between the inclined and vertical sections (40 mm $\leq$ R2 $\leq$ 60 mm).

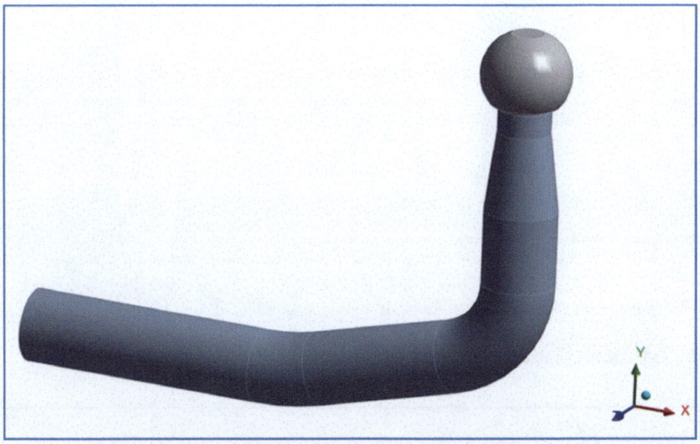

Fig. 4. The 3D model of the towing hook

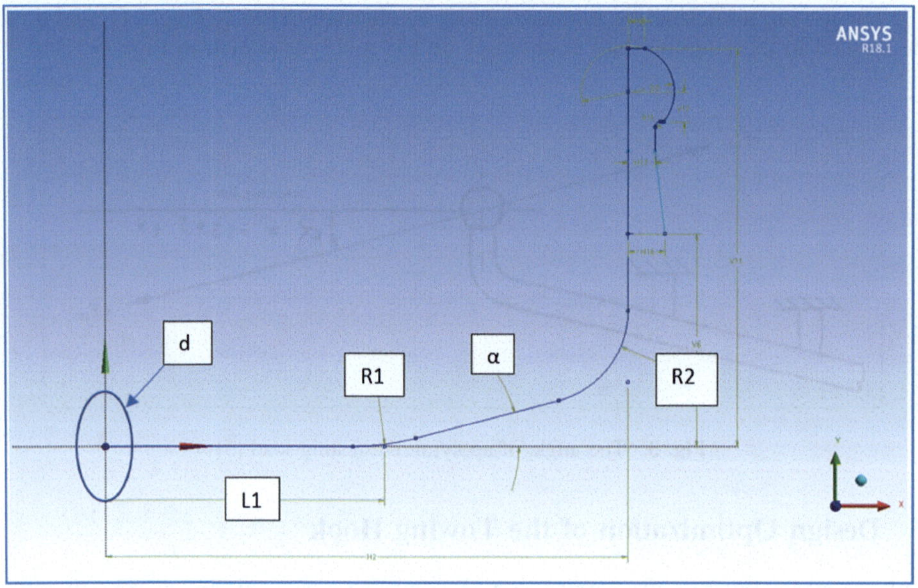

Fig. 5. The design variables for the geometry of the towing hook

Setting the design variables to their minimum and maximum values, the "minimal" and "maximal" designs of the towing hook are presented in Figs. 6 and 7.

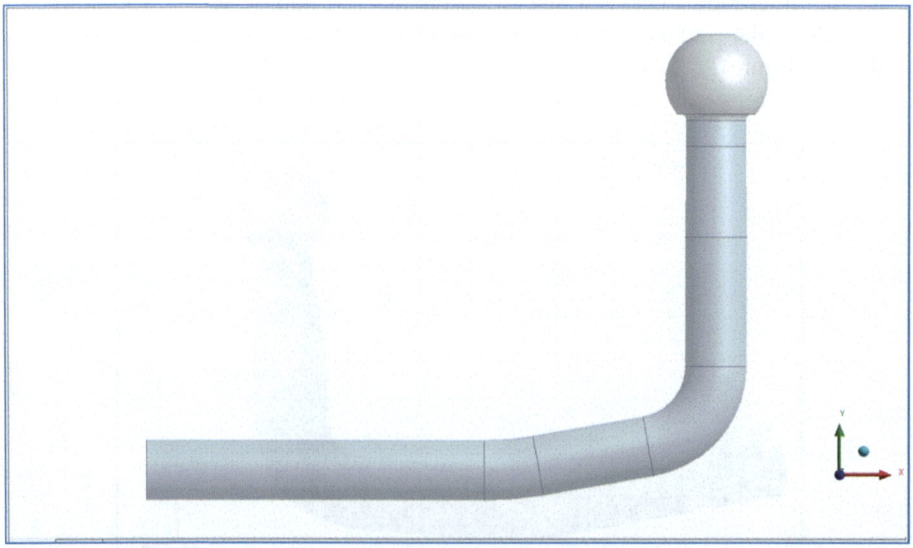

Fig. 6. The "minimal" design of the towing hook

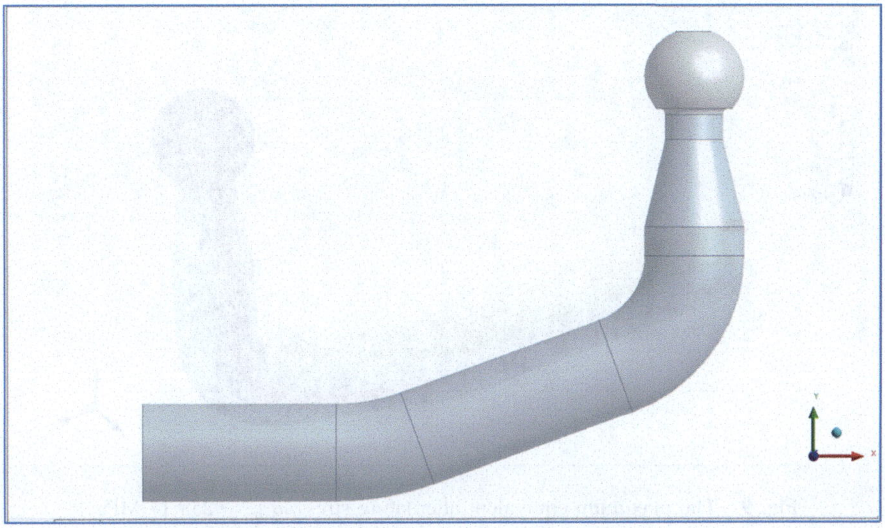

Fig. 7. The "maximal" design of the towing hook

The material used for the towing hook is SMn438 (JIS G 4106) with the yield stress of 730 MPa and ultimate stress of 844 MPa [6].

Prior to design optimization analysis a fatigue analysis of the towing hook was performed with the uniaxial fatigue test data sheet from [6]. The Soderberg mean stress theory was used with an alternating load applied as shown in Fig. 8. The direction of the loading force (Eq. 1) was established according to Fig. 3.

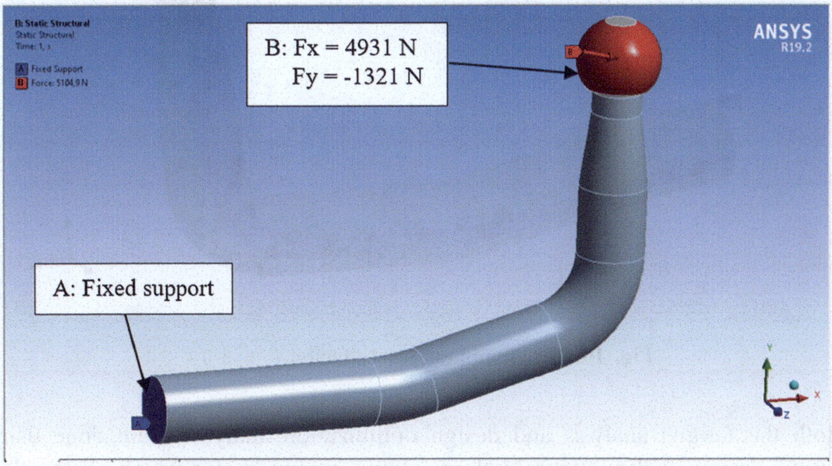

Fig. 8. Boundary condition of the model: A – fixed support, B – loading force components applied on the towing hook sphere

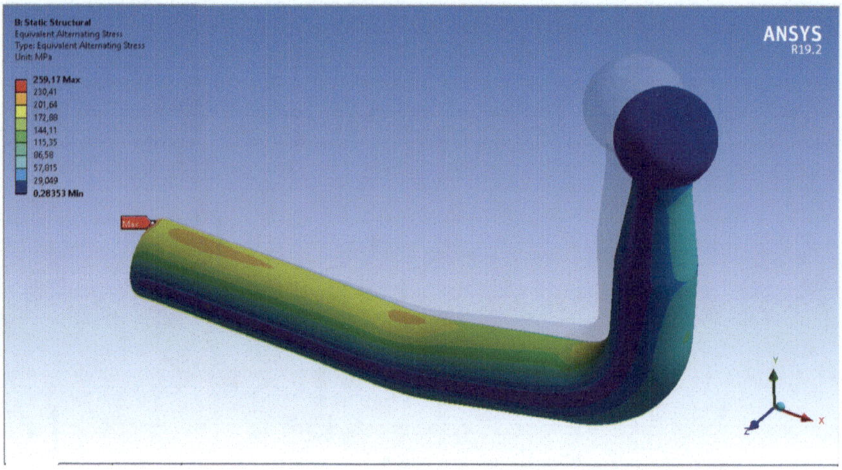

Fig. 9. The maximum equivalent alternating stress, $\sigma_{eqv} = 259.17$ MPa

The maximum equivalent alternating stress (as shown in Fig. 9) was $\sigma_{eqv} = 259.17$ MPa, in the fixed support area. The safety factor was $c_s = 1.74$, as shown in Fig. 10.

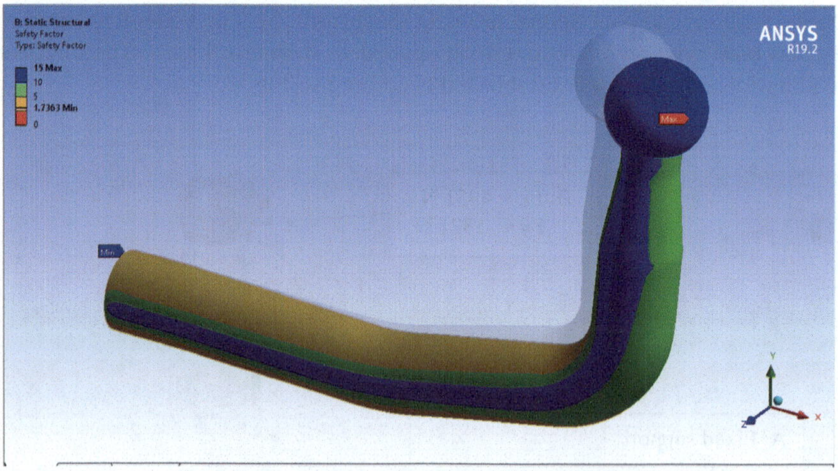

Fig. 10. The minimum safety factor, $c_s = 1.74$

Both the fatigue analysis and design optimization analyses were done using a "medium" design of the towing hook, as shown in Fig. 4, for which all the design variables were set on their median values.

For the design optimization problem, the design objective was to find the design with minimum mass subjected to the following constraints (see the design variables):

$$\begin{cases} c_s \geq 2 \\ 30\ \text{mm} \leq d \leq 50\ \text{mm} \\ 120\ \text{mm} \leq L1 \leq 180\ \text{mm} \\ 10^\circ \leq \alpha \leq 20^\circ \\ 120\ \text{mm} \leq R1 \leq 150\ \text{mm} \\ 40\ \text{mm} \leq R2 \leq 60\ \text{mm} \end{cases} \tag{2}$$

Based on a response surface optimization, the design optimization analysis found the best three candidate points shown in Fig. 11. All three candidate points were verified on the model. The candidate point 3 was chosen to be the optimum solution, having the verified solution closest to the response surface solution.

Candidate Points	Candidate Point 1	Candidate Point 1 (verified)	Candidate Point 2	Candidate Point 2 (verified)	Candidate Point 3	Candidate Point 3 (verified)
P1 - d (mm)	42,17		41,81		42,41	
P2 - L1 (mm)	121,49		146,81		132,74	
P3 - alpha (degree)	18,345		18,592		19,827	
P4 - R1 (mm)	140,37		124,53		125,97	
P5 - R2 (mm)	58,269		46,024		53,371	
P6 - Geometry Mass (tonne)	0,004627	0,0046251	0,0046572	0,0046563	0,0046915	0,0046906
P7 - Safety Factor Minimum	2,0203	2,1095	2,0024	× 1,9494	2,0395	2,0246

Fig. 11. The design optimization result

For the candidate point 3, the optimized design variables values were:

$$\begin{cases} d = 42.4\ \text{mm} \\ L1 = 132.7\ \text{mm} \\ \alpha = 19.8^\circ \\ R1 = 126\ \text{mm} \\ R2 = 53.4\ \text{mm} \end{cases} \tag{3}$$

which satisfied the minimum safety factor constrain. The mass of the towing hook (4.69 kg) was 7.32% bigger than the mass of the initial design, but the safety factor was 16.67% higher.

Using the optimum solution, a new design of the towing hook was built (see Fig. 12) and a new fatigue analysis was performed on the optimal design. Figures 13 and 14 show the results for the optimal design.

The equivalent alternating stress has the same distribution over the towing hook body with a maximum value in the fixing area, $\sigma_{eqv} = 219.64$ MPa (see Fig. 13).

The minimum safety factor, corresponding to the maximum value of the equivalent alternating stress, was $c_s = 2.05$ (see Fig. 14).

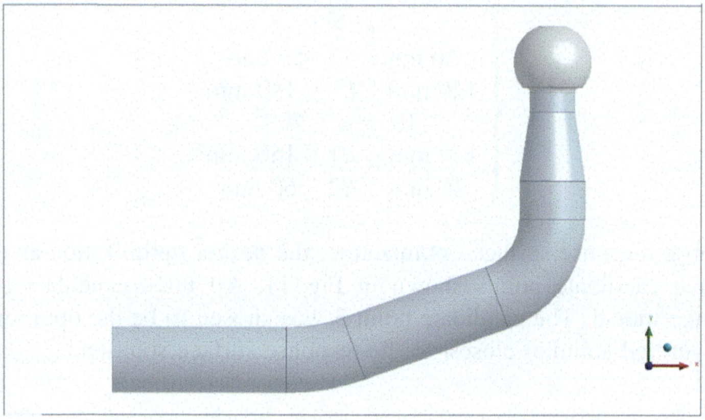

Fig. 12. The optimized design of the towing hook

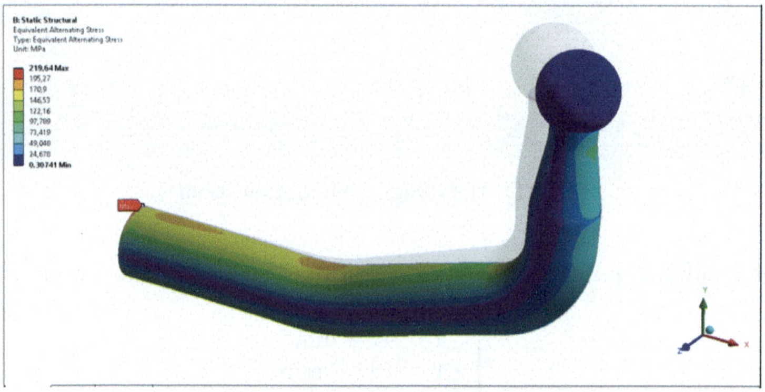

Fig. 13. The maximum equivalent alternating stress for the optimized design, $\sigma_{eqv} = 219.64$ MPa

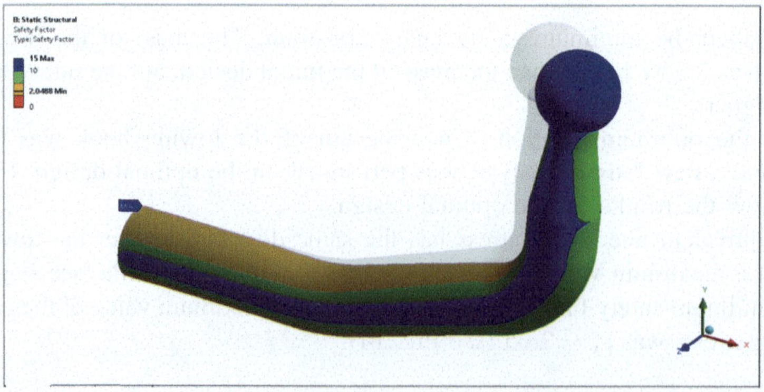

Fig. 14. The minimum safety factor for the optimized design, $c_s = 2.05$

4 Conclusions

Using ANSYS software, a parameterized design can be easily modified so the best design can be found using a design objective analysis. In this paper, the design of a towing hook was parameterized using 5 parameters in order to find the design with a minimum mass for which the minimum safety factor meets a certain value. An initial design of the towing hook was built, with all the parameters set to their median values. The fatigue analysis for this design revealed a minimum safety factor of 1.74 which was 13% lower than the imposed constrain for safety factor. After the design optimization analysis of the towing hook, the minimum safety factor meets the requirement of the optimal problem, but the mass of the new design was 7.32% higher than the mass of the initial design.

The fatigue analysis of the towing hook was done according to the 94/20/CE Direction of the European Parliament and the Council of Europe, regarding mechanical towing devices of vehicles and their trailers, as well as their safe attachment to these vehicles.

References

1. Palma, E.S., Petracconi, C.L., Ferreira, S.E.: Fatigue behavior analysis of a rear tow hook pin of a passenger vehicle. Eng. Fail. Anal. **16**(7), 2408–2416 (2009)
2. Petrici, A.V., Radu, G.: Studies on the behavior of the towbars in static applications. Adv. Mater. Methods **822**, 100–111 (2016)
3. Pruthvi, R.G., Ganesha, B.: Optimization study on tow hook of a commercial vehicle to improve the performance at different loading conditions. Int. J. Res. Eng. Technol. **04**(08), 118–123 (2015). eISSN 2319-1163, pISSN 2321-7308
4. Wang, Z.: Car front towing hook analysis and structural improvements based on CAE. In: International Industrial Informatics and Computer Engineering Conference (IIICEC). Published by Atlantis Press (2015)
5. Directive 94/20/EC of the European Parliament and of the Council of 30 May 1994 relating to the mechanical coupling devices of motor vehicles and their trailers and their attachment to those vehicles. https://eur-lex.europa.eu/legal-content/RO/ALL/?uri=CELEX%3A31994L0020. Accessed 25 July 2019
6. Data sheets on fatigue properties of SMn438 (0.38C–1.5Mn) steel. https://smds.nims.go.jp/MSDS/pdf/sheet/F16J.pdf. Accessed 25 July 2019

A Study of the Mechanical Behavior of a Car Part Realized from Epoxy Resin Reinforced with Cotton Woven

Alexandru Bolcu[1], Nicolae Dumitru[1], Dumitru Bolcu[1(✉)], Cosmin Mihai Miritoiu[1(✉)], and Marius Marinel Stanescu[2(✉)]

[1] Faculty de Mechanics, University of Craiova, 200512 Craiova, Romania
bolcualexandru@yahoo.com, dbolcu@yahoo.com, miritoiucosmin@yahoo.com
[2] Applied Mathematics Department, University of Craiova, 200585 Craiova, Romania
mamas1967@gmail.com

Abstract. In this article we were looking for the production of a physical car body part (the left side fender of a Peugeot 206) made from composite materials based on epoxy resin and reinforced cotton woven. On the newly created fender model, a modal analysis has been made in order to determine its own frequencies on a certain direction, and an harmonic analysis with Finite Element Method.

Keywords: Composite materials · Modal analysis · Own frequencies · Harmonic analysis

1 Introduction

Composite materials that are reinforced with fibers have a good market place because of their structural applications but they have a limited hardness. Until now, the use of a natural reinforcement and a synthetic resin has been considered to generate a bio-composite (see the papers [1] and [3]).

The study of the mechanical behavior of laminated composite materials is an important research direction in the mechanical field. Thus, in the works [5] and [2] are presented studies regarding the vibrational and damping behaviour of laminated composite materials that are reinforced with different types of fibers. The connection between the elasticity modulus and the damping capacity of composite materials reinforced with different types of fibers has been studied in [7].

The interest of eco-durable materials in industrial sector (cars, constructions etc.) has lead to a growing number of scientific works regarding bio-composite materials. [12] proposes a fabrication process that permits the obtaining of unidirectional bio-composites of high quality with a fraction of fiber volume of only 70%.

In [4] is evaluating the development of bio-composite specific regarding the mechanical properties.

I. Dumitru et al. (Eds.): SMAT 2019, *The 30th SIAR International Congress of Automotive and Transport Engineering*, pp. 550–558, 2020.
https://doi.org/10.1007/978-3-030-32564-0_64

The properties of elongation, bending and impact for fiber bio-composites made from polylactic acid and nettle fibers, made at high pressure have been analyzed in [6]. In [8] is presented a complete analysis of polymeric composite materials reinforced with natural fibers.

In [11] bio-composite materials have used unidirectional arranged fibers made of kenaf, bamboo, coconut to reinforce the polymeric matrix of the polylactic acid. In [10] the traction properties, bending, impact resistance have been studied for bio-composite materials such as synthetic resins reinforced with sisal fibers and ananas fibers unidirectional layered. [9] presents the results of experimental investigation of mechanical properties for composite hybrid polyester reinforced with cotton and sisal fibers.

2 Experimental Measurements

The experimental analysis for determining the own frequencies was made on a fender made from composite material of epoxy resin reinforced with cotton woven. The ratio used for the matrix is of 100 volumetric parts of Resoltech resin 1050 with 30 volumetric parts of hardener Resoltech 1058 S. The fender was hand made by pouring layer after layer (5 layers have been used) (Fig. 1).

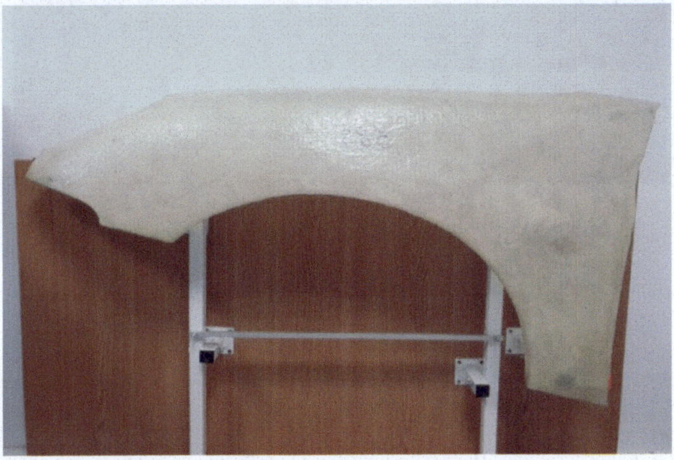

Fig. 1. Fender made of epoxy resin and reinforced with cotton woven.

Experimental measurements were performed using the SPIDER 8 data acquisition system, the NEXUS 2692-A-0I4 signal conditioner, and the accelerometer with a sensitivity of 0.04 $\mathrm{pc/ms^{-2}}$. Measurements were made in two points as shown in Figs. 2 and 3, the excitement was done with the help of an impact hammer.

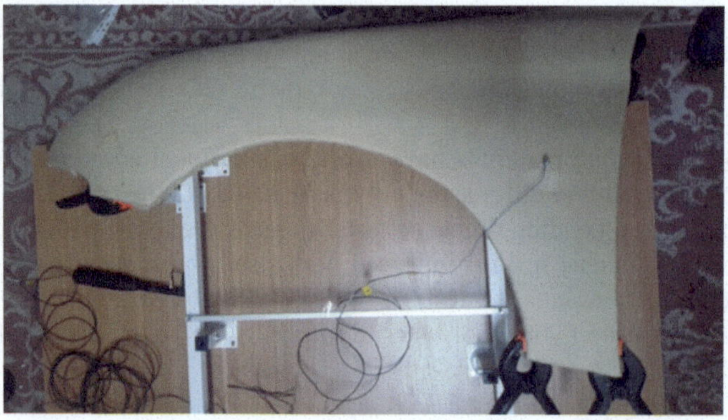

Fig. 2. Positioning of the accelerometer (lateral point)

Fig. 3. Positioning of the accelerometer (central point)

In Fig. 4 is represented the experimental measurement done on the side for a fender made from cotton.

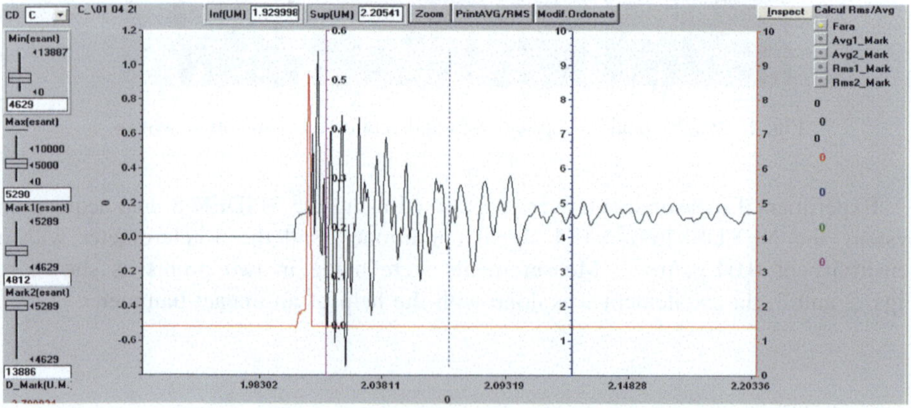

Fig. 4. Experimental measurement (lateral point)

In Fig. 5 is represented Frequency Response Function (FRF) done on the side for a fender made from cotton.

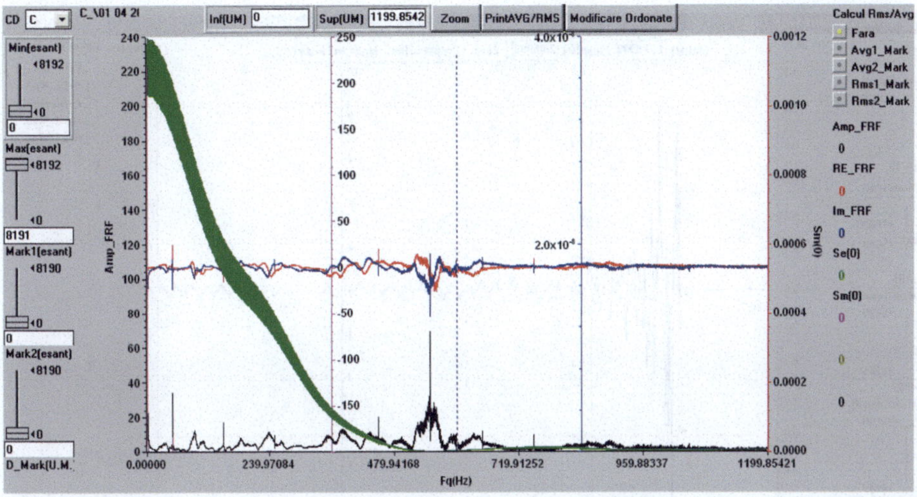

Fig. 5. Frequency Response Function (FRF)

In Fig. 6 is represented Own Frequency Modes (OFM) done on the side for a fender made from cotton.

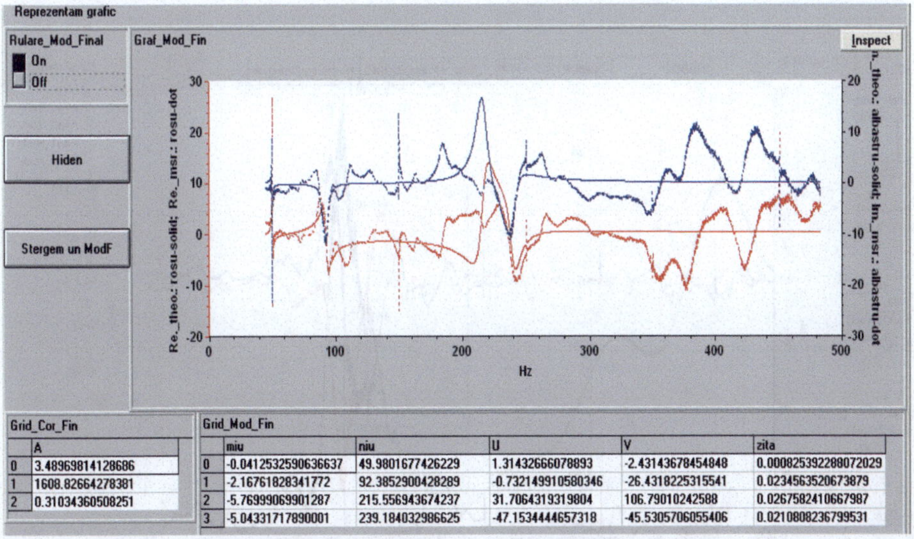

Fig. 6. OFM final panel (lateral point)

In Fig. 7 is represented the experimental measurement done on the central point for a fender made from cotton.

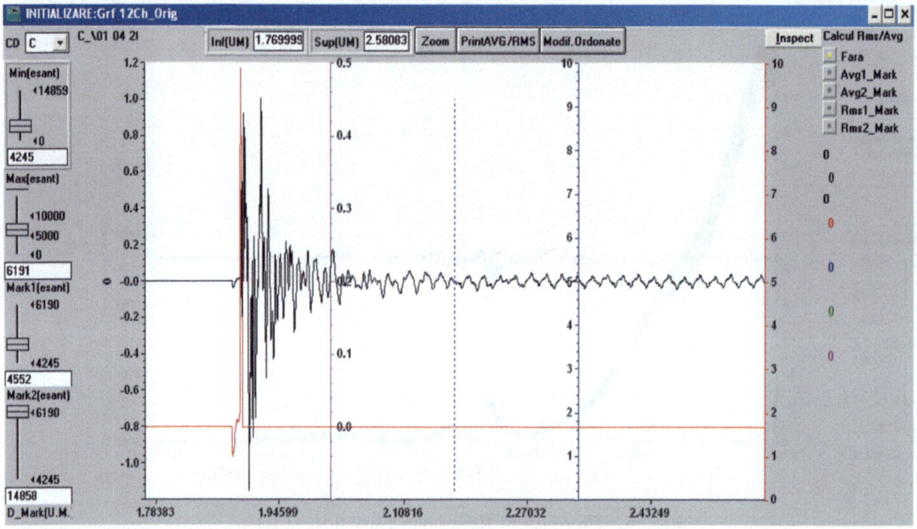

Fig. 7. Experimental measurement (central point)

In Fig. 8 is represented Frequency Response Function (FRF) done on the central point for a fender made from cotton.

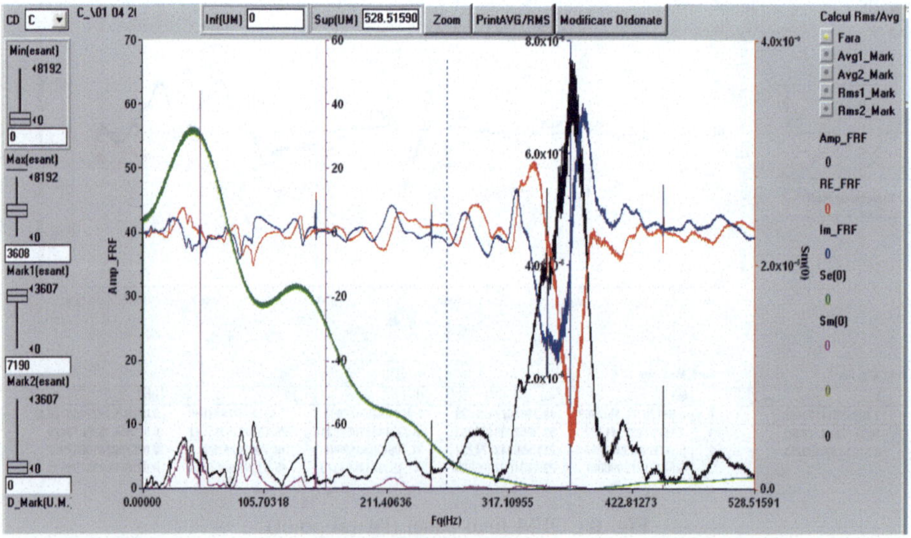

Fig. 8. Frequency Response Function (FRF)

In Fig. 9 is represented Own Frequency Modes (OFM) done on the central point for a fender made from cotton.

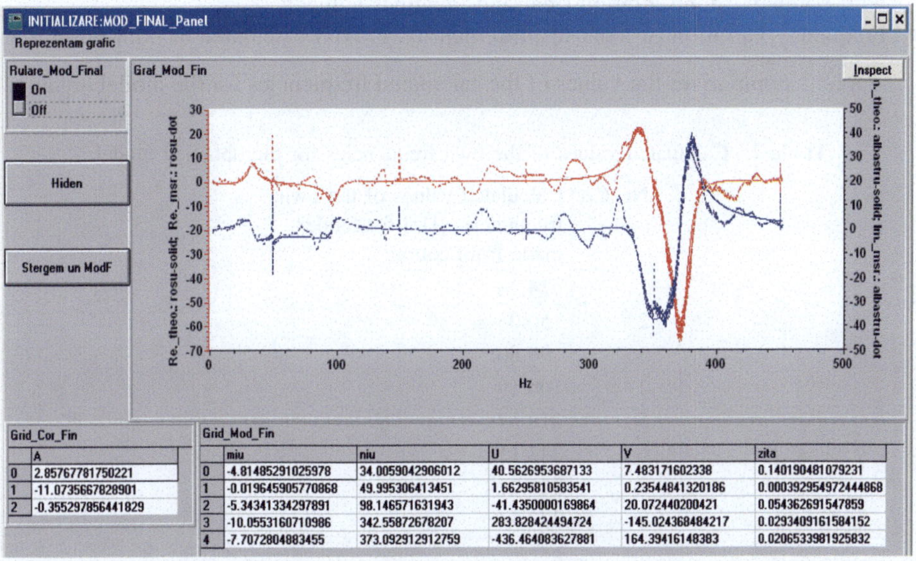

Fig. 9. OFM final panel (central point)

In Table 1 is represented the own frequencies obtained experimentally at the two measuring points, for the fender made from composite material reinforced with cotton woven.

Table 1. Own frequencies obtained from experimental modal analysis.

Fender material	Own frequencies obtained for experimental modal analysis [Hz]	
	For lateral point	For central point
Cotton	49.980	34.005
	92.385	49.995
	215.556	98.146
	239.184	342.558
	–	373.092

3 Results

We have obtained a mathematical model on the basis of witch we have determined (and highlighted) the corresponding own frequencies and vibration modes in the form of color maps and graphs. Three variation graphs from harmonic analysis have been highlighted:

- Frequency variation of axial displacement (OY axis) calculated at the lateral point, with complex representation (real and imaginar values);
- Frequency variation of axial displacement (OY axis) calculated at the central point, with complex representation (real and imaginar values);
- Frequency variation of axial displacement (OY axis) calculated at both point.

Table 2 summarizes the values of the calculated frequencies for the model obtained.

Table 2. Calculated values of the own frequencies for the obtained model.

Nr. Crt	Calculated values of the own frequencies [Hz] for fender made from cotton
1	34.26
2	50.05
3	94.80
4	100.19
5	219.54
6	247.11
7	346.27
8	377.54

For example, we present in Fig. 10 the image with its own vibration mode corresponding to 100.19 [Hz].

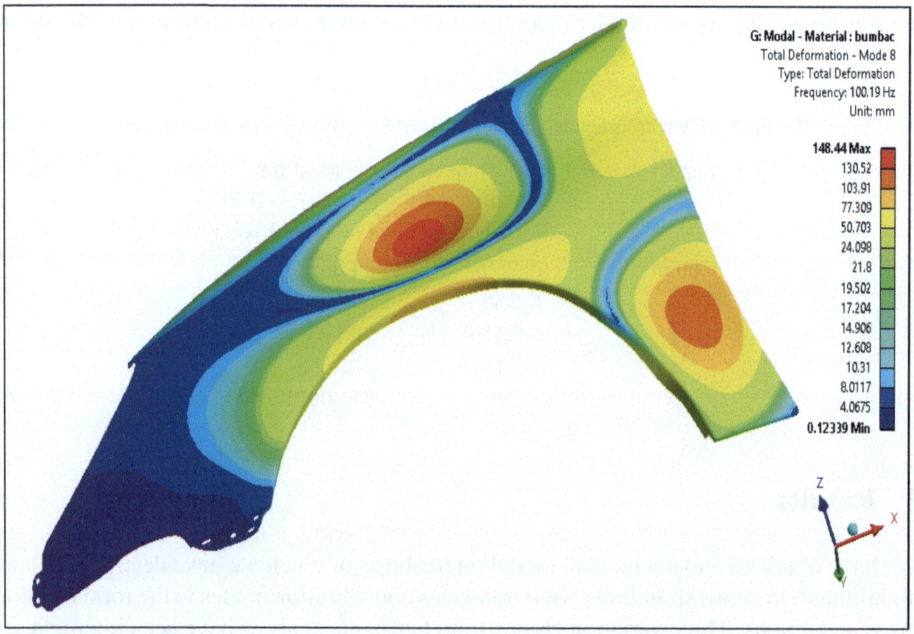

Fig. 10. Own vibration mode corresponding to 100.19 [Hz]

In Fig. 11 we present the frequency response, the amplitude of the axial displacement measured at both points.

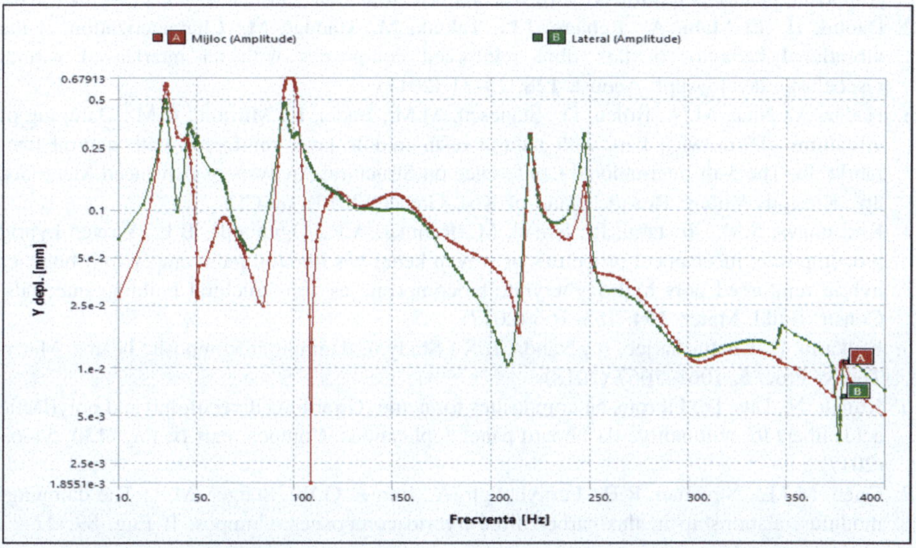

Fig. 11. Frequency Response (material: cotton). Amplitude of axial displacement [mm] (OY axis), measured at both points

Table 3 shows the comparison (as a percentage error) between experimental and calculated values

Table 3. Frequency values measured experimentally and calculated with ANSYS

Fender material	Value	Frequency [Hz]							
		1	2	3	4	5	6	7	8
Cotton	Test	34.01	49.98	92.39	98.15	215.56	239.18	342.56	379.09
	ANSYS	34.26	50.05	94.80	100.19	219.54	247.11	346.27	377.54
	Error [%]	0.75	0.14	2.61	2.08	1.85	**3.31**	1.08	1.19

4 Conclusions

- By analyzing the comparative data from Table 3, it can be seen that the values of own frequencies determined by Finite Element Method are in compliance with the ones determined by experiments, the maximum error being of 3,31%.
- This small error (less than 10%) guarantees that the geometrical model of the fender, also the model made with finite elements method are in compliance with the physical ones and can be used for future studies.

References

1. Bolcu, D., Stănescu, M.M., Ciucă, I., Trante, O., Dinu, G.: Mechanical properties of composite materials reinforced with wheat straw. Materiale Plastice **47**(2), 219–224 (2010)
2. Daoud, H., El Mahi, A., Rebiere, J.L., Taktak, M., Haddar, M.: Characterization of the vibrational behavior of flax fibre reinforced composites with an interleaved natural viscoelastic layer. Appl. Acoust. **128**, 23–31 (2017)
3. Hadăr, A., Nica, M.N., Bolcu, D., Stănescu, M.M., Baciu, F., Miriţoiu, C.M.: Damping of vibrations of sandwich bars with natural resin middle layer reinforced with natural fiber fabric. In: The 5-th International Conference on Structural Analysis of Advanced Materials, 8p., Kipriotis Village Resort, Island of Kos, Greece (2013). on CD
4. Khoshnava, S.M., Rostami, R., Ismail, M., Rahmat, A.R., Ogunbode, B.E.: Woven hybrid biocomposite: mechanical properties of woven kenaf bas fibre/oil palm empty fruit bunches hybrid reinforced poly hydroxybutyrate biocomposite as non-structural building materials. Constr. Build. Mater. **154**, 155–166 (2017)
5. Kulkarni, P., Bhattacharjee, A., Nanda, B.K.: Study of damping in composite beams. Mater. Today: Proc. **5**, 7061–7067 (2018)
6. Kumar, N., Das, D.: Fibrous biocomposites for nettle (Girardinia diversifolia) and poly(lactic acid) fibers for automotive dashboard panel applications. Compos. Part B: Eng. **130**, 54–63 (2017)
7. Guen, M.J.L., Newman, R.H., Fernyhough, A., Emms, G.W., Staiger, M.P.: The damping-modulus relationship in flax-carbon fibre hybrid composites. Compos. B Eng. **89**, 27–33 (2016)
8. Al-Oqla, F.M., Sapuan, S.M.: Natural fiber reinforced polymer composites in industrial applications: feasibility of date palm fibers for sustainable automotive industry. J. Clean. Prod. **66**, 347–354 (2014)
9. Sathishkumar, T.P., Naveen, J., Navaneethakrishnan, P., Satheeshkumar, S., Rajini, N.: Characterization of sisal/cotton fibre woven mat reinforced polymer hybrid composites. J. Ind. Text. **47**, 429–452 (2016)
10. Vikas, S., Keshav, S.B., Krishna, M.: Mechanical properties of sisal and pineapple fiber hybrid composites reinforced with epoxi resin. Int. J. Mod. Eng. Res. **5**, 32–38 (2015)
11. Yusoff, R.B., Takagi, H., Nakagaito, A.N.: Tensile and flexural properties of polylactic acid-based hybrid green composites reinforced by kenaf, bamboo and coir fibers. Ind. Crops Prod. **94**, 562–573 (2016)
12. Zuccarello, B., Marannano, G., Mancino, A.: Optimal manufacturing and mechanical characterization of high performance composites reinforced by sisal fibers. Compos. Struct. **194**, 575–583 (2018)

Modeling of Dynamic Behavior of a Spur Gear

Ionut Geonea(✉), Laurentiu Racila, Ionica Valeriu, and Cosmin Mihai Miritoiu

University of Craiova, Craiova, Romania
igeonea@yahoo.com

Abstract. The purpose to optimize gears requires a very good understanding of their dynamic behavior. Very important is the influence of vibrations on the operation of the gears. In this paper it is proposed a theoretical approach to vibration behavior of gears using a mathematical model with linear behavior, a model that allowed the application of a modern calculation method, such as the use of integral transformations.

Keywords: Gears · Dynamic response · Vibrations · Laplace transform

1 Introduction

Gears occupy a special place in transmitting the rotation motion in a uniform manner, but being the object of many researches. The gearmotor is most used to perform the rotation movement between parallel axes.

As a result of environmental constraints such as acoustics, gears are increasingly subject to stringent requirements in terms of load capacity, efficiency, noise and vibrations [1–5].

Here we presented a complex mechanical model with a large number of degrees of freedom. Then, the mathematical model solved it in an original way, using the unilateral Laplace transform in relation to the time, for that the inversion was computed [6–10]. The dynamic response it is represented in a graphical form, based on a real numerical application.

2 Mechanical Model

Figure 1 presents the mechanical model with eight degrees of freedom, obtained in the paper [1], the meaning of the notations being given below:

$$m_g - \text{output gear mass } [\text{Kg}]; m_p - \text{input gear mass } [\text{Kg}];$$

$$J_g - \text{wheel inertia moment } \left[\text{Kg m}^2\right]; J_p - \text{sprocket inertia moment } \left[\text{Kg m}^2\right];$$

$$c_{xg} - \text{wheel 1 damping coefficient } \left[\frac{\text{Ns}}{\text{m}}\right]; c_{xp} - \text{wheel 2 damping coefficient } \left[\frac{\text{Ns}}{\text{m}}\right];$$

I. Dumitru et al. (Eds.): SMAT 2019, *The 30th SIAR International Congress of Automotive and Transport Engineering*, pp. 559–564, 2020.
https://doi.org/10.1007/978-3-030-32564-0_65

c_t – torsion damping coefficient $\left[\frac{Ns}{m}\right]$; $(T_1 = T_g\ T_2 = T_p)$ – torsion [Kgf];

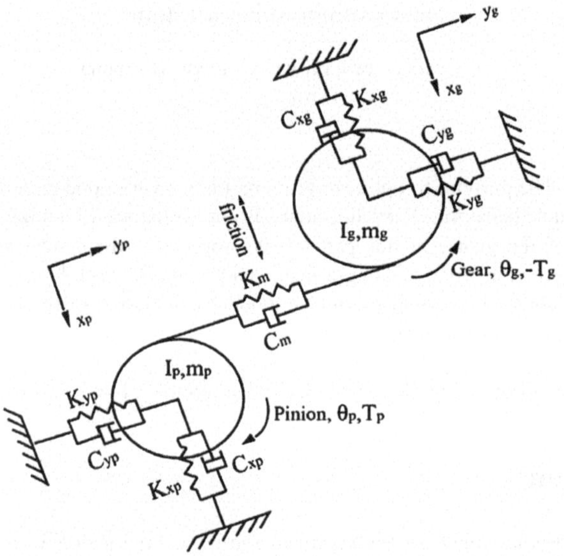

Fig. 1. Mechanical model with eight degrees of freedom.

In the matrix form, the mathematical model with eight degrees of freedom has the following aspect (see the paper [1]):

$$\left\{\ddot{q}\right\} + [A]^{-1}[B]\left\{\dot{q}\right\} + [A]^{-1}[C]\{q\} = [A]^{-1}\{F\} \tag{1}$$

Concretely, the mathematical model (1) is a system of eight differential equations with constant coefficients, obtained in Lagrange's formalism, as follows:

$$\begin{cases} m_g \ddot{x}_g + c_{xg}\dot{x}_p + k_{xg}x_g = \mu N \\ m_p \ddot{x}_p + c_{xp}\dot{x}_p + k_{xp}x_p = -\mu N \\ m_g \ddot{y}_g + c_{yg}\dot{y}_p + k_{yg}z_g = N \\ m_p \ddot{y}_p + c_{yg}\dot{y}_p + k_{yg}y_p = -N \\ m_g \ddot{z}_g + c_g\dot{z}_g + k_g z_g = 0 \\ m_p \ddot{z}_p + c_p\dot{z}_p + k_p z_p = k_p z_p \\ J_g \ddot{\theta}_g + c_{t1}\dot{\theta}_g + k_{t1}\theta_g = -R_1 N + T_1 \\ J_p \ddot{\theta}_p + c_{t2}\dot{\theta}_p + k_{t2}\theta_p = R_2 N - T_2 \end{cases} \tag{2}$$

In the absence of damping, the system (2) takes shape:

$$\begin{cases} m_g \ddot{x}_g + k_{xg}x_g = \mu N \\ m_p \ddot{x}_p + k_{xp}x_p = -\mu N \\ m_g \ddot{y}_g + k_{yg}z_g = N \\ m_p \ddot{y}_p + k_{yg}y_p = -N \\ m_g \ddot{z}_g + k_g z_g = 0 \\ m_p \ddot{z}_p + k_p z_p = k_p z_p \\ J_g \ddot{\theta}_g + k_{t1}\theta_g = -R_1 N + T_1 \\ J_p \ddot{\theta}_p + k_{t2}\theta_p = R_2 N - T_2, \end{cases} \tag{3}$$

Where:

$$\begin{aligned} N = c_m \left[\left(R_1\dot{\theta}_g - R_2\dot{\theta}_p \right) + \left(\dot{y}_p - \dot{y}_g \right)\sin\alpha + \left(\dot{x}_p - \dot{x}_g \right)\cos\alpha \right] \\ + k_m \left[(R_1\theta_g - R_2\theta_p) + \left(y_p - y_g \right)\sin\alpha + (x_p - x_g)\cos\alpha \right] \end{aligned} \tag{4}$$

3 Dynamic Response

Applying the transformed Laplace system (3), under homogeneous initial conditions, with the exception of the z coordinate, where these are taken into account, resulted in the algebraic system below in the Laplace images of the unknowns:

$$\begin{cases} (m_g s^2 + k_{sg} + \mu N_4)\widetilde{x}_g(s) + \mu N_4\widetilde{x}_p(s) - \mu N_3\widetilde{y}_g(s) + \mu N_3\widetilde{y}_p(s) \\ \quad - \mu N_1\widetilde{\theta}_g(s) + \mu N_2\widetilde{\theta}_p(s) = 0 \\ (m_p s^2 + k_{sp} - \mu N_4)\widetilde{x}_p(s) + \mu N_4\widetilde{x}_g(s) + \mu N_3\widetilde{y}_g(s) - \mu N_3\widetilde{y}_p(s) \\ \quad + \mu N_1\widetilde{\theta}_g(s) - \mu N_2\widetilde{\theta}_p(s) = 0 \\ (m_g s^2 + k_{yg} - N_3)\widetilde{y}_g(s) - N_4\widetilde{x}_g(s) + N_4\widetilde{x}_p(s) + N_3\widetilde{y}_p(s) \\ \quad - N_1\widetilde{\theta}_g(s) + N_2\widetilde{\theta}_p(s) = 0 \\ (m_g s^2 + k_{yp} - N_3)\widetilde{y}_p(s) + N_3\widetilde{y}_g(s) + N_4\widetilde{x}_g(s) \\ \quad - N_4\widetilde{x}_p(s) + N_1\widetilde{\theta}_g(s) - N_2\widetilde{\theta}_p(s) = 0 \\ (m_g s^2 + k_g)\widetilde{z}_g(s) - s z_{g_0} = 0 \\ m_p s^2\widetilde{z}_p(s) - s z_{p_0} - \dot{z}_{p_0} = 0 \\ (J_g s^2 + k_{t_1} + N_1 R_1)\widetilde{\theta}_g(s) - R_1 N_2\widetilde{\theta}_p(s) + R_1 N_4\widetilde{x}_g(s) - R_1 N_4\widetilde{x}_p(s) \\ \quad + R_1 N_3\widetilde{y}_g(s) - R_1 N_3\widetilde{y}_p(s) = T_1 \\ (J_p s^2 + k_{t_2} + N_2 R_2)\widetilde{\theta}_p(s) - R_2 N_4\widetilde{x}_g(s) + R_2 N_4\widetilde{x}_p(s) - R_2 N_3\widetilde{y}_g(s) \\ \quad + R_2 N_3\widetilde{y}_p(s) + R_2 N_1\widetilde{\theta}_g(s) = -T_2, \end{cases} \tag{5}$$

Where:

$$N_1 = k_m R_g,\ N_2 = k_m R_p,\ N_3 = k_m \sin\alpha,\ N_4 = k_m \cos\alpha \tag{6}$$

The graphical representations in Figs. 2 and 3 and were obtained based on the numerical application below, the definition laws of these coordinates being obtained after the inversion of their Laplace images obtained from solving the algebraic system (5) (Fig. 4).

$$m_g = 3,5\ [Kg];\ m_p = 0,8\ [Kg];\ J_g = 0,005\ \left[Kg\ m^2\right];\ J_p = 2,6 \bullet 10^{-3}\ \left[Kg\ m^2\right];$$

$$c_{xg} = c_{xp} = c_{yg} = c_{yp} = 10^4 \left[\frac{Ns}{m}\right];\ c_p = c_g = 0,09708 \left[\frac{Ns}{m}\right];$$

$$z_{p_0} = z_{g_0} = 0,001\ [m];\ z_{p_0} = z_p(0);\ z_{g_0} = z_g(0); c_{t_1} = c_{t_2} = 10 \left[\frac{Ns}{m}\right]; T_1 = T_2 = 2\ [Kgf];$$

$$k_m = 0,9952 \left[\frac{N}{m}\right];\ c_m = 9,9072 \left[\frac{Ns}{m}\right];$$

$$R_g = R_1 = 0,11\ [m]; R_p = R_2 = 0,05\ [m]$$

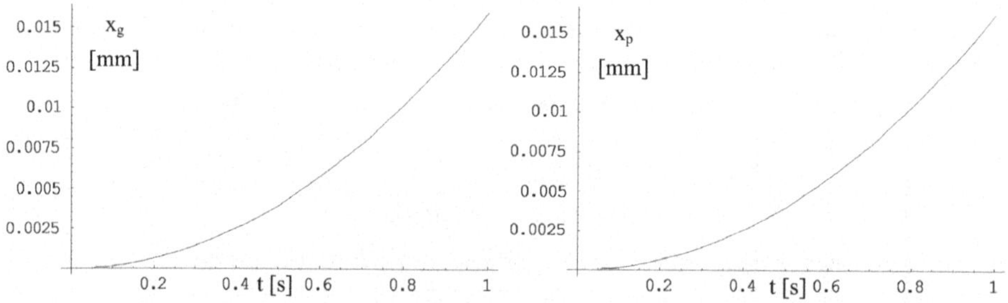

Fig. 2. Graphical representation for $x_g = x_g(t)$, $x_p = x_p(t)$.

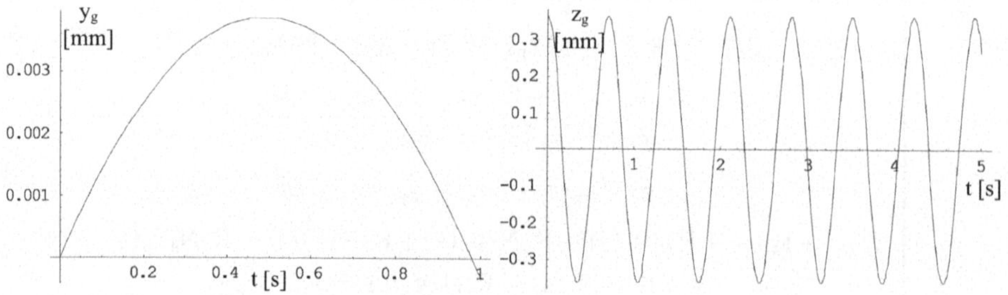

Fig. 3. Graphical representation for $y_g = y_g(t)$, $z_g = z_g(t)$.

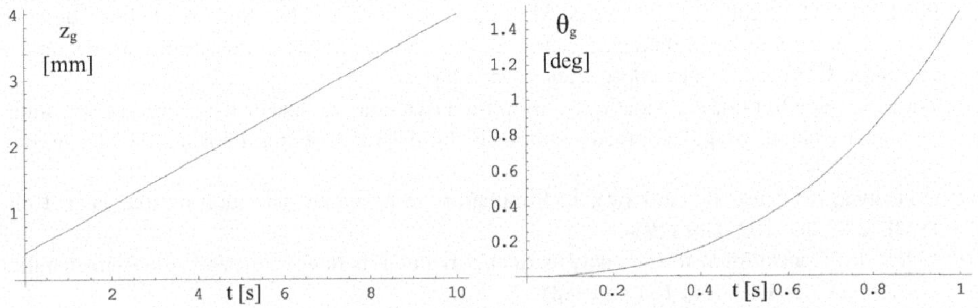

Fig. 4. Graphical representation for $z_g = z_g(t)$, $\theta_g = \theta_g(t)$.

4 Conclusions

For preventive maintenance, vibrational analysis is very useful in industrial processes.

This work, based on a mathematical model [1], simulates, following our dynamic response, the vibrational behavior of a gear subjected to deterioration.

In the event of any defects in the gearing teeth, the input and output shafts on which the wheel or the sprocket are located are twisted around the Oy axis, which could lead to their breakage.

The dynamic response we determined in the modern way of using the unilateral Laplace transform with respect to time, given that the mathematical model contains differential equations with constant coefficients.

Integration is facilitated until everything is reduced, ultimately, to a problem of elemental algebra. Then, the reversal of the unknowns Laplace's images perfectly adapted to computer work, which contains computational mathematical programs.

References

1. Ajim, B., Naguib, A., Hachim, A., Elxad, K.H.: Modeling the dynamic behavior of damaged spur gears. In: 13ème Congrès de Mècanique (2017)
2. Abousleiman, V., Velex, P.: A hybrid 3D finite element/lumped parameter model for quasi-static and dynamic analyses of planetary/epicyclic gear sets. Mech. Mach. Theory **41**(6), 725–748 (2006)
3. Băgnaru, D.Gh., Stănescu, M.M., Cuţă P.: Vibrations of the tool and tool holder unit for a shaper device. In: 8th International Conference Research and Development in Mechanical Industry, RaDMI 2008, Serbia, pp. 202–205 (2008)
4. Băgnaru, D.Gh.: On the vibration analysis of the viscoelastic bars, the constituent elements of the plane mechanisms. In: SYROM 1997, Bucureşti, pp. 25–30 (1997)
5. Băgnaru, D.Gh., Hadăr, A., Bolcu, D., Stănescu, M.M., Cuţă, P.: The vibrations influence on the field of accelerations of the linear-elastic connecting rod of a mechanism connecting rod lug. In: Annals of DAAAM for 2009 & Proceedings, vol. 20, no. 1, pp. 833–835 (2009)
6. Bard, C.: Modeling the dynamic behavior of gears transmissions, L'institut National des Sciences Appliquees de Lyon (1995)

7. Palaisi, D., Guilbault, R.L., Thomas, M., Lakis, A., Mureithi, N.: Numerical simulations of the vibration behavior of damaged gears. In: Proceedings of the 27th seminar on machinery vibration, CMVA October 2009, Vancouver (2009)
8. Thouviot, S., Omrani, A., Tawfiq, I.: Simplified dynamic model for the vibroacoustic study of a gear system, multidisciplinary approach. In: 19eme Congres Francais de Mecanique, pp. 1–6 (2009)
9. Umezawa, K., Sato, T., Ishikawa, J.: Simulation on rotational vibration of spur gear. Bull. JSME **27**(223), 102–109 (1984)
10. Velex, P.: Contribution to the analysis of the dynamic behavior of gearboxes with parallel axes, Ph.D., INSA Lyon, Lyon (1988)

Vibrations Influence on Automotive Drive

Valeriu Ionica, Dan Bagnaru, Ionut Geonea(✉), and Leonard Ciurezu

Faculty of Mechanics, University of Craiova, Craiova, Romania
igeonea@yahoo.com.com

Abstract. In this paper the influence of vibrations on the behaviour of the automotive on the road it is considered. These concern the roughness of the road, the tires, the transmission line and the engine. As a result, some behavioural patterns are addressed, whose dynamical response it is determined in the form of road functions, functions that are plotted graphically with Mathematica. To integrate differential equation systems, the Laplace unilateral transformation with report to time it is used. The coordinates layouts reversing is done also with Mathematica.

Keywords: Vehicle dynamics · Vibration · Passenger comfort · Laplace integral transformation

1 First Section

Vibration assessment is a very important criterion for judging the quality of a car. The behaviour of the vehicles in relation to the vibrations they are subjected to is closely related to the fact that they rely on the running of rubber tires under the pressure of the introduced air [1–5]. Vibration study for shafts are presented in [7], where theoretical, experimental study and Adams simulation are performed.

At the same time, it depends on the nature of the excitation and the dynamic response of the entire vehicle network, which we intend to determine here. In a succinct way, we present a schematic below.

In the paper are presented some patterns of behaviour to which we determined the dynamic response.

Since the systems of differential equations constituting these models are linear with constant coefficients, we have been able to apply a convenient integration method such as that of using the unilateral Laplace transform with time.

2 Mechanical Models

The mechanical model is presented in Fig. 1.

In Fig. 2 there is presented a mechanical model with two degrees of freedom, obtained in [6], with sprung masses and unsprung masses, the meanings of the notations being the following:

m_{us} - unsprung mass (mass of wheels);

m_s - sprung mass (machine weight);

I. Dumitru et al. (Eds.): SMAT 2019, *The 30th SIAR International Congress of Automotive and Transport Engineering*, pp. 565–572, 2020.
https://doi.org/10.1007/978-3-030-32564-0_66

k_s - the stiffness coefficient represented by a spring of the suspension;

k_{tr} - the stiffness coefficient represented by a spring associated with the wheels (tires);

c_{sh} - the damping coefficient of the suspension; c_t - damping coefficient for the tire. $z_1(t)$, $z_2(t)$ - the linear displacements of the two masses.

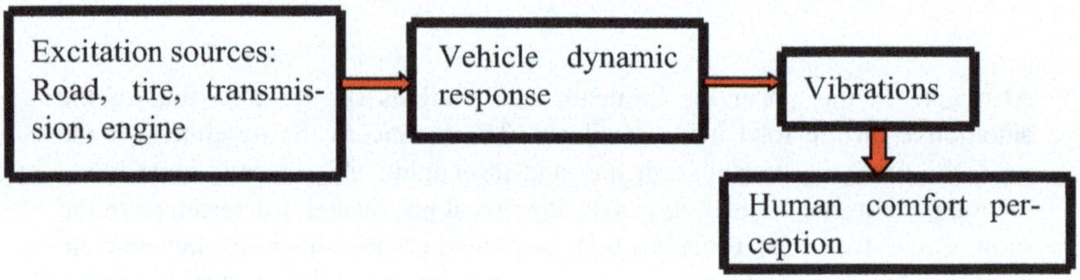

Fig. 1. Vehicle vibrations transmission flow.

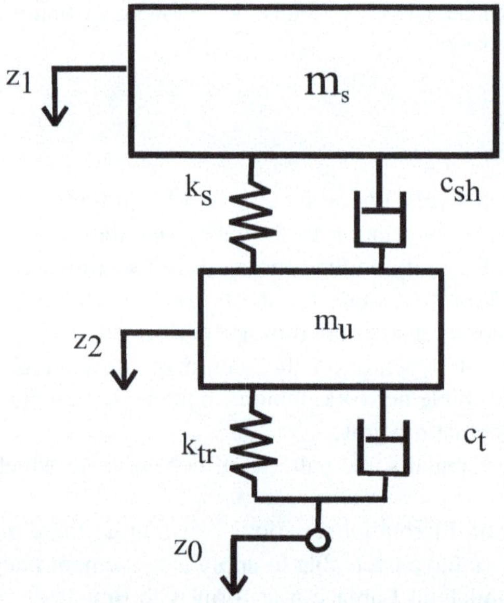

Fig. 2. The two-degree model with sprung and unsprung masses.

Figure 3 shows the pattern of pitch and roll (see [6]). Here, the damping of suspension and tires is neglected. Also, the unsprung mass is neglected, the Lagrange coordinates being the linear displacement z(t) and the rotation represented by the angular coordinate θ(t), the meanings of the notations being the following:

r_y - wheel radius;

k_f, k_r - rigidity of the suspension;

$l_1 = l_r$, $l_2 = l_f$.

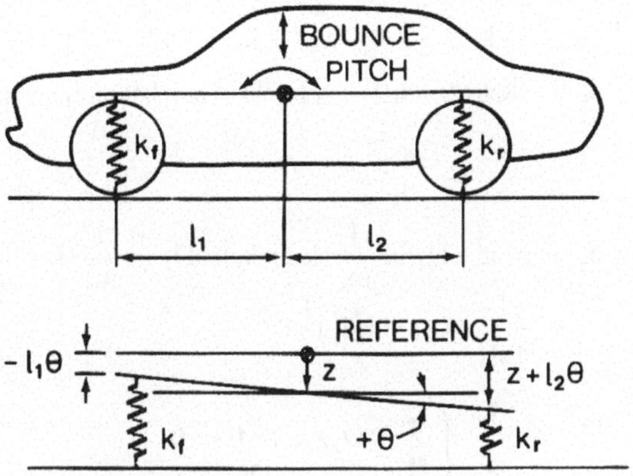

Fig. 3. The model of pitch and roll.

3 Mathematical Models

The mathematical model of the mechanical model in Fig. 2 obtained by Lagrange's formalism in the case of forced and damped vibrations is:

$$\begin{cases} m_s \ddot{z}_1 + c_{sh}\left(\dot{z}_1 - \dot{z}_2\right) + k_s(z_1 - z_2) = 0 \\ m_{us} \ddot{z}_2 + c_{sh}\left(\dot{z}_2 - \dot{z}_1\right) + k_s(z_2 - z_1) + c_t \dot{z}_2 + k_{tr} z_2 = c_t \dot{z}_0 + k_{tr} z_0, \end{cases} \tag{1}$$

In the case of forced and unamortized vibrations the model is:

$$\begin{cases} m_s \ddot{z}_1 + k_s(z_1 - z_2) = 0 \\ m_{us} \ddot{z}_2 + k_s(z_2 - z_1) + k_{tr} z_2 = k_{tr} z_0 \end{cases} \tag{2}$$

For free and damped vibrations, the mathematical model is:

$$\begin{cases} m_s \ddot{z}_1 + c_{sh}\left(\dot{z}_1 - \dot{z}_2\right) + k_s(z_1 - z_2) = 0 \\ m_{us} \ddot{z}_2 + c_{sh}\left(\dot{z}_2 - \dot{z}_1\right) + k_s(z_2 - z_1) + c_t \dot{z}_2 + k_{tr} z_2 = 0 \end{cases} \tag{3}$$

Free undamped vibrations are represented by the mathematical model:

$$\begin{cases} m_s \ddot{z}_1 + k_s(z_1 - z_2) = 0 \\ m_{us} \ddot{z}_2 + k_s(z_2 - z_1) + k_{tr} z_2 = 0 \end{cases} \tag{4}$$

The mathematical model of the mechanical model of Fig. 3, obtained through Lagrange's formalism, is:

$$\begin{cases} m_s \ddot{z} + k_f(z - l_f\theta - z_f) + k_r(z + l_r\theta - z_r) = 0 \\ I_y \ddot{\theta} - k_f l_f(z - l_f\theta - z_f) + k_r l_r(z + l_r\theta - z_r) = 0 \end{cases} \quad (5)$$

Or, with:

$$z_{f=0},\ z_{r=0};\ I_y = m_s r_y^2;\ D_1 = \frac{1}{m_s}(k_f + k_r);\ D_2 = \frac{1}{m_s}(k_r l_r - k_f l_f);\ D_3 = \frac{1}{m_s r_y^2}\left(k_f l_f + k_r l_1^2\right)$$

$$\begin{cases} \ddot{z} + D_1 z + D_2\theta = 0 \\ \ddot{\theta} + D_3\theta + \frac{D_2}{r_y^2} z = 0 \end{cases} \quad (6)$$

The equations become decoupled, if $D_2 = 0$, namely (6) becomes:

$$\begin{cases} \ddot{z} + D_1 z = 0 \\ \ddot{\theta} + D_3\theta = 0 \end{cases} \quad (7)$$

4 Dynamic Response

Applying systems of linear differential equations with constant coefficients, above, unilateral Laplace transform with respect to time, we obtain some algebraic systems that have the Laplace coordinates of the coordinates unknown.

Thus, for example, applying this integral transform to the system (4), considering that the initial displacements are null, leads to the algebraic system:

$$\begin{cases} m_s\left[s^2\tilde{z}_1(s) - z'_{10}\right] + k_s[\tilde{z}_1(s) - \tilde{z}_2(s)] = 0 \\ m_{us}\left[s^2\tilde{z}_2(s) - z'_{20}\right] + (k_s + k_{tr})\tilde{z}_2(s) - k_s\tilde{z}_1(s) = 0 \end{cases} \quad (8)$$

Similarly, applying this integral transformation to the system (6), considering that the initial displacements are null, leads to the algebraic system:

$$\begin{cases} s^2\tilde{z}(s) - z'_0 + D_1\tilde{z}(s) + D_2\tilde{\theta}(s) = 0 \\ s^2\tilde{\theta}(s) - \theta'_0 + \frac{D_2}{r_y^2}\tilde{z}(s) + D_3\tilde{\theta}(s) = 0 \end{cases} \quad (9)$$

Next, we consider a real numerical application, for both situations, for graphical representations, as follows:

$$m_s = 2120[\mathrm{Kg}]; r_y = 1,33[\mathrm{m}]; z_0' = 0,01\left[\frac{\mathrm{m}}{\mathrm{s}}\right]; \theta_0' = 0,01\left[\frac{\mathrm{rad}}{\mathrm{s}}\right]; z_{10}' = z_{20}' = 0,02\left[\frac{\mathrm{m}}{\mathrm{s}}\right]$$

$$m_{us} = 40[\mathrm{Kg}]; k_s = k_{tr} = 35\left[\frac{\mathrm{kN}}{\mathrm{m}}\right]; l_f = 1,267[\mathrm{m}]; l_r = 1,548[\mathrm{m}]; k_f = 35\left[\frac{\mathrm{kN}}{\mathrm{m}}\right];$$

$$k_r = 38\left[\frac{\mathrm{kN}}{\mathrm{m}}\right]; D_1 = 34,43\left[\mathrm{s}^{-2}\right]; D_2 = 6,83\left[\mathrm{ms}^{-2}\right]; D_3 = 39,26\left[\mathrm{s}^{-2}\right].$$

The system (8) solution is:

$$\begin{cases} \tilde{z}_1(s) = \dfrac{P_1(s)}{P(s)} \\ \tilde{z}_2(s) = \dfrac{P_2(s)}{P(s)} \end{cases} \tag{10}$$

Where:

$$\begin{aligned} P_1(s) &= m_s m_{us} z_{10}' s^2 + m_s z_{10}'(k_s + k_{tr}) + k_s m_{us} z_{20}'; \\ P_2(s) &= m_s m_{us} z_{20}' s^2 + k_s m_{us} z_{20}' + k_s m_s z_{10}'; \\ P(s) &= m_s m_{us} s^4 + [(k_s + k_{tr}) m_s + k_s m_{us}] s^2 + k_s k_{tr}. \end{aligned} \tag{11}$$

Applying the inverse of the Laplace transform, relative to time, to relations (10), using the Mathematica program, results in the dynamic vibration response of the vehicle, based on the above application, in the form of time functions:

z1(t) = cos[0.0906409 t] + cos[1.32601t] + (0.221687)sin[0.0906409t]-(0.000070 8069) Sin[1.32601 t];

z2(t) = cos [0.0906409 t] + cos[1.32601 t] + (0.111366) sin[0.0906409 t] + (0.0074703) sin[1.32601 t].

Whose graphical representations are given in Fig. 4

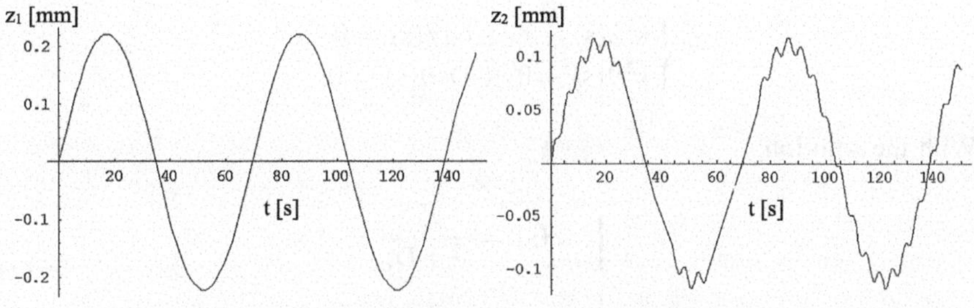

Fig. 4. Graphical results for $z_1 = z_1(t)$ and $z_2 = z_2(t)$.

The system (9) solution is:

$$\begin{cases} \widetilde{z}(s) = \dfrac{z'_0 r_y^2(s^2+D_3)-D_2 r_y^2 \theta'_0}{r_y^2(s^2+D_1)(s^2+D_3)-D_2^2} \\ \widetilde{\theta}(s) = \dfrac{\theta'_0 r_y^2(s^2+D_1)-z'_0 D_2^2}{r_y^2(s^2+D_1)(s^2+D_3)-D_2^2} \end{cases} \quad (12)$$

Reversing in (9) Laplace's transformation, the Mathematica program, based on the above application, results in the way the car behaves in pitch and roll in the form of time:

$$\begin{aligned} z(t) = {} & \mathrm{Cos}[5.58365\ t] + \mathrm{Cos}[6.52019\ t] \\ & - (0.00539967)\mathrm{Sin}[5.58365\ t] + (0.00615778)\mathrm{Sin}[6.52019\ t]; \end{aligned}$$

$$\begin{aligned} \theta(t) = {} & \mathrm{Cos}[5.58365\ t] + \mathrm{Cos}[6.52019\ t] \\ & - (0.414952)\mathrm{Sin}[5.58365\ t] + (0.356883)\mathrm{Sin}[6.52019\ t]. \end{aligned}$$

Whose graphical representations is given in Fig. 5.

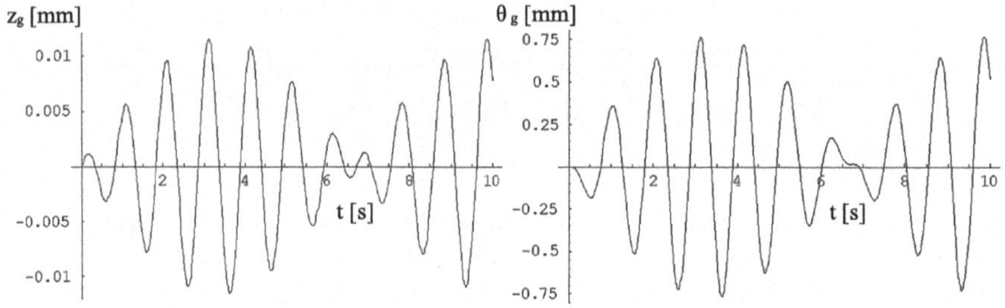

Fig. 5. Graphical results for z = z(t) and θ = θ (t).

Returning to the decoupled Eq. (7), to which we apply the transformed Laplace, we obtain, considering that the initial displacements are null, the algebraic system:

$$\begin{cases} s^2\widetilde{z}(s) - z'_0 + D_1\widetilde{z}(s) = 0 \\ s^2\widetilde{\theta}(s) - \theta'_0 + D_3\widetilde{\theta}(s) = 0, \end{cases} \quad (13)$$

With the solution:

$$\begin{cases} \widetilde{z}(s) = \dfrac{z'_0}{s^2+D_1} \\ \widetilde{\theta}(s) = \dfrac{\theta'_0}{s^2+D_3}, \end{cases} \quad (14)$$

Applying in the (14) inverse of the Laplace transform, the dynamic response in the form of time functions results:

$$\begin{cases} z(t) = \dfrac{z'_0}{\sqrt{D_1}} \sin\left(\sqrt{D_1}\ t\right) \\ \theta(t) = \dfrac{\theta'_0}{\sqrt{D_3}} \sin\left(\sqrt{D_3}\ t\right) \end{cases} \quad (15)$$

Based on the same numerical application above, the graphical representations in Fig. 6 is shown:

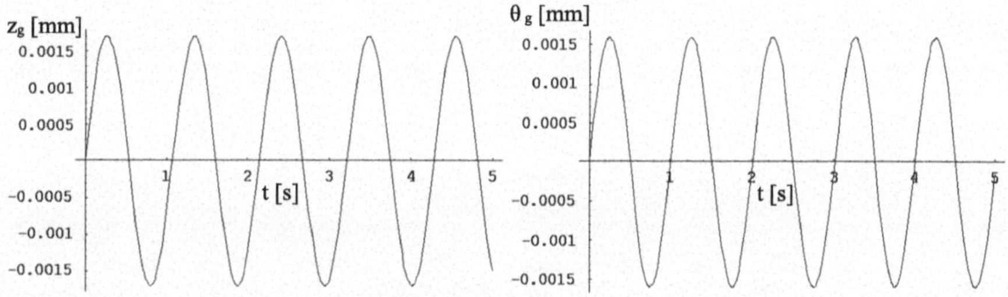

Fig. 6. Graphical results for z = z(t) and θ = θ(t).

5 Conclusions

In the paper, to determine the behaviour of the vehicle vibration, we take into account the roughness of the road and the tires. We not considered the transmission line and the engine influence, because we considered that the vibrations transmitted to the body are minimal, they are mounted on the chassis with vibration dampers (rubber pads).

Road roughness refers to asphalt pits, canal caps, and the occurrence of random deviations that are related to road surface manufacturing and repair. Taking all these into account, we present here behaviour patterns of the car at pitch and roll or vertical vibration, the dynamic response being obtained by using the Laplace transform in relation to time. The inversion of the Laplace transform was accomplished using the Mathematica program, with the same program being obtained graphical representations that are suggestive of the behaviour of the machine while traveling.

References

1. Abousleiman, V., Velex, P.: A hybrid 3D finite element/lumped parameter model for quasi-static and dynamic analyses of planetary/epicyclic gear sets. Mech. Mach. Theory **41**(6), 725–748 (2006)
2. Băgnaru, D.Gh., Stănescu, M.M., Bolcu, D., Cuţă, P.: Approximate method in order to determine the displacement fields for the kinematic elements standing in vibration. In: Second International Congress Automotive, Safety and Environment, SMAT, Craiova, pp. 9–12 (2008)
3. Lin, J., Parker, R.G.: Analytical characterization of the unique properties of planetary gear free vibration. J. Vib. Acoust. **121**(3), 316–321 (1999)

4. Parker, R.G., Agashe, V., Vijayakar, S.M.: Dynamic response of a planetary gear system using a finite element/contact mechanics model. J. Mech. Des. **122**(3), 304–310 (2000)
5. Wilk, O.: Simulation of the dynamic behavior of an automobile with aerodynamic coupling, (Simulation du comportement dynamique d'une automobile avec couplage aérodynamique, in French), Thèse de l'Ecole Centrale de Paris - Avril (1993)
6. Duysinx, P.: Comfort and handling, (Comfort et tenue de route, in French), Ingénierie des Vèhicules Terrestress, Université de Liège, Année a académique, 2015–2016
7. Geonea, I., Dumitru, N., Margine, A.: Analytical and numerical study of critical speed for right shafts. In: Acoustics and Vibration of Mechanical Structures—AVMS-2017, pp. 411–417. Springer, Cham (2018)

Analysis Methodology of Experimental Results in Case of a Cockpit Module Under Impact Load

Paul Predoiu(✉), Raluca Florentina Negrea(✉), Dinel Popa(✉), and Stefan Tabacu(✉)

Faculty of Mechanics and Technology, University of Pitesti, Pitesti, Romania
predoiu_paul@yahoo.com, Raluca.negrea@yahoo.com, {Dinel.popa, Stefan.tabacu}@upit.ro

Abstract. In this paper the authors present a method for the correlation of different datasets, captured during experimental tests performed according to ECE R21 on a cockpit module. The data were collected by specialized transducers by National Instruments data acquisition system interfaced by dedicated toolbox with MATLAB. The paper project presents the way of acquisition of data set and the test bench. According with ECE R21 it's using a gravitational pendulum consisting of a spherical rigid impactor that is supported by roll bearings to a rigid frame. By this methodology one can check a specific data set and to define consistent performance parameters set. The results obtained will be correlated with each other, so we can identify the acceleration at a certain angular position. On the other hand, in the paper project is presented the process to calculate kinetic energy and the energetic coefficient of restitution. The methodology provides the background for future investigation in designing efficient countermeasures useful for vehicle interior trimmings. Obtaining a correlation of data set we can analyze more materials with numeric simulation and we can identify new materials to minimize costs and increase performance. The paper is focused of the experimental investigation and provides the support for developing testing and data analysis procedures in the field of impact related events. The type of the material is also important thus it has an important contribution in the quality of the simulation result.

Keywords: Cockpit module · ECE R21 · Experimental investigation · Data analysis

1 Introduction

The main trend in automotive industry is to increase the performance of the products and in the same time to minimize the cost of the products. Furthermore, the most used method in automotive domain is to create a simulation at small scale or real scale for all the components which are parts of interior or exterior of the auto vehicle. Increased interest in vehicle safety has led to comprehensive research on the correlation of different data set (e.g. acceleration; angular position; video), according to ECE R21 on a cockpit module. Also in this paper are present the programs developed in MATLAB

I. Dumitru et al. (Eds.): SMAT 2019, *The 30th SIAR International Congress of Automotive and Transport Engineering*, pp. 573–580, 2020.
https://doi.org/10.1007/978-3-030-32564-0_67

which take out in front the correlation of the data. All the experimental results were processed and evaluated. According with these data set we can find new solutions in order to optimize the parts inside the vehicle regarding the structural response in impact applications. The results obtained were evaluated and processed with custom MATLAB code.

The high accelerations from inside of the compartment come from the impact of the passenger with the cockpit because of kinematic parameters recorded big variations in a short time. Obviously that the advanced technique of the experimental models, highlight results which have increased accuracy about of the studied components, but the main disadvantage is that most of the time all the pieces are destroyed after the experiment [1].

2 Experimental Procedure

The models which were simulated are built considering a homogenous distribution of material properties. In some cases, the manufacturing process alters initial performances of the initial product.

The cockpit module containing:

- A steering column (1st view from Fig. 1.)
- Air conditioning panel (2nd view and 3rd view from Fig. 1.)
- Dashboard
- Other small system

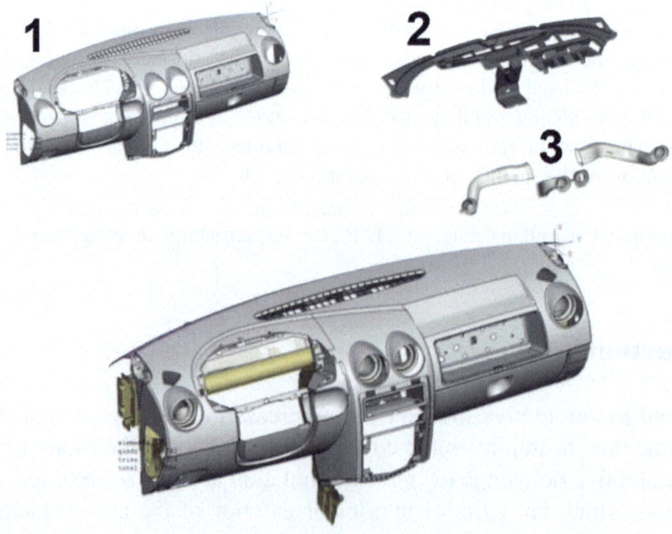

Fig. 1. Cockpit module

High performance plastics are playing an important role in the automotive domain these days. The most used materials polypropylene which is a thermoplastic polymer used in a wide variety of applications. It is rugged and unusually resistant to many cases [1].

Other advantages of high performance plastics used inside of the vehicle include:

- Substantial design freedom, allowing advanced creativity and innovation
- Flexibility in integrating components
- Safety and comfort
- Recyclability

2.1 Testing Machine

The data were collected by specialized transducers by NI data acquisition system interfaced by dedicated toolbox with MATLAB. The type of the transducer which was used is PCB 333B42 (Fig. 2).

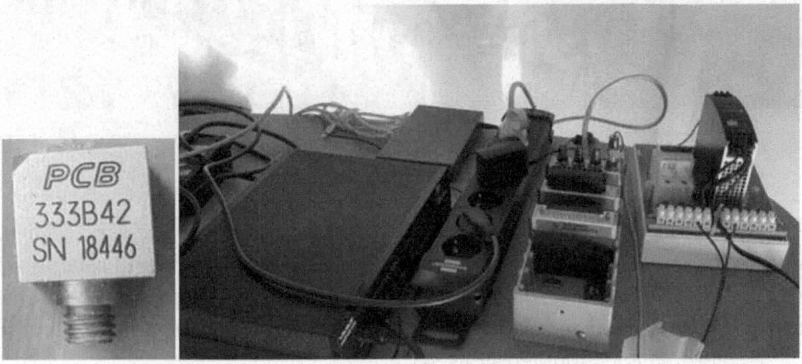

Fig. 2. PCB 333B42 Transducers and National Instruments system

These data are the impact parameters such as the impactor's acceleration, swing arm angular position and high-speed video. The transducer parameters for PCB 333B42 are presented below:

- Sensitivity (± 10%) 51 mV/(m/s^2)
- Measurement Range ±98 m/s^{2}pk
- Broadband Resolution (1 to 10000 Hz) 0.0005 m/s^{2}rms
- Resonant Frequency $\geq$ 20 kHz
- Temperature Range −18 to +66 °C
- Sensing element Ceramic

The paper project presents the way of acquisition of data set and the test bench which was designed to test a small part of vehicle interior. According with ECE R21 it's using a gravitational pendulum consisting of a spherical rigid impactor that is supported by roll bearings to a rigid frame. The swing arm is fixed with roll bearing therefore we can say that the friction forces can be neglected. Custom MATLAB codes were developed to analyze all the data and to provide a correlation measure.

The following figure represent test bench for physical testing (Fig. 3):

Fig. 3. Test bench for physical testing at 0^0 and 180^0

Special transducers were used to measure impact parameters like acceleration of the impactor and the structural response, the angular position of the arm in relation with the impact element. In the next figure, is presented how it's collected the data. The experiments were realized in accordance with ECE R21 specifications, which require having 2 accelerometers mounted (Fig. 4).

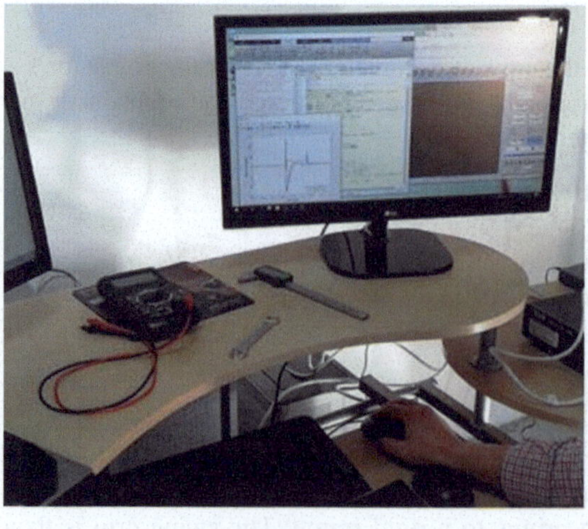

Fig. 4. Data collection

The transducer Kubler T8 5850 1281 G132 can be used between 10–30 VDC and 84 mA. By this methodology one can check a specific data set and to define consistent performance parameters set. The results obtained will be correlated with each other, so we can identify the acceleration at a certain angular position [2].

The measuring signal values are in volts and for time in seconds. The time for data acquisition is set to 10 s and it means that will be recorded all the impacts. To calculate kinetic energy and energetic coefficient of restitution we must determine the rebound height of the impactor [3]. Due to the configuration of the test bench and knowing that the kinetic energy is equal with potential energy, the restitution coefficient will result using the angular position of the arm.

$$\mathrm{I}_{\mathrm{restitution}} = \mathrm{T}_{\mathrm{max\_rebound}} = \mathrm{U}_{\mathrm{max\_rebound}} \tag{1}$$

The angles of the swing arm are determined by the formula below, only when the impactor is in contact with the vehicle's cockpit at the end of the impact and at the maximum rebound height. The signal is measured starting from 0.67 V (Fig. 5).

$$1^{\circ} = \frac{1.602 - 0.6726}{180} = 0.0051 \tag{2}$$

The maximum potential energy of the impactor after rebound is:

$$U_{max_{rebound}} = \left(\frac{m_{arm}}{2} + m_{impactor}\right) * g * h \tag{3}$$

Where:

m_{arm} = 2.157 kg mass of the arm
m_{impactor} = 5.643 kg mass of the impactor
g = 9.81 $\mathrm{m/s^2}$ gravitational acceleration
h = rebound height of the impactor

The rebound height of the impactor can be calculated with the following formula [4]:

$$h = l_{arm} * [1 - \cos(\theta_1 - \theta_3)] \tag{4}$$

$$h_0 = 2 * l_{arm} \tag{5}$$

$$m * g * h = \frac{1}{2} * m * v^2 \rightarrow v = \sqrt{2 * g * h} \tag{6}$$

$$m * g * \Delta h = \frac{1}{2} * m * (v^*)^2 \rightarrow v^* = \sqrt{2 * g * \Delta h} \tag{7}$$

The coefficient of restitution is:

$$e = \frac{v^*}{v} = 0.2815 \tag{8}$$

Where:

l_{arm} = length of the arm (1140 mm)
θ_1/θ_3 = angular position of the swing arm

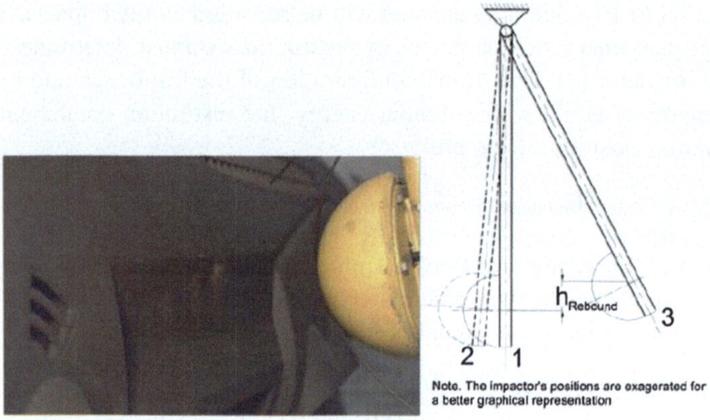

Fig. 5. Impactor's and its arm's position

3 Results and Discussion

The methodology provides the background for future investigation in designing efficient countermeasures useful for vehicle interior trimmings. Obtaining a correlation of data set we can analyze more materials with numeric simulation and we can identify new materials to minimize costs and increase performance [5] (Fig. 6).

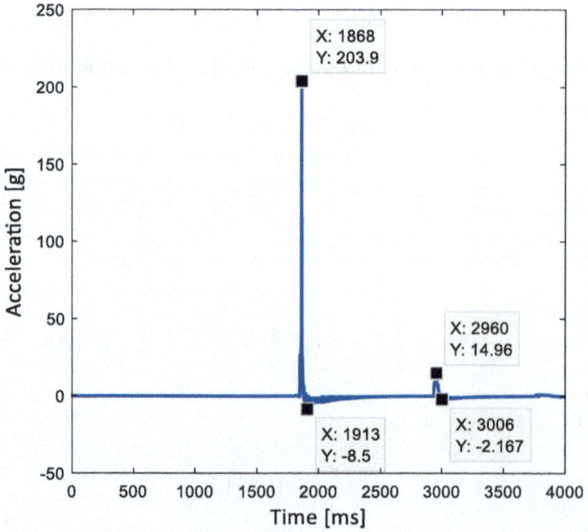

Fig. 6. Acceleration of the impactor in time (5000 Hz)

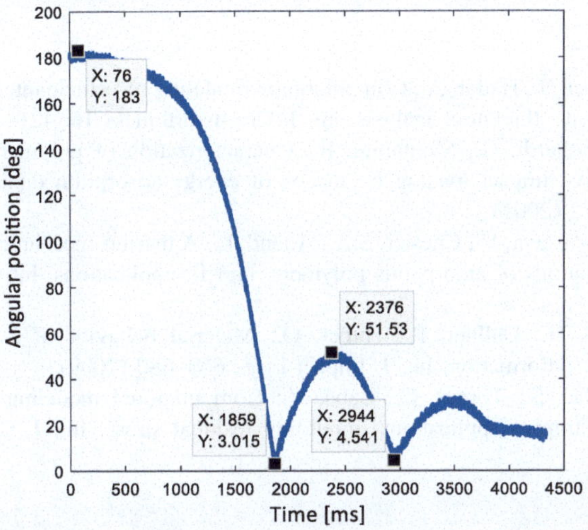

Fig. 7. Angular position of arm in time (2000 Hz)

4 Conclusion

The paper is focused of the experimental investigation and provides the support for developing testing and data analysis procedures in the field of impact related events. The type of the material is also important thus it has an important contribution in the quality of the simulation result.

According with ECE R21 it's using a gravitational pendulum consisting of a spherical rigid impactor that is supported by roll bearings to a rigid frame.

Thus, were presented the following graphics: the acceleration of the impactor and angular position of arm.

The coefficient of restitution is defined by the maximum correlated values from the Fig. 7 and the value is 0.2815. The strain energy released during restitution increases the impactor's kinetic energy. The data were collected by specialized transducers from National Instruments and it has a capacity to function from minimum value 0.6 V to maximum value 1.6 V.

The results obtained will be correlated with each other, so we can identify the acceleration at a certain angular position.

Results from these simulations can be further used to develop other analytical models for studying future frontal impact.

References

1. Tabacu, S., Tabacu, I., Hadar, A.: Computational modeling of vehicle interior components for impact applications: thickness analysis. Int. J. Crashworthiness **16**, 421–438 (2011)
2. Avalle, M., Belingardi, G., Montanini, R.: Characterization of polymeric structural foams under compressive impact loading by means of energy-absorption diagram. Int. J. Impact Eng. **25**, 455–472 (2001)
3. Ames, N.M., Srivastava, V., Chester, S.A., Anand, L.: A thermo-mechanically coupled theory for large deformations of amorphous polymers. Part II: applications. Int. J. Plast. **25**, 1495–1539 (2009)
4. Bartl, F., Klaus, H., Dallner, R., Huber, O.: Material behavior of a cellular composite undergoing large deformation. Int. J. Impact Eng., 660–680 (2009)
5. Tabacu, S., Parlac, S., Tomus, C., Baba, V.: Computational modeling of vehicle interior components for impact applications: moulding residual stress. Int. J. Crashworthiness **17**, 591–605 (2012)

Strain Analysis of a Steering Tie Rod

Ionela Alina Sandu(✉), Ştefan Tabacu, Cătălin Ducu, and Constantin Onescu

University of Piteşti, Piteşti, Romania
ionela.alina_sandu@yahoo.com

Abstract. This study is focused on developing a method to analyze the strain of the steering tie rod capable to be used for numerical investigation. The strain is determined using two experimental methods: bonded strain gauge and Digital Image Correlations (DIC). Experimental data obtained by the two methods will be compared and correlated. In conclusion, the study will develop a procedure for complex investigations of the deformation state/conditions.

Keywords: Steering tie rod · Fatigue · Strain · Variable stress · Numerical analysis · DIC method · Bonded strain gauge

1 Introduction

The steering is one of the most important system of an automobile, and under working condition the steering tie rod variable strains combined with other causes can lead sometimes to the component failure. So, it's important to develop an experimental strain analysis method to study the steering tie rod, to be used for the fatigue numerical analysis [1].

Sener [2] proposes an algorithm which uses a combination between measurements and the Finite Element Analysis to determine the fatigue life. For the measurements, the strain-gauges are applied on the steering tie rod which is assembled on the vehicle in the real conditions. For the fatigue the stress -life (S-N) method is applied.

Ikechukwu [3] defines the causes of the tie road failure as service loading, fatigue and buckling. Using the Adams Software for a McPherson suspension, within different scenarios (static, brake, cornering, bump, acceleration) the steering tie rod load is computed.

The paper uses two experimental methods, Digital Image Correlation (DIC) which is a relatively new technique used in [4] for the tensile specimens and the bounded strains gauge measurement, a well-known technique. The combination of the two methods will assure the accurate results of the stain measurements.

2 Digital Image Correlation (DIC)

For a detailed investigation the Digital Image Correlation (DIC) method has been use.

Measurements were achieved using an Aramis 3D Camera – 6 M System configuration 150, for measuring volume MV 100 with acquisition frequency 2 Hz.

The test speed was set to 2.0 mm/min (Table 1).

I. Dumitru et al. (Eds.): SMAT 2019, *The 30th SIAR International Congress of Automotive and Transport Engineering*, pp. 581–588, 2020.
https://doi.org/10.1007/978-3-030-32564-0_68

Table 1. Current calibration info

General	
Calibration date	Tue Jun 18 14:26:01 2019
Calibrated sensor	
Sensor name	ARAMIS 3D Camera/150
Measuring volume	MV100 (100 × 80× 50)
Serial number	No identifier
Calibration object	
Object type	Panel (Triple Scan)
Name	CP40-100-31268
Certification date	10.02.2016
Certification ID	160210/CP40/100/31268
Calibration points	3657 points
Certified lengths	168.511/168.511 mm
Certification temperature	22.0 °C
Expansion coefficient	3.25× 10^−6 1/K
Calibration temperature	25.5 °C
Calibration settings	
Camera lenses	35.00 mm
Snap mode	Single snap
Ellipse quality	0.4
Calibration result	
Calibration deviation	0.058 Pixels
Calibration deviation (optimized)	0.009 Pixels
Calibration deviation (check)	OK (limit value: 0.100 Pixels)
Camera angle	25.1°
Height variance	50 mm
Measuring volume	100/80/55 mm

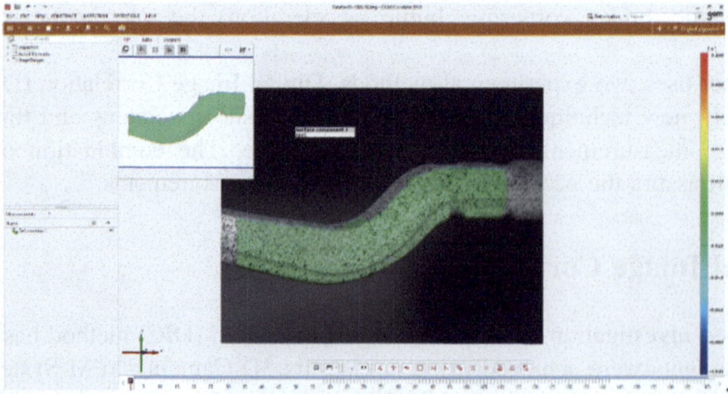

Fig. 1. Calibration step

Time step for camera to record a new frame was set for 2 frame/second.

GOM Correlate (www.gom.com) was used to process image and extract strain field distribution.

In the following sets of images are shown the measurement results (Fig. 1 – calibration step; Fig. 2 – results for a Force = ±3300 N and Fig. 3 – results for a Force = ±4200 N).

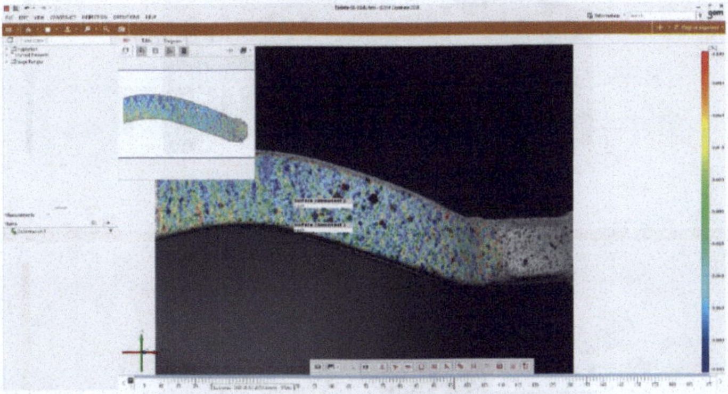

a) Front view

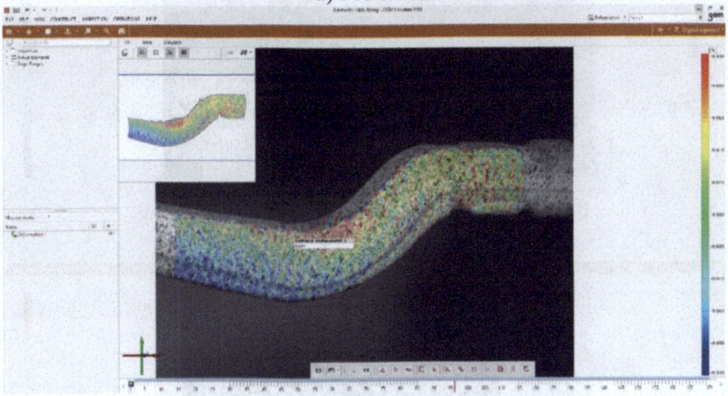

b) 90° degree rotated view from the front view

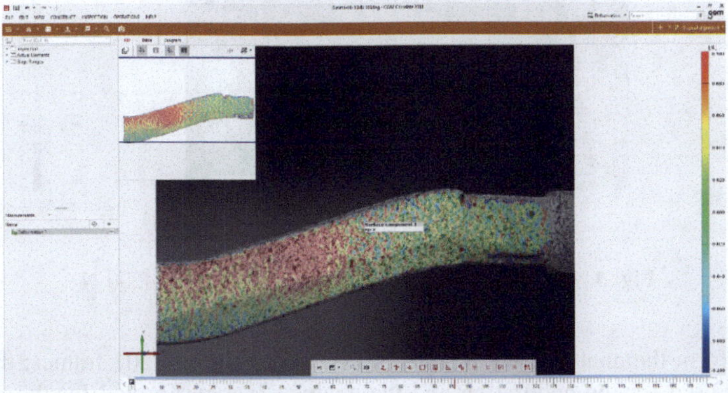

c) 180° rotated view from the front view

Fig. 2. Frame – 96 different views – Force = ±3300 N

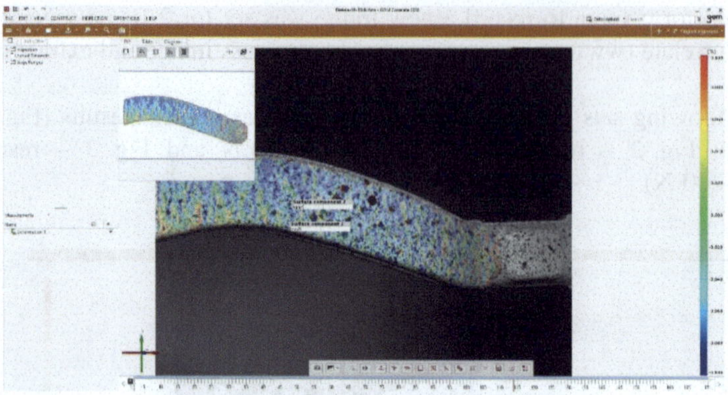

a) Front view

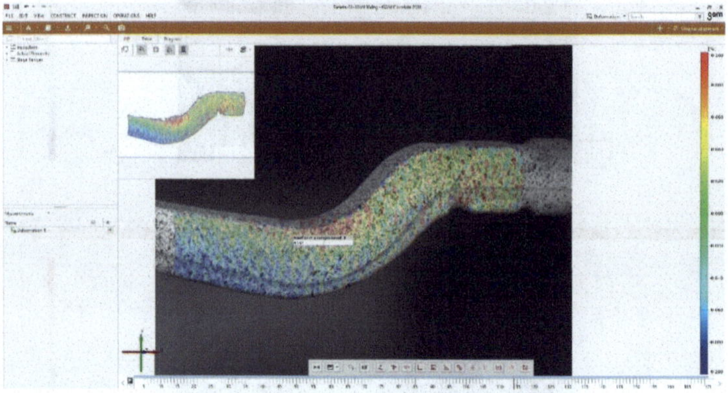

b) 90° degree rotated view from the front view

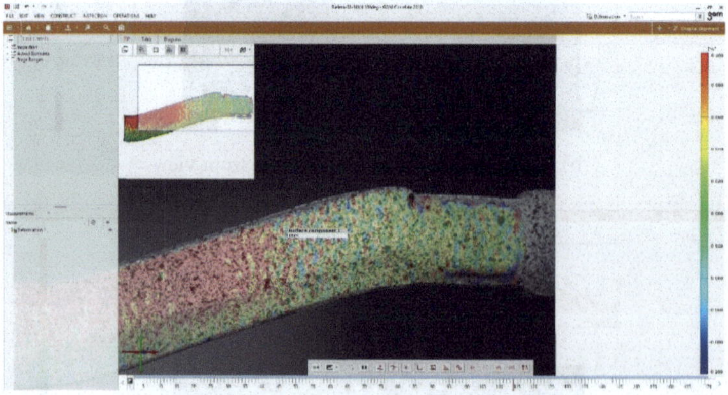

c) 180° rotated view from the front view

Fig. 3. Frame 114 – different views – Force = ±4200 N

To perform the analysis, the Aramis 3D Camera captured 201 frames: the first 30 frames are used for the calibration and the analysis was performed for the rest of the frame (171).

3 Bonded Strain Gauge

To collect data were used a C-series Wheatstone Bridge Module (NI-9237) and a Voltage and Current Input Module (NI-3207) (Fig. 4), data acquisition system.

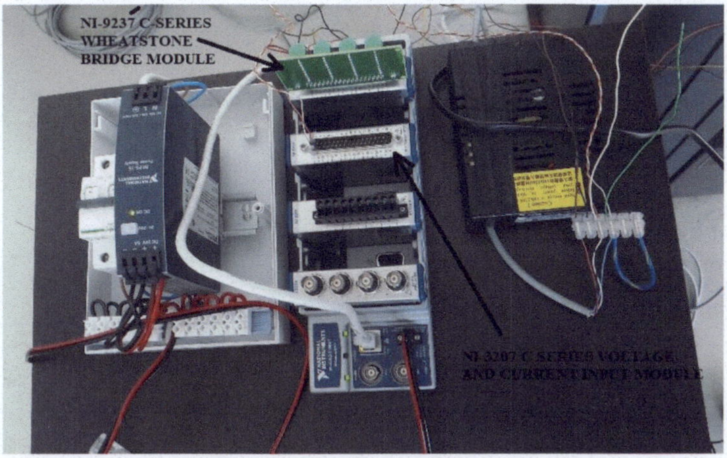

Fig. 4. Data acquisition system

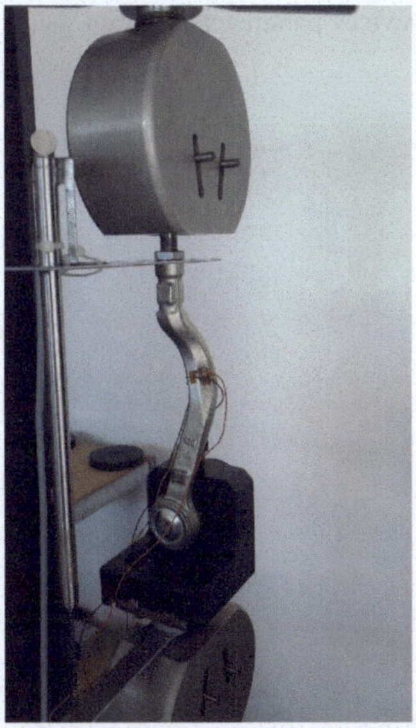

Fig. 5. Steering tie rod during the test

In Fig. 5 was presented the steering tie rod during the test. On the part can be showed the bounded strain gauge. Type of the bounded strain gauge used for analysis is linear strain gauge.

Figures 6 and 7 shows devices parameters.

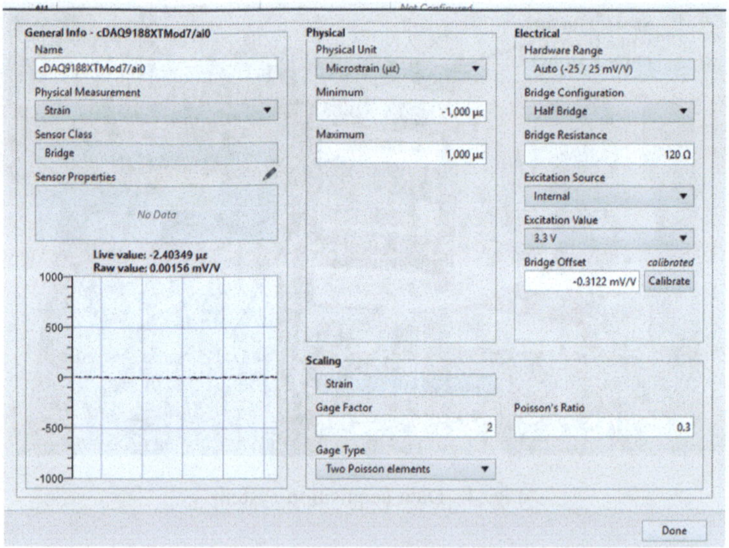

Fig. 6. Devices parameters – strain gauge analysis

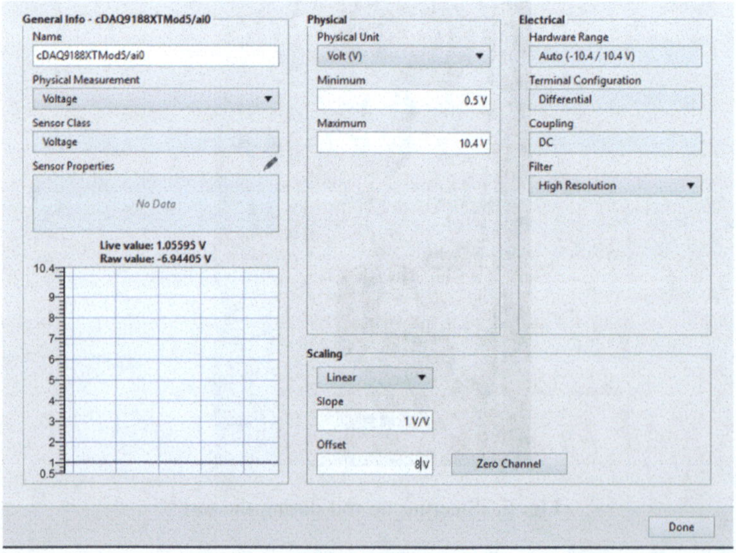

Fig. 7. Devices parameters – Voltage

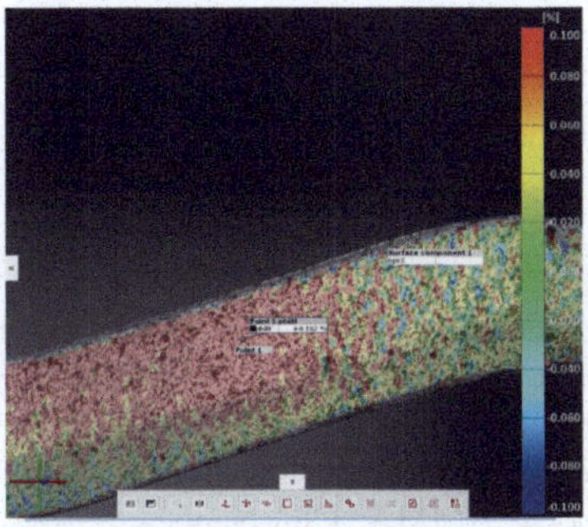

a) Point for strain measurement

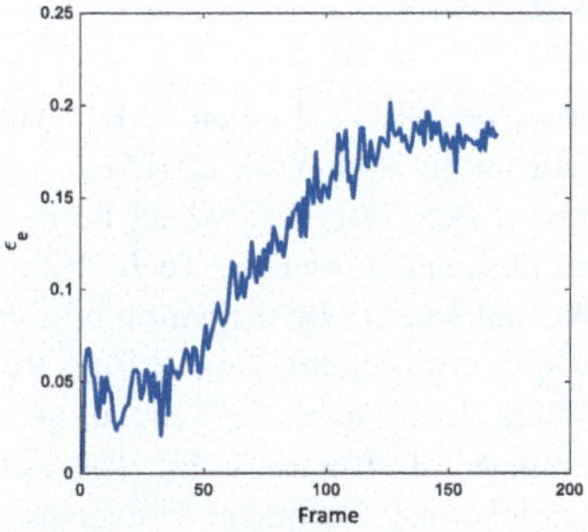

b) Strain data from GOM

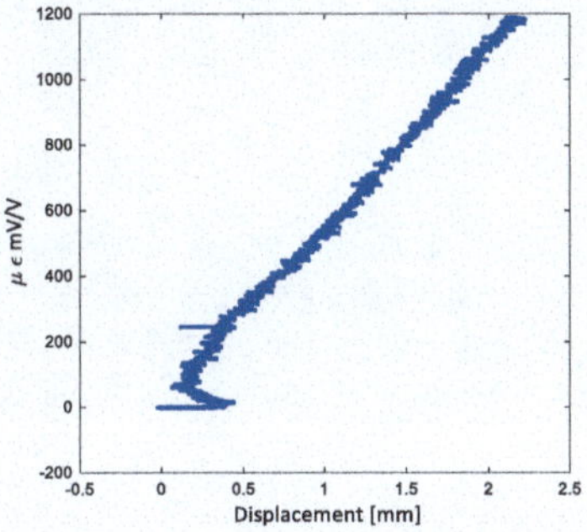

c) microstrain from strain gauge measurement

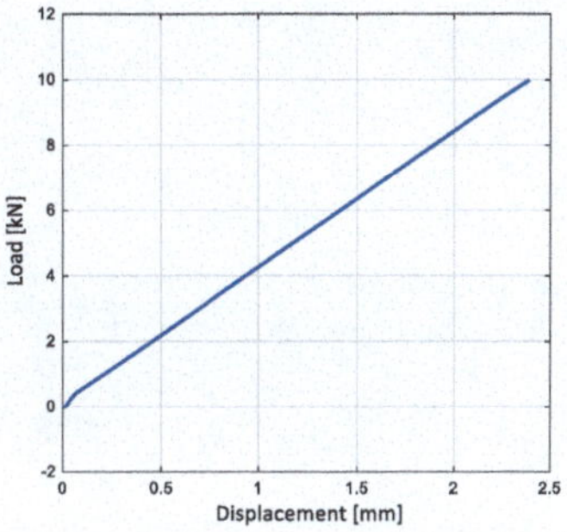

d) force – displacement data

Fig. 8. Analysis results

4 Discussions and Conclusions

Can be observed following analysis from strain measurements results (using both methods), information about a linear strain distribution.

Validation of the DIC method is achieved by using the bounded strains gauge by the similarity of the obtained results. In this study the combination of the two experimental methods, DIC and bounded strains gauge measurement, assured the accurate results of the stain.

The analysis results (Fig. 8) illustrating the complex representation of the strain. The test was performed for maximum load force of 10 kN (Fig. 8d) and field of the voltage is 9.5 V for 200 mm.

In conclusion, GOM Correlate was used to evaluate the strain and the results was used to calculate the true stress of the specimens and estimate the Poisson's ratio.

From analysis, value of point for the strain measurement resulted Poisson's ratio +0.182%.

References

1. Budynas, R.G., Nisbett, J.K.: Mechanical Engineering Design, 10th edn. McGraw-Hill Education, New York (2015)
2. Sener, A.S.: Steering wheel tie rod fatigue life determination according to Turkish mission profiles. Int. J. of Eng. Tech. **2**(2), 56–63 (2016)
3. Ikechukwu, O.: Investigation of a vehicle tie rod failure in relation to the forces acting on the suspension system. Am. J. Eng. Res. (AJER) **5**(6), 208–217 (2016)
4. Sandu, I.A., Ducu, C., Tabacu, Şt.: Development of a material model for the use in life-time estimate of structures. In: The IVth International Congress of Automotive and Transport Engineering, Technical University of Cluj-Napoca, AMMA, Cluj-Napoca, 17–19 October 2018 (2018)

Comparison Study Between a Tubular Type Camshaft and a Classic Type Using Finite Element Method

Sorin Vlad Igreț(✉) and Radu Iulian Negrilă

Faculty of Engineering, Aurel Vlaicu University, 77 Revolutiei Ave, 310130 Arad, Romania
sorin_igret_vlad@yahoo.com

Abstract. The camshaft is the main element of the timing mechanism of an internal combustion engine, it assures the fill of the cylinders, in a certain order with fuel or fresh air and evacuation of burned gases.

Because of the importance that the camshaft has in the functioning of the internal combustion engine, the authors thought is was important to study the deforming state and tension of it. In this study were analyzed two constructive ways of the camshaft. A classic solution and a new one, the camshaft being made of a tube and cams applied on it.

The study of tension state and deforming state was made on two camshafts of Renault K9 K 1,5 L diesel engine (the one that is on Dacia Logan 1.5dci). The two constructive ways are identical dimensionally speaking, the difference is made by the fact that the first one is a classic solution, the camshaft having a full circular section and is made by cast iron and the second camshaft is tubular, made of steel with the cams applied on it.

For this two camshafts were made the dimensional measurements for doing the reveal, that later was imported in the Algor program of finite elements so that the resistance calculus could be made.

Keywords: Camshaft · Tubular · Resistance · Deformation · Finite element method · Algor program

1 Introduction

The camshaft is the main element of the timing mechanism of an internal combustion engine, it assures the fill of the cylinders, in a certain order with fuel or fresh air and evacuation of burned gases.

Because of the importance that the camshaft has in the functioning of the internal combustion engine, the authors thought it was important to study the deforming state and tension of it. In this study were analyzed two constructive ways of the camshaft. A classic solution and a new one, the camshaft being made of a tube and cams applied on it.

I. Dumitru et al. (Eds.): SMAT 2019, *The 30th SIAR International Congress of Automotive and Transport Engineering*, pp. 589–596, 2020.
https://doi.org/10.1007/978-3-030-32564-0_69

2 General Presentation

The study of tension state and deforming state was made on two camshafts of Renault K9 K 1,5 L diesel engine (the one that is on Dacia Logan 1.5dci). The two constructive ways are identical dimensionally speaking, the difference is made by the fact that the first one is a classic solution, the camshaft having a full circular section and is made by cast iron and the second camshaft is tubular, made of steel with the cams applied on it.

For this two camshafts were made the dimensional measurements for doing the reveal, that later was imported in the Algor program of finite elements so that the resistance calculus could be made.

The main dimensions of the camshaft are:

- length : L = 420 mm
- shaft diameter df = 25 mm
- main hole diameter dg = 15 mm
- cam base diameter Dc = 38 mm
- cam excentricity e = 8 mm
- cam width lc = 14 mm

2.1 Calculating Theory

The reveal was made considering the important dimensions of the camshaft ignoring the dimensional variations less than 0.5 mm. Although the lubrication holes were ignored.

The counting was made in a static way, the loads and supports were applied in the counting net nods, that was obtained by keeping away the camshaft's draws.

The forces on the camshaft were considered as coming from the elastic force of the springs and the force that was obtained from the pressure that takes action on the valves. It is very important to see that the two forces are antagonists. Knowing the dimensions of the spring: spire diameter ds = 3,5 mm, medium spring diameter Dm = 37 mm, number of spires ns = 6,5, valve plate diameter and cylinder pressure value was determined maximum medium value of the sum of both forces having the value of Fmm = 2000 N on a single cam.

The force Fmm was applied to the intake-exhaust camshaft pair that can be found simultaneously in the maximum opening position [5].

2.2 The Resistance Calculating Method

The resistance calculus was made for two types of camshafts, one having the full section and the second being a tubular section [1].

Having the drawings of the camshafts with the help from Algor program, it was done in brick type finite elements on which the medium dimension of sides was imposed between 2 and 5 mm, so resulting approximately 26000 elements.

The journals of the camshaft on the combustion head was considered simply, in 8 sections, these being 4 sections between the 2 cams of each cylinder and 4 sections at the cam's end (2 sections at each end). In every section the journal area was accomplished on 36 nods, on which the removes were blocked because of the perpendicularly directions on the longitudinal axle.

The cam's load status was realized with Fmm forces appointed in 2.1 point and were applied on the camshaft in 5 nods of it [2, 3].

For load cases were considered from the cams combination that can be found in the simultaneously maximum opening position. The load cases are developed in Table 1.

Table 1. Load cases

Load case	Evacuation cam	Admision cam
1	Cylinder 1	Cylinder 2
2	Cylinder 3	Cylinder 1
3	Cylinder 4	Cylinder 3
4	Cylinder 2	Cylinder 4

For the two types of camshaft, the way of load and the journal area were considered as being the same.

2.3 Materials

First type of camshaft, having the full circular transversal section, was made of cast iron with nodular graphite having the following physical-mechanical features: hardness 220–300 HB, breaking resistance R_m = 600 N/mm^2, breaking extension: A_5 = 4–5%, longitudinal elastic modulus E = 160000 N/mm^2, transversal contraction coefficient $\mu = 0.275$.

Second variant camshaft with tubular transversal section was made in OL52 with a carbon concentration of 0.62%. Were considered the next physical-mechanical properties, breaking resistance: Rm = 610 N/mm^2, braking extension A_5 = 21%. Longitudinal elasticity module E = 208000 N/mm^2, transversal contraction coefficient $\mu = 0.3$.

3 Resistance Calculating Results

Rolling the Algor program for the two variants of the camshaft, 4 cases of considered load, the values of the deformations and tensions were obtained from calculating net nods. Maximum values of the tensions and deformations are presented in Table 2 [4].

It is obvious the fact that for the second case of load we obtain the highest tension values and deformations for both variants. For this reason it will be detailed only this case of load.

Table 2. Maximum values of tensions and deformations

Tensions [N/mm*2]				
Cam's variant	Case 1	Case 2	Case 3	Case 4
Full section	97.34	116.72	87.9	90.64
Tubular section	98.34	121.85	97.34	91.9
Deformations [mm]				
Cam's variant	Case 1	Case 2	Case 3	Case 4
Full section	0.0225	0.0292	0.0229	0.0224
Tubular section	0.0252	0.0315	0.0259	0.0246

The tension and deformation distribution kind for both variants of the shaft in second case of charge are presented in Figs. 1, 2, 3, 4, 5 and 6 [3, 6].

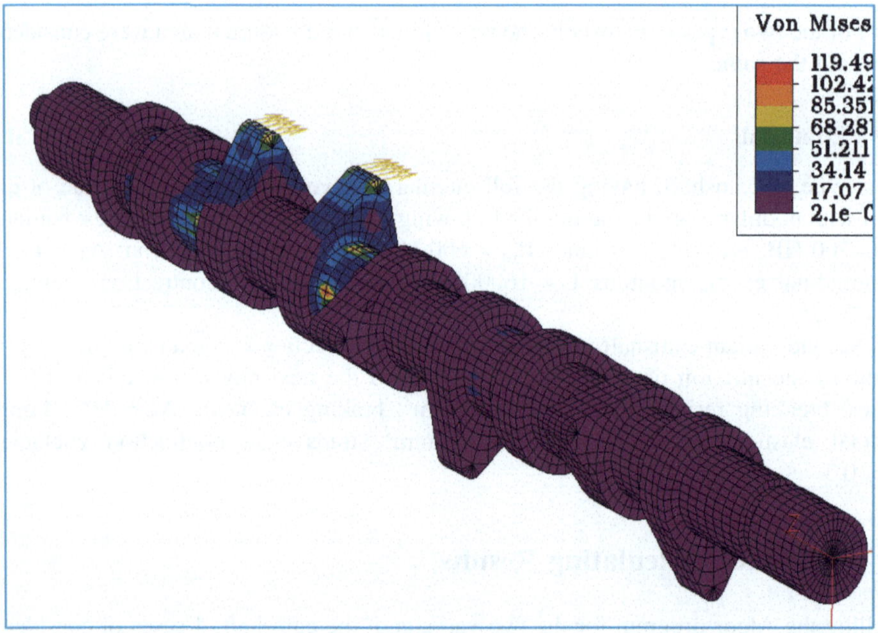

Fig. 1. Tension values for the full section camshaft in second load case

Fig. 2. Tension values for the tubular section camshaft in second load case

Fig. 3. Deformation values for the full section camshaft in second load case

Fig. 4. Deformation values for the tubular section camshaft in second load case

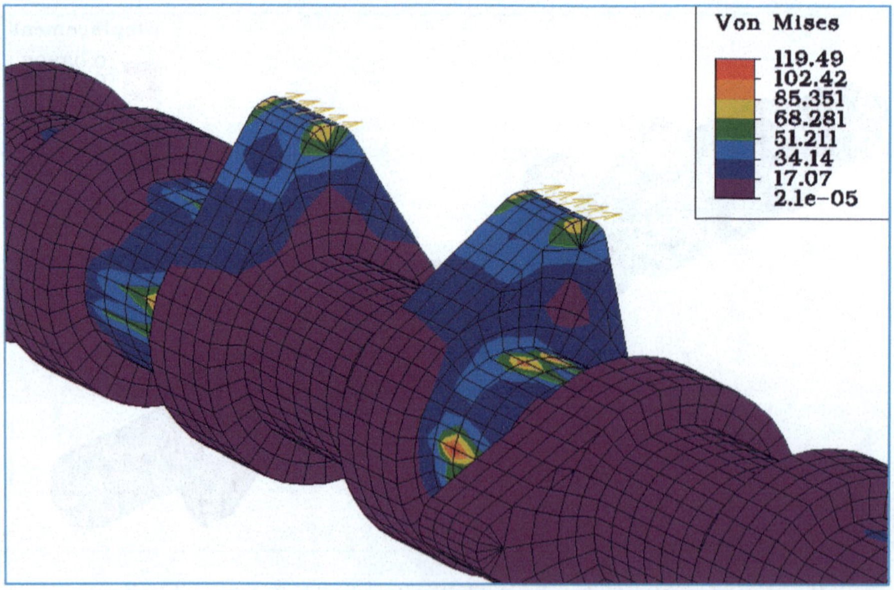

Fig. 5. Tension value detail for the full section camshaft in second load case

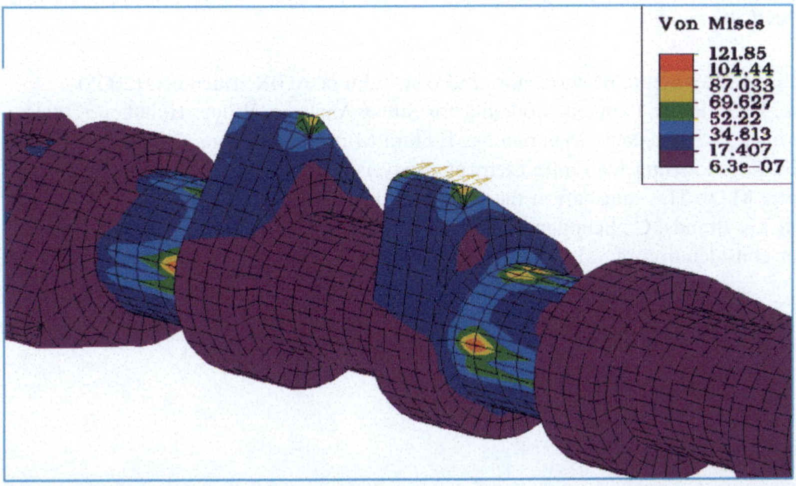

Fig. 6. Tension value detail for the tubular section camshaft in second load case

4 Conclusions

We chose to study the tubular camshaft state of charge because we considered that this item is used on engines after year 2006 and it is a quiet new technology. This technology is already used by several well known manufacturers as Renault, Peugeot, Citroen, Opel, etc., on different type of internal combustion engines.

This tubular camshaft was made to replace the conventional camshaft in order to reduce the total power plant weight and to be cheaper to built.

On the manufacturing process it is used a steel tube as main shaft and the cams are mounted on it using differentiated temperatures. This process is considerable cheaper than classic one in which the camshaft was a cast iron in a mold and after molding should be shaped on several machines(reshaped, grinded, drilled, etc.).

Comparing the tension and deformation values for the 2 variants of the camshaft, it has been ascertained that the percentage of deviation is between 0.9÷8% for tensions and for deformations is between 14.2÷20.9%.

This little difference can be explained by the fact that the value of the polar axial resistance module of a full circular section is with 30% bigger than the same calculate sizes of a tubular section by same exterior diameter.

For none of the calculation variants and for none of the charging cases there are passed the admissible tension values.

The deformation values are not significant, they are not modifying the distribution diagram.

The tubular camshaft variant is more convenient because of the weight reduction with 0,52 kg in this case.

The only think that can be a minus is the possibility of buckling of the camshaft because of the lack of resistance of the main tube witch is a common steel in comparison of a full section cast iron camshaft, and also a quicker wear of the main tube in the bearings areas.

References

1. Bejan, M.: Rezistenţa Materialelor, 2nd edn. Editura AGIR, Bucureşti (2005)
2. Cook, R.D.: Finite Element Modeling for Stress Analysis. Wiley, Hoboken (2004)
3. Gherman, G.: Rezistenţa Materialelor. Editura Mirton, Timişoara (2003)
4. ALGOR - Modeling for Finite Element Analysis. Pittsburgh (2000)
5. Rohrle, M.D.: The camshaft in the course of time. MTZ (2002)
6. Knoll, G., Brands, C., Schönen, R.: Elastohydrodynamic interaction of camshaft and bearings under consideration of valve train forces. MTZ (2003)

Basic Subsystems Reliability Influence Study on the Main Characteristics of Automotive

Radu Iulian Negrila(✉) and Sorin Vlad Igreţ

Faculty of Engineering, Aurel Vlaicu University, Arad, 77 Revolutiei Ave, 310130 Arad, Romania

radu.negrila@premiermotors.ro

Abstract. The reliability general theory is thus defined as it combines different specific chapters of reliability theory that consistently reflect all stages of vehicle construction and use. In the design and manufacturing stage, the properties that determine the reliability of the vehicles are consistently formulated and achieved in the production process. These properties are also called specific potential objectives. The notion of reliability also includes the level of potential reliability, after the introduction of prophylactic and restoration work in the use process.

In process of assessing the potential properties of motor vehicles, three groups are distinguished:

I. Group of potential properties that depend on worsening of working functions and permissible level of failure-free operation, within the set time or mileage.
II. Group of potential properties required to achieve the high level of performance and reduce the cost of manufacturing.
III. Group of potential properties required to maintain the high level of reliability during operation, such as improving the quality of diagnoses, technical maintenance and repairing; reducing all types of expenses and increasing the productivity of motor vehicles.

Operating reliability is the result of special achievement of the potential reliability of components, manufactured under identical or different operating conditions and at different times and ways.

In the case of the "automotive" subsystem, reliability can also be appreciated by making certain characteristics and indicators for its use, which, in practice, indirectly can ultimately characterize its working qualities. In this context, dynamic traction and braking characteristics are of main interest.

Keywords: Automotive · Reliability

1 Vehicle Dynamic Characteristics

The available excedentary tractive force Fe, which is used to overcome road resistances and starting resistance, characterizes the vehicle's dynamics and has the expression [1]:

I. Dumitru et al. (Eds.): SMAT 2019, *The 30th SIAR International Congress of Automotive and Transport Engineering*, pp. 597–606, 2020.
https://doi.org/10.1007/978-3-030-32564-0_70

$$F_e = F_R - R_a = F_R - k \cdot A \cdot v^2 \tag{1}$$

where FR represents tractive force (drag force), Ra air resistance force, k aerodynamic (drag) coefficient, A automotive maximum transverse section area and v vehicle speed.

From the traction equation, we can write the tractive force:

$$F_R - k \cdot A \cdot v^2 = G_a \cdot f \cdot \cos\alpha + G_a \cdot f \cdot \sin\alpha + \frac{\delta \cdot G_a}{g} \cdot \frac{dv}{dt} \tag{2}$$

where Ga vehicle weight, f rolling resistance coefficient, α the angle at which the road is inclined to the horizontal, δ rotation masses coefficient, $\frac{dv}{dt}$ vehicle acceleration at start.

In order to be used as comparison criteria for vehicles having different weight, the (2) equation is divided by its weight, thus obtaining a specific excedentary force, called dynamic factor D, which is a non-dimensional parameter and appreciates the dynamic qualities of a vehicle, namely:

$$D = f \cdot \cos\alpha + \sin\alpha + \frac{\delta}{g} \cdot \frac{dv}{dt} \tag{3}$$

The total traction force of the wheels can be determined from the engine torque, passing through the transmission to the wheel, whose dynamic radius is r:

$$F_R = \frac{M_R}{r} = \frac{M_m \cdot i_{tr} - M_f}{r} = \eta_{tr} \cdot \frac{M_m \cdot i_{tr}}{r} \tag{4}$$

where MR torque on wheel, Mm engine torque, Mf friction resistive torque, itr transmission total ratio, η_{tr} transmission efficiency.

Engine torque can also be expressed according to the effective average pressure in the cylinder, pe, and the total displacement i $\cdot$ VS and thus the Eq. (4) is:

$$F_R = \eta_{tr} \cdot \frac{k_M \cdot p_e \cdot i \cdot V_S \cdot i_{tr}}{r}, \tag{5}$$

and the dynamic factor D has the expression, according to the Eq. (1):

$$D = \frac{F_e}{G_a} = \frac{\eta_{tr} \cdot \frac{k_M \cdot p_e \cdot i \cdot V_S \cdot i_{tr}}{r} - k \cdot A \cdot v^2}{G_a} \tag{6}$$

From the Eq. (6) it results that the dynamic factor can characterize the change in the potential reliability of the vehicle by two basic parameters, namely the effective average pressure and the transmission efficiency [3]. The latter changes during vehicle operation, due to the wear increase or due to misadjustments in engine systems.

This aspect can best be emphasized by the developed form of the average pressure in the engine cylinder equation:

$$p_e = \frac{Q_i}{L_{min}\frac{\eta_i}{\lambda V_{m_0}}} \tag{7}$$

where Q_i lower calorific value of fuel, L_{min} the minimum amount of air required to burn one kg of fuel, λ the excess air coefficient, η_i engine indicated efficiency, η_m engine mechanical efficiency, η_V engine volumetric efficiency, ρ_0 air density in standard conditions.

The Eq. (7) highlights in detail the main values that can cause the dynamic factor D to decrease, by worsening the combustion process, reflected by the ratio η_i/λ, by reducing the filling capacity of the engine cylinders, due to increasing hydraulic resistances in the intake and/or exhaust manifolds, leakage of fresh air-fuel mixture reflected by η_V, or reduced mechanical efficiency η_m due to the increase in overall mechanical losses in the engine. All these changes have a negative evolution during vehicle usage and can cause the engine to exit out of the optimum power range, which compromises its potential qualities, and, finally, the overall reliability of the vehicle.

As it results from Eq. (6), the dynamic factor D may also be reduced due to the worsening of the mechanical efficiency of the transmission, which decreases due to the increase of the M_f friction lost torque (see Eq. 4), due to the wear of the gear wheels, of the transmission unit bearings etc. Thus, these changes can significantly diminish the potential qualities of the transmission aggregates and, consequently, the reliability of the vehicle.

2 Vehicle Braking Characteristics

Vehicle active safety depends largely on its braking capabilities, given that in many situations it is possible to control the movement of a car in critical conditions and thus it can eliminate the occurrence of an unintended road accident. From the point of view of the vehicle's use, braking, respectively braking system, are used for:

- avoiding inappropriate acceleration of the vehicle when driving down a slope;
- reducing the driving speed, depending on the situation, until the vehicle stops;
- avoiding any unwanted movement of a stationary vehicle.

For the first item, engine braking is used, i.e. the engine braking torque, amplified by transmission. If this is not sufficient, the service brake must also be used.

For the second, the vehicle's braking system is mainly based on the principle of dissipating kinetic energy through friction, the engine braking torque being not enough. In order to determine optimal braking characteristics and parameters, there should be analized the dependence between vehicle deceleration and braking torques, the split ratio of torques on axles and wheels, the road grip as well as the active forces on the circumference of the wheels and their interaction with the road.

For the third item, the driver will estimate the maximum ramp or slope on which the vehicle can stand. In addition to parameters such as braking time and distance, brake pedal operating effort and deceleration, the vehicle must be stable both when braking on a straight and curved track.

A number of aspects of braking are standardized in Directive 71/320/EC as well as in Regulation No. 13 UNECE, legal regulations that provide criteria for evaluating the vehicle stability as well as vehicle grip during braking, to fulfill the inequation:

$$Z = \frac{\sum_{i=1}^{n} F_{fi}}{G_a} \geq 0{,}1 + 0{,}85(\phi_\ell - 0{,}2) \tag{8}$$

where Ffi braking force on wheel i from the total of n wheels, ϕ_ℓ longitudinal adhesion coefficient.

The degree of the adhesion utilization is:

$$\varepsilon = \frac{Z}{\phi_\ell}. \tag{9}$$

If braking is achieved on all wheels of the vehicle, so that $\varepsilon = 1$, the ideal braking rate is obtained $Z = \phi_\ell$. This situation is achieved at the maximum tire slip λ_{max} [2].

The longitudinal slip λ_ℓ results from the comparison of the vehicle's driving speed v and the speed of the wheel relative to the running track $r \cdot \omega_R$, where the wheel's angular speed is ω_R. In the event of braking, the longitudinal slip becomes:

$$\lambda_\ell = \frac{v - r \cdot \omega_R}{v} \cdot 100 = \left(1 - \frac{r \cdot \omega_R}{v}\right) \cdot 100\,[\%], \tag{10}$$

The longitudinal slip is expressed as a percentage, and the dynamic radius of the wheel results from the space covered by the free wheel at one rotation $r = s_r/2\pi$.

The longitudinal slip can also be determined comparing the front and rear wheel rotation speed. For rear traction, longitudinal slip is:

$$\lambda_\ell = \left(1 - \frac{n_{fm}}{n_s}\right) \cdot 100\ [\%] \tag{11}$$

where nfm average front wheels rotation speed, ns rear wheels rotation speed.

In case of all-wheel braking, in order to determine the longitudinal slip, in addition to the wheel speed, it is necessary to measure the speed of the vehicle. The longitudinal slip, in case of individual wheel braking, is determined by the relation:

$$\lambda_{\ell i} = \left(1 - \frac{n_i}{v} \cdot \frac{\pi \cdot r}{30}\right) \cdot 100\ [\%]. \tag{12}$$

The calculations are approximate because r is considered equal on all wheels. Longitudinal adhesion coefficient ϕ_ℓ is dependent on the longitudinal slip λ_ℓ (Fig. 1). The degree of braking of actual vehicles depends on the pedaling force, the adhesion of the various wheels and the distribution of frictional forces on the axles and wheels.

The ideal braking rate can not be achieved because the point 1 in Fig. 1 is not constant and thus the value achieved will be less than the ideal one. By avoiding the wheel lock, you can get close to ideal conditions.

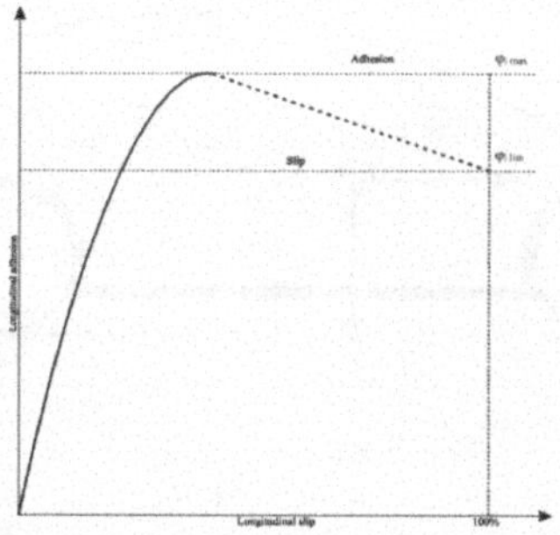

Fig. 1. Longitudinal adhesion coefficient variation according to longitudinal slip λ_ℓ [4].

The maximum adhesion utilization rate ε_{max}, is also known as the degree of worthiness.

$$\varepsilon \frac{Z_{max}}{\phi_{\ell\ max}}. \tag{13}$$

The high degree of worthiness means a reduced braking area. However, in terms of active safety, the degree of worthiness is not a sufficient criteria, since it does not consider also the stability of braking. A high degree of worthiness can lead, especially in curves, in the loss of stability and negatively reflects on the safety of the vehicle.

Vehicle behavior during braking depends primarily on the distribution of braking forces on the axles. The interdependencies can be very well represented in the so-called tangential distribution of the braking forces. First, the ideal curve of the distribution of the braking forces on the axles is theoretically determined. This curve is determined by the assumption of equal use of adhesion by the front and rear wheels. Based on relation 8, we can write:

$$F_{ff} + F_{fs} = Z \cdot G_a \tag{14}$$

where Fff braking force on front axle wheels, Ffs braking force on rear axle wheels.

The force $Z \cdot G_a$ in the relation (8) is acting in the mass center of the vehicle (Fig. 2) and causes, toward the wheel track, through the force lever arm h_{cm}, a pitch moment, which causes a dynamic front axle overload and a rear axle underload.

From the equation of moments in point B, respectively A (Fig. 2) results:

$$Z_f = G_a\left(\frac{A-a}{A} + Z\frac{h_{cm}}{A}\right);\ Z_s = G_a\left(\frac{a}{A} - Z\frac{h_{cm}}{A}\right) \tag{15}$$

The two equations assume that the position of the structure is unchanged. In modern vehicles where the braking pitch rate is balanced, this condition is accomplished.

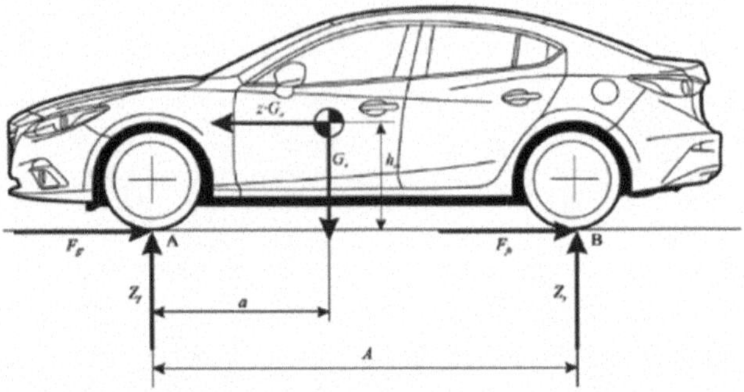

Fig. 2. Scheme of forces acting on the wheels of a car during braking

The tangential force that can be transmitted by the wheel is proportional to the Normal dynamic force. Tangential force is a friction force, the coefficient of friction at slip on the front wheels is μ_f and for the rear wheels μ_s, so we can write:

$$F_{ff} = \mu_f \cdot Z_f, \text{ respectively } F_{fs} = \mu_s \cdot Z_s \tag{16}$$

The ratio between the height of the mass center, hcm, respectively the distance, a, and the wheelbase, A, can be indicated by:

$$\chi = \frac{h_{cm}}{A}, \text{ respectively } \psi = \frac{a}{A} = \frac{Z_f}{G_a}. \tag{17}$$

By replacing Eqs. (15) and (17), the Eq. (16) can be rewritten:

$$\frac{F_{ff}}{G_a} = \mu_f(1 - \psi + Z \cdot \chi) \text{ and } \frac{F_{fs}}{G_a} = \mu_s(\psi - Z \cdot \chi). \tag{18}$$

In order to achieve an ideal distribution of braking forces, we will attept to obtain the equality $\mu_f = \mu_s = \mu$:

$$F_{ff} + F_{fs} = \mu\left(Z_f + Z_s\right) = Z \cdot G_a \text{ or } \mu = Z. \tag{19}$$

With the condition $\mu = Z$ accomplished, Eq. (18) become:

$$\frac{F_{ff}}{G_a} = Z \cdot (1 - \psi + Z \cdot \chi) \text{ and } \frac{F_{fs}}{G_a} = Z \cdot (\psi - Z \cdot \chi) \tag{20}$$

which represents, in parametric description, the braking forces balance ideal curve.

In order to graphic representing the equation $\frac{F_{ff}}{G_a} = f\left(\frac{F_{fs}}{G_a}\right)$, we determine Z from the first relation (20):

$$Z^2 \cdot \chi + Z(1-\psi) - \frac{F_{ff}}{G_a} = 0. \tag{21}$$

The equation has two solutions:

$$Z_{1,2} = -\frac{1-\psi}{2\chi} \pm \sqrt{\left(\frac{1-\psi}{2\chi}\right)^2 + \frac{1}{\chi} \cdot \frac{F_{ff}}{G_a}}. \tag{22}$$

According to the relation (19), the ratio between the rear braking force and the weight of the vehicle is:

$$\frac{F_{fs}}{G_a} = -\frac{F_{ff}}{G_a} - \frac{1-\psi}{2\chi} \pm \sqrt{\left(\frac{1-\psi}{2\chi}\right)^2 + \frac{1}{\chi} \cdot \frac{F_{ff}}{G_a}} \tag{23}$$

In order to facilitate graphic representation, the characteristic points are determined: the intersections with the coordinate axes and the position and size of the maxims.

For $\frac{F_{fs}}{G_a} = 0$ there result two points: $\frac{F_{ff}}{G_a} = 0$ and $\frac{F_{ff}}{G_a} = \frac{\psi}{\chi}$.

For $\frac{F_{ff}}{G_a} = 0$ there result also two points: $\frac{F_{fs}}{G_a} = 0$ and $\frac{F_{fs}}{G_a} = -\frac{1-\psi}{\chi}$.

By canceling the first degree derived function according to Fff/Ga, we obtain the curve maximum point:

$$\frac{d\frac{F_{fs}}{G_a}}{d\frac{F_{ff}}{G_a}} = \frac{1}{2\chi \cdot \sqrt{\left(\frac{1-\psi}{2\chi}\right)^2 + \frac{1}{\chi} \cdot \frac{F_{ff}}{G_a}}} - 1. \tag{24}$$

The maximal value $\left(\frac{F_{ff}}{G_a}\right)_{max}$ can be determined by equating the relation (24) with 0 and the maximal value $\left(\frac{F_{fs}}{G_a}\right)_{max}$ by equating the same relation with ∞. It results:

$$\left(\frac{F_{ff}}{G_a}\right)\frac{(1-\psi)^2}{4\chi}_{max} \text{ for abscissa } \frac{F_{fs}}{G_a} = \frac{(1-\psi)^2}{4\chi}, \tag{25}$$

$$\left(\frac{F_{fs}}{G_a}\right)\frac{\psi^2}{4\chi_{max}} \text{ for ordinate } \frac{F_{ff}}{G_a} = \psi \cdot \frac{2-\psi}{4\chi}. \tag{26}$$

For a more accurate image on the curve, the tangent in origin can be determined from the condition: $\frac{F_{ff}}{G_a} = 0$, and from relation (24) results $tg\alpha = \frac{\psi}{1-\psi}$.

In order to evaluate the braking behavior and stability of the braking, Fig. 3 shows the parabola of the ideal distribution of the tangential forces of the wheels in the event of braking for a medium-capacity truck having the following technical characteristics: wheelbase A = 4,5 m, mass center height hcm = 0,95 m, distance from front axle and mass center a = 2,8 m, actual mass of the vehicle m0 = 8120 kg, technically permissible maximum laden mass ma = 16000 kg, technically permissible maximum laden mass on front axle ma = 16000 kg.

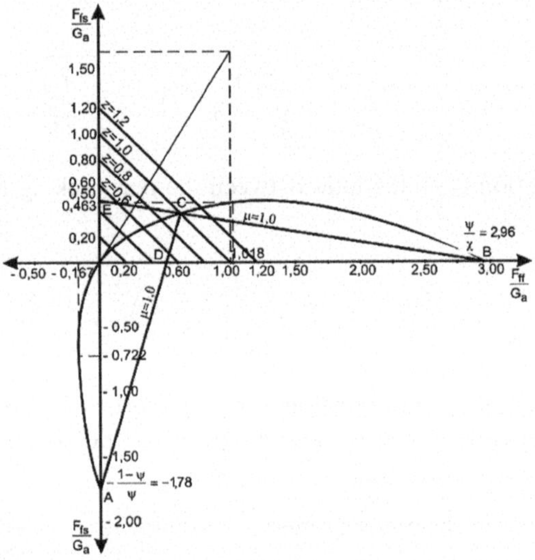

Fig. 3. Parabola of the ideal distribution of the tangential forces of the wheels in the event of braking [4].

The real operating area of the brakes is located in the first quadrant, from where the ideal correlation between the braking forces at the front and rear axles can be established, in relation to the curve described by the ideal distribution parabola. Thus, for a load to a specific tangential force of braking Fff/Ga = 0.60 for the front wheels, it will correspond to the rear wheels Ffs/Ga = 0.375, which ensures a stable and maximum efficiency process. In order to evaluate the braking behavior and the stability of the braking, Fig. 3 has been completed with two families of straight features, namely the straight-line braking rates and the constant friction coefficients.

Constant braking straights are 45° inclined to the coordinate axes and intersect these axes in Z-points. Constant friction coefficients straights start at the intersection points of the parabola with coordinate axes μ_f and μ_s. On the parabola, the lines $Z = \mu_f = \mu_s$ must intersect at a point, for example point C, in which $\mu = 1$ (Fig. 3).

3 Conclusion

A rapid assessment of the braking system can be made using the dependence $\mu - Z$ shown in Fig. 4.

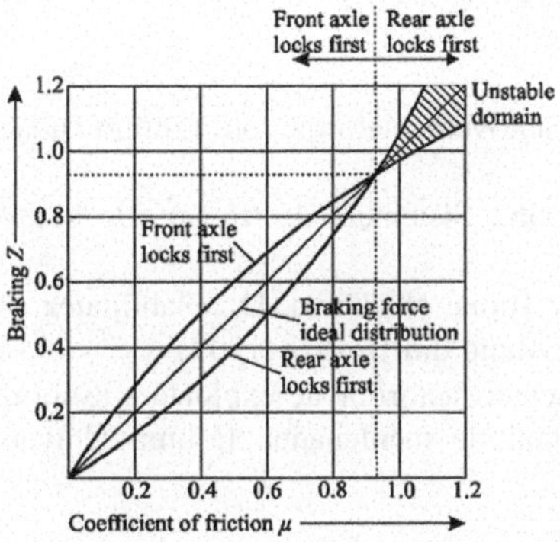

Fig. 4. Representation of the distribution of friction forces according to the coefficient of friction and the degree of braking.

The ideal distribution of braking forces is represented in Fig. 4 by a 45° inclined straight. The curves indicating the adhesion limits for the front and rear axle can be defined by:

$$Z_{fA} = \frac{\mu \cdot (1 - \psi)}{1 - \mu \cdot \chi - \varphi}, \text{ and } Z_{sA} = \frac{\mu \cdot \psi}{\mu \cdot \chi + \varphi}. \tag{27}$$

The stability condition "front axle lock before the rear axle" is respected if the curve for the front axle is above the straight line. The curves for the front axle and the rear axle intersect with the straight $\mu - Z$ right in the Zcrit point. At adhesion coefficients higher than those corresponding to the Zcrit condition, the rear axle is locked first. From the figure the stability reserves are also easily deducted.

According to Regulation no. 13 ECE - ONU, Zcrit > 0.82 for a two - axle vehicle. This critical value can be determined by the relation:

$$Z_{crit} = \frac{\psi - \varphi}{\chi}. \tag{28}$$

In order to make the car's behavior independent of the load and braking rate, it is advisable to dynamically adjust the braking forces, depending on the dynamic load on the axle, which is partly provided by the rear brake limiter.

It is clear from the above that the proper functioning of all devices and devices in the vehicle's braking system during operation is of particular importance, and the maintenance activities organized for this purpose must ensure maximum reliability at all times.

References

1. Untaru, M., ş.a.: Dinamica autovehiculelor pe roţi. Editura Didactică şi Pedagogică, Bucureşti (1981)
2. Frăţilă, Gh., Mărculescu, Gh.: Sistemele de frânare ale autovehiculelor. Editura Tehnică, Bucureşti (1986)
3. Nagy, T., Stănescu, M.A., Ţurea, N., Dima, D.: Fiabilitatea şi terotehnica autovehiculelor, vol. I. Universitatea Transilvania din Braşov (1997)
4. Negrilă, R.I.: Studiul influenţei factorilor de exploatare asupra fiabilităţii autovehiculelor în vederea optimizării activităţii de mentenanţă. Editura Universităţii "Aurel Vlaicu", Arad (2005)

Solicitations in the Rear Axle Support of a Karting Frame

Laurentiu Racila, Ilie Dumitru(✉), Lucian Matei, Mihaela Racila, Dragoş Tutunea, and Ionut Geonea

University of Craiova, Craiova, Romania
dumitru_ilie@yahoo.com

Abstract. The paper aims to determine the solicitation in the rear axle support of a karting type structure, which are used in karting competitions. The analysis was made analytically and with the help of specialized software, with a good approximation between the two methods of analysis. These methods can also be extended to compute others solicitations in areas considered dangerous for the karting type structure.

Keywords: Go kart · 3D scanning · Structure analysis

1 Introduction

The kart race is a sporty version of the "open wheels" where small body-open, four-wheeled cars called karts with or without gearboxes/gearshift, depending on the design, compete on a circuit respecting a set of rules.

Many organizations regulate these races such as IKF (International Kart Federation), WKA (World Karting Association), KART (Karters of America Racing Triad) or USPKS (United States Pro Karting Series) [1]. Race races begin at 7 or 8 years of age (5 in the US with Kid Karts) and generally run in 3-year or 15-year-olds up to the age of 15 or 16, depending on the series [2].

There are many different karting racing classes or formulations [3], classified in International Competitions (such as OKJ, OK, KZ1, KZ2 and Superkart, these being considered top classes in karting and are also racing in national championships (in the UK Super One, in the United States, Dirt Class, Australian Class Cadet, Rookie, Junior National, etc.).

The structure presented in this article is a kart structure engaged in the Kart Low Cost competition. This competition is a European student karting competition, open to all students in higher education institutions, organized by the FEEIAT (Fédération Européenne des Ecoles d'Ingénieurs d'Automobile et des Transports - European Federation of Schools in Automobile and Transport Engineering) as Challenge Kart Low Cost.

I. Dumitru et al. (Eds.): SMAT 2019, *The 30th SIAR International Congress of Automotive and Transport Engineering*, pp. 607–613, 2020.
https://doi.org/10.1007/978-3-030-32564-0_71

2 3D Frame Scanning

The 3D technology was applied in the context of the existence of mono-post structures already built at previous editions, which required the ultimate objective to detect deviations from form and to obtain an optimized solution.

3D scanning technology is the application of scientific knowledge to analyze an object (or environment) by collecting data about its shape and sometimes its appearance (color, texture) to build a 3D digital model. The device used to perform this type of analysis is a 3D scanner.

Frame scanning was done with a HandyScan 3D [4]. In this sense, the kart frame has been cleaned and for better digitization, any object that is bright, black, transparent or reflective (like bright or metallic surfaces) has been covered with a white powder to obtain a matte finish for an accurate results (Fig. 1).

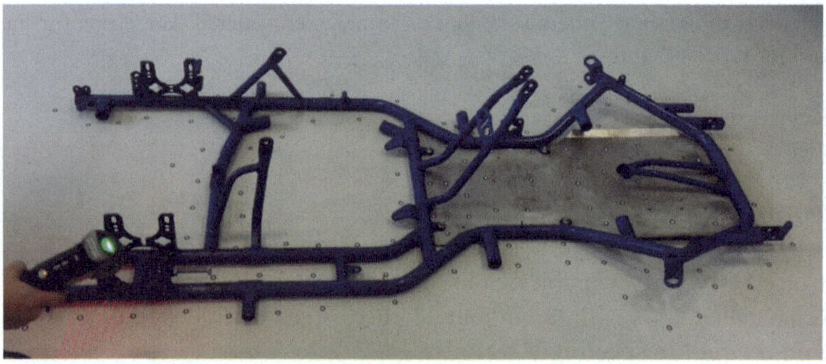

Fig. 1. Scanning position for the upper structure

To create a 3D object model, the object must be measured using 3D scanning technologies (CMM, laser scanners, structured light digitizers, etc.). After the scan operation is complete, it is possible to rebuild the 3D model using 3D CAD, CAM, CAE or other software.

For the first part of 3D rebuilding of the kart frame, a special 3D reconstruction software was used (Fig. 2).

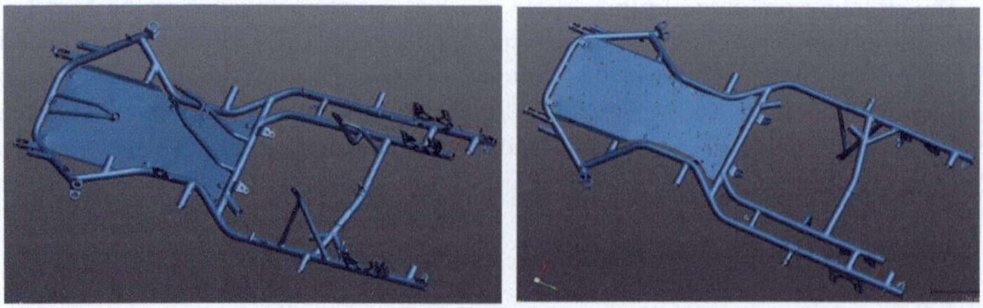

Fig. 2. Upper and lower view of the structure

3 Analytical Analysis of the Structure

The structure of the kart frame is considered symmetrical with respect to the longitudinal axis, which allows for a simplified model for analyze. Only loads due to the pilot and the seat are considered. Thus, half the weight of the assembly, the pilot and the seat, is distributed at the top of the seat fork (force F_2) and half on the front seat attachment (force F_1) (Fig. 3).

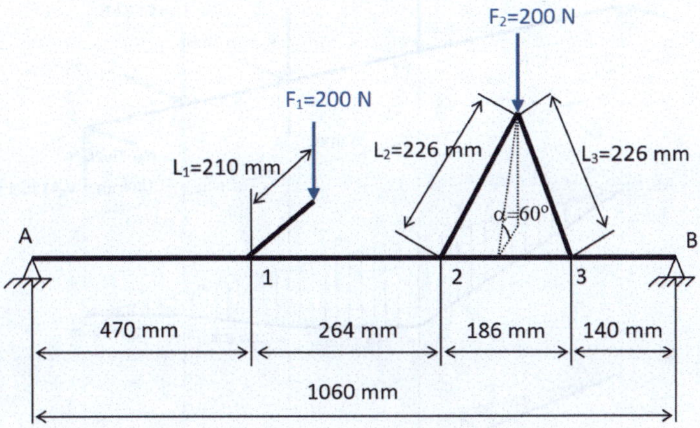

Fig. 3. Simplified model of the frame

Force F_1 and F_2 will cause vertical, horizontal reactions and twisting and bending moments. By breaking down the two forces in the support arm directions, the frame structure is required to bending in two planes [5].

The maximum bending moment will be:

$$M_{imax} = 61470{,}18\,\mathrm{N} \cdot \mathrm{mm} \tag{1}$$

The torsion moment applied to the AB beam is:

$$M_r = 81098\,\mathrm{N} \cdot \mathrm{mm} \tag{2}$$

The normal and shear stress are:

$$\sigma = \frac{M_{imax}}{W_z} = \frac{61470{,}18}{1152{,}6} = 53{,}3\,\mathrm{MPa} \tag{3}$$

$$\tau = \frac{M_r}{W_o} = \frac{81098}{2310{,}53} = 35{,}1\,\mathrm{MPa} \tag{4}$$

Because the beam is solicitate at bending and torsion it is possible to write the equivalent stress equal to:

$$\sigma_{ech} = \sqrt{\sigma^2 + 3 \cdot \tau^2} = \sqrt{53{,}3^2 + 35{,}1^2} = 80{,}8\,\text{MPa} \tag{5}$$

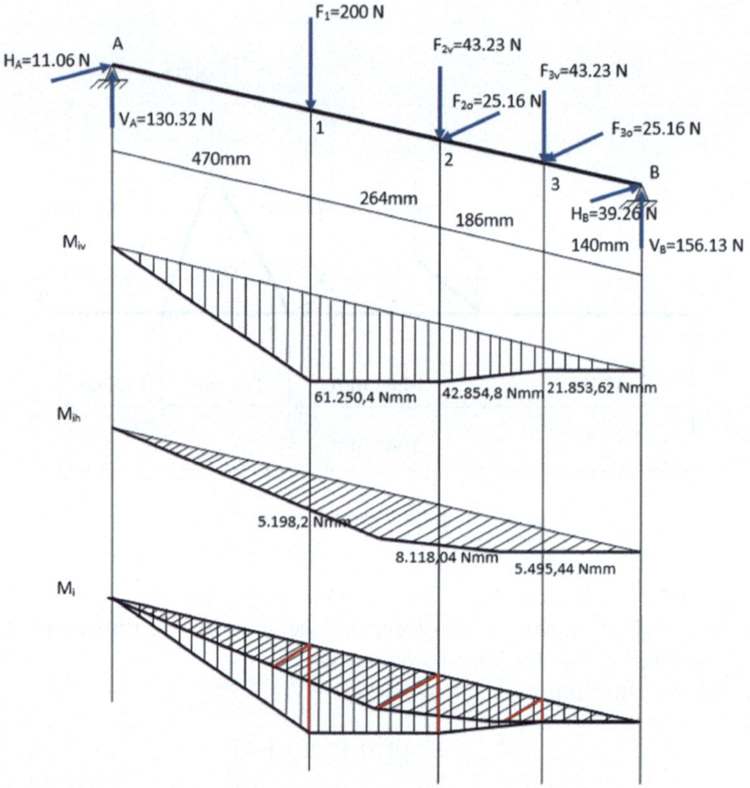

Fig. 4. Effort diagram of the simplified frame structure

This stress value is much lower than the allowable stress of the material from which the frame is made (Fig. 4).

4 Analysis of the Structure with Dedicated Software

In order to test the frame for external loads in terms of structure resistance and critical point analysis, the Solidworks software was used [6]. Only a static study was carried out on the structure of the kart.

In order to be able to perform an analysis, it is necessary to make the mesh structure so that any deformation that may occur can be accomplished at the appropriate points.

This is necessary in the later analysis of the body structure so that the resistance structure on which the functional elements of the kart will be mounted [7].

The discretization structure for static analysis can be seen in the figure below (see Fig. 5).

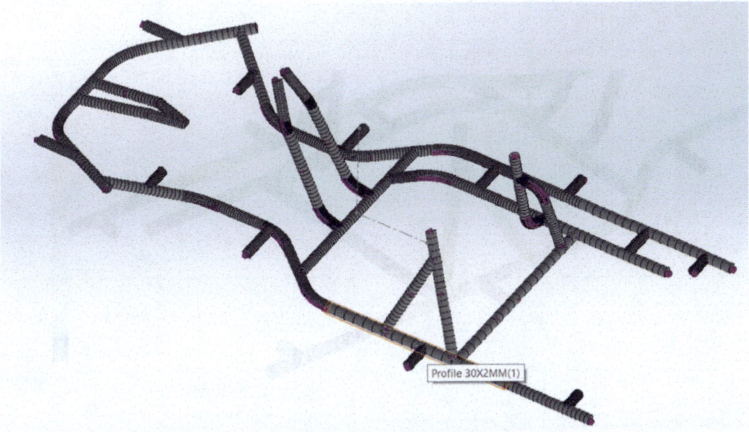

Fig. 5. Discretized structure in Solid Works

The discretized structure for static analysis can be seen in the figure above (see Fig. 6) and the pilot weight distribution in the figure below.

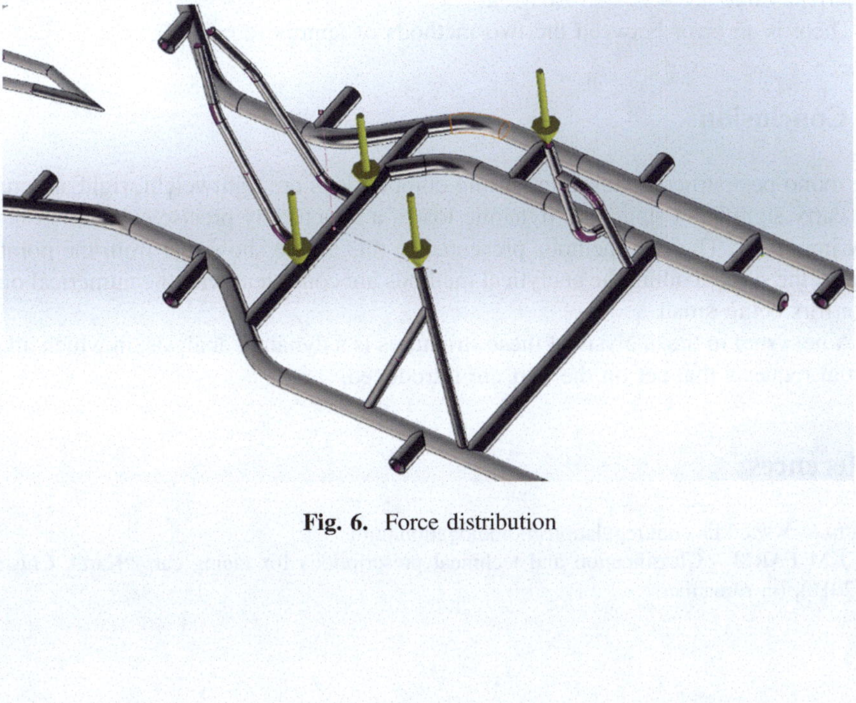

Fig. 6. Force distribution

Several maximum stress points, such as the left rear bearing area (maximum values reaching 77 MPa), and right rear bearing area, were identified as a result of the analysis, and the maximum backplanes reached 70 MPa. These right values are not symmetrical with those of the left values because of the additional reinforcements in this area. These reinforcements are added because the engine will be installed in this area [8] (Fig. 7).

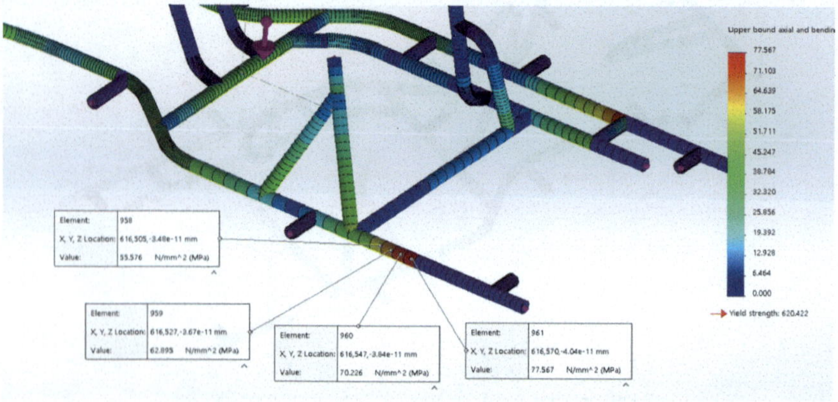

Fig. 7. Maximum stress points of the frame

It is noted that the difference between the analytical and numerical values computed with the two methods is small, the analytical value being $\sigma_{an} = 80{,}8\,\text{MPa}$ and the numerical value are $\sigma_{num} = 77{,}6\,\text{MPa}$

There is an error between the two methods of approx. $\varepsilon = 3{,}9\%$.

5 Conclusion

The mono-post structures used in karting competitions are lightweight, rigid structures that carry significant static and dynamic loads, a structurally precise calculation being very important. The two methods presented in this article show that from the point of view of the static loading the analytical methods are consistent with the numerical ones, the errors being small.

A next step in the analysis of these structures is a dynamic analysis, in which all the inertial requests that act on the kart are introduced.

References

1. http://www.cikfia.com/regulations/homologation.html
2. CKM FARM - Classification and technical prescriptions for racing cars “Kart”, Chisinau (2018). (in romanian)

3. The Romanian Federation of karting - Classification and technical prescriptions for racing cars "Kart" (2019)
4. https://www.cadworks.ro/solutii/scan-3d/handyscan-3d/
5. Racila, L.: The base of Strength of Materials: Theory and Applications (in romanian), Ed. Universitaria, Craiova (2011)
6. ***- https://www.cadworks.ro/solutii/design/solidworks-simulation/
7. Roşca, I. Dumitru, A.S.: CAD Applications in Automotive Engineering. Ed. Universitaria, Craiova (2016). (in romanian)
8. Matei, L.: Design and optimisation for a KLC karting frame, final engineering stage thesis (2018). (in romanian)

Aspects Regarding the Stability Performances Calculation of the Wheeled Vehicles While Turning

Ion Preda(✉), Gheorghe Ciolan, and Dinu Covaciu

Transilvania University of Brasov, Brasov, Romania
pion@unitbv.ro

Abstract. The stability performances are very important to appreciate the vehicle ability to be driven in different conditions.

Starting for a simple planar model of one axle, the motion equations characterizing the rollover or skidding possibility are obtained. These are interpreted as different concerns for the driver, vehicle design engineer or road design engineer.

The influence over the vehicle stability is analyzed for different driving conditions as speed, grip coefficient, lateral slope of the ground, position of the vehicle center of gravity, braking or traction intensity, suspension characteristics.

Based on the axles stability characteristics, some conclusion were drawn about the steering behavior of the full vehicle. Also, some recommendations for drivers are presented.

The procedure presented here can also be used to establish some working algorithms for potential on-board safety systems able to prevent the stability loss.

Keywords: Vehicle dynamics · Stability · Rollover · Skidding · Algorithm

1 Introduction

The concept of *stability* has a wide palette of meanings. From the definitions in dictionary (such [1] and [2]) here are retained only two: the ability to adjust to different changes and the property of a system or element by virtue of which its output will ultimately attain a steady state.

Limiting the domain of applicability, the stability of a mechanical system represents its ability to develop restoring forces equal to or greater than the disturbing forces so as to maintain a state of equilibrium or, in other words, its ability to maintain its position or to come back in its initial state of rest or movement, after having undergone a disturbing action [2]. Starting from the search of mechanical system equilibrium, the notion of stability is now largely used in mathematics, in the field of differential equation.

Particularizing more, the *stability of a vehicle* is its quality or state to remain controllable. Today is almost generally accepted [3–9] that, for a good stability, the vehicle must present the next qualities (indicated further in descending order of the associated danger probability):

I. Dumitru et al. (Eds.): SMAT 2019, *The 30th SIAR International Congress of Automotive and Transport Engineering*, pp. 614–623, 2020.
https://doi.org/10.1007/978-3-030-32564-0_72

- not to rollover;
- not to slip evidently;
- to correlate the yaw angle with the trajectory (maintaining an almost neutral steering behavior);
- to maintain the stability of the movement;
- to be predictable during driving or working.

As can be deduced, the need for vehicle stability appears not only during travelling, but also at rest, when the vehicle works stationary (for example, in the case of a dump truck, a crane, a front loader or an excavator). In that last case, the vehicle must not to rollover, slip or behave unpredictable.

The *rollover*, the situation when the wheels or tracks lose the contact with the ground and the bodywork may hit the ground, can be potentially the most dangerous case of losing stability.

The *evident slip* of the wheels or tracks with respect to the ground may present also a high level of accident risk. The important slippage is normally named more precisely *spin* and *slide* (when the slip occurs in the longitudinal direction at traction, and respectively at braking) or *skid*, when the vehicle slips laterally. It says a vehicle (or wheel, or axle, or track) skids when it slips on a laterally inclined ground or when, during turning, the vehicle is not able to inscribe into a curve, realizing a trajectory with a larger turning radius as the one intended (and commanded) by its driver.

The third situation of stability lose means that the vehicle yawing angle cannot be well correlated with its trajectory, i.e. the driver's heading deviates more from the tangent at the trajectory. That means the vehicle presents a *strong understeering or oversteering attitude*.

The fourth requirement for the vehicle stability consists in ensuring the *stability of the vehicle movement*. The movement of the vehicle can be considered stable in many ways, for example in the Lyapunov sense [10]. Accordingly, the directional stability represents the ability of a vehicle to move forward in a straight line with a minimum of driver control. Also, a vehicle turning on circle with constant speed and subjected to different disturbances must to come back to the same type of movement if the perturbations disappear. A vehicle with a stable movement (good directional stability) will not be unduly affected by side wind, lateral road inclinations, road irregularities, changes of grip coefficient, etc.

The last stability requirement, the *predictability*, means the vehicle needs to have similar behavior in different driving or working conditions and not to surprise the driver by sudden modification of its response.

Objective analyses of operating behavior are based on various experimental maneuvers, many executed at different speed and road inclination, such as steady turn on the skidpad, slalom, frequency response, steering impulse, wag (a single rapid movement from side to side), lane change, obstacle avoidance (double lane change), or braking and acceleration in a curve. For the last five maneuvers, keeping the stability is very important mainly for the first period of time (one second, for example) because this interval is comparable with the normal driver's reaction time.

Of course, obtaining theoretically the main stability characteristics of a vehicle presents a big interest, due to many advantages if compared with the experiments: cost, time, possibilities to change conditions, repeatability of disturbances and so on. The aim of this paper is to offer a simple procedure to estimate by calculation the stability performances of the wheeled vehicles while cornering.

2 Vehicle Axle Model for Turning Stability Study

The easiest way to start the study of the vehicle stability during turning consists in adopting a simple model and simplifying hypotheses. Different possibilities exist: to analyze a full (singular) vehicle [3] or only individual axles [6, 11]. This second approach was preferred here, because it permits also the study of understeering or oversteering attitude and permits to consider different grip coefficients for the wheels.

In the Fig. 1 it is presented the planar model of an axle belonging to a cornering vehicle on a laterally inclined road. The vehicle is considered as a rigid body (with no deflections at the tires and suspension springs).

The ground is banked inward with the angle β. The weight W_a exerted on the axle corresponds to the static load on the axle tires. The weight acts in the center of gravity C_g, which is placed at the height h_g and at the distance d_o to the outer wheel center plane on an axis parallel to the ground. The track width of the axle is E.

The aerodynamic force acts in the center of pressure C_p. Even in this study the lateral and lift components of the aerodynamic force are neglected (due to the small to medium speeds taken into account), these must be considered for the race vehicles. The turning radius R is the distance from the (instant) turning center O to the center of gravity C_g.

At the axle outer and inner tires act also the normal reactions of the ground Z_o and Z_i and the lateral components Y_o and Y_i of the grip forces F_o and F_i. Also, a torque M_{ar} transferred to the studied axle from the other vehicle axle (or axles) it is considered. A positive value of M_{ar} rotation moment will have a stabilizing effect on the axle (an anti-roll action against the vehicle superstructure supported by the axle) and, of course, a destabilizing influence over the other axle (or axles).

Due to the action of all the external forces (W_a, Z_o, Y_o, Z_i, Y_i) and torques (M_{ar}), the axle experiences the centripetal acceleration a_{cp}, oriented from the center of gravity C_g through the turning center O and generating the vehicle cornering (changing the orientation of the velocity vector $\vec{v}$).

Considering the three degrees of freedom of the axle planar model (translation on the normal to the ground z, translation parallel to the ground y and rotation on the vehicle longitudinal axis x) and applying the Newton second law, it results:

$$\begin{aligned} Z_o + Z_i - W_a \cos\beta &= m_a\, a_z \\ Y_o + Y_i - W_a \sin\beta &= m_a\, a_y \\ -Z_o d_o + Z_i(E - d_o) - (Y_o + Y_i)h_g - M_{ar} &= J_x\, \varepsilon_x \end{aligned} \tag{1}$$

where: $m_a = W_a/g$ is the vehicle mass distributed on the considered axle; J_x is the moment of inertia against the x axis that corresponds to the mass distributed on the axle;

a_z and a_y are the translation accelerations on the normal to the ground and parallel to the ground directions and ε_x is the angular acceleration against the longitudinal axis x.

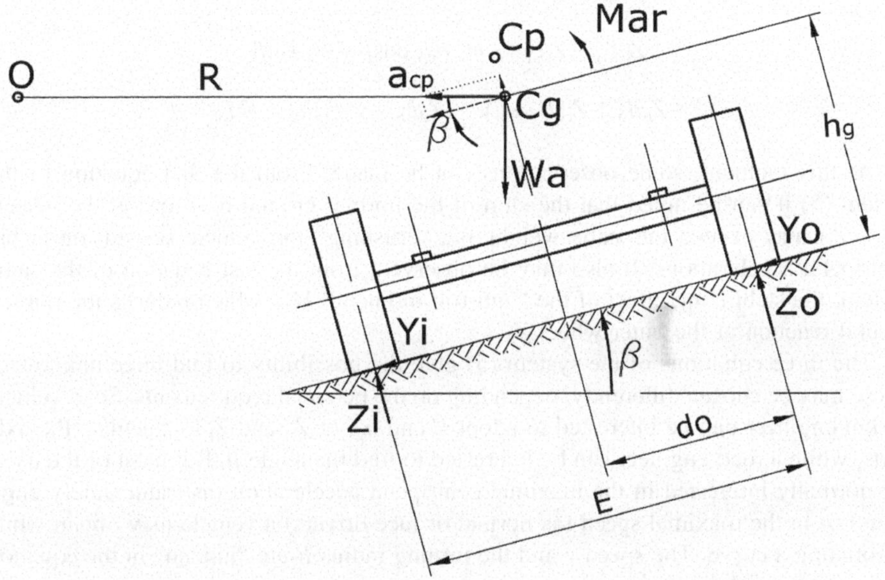

Fig. 1. Schematic of the external forces acting on a vehicle axle during cornering.

From the figure, it observes that the accelerations a_z and a_y are components of the centripetal acceleration a_{cp}:

$$a_z = a_{cp}\ \sin\beta \quad \text{and} \quad a_y = a_{cp}\ \cos\beta \tag{2}$$

Also, the lateral components Y_o and Y_i of the outer and inner grip forces can be expressed as functions of the normal reactions of the ground against the tires, Z_o and Z_i:

$$Y_o = Z_o\xi_o \quad \text{and} \quad Y_i = Z_i\xi_i \tag{3}$$

where: ξ_o and ξ_i are the coefficients of the "used" grip in lateral direction. Notating with $\mu_j = F_j/Z_j$ the grip coefficient and with X_j the longitudinal component of the grip force at a certain wheel j, it results the obvious inequalities [12]:

$$\sqrt{X_j^2 + Y_j^2} \leq \mu_j \cdot Z_j = F_j, \qquad \xi_j \leq \sqrt{\mu_j^2 - \left(\frac{X_j}{Z_j}\right)^2} = \mu_{yj}, \quad Z_j \geq 0 \tag{4}$$

Now, introducing the Eqs. (2) and (3) in the equations system (1) and considering that the rollover process didn't started ($\varepsilon_x = 0$), it results:

$$\begin{aligned} Z_o + Z_i &= m_a\, a_{cp}\, \sin\beta + W_a \cos\beta \\ Z_o\xi_o + Z_i\xi_i &= m_a\, a_{cp}\, \cos\beta - W_a \sin\beta \\ -Z_o d_o + Z_i(E - d_o) - (Z_o\xi_o + Z_i\xi_i)h_g &= M_{ar} \end{aligned} \tag{5}$$

At this moment, some observations can be made. From the first equation of the system (5) it can be noted that the sum of the normal ground reactions at the wheels ($Z_o + Z_i$) may exceed the static weight W_a, "crushing" the vehicle suspension at big centripetal accelerations. It also may be observed, from the last equation of the same system, the stabilizing effect of the "anti-roll moment" M_{ar}, which reduces the normal ground reaction at the outer wheel Z_o.

The three equations of the system (5) give the possibility to find three unknowns. These may be chosen differently, depending on the personal requirements. So, a vehicle design engineer can be interested to adopt E and h_g, or Z_o and Z_i to calculate the axle parts, while a road engineer can be interested to find the angle β. But most of the users are normally interested in the maximal centripetal acceleration (as traffic safety engineers) or in the maximal speed (as normal or race drivers) a vehicle may obtain while negotiating a curve. The speed v and the turning radius R are "hidden" in the equation of the centripetal acceleration:

$$a_{cp} = \frac{v^2}{R} = v\omega \tag{6}$$

where ω is the yawing speed of the vehicle.

Most often, the unknowns presenting the biggest interest are a_{cp}, Z_o and Z_i. Solving the equations system (5) it obtains:

$$\begin{aligned} Z_o &= W_a \cdot \frac{E - d_o + h_g \cdot \xi_i - \frac{M_{ar}}{W_a} \cdot (cos\beta - \xi_i \cdot sin\beta)}{[E + h_g \cdot (\xi_i - \xi_o)] \cdot cos\beta - [E \cdot \xi_o + d_o \cdot (\xi_i - \xi_o)] \cdot sin\beta} \\ Z_i &= W_a \cdot \frac{d_o - h_g \cdot \xi_o + \frac{M_{ar}}{W_a} \cdot (cos\beta - \xi_o \cdot sin\beta)}{[E + h_g \cdot (\xi_i - \xi_o)] \cdot cos\beta - [E \cdot \xi_o + d_o \cdot (\xi_i - \xi_o)] \cdot sin\beta} \\ a_{cp} &= g \cdot \frac{[(E - d_o) \cdot \xi_o + d_o \cdot \xi_i] \cdot cos\beta + [(E + h_g \cdot (\xi_i - \xi_o)] \cdot sin\beta + \frac{M_{ar}}{W_a} \cdot (\xi_i - \xi_o)}{[E + h_g \cdot (\xi_i - \xi_o)] \cdot cos\beta - [E \cdot \xi_o + d_o \cdot (\xi_i - \xi_o)] \cdot sin\beta} \end{aligned} \tag{7}$$

These movement equations describe the vehicle steering, applicable up to the start of rollover.

Both the rollover and skidding means the vehicle attains limit situations, which will be described further.

3 Rollover

The rollover is one of the most important and dangerous type of traffic accident. For example, only in the United States of America there are registered each year over 15000 rollovers of commercial trucks [13]. These rollovers represent about 4% of all truck accidents, but theirs consequences are even more severe: 12% of fatalities in truck accidents. That indicates the commercial truck rollover is strongly associated with severe injury and fatalities in highway accidents. From the trucks, the tractor-semitrailers combinations are the most exposed to the risk of rollover, mainly due to the height of the center of gravity and to the modality of articulation.

The rollover process starts with the lift-off of the inner wheel, Fig. 2. The rotation against a longitudinal axis passing through the outer wheel contact patch can continue to the exterior of the curve due to the destabilizing effect of the lateral grip force Y_o if that exceeds the contrary effect of the normal force Z_o and of the torque M_{ar} (if it is positive). At a certain moment, the vehicle will roll in the position in which the center of gravity comes at the vertical of the point where the forces Z_o and Y_o apply (near to the center of the outer wheel contact patch). From that moment, if the vehicle condition remains the same (speed, turning radius and lateral inclination angle), the vehicle rolls over.

However, it is still possible the driver to recover the vehicle control and to avoid the rollover if counter-steers, thus reducing or even inversing the lateral grip force Y_o. But that manoeuver must be done with a lot of skill and precaution in order not to determine the rollover in the opposite side.

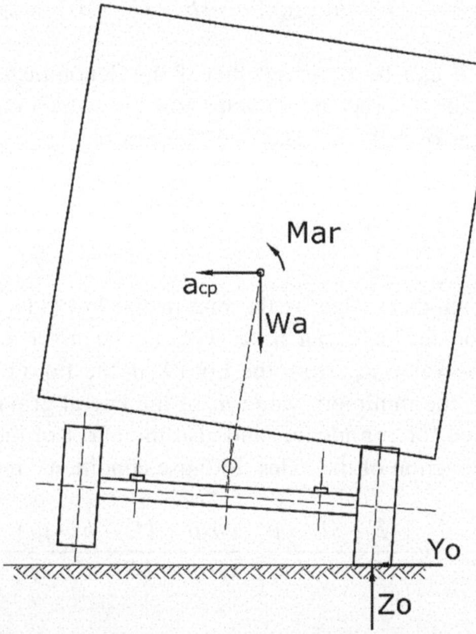

Fig. 2. Schematic of the rollover process initiation during vehicle cornering, with the indication of the forces acting on axle.

Based on these explanations, the lift-off of the inner wheel (when Z_i becomes zero) is generally considered the physical condition of the rollover imminent danger, even that this loss of wheel-ground contact doesn't necessarily means the vehicle overturns. Making $Z_i = 0$ in the second equation of the system (7), which implies also $\xi_i = 0$ due to the lack of wheel contact, and considering the maximal value of the lateral grip coefficient $\xi_{omax} = \mu_{yo}$, it results the condition to start the axle rollover:

$$h_g \cdot \mu_{yo} = d_o + \frac{M_{ar}}{W_a} \cdot (cos\beta - \mu_{yo} \cdot sin\beta) \tag{8}$$

From the five variables (h_g, d_o, μ_{yo}, β and the ratio M_{ar}/W_a) appearing in the previous equation, one can be calculated if the other four are known. All five situations may present practical interest in avoiding vehicle rollover.

For example, the maximal lateral grip coefficient at which the axle inner wheel will not lift-off is:

$$\mu_{yo} = \frac{d_o + \frac{M_{ar}}{W_a} \cdot cos\beta}{h_g + \frac{M_{ar}}{W_a} \cdot sin\beta} \tag{9}$$

Taking now the third equation of the system (7), it obtains the maximal lateral acceleration the axle can have in the considered conditions without tipping over:

$$a_{cp} = \frac{v^2}{R} = g \cdot \frac{(E - d_o) \cdot \mu_{yo} \cdot cos\beta + (E - h_g \cdot \mu_{yo}) \cdot sin\beta - \frac{M_{ar}}{W_a} \cdot \mu_{yo}}{(E - h_g \cdot \mu_{yo}) \cdot cos\beta - (E - d_o) \cdot \mu_{yo} \cdot sin\beta} \tag{10}$$

From the Eq. (10) it can be observed that if the denominator become zero, the lateral acceleration may be theoretical no matter how big, which means that the vehicle will not be in the danger of rollover. That will happen if

$$\tan\beta = g \cdot \frac{(E - h_g \cdot \mu_{yo})}{(E - d_o) \cdot \mu_{yo}} \tag{11}$$

If the angle β has another value as the one in the Eq. (11), the maximal lateral (centripetal) acceleration the axle can have without the inner wheel lift-off can be obtained introducing the value μ_{yo} from the Eq. (9) in the Eq. (10).

For the full vehicle, the minimum value μ_y of the lateral grip coefficient of all the axles must be considered for confidence, and also the effect of the torque M_{ar} will be canceled due to the interaction of the axles. In those conditions, the Eq. (10) becomes:

$$a_{cp} = \frac{v^2}{R} = g \cdot \frac{(E - d_o) \cdot \mu_y \cdot cos\beta + (E - h_g \cdot \mu_y) \cdot sin\beta}{(E - h_g \cdot \mu_y) \cdot cos\beta - (E - d_o) \cdot \mu_y \cdot sin\beta} \tag{12}$$

where

$$\mu_y = d_{ov}/h_{gv} \tag{13}$$

can be derived from the Eq. (9), considering d_{ov} and h_{gv} for the entire vehicle.

Annulling the centripetal acceleration (a_{cp} = 0) in the Eq. (12) and combining with the Eq. (13) it is obtained the limit of the lateral slope at which the vehicle still doesn't overturns if it moves in straight line (or is tested on a tilting platform):

$$\beta = atan\frac{E}{2h_g} \tag{14}$$

The basic measure of roll stability in a curve is the *static rollover threshold*, expressed in gravitational units (*g*), which is in fact the maximal lateral acceleration the vehicle can have without appearing the rollover. Starting from the Eqs. (12) and (13) and considering a flat ground (β = 0) and a symmetrically placed center of gravity in the lateral plane (d_{ov} = *E*/2), that threshold results as:

$$a_{cp} = g \cdot \frac{E}{2h_g} \tag{15}$$

Most passenger cars have rollover thresholds greater than 1 g, while light trucks, vans, and SUVs range from 0.8 to 1.2 g [13]. However, the rollover threshold of a loaded heavy truck often lies well below 0.5 g, mainly due to influence of the suspension and tires compliance.

The Eq. (12) also leads to the well-known relation giving the maximal speed a vehicle can negotiate a curve with the radius *R* without overturning [3]:

$$v_r = \sqrt{g \cdot R \cdot \frac{\frac{0.5 \cdot E}{h_g} + \tan\beta}{1 - \frac{0.5 \cdot E}{h_g} \cdot \tan\beta}} \tag{16}$$

4 Skidding

To estimate the skidding avoidance performance of an axle it is necessary to determine the maximal lateral (centripetal) acceleration the axle can have in given conditions. This is possible using the third equation in the system (7) and replacing the actual coefficients of the "used" lateral grip with theirs maximal values realizable in the considered conditions:

$$a_{cp} = \frac{v^2}{R} = g \cdot \frac{[(E - d_o) \cdot \mu_{yo} + d_o \cdot \mu_{yi}] \cdot cos\beta + [(E + h_g \cdot (\mu_{yi} - \mu_{yo})] \cdot sin\beta + \frac{M_{ar}}{W_a} \cdot (\mu_{yi} - \mu_{yo})}{[E + h_g \cdot (\mu_{yi} - \mu_{yo})] \cdot cos\beta - [E \cdot \mu_{yo} + d_o \cdot (\mu_{yi} - \mu_{yo})] \cdot sin\beta} \tag{17}$$

For the full vehicle the effect of the torque M_{ar} will be canceled due to the interaction of the axles. If also it considers the lateral grip coefficients are identic $\mu_{yo} = \mu_{yi} = \mu_y$, the Eq. (17) becomes

$$a_{cp} = \frac{v^2}{R} = g \cdot \frac{\mu_y \cdot cos\beta + sin\beta}{cos\beta - \mu_y \cdot sin\beta} = g \cdot \frac{\mu_y + tan\beta}{1 - \mu_y \cdot tan\beta} \tag{18}$$

The Eq. (18) also leads to the well-known relation giving the maximal speed a vehicle can negotiate a curve with the radius R without skidding [3]:

$$v_s = \sqrt{g \cdot R \cdot \frac{\mu_y + \tan\beta}{1 - \mu_y \cdot \tan\beta}} \tag{19}$$

From the previous equation it observes that the increase of the transversal inclination β leads to a highest limit speed for skidding v_s, and for the angle

$$\beta = atan \frac{1}{\mu_y} \tag{20}$$

the skidding will not appear, doesn't matter the speed.

The same value as in Eq. (20) it is obtained as limit of the lateral slope at which the vehicle still doesn't skids if move in straight line (or is tested on a tilting platform). That results if in the Eq. (18) the centripetal acceleration is annulled ($a_{cp} = 0$).

5 Conclusions

This paper presented one way to establish the necessary equations to estimate the rollover or skidding situations which may appear when a vehicle negotiates a curve.

The demonstration presented here exceeds the normal approaches found in the literature. Many parameters influencing the vehicle stability were taken into account.

More precise results can be obtained if supplementary parameters (as compliance of the tires, springs and anti-roll bars or axle roll centers) will be considered.

The procedure presented here may be used to establish some working algorithms for potential on-board safety systems able to prevent the stability loss.

References

1. Motorera Homepage: Dictionary of automotive terms. Illustrated glossary for automobiles, motorcycles, bicycles, and motors. https://www.motorera.com/dictionary/. Accessed 22 July 2019
2. Frățilă, G., Andreescu, C., Vladu, C., Moisescu, R., Stan, C., Toma, M., Cristea, G.: Dicționar explicativ pentru știință și tehnologie Român/Englez/Francez/German. Autovehicule Rutiere. Editura AGIR, București (2015)

3. Untaru, M., Poţincu, G., Stoicescu, A., Pereş, G., Tabacu, I.: Dinamica autovehiculelor pe roţi. Editura Didactică şi Pedagogică, Bucureşti (1981)
4. d'Entremont, K.L.: Light-Duty Vehicles in Tripped-Rollover Situations, vol. 1, no. 4. Safety Bulletin, Triodyne Environmental Engineering, Inc., August 1995
5. Gillespie, T.D.: Fundamentals of Vehicle Dynamics. SAE Inc., Warrendale (1992)
6. Verschoore, R.: The overturning stability of heavy vehicles. In: The IXth International Conference CONAT 1999, Transilvania University of Braşov, Braşov, Romania (1999)
7. Rill, G.: Road Vehicle Dynamics - Fundamentals and Modeling. CRC Press, Boca Raton (2012)
8. Guiggiani, M.: The Science of Vehicle Dynamics – Handling, Braking, and Ride of Road and Race Cars. Springer, London (2014)
9. Wiegand, B.P.: Automotive rollover equation. In: Future Annual International Conference of the Society of Allied Weight Engineers, Los Angeles (2015)
10. Andrzejewski, R., Awrejcewicz, J.: Nonlinear Dynamics of a Wheeled Vehicle. Springer, New York (2005)
11. Preda, I., Flueraş, A., Câmpian, O.V.: Bending torques on solid beam axle during cornering. In: ESFA Conference, Paper S17, Bucureşti, Romania (2003)
12. Preda, I., Ciolan, G.: Vehicle mathematical model for the study of cornering. In: Annals of the Oradea University, Fascicle of Management and Technological Engineering, vol. XI (XXI), no. 2/1, pp. 22–32 (2012)
13. Winkler, C.: Rollover of heavy commercial vehicles. UMTRI Res. Rev. **31** (4) (2000)

Finite Element Analyses of the Influence of Tyre Functional Parameters on Contact Patch Stresses

Alexandra-Raluca Moisescu(✉) and Gabriel Anghelache

University Politehnica of Bucharest, Splaiul Independentei 313, 060042 Bucharest, Romania
raluca.moisescu@upb.ro

Abstract. The study of contact patch stresses represents an important field of research with respect to automotive safety, as well as tyre and road durability, but also concerning vehicle comfort. The finite element method is an important tool for the investigation of tyre-road contact that provides valuable information regarding the generation of forces in the contact patch. The current paper presents an investigation performed on the distribution of stresses developed in the contact patch, using the finite element model of a 11R22.5 radial truck tyre. The analyses performed on the tyre model have emphasized the influence of various functional parameters, such as inflation pressure, camber angle and rolling speed, on the distributions of the contact patch stresses. The inflation pressure has an important effect on contact patch size and normal stress magnitudes. At lower inflation pressure the higher values of shear stress are applied on larger areas, suggesting an increase in the intensity of slip phenomena. The camber angle has an influence on the contact patch shape, as well as on the asymmetry of contact stress distributions with respect to the longitudinal plane. The differences between the distributions of contact patch stresses at different speed values are very small, possibly due to the relatively low speed values. The comparison between the results obtained in static conditions and in rolling conditions has showed insignificant changes in the distribution of normal stresses, but it has revealed major differences concerning the magnitudes and distributions of longitudinal and lateral stresses.

Keywords: Truck tyre · Contact patch · Finite element analysis

1 Introduction

The study of contact patch stresses represents an important field of research with respect to automotive safety [1–3], as well as tyre and road durability [4–6], but also concerning vehicle comfort [7, 8]. Among the numerical methods used for the investigation of tyre-road contact, the finite element method is an important tool that provides valuable information regarding the generation of forces in the contact patch [9–14]. It is also used in the most recent research focusing on the development of intelligent tyres [15–17].

I. Dumitru et al. (Eds.): SMAT 2019, *The 30th SIAR International Congress of Automotive and Transport Engineering*, pp. 624–635, 2020.
https://doi.org/10.1007/978-3-030-32564-0_73

The current paper presents an investigation performed using the finite element model of a 11R22.5 radial truck tyre, on the distribution of stresses developed in the contact patch, for various tyre functional parameters. Finite element analyses have included static conditions and rolling conditions, at different values of inflation pressure and camber angle.

2 Finite Element Model of the 11R22.5 Truck Tyre

The finite element model has been developed using the coordinates measured on a real 11R22.5 truck tyre, before and after it has been sectioned, so that the location and structure of different rubber and reinforcement layers could also be determined and applied in the model. The tyre model takes into consideration the tread layers, radial carcass and beads corresponding to the real tyre construction, as well as nonlinear behaviour of rubber, using a hyperelastic neo-Hookean model [18, 19]. Meshing has been defined starting from a two-dimensional model of the tyre section, with different types of elements that correspond to the structure nonuniformity and anisotropy, and then developing into the tyre three-dimensional model with circumferential grooves and with non-uniform circumferential meshing density. Boundary conditions have been applied to achieve rim mounting, tyre inflation and contact with the road surface, including the friction coefficient [20]. Thorough information regarding the finite element model has been presented in [21–23].

The finite element analyses with inflation pressure applied on the 11R22.5 truck tyre model provided the section shape and width of inflated tyre, allowing partial validation of finite element model with respect to experimental results. These analyses have been performed at inflation pressures of 800 kPa, 700 kPa and 400 kPa corresponding to the conditions in which the real tyre dimensions have been measured.

Figure 1 shows the displacements of the two-dimensional model of truck tyre obtained from finite element analysis performed in Abaqus at 800 kPa inflation pressure without road contact. The undeformed contour of the tyre model is superimposed with dotted line on the deformed contour, so that the change in section shape and dimensions can be observed. Figure 2 shows the displacements of the three-dimensional model of truck tyre obtained from finite element analysis performed also at 800 kPa inflation pressure without road contact.

When inflation pressure is applied, the section shape of the finite element tyre model becomes wider, simultaneously with a slight increase of outer diameter. The increase in section width is more noticeable in the area near the beads, where a slight rotation also takes place. The increase of tyre radius occurs predominantly on the central rib. The outer ribs have very small displacements, oriented outwards for the inner edge of the rib, and respectively inwards for the outer edge of the rib, so that the curvature of tread layers tends to increase under the effect of inflation pressure.

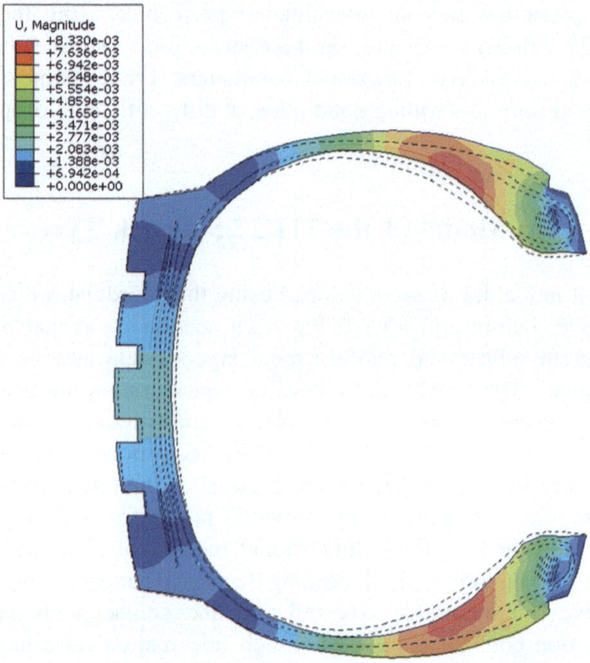

Fig. 1. Two-dimensional model of the 11R22.5 truck tyre: displacements at 800 kPa inflation pressure.

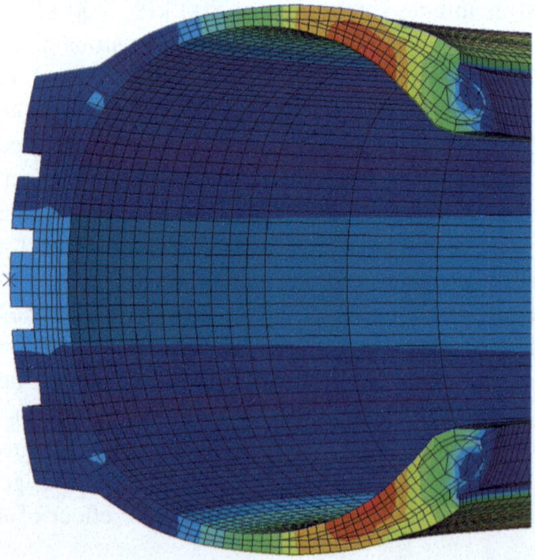

Fig. 2. Partial view of the three-dimensional model of 11R22.5 truck tyre with non-uniform circumferential meshing: displacements at 800 kPa inflation pressure.

Table 1 shows a comparison between the results of analyses performed on the finite element model of 11R22.5 truck tyre and the experimental data obtained at inflation pressures of 400 kPa, 700 kPa and 800 kPa.

Table 1. Comparison between inflated shape of FE truck tyre model and experimental results.

Inflation pressure [kPa]		400	700	800
Variation of tyre radius [mm]	FE model	0.97	1.72	1.90
	Experiment	0.95	1.11	1.20
Variation of circumference [mm]	FE model	6.1	10.8	11.9
	Experiment	6.0	7.0	7.5
Variation of tyre width [mm]	FE model	8.0	10.5	11.3
	Experiment	–	–	8.0

When the 800 kPa inflation pressure is applied, the outer radius of the tyre model increases by 1.9 mm and therefore the circumference on the tyre central rib increases by 11.9 mm, while the deformations measured on the real tyre are 1.2 mm and respectively 7.5 mm. The difference can be caused by a lower stiffness of the finite element model due to the material properties which were not available for the same truck tyre used for measurements. When the 700 kPa inflation pressure is applied, the outer radius of the tyre model augments by 1.72 mm and therefore the circumference on the tyre central rib augments by 10.8 mm, while the variations measured on the real tyre are 1.11 mm and respectively 7 mm. The best correlation is obtained when the inflation pressure of 400 kPa is applied, resulting in a 0.97 mm increase of outer radius, which represents a 2.06% difference between the results of analyses performed on the finite element model and the experimental data.

From the comparison, it can be noticed that when the 800 kPa inflation pressure is applied, the width of the tyre model increases by 11.3 mm in the sidewall area, from the value of 268.8 mm for the uninflated tyre to 280.1 mm for the inflated tyre. The experimental results show that the width increases by 8 mm when the real tyre is inflated at 800 kPa. The finite element model has larger sidewall deflections because of the slightly lower stiffness, as explained above.

3 Influence of Inflation Pressure

Finite element analyses have been performed on the model of the 11R22.5 radial truck tyre in contact with the road at different values of inflation pressure. Figure 3 shows the distributions of stresses in the contact patch of the tyre model in static conditions, with 1° camber angle and 20445 N vertical load, at 650 kPa and 780 kPa inflation pressure. These values are equivalent to the real conditions in which contact stress distributions have been measured using the complex road transducer [21, 24].

In these images only a sector of the tyre model is shown, representing the region closest to the contact with road surface. For each pair of compared images, the same limits have been used for the corresponding contour maps. The frames drawn on the distributions of normal stresses are intended to facilitate the comparison of contact patch shapes and sizes.

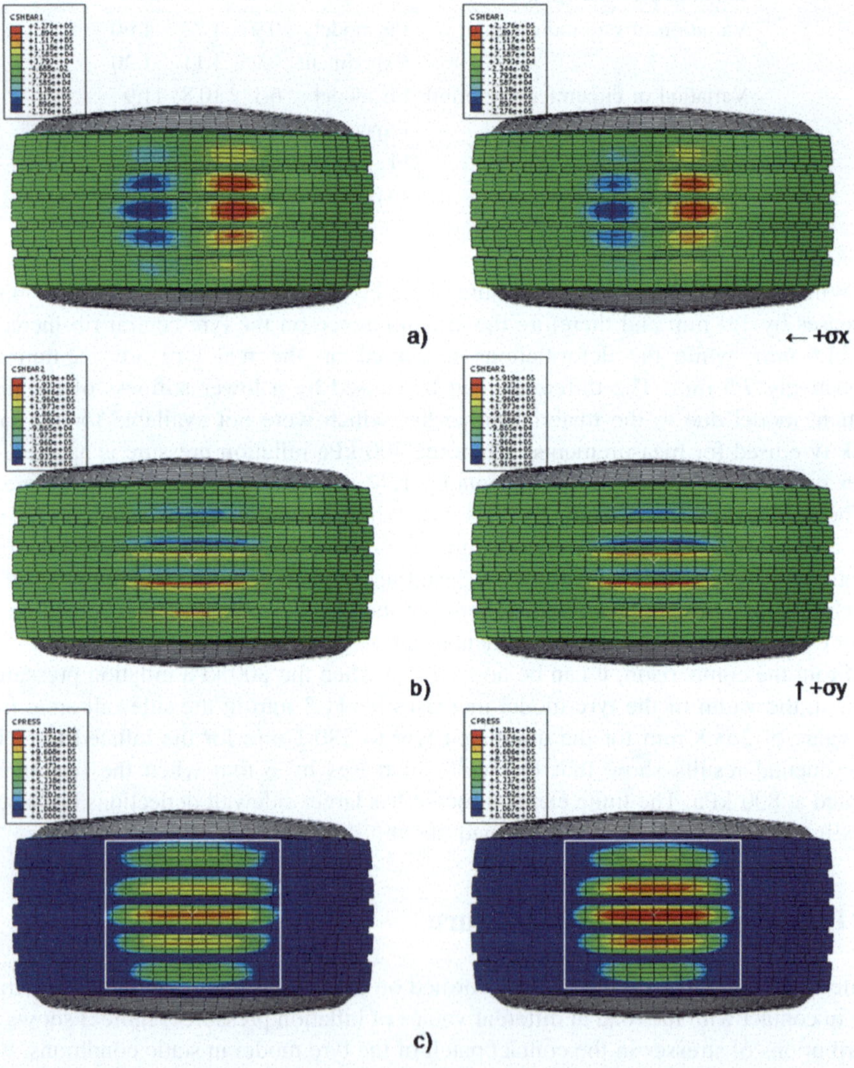

Fig. 3. Distributions of contact stresses in [N/m^2] for finite element tyre model with 650 kPa inflation pressure (left) and 780 kPa inflation pressure (right): (a) longitudinal stresses, (b) lateral stresses, (c) normal stresses.

The distributions of longitudinal stresses are antisymmetrical with respect to the lateral plane in the centre of the contact patch, while the distributions of lateral and normal stresses are symmetrical with respect to this plane. The orientation of longitudinal stresses is towards the lateral plane in the centre of the contact patch, while the overall orientation of lateral stresses is towards the longitudinal plane in the centre of the contact patch, resulting in distributions of shear stresses that converge towards the contact patch centre. However, the local orientation of lateral stresses on each rib is divergent, tending towards the nearest groove in the tyre tread.

It can be clearly noticed that when the inflation pressure is lower, the area of the contact patch is larger and the magnitude of normal stresses is significantly lower. In each of the two cases, the magnitude of normal stresses exceeds the value of the inflation pressure, as mentioned in [25].

In the case of lower inflation pressure, the high values of longitudinal stresses appear on larger areas. The distributions of lateral stresses are only slightly different in the two cases, with apparently lower stress magnitudes for the tyre model at 650 kPa.

The symmetry of stress distributions with respect to the longitudinal plane in the centre of the contact patch is affected by the presence of camber angle, whose influence will be discussed in the following section.

4 Influence of Camber Angle

The influence of camber angle on the distributions of contact patch stresses has been investigated through finite element analyses performed on the model of the 11R22.5 radial truck tyre in contact with the road. Figure 4 shows the distributions of stresses in the contact patch of the tyre model in static conditions, with 650 kPa inflation pressure and 20445 N vertical load, at 0° and 1° camber angle.

It can be noticed that the contact patch shape is influenced by the camber angle, which induces a slight asymmetry with respect to the longitudinal plane in the centre of the contact patch. The rib located on the side of the contact patch towards which the tyre inclines has higher values of normal stresses and longer contact area than the rib located on the opposite side of the contact patch. The distribution of normal stresses for the tyre without camber angle is symmetrical with respect to the longitudinal and the lateral plane, but for the tyre with camber angle it is symmetrical only with respect to the lateral plane.

The influence of camber angle is also seen on the distribution of longitudinal stresses which are higher on the rib located on the side of the contact patch towards which the tyre has camber angle. In both cases, with and without camber angle, the distribution of longitudinal stresses is symmetrical with respect to the lateral plane.

The distribution of lateral stresses in the case of tyre with 1° camber angle is characterized by the lack of symmetry with respect to the longitudinal plane and by lower stress values on the rib located on the contact patch side opposite to the camber angle.

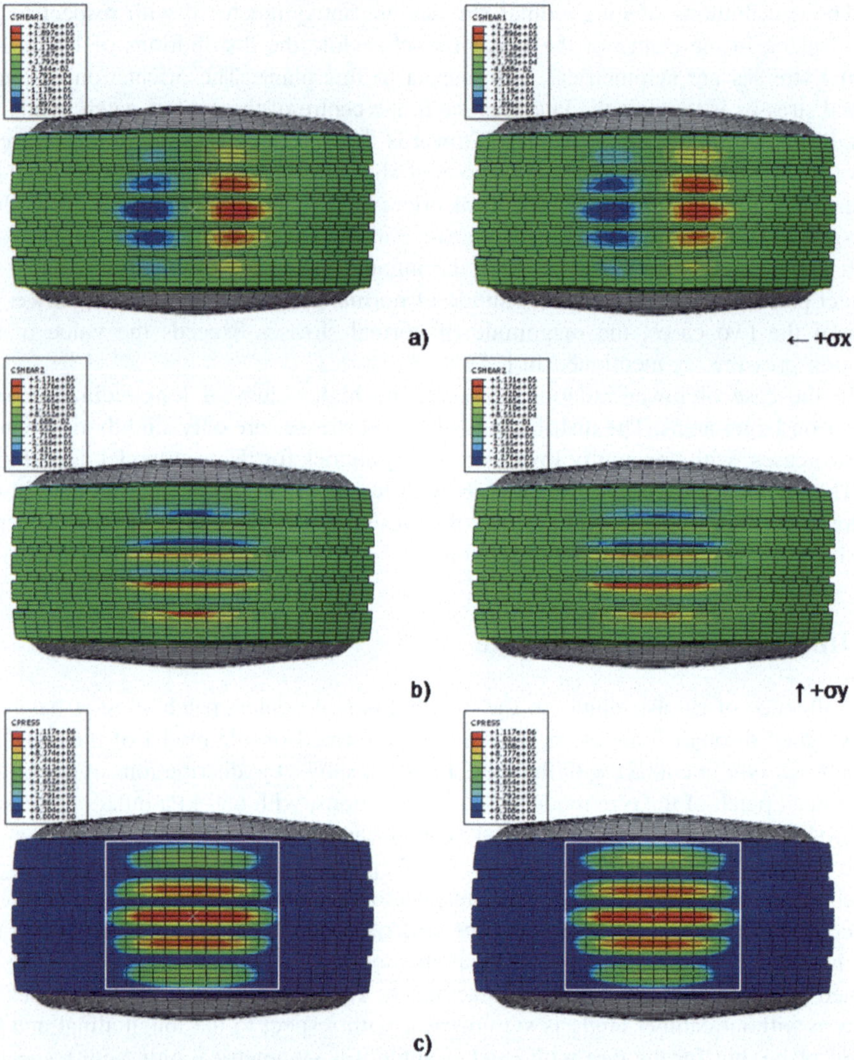

Fig. 4. Distributions of contact stresses in [N/m^2] for finite element tyre model with 0° camber angle (left) and 1° camber angle (right): (a) longitudinal stresses, (b) lateral stresses, (c) normal stresses.

5 Influence of Rolling Speed

Finite element analyses have also been performed on the model of the 11R22.5 radial truck tyre in contact with the road at different values of rolling speed. In the case of tyre in free rolling conditions, with 1° camber angle, 780 kPa inflation pressure and 20445 N vertical load, the values of 3.2 km/h and 14.5 km/h have been considered for rolling speed, corresponding to the real conditions in which contact stress distributions have been measured using the complex transducer [21, 24].

Figure 5 presents the distributions of stresses in the contact patch of the tyre model with 3.2 km/h and 14.5 km/h rolling speed, while Fig. 6 shows a comparison between distributions of stresses in static conditions and in free rolling conditions at 14.5 km/h.

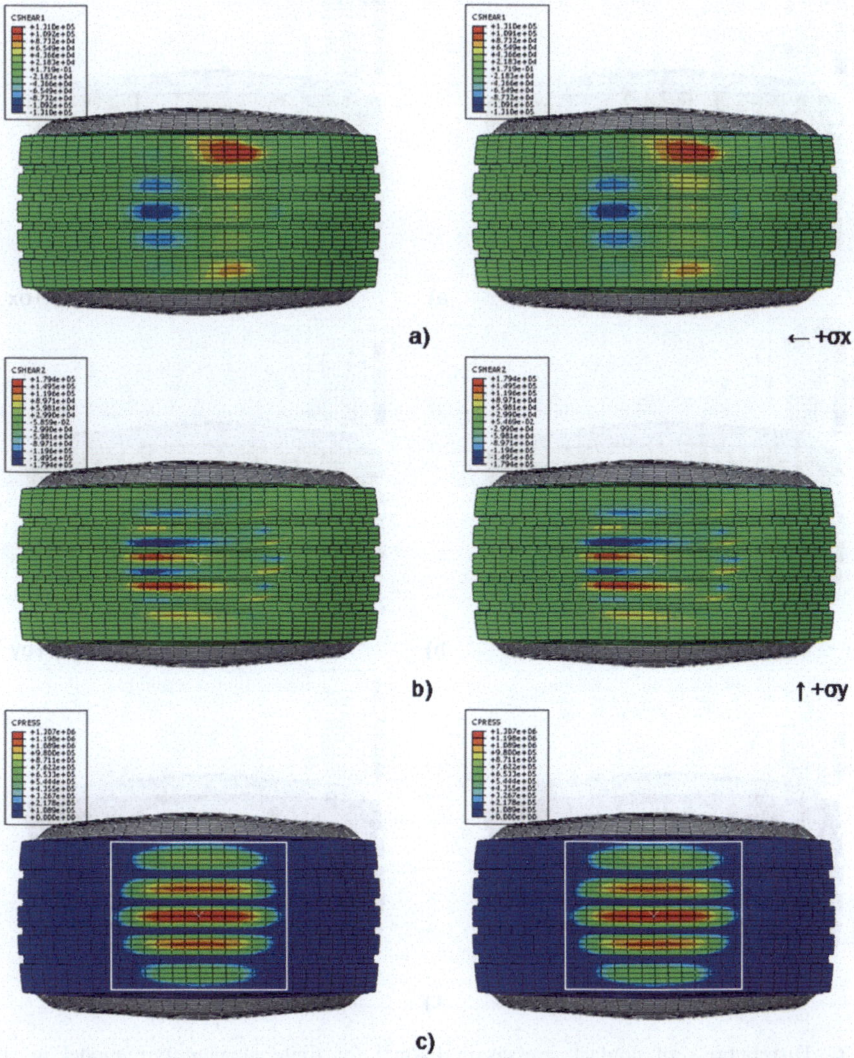

Fig. 5. Distributions of contact stresses in [N/m^2] for finite element tyre model with 3.2 km/h rolling speed (left) and 14.5 km/h rolling speed (right): (a) longitudinal stresses, (b) lateral stresses, (c) normal stresses.

It can be noticed that in the case of the two different values of rolling speed the distributions of longitudinal, lateral and normal stresses are not visibly influenced, however there are significant differences between the case of the tyre in static conditions and the case of the tyre in free rolling conditions.

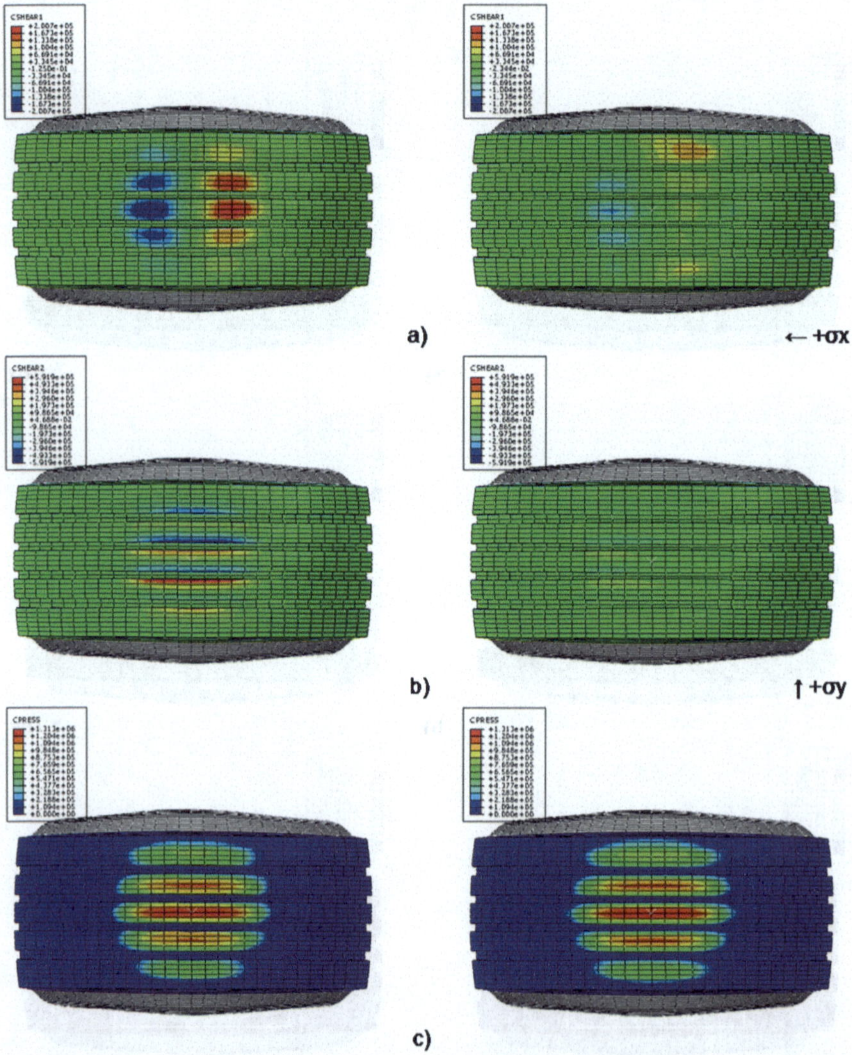

Fig. 6. Distributions of contact stresses in [N/m^2] for finite element tyre model in static conditions (left) and with 14.5 km/h rolling speed (right): (a) longitudinal stresses, (b) lateral stresses, (c) normal stresses.

The distribution of longitudinal stresses in free rolling conditions is not antisymmetrical with respect to the lateral plane in the centre of the contact patch, as in the case of static conditions. It can be noticed that in rolling conditions, the longitudinal stresses near the leading edge of the contact patch are higher on the central rib and intermediate ribs than on the outer ribs. Moreover, in rolling conditions, in the rear half of the contact patch positive values of longitudinal stress are predominant but some negative values appear near the trailing edge, completely different from the case of static conditions. Furthermore, in rolling conditions, the positive values of longitudinal stresses are higher on the outer ribs than on the central rib and intermediate ribs, unlike in the case of static conditions.

The distribution of lateral stresses in free rolling conditions is not symmetrical with respect to the lateral plane in the centre of the contact patch, as in the case of static conditions. It can be noticed that in rolling conditions, the lateral stresses in the front half of the contact patch are significantly higher than in the rear half of the contact patch, in which only some local peaks occur near the trailing edge. In static conditions, the peaks of lateral stresses are much higher and they are applied only on the rib edges near the tread grooves, while in rolling conditions the peaks are very low and they are located across the entire width of the ribs [26].

The distribution of normal stresses in free rolling conditions is not significantly different with respect to the case of static conditions, from the point of view of the magnitudes as well as the distribution across the contact patch.

6 Conclusions

The finite element analyses performed on the radial truck tyre have emphasized the influence of inflation pressure, camber angle and rolling speed on the distributions of the contact patch stresses. Thus, the effect of tyre inflation pressure is clearly visible from the point of view of contact patch size and normal stress magnitudes. Also, it has been confirmed that the normal stress peaks exceed the value of the inflation pressure. The influence of this parameter on longitudinal and lateral stresses is rather small, but at lower inflation pressure the higher values of shear stress seem to be distributed on larger areas, suggesting an increase in the intensity of slip phenomena, even in static conditions.

The influence of camber angle on the contact patch shape is clearly noticeable, as well as the effect on the asymmetry of contact stress distributions with respect to the longitudinal plane in the centre of the contact patch.

The differences between the distributions of contact patch stresses at different values of rolling speed are very small, probably due to the relatively low speed values. The comparison between the distributions obtained in static conditions and in rolling conditions has showed insignificant changes in the distribution of normal stresses, but it has revealed major differences concerning the magnitudes and distributions of longitudinal and lateral stresses.

Acknowledgement. The research activities presented in this paper were performed within the scientific research contract "Experimental and Numerical Research on Tyre-Road Interaction in View of Increasing Automotive Safety and Road Transport Efficiency", in the National Research Development and Innovation Plan – IDEAS Program, financed by National Authority for Scientific Research of the Romanian Ministry of Education, Research and Innovation.

References

1. Pacejka, H.: Tire and Vehicle Dynamics, 3rd edn. Butterworth-Heinemann, Oxford (2012)
2. Trinko, M.J.: Tire stress analysis. In: Gent, A.N., Walter, J.D. (eds.), The Pneumatic Tire, pp. 207–230, chap. 6, The University of Akron, U.S. Department of Transportation - National Highway Traffic Safety Administration, Washington DC (2005)
3. Anghelache, G.: Tyre-road interaction at high passenger car rolling speed. Ph.D. thesis, University Politehnica of Bucharest, Romania (1999)
4. De Beer, M., Fisher, C.: Stress-In-Motion (SIM) system for capturing tri-axial tyre–road interaction in the contact patch. Measurement **46**(7), 2155–2173 (2013)
5. Huang, H., Chiu, Y., Wang, C., Jin, X.: Three-dimensional global pattern prediction for tyre tread wear. Proc. Inst. Mech. Eng. Part D: J. Automob. Eng. **229**(2), 197–213 (2015)
6. Wang, C., Huang, H., Chen, X., Liu, J.: The influence of the contact features on the tyre wear in steady-state conditions. Proc. Inst. Mech. Eng. Part D: J. Automob. Eng. **231**(10), 1326–1339 (2017)
7. Negrus, E., Anghelache, G., Stanescu, A.: Finite element analysis and experimental analysis of natural frequencies and mode shapes for a non-rotating tyre. Veh. Syst. Dyn. **27**(S1), 221–224 (1997)
8. Anghelache, G., Negrus, E.M., Sorohan, S.: Radial vibrations of tire tread-band at different rolling speeds. In: FISITA World Automotive Congress 2000, F2000G354, Seoul, Korea (2000)
9. Anghelache, G., Moisescu, A.R.: Numerical and experimental research on truck tire-road contact stress. In: FISITA World Automotive Congress 2006, Paper No. F2006V162, Session 3, Yokohama, Japan (2006)
10. Hall, W., Mottram, J.T., Jones, R.P.: Tire modeling methodology with the explicit finite element code LS-DYNA. Tire Sci. Technol. **32**(4), 236–261 (2004)
11. Olatunbosun, O.A., Bolarinwa, O.: FE simulation of the effect of tire design parameters on lateral forces and moments. Tire Sci. Technol. **32**(3), 146–163 (2004)
12. Ghoreishy, M.H.R.: A state of the art review of the finite element modelling of rolling tyres. Iran. Polym. J. **17**(8), 571–597 (2008)
13. Wang, H., Al-Qadi, I.L., Stanciulescu, I.: Simulation of tyre–pavement interaction for predicting contact stresses at static and various rolling conditions. Int. J. Pavement Eng. **13**(4), 310–321 (2012)
14. Wei, C., Olatunbosun, O.A., Yang, X.: A finite-element-based approach to characterising FTire model for extended range of operation conditions. Veh. Syst. Dyn. **55**(3), 295–312 (2017)
15. Yang, X.: Finite element analysis and experimental investigation of tyre characteristics for developing strain-based intelligent tyre system. Ph.D. thesis, University of Birmingham (2011)
16. Yang, X., Olatunbosun, O., Garcia-Pozuelo, D., Bolarinwa, E.: FE-based tire loading estimation for developing strain-based intelligent tire system. SAE Technical Paper, No. 2015-01-0627 (2015)
17. Garcia-Pozuelo, D., Olatunbosun, O., Yunta, J., Yang, X., Diaz, V.: A novel strain-based method to estimate tire conditions using fuzzy logic for intelligent tires. Sensors **17**(2), 350 (2017)
18. Anghelache, G., Moisescu, R.: Analysis of rubber elastic behaviour and influence on modal properties of truck tyre. Mater. Plastice **45**(2), 143–148 (2008)
19. Hernandez, J.A., Al-Qadi, I.L.: Tire–pavement interaction modelling: hyperelastic tire and elastic pavement. Road Mater. Pavement Des. **18**(5), 1067–1083 (2017)

20. Svendenius, J.: Tire modeling and friction estimation. Ph.D. thesis, Department of Automatic Control, Lund Institute of Technology, Lund University, Sweden (2007)
21. Anghelache, G., Moisescu, R.: Truck tyre-road contact stress measurement and modelling. In: Gruber, P., Sharp, R.S. (eds.) Proceedings of the 4th International Tyre Colloquium: Tyre Models for Vehicle Dynamics Analysis, pp. 218–227, WOS 000375226400026, University of Surrey, Guildford, United Kingdom (2015). ISBN 978-1-84469-032-9
22. Moisescu, R., Anghelache, G.: Investigation of the influence of vertical force on the contact between truck tyre and road using finite element analyses. In: Proceedings of International Congress of Automotive and Transport Engineering - Mobility Engineering and Environment (CAR2017), IOP Conference Series: Materials Science and Engineering, vol. 252, no. 1, UNSP 012015. IOP Publishing, Pitesti (2017). ISSN 1757-8981
23. Moisescu, R., Anghelache, G.: Study on the modal behaviour of contactless 11R22.5 truck tyre. In: Nedelcu, D., et al. (eds.) Proceedings of ModTech International Conference - Modern Technologies in Industrial Engineering VI (ModTech 2018), IOP Conference Series: Materials Science and Engineering, vol. 400, no. 042040. IOP Publishing, Constanta, (2018). ISSN 1757-8981
24. Anghelache, G., Moisescu, R., Sorohan, S., Buretea, D.: Measuring system for investigation of tri-axial stress distribution across the tyre–road contact patch. Measurement **44**(3), 559–568 (2011)
25. Browne, A., Ludema, K.C., Clark, S.K.: Contact between the tire and roadway. In: Clark, S. K. (ed.) Mechanics of Pneumatic Tires, pp. 271–307. U.S. Government Printing Office, Washington DC (1980)
26. Moisescu, R.: Research on automotive active safety – the tyre as an element of active safety. Ph.D. thesis, University Politehnica of Bucharest, Romania (2010)

Construction of a Kart Chassis Through 3D Reconstruction Methods – Part 1 – Scanning and Alignment

Lucian Matei, Ilie Dumitru(✉), Laurentiu Racila, Mihaela Racila, Dragoş Tutunea, and Oana Otat

University of Craiova, Craiova, Romania
dumitru_ilie@yahoo.com

Abstract. The purpose of the paper is to build a 3D frame with of a kart with the opportunity to make it physically. The starting point, consists in analyzing an already existing kart chassis so that the key points and elements can be found in the final model. The methods in which this can be done are multiple, but the most efficient in terms of time, is the 3D scanning of the chassis and passing through the re-verse engineering methods, then following the 3D reconstruction with some specialized software.

Keywords: Kart · 3D scanning · Reverse engineering · 3D modelling

1 Introduction

Kart racing is a sports variant considered "open wheels" where small, open body vehicles, with four wheels called kart with or without gearboxes/gearbox, depending on the design, compete on a circuit following a set of rules [1]. Typically, these circuits are made on a smaller scale than normal racing circuits.

Kart racing is generally accepted as the most economical form of four-wheel motor sports available. As a free activity, it can be done by almost anyone, and as a sport itself, it is one of the sports regulated by the FIA (under the name of CIK), allowing licensed races for anyone from the age of 8 years.

The chassis is made of tube profiles (often chromed, aesthetic tendency). There is no suspension, so the chassis must be flexible enough to operate as a suspension and rigid enough not to break or yield upon return [2]. All chassis approved by the International Karting Commission - International Automobile Federation or CIK-FIA are "straight" and "open".

Karting chassis are generally classified into 4 broad categories:

- Open – karts do not have a protective cage;
- Closed – karts have a cage around the pilot; they are mostly used on other types of circuits;
- Right – the pilot is in the center. The straight chassis are used for sprint racing;
- Unbalanced – the driver is on the left side [3]. This type of chassis is generally used for races that have curves predominantly to the left.

I. Dumitru et al. (Eds.): SMAT 2019, *The 30th SIAR International Congress of Automotive and Transport Engineering*, pp. 636–642, 2020.
https://doi.org/10.1007/978-3-030-32564-0_74

2 3D Scanning of the Art Chassis

3D scanning technology is the application of scientific knowledge for analyzing an object (or the environment) by collecting data on its shape and sometimes its appearance (color, texture) to build a 3D digital model. The device used to perform this type of analysis is a 3D scanner.

There are many types of equipment used to measure 3D objects. 3D scanners, dedicated to the acquisition of digital forms, can be classified into two types of equipment: 3D contact and non-contact scanners [4]. However, non-contact 3D scanners are also divided into two categories: passive and active. The scanner used for the 3D reconstruction of the current chassis is a non-contact, active scanner.

In order to probe and scan objects or the environment, active scanners emit certain types of radiation (X-ray, ultrasound) or light and detect reflection or radiation passing through the object. The light or radiation emitted by the scanner is reflected by the scanned object and sends the data back to the scanner, recording the surface distance.

2.1 Preparation of the Assembly

The frame was scanned using a 3D HandyScan. 3D scanning means getting the shape of an object so that the object to be scanned should be as clean as possible, as all defects/dirt/dust will be seen on the resulting scan data. Good preparation is essential for a good scan result (Fig. 1).

In this sense, the kart frame has been cleaned and for better digitization, any object that is shiny, black, transparent or reflective (such as mirrors and glossy or metallic surfaces) has been covered with a white powder to obtain a matte finish, achieving beautiful and accurate results.

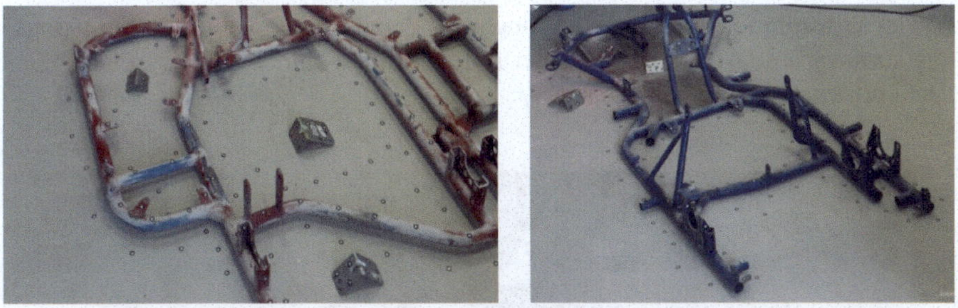

Fig. 1. Preparation of glossy surfaces for the frame scanning step

The scanner is positioned relative to the scanned object with the positioning targets. These targets have been positioned in an organized manner to ensure a surface free of empty areas and that the scanner will be well positioned at any time or will find markings at any point on the scanned surface. This is true even if randomization is required to create unique feature patterns (Fig. 2).

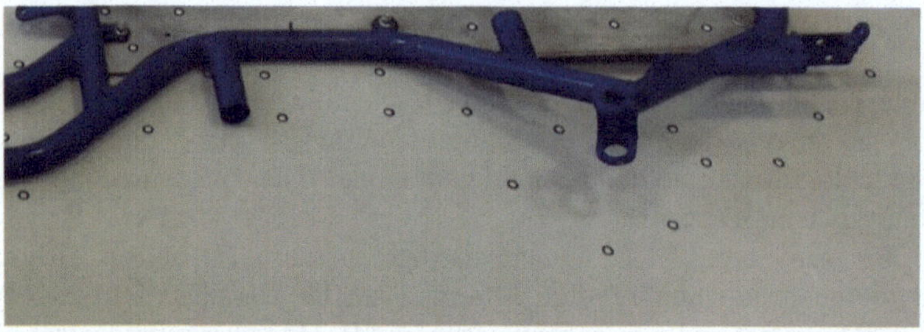

Fig. 2. Positioning of the targets for triangulation of the scanned position

2.2 Scanning the Chassis

After positioning the scan targets on the kart, the first thing that has been achieved is the positioning of the targets in space [5]. This is achieved by positioning and fixing the frame in the scanning area and simple scanning the targets (Fig. 3).

The positional targets were placed both on the kart frame and on related surfaces (floor, etc.). This was achieved because, when the angle of incidence is acute (closer to 45 °). The influence of the error in calculating the center of the ellipse is greater than in the case of a circle. Therefore, the accuracy of the positioning model will be higher with higher angles of incidence (Fig. 4).

The scanning of the frame was performed in two distinct stages:

- Scanning the upper section of the kart frame
- Scanning the lower section of the kart frame

For both scanned sections, the positional targets were scanned in advance. At the same time, the following details were considered:

- The scan resolution is 1 mm (the roughness details or the natural deformations of the surfaces are taken by the scanner starting with 1 mm)
- Data retrieval speed is 2.42 ms

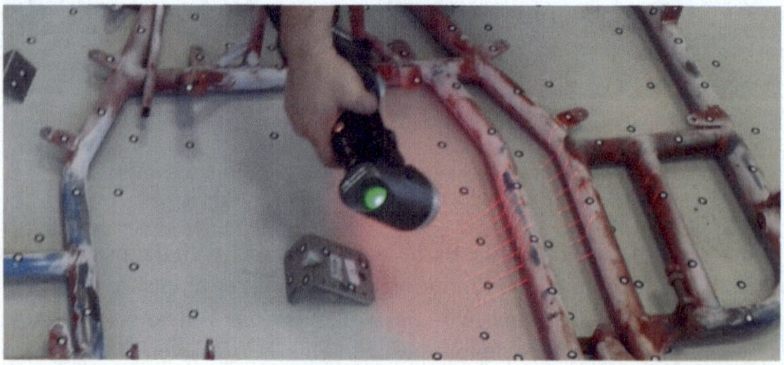

Fig. 3. Positioning of laser lines on the scanned surface

In order to have all the details needed for further processing (creating a 3D model based on scanning), it was necessary to scan in two sections:

- Scanning at the top of the kart chassis

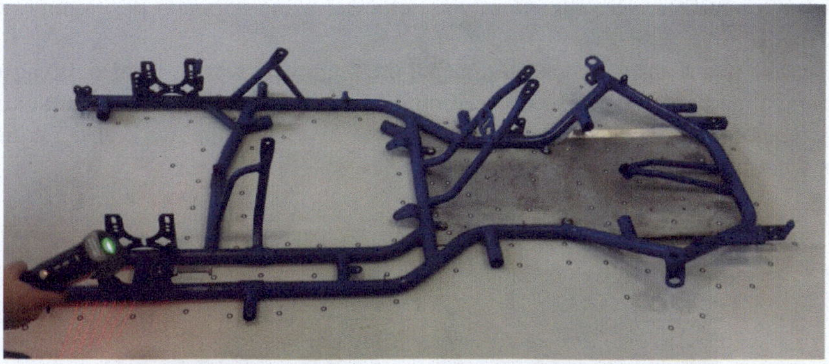

Fig. 4. The scan position for the top of the chassis

- Scanning at the bottom of the kart chassis

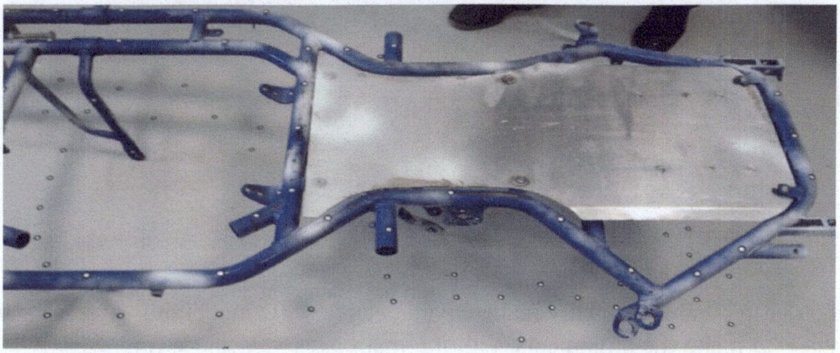

Fig. 5. The scan position for the bottom of the chassis

The results of the two scans being a discretized object for each scanned model as in the figures below (Fig. 5).

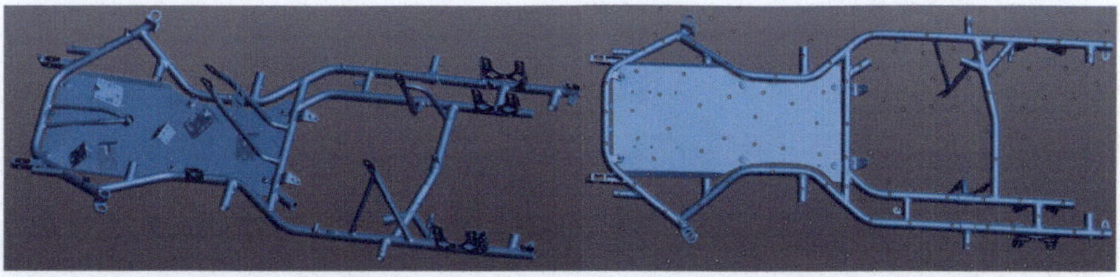

Fig. 6. Capture from VxElements of the discretized object

2.3 Merge the Scanned Sections and Align the Scans

Because a very large area scan was performed, it was necessary to scan multiple sessions. Following these scans or created two scanning files (discretization's) thus creating the need to combine scans into a single session, thus generating a single mesh for the entire surface. The alignment was achieved using common positioning objectives (Fig. 6).

After the two scans were loaded, manual pre-alignment was performed, because the automatic alignment is performed based on common targets, these targets not existing in this type of scan [6]. In this regard, for a better overlap of the two scans, 5 common points were taken (Fig. 7).

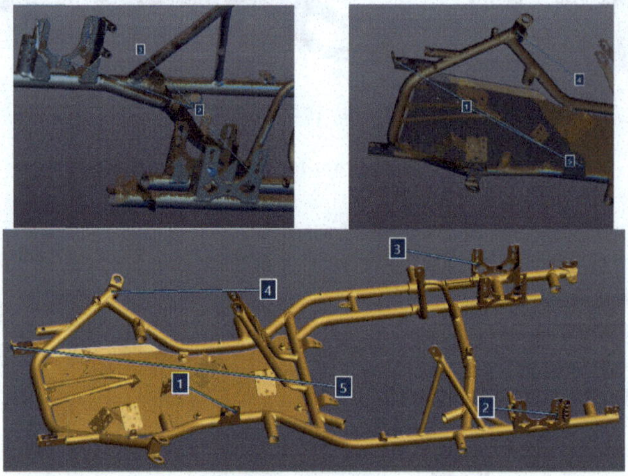

Fig. 7. Overview of the alignment points

In order to be able to finally make a parameterized 3D model, it is necessary in the first phase to make the alignments with the main axes x, y and z (Fig. 8).

This alignment was achieved by creating geometries based on the scanning construction, i.e. the discretized structure. These geometries are represented by 4 flat geometric shapes and 3 cylindrical geometric shapes, each of which can be seen in the images below.

Fig. 8. Creating geometric shapes for aligning the frame with the x, y and z coordinate axes

Following the creation of the basic geometries for alignment, it was possible to generate an alignment for the x, y and z axes, so that in the next stage, the 3D reconstruction of the frame, the bases of the construction plans can be realized according to the geometric positions in space (Fig. 9).

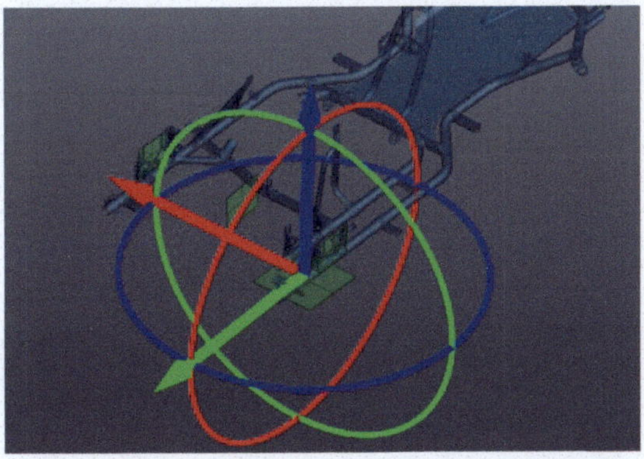

Fig. 9. Capture of the final alignment

3 Conclusion

3D scanning offers many returns, such as: fast and easy to use, comparing results with initial ones, precise and non-destructive, very useful for equipment's that cannot be measured by classical or adaptive methods.

The 3D scanning method used in this paper offers the opportunity to preleverage the geometrical features, in a discretized format, of a professional kart chassis.

The advantage of this method is that, through traditional methods, certain geometries cannot be measured or calculated so that the only way to design a chassis based on a professional one is to perform several comparative tests and assessments that are generally expensive in terms of technicality as well as time.

Acknowledgements. We thank CAD WORKS International SRL for the support and help given in the elaboration of this work.

References

1. Roşca, V., Geonea, I.D.: Romanescu Alina, Rezistenţa materialelor, vol.1, Solicitări simple
2. Teorie şi aplicaţii rezolvate, Editura SITECH, Craiova (2015)
3. Roşca, A.S., Dumitru, I.: CAD Applications in Automotive Engineering (in Romanian), Ed. Universitaria, Craiova (2016)
4. https://www.cadworks.ro/solutii/scan-3d/handyscan-3d/
5. Racila, L.: The Base of Strength of Materials: Theory and Applications (in Romanian), Ed. Universitaria, Craiova (2011)

6. Jovanov, D., Vollpracht, H. J., Beles, H., Popa, V., Tolea, B.A.: Most common road safety engineering deficiencies in South Eastern Europe as a part of safe system approach. In: International Congress of Automotive and Transport Engineering - Mobility Engineering and Environment (CAR2017). IOP Conference Series-Materials Science and Engineering, vol. 252, pp. 12–21. IOP Publishing LTD., England (2017). Article Number: UNSP 012021, ISSN: 1757-8981. https://doi.org/10.1088/1757-899x/252/1/012021

Construction of a Kart Chassis Through 3D Reconstruction Methods – Part 2 – Reverse Engineering

Lucian Matei, Ilie Dumitru, Laurentiu Racila(✉), Mihaela Racila, Dragoş Tutunea, and Oana Otat

University of Craiova, Craiova, Romania
racila_laurentiu@yahoo.com

Abstract. The second part of the paper aims to resolve the reconstruction process of the chassis. The method by which the 3D scanned chassis will be reconstructed is through the utilization of the reverse engineering methods with the help of specialized software. The base point of the reverse engineering is to utilize a discretized 3D model, which in this case was created by 3D scanning. Typically, reverse engineering it is used to analyzing in detail the model, in order to create a new device or similar system, which does not copy anything from the original one.

Keywords: Kart · 3D scanning · Reverse engineering · 3D modelling

1 Introduction

Reverse engineering is the process that identifies the technological properties of the object, device or system, performing a comprehensive analysis of its structure, functions and operations [1]. In mechanical engineering, this process aims to create a virtual 3D model from an existing physical object to duplicate or enhance it.

There are many reasons to use reverse engineering of physical objects. For example, the reverse engineering process will be used if the original design is not supported by sufficient or adequate documentation or if the original CAD model is not adequate to support standard production modifications and/or methods. In some cases, the original manufacturer no longer exists and does not produce a product, but certain requirements remain for this product.

To create a 3D model of the object, the object must be measured using 3D scanning technologies (CMM, laser scanners, structured light digitizers, etc.). After the scan operation is completed, it is possible to rebuild the 3D model using 3D CAD, CAM, CAE or other software.

For the first part of the 3D reconstruction of the kart chassis, a special 3D reconstruction software was used. In order to be able to reconstruct this framework, several stages have been accomplished [2]. The purpose of this paper in the realization of the 3D chasis involves finding the vector axes of construction of each element of the frame. These vector axes are subsequently used as construction axes in Solidworks and as geometric interlocking and parametric points of the construction lines.

I. Dumitru et al. (Eds.): SMAT 2019, *The 30th SIAR International Congress of Automotive and Transport Engineering*, pp. 643–649, 2020.
https://doi.org/10.1007/978-3-030-32564-0_75

2 Reverse Engineering

2.1 Vectorized Axes of the Chassis Frame

The first step in the realization of the parameterized 3D model consists in dividing the initial discretization into areas with the same geometry as well as into the corresponding geometries along the segments as can be see o the figure below [3] (Fig. 1).

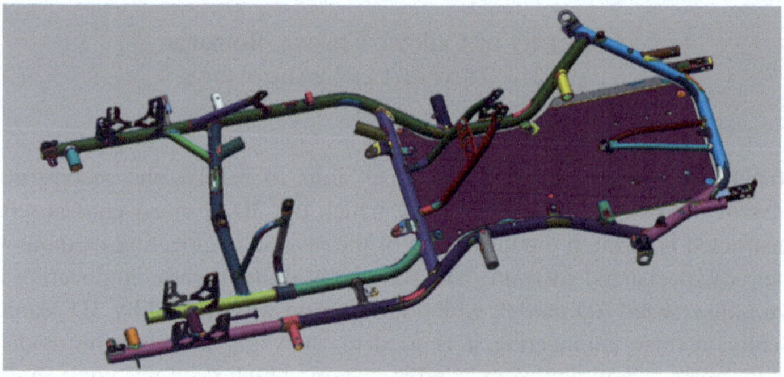

Fig. 1. Zoning similar geometries

The second stage consists in the creation of the construction plans based on which the axes of construction of the cylinders will be created, the pipes from which the metal frame is actually built. The construction plans were made for each geometric section so that more than 60 construction plans were built (Fig. 2).

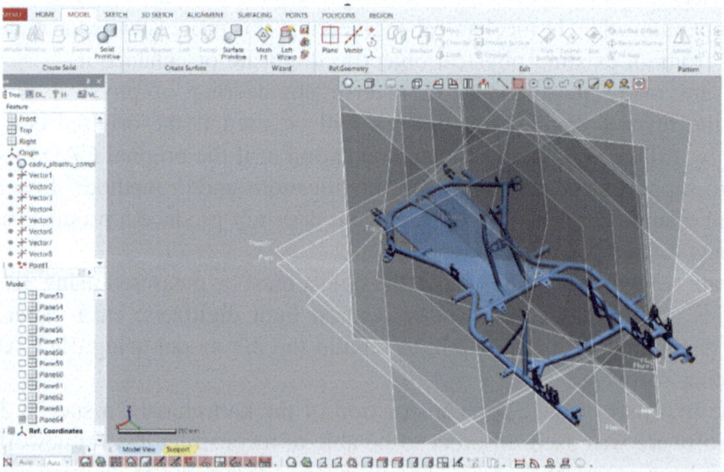

Fig. 2. Construction planes

Based on the symmetry planes built in the previous stage, the central axes of the cylinders were realized, automatically already being able to find out from this intermediate stage values of the spatial angles where the pipes are bent in order to weld the rocket support, of the support on which the pedal is attached and the front spoiler, the welding angles of the pipes for securing the pilot's seat, etc. [4] (Fig. 3).

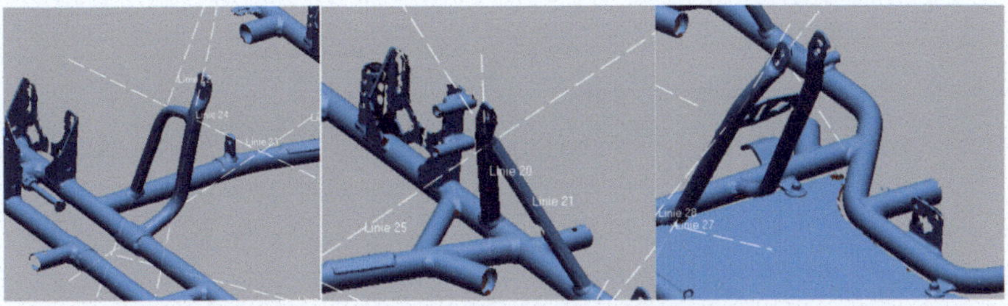

Fig. 3. Views of individual vector elements

As can be seen from the previous figures, there is a set of construction axes for each pipe profile, namely 30 × 2 mm and 16 × 2 mm [5]. At the same time, it can be observed that there are a number of 30 axes of construction of the pipes being able to build on the basis of these axes all the basic geometries of the kart frame (Fig. 4).

Fig. 4. General capture of construction vectors

2.2 Parameterized Reconstruction

Following the construction of the axes for the construction of the pipes that form the basis of the kart frame, we proceeded to the next stage of 3D reconstruction, namely to the Solidworks model [6].

Weldments elements will be used to make 3D reconstruction as easy as possible. In order to achieve this, a 3D sketch was made in which it is necessary to draw each section as simple lines. These lines were created based on the geometric axes built in the previous step (Fig. 5).

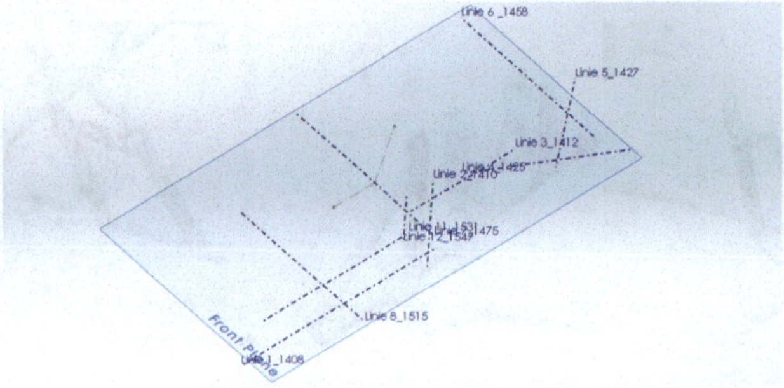

Fig. 5. Capture with the lines of construction of the 3D sketch in Solidworks

Moving on to the next step, a 3D sketch was created to start rebuilding the construction axes of the main frame pipes (Fig. 6).

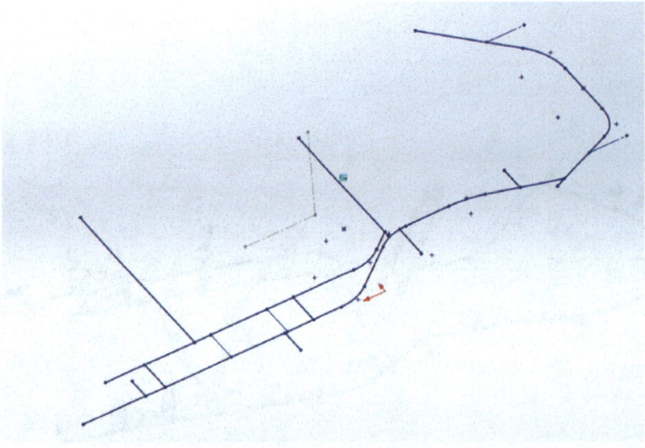

Fig. 6. The first phase of construction of the frame elements

The frame construction is further realized by creating the constructions of the elements that support the pilot's seat as well as the elements that support the steering (Fig. 7).

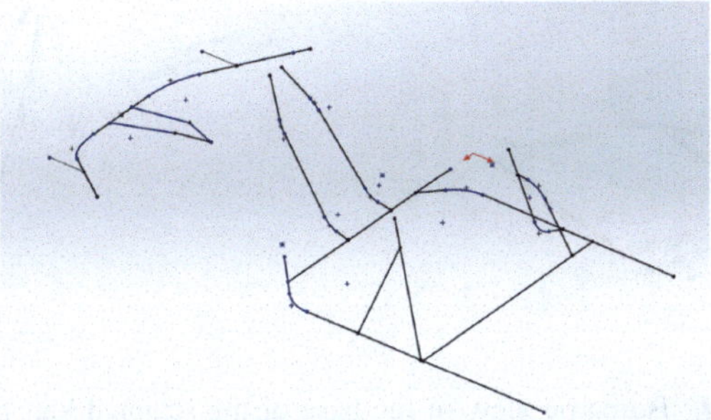

Fig. 7. Construction of the supporting elements of the pilot's seat and steering

After realizing and parameterizing the frame construction elements, the basic sketch looks like in the following figure (Fig. 8).

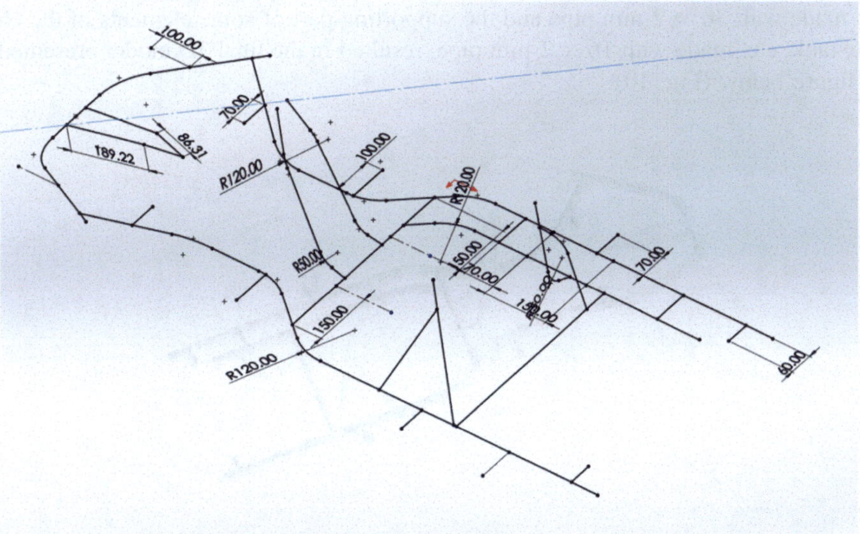

Fig. 8. Parameterizing the elements of the frame sketch

The kart framework was created with the help of the automatic pipe generation functions within the Solidworks platform. The first phase of this stage is the realization of the frame elements made of pipe type profile with dimensions 30 × 2 mm. The constructive elements and the base of the frame are in the following figures.

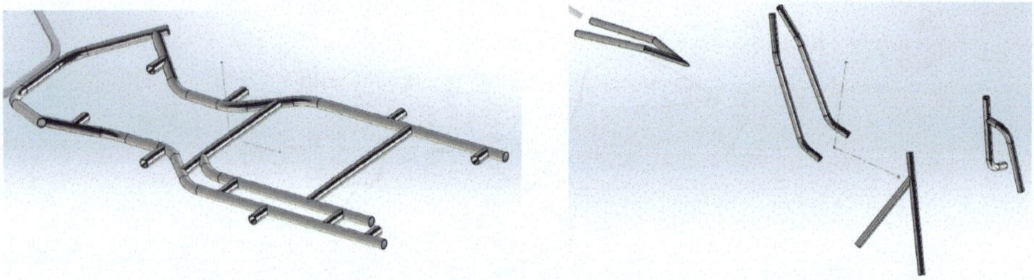

Fig. 9. Isometric view of the base of the scanned kart frame

The next step is to generate and unite the elements that support the pilot's seat and the frame's structural strength elements (Fig. 9).

At the same time, the elements related to supporting the steering column based on the frame will be joined. These elements are made of 16 × 2 mm pipe and are welded to the frame and stiffener elements.

Following the concatenation of the two distinct elements of the frame, the stiffening part made with 30 × 2 mm pipe and the supporting part of some elements of the chair type, tank, etc. made with 16 × 2 mm pipe, resulted in the final 3D model presented in the figure below (Fig. 10).

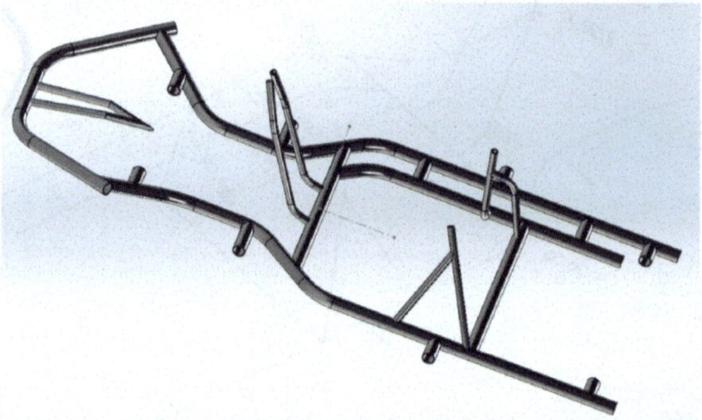

Fig. 10. Final view of the 3D model made for the kart

3 Conclusion

As computer-aided design (CAD) becomes more popular, reverse engineering is used to create three-dimensional virtual models of existing parts and subassemblies. 3D models are analyzed in CAE (computer-aided engineering) software, and finally CAM software (computer aided manufacturing).

In order to determine the dimensions of the physical object, 3D scanning technologies such as coordinate measuring machines (CMM), laser triangulation scanner, structured light scanner and industrial computed tomography are used. The measured data are usually represented as point clouds.

Acknowledgements. We thank CAD WORKS International SRL for the support and help given in the elaboration of this work.

References

1. Roşca, V., Geonea, I.D.: Romanescu Alina, Rezistenţa materialelor, vol.1, Solicitări simple
2. Teorie şi aplicaţii rezolvate, Editura SITECH, Craiova (2015)
3. Roşca, A.S., Dumitru, I.: CAD Applications in Automotive Engineering (in Romanian), Ed. Universitaria, Craiova (2016)
4. https://www.cadworks.ro/solutii/scan-3d/handyscan-3d/
5. Racila, L.: The Base of Strength of Materials : Theory and Applications (in Romanian), Ed. Universitaria, Craiova (2011)
6. Jovanov, D., Vollpracht, H.J., Beles, H., Popa, V., Tolea, B.A.: Most common road safety engineering deficiencies in South Eastern Europe as a part of safe system approach. In: International Congress of Automotive and Transport Engineering - Mobility Engineering and Environment (CAR2017). IOP Conference Series-Materials Science and Engineering, vol. 252, pp. 12–21. IOP Publishing LTD., England (2017). Article Number: UNSP 012021, ISSN: 1757-8981. https://doi.org/10.1088/1757-899x/252/1/012021

Materials and Technologies

Generating Loci Through the Theory of Mechanisms

Iulian Popescu, Alina Elena Romanescu(✉), and Ludmila Sass

Faculty of Mechanical Engineering, University of Craiova, Craiova, Romania
romanescualinaelena@yahoo.com

Abstract. The problems focused on loci are usually solved by using Geometry, the approaching being difficult. Moreover, due to difficulties related to geometric aspects, ordinary solutions of this type are associated only to problems generating simple loci: line segments, circle arcs or conical circle arcs. In order to find loci with complicated shapes, the geometric methods are hard to apply. Moreover, there is no clear sequence of steps (procedures) applicable for solving problems of this kind. A common approach consists in trying to apply different theorems from the classical Geometry. On the other hand, if the locus can lead to finding an equivalent mechanism (which in most of the cases is possible), then the methods from the Theory of Mechanisms can be applied and consequently the loci become trajectories (rod curves) which can be determined as full curves. The paper's starting point consists in a loci problem which is solved by using methods from the Theory of Mechanisms. Firstly the synthesis of the equivalent mechanism is made. It contains 5 dyads. By using projections one obtained the relations defining the positions of 4 points belonging to the mechanism for which loci were found. The curves associated to the found loci were particular for each distinct point. The lengths of certain elements were modified and the consequences of these actions over the loci were studied.

Keywords: Loci · Rod curves · Mechanisms

1 Introduction

The geometric methods are hard to apply when finding loci with complicated shapes is intended. The plotting of several mathematic curves by using geometrical aspects is made available by means of [1]. Numerous plane curves are presented in [2], few of them being solved as loci. A study of loci was performed by means of the Theory of Mechanisms in [3]. Many examples of loci are provided in [4]. They address centers of circles tangent to two circles, the intersection of two circles, the middle of a segment, an orthogonal projection, places in a triangle and so on. Other examples are reported by [5]. Books concerned with Geometry present solutions for many loci problems. The solution for a loci problem, involving the analysis of a mechanism with 5 dyads is provided in the below section.

I. Dumitru et al. (Eds.): SMAT 2019, *The 30th SIAR International Congress of Automotive and Transport Engineering*, pp. 653–659, 2020.
https://doi.org/10.1007/978-3-030-32564-0_76

2 The Loci Problem and Equivalent Mechanism

Given the triangle ACD, a line is traced between C and a point B belonging to AD. Two perpendicular lines are plotted: BG is perpendicular to CD and CE is perpendicular to AD. The loci corresponding to the points D, G, E and F (Fig. 1) are to be determined, considering that C slips along AC. Three rectangular triangles can be noticed in Fig. 1.

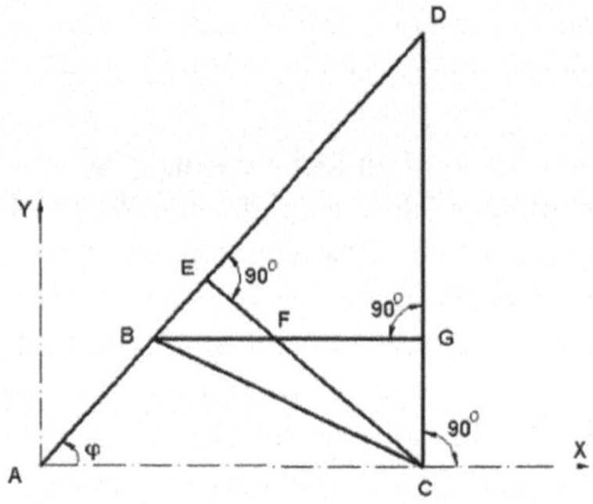

Fig. 1. The locus.

Below are given the steps performed by the mechanism's synthesis (Fig. 2): the rectangular triangle ADC was built; the line ABED becomes leading element with variable length and therefore in D one placed the coulisse 5, which is linked to the coulisse 4 by means of a rotation coulisse (4 slides along DC). The point C hosts the coulisse 3, welded at 90° on the line DC, such as to have the point sliding long the axis Ox. In order to have an angle of 90°, the double coulisse 11 was placed. It allows for a variable length of the segment 10 along with the mobility of G along DC. Similar, the double coulisse 9 was placed in order to have an angle of 90° in E. Therefore the element no. 6 (of variable length) can slide through 9. The point representing the intersection between CE and BG was represented by two coulisses able to rotate one around the other.

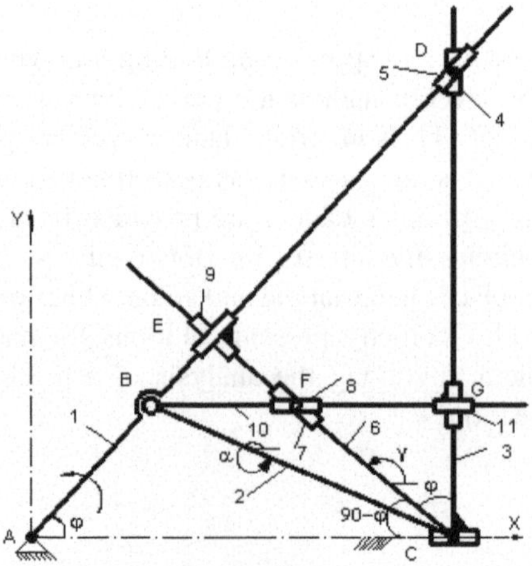

Fig. 2. The equivalent mechanism.

The mechanism has 11 mobile elements with 16 couples of class V and therefore the mobility degree is:

$$M = 3 \times 11 - 2 \times 16 = 1 \tag{1}$$

The structural schematic and its decomposition in kinematic groups is given in Fig. 3. The mechanism has a leading element (1) and 5 dyads: BCC (of type RRP), BGG (RPP), CEE (RPP), FFF (PRP), DDD (PRP).

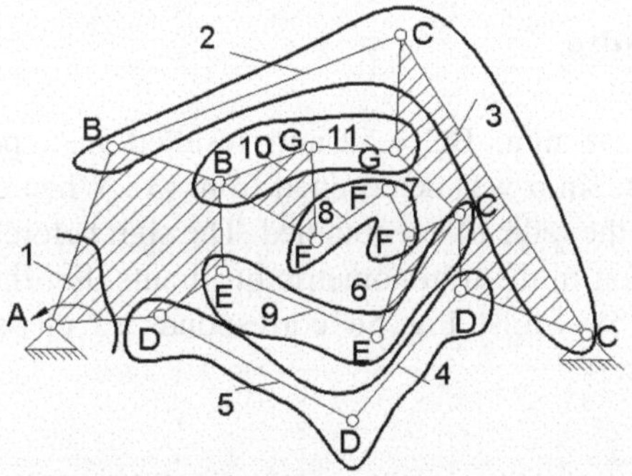

Fig. 3. The mechanism structure.

3 Relations

The following relations can be written based on Fig. 2:

$$\begin{cases} x_B = AB cos\varphi \\ y_B = AB sin\varphi \end{cases} \tag{2}$$

$$\begin{cases} x_C = x_B + BC cos\alpha \\ y_C = y_B + BC sin\alpha = 0 \end{cases} \tag{3}$$

$$\begin{cases} x_D = AD\, cos\varphi = x_C \\ y_D = AD sin\varphi \end{cases} \tag{4}$$

$$\begin{cases} x_G = x_C \\ y_G = y_B \end{cases} \tag{5}$$

$$\begin{cases} x_E = AE cos\varphi = x_C + CE cos\gamma \\ y_E = AE sin\varphi = CE sin\gamma \end{cases} \tag{6}$$

$$\gamma = 90^0 + \varphi \tag{7}$$

$$\begin{cases} x_F = x_B + BF = x_C + CF\cos\gamma \\ y_F = y_B = CF\sin\gamma \end{cases} \quad (8)$$

The position of B can be deduced from (2), the angle α can be computed based on (2) whilst x_C can be derived from the Eq. (2). AD and y_D can be determined from the Eq. (4). The position of G can be deduced from (5) and the Eq. (6) provide the formula to determine AE, CE, x_E, y_E, where γ is determined in a geometric way from (7). The Eq. (8) allow for the determination of BF, CF, x_F, y_F.

4 Graphical Results

The initial data AB = 48 mm, BC = 52 mm resulted in a position of the control mechanism from Fig. 4. sin α was determined from (3). When determining cos α, two solutions (for ± before the radical) are obtained. The sign before the radical determined the mechanism's position in the trigonometric quadrants. For the sign (+) one gets the mechanism from Fig. 5a whilst Fig. 5b corresponds to the sign (−). Figure 5c is obtained for both signs.

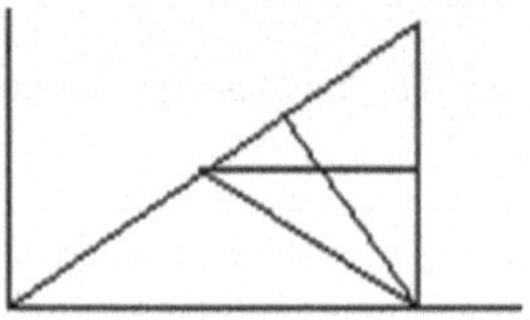

Fig. 4. The mechanism in a position.

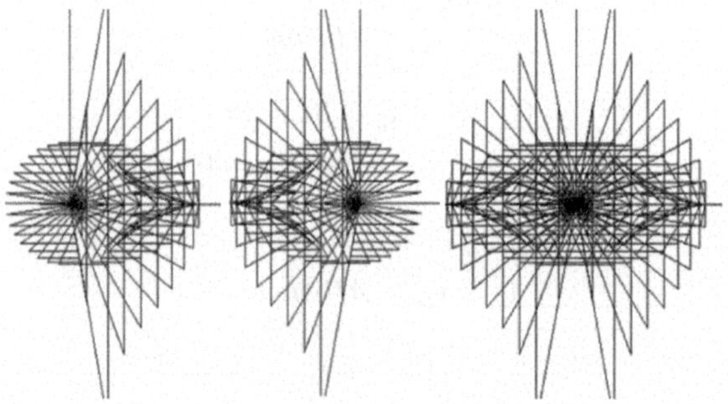

a. b. c.

Fig. 5. Mechanism successive positions. a. For the sign (+); b. For the sign (−); c. All the successive positions

The locus for the point F is given in Fig. 6a (full curve). The locus (with a shape similar to spectacles) of the point G is the curve from Fig. 6b. It is interesting to notice that it includes segments almost straight. Therefore the mechanism can be considered as generating line segments as well.

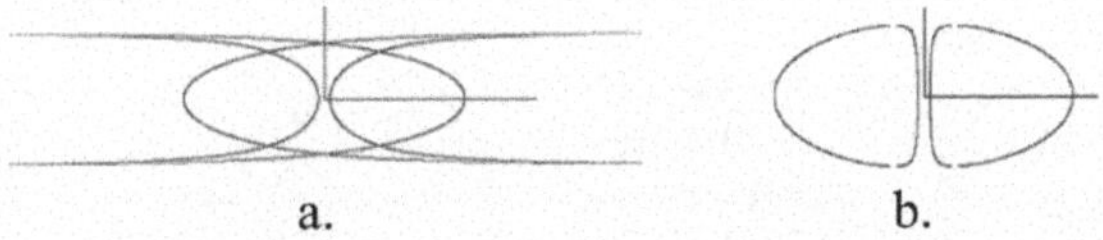

Fig. 6. Loci. a. Locus of the point F; b. Locus of the point G.

The curves obtained as loci of different points are also affected by signs. For example the curve described by the point D and obtained for the sign (+) is depicted in Fig. 7a whilst that corresponding to the sign (−) is depicted in Fig. 7b. The full curve is represented in Fig. 7c. The curves are symmetric relativ to the oy axis.

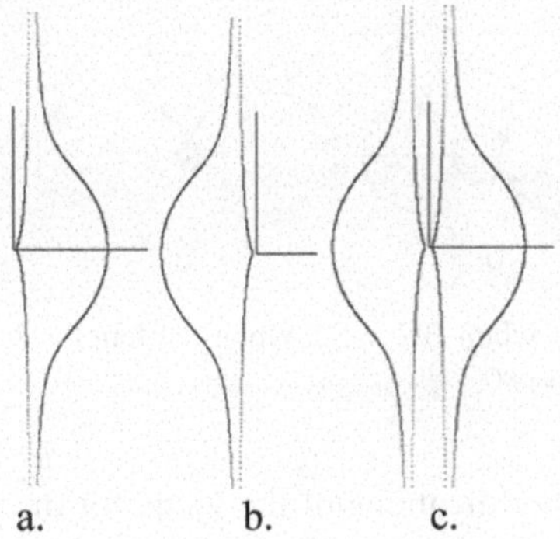

Fig. 7. Curves described by the point D. a. For the sign (+); b. For the sign (−); c. Full curve.

The locus of E is represented in Fig. 8a. This presents interest in its middle area, where a shape very similar to a rectangle can be noticed. Figure 8b represents the locus for the sign (+) (a half of the above mentioned rectangle from the interior of the right curve can be seen).

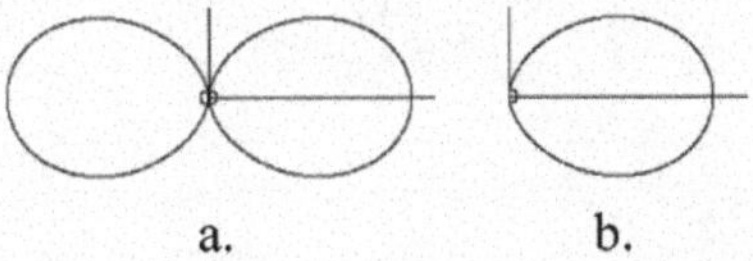

Fig. 8. Locus for the point E: a. Locus with initial data; b. Locus for the sign (+).

In order to follow the middle area, the length of the rod BC was modified. The resulting loci are given in Fig. 9. One can see that for small values of BC both branches of the curve are apart. Theu become closer once BC is increased and the rectangle from the middle is no longer visible. Two inner incomplete curves appear instead, similar to the sign ∞.

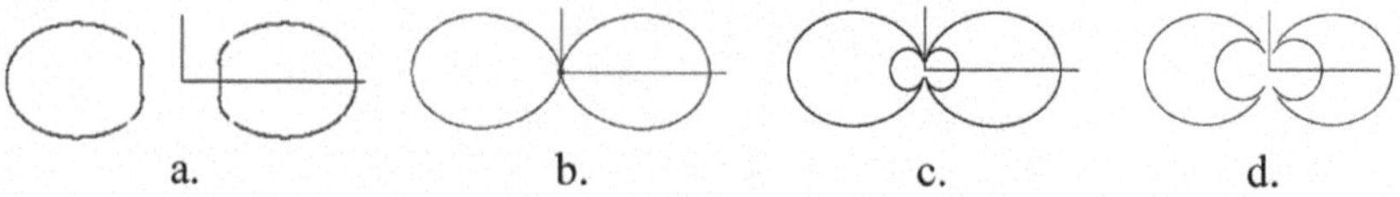

Fig. 9. Locus for the point E when BC = 52, when the length of AB is modified. BC = 30; b. BC = 50; c. BC = 80; d. BC = 120.

The influence of the rod length (AB) was also studied. The loci for E when BC = 52 mm are depicted in Fig. 10. One can see that the curves have shapes variable with AB. When AB is increased, the curves have rectangular shapes, tend to shapes which includes side cutes and finally become almost straight lines.

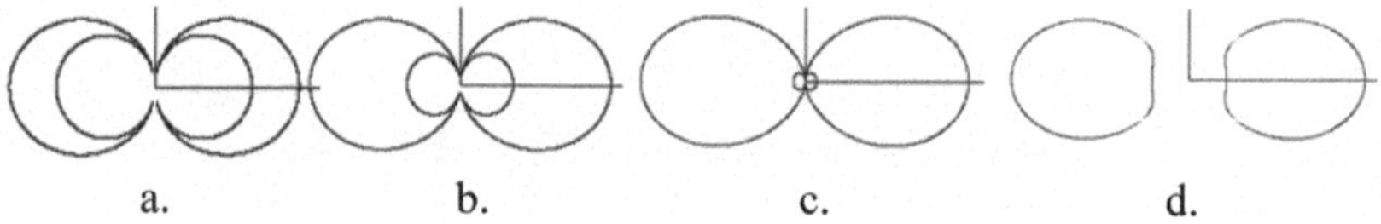

Fig. 10. Locus for the point E when BC = 52, when the length of AB is modified. AB = 10; b. AB = 25; c. AB = 45; d. AB = 80.

When modifying AB, modifications of the locus for the point F appear as well, as one can see in Fig. 11.

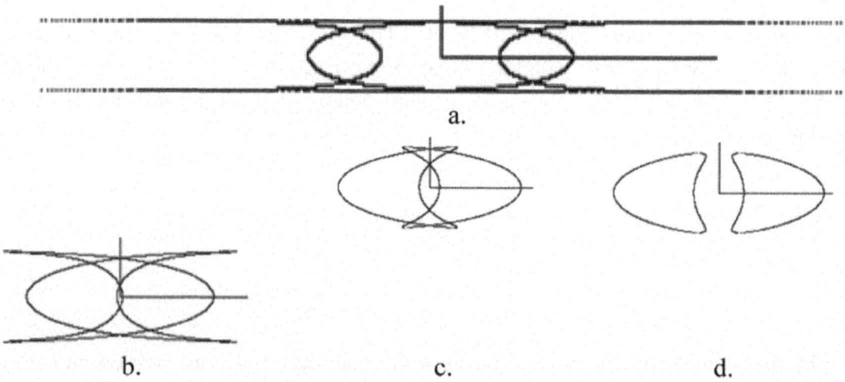

Fig. 11. Locus for the point F when BC = 52, when AB is varied a. AB = 20; b. AB = 55; c. AB = 65; d. AB = 85.

5 Conclusions

A problem with loci was solved by using methods from the Theory of Mechanisms.

The synthesis of the equivalent mechanism was made. It has a leading element and 5 dyads. The mechanism's analysis was performed. The loci for 4 points belonging to the mechanism were determined. Different resulting curves (loci) were obtained for different points. Afterward the influence of sizes corresponding to various elements from the mechanism over the loci was studied.

References

1. Artobolevskii, I.I.: Teoria mehanismov dlia vosproizvedenia ploskih krivâh. Izd. Akademii Nauk SSSR, Moskva (1959)
2. Ferreol, R.: Mathcurve. https://www.mathcurve.com/courbes2d/courbes2d.shtml. Accessed 15 Mar 2019
3. Popescu, I.: Curbe de bielă în diferite planuri şi locuri geometrice generate de mecanisme. Editura Sitech, Craiova (2017)
4. http://debart.pagesperso-orange.fr/geoplan/lieux_geometriques_classique.html. Accessed 10 Feb 2019
5. https://fr.wikipedia.org/wiki/Lieu_g%C3%A9om%C3%A9trique. Accessed 11 Mar 2019

Stereometric Analysis of Pt Schottky Contacts on Quaternary n-$Al_{0.08}In_{0.08}Ga_{0.84}N$ Thin Film

Ştefan Ţălu[1(✉)], Mihai Ţălu[2], Daniela Vintilă[2], Laura Diana Grigorie[2], and Alina Romanescu[2]

[1] The Directorate of Research, Development and Innovation Management (DMCDI), Technical University of Cluj-Napoca, Constantin Daicoviciu Street, no. 15, 400020 Cluj-Napoca, Cluj County, Romania
stefan_ta@yahoo.com

[2] Faculty of Mechanics, University of Craiova, 107 Calea Bucureşti Street, 200512 Craiova, Romania

Abstract. The purpose of this work is a stereometric analysis of Pt Schottky contacts on the quaternary n-$Al_{0.08}In_{0.08}Ga_{0.84}N$ thin film to obtain the three-dimensional (3-D) micro-texture surface based on the atomic force microscopy (AFM). The thermal annealing processes were performed in the thermal ranges 300–500 °C on the Pt Schottky contacts grown on nanostructure $Al_{0.08}In_{0.08}$-$Ga_{0.84}N$ thin film grown on a sapphire substrate using molecular beam epitaxy (MBE) technique. The samples were divided into three analysis groups, in correlation with the thermal annealing treatments applied, as follows: group I (as deposited samples); group II (samples after thermal annealing at 400 °C); group III (samples after thermal annealing at 500 °C). AFM images with a size of 9.96 μm × 9.96 μm were recorded with a scanning resolution of 256 × 256 pixels in contact mode. The AFM data of the samples were analyzed stereometrically, and the surface microtexture was characterized according to the definition of the related parameters in the ISO 25178-2: 2012 and ASME B46.1-2009 standards. The surface of group I was characterized by the highest value of the fractal dimension (D = 2.20 ± 0.01) and the lowest value of root mean square height Sq (Sq = 6.7 ± 0.1 nm); while the smallest value of the fractal dimension (D = 2.13 ± 0.01) was determined in group II. Furthermore, the highest value of Sq (Sq = 10.2 ± 0.1 nm) was determined in group III. All the characteristics of the micro-texture can be implemented in numerical programs to simulate advanced micro-texture models under specific composition and microstructure conditions.

Keywords: Fractal dimension · Quaternary AlInGaN · Stereometric analysis · Surface microtexture

1 Introduction

In the last decades the quaternary AlInGaN alloy, based on a series of advanced optical and electrical properties, they have been produced and integrated into a variety of opto-electronic devices which are used in various fields of activity, such as the military

I. Dumitru et al. (Eds.): SMAT 2019, *The 30th SIAR International Congress of Automotive and Transport Engineering*, pp. 660–669, 2020.
https://doi.org/10.1007/978-3-030-32564-0_77

industry, the telecommunications industry, the aeronautical industry, the automotive industry, automatic access control systems, and the medical industry [1–4].

In the advanced researches in this field, important experimental and theoretical results regarding the effect of thermal recovery of the Pt contact on binary GaN and ternary AlGaN have been obtained and validated [5–7].

As shown by the studies of several authors in the literature, it was found that only a small number of experiments were performed on the thermal annealing effect on quaternary AlInGaN alloy and the optimal methods for quantitative and qualitative characterization of the associated 3-D surface microtexture of the samples [8, 9].

In the current context of rapid scientific and technological progress in nanotechnology, atomic force microscopy (AFM) is the most powerful method of study and morphological characterization at the micro and nanoscale [10–14].

Atomic force microscopy as an advanced non-destructive technique used mainly in the industry as well as in surface physics research, offers the advantage over optical and electron microscopy, to present the direct measurement of the dimensions in the x, y, z-direction. Compared to other microscopy techniques, AFM can be applied in the research of samples arranged in air, liquid or vacuum which has certain advantages such as simplicity in operation, imaging ability of atomic resolution and relatively low costs [15–17].

Furthermore, stereometric [17–19] and fractal/multifractal analyses [14–16, 20–22] are new reliable modern mathematical tools used to characterize and to give a basic understanding of the nano-scale patterns in 3-D complex surfaces of thin films.

An in-depth understanding of the technological applications of these materials were correlated with their 3-D surface micromorphology features [8, 9].

To achieve the objectives of this investigation of the 3-D surface microtexture of Pt Schottky contacts on the thin film n-$Al_{0.08}In_{0.08}Ga_{0.84}N$, the AFM analysis, and the stereometric analysis were used.

2 Materials and Methods

2.1 Experimental Section

The experimental procedure to obtain Pt Schottky contacts grown on nanostructure $Al_{0.08}In_{0.08}Ga_{0.84}N$ thin film grown on sapphire substrate by molecular beam epitaxy technique has been published in detail in our previous work [9].

The morphological study of the samples was performed using an AFM Nanosurf Easyscan 2 system [23] in contact mode and in static force, at a temperature of (23 ± 1 °C) and a relative humidity (50 ± 1%), on square areas 9.96 μm × 9.96 μm squares with a CL-10 cantilever [24]. All images were acquired at 256 pixels × 256 pixels, by repeating the same experiment three times over different square scanning areas for the samples: (a) group I (as deposited); (b) group II (after thermal annealing at 400 °C); (c) group III (after thermal annealing at 500 °C).

Relevant 2-D and 3-AF AFM images of the samples taken and those obtained by annealing at temperatures of 400 °C and 500 °C, for 30 min, for 9.96 × 9.96 μm^2 scanning square areas, are shown in Fig. 1.

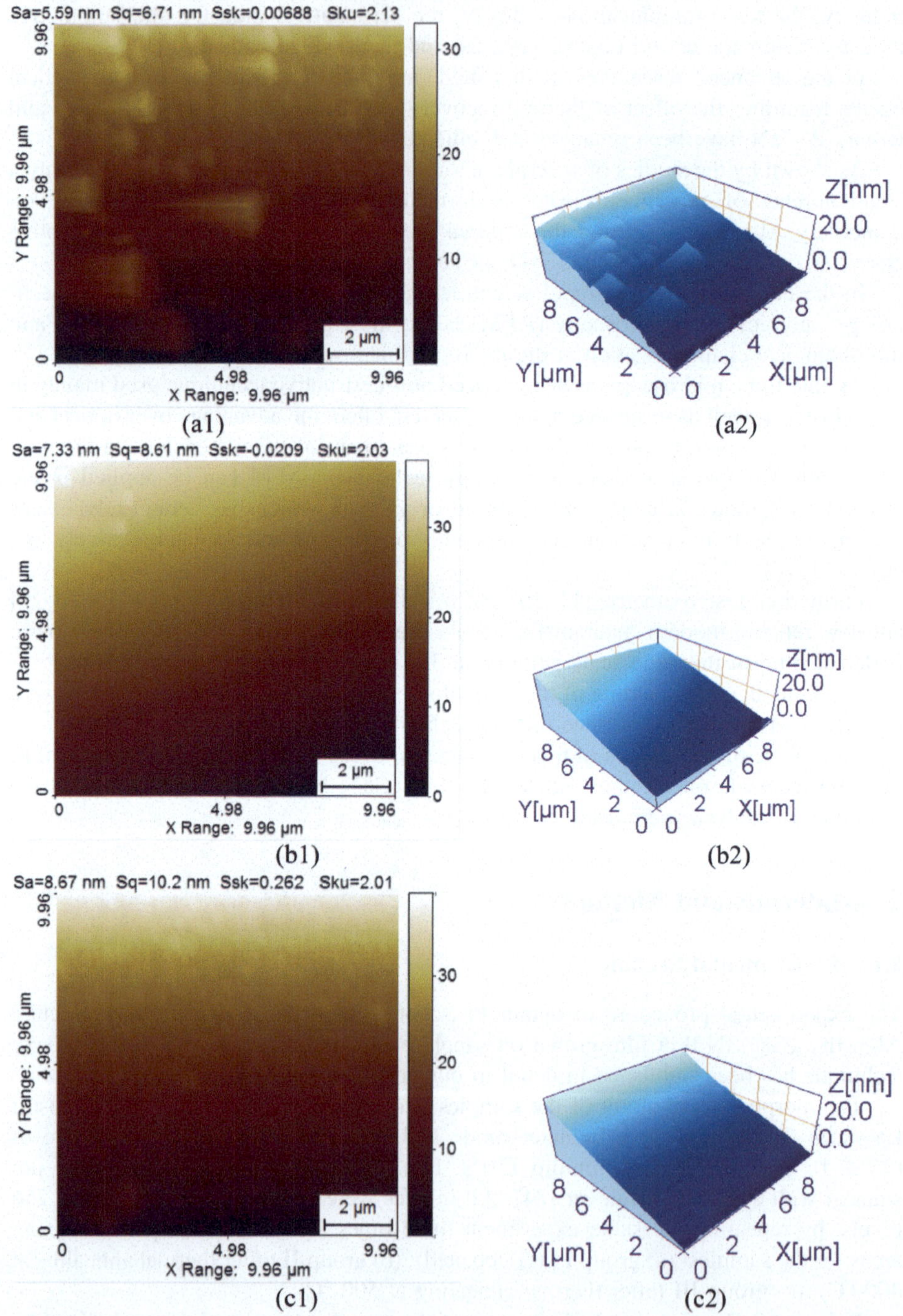

Fig. 1. Relevant 2-D and 3-D AFM micrographs of the samples: (a) group I; (b) group II; (c) group III.

3 Results and Discussion

Stereometric analysis of the samples based on the data provided by AFM software was evaluated and described using the appropriate parameters using SPIP^{TM} 6.7.4 software [25], according to ISO 25178-2: 2012 [26] and ASME B46.1-2009 [27].

The histogram of the height distribution on which the integration curve of the histogram for samples is calculated and plotted is shown in Fig. 2. The height distribution histogram gives precise data on the flatness of the surface.

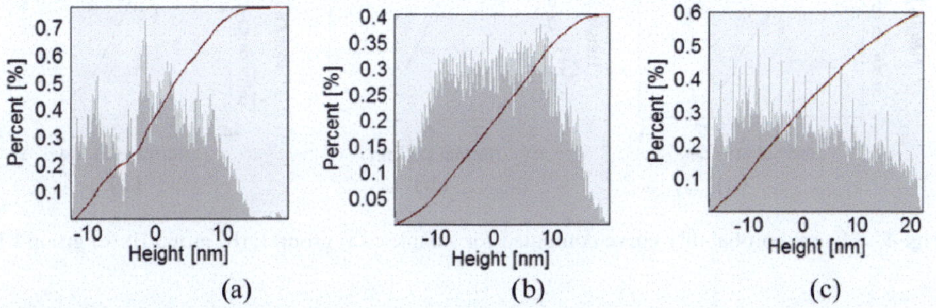

Fig. 2. The height distribution histogram with the integration curve of the histogram computed for samples: (a) group I; (b) group II; (c) group III.

Graphical representations of Abbott–Firestone curve computed for samples are shown in Fig. 3. Mathematically, the bearing ratio curve can be defined as a density function of the cumulative probability of the height of the surface profile and is obtained by mathematically integrating the profile traces [25].

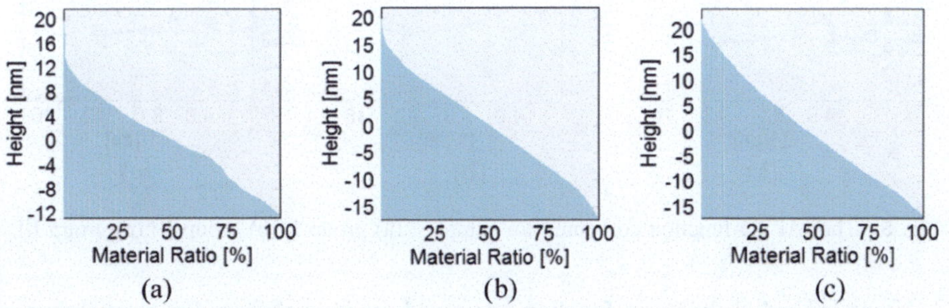

Fig. 3. The Abbott–Firestone curve computed for samples: (a) group I; (b) group II; (c) group III.

Graphical representations of material probability curve computed for samples is shown in Fig. 4. The material probability curve is defined as a graphical representation of the material ratio curve in which the heights are represented in units of standard deviations of a normal (Gaussian) distribution with the same mean and standard deviation as the height data.

It is known that a normal distribution in this type of plot is represented by a straight line. The useful of this chart is in studying of how close the distribution of heights from the analyzed surface is to a normal distribution [25].

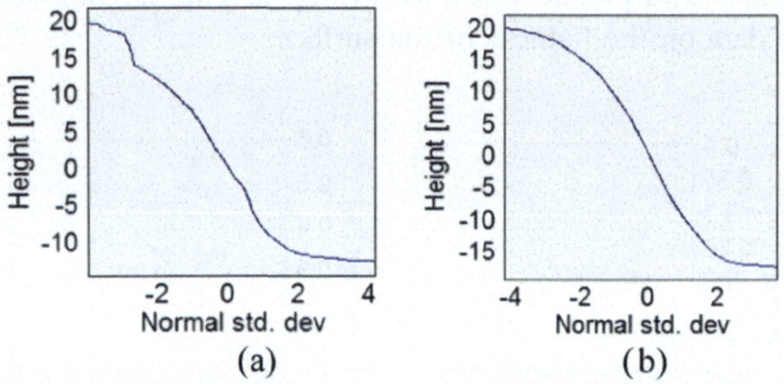

Fig. 4. Material probability curve computed for samples: (a) group I; (b) group II; (c) group III.

Isotropic Area Power Spectral Density (IAPSD) function computed for samples is shown in Fig. 5. The IAPSD function is the APSD (Area Power Spectral Density) function integrated for all angles with the given frequencies (considering inverse wavelength) [25].

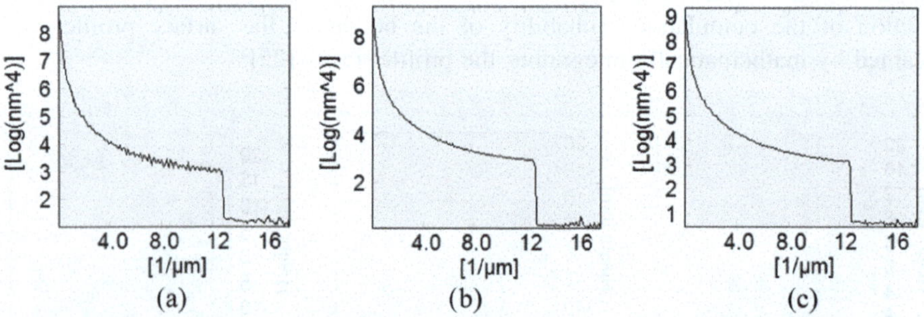

Fig. 5. The IAPSD function computed for samples: (a) group I; (b) group II; (c) group III.

Integrated radial spectrum function computed for images from Fig. 1 is shown in Fig. 6.

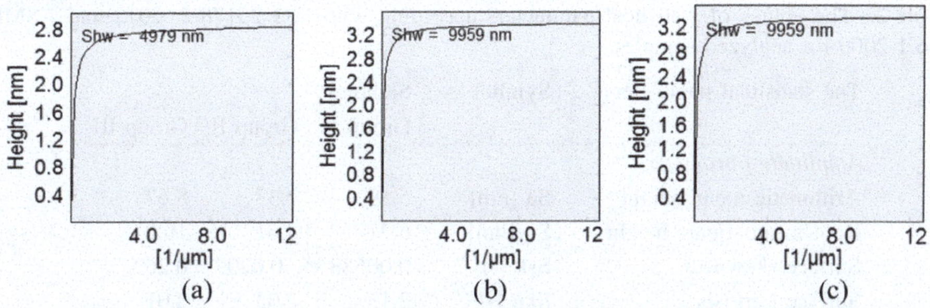

Fig. 6. The integrated radial spectrum function computed for samples: (a) group I; (b) group II; (c) group III.

The fractal dimension, S_{fd}, is calculated by considering each individual direction with the mathematical formula $S_{fd} = (6 + s)/2$, where the parameter s is associated with the (negative) slope of the log-log curves (Fig. 7) [25]. Following the average calculation for the fractal dimension for all directions, the values given in Table 1 are obtained.

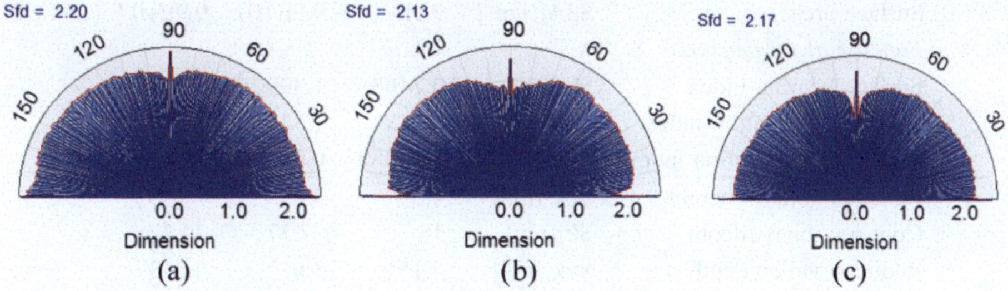

Fig. 7. The fractal dimension, S_{fd}, computed for samples: (a) group I; (b) group II; (c) group III.

Table 1. The results of fractal dimension, S_{fd}.

The fractal dimension, S_{fd}	Samples		
	Group I	Group II	Group III
Values	2.20 ± 0.01	2.13 ± 0.01	2.17 ± 0.01

*Statistically significant difference: $P < 0.05$.

In Table 2 are shown the values of statistical parameters of the AFM micrographs, according with ISO 25178-2: 2012 and ASME B46.1-2009.

Table 2. The values of statistical parameters according with ISO 25178-2: 2012 and ASME B46.1-2009 for analyzed samples.

The statistical parameters	Symbol	Samples		
		Group I	Group II	Group III
Amplitude parameters				
Arithmetic mean height	Sa [nm]	5.59	7.33	8.67
Root mean square height	Sq [nm]	6.71	8.61	10.2
Surface skewness	Ssk [-]	0.00688	−0.0209	0.262
Surface kurtosis	Sku [-]	2.13	2.03	2.01
Maximum height	Sz [nm]	32.5	37.5	39.5
Ten point height	S10z [nm]	25.88	30.88	32.78
Maximum valley depth	Sv [nm]	12.8	17.4	17.5
Maximum peak height	Sp [nm]	19.7	20.0	22.2
Hybrid parameters				
Mean summit curvature	Ssc 1/nm]	0.000612	0.000292	0.000202
Root mean square gradient	Sdq [-]	0.0125	0.00376	0.0061
Area root mean square slope	Sdq6	0.009527	0.003525	0.005794
Surface area ratio	Sdr [%]	0.00785	0.000708	0.00186
Projected area	S2A [nm^2]	9.9E+07	9.9E+07	9.9E+07
Surface area	S3A [nm^2]	9.9E+07	9.9E+07	9.9E+07
Functional parameters				
Surface bearing index	Sbi	0.7468	1.399	2.383
Core fluid retention index	Sci	1.548	1.543	1.716
Valley fluid retention index	Svi	0.07688	0.08483	0.06133
Reduced summit height	Spk [nm]	4.54	3.635	7.108
Core roughness depth	Sk [nm]	18	27.37	29.77
Reduced valley depth	Svk [nm]	3.151	3.6	0.8537
Spatial parameters				
Density of summits	Sds [1/μm^2]	12.93	0.1008	4.295
Texture direction	Std [°]	89.65°	89.65°	89.65°
Texture direction index	Stdi [-]	0.1465	0.1603	0.1209
Dominant radial wavelength	Srw [nm]	9959	9959	9959
Radial wave index	Srwi [-]	0.01798	0.01357	0.01444
Mean half wavelength	Shw [nm]	4979	9959	9959
Correlation length at 20%	Scl20 [nm]	1445	1836	1562
Correlation length at 37%	Scl37 [nm]	1093	1406	1133

*Statistically significant difference for all values: $P < 0.05$.

The qualitative values of amplitude parameters of the 3-D surface texture revealed that the smoothest surface is group I (Sa, Sq, Sz, S10z, Sv, Sp have the lower values), while the roughest surface is group III (Sa, Sq, Sz, S10z, Sv, Sp have the higher values). The surface of group I was characterized by the highest value of the fractal dimension ($D = 2.20 \pm 0.01$) and the lowest value of root mean square height

Sq (Sq = 6.7 ± 0.1 nm); while the smallest value of the fractal dimension (D = 2.13 ± 0.01) was determined in group II. Furthermore, the highest value of Sq (Sq = 10.2 ± 0.1 nm) was determined in group III.

4 Conclusions

Based on the data obtained in this study, regarding the nanometric changes in the geometry and the spatial distribution of the microtextures of the analyzed sample surfaces, it is obvious that the topography geometric features influence relation set between surface and structure, in thin film in different experimental conditions at the micrometer- and nanometer-scale. The surface topography measurement and characterization can lead to a correct understanding of manufacturing process in correlation with physical properties from both theoretically and experimentally.

Acknowledgements. The authors would like to express their appreciation to Asst. Prof. Dr. Alaa Jabbar Ghazai, Al-Nahrain University, Science College, Physics Dept., Iraq, for providing AFM data.

Competing Interests. The authors declare that they have no competing interests.

References

1. Ghazai, A.J., Thahab, S.M., Abu Hassan, H., Hassan, Z.: A study of the operating parameters and barrier thickness of $Al_{0.08}In_{0.08}Ga_{0.84}N/Al_xIn_yGa_{1-x-y}$ N double quantum well laser diodes. Sci. China Technol. Sci. **54**(1), 47–51 (2011)
2. Ghazai, A.J., Thahab, S.M., Abu Hassan, H., Hassan, Z.: Quaternary ultraviolet AlInGaN MQW laser diode performance using quaternary AlInGaN electron blocking layer. Opt. Express **19**(10), 9245–9254 (2011). https://doi.org/10.1364/oe.19.009245
3. Ghazai, A.J., Aziz, W.J., Abu Hassan, H., Hassan, Z.: Quaternary n-$Al_{0.08}In_{0.08}Ga_{0.84}N$/p-Si-based solar cell. Superlattices Microstruct. **51**(4), 480–485 (2012). https://doi.org/10.1016/j.spmi.2012.01.011
4. Ghazai, A.J., Abu Hassan, H., Hassan, Z.B.: Structural and optical properties of Si-doped $Al_{0.08}In_{0.08}Ga_{0.84}N$ thin films grown on different substrates for optoelectronic devices. Superlattices Microstruct. **95**, 95–107 (2016)
5. Rajagopal, V., Ravinandan, M., Koteswara, P., Choi, C.J.: Effects of thermal annealing on the electrical and structural properties of Pt/Mo Schottky contacts on n-type GaN. J. Mater. Sci.: Mater. Electron. **20**(10), 1018–1025 (2009)
6. Reddy, V.R., Ravinandan, M., Rao, P.K., Choi, C.J.: Electrical and structural properties of rapidly annealed Pd/Mo Schottky contacts on n-type GaN. Semicond. Sci. Technol. **23**(9), 095026 (2008)
7. Wang, J., Zhao, D.G., Sun, Y.P., Duan, L.H., Wang, Y.T., Zhang, S.M., Yang, H., Zhou, S., Wu, M.: Thermal annealing behaviour of Pt on n-GaN Schottky contacts. J. Phys. D Appl. Phys. **36**(8), 1018–1022 (2003)
8. Ghazai, A.J., Abu Hassan, H., Hassan, Z., Hussein, A.Sh.: Effects of thermal annealing of Pt Schottky contacts on quaternary n-$Al_{0.08}In_{0.08}Ga_{0.84}N$ thin film. J. Optoelectron. Adv. M. RC **6**(1–2), 324–326 (2012)

9. Ţălu, Ş., Ghazai, A.J., Stach, S., Abu Hassan, H., Hassan, Z., Ţălu, M.: Characterization of surface roughness of Pt Schottky contacts on quaternary n-$Al_{0.08}In_{0.08}Ga_{0.84}N$ thin film assessed by atomic force microscopy and fractal analysis. J. Mater. Sci. Mater. Electron. **25**(1), 466–477 (2014). https://doi.org/10.1007/s10854-013-1611-6
10. Ţălu, Ş.: Micro and nanoscale characterization of three dimensional surfaces. Basics and applications. Napoca Star Publishing House, Cluj-Napoca, Romania (2015)
11. Naseri, N., Solaymani, S., Ghaderi, A., Bramowicz, M., Kulesza, S., Ţălu, Ş., Pourreza, M., Ghasemi, S.: Microstructure, morphology and electrochemical properties of Co nanoflake water oxidation electrocatalyst at micro- and nanoscale. RSC Adv. **7**(21), 12923–12930 (2017). https://doi.org/10.1039/C6RA28795F
12. Ţălu, Ş., Stach, S., Valedbagi, S., Elahi, S.M., Bavadi, R.: Surface morphology of titanium nitride thin films synthesised by DC reactive magnetron sputtering. Mater. Sci.-Poland **33**(1), 137–143 (2015). https://doi.org/10.1515/msp-2015-0010
13. Ţălu, Ş., Solaymani, S., Bramowicz, M., Kulesza, S., Ghaderi, A., Shahpouri, S., Elahi, S. M.: Effect of electric field direction and substrate roughness on three-dimensional self-assembly growth of copper oxide nanowires. J. Mater. Sci.: Mater. Electron. **27**(9), 9272–9277 (2016). https://doi.org/10.1007/s10854-016-4965-8
14. Ţălu, Ş., Bramowicz, M., Kulesza, S., Dalouji, V., Solaymani, S., Valedbagi, S.: Fractal features of carbon–nickel composite thin films. Microsc. Res. Tech. **79**(12), 1208–1213 (2016). https://doi.org/10.1002/jemt.22779
15. Ţălu, Ş., Stach, S., Zaharieva, J., Milanova, M., Todorovsky, D., Giovanzana, S.: Surface roughness characterization of poly(methylmethacrylate) films with immobilized Eu(III) β-Diketonates by fractal analysis. Int. J. Polym. Anal. Charact. **19**, 1–18 (2014). https://doi.org/10.1080/1023666X.2014.904149
16. Elenkova, D., Zaharieva, J., Getsova, M., Manolov, I., Milanova, M., Stach, S., Ţălu, Ş.: Morphology and optical properties of SiO_2-based composite thin films with immobilized terbium(III) complex with a biscoumarin derivative. Int. J. Polym. Anal. Charact. **20**(1), 42–56 (2015). https://doi.org/10.1080/1023666X.2014.955400
17. Arman, A., Ţălu, Ş., Luna, C., Ahmadpourian, A., Naseri, M., Molamohammadi, M.: Micromorphology characterization of copper thin films by AFM and fractal analysis. J. Mater. Sci.: Mater. Electron. **26**, 9630–9639 (2015)
18. Ţălu, Ş., Stach, S., Mahajan, A., Pathak, D., Wagner, T., Kumar, A., Bedi, R.K.: Multifractal analysis of drop-casted copper (II) tetrasulfophthalocyanine film surfaces on the indium tin oxide substrates. Surf. Interface Anal. **46**(6), 393–398 (2014). https://doi.org/10.1002/sia.5492
19. Stach, S., Sapota, W., Ţălu, Ş., Ahmadpourian, A., Luna, C., Ghobadi, N., Arman, A., Ganji, M.: 3D Surface stereometry studies of sputtered TiN thin films obtained at different substrate temperatures. J. Mater. Sci.: Mater. Electron. **28**(2), 2113–2122 (2017). https://doi.org/10.1007/s10854-016-5774-9
20. Ţălu, Ş., Stach, S., Valedbagi, S., Bavadi, R., Elahi, S.M., Ţălu, M.: Multifractal characteristics of titanium nitride thin films. Mater. Sci.-Poland **33**(3), 541–548 (2015). https://doi.org/10.1515/msp-2015-0086
21. Ţălu, Ş., Stach, S., Mendez, A., Trejo, G., Talu, M.: Multifractal characterization of nanostructure surfaces of electrodeposited Ni-P coatings. J. Electrochem. Soc. **161**, D44–D47 (2013). https://doi.org/10.1149/2.039401jes
22. Ţălu, Ş., Morozov, I.A., Yadav, R.P.: Multifractal analysis of sputtered indium tin oxide thin film surfaces. Appl. Surf. Sci. **484**, 892–898 (2019). https://doi.org/10.1016/j.apsusc.2019.04.170
23. Product catalog. https://www.nanosurf.com/en/software/easyscan-2. Accessed 10 Aug 2019

24. Silicon AFM Probes. https://www.ballymore.com/cantilever-series-specification. Accessed 10 Aug 2019
25. SPIP™ 6.7.4 software (Copyright © 1998–2018 Image Metrology A/S). http://www.imagemet.com. Accessed 10 Aug 2019
26. ISO 25178-2: Geometrical product specifications (GPS) - Surface texture: Areal - Part 2: Terms, definitions and surface texture parameters (2012). http://www.iso.org. Accessed 10 Aug 2019
27. American Society of Mechanical Engineers 2009: ASME B46.1-2009, Surface Texture (Surface Roughness, Waviness and Lay), ASME. New York, NY, USA

Surface Roughness and Temperature in Dry Milling of an Austenitic Stainless Steel

Nicolae Craciunoiu(✉), Daniela Tarata, Adrian Sorin Rosca, and Ionut Geonea

Faculty of Mechanics, University of Craiova, 200512 Craiova, Romania
ncraciunoiu@yahoo.com

Abstract. In this paper some results concerning the cutting temperature and surface roughness using different milling conditions (tool material, cutting speed, speed feed and depth of cut) are presented.

In order to measure the milling temperature an experimental set-up with six thermocouples was used. All the thermocouples are adequate placed along the workpiece. After the each pass of the tool, the surface roughness was measured with a portable surface roughness tester.

The results are graphically represented and conclusions on the performed experiments are presented.

Keywords: Austenitic stainless steel · Millling · Surface roughness · Thermocouples · Temperature

1 Introduction

Austenitic stainless steel became very useful in automotive industry, many parts of the vehicles being manufactured using these types of steel.

Low conductivity and pronounced tendencies for built-up edge affect the machinability of this steel type, and from this point of view these materials are very difficult to machine.

For this reasons many authors, in their works, present different aspects regarding the cutting temperature, surface roughness, chips type, the forces magnitude developed during cutting process, residual stress, types of cutting tools etc.

In [1] the authors develop a comparative study between a heat-treated carbon steel and a stainless steel, regarding the effect of material treatment on tool temperature.

A review regarding the metallurgy, applications, machining and future research areas of austenitic stainless steel are presented in [2].

Studies regarding the tool material selection for milling of a type of stainless steel, from point of surface roughness during finishing process of these materials are presented in [3].

In [4] the authors study the surface integrity in dry milling of an austenitic stainless steel using a statistical model with three different parameters and their interaction.

A method to optimize of the cutting parameters (cutting speed, feed rate and depth of cut) in milling of a stainless steel, using a tungsten carbide end mill, is presented in [5].

I. Dumitru et al. (Eds.): SMAT 2019, *The 30th SIAR International Congress of Automotive and Transport Engineering*, pp. 670–677, 2020.
https://doi.org/10.1007/978-3-030-32564-0_78

An analysis regarding the flatness deviation and roughness parameters for austenitic stainless steel after three operations (milling, grinding, smoothing) is presented in the paper [6].

In order to obtain an optimum set of input parameters, a numerical modeling and optimization of machining duplex stainless steel is presented in [7].

Study regarding the cutting force and temperature for up-milling and down-milling for an austenitic stainless steel is in [8].

Researches about surface roughness obtained in machining of stainless steel are presented, also, in [9] and [10].

The mains purpose of this paper is to measuring and analyze, first, the temperature during milling process of an austenitic stainless steel, using a set of thermocouples, mounted along the workpiece, close to cutting area, and then, the surface roughness.

2 Experimental Details

2.1 Workpiece Material and Milling Conditions

The material used in present experiments is an austenitic stainless steel with the main mechanical properties presented in Table 1.

This material was obtained using a special technology, detailed presented in [11, 12] and [13].

Table 1. Mechanical properties of austenitic stainless steel

Mechanical properties				
$Rp_{0,2}$ MPa	Rm, MPa	A, %	Z, %	HV, MPa
538	729	61,7	70,5	213.8

The temperature during milling process of austenitic stainless steel and surface roughness after each pass of the tool, was measured for different cutting speed (v = 65.973 m/min, 32.986 m/min and 21.865 m/min). The feed speed and depth of cut was kept constant, at the values of v_f = 100 mm/min, and a_p = 1.5 mm, respectively.

A carbide end mill tool, Fig. 1, having 4 teeth and ϕ12 mm diameter, [14] was used for experimental determinations.

The part, having the dimensions 26 mm · 26 mm · 95 mm, is presented in Fig. 2. In order to connect the thermocouples with the part, six holes with ϕ2 mm diameter and 2 mm depth was machining, keeping an equal distance, 15 mm, between the holes. The distance between the center of the holes and edge part was 3.5 mm.

Machine tool used in experiments was a milling machine UWF 80 E type.

2.2 Experimental Set-Up

As it is presented in Fig. 3, to measure and analyze the temperature, six thermocouples connected to an acquisition system with six channels, and, then, a computer for storage and data processing was used.

In order to measure the surface roughness, a Portable Surface Roughness Tester – TR200 was used, [15].

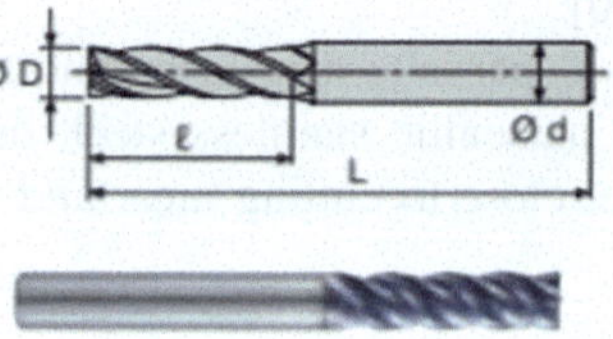

unit: mm

D(1)	ℓ	L	d	N	STOCK
2	10	50	4	4	o
3	15	50	6	4	o
4	17	50	6	4	o
6	20	60	6	4	o
8	30	60	8	4	o
10	34	90	10	4	o
12	40	90	12	4	o
16	50	115	16	4	o
20	56	125	20	4	o

N: Number of flutes Zähnezahl Nombre de dents Número de dientes

Fig. 1. Carbide end mill used in experimental determination

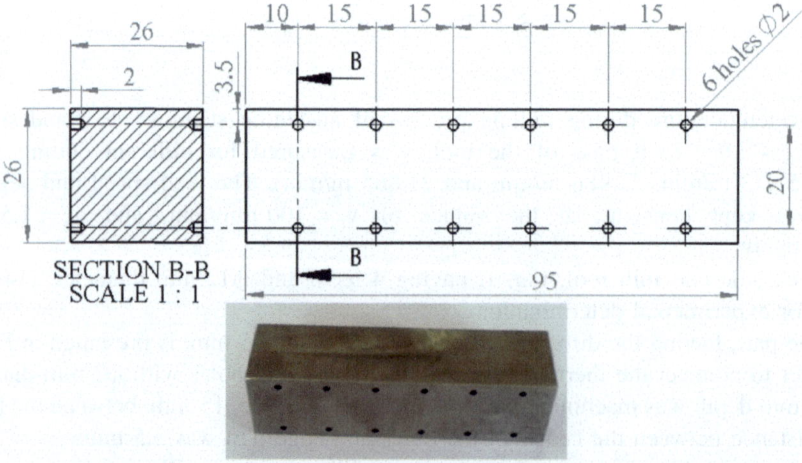

Fig. 2. The dimensions of the part used in experimental determinations

3 Results

Using an adequate acquisition system with six channels, one for each of the six thermocouples, the temperature was registered, and then, using Excel program, the graphical representations, for each case, are presented.

3.1 Milling Temperature

After each pass of the carbide end mill, the all six thermocouple recorded the milling temperature, so that the Figs. 5, 6 and 7 show the variation of the temperature, for cutting speed values: v = 65.973 m/min, v = 32.986 m/min and v = 21.865 m/min, respectively.

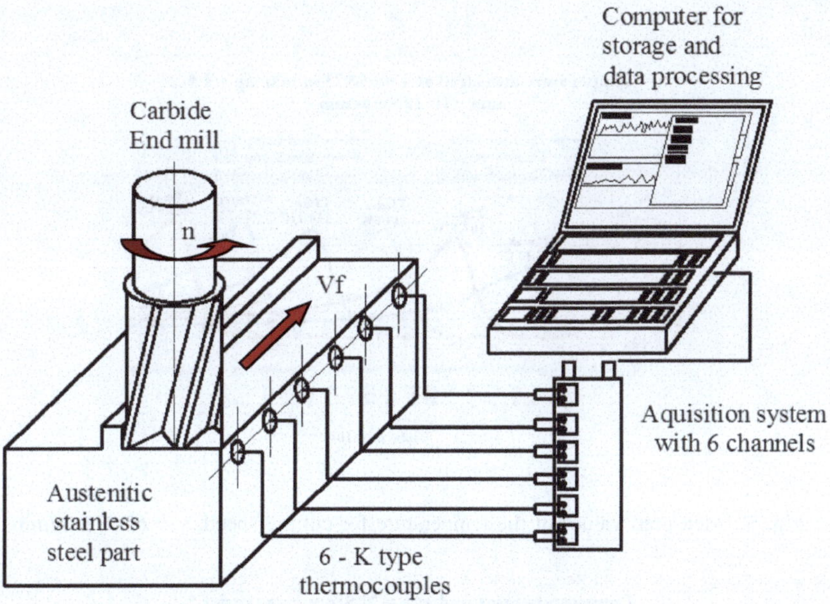

Fig. 3. Experimental set-up for measuring of the milling temperature

3.2 Surface Roughness

The surface roughness was measure after each pass of the carbide end mill using portable surface roughness, as it is shown in Fig. 4.

Figures 8, 9 and 10, present the roughness profile of the surfaces, obtained as a capture of the roughness tester software.

Fig. 4. Measuring of the surface roughness of the part after milling operation

After surface roughness measurements, graphical variation of Ra (arithmetical mean deviation of the assessed profile) for the three values of cutting speed is presented in Fig. 11.

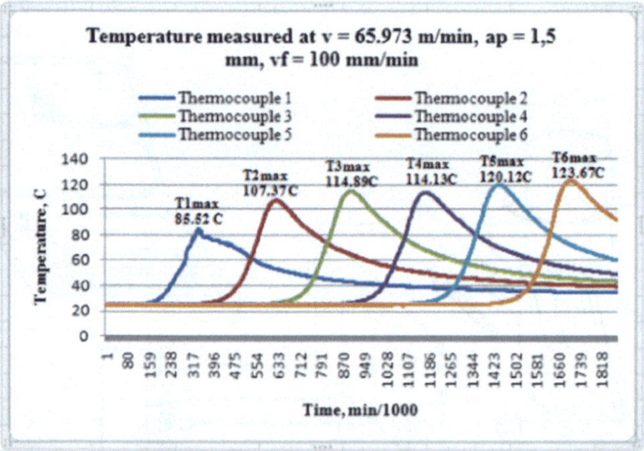

Fig. 5. Measured values of the temperature for cutting speed, v = 65.973 m/min

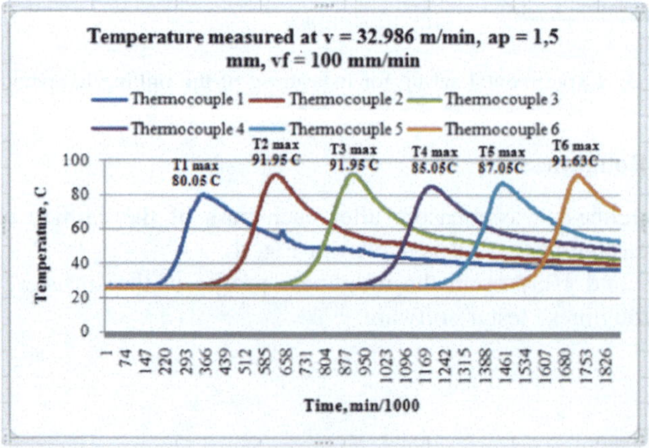

Fig. 6. Measured values of the temperature for cutting speed, v = 32.986 m/min

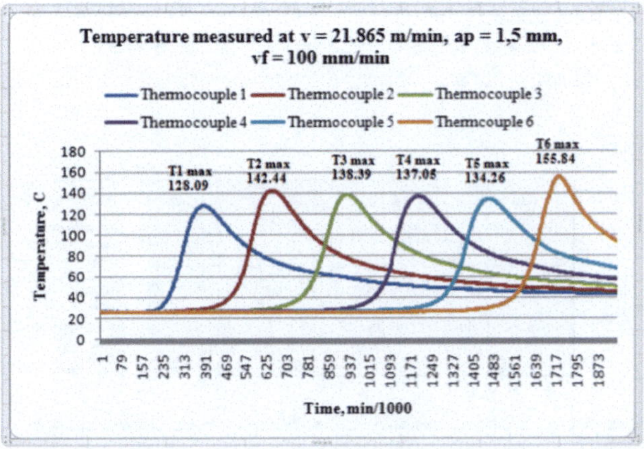

Fig. 7. Measured values of the temperature for cutting speed, v = 21.865 m/min

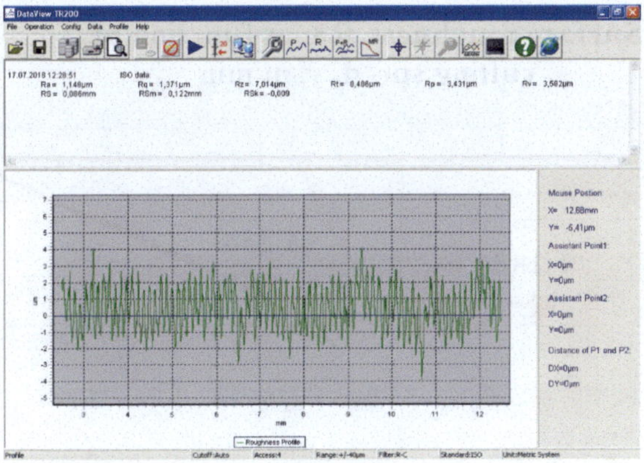

Fig. 8. The roughness profile of the milling surface for v = 65.973 m/min, (R_a = 1.148 μm)

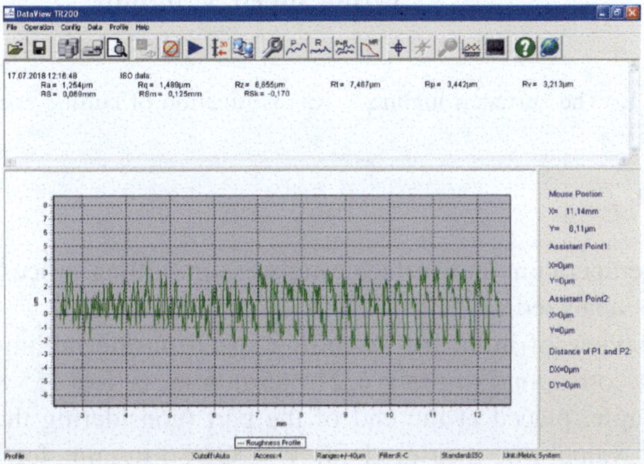

Fig. 9. The roughness profile of the milling surface for v = 32.986 m/min, (R_a = 1.254 μm)

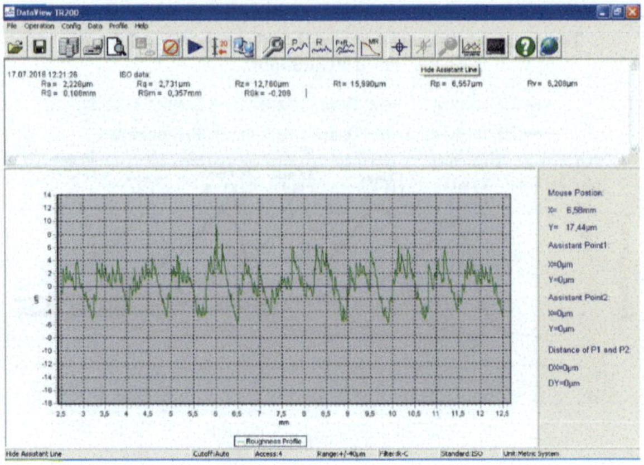

Fig. 10. The roughness profile of the milling surface for v = 21.865 m/min, (R_a = 2.228 μm)

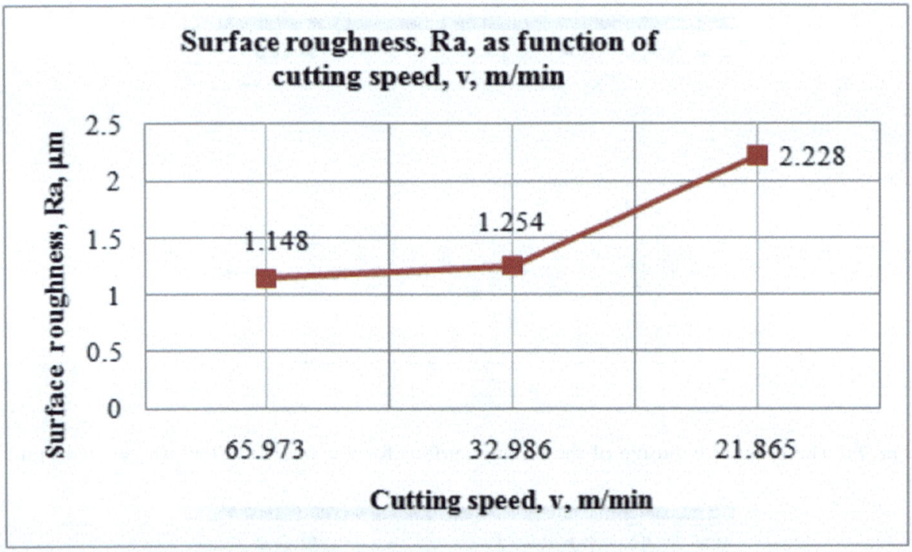

Fig. 11. The surface roughness, Ra, as function of cutting speed, v

4 Conclusion

In this paper the temperature and surface roughness in milling process of an austenitic stainless steel was analyzed.

From point of view of influence of cutting speed on the milling temperature the greater value, that correspond to the v = 21.865 m/min, is T = 155.84 °C, Fig. 7, for the last thermocouple, placed at the end of the part (considering the feed direction), while the smaller value was measured for v = 32.986 m/min, for the first thermocouple, T = 80.05 °C.

The value of surface roughness, Ra, increases with the decreasing of the values of rotational speed, and graphical representation from Figs. 8, 9, 10 and 11, show that.

In order to validate all this data presented in this paper (both for milling temperature of austenitic stainless steel and surface roughness) our main goal is to continue this research.

References

1. M'Saoubi, R., Chandrasekaran, H.: Experimental study and modelling of tool temperature distribution in orthogonal cutting of AISI 316L and AISI 3115 steels. Int. J. Adv. Manuf. Technol. **56**, 865–877 (2011). https://doi.org/10.1007/s00170-011-3257-y
2. Kaladhar, M., Venkata Subbaiah, K., Srinivasa Rao, C.H.: Machining of austenitic stainless steels – a review. Int. J. Mach. Mach. Mater. **12**(1/2), 178–192 (2012)
3. Fulemova, J., Janda, Z.: Selection of cutting material for milling of ferritic – martensitic stainless steel tool wear vs. roughness of machined surface. In: Annals of DAAAM for 2012 & Proceedings of the 23rd International DAAAM Symposium, vol. 23, no. 1, pp. 0361–0364 (2012). ISSN 2304-1382, ISBN 978-3-901509-91-9
4. Maurotto, A., Tsivoulas, D., Burke, M.G.: Surface integrity in dry milling of 304L steel: a parametric study. Procedia CIRP **13**, 156–162 (2014)
5. Naresh, N., Rajasekhar, K.: Multi-response optimization for milling AISI 304 Stainless steel using GRA and DFA. Adv. Mater. Res. **5**(2), 67–80 (2016). https://doi.org/10.12989/amr.2016.5.2.067
6. Nadolny, K., Kapłonek, W.: Analysis of flatness deviations for austenitic stainless steel workpieces after efficient surface machining. Measur. Sci. Rev. **14**(4), 204–212 (2014)
7. Koyee, R.D., Schmauder, S., Heisel, U., Eisseler, R.: Numerical modeling and optimization of machining duplex stainless steels. Prod. Manuf. Res. **3**(1), 36–83 (2015). https://doi.org/10.1080/21693277.2014.990539
8. San-Juan, M., Martín, Ó., de Tiedra, M.D.P., Santos, F.J., López, R., Cebrián, J.A.: Study of cutting forces and temperatures in milling of AISI 316L. In: The Manufacturing Engineering Society International Conference, MESIC 2015, Procedia Engineering, vol. 132, pp. 500–506 (2015)
9. Mahdavinejad, R.A., Saeedy, S.: Investigation of the influential parameters of machining of AISI 304 stainless steel. Sādhanā, vol. 36, Part 6, pp. 963–970, December 2011
10. Philip Selvaraj, D., Chandramohan, P.: Optimization of surface roughness of AISI 304 austenitic stainless steel in dry turning operation using Taguchi design method. J. Eng. Sci. Technol. **5**(3), 293–301 (2010)
11. Craciunoiu, N., Tarata, D., Miritoiu, C., Patru, E.N., Panduru, D.: On the temperature and chips types in milling of an austenitic stainless steel. In: Proceedings of the 4th International Congress of Automotive and Transport Engineering (AMMA 2018), AMMA2018 Proceedings in Automotive Engineering, pp. 455–463. Springer, Cham (2018)
12. Tarata, D.: Research on the properties of the austenitic steel allied with manganese. In: 10th International Metallurgy and Materials Congress and Trade Fair, Metallurgy and Materials, Istanbul, 24–28 May 2000, vol. 3, pp. 2015–2022 (2000)
13. Tarata, D.: Experimental research regarding the austenitic stainless steel with cryogenic resistance. In: International Conference of Mechanical Engineering, ICOME 2010, vol. 2, pp. 1009–1012 (2010)
14. http://www.magistor.nl/upload/content/documents/productbrochure_nachi_frezen.pdf
15. Portable Surface Roughness Tester - TR200. Innovatest, Manual TR200-Surface roughness tester

Complex Characterization of the Material

Ionela Alina Sandu(✉), Constantin Onescu(✉), Ştefan Tabacu(✉), and Cătălin Ducu(✉)

University of Piteşti, Piteşti, Romania
ionela.alina_sandu@yahoo.com,
constantin.onescu@gmail.com, stefan.tabacu@gmail.com,
catalinducu@yahoo.com

Abstract. Developing a complex characterization method for analyzed material represent the focus of this study. We aimed to develop a method that can be used for numerical investigation of mechanical structures. The first test performed was tensile test (classical test), the material analysis continued with shear test and triaxiality. The measurements are achieved by Digital Image Correlation (DIC), a new technique for an for improved analysis of the strain field.

It is known that if the triaxiality factor is known the state of stress in any point of strained object can be determined.

Keywords: Material characterization · Tensile test · Compression test · Shear test · Triaxiality · Bonded strain gauge

1 Introduction

In the presented study, a complex characterization method of materials, suitable for using in numerical investigation of mechanical structures, has been developed.

Using experimental tests was developed a material model for lifetime estimation, on different metallic specimens, used for tensile test and shear test, designed to provide the most accurate results [1, 2].

The tensile test is the well-known method to obtain information about the material, and the article [3] presents several new techniques as DIC method and X-Ray Diffraction (XRD) to analyze stresses and strains within the material. Dedicated specimens are used [4] for the uniaxial tension tests and plane stress tension.

For shear tests various specimens could be used, with one shear test zones as ASTM B831-05 [2] or modified ASTM proposed by Merklein and Biasutti and two shear test zones proposed by Miyauchi [5]. A comparison between experimental tests performed by Digital Image Correlation (DIC) and numerical simulation is presented in [4].

DIC, a non-contact optical technique, is an ideal technique suited to study the crack propagation and material deformation [6]. The grey value digital images represent the measurement principle of the DIC method, which can determine in three dimensions the displacements and the contour of an object under load [7].

Stress state is often characterized in terms of stress triaxiality level [8]. The triaxiality is known to greatly influence the amount of plastic strain which a material may support before failure occurs.

I. Dumitru et al. (Eds.): SMAT 2019, *The 30th SIAR International Congress of Automotive and Transport Engineering*, pp. 678–683, 2020.
https://doi.org/10.1007/978-3-030-32564-0_79

2 Stress Triaxiality Factor

Stress state is often characterized in terms of stress triaxiality factor, η, which is defined as ration between mean stress (σ_m) and equivalent von Misses stress (σ_{eq}) [9]:

$$\eta = \frac{\sigma_m}{\sigma_{eq}} \tag{1}$$

If this factor is known, the state of stress in any point of strained is determined: if $\eta = 0$ means simple shear; $\eta = 0,33$ means uniaxial tension; $\eta = 0,66$ means biaxial tension and $\eta = -0,33$ means uniaxial compression.

Within the present study the following specimens are used to characterize the material: (a) uniaxial tensile; (b) notched tensile; (c) combined tensile-shear and (d) simple shear (ASTM B831) [2] (Fig. 1).

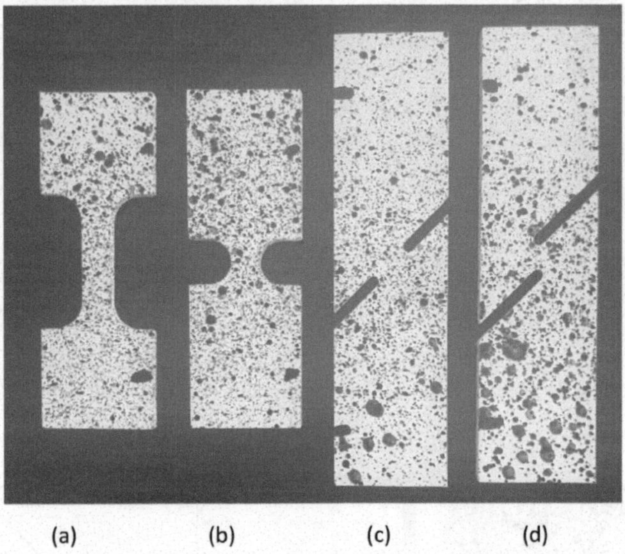

Fig. 1. Test specimens (prepared for DIC tests).

The material used for the present study was 30MnVS6 (raw material: round bar).

Flat specimens were manufactured from steel bar, the final shape was obtained by milling up to 3 mm thickness and by electro-erosion with wire. These processes were selected to not thermally influenced the structure of the base material.

3 Experimental Results

For all types of test specimens, a universal mechanical testing machine has been used with a test speed set to 2.0 mm/min.

DIC method has been used for a detailed investigation of the strain field. Capturing images during tensile test was done with a Nikon digital camera equipped with macro lenses. To drive the camera to record a new frame at a defined time step (10 s) was used DigiCamControl software and to process image and extract strain field distribution was used GOM Correlate.

3.1 Tensile Stress

The samples were tensile tested to obtain the strain state and the displacement-force characteristics of the material. The critical area is the area with maximum strain.

The following two figures (Figs. 2 and 3) presents the experimental results of tensile specimens, strain is given by DIC and force-displacement curves given by the testing machine.

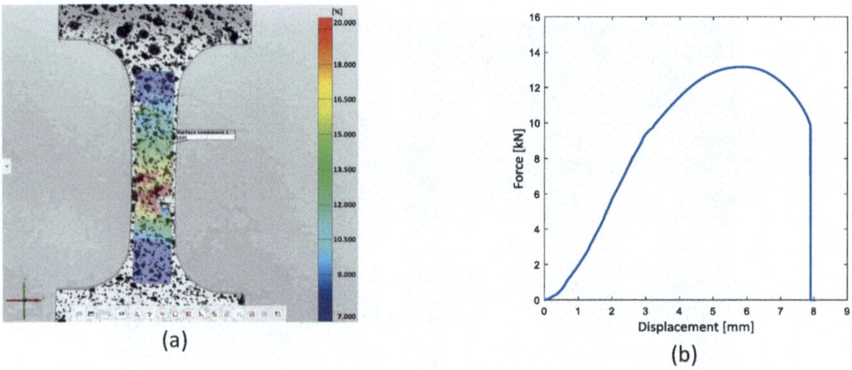

Fig. 2. Uniaxial tensile specimen: (a) DIC measurement result; (b) curve force – displacement.

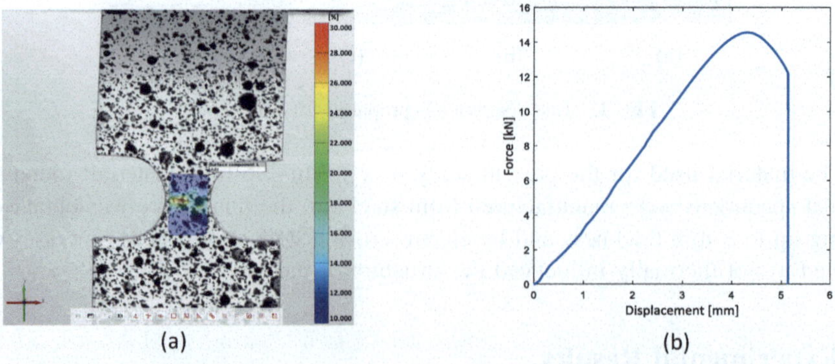

Fig. 3. Notched tensile specimen: (a) DIC measurement result; (b) curve force – displacement.

3.2 Shear Stress

Simple shear tests are widely used to characterization of materials, mainly for metal sheets to obtain an ample deformation without plastic instability.

The Figs. 4 and 5 presents the experimental results of the shear specimens, strain given by DIC and force-displacement curve given by the testing machine.

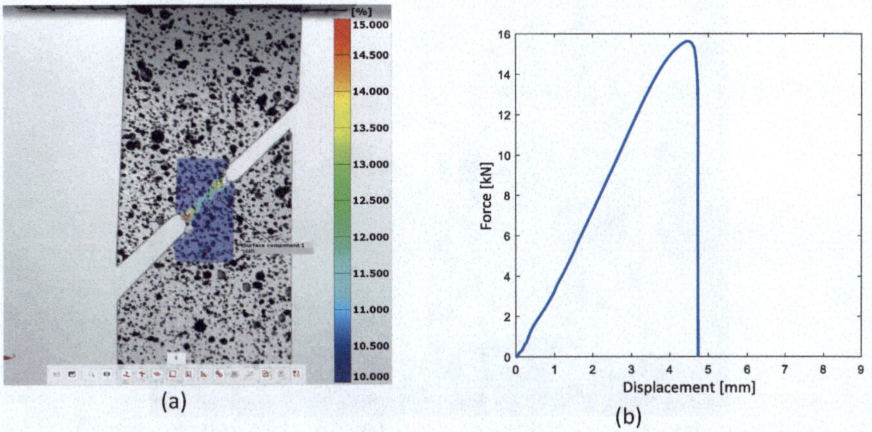

Fig. 4. Combined tensile-shear specimen: (a) DIC measurement result; (b) curve force – displacement.

3.3 Stress Triaxiality Factor

The specimens were tested until rupture (Fig. 6). From the Figs. 2, 3, 4 and 5 on deduce that the stress state of critical areas (red color), corresponds which that defined by the triaxiality factor.

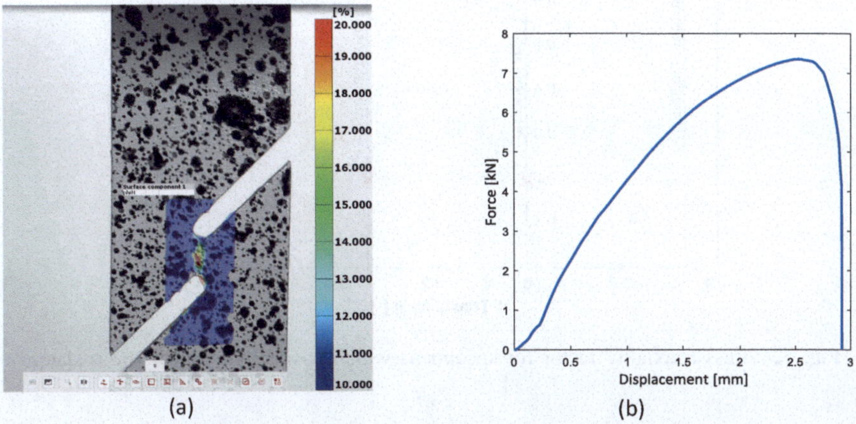

Fig. 5. Shear tensile specimen: (a) DIC measurement result; (b) curve force – displacement

The experimental triaxiality factor is presented in the Fig. 7. The main advantage is the achieving a stress shear using the specimen shape and tensile test.

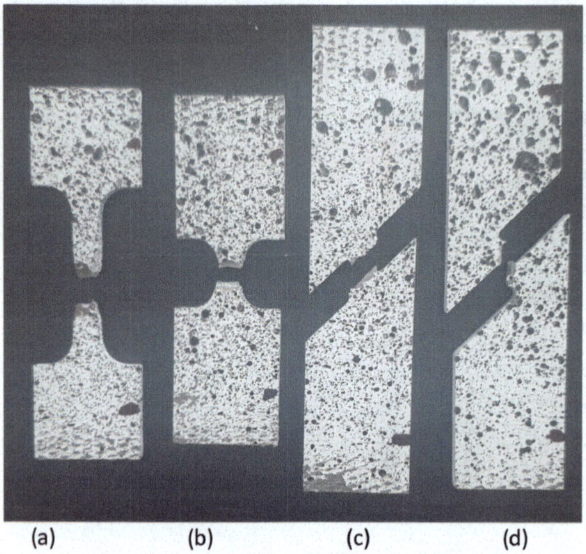

Fig. 6. Tested specimens (prepared for DIC tests).

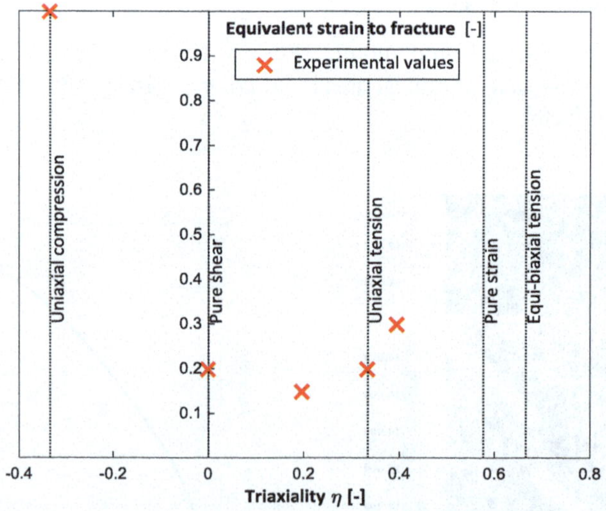

Fig. 7. Stress triaxiality factor for specimens which cover a variety of stress states.

4 Conclusion

In this study four different specimen for tensile and shear test were measured using non-contact Digital Image Correlation technique. The DIC results were captured in the same period with those given by the universal mechanical testing machine acquisition system.

Using stress triaxiality fact b or, it's possible to create new specimens for the tensile test in order to obtain a various types of the stress states.

In this study we discussed about a concept, triaxiality, a method of customizing the tensile state and some parameters that can determine the conditions of use of a part.

Specific measurements (elongation at break, also known as fracture strain), are specific for a material (for this study 30MnVS6).

The aim of the study is not to establish and transmit absolute values.

In this study we discussed about the DIC method, we have used a series of tests that should induce particular states.

In conclusion, the obtained diagram is in concordance with the diagrams obtained in the specialized literature.

References

1. ASTM International: Standard Test Methods for Tension Testing of Metallic Materials 1, Astm, no. C, pp. 1–27 (2009)
2. ASTM International: Standard Test Method for Shear Testing of Thin Aluminum Alloy Products 1, pp. 5–8 (2005)
3. Sandu, I.A., Ducu, C., Moga, S., Tabacu, Ş.: Development of a material model for the use in life-time estimate of structures
4. Zou, D.Q., Li, S.H., He, J., Gu, B., Li, Y.F.: The deformation induced martensitic transformation and mechanical behavior of quenching and partitioning steels under complex loading process. Mater. Sci. Eng.: A **715**, 243–256 (2018)
5. Yin, Q., Soyarslan, C., Isik, K., Tekkaya, A.E.: A grooved in-plane torsion test for the investigation of shear fracture in sheet materials. Int. J. Solids Struct. **66**, 121–132 (2015). November 2013
6. McCormick, N., Lord, J.: Digital image correlation. Mater. Today **13**(12), 52–54 (2010)
7. Dantec Dynamics: Measurement Principles - Digital Image Correlation (DIC) (2013). https://www.dantecdynamics.com/measurement-principles-of-dic. Accessed 12 Apr 2018
8. Kut, S.: A simple method to determine ductile fracture strain in a tensile test of plane specimen's. Metalurgija **49**(4), 295–299 (2010)
9. Brunig, M.: From Creep Damage Mechanics to Homogenization Methods. Springer International Publishing, Switzerland (2015)

Study About the Roughness Surface Quality for a Duralumin Alloy After Cutting Process

Ileana Cristina Pascu(✉) and Iulian Popescu

Faculty of Mechanics, University of Craiova, Calea Bucuresti str. 107, 200478 Craiova, Romania
i_pascu@yahoo.com

Abstract. In this paper results of researches regarding the aspects of the micro-geometry of the chips and the final quality of the surface of a duralumin Al Cu4 Mg1Mn pipe type are presented. The piece was inner turned, and after processing, it was cut in two in order to determine the roughness and preview the shapes of cutting surfaces. For the cutting regime, a speed of 500 rot/min and a cutting depth of 0.5 mm were used, these values being constant, and using different values of the feed with values ranging from 0.096 mm/rot until 0.584 mm/rot. The Ra, Rz and Rq roughness parameters of the surface were obtained by the profiles micro-unevenness measured by an electronic roughness tester Mitutoyo type, a SJ-201 P model. We have also studied the aspect of the processed surfaces, both by enlarging them by microscopy and by studying the chips sections resulting during the cutting process, giving the values of the theoretical and real dimensions, also interpreted on the basis of deformations in the area cutting. The diagram shows the curves on chip sections and the distribution parameters of the cutting parameters.

Keywords: Duralumin · Roughness · Inner turning

1 Introduction

Due to its lightweight properties, aluminum and its alloys are mainly used in the field of automotive components, aircraft engines, fuselage and energy saving strategies [1].

Between of Al based alloys, duralumin (aluminum-copper-magnesium components) is characterized by a high strength, heat-treatable, anticorrosion properties and low specific weight allowing the development of many complicated shapes [2]. Because of these properties, processing of duralumin aroused interest, there are studies on its workability on processing by different methods [3–6].

Dogra and all [7] a study about the effect of the parameter process on the surface roughness for duralumin turning and the effect is presented. In [8] the almost pure aluminum turning with 99.39% aluminum are studied. The shapes of chips resulting from different cutting regimes and the depositions on the edge, which are very pronounced have been studied. In [9] an aluminum alloy ATSi7 Mg processed by turning was presented, with studying the chips shapes with different cutting speeds regimes. The dimensions of the cuttings at different cutting speeds have been analyzed.

I. Dumitru et al. (Eds.): SMAT 2019, *The 30th SIAR International Congress of Automotive and Transport Engineering*, pp. 684–690, 2020.
https://doi.org/10.1007/978-3-030-32564-0_80

2 Initial Data

For this study, a duralumin Al Cu4 Mg1 Mn alloy part tube was inner turned with different values of feeds, without cutting fluid.

After inner turning, the pipe was clipped in two for determining the roughness and observing the processed surfaces (see Fig. 1). For the cutting regime, a speed of 500 rot/min and a cutting depth of 0.5 mm have been settled. The 6 values for cutting feeds were chosen: 0.096 mm/rot, 0.208 mm/rot, 0.302 mm/rot, 0.416 mm/rot, 0.5 mm/rot and 0.584 mm/rot.

Fig. 1. Inner turned part

For processing, an inside boring tool from high-speed steel (see Fig. 2), with the following geometry: a channel on the front face with 11.46 mm × 3.36 mm dimensions and depth of 3.5 mm, $\alpha_n = 18°$, $\gamma_n = 30°$, $\kappa_r = 110°$, $\kappa_r' = 8°$, has been used.

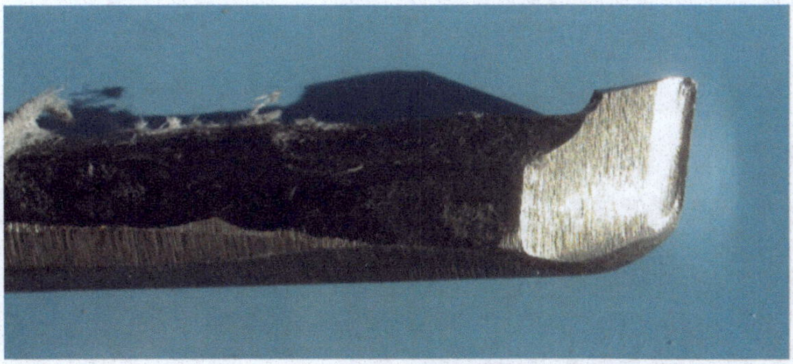

Fig. 2. The inside boring tool

The Ra, Rz and Rq surface roughness parameters were measured with an electronic roughness tester Mitutoyo type, a SJ-201 P model.

3 Experimental Results

The surface roughness for the cutting feed of f = 0.096 mm/rotis given by the profile micro-unevenness is presented in Fig. 3.

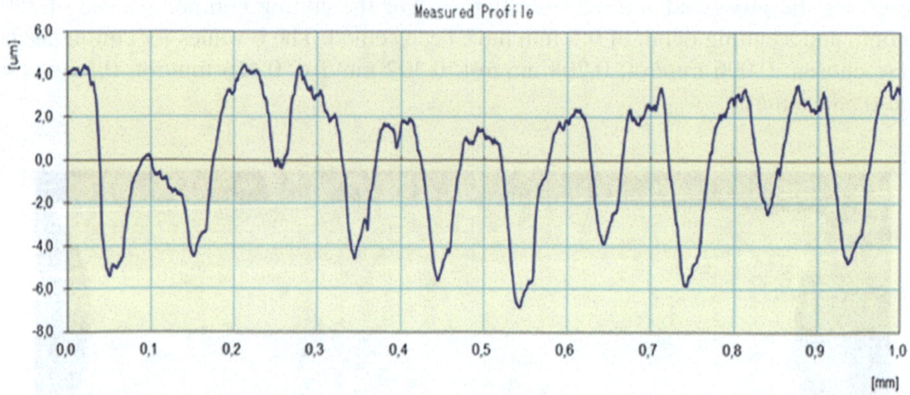

Fig. 3. Roughness parameters for the cutting feed of 0.096 mm/rot

The forms of chips resulted after inside turning with f = 0.208 mm/rot feed and the chips microscopy are presented in Fig. 4. From the Fig. 4 it can be observed that the chips are white-silver colour, some of them having helical shapes with uneven steps between 11 and 14 mm, with a pitch of 8.25 mm. There are splinters in the form of uneven spatial curves, with a gauge of 44 × 28 mm.

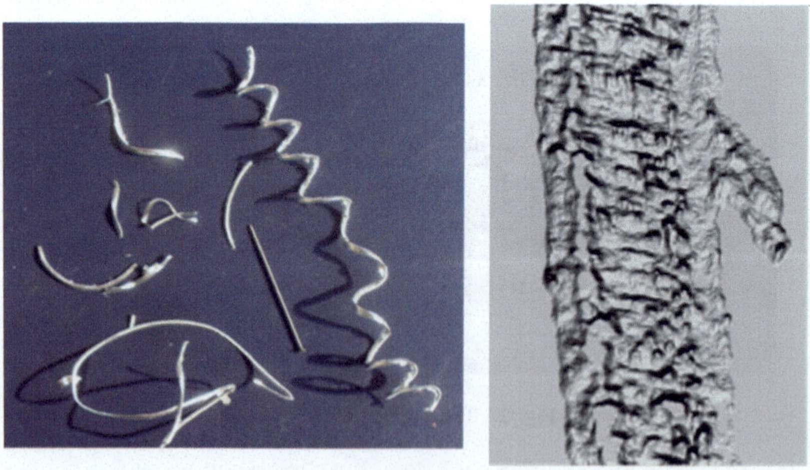

Fig. 4. Resulted chips after turning with a feed f = 0.208 mm/rot

In Fig. 5 the forms and microscopy of chips obtained after inside turning with f = 0.208 mm/rot feed are shown.

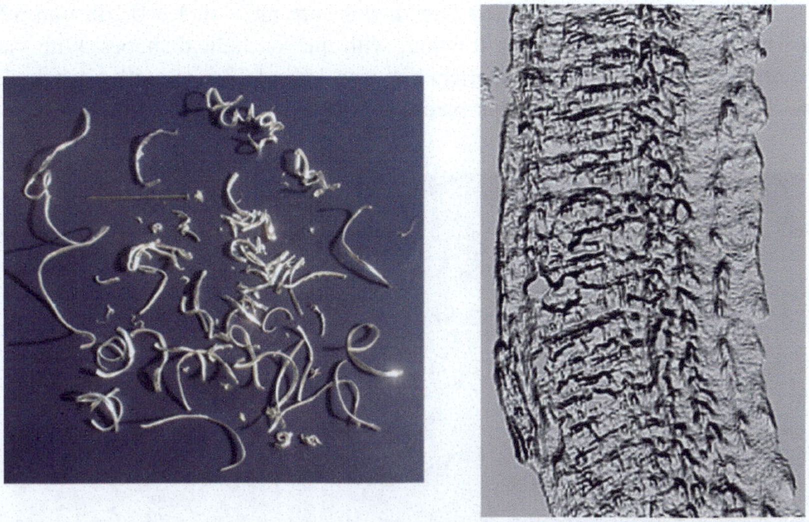

Fig. 5. Resulted chips and micro-geometry after turning with a feed f = 0.302 mm/rot

From the Fig. 5 it can be noted that the chips have very varied, random shapes. They start from the cutting area in the form of a propeller, but immediately deform, clinging. There are some helical spires (at some) with a diameter of 5.38 mm and a pitch of 5.68 mm, and on others some spirals with a diameter of 17 mm and a pitch of 27 mm appear. There are also pieces of propeller, broken, deformed. Also, it can be observed that the chip is wider than in the previous cases, the scales forming parallel transverse rows to the chip - in the left area and longitudinal rows to the right, which highlights the large plastic deformations of the chip during forming and the friction on the groove on the opening face of the inside boring tool.

The surface roughness is given by the profile micro-unevenness (see in Fig. 6).

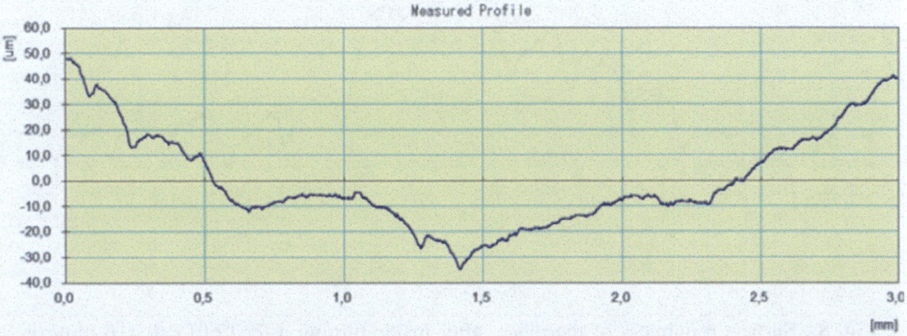

Fig. 6. Surface roughness of the piece, after turning with f = 0.302 mm/rot feed

The ruggedness of chips resulted after inside turning with f = 0.208 mm/rot feed are shown in Fig. 6. The chips are flowing, with uneven helical shapes, with variable pitch, with the diameters of the propellers between 15 and 37 mm, with pitches of 30 to 56 mm. There are also broken chips pieces of those mentioned.

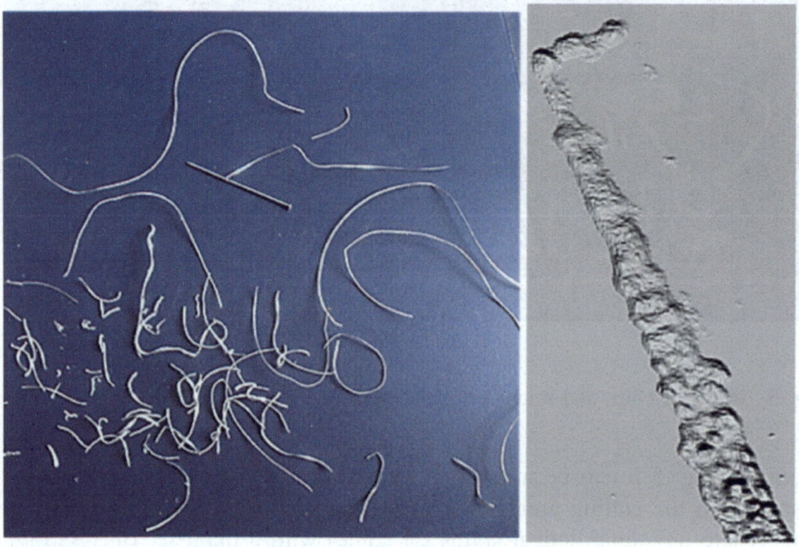

Fig. 7. Chip micro-geometry for a f = 0.416 mm/rot feed

From the Fig. 7, it can be observed that the chip has lateral deformations, too, not uniform, the elements are generally transversal and the scales are large, uneven.

The profile micro-unevenness presented in Fig. 8 shows some symmetries, a sign that the process has gone orderly.

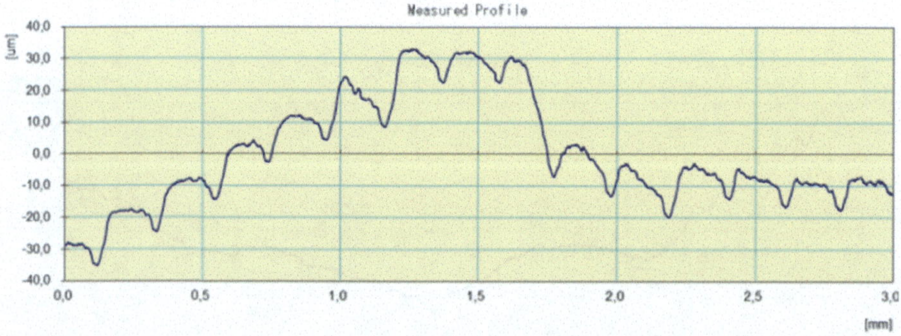

Fig. 8. Surface roughness of the piece, after inside turning a feed of f = 0.416 mm/rot

The forms and micro-geometry of chips obtained after inside turning with f = 0.5 mm/rot feed are given in Fig. 9. It can be noted that the chips are in the form of a propeller with a height of 70–100 mm, diameter between 20 and 30 mm, pitches between 21 to 33 mm. No small or broken splinters. It can be shown a longitudinal channel approximately halfway through the chip width, cross cuts in the left area and large chamfers of the chip elements in the right area, where the friction on the groove on the front face of the inside boring tool was high.

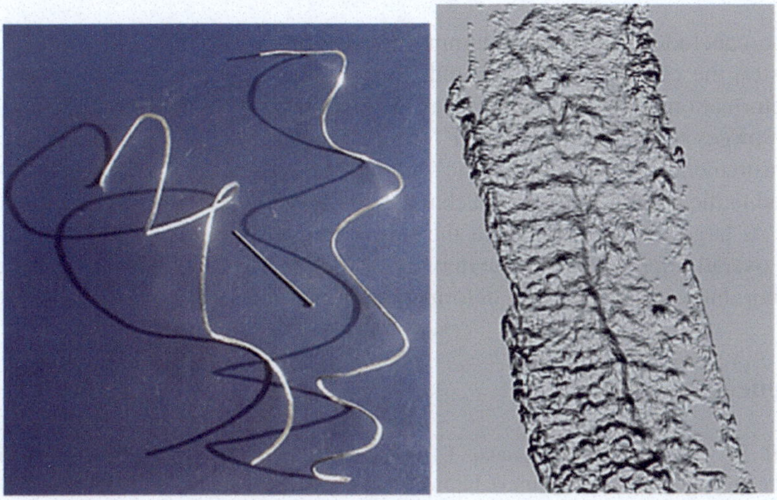

Fig. 9. Resulted chips after turning with a feed f = 0.5 mm/rot

In Fig. 10 the measured values of the roughness parameter for piece processed surfaces are presented.

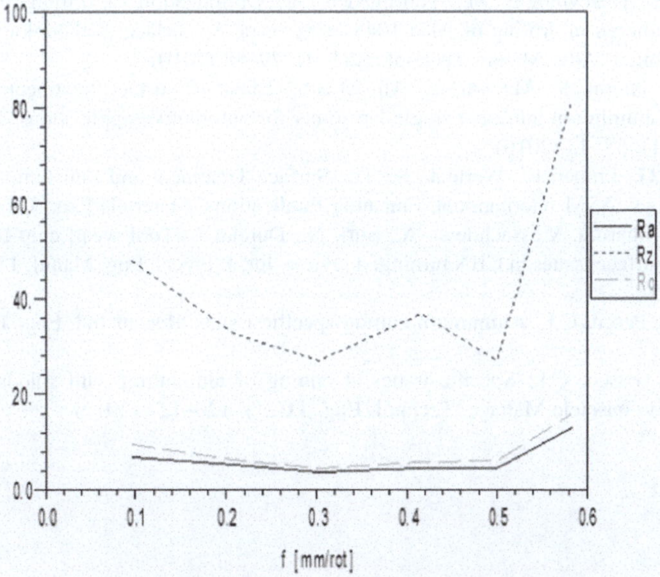

Fig. 10. The roughness parameters (Ra, Rz, Rq)

From the Fig. 10, it can be observed that there are normal variations for Rq and Ra and with jumps for Rz, which proves large deformations in the cutting area, which influence the quality of the processed surfaces.

4 Conclusions

It can be concluded that the duralumin alloy is easy to machining, and there are no deposits on the cutting edge that could disrupt the cutting process. However, appear large deformations of the chipping faces, so that the process does not flow uniformly, the chip pieces overlap on each other forming the scales.

Measurements have shown that the sections of the chips are uneven, the chips being deformed in the cutting area and the channel on the front face, where the long chips are broken. At large feed values cut off the side cutting edge, too.

The overall inference is that the duralumin Al Cu4 Mg1Mn alloy material is easy to machining, but can appear large deformations in the cutting area.

References

1. Sayuti, M., Sarhan, A.A.D., Tanaka, T., Hamdi, M., Saito, Y.: Cutting force reduction and surface quality improvement in machining of aerospace duralumin AL-2017-T4 using carbon onion nanolubrication system. Int. J. Adv. Manuf. Technol. **65**(9-12), 1493–1500 (2012)
2. Kadlečková, M., Minařík, A., Smolka, P., Mráček, A., Wrzecionko, E., Novák, L., Musilová, L., Gajdošík, R.: Preparation of textured surfaces on aluminum-alloy substrates. Materials **12** (1), 109–121 (2018)
3. Aniket, A., Kumar, S.: Laser cutting of AISI 304 material: an experimental investigation on surface roughness. Adv. Mater. Process. Technol. **5**(3), 429–437 (2019)
4. Kilickap, E., Huseyinoglu, M., Yardimeden, A.: Optimization of drilling parameters on surface roughness in drilling of AISI 1045 using response surface methodology and genetic algorithm. Int. J. Adv. Manuf. Technol. **52**(1–4), 79–88 (2010)
5. Boutar, Y., Naïmi, S., Mezlini, S., Ali, M.B.S.: Effect of surface treatment on the shear strength of aluminium adhesive single-lap joints for automotive applications. Int. J. Adhes. Adhes. **67**(1), 38–43 (2016)
6. Sheasby, P.G., Pinner, R., Wernick, S.: The Surface Treatment and Finishing of Aluminum and Its Alloys. ASM International, Finishing Publications, Materials Park, USA (2001)
7. Dogra, M., Sharma, V., Sachdeva, A., Suri, N., Dureja, J.: Tool wear, chip formation and workpiece surface issues in CBN turning: a review. Int. J. Precis. Eng. Manuf. **11**(1), 341–349 (2010)
8. Popescu, I., Pascu, C.I.: Aluminium cutting specific issues. Recent Ind. Eng. J. **12**(1), 384–392 (2011)
9. Popescu, I., Pascu, C.I.: Specific issues at cutting of aluminium with silicon alloy. Ann. Oradea Univ. Fascicle Manage. Technol. Eng. **11**(21), 120–127 (2012)

Valves Processing of Internal Power Engines by Ultrasonic Vibro-Cutting

Bebe Tică[1(✉)] and C. Klement[2]

[1] Department ATII, Faculty of Mechanical Engineering, University of Craiova, Calea Bucuresti Street 107, Craiova, Romania
ticabebe@yahoo.com

[2] Faculty of Mechanical Engineering, Transilvania University of Brasov, Politehnicii Street, no.1, Brasov, Romania

Abstract. The processing of materials with a hardness greater than 55 HRC has always required the use of unconventional or rectifying methods to obtain the quality and precision required by technical conditions. For these reasons, the research of the past decades has forced an unconventional method of processing, with high precision and accuracy, by assisting with classical processing techniques, forced, un-amorphized ultrasonic vibrations of low amplitude. The present paper presents a modern technology for the processing of internal combustion engine valves, namely ultrasonic vibrating. The valves are very well suited to vibration machining because they are shaft pieces with a large ratio between the length and the diameter of the rod and can easily be deformed by conventional machining. The technological equipment used is simple and is very quickly depreciated in serial or mass production.

Keywords: Ultrasonic vibrations · Process · Valve

1 Introduction

Typically, the vibrations occurring in the mechanical machining process are undesirable, leading to worsening of the conditions in which the mechanical processing takes place with unfavorable consequences on the quality of the parts executed. The continuous development of materials with special properties has forced the use of highly advanced and increasingly sophisticated processing technologies. It has thus been found that, in a new sense, paradoxically, the introduction of unstimulated forced vibrations of defined, controlled and controlled shape, amplitude and frequency in order to obtain mechanical processing benefits.

The great Romanian inventor, George Constantinescu, was the one who first used sonic acoustic vibrations for processing glass, quartz and extremely hard stones, providing the possibility of using vibration ultrasonic vibrations in the 20's of the 20th century [1]. As has been shown in the field, all research uses the tool's vibration direction, the axial direction, considered to be the most advantageous. All this until the Japanese researcher J. Kumabe, in his work "Vibration Cutting", imposed a new theory that would open new perspectives for the process of vibrating. Thus, he established that

I. Dumitru et al. (Eds.): SMAT 2019, *The 30th SIAR International Congress of Automotive and Transport Engineering*, pp. 691–698, 2020.
https://doi.org/10.1007/978-3-030-32564-0_81

the application of the vibrating machining process must comply with two basic conditions, namely:

1. ultrasonic vibrations must be oriented along the direction of the main component of the cutting force;
2. cutting speed (v) must be less than critical speed V_c:

$$V < V_c = 2\pi fa \tag{1}$$

where: V - cut speed; V_c - critical cutting speed; f - frequency; a - the amplitude of the ultrasonic oscillation.

After Kumabe [3], if $V \geq V_c$, there is no distinction between ordinary and vibrational splitting. From now on, all the research done by all the specialists was based on Kumabe's theory [2–4, 6]. The results of the research were among the most optimistic, the advantages of the new theory being very competitive.

The general conclusion resulting from industrial and laboratory research shows that vibro-cutting is not a chip-making mechanism, but a way of perfecting traditional shredding. These conclusions are based on the following peculiarities of vibration:

- in vibration scrubbing, the apparent stiffness increases by 3 to 10 times the actual thickness, allowing the processing of thin-walled parts, cylinders or thin-walled plates, and the use of low-rigidity tools (drills, reamers, boring bars);
- vibration-proof surfaces have a uniform appearance, with no deviations in shape and small roughness, in many cases smaller than the theoretical roughness [5, 6];
- the tool's durability increases several times, the tool wear is very low and the cutting forces drop to 3–4 times compared to conventional machining.

2 Trends and Perspectives

Due to the high technical performance achieved with vibration machining, it is foreseen to extend this process by making new types of machinery, both universal and specialized installations. There is a continuing concern for the improvement of ultrasonic processing equipment, through the development of high-tech installations, including the latest achievements in the technology: program control and adaptive control.

One of the main directions for the development of ultrasonic vibration processing processes is related to the improvement of the respective installations [6]. This refinement pursues two main directions, namely:

- the over-dimensional improvement of the ultrasonic installations in order to fully satisfy the stresses in assisting the different mechanical processing processes;
- improvement of the ultrasonic equipment subassemblies in order to achieve superior technical characteristics, with the cost reduction.

Starting from the above, it is evident that the method of vibrating machining, presents great advantages compared to conventional chipping, and opens new possibilities in the field of extrusion material processing, where classical chipping has great difficulty sometimes even impossibilities.

3 Mechanical Machining Technology of Internal Combustion Engine Valves

The constructive form, as well as the character of mass production, make it possible to automate the mechanical processing process.

Modern technologies require the production of very precise semi-finished products, with small additions, so that the processing can only be done with a very fine processing operation.

In the execution of the valves, strict conditions are imposed regarding the relative position of the surfaces of the oblique and the tail of the valve in relation to the rod, as well as on the straightness of the rod.

The main condition to be met when working the valve is to obtain the most perfect concentricity of the valve head cone with the guide portion of the rod (Fig. 1).

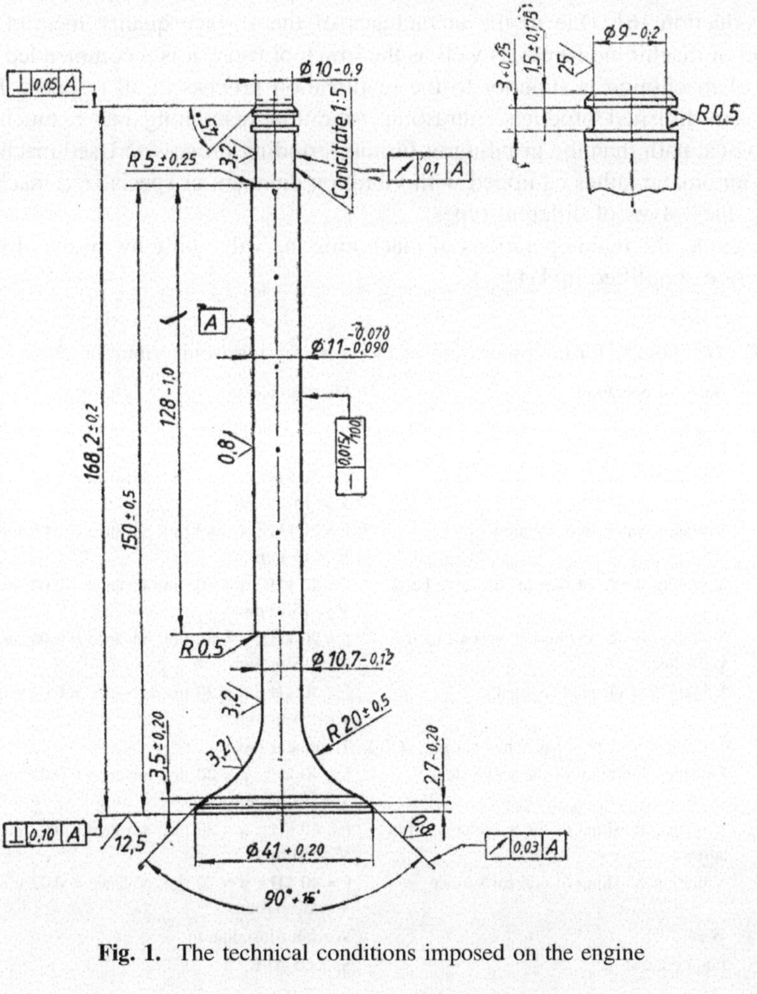

Fig. 1. The technical conditions imposed on the engine

In general, valves are processed by turning and grinding. Modern technologies that have emerged in the last decade offer new methods of valve processing, one of them being ultrasonic vibration, with its advantages, as shown above.

The valves are very suitable for processing by vibro-cutting, because they are pieces in the form of a shaft, with a large ratio between the length and the diameter of the rod, which deforms slightly by conventional processing (Fig. 1).

In the execution of the valves, strict conditions are imposed regarding the relative position of the conical surfaces and the tail of the valve in relation to the rod, as well as the rectilinearity of the rod.

A general type of conventional valve processing technology ensures the execution by turning and fine abrasion, but when calibrating the blanks, they can be executed only by fine abrasion.

The present paper proposes a manufacturing technology only by ultrasonic vibrating.

The technological equipment used is fairly simple and is very fast in production or mass production [6]. Due to the advantages of the surface quality method and the reduction of machining forces as well as the low tool wear, it is recommended that this method of machining is superior to the rectification process in all respects. For very precise semi-finished products, ultrasonic vibration machining has a much shorter technological path than the grinding or turning-grinding processes. Used machines can be semi-automatic lathes equipped with vibration dampers or specialized machines for vibrating the valves of different types.

As a guide, the main operations of machining the valve only by means of vibration cutting are exemplified in Table 1.

Table 1. Technology for the processing of the valve by ultrasonic vibration-cutting (Fig. 1)

Nr. operation	Name of operation	Machine tools
1	Supply of the technological line	
2	Vibration of front valve stem surface	f = 20 kHz; a = 40 μm; advance = 0.05 mm/rot; V = 15 m/min
3	Vibrating valve stem rotation	f = 20 kHz; a = 40 μm; advance = 0.05 mm/rot; V = 15 m/min
4	Vibrating the front face of the valve head	f = 20 kHz; a = 40 μm; advance = 0.04 mm/rot; V = 15 m/min
5	Vibration of the cylindrical surface of the valve head	f = 20 kHz; a = 30 μm; advance = 0.04 mm/rot; V = 15 m/min
6	Vibration of channel segments	f = 40 kHz; a = 20 μm; advance = 0.04 mm/rot; V = 15 m/min
7	Treatment with high frequency currents (CIF)	Treatment facility
8	Finishing vibration of the valve stem	f = 40 kHz; a = 20 μm; advance = 0.02 mm/rot; V = 30 m/min
9	Vibrating finishing of the valve stem front surface	f = 40 kHz; a = 20 μm; advance = 0.02 mm/rot; V = 30 m/min
10	Vibration finishing of conical surface	f = 40 kHz; a = 20 μm; advance = 0.02 mm/rot; V = 30 m/min
11	Wash	Washing installation
12	Final control	Special AMC

Some operations in the presented technological process can be performed by classical processing, on the same machine-tool or machine, with the decoupling of the ultrasonic vibration installation, resulting in a combined vibrational sharpening technology. It is, however, preferable to a vibrant processing technology, because the processing is obtained in a higher precision class, with very high surface qualities, with implicit operating advantages.

Following the experiments, the following results regarding the quality of the valve surfaces (Ra) and the wear of the tools used for vibration processing were revealed. Thus, in Figs. 2, 3 and 4, the roughness values Ra are given, depending on the parameters of the cutting regime, obtained under the ultrasonic vibration conditions of the cutting edge of the tool, at the regime of Table 1.

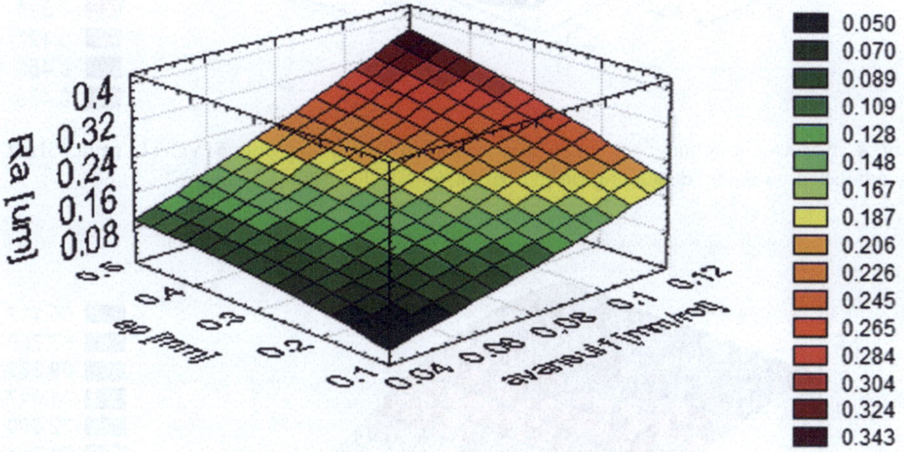

Fig. 2. Surface roughness at ultrasonic vibration machining of 51VCr11 (min.65HRC) semifinished products, depending on the parameters of the cutting regime

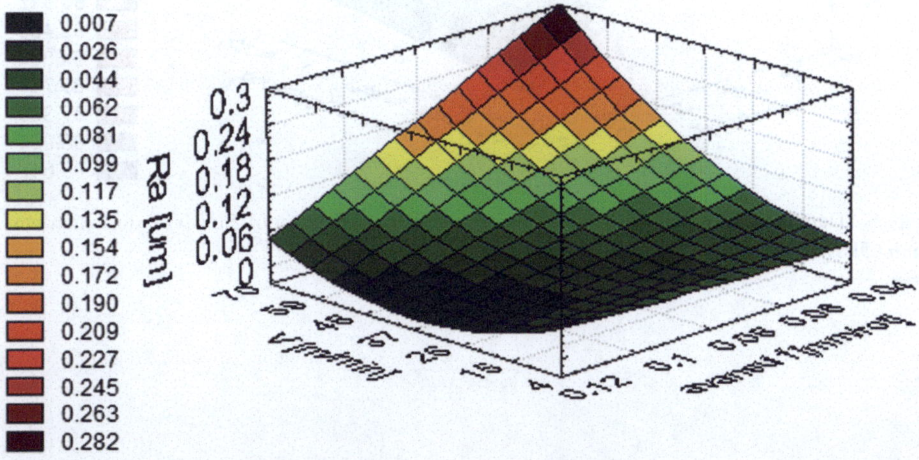

Fig. 3. Surface roughness at ultrasonic vibration machining of 51VCr11 (min.65HRC) semifinished products, depending on the parameters of the cutting regime

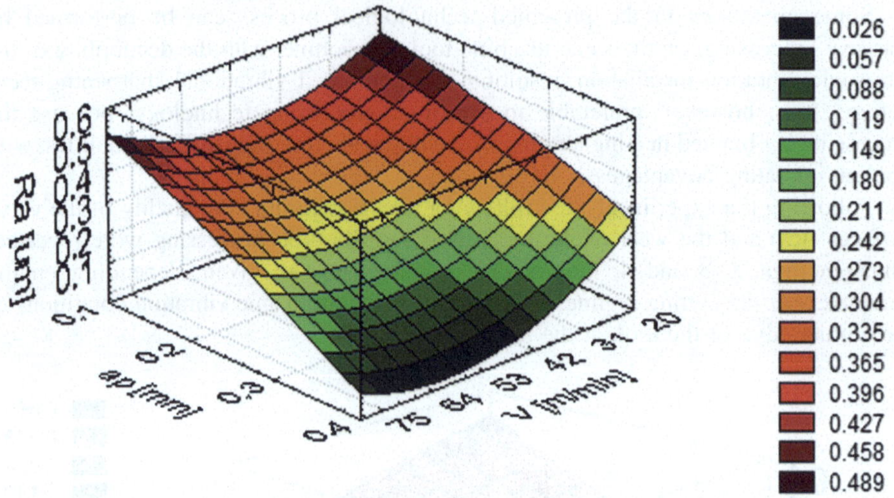

Fig. 4. Surface roughness at ultrasonic vibration machining of 51VCr11 (min.65HRC) semifinished products, depending on the parameters of the cutting regime

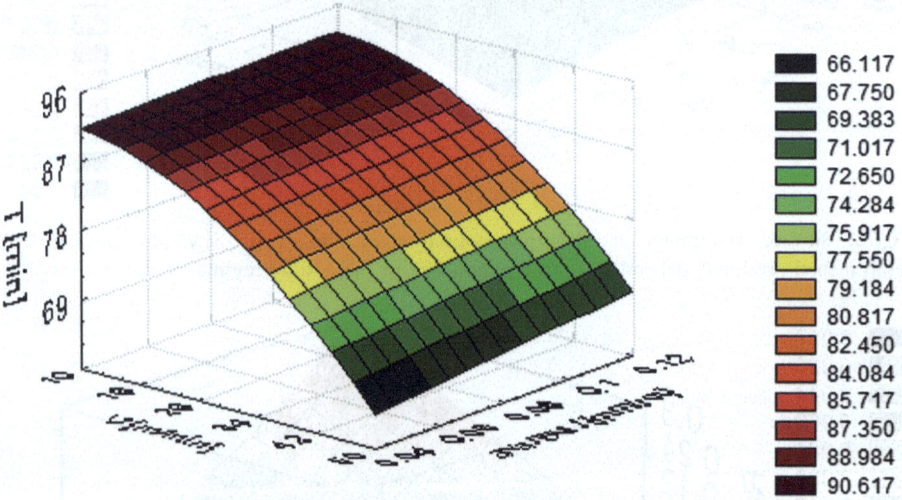

Fig. 5. Durability variation of diamond tool to ultrasonic vibration of 51VCr11 steel (hardness min.65HRC), depending on feed rate and cutting speed

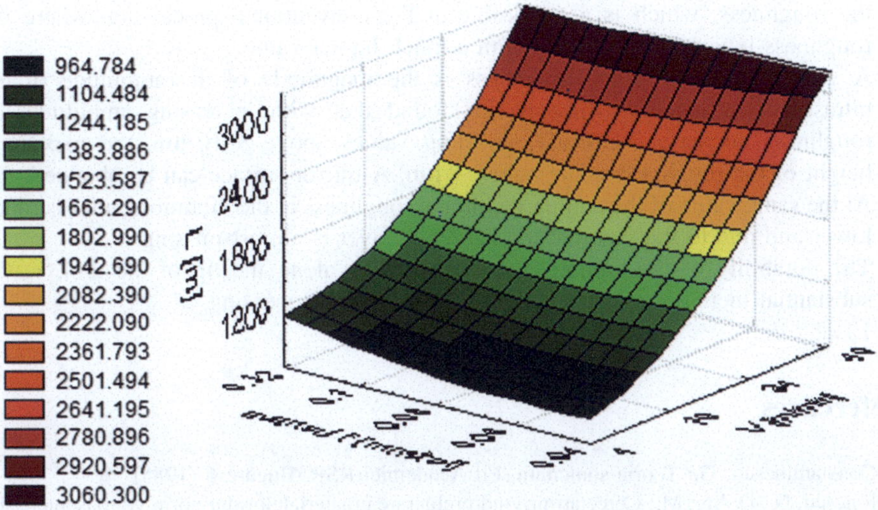

Fig. 6. The variation of the road from the top of the tool to the surface processed by ultrasonic vibration, 51VCr11 steel

In Figs. 5 and 6, the durability values of the cutting tool and the path of the diamond peak, are presented under the conditions of the valve processing in Fig. 1, by vibrating with the vibration parameters in Table 1.

By default, it can be determined with what parameters of the cutting regime and under what conditions of ultrasonic vibration of the tool tip can be obtained certain desired values, roughness Ra, durability T and the path of the diamond peak L.

4 Conclusions

From the analysis of the obtained results it was found that the fields of variation of the roughness parameter Ra depend significantly on the parameters of the cutting regime, as follows:

- The increase of the cutting speed has no influence on the roughness, except in the field of the values in the critical speed area, where the roughness becomes irregular, increasing and becoming equal to that of the classic cutting, at values above the critical speed.
- At increasingly lower values of speed, the roughness is almost equal to the theoretical roughness.
- It was also found that at any value of the speed below the critical speed, no deposits were formed on the cutting edge of the tool, which determines a roughness much better than the classic cutting.
- The increase of the working advance determines the increase of the surface roughness, but not over the geometrical factor, which gives the maximum value of

the roughness, which is not possible at the conventional processing, where the roughness is very irregular and with a much higher value.

- A great influence on the roughness is the magnitude of the amplitude of the ultrasonic oscillations. Thus, it was found that with increasing amplitude the roughness decreases vertically, reaching values above 8–10 μm, the maximum height of the roughness being below 1 μm. A mirror surface can be obtained.
- At the same value of the cutting depth, the roughness at the vibration cutter is much lower and due to the fact that the hardened layer is almost missing.
- The durability of the cutting tool and the way along the tip of the tool, was a substantial increase compared to the conventional processing.

References

1. Constantinescu, G.: Teoria sonicitatii. Ed Academiei RSR, Bucuresti (1985)
2. Ionescu, N., Dobre, M.: Cercetari privind prelucrarea materialelor dure prin vibroaschiere. In: 1994 Simpozion de tehnologii speciale si neconventionale, Galati, October 1994
3. Kumabe, J.: Vibration Cutting. Jikkyou Publishing Co., Tokyo (1979)
4. Kumabe, J.: Aplication of ultrasonic vibratory Machining. Tool Eng. **30**(1), 58–66 (1986)
5. Moriwaki, T., Shamoto, E.: Ultraprecision diamond turning of stainless steel by applying ultrasonic vibration. Ann. CIRP **40**, 559–562 (1991)
6. Tică, B.: Researches on vibrating processing of superhard materials. Ph.D. thesis (1998)

Experimental Testing of Energy Absorbing Foams Used as Safety Structure Within Automobile's Cockpit Module

Raluca Florentina Negrea[1(✉)], Paul Predoiu[1], Stefan Tabacu[1], Catalin Ducu[2], and Denis Negrea[2]

[1] Faculty of Mechanics and Technology, University of Pitesti, Pitești, Romania
raluca.negrea@yahoo.ro, predoiu_paul@yahoo.com, stefan.tabacu@upit.ro

[2] Regional Center of Research & Development for Materials, Processes and Innovative Products Dedicated to the Automotive Industry (CRC&D-Auto), University of Pitesti, Pitești, Romania
{catalin.ducu,denis.negrea}@upit.ro

Abstract. The main aim of this scientific paper is to present the experimental testing performed on three foam materials: expanded polypropylene foams (EPP), Expanded Polystyrene (EPS) and Piocelan (60 g/l), in order to evaluate their performance as add on safety structures within automobile's cockpit module. The performance analysis of these foam materials used as add on safety structures was done using two methods of in order to estimate the performance as safety devices for interior trimmings: material's structural response to impact (high variation vs material) and Quasi-static tests to dynamic loadings for studying the strain rate effect on the foam response using a conventional testing machine for quasi-static tests at the strain rate of 0.01 and 1s − 1. Test showed an increasing in strength proportionally to foams density, bead thickness and cell walls as a consequence of the increase of the bulk material in the foam. Also, Expanded Polystyrene (EPS) seems to manifest best properties regarding energy absorption. For the experimental data analysis, custom MATLAB codes have been developed. All the acquisition data will be integrated into a wider database which could be further used for comparing the results, data interpretation and appropriate material selection according to a specific application.

Keywords: Foam materials · Cockpit module · Impact performance · ECE R21

1 Introduction

The objective of the work presented in this article was the identification of the polypropylene foam behavior under dynamic stresses and to test a number of foam materials under static and dynamic loading conditions in order to estimate the performance as safety devices for interior trimmings. From the stresses which are imposed to it during an impact – mainly of compression -, it is natural to identify the response of this material in dynamic compression [8]. The expanded polypropylene foams

I. Dumitru et al. (Eds.): SMAT 2019, *The 30th SIAR International Congress of Automotive and Transport Engineering*, pp. 699–705, 2020.
https://doi.org/10.1007/978-3-030-32564-0_82

(EPP) are often used for packaging (aeronautical structures, electronic components, food, etc.), for personal safety (knee-pads, helmets, etc.) and also as core material for sandwich composite structures used in automotive, aircraft or naval industries on the strength of their great capacity to absorb energy and their lightness [2].

In order to bear severe loadings which imply high strain rates, the above mentioned foams are designed to absorb maximum energy. Thus, it is most important that before starting to optimize the design of this structures, to observe and understand first the behavior of these materials subjected to actual condition of use.

One of the main candidates to fulfill the above mentioned criteria is the expanded polypropylene (EPP) due to its low density, high elasticity, resistance to acid, alkali chemicals, oils and solvents and does not readily absorb water. It's not toxic and can be easily recycled. All these properties make EPP suitable for packaging and shipping products where shock protection during handling is necessary (electronics and medical devices). Most important, it can serve as core material for bumpers within car parts manufacturing industry due to the fact that it's strong, light, shock resistant, insulates heat and absorbs noise [4].

Another candidate is a rigid and tough closed-cell foam: the expanded polystyrene (EPS) which has the advantage of being able to produce products with different densities, it can easily be molded into sheets of varying thicknesses or flexible design as necessary. The versatility is due to the production process in which the polystyrene beads are placed inside the mold and expanded (by around 4000%) using steam and a blowing agent until the cavity in the mold is filled to produce the required shape. EPS is light, has good thermal insulation properties, moisture resistance, durability, acoustic absorption and excellent energy dissipation properties [5]. The EPS foam is a good energy absorber which can undergo large compressive deformation due to energy dissipation through cell bending, buckling or fracture, the stress being generally limited by the long and flat plateau of the stress-strain curve. This behavior explains the high energy efficiency that can be obtained with foamed materials compared to a corresponding solid specimen of equal volume made of the same material. Besides that, the associated EPS properties are relatively insensitive to temperature changes as compared to most other closed cell foams. The EPS application includes packaging and refrigeration, but is primarily employed as a structural insulator to improve the energy efficiency of buildings walls and floors [7].

The third candidate is Piocelan - a highly versatile type of polystyrene foam obtained by adding polyolefin to conventional polystyrene. Thus, Piocelan keeps the qualities of polystyrene (high level of rigidity and low bulk weight), while at the same time improving shock absorption, flexibility, resistance to chemicals and mechanical properties. This makes it more suitable for packing and load carriers of electronics and precision components and robust moldings [9].

Piocelan hybrid foam provides better impact strength and resiliency compared to EPP and EPS foams. Offers cost effective part production because it can be molded at lower densities than EPP and standard EPS equipment can be utilized [6].

2 Experimental Procedure

2.1 Material and Specimens

The mechanical properties of these types of industrial cellular materials are well known under quasi-static compression and also under large strain. We use for the two tests, tree foam materials: expanded polypropylene foams (EPP), Expanded Polystyrene (EPS) and Piocelan (60 g/l). Specimens are extracted from the large foam blocks at 30 mm height and 36 × 33 mm area and for the second test blocks at 30 mm height and 140 × 33 mm area (Fig. 1).

(a)

(b)

Fig. 1. The materials that we use for the universal testing machine (a) and experimental equipment physical test bench (b)

2.2 Testing Machines

Experiments were performed with a test bench designed for full-scale tests which used special transducers such as the impactor's acceleration during the event, its structural response (acceleration) and the angular position of the impactor's arm [1].

For EU (European Union) countries, experiments are performed according to ECE 21 specifications. Quasi-static to dynamic loadings is then worked out to study the strain rate effect on the foam response using a conventional testing machine for quasi-static tests at the strain rate of 0.01 and 1 s^{-1}.

The microcomputer controlled electronic universal testing machine is capable to developed a displacement rate of 500 mm/min at 1000 Hz and speed 0,01 mm/min. Data acquisition was made in MATLAB with National Instruments data acquisition system [3] (Fig. 2).

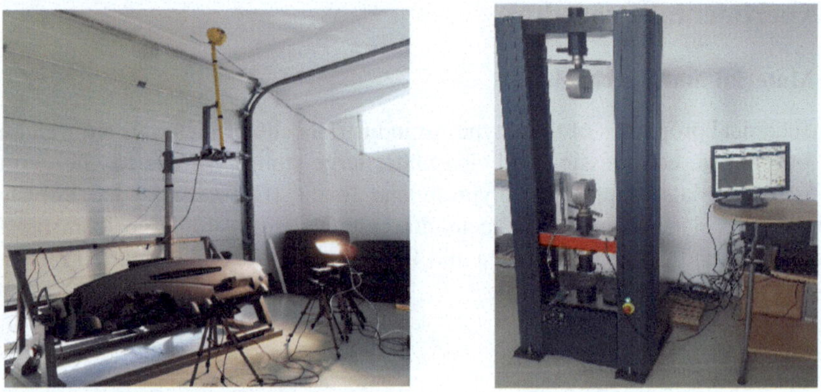

(a) **(b)**

Fig. 2. The experimental equipment physical test bench (a) and the universal testing machine (b)

3 Results and Discussion

The resulting curves from the experimental data, which were obtained with Universal testing machine tests for expanded polypropylene foams (EPP), Expanded Polystyrene (EPS) and Piocelan are (Fig. 3):

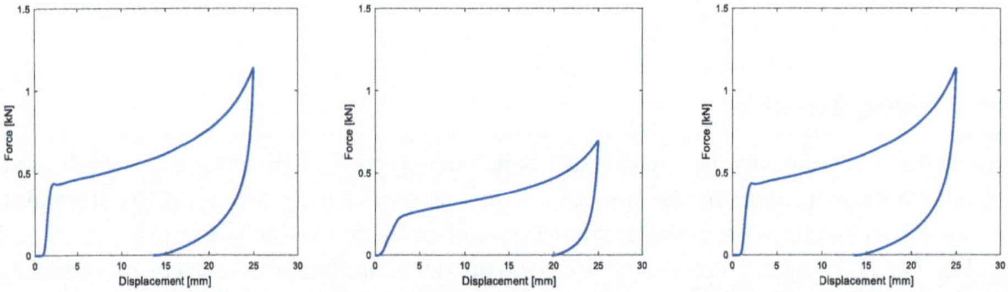

(a) **(b)** **(c)**

Fig. 3. EPP material (a), EPS material (b) and Piocelan (c)

In Fig. 4 is presented the variation of the height samples tested before and after the compression test.

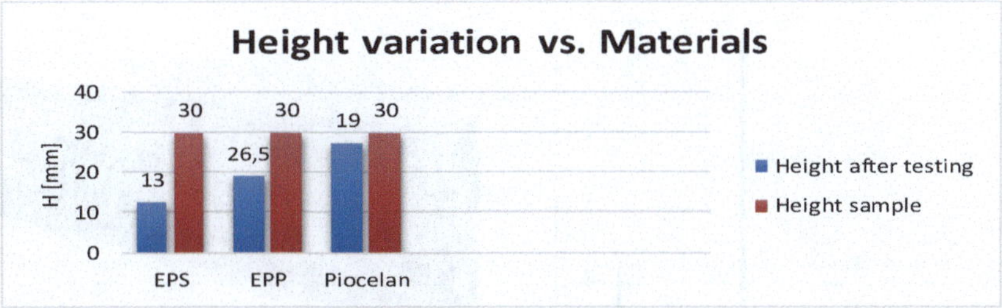

Fig. 4. The variation of the height samples.

The resulting curves of the acceleration from the experimental data of the tree materials EPS, EPP, Piocelan, which were obtained with experimental testing and experimental equipment physical test bench are (Figs. 5, 6 and 7):

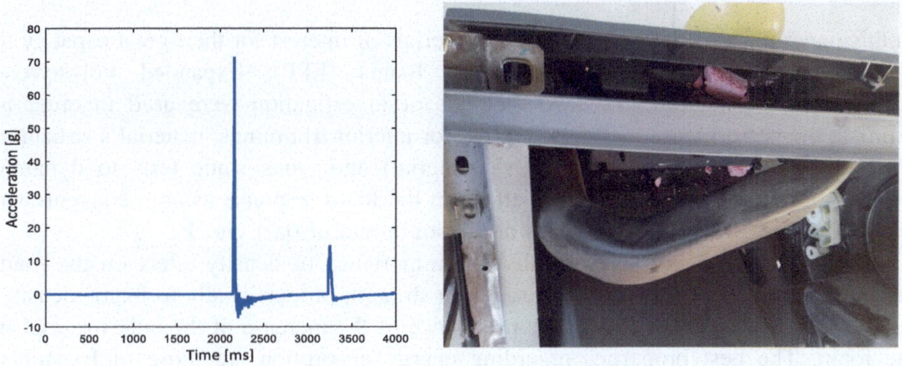

Fig. 5. EPS material after the experiment and its result of curve acceleration

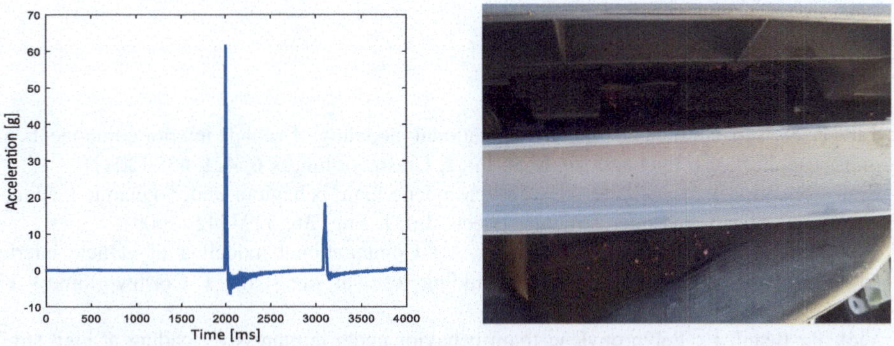

Fig. 6. EPP material after the experiment and its result of curve acceleration

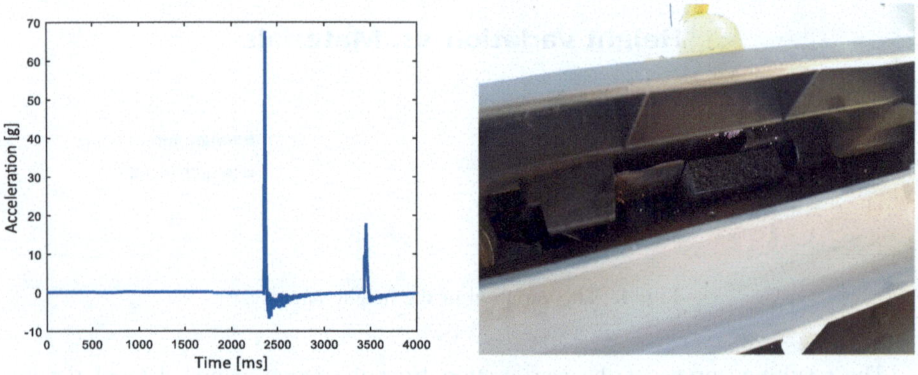

Fig. 7. Piocelan material after the experiment and its result of curve acceleration

4 Conclusion

In this paper we have investigated three materials of interest for their great capacity to absorb energy: expanded polypropylene foams (EPP), Expanded Polystyrene (EPS) and Piocelan (60 g/l). Two methods of investigation were used in order to estimate the performance as safety devices for interior trimmings: material's structural response to impact (high variation vs material) and Quasi-static tests to dynamic loadings for studying the strain rate effect on the foam response using a conventional testing machine for quasi-static tests at the strain rate of 0.01 and 1 s^{-1}.

The tests performed have revealed the importance of density effect on the foam behavior. Thus, test showed an increasing in strength proportionally to foams density, bead thickness and cell walls as a consequence of the increase of the bulk material in the foam. The best properties regarding energy absorption are those of Expanded Polystyrene (EPS).

Acknowledgements. All authors made an equal contribution to the present work.

References

1. Tabacu, S., Tabacu, I., Hadar, A.: Computational modeling of vehicle interior components for impact applications: thickness analysis. Int. J. Crashworthiness **6**, 421–438 (2011)
2. Bouix, R., Viot, P., Lataillade, J.L.: Polypropylene foam behaviour under dynamic loadings: strain rate, density and microstructure effects. Int. J. Eng. **36**, 329–342 (2009)
3. Tabacu, S., Parlac, S., Tomus, C., Baba, V.: Computational modeling of vehicle interior components for impact applications: moulding residual stress. Int. J. Crashworthiness **17**, 591–605 (2012)
4. Viot, P., Beani, F.: Polypropylene foam behavior under compressive loading at high strain rate. Struct. Shock. Impact **VIII**, 2–9 (2004)
5. Avalle, M., Belingardi, G., Montanini, R.: Characterization of polymeric structural foams under compressive impact loading by means of energy-absorption diagram. Int. J. Impact Eng **25**, 455–472 (2001)

6. Nising, P.: Flexible foam production. Sulzer Technical Review, February 2012
7. Matthew Barsotti, M.S.: Comparison of FEM and SPH for modeling a crushable foam aircraft arrestor bed. In: 11th International LS-DYNA Users Conference, vol. 16, pp. 37–54 (2010)
8. Slik, G., Vogen, G., Chawda, V.: Material model validation of a high efficient energy absorbing foam. In: 5th LS-DYNA Forum, Material III – Foam/Composites (2006)
9. Chen, W., Hao, H., Hughes, D., Shi, Y., Cui, J., Li, Z.X.: Static and dynamic mechanical properties of expanded polystyrene. Mater. Des. **69**, 170–180 (2015)

Virtual Modeling and Numerical Simulations of the Latitude Prosthesis - Human Elbow Assembly

Daniela Tarnita[1(✉)], C. Boborelu[2], D. Popa[1], R. Malciu[1], and D.-N. Tarnita[3]

[1] University of Craiova, Craiova, Romania
tarnita.daniela@gmail.com
[2] Emergency Hospital of Dolj County, Craiova, Romania
[3] University of Medicine and Pharmacy, Craiova, Romania

Abstract. The Latitude Prosthesis is a complex implant for human elbow joint with a great constructional flexibility, providing the possibility, depending on the intraoperative situation and the residual instability, to “move” intra-operatively from a non-constrained prosthesis to a constrained prosthesis by a piece attached to the component ulnar. This study aims to present the virtual modeling of the Latitude elbow prosthesis and of the human elbow joint-Latitude prosthesis assembly. The kinematic simulation of the flexion-extension and pronation-supination movements of the elbow joint-Latitude prosthesis–assembly are performed.

Keywords: Virtual elbow · Latitude prosthesis · Flexion-extension · Pronation-supination

1 Introduction

Arthritis elbow last phase is managed by resecting or arthroplasty. The total and partial elbow arthroplasty was studied in different papers [1–5]. Contemporary half-constrained prostheses have a mechanical link that accurately reproduces the kinematics of the native elbow. Physiological rotation and varus/valgus forces occurring during flexion and extension are dissipated into soft tissues more than in the mechanical joint or stem fixation. When the prostheses are properly positioned and balanced, such an arthroplasty can almost completely restore the kinematics of the human elbow. Some prosthetic implants have used a radial head prosthesis with the theoretical advantage of improving the stability of the elbow, as well as of reducing the forces on the radio-humeral articulation by balancing the distribution of forces along the elbow joint [5].

Good results were reported after total arthroplasty for both constrained and non-constrained prostheses. The success of the elbow arthroplasty and advances in surgical technique and prosthetic design have expanded indications for use in younger patients but also for other diagnoses. In an attempt to allow the intra-operative decision to choose between constrained or unconstrained prostheses, the concept of prosthetic convertibility was approached by two systems: Latitude (see Fig. 1) and DePuy Acclaim [5].

I. Dumitru et al. (Eds.): SMAT 2019, *The 30th SIAR International Congress of Automotive and Transport Engineering*, pp. 706–712, 2020.
https://doi.org/10.1007/978-3-030-32564-0_83

These implants allow the post-operative conversion of unstable devices into a stable joint without requiring extensive revision. Aligned devices and modular components can help replicate the native rotation center more accurately, but also the soft tissue balance needed for a good tracking component. The Latitude prosthesis system is a complex implant with a great constructional flexibility, providing the possibility to "intraoperatively" turn from a non-constrained prosthesis to a constrained prosthesis through a piece attached to the ulnar component, depending on the intraoperative situation and the residual instability. It also contains a bipolar radial head prosthesis that allows the mechanical balance of the radio-humeral joint when the alignment is correct.

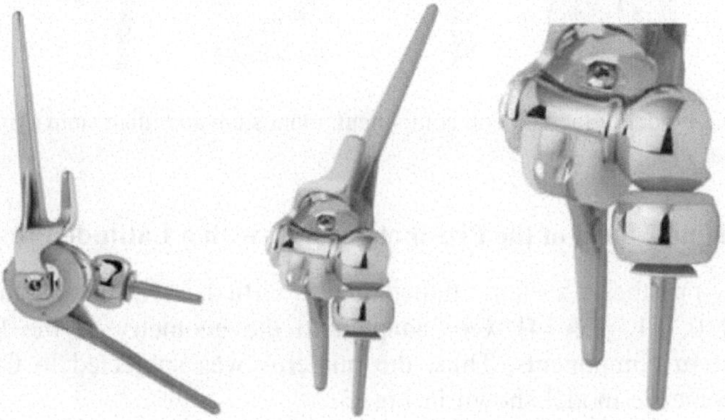

Fig. 1. Latitude total elbow prosthesis system [5]

1.1 Three-Dimensional Modeling of Latitude Elbow Total Prosthesis

Virtual prototyping, modeling and simulation are very important tools used in biomechanics and biomedical engineering, in order to study the biomechanical behavior of the bones, of the healthy joints, damaged joints, or prosthetic and orthotic joints. [6–16]. In order to determine the geometric parameters of the Latitude elbow prosthesis components, the direct measurement method combined with measurement on the photographic images was used. Certain geometrically directly measurable parameters were determined by direct measurement on other similar elbow prostheses. Also, the primary forms of these components, have been identified. The geometric and dimensional elements were determined by measuring the components. The geometric elements of the total elbow prosthesis components were fully dimensionally determined and then the three-dimensional modeling was performed using the SolidWorks software. In the following, part of the humeral component of the studied prosthesis is partially detailed, the model being designed to allow easy transformation into a different size model by geometrically complete parameterization. The final model of the humeral component and the other components of the elbow prosthesis were modeled (see Fig. 2). The model of the Latitude total elbow prosthesis was obtained by using the Assembly module of the software and specific commands that canceled certain degrees of freedom.

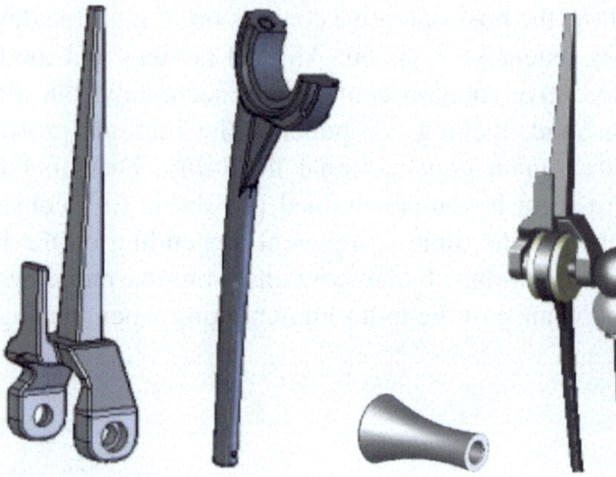

Fig. 2. The final model of the humeral component, ulnar stem and ulnar stem axis. The model of the Latitude total elbow prosthesis

1.2 The Virtual Model of the Prosthetic Elbow with a Latitude Prosthesis

For the elbow-prosthesis system studied in the virtual environment, the three bone components [10, 11, 13, 14] were adapted to the geometry of the Latitude total prosthetic system components. Thus, the humerus was subjected to Cut operations having the geometric model shown in Fig. 3.

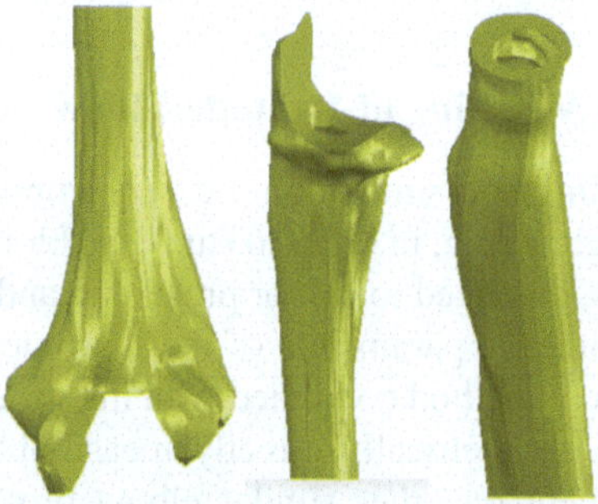

Fig. 3. The humerus, cubitus and radius resected models

The modified and adapted bone components were loaded into the Assembly module and attached to the Latitude total elbow prosthesis stems. The final model of the assembly is shown in Fig. 4.

1.3 Kinematic and Dynamic Simulation of the Prosthetic Elbow Joint Assembly for Flexion-Extension Movements

This model was exported to the kinematic simulation and finite element analysis software where the initial parameters of the simulation were defined:

- full length of simulation T = 1 s;
- the force F was 100 N having vertical permanent action and the point of application at the extremities of ulna and radius;

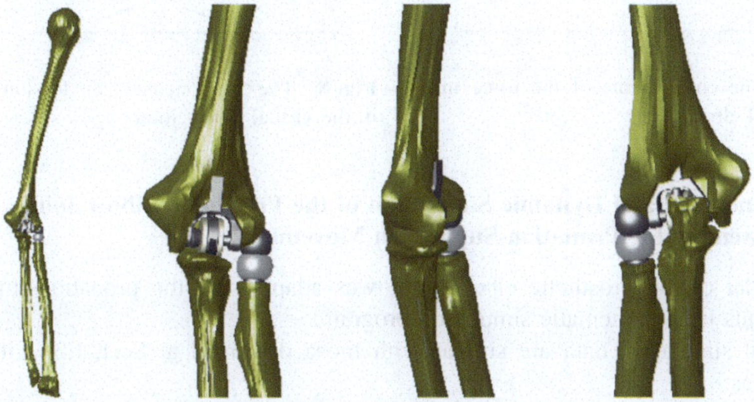

Fig. 4. The prosthetic model of the elbow joint

- the humerus was the fixed component, the other elements were considered mobile;
- in the revolute pair simulating the humero-ulnar and humero-radial joint a driving pair was defined having an angular velocity $\omega = -450 \sin(2\pi t)$ grd/s;
- the simulated situation corresponded to lifting and lowering of a weight of about 10 kg by flexion and extension movements.

It is possible to extract important kinematic parameters of the prosthetic elbow:

- the positions of mobile components: cubitus (see Fig. 5) and radius;
- the velocities and accelerations with respect to the fixed global system of the model;
- the angular positions of mobile components (Fig. 6);
- the angular velocities and accelerations of the mobile elements;
- the force components in the joint of the elbow (Fig. 7);
- the torsion moment in the elbow joint (Fig. 8).

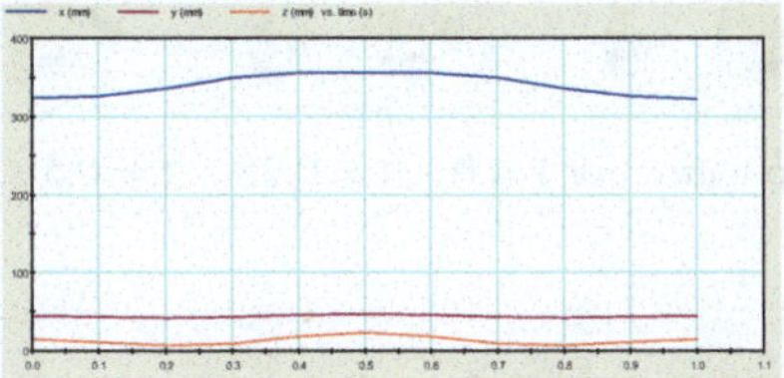

Fig. 5. The cubitus center of mass position (in x, y, z coordinates)

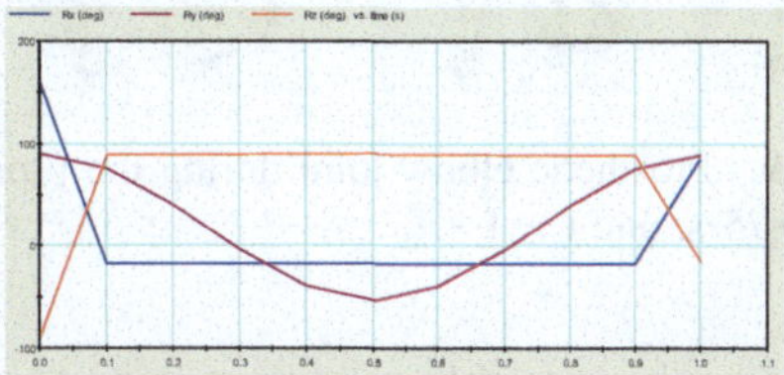

Fig. 6. The angular position (in angles Rx, Ry, Rz) of the cubitus center of mass

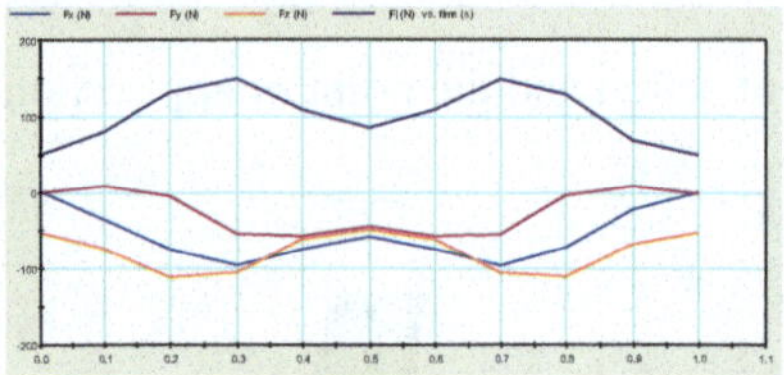

Fig. 7. The components of the force in the virtual elbow joint

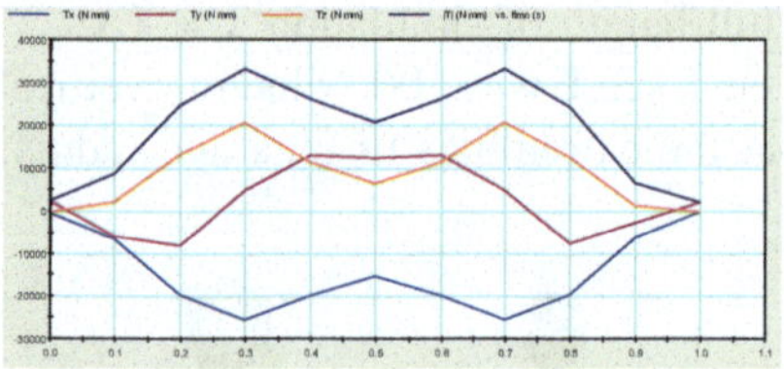

Fig. 8. The components of the torsion moment in the virtual elbow joint

1.4 Kinematic and Dynamic Simulation of the Prosthetic Elbow Joint Assembly for Pronation-Supination Movements

The model of the prosthetic elbow joint was adapted for the pronation-supination movements in the kinematic simulation program:

Initial simulation data are similar with those described at Sect. 1.3, with some differences:

– in the revolute pair simulating the humero-radial joint a driving pair was defined having the angle of the articulation defined by the relation: $\alpha = -47 \sin(\pi t)$ grd;

Figure 9 shows five frames of the simulation in which the position of the bone components and the prosthesis is highlighted over the studied length of time.

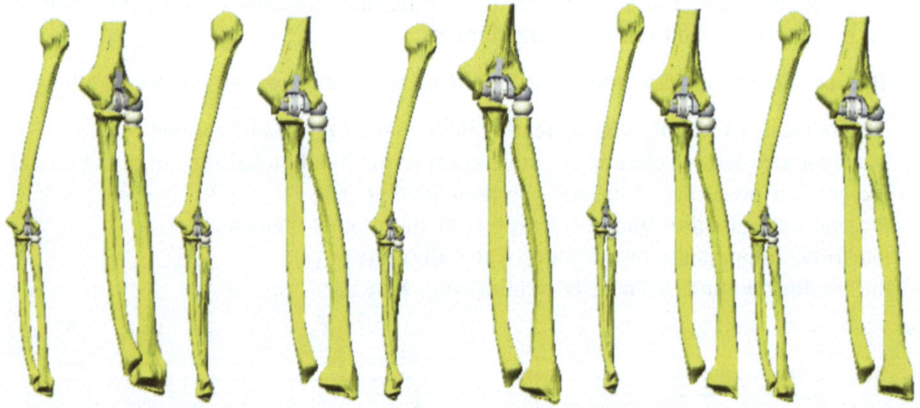

Fig. 9. Prosthetic elbow joint during the pronation-supination for t = 0 s, t = 0.25 s, t = 0.5 s, t = 0.75 s, and t = 1 s.

It is also possible to extract important kinematic parameters, such as:

– Rx, Ry, Rz – the variations of the angles between the straight line joining the origin of the reference system and the radius center of mass, on the one hand, with the axes of the reference system, on the other hand, occurred during the motion (Fig. 10);

– the angular velocities and the angular accelerations of Rx, Ry, Rz;
– the torsion moment in the elbow joint (Fig. 11).

It is noted that the kinematic parameters diagrams are similar to the diagrams obtained during the kinematic simulation of the non-prosthetic elbow.

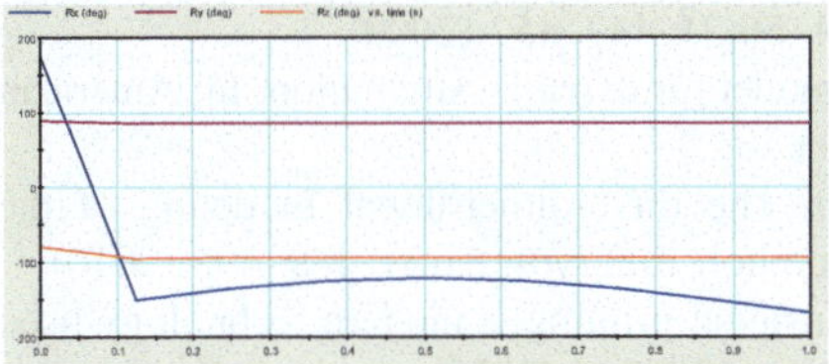

Fig. 10. The angular variations Rx, Ry, Rz of the radius center of mass

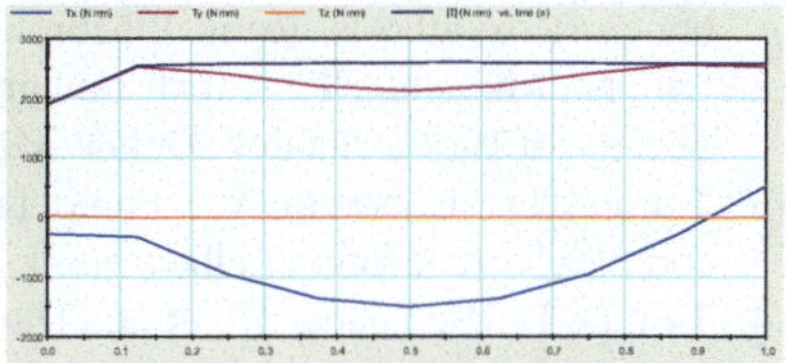

Fig. 11. The components of the torsion moment in the virtual elbow joint

2 Conclusions

A convertible implant allows the surgeon the versatility to choose to perform a hemi-arthroplasty or an unlinked or linked total elbow arthroplasty with assurance that later revision can be performed without the need to remove well-fixed stems [5]. Conversion from an unlinked to a linked prosthesis can be performed at the time of the initial implantation or post-operatively in a minimally invasive fashion [5]. The Latitude (Tornier, Stafford, TX) total elbow system currently offers these options. This paper presents the 3D modeling of Latitude modular elbow prosthesis using the SolidWorks parameterized modelling environment, the 3D modelling of the prosthetic elbow joint and the results of kinematic and dynamic simulation of the flexion-extension and pronation-supination movements of the Latitude prosthesis-joint assembly.

References

1. Morrey, B.F., An, K.N., Chao, E.Y.S.: Functional evaluation of the elbow. In: Morrey, B.F. (ed.) The Elbow and Its Disorders, 2nd edn, Philadelphia, pp. 86–89 (1993)
2. Van der Lugt, J.C., Rozing, P.M.: Systematic review of primary total elbow prostheses used for the rheumatoid elbow. Clin. Rheumatol. **23**(4), 291–298 (2004)
3. Kudo, H., Iwano, K., Watanabe, S.: Total replacement of the rheumatoid elbow with a hingeless prosthesis. J. Bone Joint Surg. **62-A**, 277–285 (1980)
4. Ewald, F.C., Jacobs, M.A.: Total elbow arthroplasty. Clin. Orthop. **182**, 137–142 (1984)
5. King, G.J.W., Yamaguchi, K., O'Driscoll, K.: Latitude convertible total elbow prosthesis. In: Treatment of Elbow Lesions, pp. 251–261. Springer (2008)
6. Barbu, I., Alexandru, C.: Virtual prototyping tools applied in mechanical engineering. Mecatronica **3**, 9–12 (2004)
7. Alexandru, C.: Modeling and Simulation in Engineering, Intech Publishing House (2012)
8. Tarnita, D., Tarnita, D.N.: Numerical simulations of human tibia osteosynthesis using modular plates based on Nitinol staples. Rom. J. Morphol. Embryol. **51**(1), 145–150 (2010)

9. Zhang, L.L., Zhang, X.A., Wang, C.T.: Upper limb musculoskeletal model for biomechanical analysis-investigation of elbow flexion movement. The National Natural Science Foundation of China (No. 30530230) (2008)
10. Tarnita, D., Tarnita, D.N., et al.: In vitro experiment of the modular orthopedic plate based on Niti used for human radius bone fractures. Rom. J. Morphol. Embryol. **51**(2), 315–320 (2010)
11. Tarnita, D., Tarnita, D.N., Bizdoaca, N., et al.: Modular adaptive bone plate for humerus bone osteosynthesis. Rom. J. Morphol. Embryol. **50**(3), 447–452 (2009)
12. Lan, N., Murakata, T.: A realistic human elbow model for dynamic simulation. In: American Society of Biomechanics Annual Meeting (2001)
13. Tarnita, D., Boborelu, C., Popa, D., Rusu, L.: The three-dimensional modeling of the complex virtual human elbow joint. Rom. J. Morphol. Embryol. **51**(3), 489–495 (2010)
14. Tarnita, D., Berceanu, C., et al.: The three-dimensional printing–a modern technology used for biomedical prototypes. Materiale Plastice **47**(3), 328–334 (2010)
15. Vaida, C., Plitea, N., Pisla, D., Gherman, B.: Orientation module for surgical instruments—a systematical approach. Meccanica **48**(1), 145–158 (2013)
16. Tarnita, D., Tarnita, D.N., Bolcu, D.: Orthopedic implants based on shape memory alloys. In: Biomedical Engineering – From Theory to Applications. InTechOpen, pp. 431–468 (2011)

Author Index

A
Abdullah, Oday I., 68
Allah, Fazal Um Min, 113
Anghelache, Gabriel, 624

B
Bagnaru, Dan, 565
Balcau, Monica, 502
Băldean, Doru-Laurean, 82, 160, 203, 209, 468
Balint, Richard János, 153
Bancă, Gheorghe, 31
Barabás, István, 195
Barothi, Laszlo, 165, 451
Bățăuş, Marius, 103
Beles, Horia, 375
Benea, Bogdan Cornel, 359
Bilal, Levent, 525
Blaga, Vasile, 3
Boborelu, C., 706
Bolcu, Alexandru, 550
Bolcu, Dumitru, 550
Boroiu, Alexandru, 349
Boroiu, Andrei Alexandru, 349
Borza, Emilian Vasile, 203, 485
Borzan, Adela-Ioana, 82, 494
Brabie, Gheorghe, 533
Burnete, Nicolae, 95, 406
Burnete, Nicolae Vlad, 153

C
Calangiu, Adrian, 15
Capătă, Marius S. D., 95, 406
Cerbu, Camelia, 231
Chereches, Ioan-Aurel, 82
Chioreanu, Catalin, 3
Chiriac, Rareş Lucian, 131, 289
Chiru, Anghel, 131, 221, 273, 281, 289
Ciolan, Gheorghe, 614
Cipleu, Adrian, 366, 400
Ciunel, Stefanita, 54, 332, 341
Ciurezu, Leonard, 565
Clenci, Adrian, 58
Clinciu, Mihai, 332, 341
Cofaru, Corneliu, 251
Condrea, Ovidiu Andrei, 131, 273, 281, 289
Constantinescu, Augustin, 54, 540
Copae, Ion, 165, 177, 477
Copilusi, Cristian, 15
Cordoş, Nicolae, 195
Coropeţchi, Iulian Constantin, 165
Covaciu, Dinu, 392, 614
Covaciu, Florin, 468
Craciunoiu, Nicolae, 670
Cucu, Leonard-Iulian, 165

D
Darie, Cătălin-Dumitru, 165
Diaconescu, Eugen, 508
Dima, Dragos Sorin, 265, 312
Dragomir, George, 375
Drosescu, Radu, 426
Drosu, Alin, 251
Ducu, Cătălin, 581, 678, 699
Dumitru, Ilie, 26, 383, 607, 636, 643
Dumitru, Nicolae, 15, 550

F
Filip, Nicolae, 297, 304
Florea, Daniela, 392
Frățilă, Gheorghe, 103, 214

I. Dumitru et al. (Eds.): SMAT 2019, *The 30th SIAR International Congress of Automotive and Transport Engineering*, pp. 713–715, 2020.
https://doi.org/10.1007/978-3-030-32564-0

G
Gallo, Waldyr Luiz Ribeiro, 113
Geonea, Ionut, 15, 26, 559, 565, 607, 670
Gheorghe, Carmen, 297, 304
Gönczi, Attila Iuliu, 366, 400
Grigorie, Laura Diana, 265, 660
Grujic, Ivan, 68

I
Iclodean, Călin, 195
Igreţ, Sorin Vlad, 589, 597
Ilea, Laurentiu, 124
Ionica, Valeriu, 565
Iorga-Simăn, Victor, 58
Iozsa, Daniel, 31, 124, 214
Ispas, Nicolae, 439
Ivan, Florian, 31

J
Ji, Liu, 138
Jurchiş, Bogdan-Manolin, 160, 203
Jurco, Ancuta, 485

K
Karaca, Merve, 525
Klement, C., 691

L
Leahu, Cristian-Ioan, 42, 88, 188, 241
Lehel, Csokmai, 458
Lespezeanu, Ion, 165, 177, 477
Liang, Chen, 138
Luca-Motoc, Dana, 188

M
Măgherusan, Corneliu Adrian, 153
Maican, Sorin, 103
Malciu, R., 706
Manea, Adriana, 312
Manea, Ana Maria, 214
Manea, Laurentiu Claudiu, 312
Mariasiu, Florin, 485, 494
Marin, Mihnea, 265
Matei, Lucian, 26, 383, 607, 636, 643
Miloraddovic, Danijela, 68
Miritoiu, Cosmin Mihai, 550, 559
Mitran, Tudor, 3, 458
Mitroi, Marian Florin, 221, 231, 322
Moca, Sorin, 3, 375, 458
Moisescu, Alexandra-Raluca, 624
Moldovanu, Dan, 153
Morariu, Silvan, 485, 494
Mousavi, Hoda, 138
Muresan, Georgiana Silvia, 304

N
Năstăsoiu, Mircea, 88, 439
Neagu, Elena, 349
Negrea, Denis, 699
Negrea, Raluca Florentina, 573, 699
Negrilă, Radu Iulian, 589, 597
Niculae, Mihai, 58
Niculescu, Rodica, 58
Nişulescu, Valentin, 31

O
Oncescu, Alin Teofil, 518
Onescu, Constantin, 581, 678
Oprean, Mircea, 103
Oţăt, Oana, 26, 383, 636, 643
Oţăt, Victor, 383

P
Pântece, Constantin-Iulian, 477
Parlac, Sebastian, 349
Pascu, Ileana Cristina, 684
Popa, Dinel, 573, 706
Popa, Mădălin Florin, 95, 406
Popescu, Iulian, 653, 684
Popescu, Mihaela Virginia, 251
Preda, Ion, 614
Predoiu, Paul, 573, 699
Prutica, Dan George, 533

R
Racila, Laurentiu, 26, 559, 607, 636, 643
Racila, Mihaela, 607, 636, 643
Radu, Alexandru Ionut, 312, 359, 375
Renţea, Cristian, 103
Romanescu, Alina, 660
Romanescu, Alina Elena, 265, 653
Rosca, Adrian Sorin, 670
Rotea, Cristina, 26
Rufino, Caio Henrique, 113
Rus, Alexandru, 458

S
Sandu, Corina, 138
Sandu, Ionela Alina, 581, 678
Sass, Ludmila, 653
Sava, Daniel, 177
Schlattmann, Josef, 68
Scurtu, Liviu, 485, 494, 502
Sebesan, Ioan, 414
Simniceanu, Loreta, 50, 540
Simniceanu, Robert Emil, 50, 54
Spirleanu, Cristian, 508
Stan, Cornelia, 124, 214
Stanescu, Marius Marinel, 550

Ştefănescu, Werner, 400
Stoica, Ramona-Monica, 177, 451
Stojanovic, Nadica, 68

T
Tabacu, Ştefan, 573, 581, 678, 699
Ţălu, Mihai, 660
Ţălu, Ştefan, 660
Tarata, Daniela, 670
Tarnita, D.-N., 706
Tarnita, Daniela, 518, 706
Ţârulescu, Radu, 88, 188, 241
Ţârulescu, Stelian, 188, 241
Teodorescu, Claudiu, 124, 214
Tică, Bebe, 332, 691
Timar, Janos, 392
Todoruţ, Adrian, 195
Togănel, George Radu, 273, 281, 289, 359
Tolea, Bogdan Adrian, 375
Topaç, Mehmet Murat, 525
Trotea, Mario, 54, 540
Trusca, Daniel Dragos, 273, 281, 289, 359
Turcanu, Iulian, 322, 414
Tutunea, Dragoş, 26, 383, 607, 636, 643

V
Valeriu, Ionica, 559
Vandici, Ionel, 400
Varga, Bogdan, 494
Vasile, Alexandru, 477
Vesselenyi, Tiberiu, 458
Vilău, Radu, 177, 477
Vintilă, Daniela, 660
Vlad, Nicolae, 485, 494
Voicu, Daniela, 451

Z
Zabeu, Clayton Barcelos, 113
Zamfir, Silviu, 426

The manufacturer's authorised representative in the EU is Springer Nature Customer Service Centre GmbH, Europaplatz 3, 69115 Heidelberg, Germany. If you have any concerns regarding our products, please contact ProductSafety@springernature.com

Printed and bound by CPI Group (UK) Ltd, Croydon, CR0 4YY
05/10/2026
02228339-0009